Alice and Bob Meet Banach

The Interface of Asymptotic Geometric Analysis and Quantum Information Theory

Mathematical Surveys and Monographs

Volume 223

Alice and Bob Meet Banach

The Interface of Asymptotic Geometric Analysis and Quantum Information Theory

Guillaume Aubrun
Stanisław J. Szarek

AMERICAN MATHEMATICAL SOCIETY
Providence, Rhode Island

2010 *Mathematics Subject Classification.* Primary 46Bxx, 52Axx, 81Pxx, 46B07, 46B09, 52C17, 60B20, 81P40.

For additional information and updates on this book, visit
www.ams.org/bookpages/surv-223

Library of Congress Cataloging-in-Publication Data

Names: Aubrun, Guillaume, 1981- author. | Szarek, Stanisław J., author.
Title: Alice and Bob Meet Banach: The interface of asymptotic geometric analysis and quantum information theory / Guillaume Aubrun, Stanisław J. Szarek.
Description: Providence, Rhode Island : American Mathematical Society, [2017] | Series: Mathematical surveys and monographs ; volume 223 | Includes bibliographical references and index.
Identifiers: LCCN 2017010894 | ISBN 9781470434687 (alk. paper)
Subjects: LCSH: Geometric analysis. | Quantum theory. | AMS: Functional analysis – Normed linear spaces and Banach spaces; Banach lattices – Normed linear spaces and Banach spaces; Banach lattices. msc | Convex and discrete geometry – General convexity – General convexity. msc | Quantum theory – Axiomatics, foundations, philosophy – Axiomatics, foundations, philosophy. msc | Functional analysis – Normed linear spaces and Banach spaces; Banach lattices – Local theory of Banach spaces. msc | Functional analysis – Normed linear spaces and Banach spaces; Banach lattices – Probabilistic methods in Banach space theory. msc | Convex and discrete geometry – Discrete geometry – Packing and covering in n dimensions. msc | Probability theory and stochastic processes – Probability theory on algebraic and topological structures – Random matrices (probabilistic aspects; for algebraic aspects see 15B52). msc | Quantum theory – Axiomatics, foundations, philosophy – Quantum coherence, entanglement, quantum correlations. msc
Classification: LCC QA360 .A83 2017 | DDC 515/.732–dc23 LC record available at https://lccn.loc.gov/2017010894

Softcover ISBN: 978-1-4704-7796-7

Reprinted by the American Mathematical Society, 2024.

∞ The paper used in this book is acid-free and falls within the guidelines
established to ensure permanence and durability.
Visit the AMS home page at `http://www.ams.org/`

11 10 9 8 7 6 5 4 3 2   29 28 27 26 25 24

To Aurélie and Margaretmary

Contents

List of Tables xiii

List of Figures xv

Preface xix

Part 1. Alice and Bob: Mathematical Aspects of Quantum Information Theory 1

Chapter 0. Notation and basic concepts 3
- 0.1. Asymptotic and nonasymptotic notation 3
- 0.2. Euclidean and Hilbert spaces 3
- 0.3. Bra-ket notation 4
- 0.4. Tensor products 6
- 0.5. Complexification 6
- 0.6. Matrices vs. operators 7
- 0.7. Block matrices vs. operators on bipartite spaces 8
- 0.8. Operators vs. tensors 8
- 0.9. Operators vs. superoperators 8
- 0.10. States, classical and quantum 8

Chapter 1. Elementary convex analysis 11
- 1.1. Normed spaces and convex sets 11
  - 1.1.1. Gauges 11
  - 1.1.2. First examples: ℓ_p-balls, simplices, polytopes, and convex hulls 12
  - 1.1.3. Extreme points, faces 13
  - 1.1.4. Polarity 15
  - 1.1.5. Polarity and the facial structure 17
  - 1.1.6. Ellipsoids 18
- 1.2. Cones 18
  - 1.2.1. Cone duality 19
  - 1.2.2. Nondegenerate cones and facial structure 21
- 1.3. Majorization and Schatten norms 22
  - 1.3.1. Majorization 22
  - 1.3.2. Schatten norms 23
  - 1.3.3. Von Neumann and Rényi entropies 27
- Notes and Remarks 29

Chapter 2. The mathematics of quantum information theory 31
- 2.1. On the geometry of the set of quantum states 31
  - 2.1.1. Pure and mixed states 31

2.1.2. The Bloch ball $D(\mathbb{C}^2)$ 32
2.1.3. Facial structure 33
2.1.4. Symmetries 34
2.2. States on multipartite Hilbert spaces 35
2.2.1. Partial trace 35
2.2.2. Schmidt decomposition 36
2.2.3. A fundamental dichotomy: Separability vs. entanglement 37
2.2.4. Some examples of bipartite states 39
2.2.5. Entanglement hierarchies 41
2.2.6. Partial transposition 41
2.2.7. PPT states 43
2.2.8. Local unitaries and symmetries of Sep 46
2.3. Superoperators and quantum channels 47
2.3.1. The Choi and Jamiołkowski isomorphisms 47
2.3.2. Positive and completely positive maps 48
2.3.3. Quantum channels and Stinespring representation 50
2.3.4. Some examples of channels 52
2.4. Cones of QIT 55
2.4.1. Cones of operators 55
2.4.2. Cones of superoperators 56
2.4.3. Symmetries of the $\mathcal{PSD}$ cone 58
2.4.4. Entanglement witnesses 60
2.4.5. Proofs of Størmer's theorem 62
Notes and Remarks 63

Chapter 3. Quantum mechanics for mathematicians 67
3.1. Simple-minded quantum mechanics 67
3.2. Finite vs. infinite dimension, projective spaces, and matrices 68
3.3. Composite systems and quantum marginals: Mixed states 68
3.4. The partial trace: Purification of mixed states 70
3.5. Unitary evolution and quantum operations: The completely positive maps 71
3.6. Other measurement schemes 73
3.7. Local operations 74
3.8. Spooky action at a distance 75
Notes and Remarks 75

Part 2. Banach and His Spaces: Asymptotic Geometric Analysis Miscellany 77

Chapter 4. More convexity 79
4.1. Basic notions and operations 79
4.1.1. Distances between convex sets 79
4.1.2. Symmetrization 80
4.1.3. Zonotopes and zonoids 81
4.1.4. Projective tensor product 82
4.2. John and Löwner ellipsoids 84
4.2.1. Definition and characterization 84
4.2.2. Convex bodies with enough symmetries 89

4.2.3. Ellipsoids and tensor products 91
4.3. Classical inequalities for convex bodies 91
4.3.1. The Brunn–Minkowski inequality 91
4.3.2. log-concave measures 93
4.3.3. Mean width and the Urysohn inequality 94
4.3.4. The Santaló and the reverse Santaló inequalities 98
4.3.5. Symmetrization inequalities 98
4.3.6. Functional inequalities 101
4.4. Volume of central sections and the isotropic position 101
Notes and Remarks 103

Chapter 5. Metric entropy and concentration of measure in classical spaces 107
5.1. Nets and packings 107
5.1.1. Definitions 107
5.1.2. Nets and packings on the Euclidean sphere 108
5.1.3. Nets and packings in the discrete cube 113
5.1.4. Metric entropy for convex bodies 114
5.1.5. Nets in Grassmann manifolds, orthogonal and unitary groups 116
5.2. Concentration of measure 117
5.2.1. A prime example: concentration on the sphere 119
5.2.2. Gaussian concentration 121
5.2.3. Concentration tricks and treats 124
5.2.4. Geometric and analytic methods. Classical examples 129
5.2.5. Some discrete settings 136
5.2.6. Deviation inequalities for sums of independent random variables 139
Notes and Remarks 142

Chapter 6. Gaussian processes and random matrices 149
6.1. Gaussian processes 149
6.1.1. Key example and basic estimates 150
6.1.2. Comparison inequalities for Gaussian processes 152
6.1.3. Sudakov and dual Sudakov inequalities 154
6.1.4. Dudley's inequality and the generic chaining 157
6.2. Random matrices 160
6.2.1. ∞-Wasserstein distance 161
6.2.2. The Gaussian Unitary Ensemble (GUE) 162
6.2.3. Wishart matrices 166
6.2.4. Real RMT models and Chevet–Gordon inequalities 173
6.2.5. A quick initiation to free probability 176
Notes and Remarks 178

Chapter 7. Some tools from asymptotic geometric analysis 181
7.1. ℓ-position, K-convexity and the MM^*-estimate 181
7.1.1. ℓ-norm and ℓ-position 181
7.1.2. K-convexity and the MM^*-estimate 182
7.2. Sections of convex bodies 186
7.2.1. Dvoretzky's theorem for Lipschitz functions 186
7.2.2. The Dvoretzky dimension 189
7.2.3. The Figiel–Lindenstrauss–Milman inequality 193

7.2.4. The Dvoretzky dimension of standard spaces 195
7.2.5. Dvoretzky's theorem for general convex bodies 200
7.2.6. Related results 201
7.2.7. Constructivity 205
Notes and Remarks 207

Part 3. The Meeting: AGA and QIT 211

Chapter 8. Entanglement of pure states in high dimensions 213
8.1. Entangled subspaces: Qualitative approach 213
8.2. Entropies of entanglement and additivity questions 215
8.2.1. Quantifying entanglement for pure states 215
8.2.2. Channels as subspaces 216
8.2.3. Minimal output entropy and additivity problems 216
8.2.4. On the $1 \to p$ norm of quantum channels 217
8.3. Concentration of E_p for $p > 1$ and applications 218
8.3.1. Counterexamples to the multiplicativity problem 218
8.3.2. Almost randomizing channels 220
8.4. Concentration of von Neumann entropy and applications 222
8.4.1. The basic concentration argument 222
8.4.2. Entangled subspaces of small codimension 224
8.4.3. Extremely entangled subspaces 224
8.4.4. Counterexamples to the additivity problem 228
8.5. Entangled pure states in multipartite systems 229
8.5.1. Geometric measure of entanglement 229
8.5.2. The case of many qubits 230
8.5.3. Multipartite entanglement in real Hilbert spaces 231
Notes and Remarks 232

Chapter 9. Geometry of the set of mixed states 235
9.1. Volume and mean width estimates 236
9.1.1. Symmetrization 236
9.1.2. The set of all quantum states 236
9.1.3. The set of separable states (the bipartite case) 238
9.1.4. The set of block-positive matrices 240
9.1.5. The set of separable states (multipartite case) 242
9.1.6. The set of PPT states 244
9.2. Distance estimates 245
9.2.1. The Gurvits–Barnum theorem 246
9.2.2. Robustness in the bipartite case 247
9.2.3. Distances involving the set of PPT states 248
9.2.4. Distance estimates in the multipartite case 249
9.3. The super-picture: Classes of maps 250
9.4. Approximation by polytopes 252
9.4.1. Approximating the set of all quantum states 252
9.4.2. Approximating the set of separable states 256
9.4.3. Exponentially many entanglement witnesses are necessary 258
Notes and Remarks 260

Chapter 10. Random quantum states 263
10.1. Miscellaneous tools 263
10.1.1. Majorization inequalities 263
10.1.2. Spectra and norms of unitarily invariant random matrices 264
10.1.3. Gaussian approximation to induced states 266
10.1.4. Concentration for gauges of induced states 267
10.2. Separability of random states 268
10.2.1. Almost sure entanglement for low-dimensional environments 268
10.2.2. The threshold theorem 269
10.3. Other thresholds 271
10.3.1. Entanglement of formation 271
10.3.2. Threshold for PPT 272
Notes and Remarks 272

Chapter 11. Bell inequalities and the Grothendieck–Tsirelson inequality 275
11.1. Isometrically Euclidean subspaces via Clifford algebras 275
11.2. Local vs. quantum correlations 276
11.2.1. Correlation matrices 277
11.2.2. Bell correlation inequalities and the Grothendieck constant 280
11.3. Boxes and games 283
11.3.1. Bell inequalities as games 284
11.3.2. Boxes and the nonsignaling principle 285
11.3.3. Bell violations 289
Notes and Remarks 294

Chapter 12. POVMs and the distillability problem 299
12.1. POVMs and zonoids 299
12.1.1. Quantum state discrimination 299
12.1.2. Zonotope associated to a POVM 300
12.1.3. Sparsification of POVMs 300
12.2. The distillability problem 301
12.2.1. State manipulation via LOCC channels 301
12.2.2. Distillable states 302
12.2.3. The case of two qubits 302
12.2.4. Some reformulations of distillability 304
Notes and Remarks 305

Appendix A. Gaussian measures and Gaussian variables 307
A.1. Gaussian random variables 307
A.2. Gaussian vectors 308
Notes and Remarks 309

Appendix B. Classical groups and manifolds 311
B.1. The unit sphere S^{n-1} or $S_{\mathbb{C}^d}$ 311
B.2. The projective space 312
B.3. The orthogonal and unitary groups $\mathsf{O}(n)$, $\mathsf{U}(n)$ 312
B.4. The Grassmann manifolds $\mathsf{Gr}(k, \mathbb{R}^n)$, $\mathsf{Gr}(k, \mathbb{C}^n)$ 314
B.5. The Lorentz group $\mathsf{O}(1, n-1)$ 318
Notes and Remarks 319

Appendix C. Extreme maps between Lorentz cones and the S-lemma 321
Notes and Remarks 324

Appendix D. Polarity and the Santaló point via duality of cones 325

Appendix E. Hints to exercises 329

Appendix F. Notation 375
General notation 375
Convex geometry 375
Linear algebra 376
Probability 377
Geometry and asymptotic geometric analysis 378
Quantum information theory 379

Bibliography 381
Websites 408

Index 409

List of Tables

2.1 Cones of operators and their duals 55

2.2 Cones of superoperators 57

3.1 Spooky action at a distance: outcome distribution for a 2-qubit measurement experiment 75

4.1 Radii, volume radii, and widths for standard convex bodies in $\mathbb{R}^n$ 96

5.1 Covering numbers of classical manifolds 116

5.2 Constants and exponents in subgaussian concentration inequalities 118

5.3 Optimal bounds on Ricci curvature of classical manifolds 131

5.4 log-Sobolev and Poincaré constants for classical manifolds 134

7.1 Derandomization/randomness reduction for Euclidean sections of B_1^n 207

9.1 Radii, volume radii, and widths for sets of quantum states 235

9.2 References for proofs of the results from Table 9.1 236

9.3 Volume estimates for bases of cones of superoperators 251

9.4 Verticial and facial dimensions for sets of quantum states 253

11.1 The magic square game 294

List of Figures

1.1 Gauge of a convex body 12
1.2 A polytope and its polar 17
1.3 A cone and its dual cone 20

2.1 The set of quantum states and the set of separable states 38
2.2 The set of PPT states 44

4.1 Symmetrizations of a convex body 80
4.2 An equilateral triangle in Löwner position 85
4.3 Width and half-width of a convex body 94

5.1 A net and a packing for an equilateral triangle 108
5.2 Upper-bounding the volume of a spherical cap 109
5.3 Volume growth on the sphere S^2 as a function of geodesic distance 130

6.1 Empirical eigenvalue distribution of a GUE matrix 164
6.2 Marčenko–Pastur densities 167

7.1 Low-dimensional illustration of Dvoretzky's theorem 191

11.1 Diagrammatic representation of a quantum game 284

D.1 Changing the center of polarity and duality of cones 326
D.2 The Santaló point via duality of cones 327

E.1 An example of an extreme point which is not exposed. 330
E.2 Schatten unit balls in 2×2 real self-adjoint matrices. 333
E.3 Sharper upper and lower bounds of the volume of spherical caps 345

Credit: Aurélie Garnier

Preface

The quest to build a quantum computer is arguably one of the major scientific and technological challenges of the twenty-first century, and *quantum information theory* (QIT) provides the mathematical framework for that quest. Over the last dozen or so years, it has become clear that quantum information theory is closely linked to geometric functional analysis (Banach space theory, operator spaces, high-dimensional probability), a field also known as *asymptotic geometric analysis* (AGA). In a nutshell, asymptotic geometric analysis investigates quantitative properties of convex sets, or other geometric structures, and their approximate symmetries as the dimension becomes large. This makes it especially relevant to quantum theory, where systems consisting of just a few particles naturally lead to models whose dimension is in the thousands, or even in the billions.

While the idea for this book materialized after we independently taught graduate courses directed primarily at students interested in functional analysis (at the University Lyon 1 and at the University Pierre et Marie Curie-Paris 6 in the spring of 2010), the final product goes well beyond enhanced lecture notes. This book is aimed at multiple audiences connected through their interest in the interface of QIT and AGA: at quantum information researchers who want to learn AGA or apply its tools; at mathematicians interested in learning QIT, or at least the part of QIT that is relevant to functional analysis/convex geometry/random matrix theory and related areas; and at beginning researchers in either field. We have tried to make the book as user-friendly as possible, with numerous tables, explicit estimates, and reasonable constants when possible, so as to make it a useful reference even for established mathematicians generally familiar with the subject.

The first four chapters are of introductory nature. Chapter 0 outlines the basic notation and conventions with emphasis on those that are field-specific to AGA or to physics and may therefore need to be clarified for readers that were educated in other areas. It should be read lightly and used later as a reference. Chapter 1 introduces basic notions from convexity theory that are used throughout the book, notably duality of convex bodies or of convex cones and Schatten norms. Chapter 2 goes over a selection of mathematical concepts and elementary results that are relevant to quantum theory. It is aimed primarily at newcomers to the area, but other readers may find it useful to read it lightly and selectively to familiarize themselves with the "spirit" of the book. Chapter 3 may be helpful to mathematicians with limited background in physics; it shows why various mathematical concepts appear in quantum theory. It could also help in understanding physicists' discussions of the subject and in seeing the motivation behind their enquiries. The choice of topics largely reflects the aspects of the field that we ourselves found not immediately obvious when encountering them for the first time.

Chapters 4 through 7 include the background material from the widely understood AGA that is either already established to be directly or indirectly relevant to QIT, or that we consider to be worthwhile making available to the QIT community. Even though most of this material can be found in existing books or surveys, many items are difficult to locate in the literature and/or are not readily accessible to outsiders. Here we have organized our exposition of AGA so that the applications follow as seamlessly as possible. Our presentation of some aspects of the theory is nonstandard. For example, we exploit the interplay between polarity and cone duality (outlined in Chapter 1 and with a sample application in Appendix D) to give novel and potentially useful insights. Chapters 4 (More convexity) and 5 (Metric entropy and concentration of measure) can be read independently of each other, but Chapters 6 and 7 depend on the preceding ones.

Chapters 8 through 12 discuss topics from the QIT proper, mostly via application of tools from the prior chapters. These chapters can largely be read independently of each other. For the most part, they present results previously published in journal articles, often (but not always) by the authors and their collaborators, most notably Cécilia Lancien, Elisabeth Werner, Deping Ye, Karol Życzkowski, and The Horodecki Group. A few results are byproducts of the work on this book (e.g., those in Section 9.4). This book also contains several new proofs. Some of them could arguably qualify as "proofs from *The Book*," for example the first proof of Størmer's Theorem 2.36 (Section 2.4.5) or the derivation of the sharp upper bound for the expected value of the norm of the complex Wishart matrix (Proposition 6.31).

Some statements are explicitly marked as "not proved here"; in that case the references (to the original source and/or to a more accessible presentation) are indicated in the "Notes and Remarks" section at the end of each chapter. Otherwise, the proof can be found either in the main text or in the exercises. There are over 400 exercises that form an important part of the book. They are diverse and aim at multiple audiences. Some are simple and elementary complements to the text, while others allow the reader to explore more advanced topics at their own pace. Still others explore details of the arguments that we judged to be too technical to include in the main text, but worthwhile to be outlined for those who may need sharp versions of the results and/or to "reverse engineer" the proofs. All but the simplest exercises come with hints, collected in Appendix E. Appendices A to D contain material, generally of reference character, that would disrupt the narrative if included in the main text.

The back matter of the book contains material designed to simplify the task of readers wanting to use the book as a reference: a guide to notation and a keyword index. The bibliography likewise contains back-references displaying page(s) where a given item is cited. For additional information and updates on or corrections to this book, we refer the reader to the associated blog at `https://aliceandbobmeetbanach.wordpress.com`. At the same time, we encourage—or even beg—readers to report typos, errors, improvements, solutions to problems and the like to the blog. (An alternative path to the online post-publication material is by following the link given on the back cover of the book.)

While the initial impulse for the book was a teaching experience, it has not been designed, in its ultimate form, with a specific course or courses in mind. For starters, the quantity of material exceeds by far what can be covered in a single semester. However, a graduate course centered on the main theme of the book—the interface of QIT and AGA—can be easily designed around selected topics from Chapters 4–7, followed by selected applications from Chapters 8–12. While we assume at least a cursory familiarity with functional analysis (normed and inner product spaces and operators on them, duality, Hahn–Banach-type separation theorems, etc.), real analysis (L_p-spaces), and probability, deep results from these fields appear only occasionally and—when they do—an attempt is made to soften the blow by presenting some background via appropriately chosen exercises. Alternatively, most chapters could serve as a core for an independent study course. Again, this would be greatly facilitated by the numerous exercises and—mathematical maturity being more critical than extensive knowledge—the text will be accessible to sufficiently motivated advanced undergraduates.

Acknowledgements. This book has been written over several years. During this period the project benefited greatly from the joint stays of the authors at the Isaac Newton Institute in Cambridge, Mathematisches Forschungsinstitut Oberwolfach (within the framework of its Research in Pairs program), and the Instituto de Ciencias Matemáticas in Madrid. We are grateful to these institutions and their staff for their support and hospitality. We are indebted to the many colleagues and students who helped us bring this book into being, either by reading and commenting on specific chapters, or by sharing with us their expertise and/or providing us with references. We thank in particular Dominique Bakry, Andrew Blasius, Michał Horodecki, Cécilia Lancien, Imre Leader, Ben Li, Harsh Mathur, Mark Meckes, Emanuel Milman, Ion Nechita, David Reeb, and Quanhua Xu. We also thank the anonymous referees for many suggestions which helped to improve the quality of the text. We are especially grateful to Gaëlle Jardine for careful proofreading of parts of the manuscript. We acknowledge Aurélie Garnier, who created the comic strip. Thanks are also due to Sergei Gelfand of the American Mathematical Society's Editorial Division, who guided this project from conception to its conclusion and whose advice and prodding were invaluable. Finally, we would like to thank our families for their support, care, and patience throughout the years.

While working on the book the authors benefited from partial support of the Agence Nationale de la Recherche (France), grants OSQPI (2011-BS01-008-02, GA and SJS) and StoQ (2014-CE25-0003, GA), and of the National Science Foundation (U.S.A.), awards DMS-0801275, DMS-1246497, and DMS-1600124 (all SJS).

Part 1

Alice and Bob

Mathematical Aspects of Quantum Information Theory

CHAPTER 0

Notation and basic concepts

0.1. Asymptotic and nonasymptotic notation

The letters $C, c, c', c_0, \ldots$ denote absolute numerical constants, independent of the instance of the problem at hand. However, the actual values corresponding to the same symbol may change from place to place. Such constants are always assumed to be positive. Usually C or C' stands for a large (but finite) number, while c or c_0 denotes a small (but nonzero) number. If a constant is allowed to depend on a parameter (say n or ε), we use expressions such as C_n or $c(\varepsilon)$.

When A, B are quantities depending on the dimension (and/or perhaps on some other parameters), the notation $A = O(B)$ means that there exists an *absolute* constant $C > 0$ such that the inequality $A \leqslant CB$ holds in every dimension. Similarly, $A = \Omega(B)$ means that $B = O(A)$, and $A = \Theta(B)$ means both $A = O(B)$ and $B = O(A)$. We emphasize that these are *nonasymptotic* relations; they are supposed to hold *universally*, in every instance of the problem, independently of any other parameters that may be involved, and not just in the limit. We also write $A \lesssim B$, $A \gtrsim B$ and $A \simeq B$ as alternative notation for $A = O(B)$, $A = \Omega(B)$ and $A = \Theta(B)$, respectively. However, sometimes we will want to indicate relations that have an asymptotic flavor. For example, $A \sim B$ will mean that $A/B \to 1$ as the dimension tends to ∞ (or as some other relevant parameter tends to its limiting value), and both $A = o(B)$ and $A \ll B$ mean that $A/B \to 0$. If we want to indicate or emphasize that a dependence (of either kind) is not necessarily uniform in some of the parameters, we may write, for example, $c(\alpha)$ or $A = O_\varepsilon(B)$ to identify the parameter(s) on which the relation in question *does* or *may* depend, and similarly for $A \sim_p B$ (asymptotic equivalence for fixed p). Note that if there is only one parameter involved (say, the dimension n), then $A \sim B$ implies $A \simeq B$; however, $A \sim_p B$ does not necessarily entail $A \simeq B$.

0.2. Euclidean and Hilbert spaces

Throughout this book, virtually all the normed spaces we consider will be finite-dimensional (most concepts *do* extend to infinite-dimensional spaces, but we do not dwell on this). In the case of real or complex Hilbert spaces, we denote by $\langle \psi, \chi \rangle$ the inner product of two vectors ψ, χ, and by $|\psi| = \sqrt{\langle \psi, \psi \rangle}$ the corresponding Hilbert space norm. For a complex Hilbert space $\mathcal{H}$, we use the convention that the inner product is conjugate linear in the first argument and linear in the second argument: if $\psi, \chi \in \mathcal{H}$ and $\lambda \in \mathbb{C}$, then

$$\langle \lambda\psi, \chi \rangle = \bar{\lambda} \langle \psi, \chi \rangle \quad \text{and} \quad \langle \psi, \lambda\chi \rangle = \lambda \langle \psi, \chi \rangle.$$

This convention is common in physics literature, but differs from the one usually employed in mathematics.

When $\mathcal{H}, \mathcal{H}'$ are (real or complex) finite-dimensional Hilbert spaces, we denote by $B(\mathcal{H}', \mathcal{H})$ the space of operators (another name for linear maps) from $\mathcal{H}'$ to $\mathcal{H}$, and $B(\mathcal{H}) = B(\mathcal{H}, \mathcal{H})$. The *adjoint* of an operator $A \in B(\mathcal{H}', \mathcal{H})$ is the unique operator $A^\dagger \in B(\mathcal{H}, \mathcal{H}')$ satisfying the property

$$\langle \psi, A\psi' \rangle = \langle A^\dagger \psi, \psi' \rangle \tag{0.1}$$

for any $\psi \in \mathcal{H}, \psi' \in \mathcal{H}'$. We denote by $B^{\mathrm{sa}}(\mathcal{H})$ the space of self-adjoint operators satisfying $A^\dagger = A$; $B^{\mathrm{sa}}(\mathcal{H})$ is a *real* (but not complex) vector subspace of $B(\mathcal{H})$.

The dependence $A \mapsto A^\dagger$ is conjugate linear. A simple but important instance of this operation is when $\mathcal{H}' = \mathbb{C}$: if we identify $\varphi \in \mathcal{H}$ with an operator $z \mapsto z\varphi$ belonging to $B(\mathbb{C}, \mathcal{H})$, then the adjoint of that operator is $\varphi^\dagger = \langle \varphi, \cdot \rangle \in B(\mathcal{H}, \mathbb{C}) = \mathcal{H}^*$.

The notation $B(\cdot, \cdot)$ will be occasionally used for the corresponding concepts in the category of normed (or just vector) spaces. Note that while B stands for "bounded," in the finite-dimensional setting all linear operators are bounded and so—if minimal care is exercised—this will not introduce ambiguity. On the other hand, the notation † will be reserved for operators acting between Hilbert spaces; in other contexts we will use the usual functional analytic notation T^* for the adjoint of a linear map T.

If $\mathcal{H}$ is a complex Hilbert space, we denote by $\overline{\mathcal{H}}$ the Hilbert space which coincides with $\mathcal{H}$ as far as the additive structure is concerned, but with multiplication defined as $(\lambda, x) \mapsto \overline{\lambda} x$. Again, the identity map $\mathcal{H} \ni \psi \mapsto \psi \in \overline{\mathcal{H}}$ is $\mathbb{R}$-linear, but not $\mathbb{C}$-linear. Still, the Hilbert spaces $\mathcal{H}$ and $\overline{\mathcal{H}}$ are isomorphic. Explicit isomorphisms can be constructed as follows: if (e_j) is an orthonormal basis in $\mathcal{H}$ and $\psi = \sum \lambda_j e_j \in \mathcal{H}$, we denote by $\overline{\psi}$ the vector $\sum \overline{\lambda_j} e_j$; then the map $\psi \mapsto \overline{\psi}$ is a Hilbert-space isomorphism between $\mathcal{H}$ and $\overline{\mathcal{H}}$. However, this identification between $\mathcal{H}$ and $\overline{\mathcal{H}}$ is not *canonical* since it depends on the choice of a basis. (In general, a mathematical procedure/construction/morphism is said to be canonical when it depends only on the underlying structure of the object(s) at hand and does not involve any additional arbitrary choices. An identification between two spaces is canonical when there is only one natural candidate for an isomorphism. In the setting of vector spaces, "canonical" is roughly the same as "can be defined in a coordinate-free way.") In our context, it is the dual space $\mathcal{H}^* = B(\mathcal{H}, \mathbb{C})$ which identifies canonically with $\overline{\mathcal{H}} = \overline{B(\mathbb{C}, \mathcal{H})}$ via the map $\mathcal{H}^* \ni \psi^\dagger \leftrightarrow \psi \in \overline{\mathcal{H}}$. This subtlety does not arise in the real case since the map $\psi \mapsto \psi^\dagger$ is $\mathbb{R}$-linear and so the dual space $\mathcal{H}^* = B(\mathcal{H}, \mathbb{R})$ identifies canonically with $\mathcal{H}$.

Here is some more notation: $S_{\mathcal{H}}$ is the sphere of a real or complex Hilbert space $\mathcal{H}$, and $S^{n-1} = S_{\mathbb{R}^n}$. We denote by vol the Lebesgue measure on a finite-dimensional Euclidean space, and occasionally by vol_n the Lebesgue measure on $\mathbb{R}^n$ if we want to emphasize the dimension. If H is a linear or affine subspace, we denote by vol_H the Lebesgue measure on H. We also denote by σ the Lebesgue measure on S^{n-1}, normalized so that $\sigma(S^{n-1}) = 1$ (see Appendix B.1).

0.3. Bra-ket notation

When working with objects related to Hilbert spaces, particularly the complex ones, we use throughout the book Dirac's *bra-ket notation.* It resembles the convention, which may be familiar to some readers and is commonly used, usually in the real setting, in linear programming/optimization. In that convention, $x \in \mathbb{R}^m$

is a *column vector* (an $m \times 1$ matrix, which can also be identified with an operator from $\mathbb{R}$ to $\mathbb{R}^m$); the transposition x^T is a *row vector*, or a linear functional on $\mathbb{R}^m$; xy^T is the outer product of column vectors x and y, while $x^T y$ is their inner (scalar) product, defined if x and y have the same dimension.

The Dirac notation has a very similar structure, the differences being that it is (at least *a priori*) coordinate-free, that the primary operation is † rather than ^{T}, and that the identification of a given object as a vector or as a functional is intrinsic in the notation. "Standard" vectors in $\mathcal{H}$ are written as $|\psi\rangle$ (a *ket* vector). The same vector, but thought of as an element of $\mathcal{H}^* \leftrightarrow \overline{\mathcal{H}}$, is identified with $|\psi\rangle^\dagger$ and written as $\langle\psi|$ (a *bra* vector). The bra-ket notation works seamlessly with standard operations on Hilbert spaces. The action of a functional $\langle\psi|$ on a vector $|\chi\rangle$ is $\langle\psi|\chi\rangle$, an alternative notation for the scalar product $\langle\psi, \chi\rangle$. If $A \in B(\mathcal{H})$ and $\psi \in \mathcal{H}$, then we have $A|\psi\rangle = |A\psi\rangle$ and $\langle A\psi| = (A|\psi\rangle)^\dagger = \langle\psi|A^\dagger$. Consequently, the quantity $\langle\psi'|A|\psi\rangle$ can be read as $\langle\psi', A\psi\rangle$ or as $\langle A^\dagger\psi', \psi\rangle$, the equality of which is a restatement of the definition (0.1).

Let $\mathcal{H}_1, \mathcal{H}_2$ be real or complex Hilbert spaces, and let ψ_1, ψ_2 be vectors in $\mathcal{H}_1, \mathcal{H}_2$ respectively. Then the operator $|\psi_1\rangle\langle\psi_2| : \mathcal{H}_2 \to \mathcal{H}_1$ acts on $\chi \in \mathcal{H}_2$ as follows:

$$|\chi\rangle \mapsto |\psi_1\rangle\langle\psi_2|\chi\rangle = \langle\psi_2|\chi\rangle|\psi_1\rangle$$

or, in the standard notation, $\chi \mapsto \langle\psi_2, \chi\rangle\psi_1$. This operator has rank one unless one of the vectors ψ_1, ψ_2 is zero.

In some mathematical circles, the operator $|\psi_1\rangle\langle\psi_2|$ is sometimes denoted $\psi_1 \otimes \overline{\psi_2}$ or $\overline{\psi_2} \otimes \psi_1$, or even $\psi_1 \otimes \psi_2$. However, such notation is inconvenient and often ambiguous, and it becomes unmanageable when the Hilbert spaces, in which ψ_1 and ψ_2 live, are themselves equipped with a tensor product structure.

When $E \subset \mathcal{H}$ is a linear subspace, we denote by P_E the orthogonal projection onto E. When E is 1-dimensional, we have $P_E = |x\rangle\langle x|$ for any unit vector $x \in E$.

We denote the standard basis of $\mathbb{C}^d$ by $(|1\rangle, \ldots, |d\rangle)$. (Note that while $(|j\rangle)$ is just one of many orthonormal bases of $\mathbb{C}^d$, it becomes canonical if we take into account the lattice structure.) However, sometimes we will employ the enumeration $(|0\rangle, |1\rangle, \ldots, |d-1\rangle)$, particularly for $d = 2$, where we will follow the traditional convention from computer science and use $(|0\rangle, |1\rangle)$. Either way, we will refer to this basis as the *computational basis*. (As explained in Section 3.1, the designation "computational basis" may have an operational meaning, but such subtleties will be normally beyond the scope of our analysis.) Nevertheless, in some cases, particularly in the real context, we will use the notation $e_1, e_2, \ldots, e_d$ that is more common in the mathematical literature.

EXERCISE 0.1. Check the following properties, where $\psi_1, \chi_1 \in \mathcal{H}_1$, $\psi_2, \chi_2 \in \mathcal{H}_2$, $\chi_3 \in \mathcal{H}_3$, and $A \in B(\mathcal{H}_1, \mathcal{H}_2)$.
(i) Product/composition: $|\psi_1\rangle\langle\psi_2| \circ |\chi_2\rangle\langle\chi_3| = \langle\psi_2, \chi_2\rangle|\psi_1\rangle\langle\chi_3|$.
(ii) Adjoint: $(|\psi_1\rangle\langle\psi_2|)^\dagger = |\psi_2\rangle\langle\psi_1|$.
(iii) Trace: $\operatorname{Tr}|\psi_1\rangle\langle\chi_1| = \langle\chi_1, \psi_1\rangle$, $\operatorname{Tr}\left(A|\psi_1\rangle\langle\psi_2|\right) = \langle\psi_2|A|\psi_1\rangle$.

0.4. Tensor products

Whenever $(\mathcal{H}_i)_{1\leqslant i\leqslant k}$ are real or complex finite-dimensional Hilbert spaces, we consider the tensor product over the real or complex field, respectively,

$$\mathcal{H} = \bigotimes_{i=1}^{k} \mathcal{H}_i = \mathcal{H}_1 \otimes \mathcal{H}_2 \otimes \cdots \otimes \mathcal{H}_k, \tag{0.2}$$

which is often called a *multipartite* Hilbert space (or *bipartite* when $k = 2$). The space $\mathcal{H}$ carries a natural Hilbert space structure given by the inner product defined for product vectors by

$$\langle \psi_1 \otimes \cdots \otimes \psi_k, \chi_1 \otimes \cdots \otimes \chi_k \rangle = \prod_{i=1}^{k} \langle \psi_i, \chi_i \rangle$$

and extended to $\mathcal{H}$ by multilinearity. There are canonical identifications

$$B\left(\bigotimes_{i=1}^{k} \mathcal{H}_i\right) \longleftrightarrow \bigotimes_{i=1}^{k} B(\mathcal{H}_i),$$

where the tensor products are over the real or complex field, respectively. *In the complex case only*, another canonical identification is

$$B^{\mathrm{sa}}\left(\bigotimes_{i=1}^{k} \mathcal{H}_i\right) \longleftrightarrow \bigotimes_{i=1}^{k} B^{\mathrm{sa}}(\mathcal{H}_i), \tag{0.3}$$

where the tensor products are over the complex field on the left-hand side and over the real field on the right-hand side. Except in the trivial cases, the analogue of (0.3) is false in the setting of real Hilbert spaces: e.g., $B^{\mathrm{sa}}(\mathbb{R}^2) \otimes B^{\mathrm{sa}}(\mathbb{R}^2)$ is a proper subspace of $B^{\mathrm{sa}}(\mathbb{R}^2 \otimes \mathbb{R}^2)$, which can be easily seen by comparing the dimensions.

While it is occasionally computationally convenient to allow some of the factors in (0.2) to be 1-dimensional, such factors may be just dropped and so, when referring to a multipartite Hilbert space, we will normally assume that all the factors are of dimension at least 2.

We often work with concrete spaces such as $(\mathbb{C}^2)^{\otimes k}$, which corresponds to k *qubits*. In that case the computational basis is obtained by the 2^k vectors of the form $|i_1\rangle \otimes \cdots \otimes |i_k\rangle$, where $(i_1, \ldots, i_k) \in \{0,1\}^k$. It is customary to drop the tensor product sign: for example the computational basis of $\mathbb{C}^2 \otimes \mathbb{C}^2$ consists of the four vectors $|00\rangle, |01\rangle, |10\rangle, |11\rangle$.

We also point out that tensor products commute with the operation of taking dual, i.e., there is a canonical identification

$$(\mathcal{H}_1 \otimes \mathcal{H}_2)^* \leftrightarrow \mathcal{H}_1^* \otimes \mathcal{H}_2^*.$$

EXERCISE 0.2. Let $\mathcal{H}_1, \mathcal{H}_2$ be *complex* Hilbert spaces, and consider vectors $x_1, y_1 \in \mathcal{H}_1$ and $x_2, y_2 \in \mathcal{H}_2$. Write explicitly the operator $|x_1 \otimes x_2 + y_1 \otimes y_2\rangle\langle x_1 \otimes x_2 + y_1 \otimes y_2| \in B^{\mathrm{sa}}(\mathcal{H}_1 \otimes \mathcal{H}_2)$ as a linear combination of operators of the form $|z\rangle\langle z| \otimes |z'\rangle\langle z'|$, with $z \in \mathcal{H}_1$ and $z' \in \mathcal{H}_2$.

0.5. Complexification

Let V be a real vector space. The *complexification* of V is the vector space $V^{\mathbb{C}} = V \otimes \mathbb{C}$ (the tensor product is over the reals). Elements of $V^{\mathbb{C}}$ are of the form $x \otimes 1 + y \otimes i$ (for $x, y \in V$), which we write $x + iy$ for short.

Note that the complexification of $B^{\mathrm{sa}}(\mathbb{C}^n)$ is canonically isomorphic to $B(\mathbb{C}^n)$. Note also that for real spaces V, W, $(V \otimes_{\mathbb{R}} W)^{\mathbb{C}}$ and $V^{\mathbb{C}} \otimes_{\mathbb{C}} W^{\mathbb{C}}$ are canonically isomorphic.

Similarly, if $f : V \to W$ is a linear map between real vector spaces, the map $x + iy \mapsto f(x) + if(y)$ defines canonically a $\mathbb{C}$-linear map (the complexification of f) from $V^{\mathbb{C}}$ to $W^{\mathbb{C}}$.

An operation that goes in the opposite direction to complexification is that of *dropping the complex structure*, i.e., considering a complex space as a real space, so that for example $\mathbb{C}^n$ is treated as $\mathbb{R}^{2n}$. In the abstract setting, if the original complex space was endowed with a scalar product $\langle \cdot, \cdot \rangle$, the corresponding *real* scalar product is $\mathrm{Re}\langle \cdot, \cdot \rangle$. While this is frequently a useful point of view, particularly in geometric considerations (see Section 1.1), some caution is needed as this operation is not as sound functorially as complexification. For example, $\mathbb{C} \otimes_{\mathbb{C}} \mathbb{C} = \mathbb{C}$ identifies this way with $\mathbb{R}^2$, even though $\mathbb{R}^2 \otimes_{\mathbb{R}} \mathbb{R}^2$ is 4-dimensional.

0.6. Matrices vs. operators

We denote by $\mathsf{M}_{m,n}$ the space of $m \times n$ matrices, either real or complex, and by M_n if $m = n$. The entries of a matrix $M \in \mathsf{M}_{m,n}$ are denoted by $(m_{ij})_{1 \leqslant i \leqslant m, 1 \leqslant j \leqslant n}$. We denote by $M^\dagger$ the *Hermitian conjugate* of M, i.e., $(m_{ij})^\dagger = (\overline{m_{ji}})$. We will denote by $\mathsf{M}_m^{\mathrm{sa}} := \{M \in \mathsf{M}_m : M = M^\dagger\}$, the subspace of M_m consisting of *Hermitian* (or *self-adjoint*) matrices. For matrices with real entries, "self-adjoint" simply means "symmetric".

As a default, we identify complex $m \times n$ matrices with operators from $\mathbb{C}^n$ to $\mathbb{C}^m$ and write $\mathsf{M}_{m,n} = B(\mathbb{C}^n, \mathbb{C}^m)$, and similarly $\mathsf{M}_n = B(\mathbb{C}^n)$, $\mathsf{M}_n^{\mathrm{sa}} = B^{\mathrm{sa}}(\mathbb{C}^n)$. The preceding definitions ensure that the above notion of † is consistent with that introduced in Section 0.2, and that the operator composition is consistent with matrix multiplication. Again, this is fully parallel to the conventions in linear analysis/optimization in the real setting.

More generally, $\mathsf{M}_{m,n}$ and M_n can (and often will) be identified with operators on/between *any* Hilbert spaces of the appropriate dimensions. However, such identification requires specifying bases in the spaces in question and, consequently, is not canonical.

In the real case, M_n is a vector space of dimension n^2, and $\mathsf{M}_n^{\mathrm{sa}}$ is a subspace of dimension $n(n+1)/2$. In the complex case, M_n is a complex vector space of complex dimension n^2, while $\mathsf{M}_n^{\mathrm{sa}}$ is a *real* vector space of real dimension n^2.

A natural inner product on $\mathsf{M}_{m,n}$ is given by the trace duality: if $M, N \in \mathsf{M}_{m,n}$, then

$$\langle M, N \rangle = \operatorname{Tr} M^\dagger N. \tag{0.4}$$

(Recall that we use the "physics" convention for sesquilinear forms, as explained in Section 0.3.) The Euclidean structure on $\mathsf{M}_{m,n}$ induced by this inner product is called the *Hilbert–Schmidt Euclidean structure*, and the corresponding norm is the *Hilbert–Schmidt norm* $\|M\|_{\mathrm{HS}} = \sqrt{\operatorname{Tr} M^\dagger M}$. (In linear algebra the more commonly used name is *Frobenius*.) Note that in the complex case the inner product will, in general, not be real. However, if $M, N \in \mathsf{M}_m^{\mathrm{sa}}$, then $\langle M, N \rangle = \operatorname{Tr} MN$ is real (even if some of the entries of M, N are complex).

0.7. Block matrices vs. operators on bipartite spaces

It is convenient to identify operators on $\mathbb{C}^m \otimes \mathbb{C}^n$ with elements of M_{mn} having a block structure. More precisely, to each operator $A \in B(\mathbb{C}^m \otimes \mathbb{C}^n)$ there corresponds the block matrix

$$M = \begin{bmatrix} M_{11} & \cdots & M_{1m} \\ \vdots & & \vdots \\ M_{m1} & \cdots & M_{mm} \end{bmatrix} \tag{0.5}$$

where, for each $i, j \in \{1, \ldots, m\}$, the matrix $M_{ij} \in \mathsf{M}_n$ is defined as

$$M_{ij} = \begin{bmatrix} (\langle i| \otimes \langle 1|)A(|j\rangle \otimes |1\rangle) & \cdots & (\langle i| \otimes \langle 1|)A(|j\rangle \otimes |n\rangle) \\ \vdots & & \vdots \\ (\langle i| \otimes \langle n|)A(|j\rangle \otimes |1\rangle) & \cdots & (\langle i| \otimes \langle n|)A(|j\rangle \otimes |n\rangle) \end{bmatrix}. \tag{0.6}$$

0.8. Operators vs. tensors

Let $\mathcal{H}_1, \mathcal{H}_2$ be complex Hilbert spaces. The map $u \otimes v \mapsto |v\rangle\langle u|$ induces a canonical identification between the spaces $\overline{\mathcal{H}_1} \otimes \mathcal{H}_2$ and $B(\mathcal{H}_1, \mathcal{H}_2)$. Recall from Section 0.2 that $\overline{\mathcal{H}_1}$ identifies canonically with $\mathcal{H}_1^*$.

As explained in Section 0.2, the use of the complex conjugacy can be avoided if we agree to work with specified bases. Fix bases $(e_i)_{i \in I}$ in $\mathcal{H}_1$ and $(f_j)_{j \in J}$ in $\mathcal{H}_2$. Define a map $\mathrm{vec} : B(\mathcal{H}_1, \mathcal{H}_2) \to \mathcal{H}_2 \otimes \mathcal{H}_1$ as follows: for $i \in I$ and $j \in J$, set $\mathrm{vec}(|f_j\rangle\langle e_i|) = f_j \otimes e_i$ and extend the definition by $\mathbb{C}$-linearity. In other words, for $\psi_1 \in \mathcal{H}_1, \psi_2 \in \mathcal{H}_2$ we have $\mathrm{vec}\,|\psi_2\rangle\langle\psi_1| = \psi_2 \otimes \overline{\psi_1}$ where conjugacy is taken with respect to the basis (e_i).

0.9. Operators vs. superoperators

It is convenient to use the term *superoperator* to denote maps acting between spaces of operators, or between spaces of matrices. The distinction between operators and superoperators may seem rather arbitrary since, as we noted earlier, $B(\mathcal{H})$ and $\mathsf{M}_{m,n}$ carry a natural Hilbert space structure. However, it helps to organize one's thinking and is widely used in quantum information theory.

Accordingly, we use two different types of notation to denote the identity map: the identity operator on a Hilbert space $\mathcal{H}$ is denoted by $\mathrm{I}_{\mathcal{H}}$ (or I_n if $\mathcal{H} = \mathbb{C}^n$ or $\mathbb{R}^n$, or even simply I if there is no ambiguity), while the identity superoperator on $B(\mathcal{H})$ is denoted by $\mathrm{Id}_{B(\mathcal{H})}$ (or simply Id).

0.10. States, classical and quantum

The concept that plays a central role throughout this book is that of a quantum state.

We start by introducing the classical analogue: given a finite set S, a *classical state* on S is simply a probability measure on S (or, equivalently, a probability mass function indexed by $s \in S$). We denote by Δ_n the set of classical states on $\{0, 1, \ldots, n\}$. Geometrically, Δ_n is an n-dimensional simplex; we shall return to this circle of ideas in Chapter 1.

Let $\mathcal{H}$ be a complex finite-dimensional Hilbert space. A *quantum state* (or simply a state) on $\mathcal{H}$ is a positive self-adjoint operator of trace one. We denote by

$\mathrm{D}(\mathcal{H})$ the set of states on $\mathcal{H}$ (the letter D stands for *density matrix*, which is an alternative terminology for states). If $\mathcal{H} = \mathbb{C}^{n+1}$, the subset of $\mathrm{D}(\mathcal{H})$ consisting of diagonal operators identifies naturally with Δ_n (and similarly for operators diagonal with respect to any *fixed* basis in any finite-dimensional Hilbert space).

In functional analysis, a state on a C^*-algebra is—by definition—a positive linear functional of norm 1. This is consistent with the definitions of classical and quantum states introduced above. Indeed, given a finite set S, states on the commutative C^*-algebra $\mathbb{C}^S$ correspond to classical states on S. Similarly, given a finite-dimensional complex Hilbert space $\mathcal{H}$, the states on the C^*-algebra $B(\mathcal{H})$ can be identified with elements of $\mathrm{D}(\mathcal{H})$ via trace duality (0.4).

CHAPTER 1

Elementary convex analysis

In this chapter we present an overview of basic properties of convex sets and convex cones. Unless stated explicitly otherwise, we shall assume that the base field is $\mathbb{R}$ and that all the objects involved are finite-dimensional. However, notions for complex spaces will be important and even indispensable in some settings. They are typically introduced by repeating *mutatis mutandis* the definitions of their real counterparts. At the same time, one can always consider them as real spaces by ignoring the complex structure.

If V is an n-dimensional vector space over $\mathbb{R}$, we will usually assume that V is identified with $\mathbb{R}^n$. This implies in particular that there is a distinguished *Euclidean structure* (i.e., a scalar product) in V, so that V is also identified with its dual V^*.

1.1. Normed spaces and convex sets

1.1.1. Gauges. We start with a simple proposition which characterizes the subsets of $\mathbb{R}^n$ that can be the unit balls for some norm. A subset $K \subset \mathbb{R}^n$ is a *convex body* if it is convex, compact, and with non-empty interior. We similarly define convex bodies in linear (or affine) subspaces of $\mathbb{R}^n$. We will call K *symmetric* (or 0-symmetric if there is an ambiguity) if it is centrally symmetric with respect to the origin, i.e., $K = -K$.

PROPOSITION 1.1 (Easy). *Let K be a subset of $\mathbb{R}^n$. The following are equivalent*

(1) K is a symmetric convex body.

(2) There is a norm on $\mathbb{R}^n$ for which K is the unit ball.

Given K, the corresponding norm can be retrieved by considering the *gauge* of K, also called the *Minkowski functional* of K, which is defined for $x \in \mathbb{R}^n$ by

$$\|x\|_K := \inf\{t \geqslant 0 \ : \ x \in tK\}, \tag{1.1}$$

where $tK = \{tx \ : \ x \in K\}$ (see Figure 1.1). If X is a normed space (most often, $X = (\mathbb{R}^n, \|\cdot\|)$), we will denote its unit ball $\{x : \|x\| \leqslant 1\}$ by B_X. (However, to lighten the notation, we will use specialized symbols for various "common" spaces.) The correspondence $X \mapsto B_X$ is the inverse of the correspondence $K \mapsto \|\cdot\|_K$.

In the complex case, the analogue of symmetry is circledness. A convex body $K \subset \mathbb{C}^n$ is said to be *circled* if for every $\theta \in \mathbb{R}$ and $x \in K$ we have $e^{i\theta}x \in K$. Circled convex bodies are exactly the unit balls of norms in $\mathbb{C}^n$.

Equation (1.1) will also be used to define the gauge of a non-necessarily symmetric convex set K. However, in order for the gauge to take only finite values and to avoid other degeneracies, we will usually insist that K contain the origin in its interior and that K be closed. We will still denote by $\|\cdot\|_K$ the gauge of such a convex set, and we will still have the (essentially tautological) relation

$$K = \{x \ : \ \|x\|_K \leqslant 1\}. \tag{1.2}$$

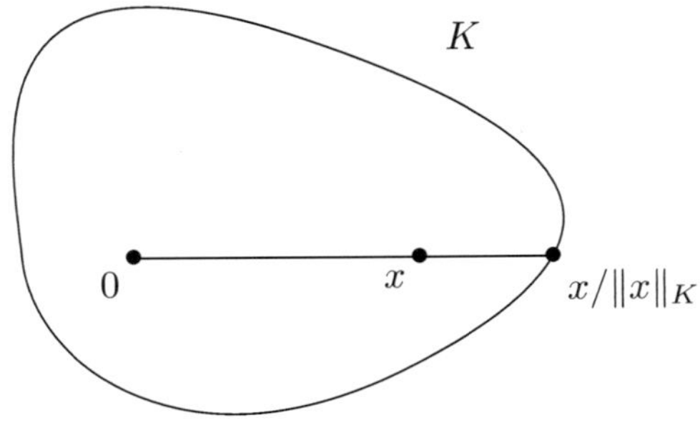

FIGURE 1.1. Gauge of a convex body.

(Observe that if K is closed, the infimum in (1.1) is always attained.) However, if K is not assumed to be symmetric, we should note that in general $\|x\|_K \neq \|-x\|_K$.

We point out that the correspondence between convex bodies and their gauges is order-reversing: $K \subset L$ if and only if $\|\cdot\|_K \geqslant \|\cdot\|_L$. In the same vein, we have $\|\cdot\|_{tK} = t^{-1}\|\cdot\|_K$ for $t > 0$.

1.1.2. First examples: ℓ_p-balls, simplices, polytopes, and convex hulls. For $1 \leqslant p \leqslant +\infty$, we denote by $\|\cdot\|_p$ the ℓ_p-*norm*, defined for $x \in \mathbb{R}^n$ via

$$\|x\|_p = \left(\sum_{k=1}^n |x_k|^p\right)^{1/p}, \tag{1.3}$$

where the limit case $p = +\infty$ should be understood as $\|x\|_\infty = \max\{|x_k| : 1 \leqslant k \leqslant n\}$. Recall also that $\|\cdot\|_2$ will be usually denoted by $|\cdot|$. The ℓ_p-norms satisfy the following inequalities: if $1 \leqslant p \leqslant q \leqslant \infty$ and $x \in \mathbb{R}^n$, then

$$\|x\|_q \leqslant \|x\|_p \leqslant n^{1/p-1/q}\|x\|_p. \tag{1.4}$$

The normed space $(\mathbb{R}^n, \|\cdot\|_p)$ is denoted by ℓ_p^n and its unit ball by B_p^n.

If $A \subset \mathbb{R}^n$, we denote by $\operatorname{conv} A$ the *convex hull* of A, i.e., the set of all convex combinations of elements of A, which is also the smallest convex set containing A. The following theorem bounds the length of convex combinations needed to generate the convex hull.

THEOREM 1.2 (Carathéodory's theorem; see Exercise 1.1). *Let $A \subset \mathbb{R}^n$. Then $\operatorname{conv}(A)$ is the set of all convex combinations of at most $n+1$ elements of A. The same assertion holds if $A \subset H$, where H is an n-dimensional affine subspace of $\mathbb{R}^m$ for some $m > n$.*

A convex body is a *polytope* if it is the convex hull of finitely many points. The simplest polytope is the *simplex*, which is the convex hull of $n+1$ affinely independent points in $\mathbb{R}^n$. This is the prototypical example of a non-symmetric convex body (for $n \geqslant 2$). Note that Carathéodory's theorem implies that when $K = \operatorname{conv} A$, then K is the union of all simplices with vertices in A (the dimension of each simplex being equal to $\dim K$).

A simplex is *regular* if all the pairwise distances between the $n+1$ vertices are equal. A convenient representation of a regular simplex is as follows: consider the affine hyperplane $H \subset \mathbb{R}^{n+1}$ formed by all vectors whose coordinates add up to 1, and denote by Δ_n the convex hull of the vectors from the canonical basis in

$\mathbb{R}^{n+1}$. Note that Δ_n is a convex body in H, but only a convex subset of $\mathbb{R}^{n+1}$. The simplex Δ_n corresponds to the set of *classical* states, i.e., probability measures on $\{0, \ldots, n\}$.

EXERCISE 1.1 (Carathéodory's theorem). Let $A \subset \mathbb{R}^n$, $x \in \operatorname{conv} A$ and consider a decomposition $x = \sum_{i=1}^N \lambda_i x_i$ (where (λ_i) is a convex combination and $x_i \in A$) of minimal length N. Show that the points (x_i) must be affinely independent, and conclude that $N \leqslant n + 1$.

EXERCISE 1.2. Let $A \subset \mathbb{R}^n$ be a compact set. Show that $\operatorname{conv} A$ is compact.

1.1.3. Extreme points, faces. Let $K \subset \mathbb{R}^n$ be a convex set. A point $x \in K$ is said to be *extreme* if it cannot be written in a nontrivial way as a convex combination of points of K, i.e., if the equality $x = ty + (1-t)z$ for $t \in (0, 1)$ and $y, z \in K$ implies that $x = y = z$. The following fundamental theorem asserts that, in a sense, all information about a convex body is contained in its extreme points.

THEOREM 1.3 (Krein–Milman theorem; see Exercise 1.6). *Let $K \subset \mathbb{R}^n$ be a convex body. Then K is the convex hull of its extreme points.*

Let F, K be closed convex sets with $F \subset K$. Then F is called a *face* of K if every segment contained in K whose (relative) interior intersects F is entirely contained in F. If $F \neq \varnothing$ and $F \neq K$, F is said to be a *proper face*. Note that a singleton $\{x\}$ is a face if and only if x is an extreme point. If F is a face of K with $\dim F = \dim K - 1$, then F is called a *facet*.

A frequently encountered setting in convex or functional analysis is that of two convex sets K, L and a linear or affine map u such that $u(L) \subset K$. For example, if X, Y are normed spaces, and $u : X \to Y$ a linear operator, then u is a contraction iff $u(B_X) \subset B_Y$. The following elementary observation makes it possible to use the facial structure of the sets in question to study these kinds of situations.

PROPOSITION 1.4 (Affine maps preserve faces; see Exercise 1.4). *Let K, L be closed convex sets, let x be a point in the relative interior of L, and let $u : L \to K$ be an affine map. If F is a face of K such that $u(x) \in F$, then $u(L) \subset F$.*

Finally, we introduce some more vocabulary. Let $K \subset \mathbb{R}^n$ be a closed convex set. An affine hyperplane $H \subset \mathbb{R}^n$ is said to be a *supporting hyperplane* for K if $H \cap \partial K \neq \varnothing$ and K is entirely contained in one of the closed half-spaces delimited by H. Note that for any $x \in \partial K$, there is at least one supporting hyperplane for K which contains x. A proper subset $F \subset K$ is an *exposed face* if it is the intersection of K with a supporting hyperplane. We say then that H *isolates* F (as a face of K). Similarly, a point $x \in K$ is an *exposed point* if $\{x\}$ is an exposed face, i.e., if there exists a vector $y \in \mathbb{R}^n$ such that the linear functional $\langle y, \cdot \rangle$ attains its maximum on K only at x. These notions are studied in Exercise 1.5.

EXERCISE 1.3. Show that the (relative) boundary of a closed convex set is a union of exposed faces.

EXERCISE 1.4. Prove Proposition 1.4.

EXERCISE 1.5 (Extreme vs. exposed points, faces vs. exposed faces). Let $K \subset \mathbb{R}^n$ be a closed convex set.

(a) Show that every exposed face F of a closed convex set K is indeed a face of K, which is necessarily proper (i.e., $F \neq K, \varnothing$).

(b) Show that the relation "F is a face of G" is transitive.
(c) Show that every maximal proper face of a closed convex set K is exposed. Deduce that every facet of K (i.e., a face of dimension $\dim K - 1$) is exposed.
(d) By (a), any exposed point is extreme. Give an example of a convex body $K \subset \mathbb{R}^2$ with an extreme point which is not exposed. (However, a theorem by Straszewicz states that any extreme point is a limit of exposed points; see Theorem 18.6 in [**Roc70**].) Deduce that the relation "F is an exposed face of G" is not transitive.
(e) More generally, for $k \leqslant n-2$, give an example of a convex body $L \subset \mathbb{R}^n$ with a k-dimensional face which is not exposed.
(f) Show that F is a face of K if and only if there exists a sequence $F = F_0 \subset F_1 \subset \cdots \subset F_s = K$ such that F_{i-1} is an exposed face of F_i for $i = 1, \ldots, s$.
(g) If every point in the (relative) boundary of a convex set K is extreme, K is called *strictly convex*. Show that, in that case, every point of the boundary is an exposed point.

EXERCISE 1.6. Prove the Krein–Milman Theorem 1.3 by induction with respect to n. (Start by showing that any convex body has at least one extreme point.)

EXERCISE 1.7. Show that the extreme points of the set of quantum states $\mathrm{D}(\mathcal{H})$ are operators of the form $|\psi\rangle\langle\psi|$, where $\psi \in \mathcal{H}$ is a norm one vector (i.e., rank one orthogonal projections).

EXERCISE 1.8. Show that every face of a polytope is a polytope.

EXERCISE 1.9. Show that every proper face of a polytope is exposed.

EXERCISE 1.10. Find the extreme points of B_p^n for $1 \leqslant p \leqslant \infty$.

EXERCISE 1.11 (Hanner's inequalities and uniform convexity). The goal of this exercise is to prove Hanner's inequalities about the geometry of the p-norm, which lead to precise quantitative statements about convexity and smoothness of balls in L_p-spaces.
(i) Let $p \in (1,2]$. For $t > 0$, set $\alpha(t) = (1+t)^{p-1} + |1-t|^{p-1}\mathrm{sign}(1-t)$. Show that for $a, b \in \mathbb{R}$, we have $|a+b|^p + |a-b|^p = \sup\{\alpha(t)|a|^p + \alpha(1/t)|b|^p \ : \ t > 0\}$.
(ii) Let $p \in (1,2]$. Show that for $x, y \in \mathbb{R}^n$

$$\|x+y\|_p^p + \|x-y\|_p^p \geqslant (\|x\|_p + \|y\|_p)^p + \big|\|x\|_p - \|y\|_p\big|^p . \tag{1.5}$$

Show also that, for $p \in [2,\infty)$, (1.5) holds with $\leqslant$ instead of $\geqslant$.
(iii) Let $p \in (1,2]$. Prove also that for $x, y \in \mathbb{R}$.

$$\left(\frac{\|x+y\|_p^p + \|x-y\|_p^p}{2}\right)^{2/p} \geqslant \|x\|_p^2 + (p-1)\|y\|_p^2. \tag{1.6}$$

(iv) Fix $p \in (1,\infty)$. Show that for any $\varepsilon > 0$ there exists $\delta > 0$ such that whenever $x, y \in B_p^n$ verify $\|x-y\|_p \geqslant \varepsilon$, then $\left\|\frac{x+y}{2}\right\|_p \leqslant 1-\delta$. (This property of B_p^n is a quantitative version of strict convexity and is called *uniform convexity*.)

EXERCISE 1.12 (A Borel selection theorem). Let $K \subset \mathbb{R}^n$ be a convex body. Show that there is a Borel map $\Theta : \mathbb{R}^n \to K$ with the property that for every $x \in \mathbb{R}^n$ we have $\langle\Theta(x), x\rangle = \max\{\langle z, x\rangle \ : \ z \in K\}$.

1.1.4. Polarity. This section and the next one will present elements of convex analysis. Readers not familiar with the subject are encouraged to go over the suggested exercises, which are generally simple and elementary, but often contain facts not included in standard texts.

Since norms on $\mathbb{R}^n$ are in one-to-one correspondence with symmetric convex bodies, the notion of duality between normed spaces induces a duality for convex bodies, which is called *polarity*. Its explicit definition is as follows: if $A \subset \mathbb{R}^n$, the *polar* of A is

$$A^\circ := \{y \in \mathbb{R}^n \ : \ \langle x, y\rangle \leqslant 1 \ \text{ for all } \ x \in A\}. \tag{1.7}$$

In particular (cf. (1.2) and Exercise 1.13)

$$\|y\|_{A^\circ} = \sup_{x \in A \cup \{0\}} \langle x, y\rangle. \tag{1.8}$$

The key example is $A = B_X$ (the unit ball of X); we have then $A^\circ = B_{X^*}$, the unit ball with respect to the dual norm, the duality being induced by the standard Euclidean structure. For example, duality of ℓ_p-norms translates into

$$(B_p^n)^\circ = B_q^n, \tag{1.9}$$

where $1/p + 1/q = 1$.

A larger important class of sets is that of convex bodies containing 0 in the interior; it is stable under the operation of polarity. While most of the properties of the operation $K \mapsto K^\circ$ listed below hold for more general sets, this last class is sufficient for most applications (with the notable exception of *cones*, see Section 1.2).

Because of the inequality appearing in the definition (1.7), the concept of polarity *a priori* makes sense only in the category of *real* Euclidean spaces. We exemplify adjustments needed to make it work in the complex setting in Section 1.3.2, where that setting is at times indispensable.

Since the notion of polarity appeals to the Euclidean structure on $\mathbb{R}^n$, it is not immediately canonical in the category of vector spaces. Equivalently, it depends on how we identify the vector space $\mathbb{R}^n$ with its dual. One useful way to describe this dependence is as follows: if $u \in \mathsf{GL}(n, \mathbb{R})$, then

$$(uA)^\circ = (u^T)^{-1}(A^\circ). \tag{1.10}$$

(The dependence of polarity on translation is somewhat less transparent; one promising approach to its description is explored in Appendix D.) A way to make polarity canonical is to consider the polar K° as a subset of V^*, the dual of the ambient space V containing K. Basically, all the formulas remain the same, except that if $x \in V$ and $x^* \in V^*$, then $\langle x^*, x\rangle$ needs to be understood as $x^*(x)$. This approach is occasionally useful, but is normally avoided since it requires considering twice as many spaces as the other approach.

A fundamental result from convex analysis is that if K is closed, convex, and contains the origin, then

$$(K^\circ)^\circ = K \tag{1.11}$$

(see also Exercise 1.15). This is the *bipolar theorem*, a baby version of the Hahn–Banach theorem. When K is a symmetric convex body, this is just saying that a finite-dimensional normed space is reflexive (i.e., canonically isomorphic to its double dual, see [**Fol99**]).

At the functional-analytic level, the duality exchanges the operations of taking a subspace and taking a quotient. Geometrically, this translates into the fact that polarity exchanges the projection and the section operations. Here is a more precise statement: *if $K \subset \mathbb{R}^n$, then, for every linear subspace $E \subset \mathbb{R}^n$,*

$$(P_E K)^\circ = E \cap K^\circ, \tag{1.12}$$

where P_E denotes the orthogonal projection onto E. Moreover, if K is a convex set containing 0 in the interior, then

$$(K \cap E)^\circ = P_E(K^\circ). \tag{1.13}$$

Note that in the left-hand sides in (1.12) and (1.13), the polars are taken *inside* E, equipped with the induced inner product.

Another pair of simple but useful relations involving polars is

$$(K \cup L)^\circ = K^\circ \cap L^\circ \tag{1.14}$$

for any $K, L \subset \mathbb{R}^n$ and

$$(K \cap L)^\circ = \overline{\operatorname{conv}(K^\circ \cup L^\circ)} \tag{1.15}$$

if K, L are closed, convex and contain the origin.

EXERCISE 1.13. Find a gap in the following argument. *Since $\|y\|_{A^\circ} \leqslant 1$ iff $y \in A^\circ$ iff $\sup_{x\in A}\langle x, y\rangle \leqslant 1$, it follows by homogeneity that $\|y\|_{A^\circ} = \sup_{x\in A}\langle x, y\rangle$.*

EXERCISE 1.14 (Stability properties of polarity). Show that $K \subset \mathbb{R}^n$ is bounded iff K° contains 0 in the interior. Similarly, if K is convex, then it contains 0 in its interior iff K° is bounded.

EXERCISE 1.15 (The general bipolar theorem). Show that if $K \subset \mathbb{R}^n$ is an arbitrary subset, then $(K^\circ)^\circ = \overline{\operatorname{conv}(K \cup \{0\})}$. (This holds even if $K = \varnothing$, if one applies reasonable conventions.) The bipolar theorem (1.11) is a special case of this statement.

EXERCISE 1.16 (Polar of a projection). Prove (1.12).

EXERCISE 1.17 (Polar of a section). The following argument seems to prove that $(K \cap E)^\circ \subset P_E(K^\circ)$, whenever K is an arbitrary convex body containing the origin.

We will represent any point in $\mathbb{R}^n$ as (x, x'), where $x \in E, x' \in E^\perp$. The condition $y \in (K \cap E)^\circ \subset E$ means that $\langle x, y\rangle \leqslant 1$ for $x \in K \cap E$. In other words, the functional $x \mapsto \langle x, y\rangle$ defined on E is dominated by $\|\cdot\|_K$, and so, by the Hahn–Banach theorem, it extends to a linear functional on $\mathbb{R}^n$ also dominated by $\|\cdot\|_K$. That extension must be of the form $(x, x') \mapsto \langle (x, x'), (y, y')\rangle$ for some $y' \in E^\perp$, and the domination by $\|\cdot\|_K$ means that $(y, y') \in K^\circ$. In particular, $y \in P_E(K^\circ)$.

Find an error. Fix it and complete the proof of (1.13) (under the assumptions stated there). Give an example of K with 0 on the boundary such that (1.13) fails.

EXERCISE 1.18 (Polars of unions and intersections). Prove (1.14) and (1.15). For the latter, show by examples that each of the hypotheses and the closure on the right-hand side may be needed.

EXERCISE 1.19 (Polars of polytopes). Show that the polar of a polytope $K \subset \mathbb{R}^n$ is a polytope if and only if $\dim K = n$ and 0 is an interior point of K.

1.1.5. Polarity and the facial structure. If $K \subset \mathbb{R}^n$ is a closed convex set containing 0 in the interior and F is an exposed face of K, let us define

$$\nu_K(F) := \{y \in K^\circ \ : \ \langle y, x\rangle = 1 \text{ for all } x \in F\}. \tag{1.16}$$

Then (see Exercise 1.20) $\nu_K(F)$ is an exposed face of K°. Moreover, $F \mapsto \nu_K(F)$ is an injective order-reversing (with respect to inclusion) map between the corresponding sets of exposed faces. If K is a convex body (and so ν_{K° is also defined), then $\nu_{K^\circ}\big(\nu_K(F)\big) = F$ for any exposed face F of K.

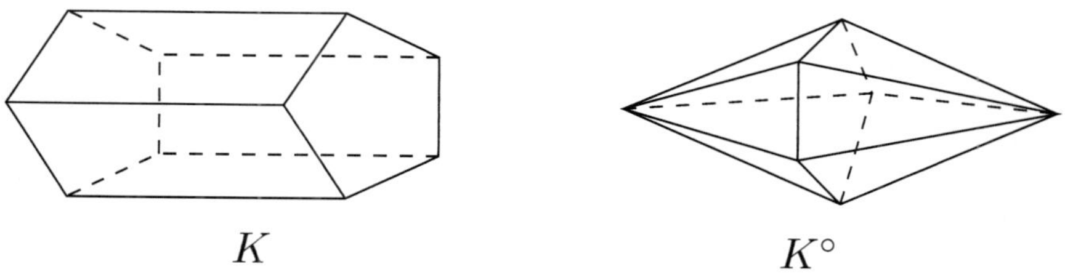

FIGURE 1.2. A polytope and its polar. The reader is encouraged to visualize the bijection ν_K between vertices (resp. edges, facets) of K and facets (resp. edges, vertices) of K°. The map ν_K is vaguely related to the Gauss map from differential geometry.

If K is a polytope, then the action of ν_K is very regular: every vertex is mapped to a facet and *vice versa*, and, more generally, every k-dimensional face is mapped to an $(n-k-1)$-dimensional face (see Figure 1.2).

The situation gets more complicated when dealing with general convex bodies: if F is a maximal face (necessarily exposed, see Exercise 1.5), then $\nu_K(F)$ is a minimal exposed face (not necessarily a minimal *face*, and certainly not necessarily an extreme point of K°). However, it is still possible to retrieve all maximal faces of K from extreme points of K°. We have

PROPOSITION 1.5. *Let $K \subset \mathbb{R}^n$ be a convex body containing 0 in the interior. For $y \in \partial K^\circ$ we define*

$$F_y := \{x \in K \ : \ \langle y, x\rangle = 1\}. \tag{1.17}$$

Then F_y is an exposed face of K. Moreover, the family

$$\{F_y : \ y \text{ is an extreme point of } \ K^\circ\}$$

contains the family of maximal faces of K.

The proof of Proposition 1.5 is outlined in Exercise 1.21 (see also Exercise 1.22).

EXERCISE 1.20. Prove the properties of ν_K listed in the paragraph following its definition in (1.16).

EXERCISE 1.21 (Extreme points and maximal faces). Prove Proposition 1.5. How does the assertion need to be modified if K is only a closed convex set containing 0 in the interior (i.e., not necessarily bounded)?

EXERCISE 1.22 (A dual Krein–Milman theorem). Let $K \subset \mathbb{R}^n$ be a closed convex set containing 0 in the interior, let F_y be defined by (1.17), and let E be the set of extreme points of K°. Show that the formula $\bigcup_{y\in E} F_y = \partial K$ is a dual restatement of the Krein–Milman theorem (Theorem 1.3).

EXERCISE 1.23. Give an example of a body $K \subset \mathbb{R}^2$ (containing 0 in the interior) with a maximal face F such that $\nu_K(F)$ is not necessarily a minimal face.

EXERCISE 1.24. Give an example of a body $K \subset \mathbb{R}^2$ (with 0 in the interior) and y, an extreme point of K°, such that the face F_y given by (1.17) is not maximal.

1.1.6. Ellipsoids. A convex body $K \subset \mathbb{R}^n$ is an *ellipsoid* if it is the image of B_2^n under an affine transformation. In particular, 0-symmetric ellipsoids are exactly the unit balls of Euclidean norms on $\mathbb{R}^n$ (i.e., norms induced by an inner product). Given a 0-symmetric ellipsoid $\mathscr{E} \subset \mathbb{R}^n$, we denote by $\langle \cdot, \cdot \rangle_{\mathscr{E}}$ the inner product associated to $\mathscr{E}$. Note also that given a 0-symmetric ellipsoid $\mathscr{E}$, there is a unique positive invertible matrix T such that $\mathscr{E} = T(B_2^n)$.

As explained in Section 0.4, there is a canonical notion of tensor products within the category of Euclidean spaces. Accordingly, given two 0-symmetric ellipsoids $\mathscr{E} \subset \mathbb{R}^n$ and $\mathscr{E}' \subset \mathbb{R}^{n'}$, we denote by $\mathscr{E} \otimes_2 \mathscr{E}' \subset \mathbb{R}^n \otimes \mathbb{R}^{n'}$ the resulting ellipsoid, which satisfies

$$\langle x \otimes x', y \otimes y' \rangle_{\mathscr{E} \otimes_2 \mathscr{E}'} = \langle x, y \rangle_{\mathscr{E}} \langle x', y' \rangle_{\mathscr{E}'}.$$

for $x, y \in \mathbb{R}^n$ and $x', y' \in \mathbb{R}^{n'}$. An alternative presentation is to say that if T (resp., T') is a linear transformation on $\mathbb{R}^n$ (resp., on $\mathbb{R}^{n'}$) such that $\mathscr{E} = T(B_2^n)$ (resp., such that $\mathscr{E}' = T'(B_2^{n'})$), then

$$\mathscr{E} \otimes_2 \mathscr{E}' = (T \otimes T')(B_2^{nn'}),$$

where we identified $\mathbb{R}^n \otimes \mathbb{R}^{n'}$ with $\mathbb{R}^{nn'}$.

EXERCISE 1.25 (Spherical sections of ellipsoids). Show that any $(2n-1)$-dimensional ellipsoid $\mathscr{E}$ admits an n-dimensional central section which is a Euclidean ball.

EXERCISE 1.26 (Polar of an ellipsoid is an ellipsoid). Follow the outline below to give an elementary proof of the fact that the polar of an ellipsoid $\mathscr{E} \subset \mathbb{R}^n$ containing 0 in its interior is again an ellipsoid, and that among translates of a given ellipsoid the volume of the polar is minimized iff the translate is 0-symmetric. (See Exercise D.3 for a computation-free proof.)
(a) Show, by direct calculation, that if $0 \leqslant a < 1$ and $D_a \subset \mathbb{R}^2$ is the disk of unit radius and center at $(a, 0)$, then D_a° is an ellipse with center at $(-\frac{a}{1-a^2}, 0)$ and principal semi-axes of length $\frac{1}{1-a^2}$ and $\frac{1}{\sqrt{1-a^2}}$. In particular the area of D_a° is minimal iff $a = 0$.
(b) Infer similar statements for the n-dimensional Euclidean ball, and then deduce the desired conclusion.

1.2. Cones

A nonempty closed convex subset $\mathcal{C}$ of $\mathbb{R}^n$ (or of any real vector space) is called a *cone* if whenever $x \in \mathcal{C}$ and $t \geqslant 0$, then $tx \in \mathcal{C}$. An equivalent definition: $\mathcal{C}$ is a closed set such that $x, x' \in \mathcal{C}$ and $t, t' \geqslant 0$ imply $tx + t'x' \in \mathcal{C}$. Examples of cones include:
(1) the cone of elements of $\mathbb{R}^n$ with nonnegative coordinates (the *positive orthant* $\mathbb{R}_+^n$),
(2) the *Lorentz cone* $\mathcal{L}_n = \{(x_0, x_1, \ldots, x_{n-1}) : x_0 \geqslant 0,\ \sum_{k=1}^{n-1} x_k^2 \leqslant x_0^2\} \subset \mathbb{R}^n$ for $n \geqslant 2$,
(3) the cone $\mathcal{PSD} = \mathcal{PSD}(\mathbb{C}^n) \subset \mathsf{M}_n^{\mathrm{sa}}$ of complex *positive semi-definite* matrices.

1.2.1. Cone duality. The *dual cone* $\mathcal{C}^*$ is defined via

$$\mathcal{C}^* := \{x \in \mathbb{R}^n \ : \ \forall\, y \in \mathcal{C}\ \langle x, y\rangle \geqslant 0\}. \tag{1.18}$$

As was the case with the polarity (see Section 1.1.4), the notion of the dual cone is not canonical in the category of vector spaces since it appeals to the scalar product. This can be again circumvented by considering $\mathcal{C}^*$ as a subset of the vector space that is dual to the one containing $\mathcal{C}$. We will present some advantages of this point of view in Appendix D, but will otherwise stick to the more familiar Euclidean setting.

It is easily checked that the cones $\mathbb{R}^n_+$, $\mathcal{L}_n$ and $\mathcal{PSD}$ defined in the preamble to Section 1.2 have the remarkable property of being *self-dual*, i.e., verify $\mathcal{C}^* = \mathcal{C}$. (For $\mathcal{C} = \mathcal{PSD}$, extend the definition (1.18) *mutatis mutandis* to the setting of arbitrary real inner product spaces and use trace duality (0.4).)

Not surprisingly, the notion of cone duality is strongly related to that of polarity. First, a simple argument shows that if $\mathcal{C}$ is a (closed convex) cone, then $\mathcal{C}^* = -\mathcal{C}^\circ$ and, therefore, by (1.11),

$$(\mathcal{C}^*)^* = \mathcal{C}. \tag{1.19}$$

Similarly, for two closed convex cones $\mathcal{C}_1, \mathcal{C}_2$,

$$(\mathcal{C}_1 \cap \mathcal{C}_2)^* = \overline{\mathcal{C}_1^* + \mathcal{C}_2^*} \tag{1.20}$$

by (1.15). However, we also have another link to polarity of convex bodies, which is less obvious. To point out that link, let us first define a *base* of a closed convex cone $\mathcal{C} \subset \mathbb{R}^n$ to be a closed convex set $K \subset \mathcal{C}$ such that (1) the affine space generated by K does not contain the origin and (2) K generates $\mathcal{C}$, i.e., $\mathcal{C} = \overline{\mathbb{R}_+ K}$. An alternative description (which is equivalent, see Exercise 1.27) is as follows: fix a distinguished nonzero vector $e \in \mathbb{R}^n$ and the corresponding affine hyperplane

$$H_e := \{x \in \mathbb{R}^n \ : \ \langle x, e\rangle = |e|^2\}, \tag{1.21}$$

in which e is the point closest to the origin. If $\mathcal{C} \subset \mathbb{R}^n$ is a closed convex cone such that $e \in \mathcal{C}^* \backslash \mathcal{C}^\perp$, the set $\mathcal{C}^{\mathrm{b}}$ defined as

$$\mathcal{C}^{\mathrm{b}} = \mathcal{C} \cap H_e \tag{1.22}$$

is then a base of $\mathcal{C}$ (that is, $\mathcal{C}$ is the smallest closed cone containing $\mathcal{C}^{\mathrm{b}}$, see Exercise 1.28). In particular, knowing $\mathcal{C}^{\mathrm{b}}$ allows to reconstruct $\mathcal{C}$.

As was to be expected, natural set-theoretic and algebraic operations on cones induce analogous operations on bases of cones. Sometimes this is as trivial as $(\mathcal{C}_1 \cap \mathcal{C}_2)^{\mathrm{b}} = \mathcal{C}_1^{\mathrm{b}} \cap \mathcal{C}_2^{\mathrm{b}}$, or as simple as $(\mathcal{C}_1 + \mathcal{C}_2)^{\mathrm{b}} = \operatorname{conv}(\mathcal{C}_1^{\mathrm{b}} \cup \mathcal{C}_2^{\mathrm{b}})$. In fact, if we want to stay in the class of *closed* cones, the more appropriate form of the latter formula would be

$$(\overline{\mathcal{C}_1 + \mathcal{C}_2})^{\mathrm{b}} = \overline{\operatorname{conv}(\mathcal{C}_1^{\mathrm{b}} \cup \mathcal{C}_2^{\mathrm{b}})} \tag{1.23}$$

(see Exercise 1.30; however, such adjustments are not needed under some natural nondegeneracy assumptions, which we will describe later in Section 1.2.2).

What is more interesting—and somewhat surprising—is that the duality of cones likewise carries over to a precise duality of bases in the following sense (see Figure 1.3; see also Lemma D.1 in Appendix D).

LEMMA 1.6. *Let $\mathcal{C} \subset \mathbb{R}^n$ be a closed convex cone and let $e \in \mathcal{C} \cap \mathcal{C}^*$ be a nonzero vector. Let $\mathcal{C}^{\mathrm{b}} = \mathcal{C} \cap H_e$ and $(\mathcal{C}^*)^{\mathrm{b}} = \mathcal{C}^* \cap H_e$ be the corresponding bases of $\mathcal{C}$ and $\mathcal{C}^*$. Then*

$$(\mathcal{C}^*)^{\mathrm{b}} = \{y \in H_e \ : \ \forall x \in \mathcal{C}^{\mathrm{b}} \ \ \langle -(y-e), x-e\rangle \leqslant |e|^2\}. \tag{1.24}$$

In other words, if we think of H_e as a vector space with the origin at e, and of $\mathcal{C}^{\mathrm{b}}$ and $(\mathcal{C}^*)^{\mathrm{b}}$ as subsets of that vector space, then $(\mathcal{C}^*)^{\mathrm{b}} = -|e|^2(\mathcal{C}^{\mathrm{b}})^\circ$.

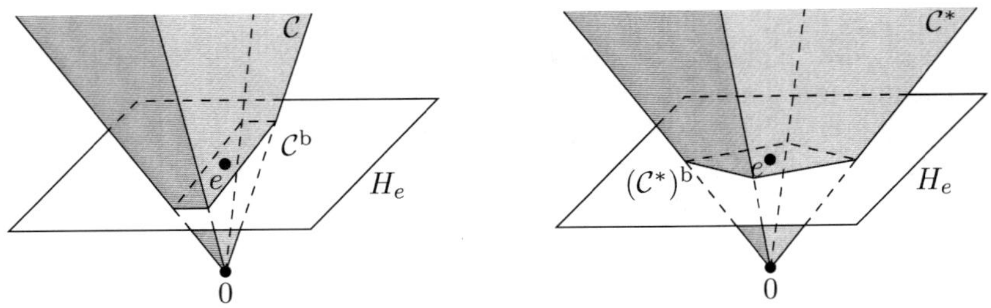

FIGURE 1.3. A cone and its dual cone. Up to a reflection, the bases $\mathcal{C}^{\mathrm{b}}$ and $(\mathcal{C}^*)^{\mathrm{b}}$ are polar to each other with respect to e.

PROOF. If $\langle x, e\rangle = \langle y, e\rangle = |e|^2$, then $\langle -(y-e), x-e\rangle = -\langle y, x\rangle + |e|^2$ and so the condition from (1.24) can be restated as "$\forall x \in \mathcal{C}^{\mathrm{b}} \ \ -\langle y, x\rangle + |e|^2 \leqslant |e|^2$" or, more simply, "$\forall x \in \mathcal{C}^{\mathrm{b}} \ \langle y, x\rangle \geqslant 0$." Since $\mathcal{C}^{\mathrm{b}}$ generates $\mathcal{C}$ (see Exercise 1.28), the latter condition is further equivalent to "$\langle y, x\rangle \geqslant 0$ for all $x \in \mathcal{C}$," i.e., to "$y \in \mathcal{C}^*$," as required. □

Here are two important classical examples where Lemma 1.6 applies.
(1) The positive orthant $\mathbb{R}_+^{n+1} \subset \mathbb{R}^{n+1}$. Take $e = (\frac{1}{n+1}, \dots, \frac{1}{n+1})$, so that H_e is given by the equation $x_0 + \cdots + x_n = 1$. Then $\left(\mathbb{R}_+^{n+1}\right)^{\mathrm{b}} = \Delta_n$, the set of classical states. Since $\mathbb{R}_+^{n+1}$ is self-dual, it follows from Lemma 1.6 that

$$\Delta_n^\circ = -(n+1)\Delta_n. \tag{1.25}$$

Note that the prefactor is $-(n+1)$ and not $-n$ because the n-dimensional ball circumscribed around Δ_n is not of unit radius.
(2) The cone $\mathcal{PSD}(\mathbb{C}^n) \subset \mathsf{M}_n^{\mathrm{sa}}$. Take $e = \mathrm{I}/n$ (the *maximally mixed state*), so that H_e is the hyperplane of trace one matrices. Then $\mathcal{PSD}^{\mathrm{b}} = \mathrm{D}(\mathbb{C}^n)$, the set of quantum states. Since $\mathcal{PSD}$ is self-dual, it follows from Lemma 1.6 that

$$\mathrm{D}(\mathbb{C}^n)^\circ = -n\mathrm{D}(\mathbb{C}^n). \tag{1.26}$$

The bases of the Lorentz cones $\mathcal{L}_n$ relative to the natural choice $e = e_0$ are Euclidean balls, so applying Lemma 1.6 just tells us that the Lorentz cone is self-dual (a property which is easy to verify directly). However, other choices of e lead to nontrivial consequences, see Exercise D.3. Another simple but important observation is that since $\mathrm{D}(\mathbb{C}^2)$ is a 3-dimensional Euclidean ball (the Bloch ball), the cone $\mathcal{PSD}(\mathbb{C}^2)$ is isomorphic (or even isometric in the appropriate sense) to the Lorentz cone $\mathcal{L}_4$ (see Section 2.1.2).

EXERCISE 1.27. Let K be a base of a closed convex cone $\mathcal{C}$, and H the affine space generated by K. Show that $K = \mathcal{C} \cap H$.

EXERCISE 1.28 (Bases generate cones). Show that if $e \in \mathbb{R}^n$ and a closed convex cone $\mathcal{C} \subset \mathbb{R}^n$ are such that $e \in \mathcal{C}^* \backslash \mathcal{C}^{\perp}$, and if $\mathcal{C}^{\mathrm{b}}$ is defined by (1.21) and (1.22), then $\overline{\mathbb{R}_+\mathcal{C}^{\mathrm{b}}} = \mathcal{C}$. Give an example showing that the closure is needed.

EXERCISE 1.29 (Nontrivial cones admit bases). Let $\mathcal{C} \subset \mathbb{R}^n$ be a closed convex cone. Show that $\mathcal{C}$ admits a base iff $\mathcal{C}$ is not a linear subspace iff $\mathcal{C} \neq -\mathcal{C}$.

EXERCISE 1.30. Give an example of closed cones $\mathcal{C}_1, \mathcal{C}_2$ in $\mathbb{R}^3$ such that the cone $\mathcal{C}_1 + \mathcal{C}_2$ is not closed.

EXERCISE 1.31 (Time dilation and the Lorentz cone). Consider the cone $\mathcal{C}_y = \{x \in \mathbb{R}^n : |x| \leqslant \langle x, y\rangle\}$ where $y \in \mathbb{R}^n$ satisfies $|y| > 1$. Show that $\mathcal{C}_y^* = \mathcal{C}_z$ for $z = y/\sqrt{|y|^2 - 1}$.

1.2.2. Nondegenerate cones and facial structure. We will be mostly dealing with (closed convex) cones $\mathcal{C} \subset \mathbb{R}^n$ verifying (i) $\mathcal{C} \cap (-\mathcal{C}) = \{0\}$ and (ii) $\mathcal{C} - \mathcal{C} = \mathbb{R}^n$; we will call such cones *nondegenerate*. The properties (i) and (ii) are often referred to as $\mathcal{C}$ being respectively *pointed* and *full*. They are dual to each other, i.e., $\mathcal{C}$ verifies (i) iff $\mathcal{C}^*$ verifies (ii), and *vice versa*; the reader may explore them further in Exercise 1.32. Here we note the following:

LEMMA 1.7. *Let $\mathcal{C} \subset \mathbb{R}^n$ be a closed convex cone. Then y is an interior point of $\mathcal{C}^*$ iff $\langle y, x\rangle > 0$ for every $x \in \mathcal{C}\backslash\{0\}$.*

PROOF. Let $x \in \mathcal{C}$. If $B(y, \varepsilon) \subset \mathcal{C}^*$ for some $\varepsilon > 0$, then

$$\langle y + u, x\rangle \geqslant 0 \quad \text{for any } \; |u| < \varepsilon. \tag{1.27}$$

Since $\inf_{|u|<\varepsilon}\langle y + u, x\rangle = \langle y, x\rangle - \varepsilon|x|$, this is only possible if either $\langle y, x\rangle > 0$ or $|x| = 0$. This proves the "only if" part (see also Exercise 1.34). For the "if" part, we note that $B(y, \varepsilon) \subset \mathcal{C}^*$ follows if (1.27) holds for x in $\mathcal{C} \cap S^{n-1} =: A$. This could be ensured by choosing $\varepsilon = \inf_{x\in A}\langle y, x\rangle$, which is strictly positive since the continuous function $\langle y, \cdot\rangle$ is pointwise positive on the compact set A. □

COROLLARY 1.8. *If $\mathcal{C}$ is a closed convex cone which is pointed, then 0 is an exposed point of $\mathcal{C}$. If, moreover, $\mathcal{C} \neq \{0\}$, then $\mathcal{C}$ admits a compact base.*

PROOF. Since $\mathcal{C}$ is pointed, $\mathcal{C}^*$ has nonempty interior. If y is any interior point of $\mathcal{C}^*$, Lemma 1.7 says that the hyperplane $H = \{x \in \mathbb{R}^n : \langle y, x\rangle = 0\}$ isolates 0 as an exposed point of $\mathcal{C}$, and it readily follows that the base of $\mathcal{C}$ induced by $e = y$ is compact. In fact, all the three properties stated in the Corollary are equivalent (see Exercise 1.32). □

We are now ready to state the main observation of this section. Once made, it is fairly straightforward to show.

PROPOSITION 1.9 (Faces of cones and faces of bases; see Exercise 1.35). *Let $\mathcal{C} \subset \mathbb{R}^n$ be a closed convex cone with a compact base $\mathcal{C}^{\mathrm{b}}$. When we exclude the exposed point 0 of $\mathcal{C}$, there is a one-to-one correspondence between faces of $\mathcal{C}^{\mathrm{b}}$ and those of $\mathcal{C}$ given by $F \mapsto \mathbb{R}_+F$. Moreover, this correspondence preserves the exposed (or non-exposed) character of each face.*

An important special case is when x is an extreme (or exposed) point of $\mathcal{C}^{\mathrm{b}}$; the corresponding face of $\mathcal{C}$ is then the *ray* $\mathbb{R}_+x$, called an *extreme ray* (or an *exposed ray*). The Krein–Milman theorem (see Section 1.1.2) implies then that $\mathcal{C}$ is the convex hull of its extreme rays. We also note for future reference the following consequence of Proposition 1.9 (for the second part, appeal to Exercise 1.7).

COROLLARY 1.10. *All extreme rays of $\mathcal{PSD}(\mathbb{C}^n)$ are of the form $\mathbb{R}_+|\psi\rangle\langle\psi|$, where $\psi \in S_{\mathbb{C}^n}$. All rays contained in the boundary of the Lorentz cone $\mathcal{L}_n$ are extreme.*

EXERCISE 1.32 (Full cones and pointed cones). Let $\mathcal{C} \subset \mathbb{R}^n$ be a closed convex cone, $\mathcal{C} \neq \{0\}$. Show that the following conditions are equivalent:
(a) $\mathcal{C}$ is pointed (i.e., $\mathcal{C} \cap (-\mathcal{C}) = \{0\}$),
(b) $\mathcal{C}^*$ is full (i.e., $\mathcal{C}^* - \mathcal{C}^* = \mathbb{R}^n$),
(c) 0 is an exposed point of $\mathcal{C}$,
(d) $\mathcal{C}$ does not contain a line,
(e) $\mathcal{C}$ admits a compact base,
(f) $\dim \mathcal{C}^* = n$,
(g) $\operatorname{span} \mathcal{C}^* = \mathbb{R}^n$.

EXERCISE 1.33 (Structure theorem for a general cone). If $\mathcal{C} \subset \mathbb{R}^n$ is a closed convex cone, then there exists a vector subspace $V \subset \mathbb{R}^n$ and a pointed cone $\mathcal{C}' \subset V^\perp$ such that $\mathcal{C} = V + \mathcal{C}'$ (a direct Minkowski sum).

EXERCISE 1.34. Deduce the "only if" part of Lemma 1.7 from Proposition 1.4.

EXERCISE 1.35. Prove Proposition 1.9, relating faces of cones to those of their bases.

EXERCISE 1.36. Show that if the cones $\mathcal{C}_1^*, \mathcal{C}_2^*$ are pointed with the same isolating hyperplane, then the closure on the right-hand side of (1.20) is not needed.

1.3. Majorization and Schatten norms

1.3.1. Majorization. If $x \in \mathbb{R}^n$, we denote by $x^\downarrow \in \mathbb{R}^n$ the non-increasing rearrangement of x, i.e., the coordinates of $x^\downarrow$ are equal to the coordinates of x up to permutation, and $x_1^\downarrow \geqslant \cdots \geqslant x_n^\downarrow$.

DEFINITION 1.11. If $x, y \in \mathbb{R}^n$ with $\sum_{i=1}^n x_i = \sum_{i=1}^n y_i$, we say that x is *majorized* by y, and write $x \prec y$, if

$$\sum_{j=1}^k x_j^\downarrow \leqslant \sum_{j=1}^k y_j^\downarrow \quad \text{for any } k \in \{1, 2, \ldots, n\}. \tag{1.28}$$

Note that, by hypothesis, (1.28) becomes an equality for $k = n$.

The majorization property will be a crucial tool in Chapter 10. As a warm-up, we will use it in the next section to prove the Davis convexity theorem and various properties of Schatten norms (noncommutative ℓ_p-norms).

There are several equivalent reformulations of the majorization property. We gather some of them in the following proposition.

PROPOSITION 1.12. *For $x, y \in \mathbb{R}^n$ with $\sum x_i = \sum y_i$, the following conditions are equivalent.*

(i) *$x \prec y$.*
(ii) *x can be written as a convex combination of coordinatewise permutations of y.*

(iii) *There is an $n \times n$ bistochastic matrix B such that $y = Bx$ (a matrix is bistochastic if its entries are non-negative, and add up to 1 in each row and each column).*

(iv) *Whenever ϕ is a permutationally invariant convex function on $\mathbb{R}^n$, then $\phi(x) \leqslant \phi(y)$.*

(v) *For every $t \in \mathbb{R}$, we have $\sum_{i=1}^n |x_i - t| \leqslant \sum_{i=1}^n |y_i - t|$.*

(vi) *For every $t \in \mathbb{R}$, we have $\sum_{i=1}^n (x_i - t)^+ \leqslant \sum_{i=1}^n (y_i - t)^+$, where $x^+ = \max(x, 0)$.*

SKETCH OF THE PROOF. Fix $y \in \mathbb{R}^n$, and consider the non-empty convex compact set

$$K_y = \{x \in \mathbb{R}^n \ : \ x \prec y\}.$$

It is easily checked that x is an extreme point of K_y if and only if $x^\downarrow = y^\downarrow$, and it follows from the Krein–Milman theorem that (i) is equivalent to (ii). Similarly, the classical Birkhoff theorem, which asserts that extreme points of the set of bistochastic matrices are exactly permutation matrices, gives the equivalence of (ii) and (iii). The implications (ii) $\Rightarrow$ (iv) $\Rightarrow$ (v) are obvious. We check that (v) and (vi) are equivalent since $|x| = 2x^+ - x$ (using the fact that $\sum x_i = \sum y_i$). Finally, for $t = y_k^\downarrow$, we compute

$$\sum_{i=1}^n (y_i - t)^+ = \sum_{i=1}^k (y_i^\downarrow - t) = \sum_{i=1}^k y_i^\downarrow - kt$$

$$\sum_{i=1}^n (x_i - t)^+ = \sum_{i=1}^n (x_i^\downarrow - t)^+ \geqslant \sum_{i=1}^k (x_i^\downarrow - t)^+ \geqslant \sum_{i=1}^k (x_i^\downarrow - t) = \sum_{i=1}^k x_i^\downarrow - kt.$$

Therefore, the inequality from (vi) implies that $\sum_{i=1}^k x_i^\downarrow \leqslant \sum_{i=1}^k y_i^\downarrow$, hence $x \prec y$. □

EXERCISE 1.37. Show that, in the statement of Proposition 1.12, we have to assume the hypothesis $\sum x_i = \sum y_i$ only in (vi); in (ii)–(v) this property follows formally.

EXERCISE 1.38 (Submajorization). Given $x, y \in \mathbb{R}^n$, we say that x is *submajorized* by y and write $x \prec_w y$ if (1.28) holds (the difference with majorization is that we do not assume $\sum x_i = \sum y_i$). Show that $x \prec_w y$ if and only if there exists $u \in \mathbb{R}^n$ such that $u \prec y$ and $x_k \leqslant u_k$ for every $1 \leqslant k \leqslant n$.

1.3.2. Schatten norms. Recall that the space $\mathsf{M}_{m,n}$ of (real or complex) $m \times n$ matrices carries a Euclidean structure given by the Hilbert–Schmidt inner product (see Section 0.6). The Hilbert–Schmidt norm is a special case of the *Schatten p-norms*, which are the noncommutative analogues of the ℓ_p-norms. If $M \in \mathsf{M}_{m,n}$, define $|M| := (M^\dagger M)^{1/2}$, and for $1 \leqslant p \leqslant \infty$,

$$\|M\|_p := \left(\operatorname{Tr} |M|^p\right)^{1/p}.$$

Note that $\|\cdot\|_{\mathrm{HS}} = \|\cdot\|_2$. The case $p = \infty$ should be interpreted as the limit $p \to \infty$ of the above, and corresponds to the usual operator norm

$$\|M\|_\infty = \|M\|_{\mathrm{op}} := \sup_{|x| \leqslant 1} |Mx|.$$

The quantity $\|M\|_1 = \operatorname{Tr}|M|$ is called the *trace norm* of M. Occasionally we will loosely refer to various matrix spaces endowed with Schatten norms as *Schatten spaces* or p-Schatten spaces.

There is ambiguity in the notation $\|\cdot\|_p$ in that it has two possible meanings: the Schatten p-norm on $\mathsf{M}_{m,n}$ (matrices) and the usual ℓ_p-norm on $\mathbb{R}^n$ or $\mathbb{C}^n$ (sequences). However, it will be always clear from the context which of the two is the intended one.

If $M \in \mathsf{M}_{m,n}$, and if we denote by $s(M) = (s_1(M), \dots, s_n(M))$ the singular values of M (i.e., the eigenvalues of $|M|$) arranged in the non-increasing order, then for any p,

$$\|M\|_p = \|s(M)\|_p. \tag{1.29}$$

The following lemma allows us to reduce the study of Schatten norms to the case of self-adjoint matrices.

LEMMA 1.13. *Let $M \in \mathsf{M}_{m,n}$, and $\tilde{M} \in \mathsf{M}_{m+n}$ be the self-adjoint matrix defined by*

$$\tilde{M} = \begin{bmatrix} 0 & M \\ M^\dagger & 0 \end{bmatrix}.$$

Then we have $\|\tilde{M}\|_p = 2^{1/p}\|M\|_p$ for $1 \leqslant p \leqslant \infty$. Similarly, if $M, N \in \mathsf{M}_{m,n}$, then $\operatorname{Tr} \tilde{M}\tilde{N} = 2 \operatorname{Re} \operatorname{Tr} M^\dagger N$.

PROOF. For the first assertion, it suffices to notice that the eigenvalues of $\tilde{M}$ are equal to $\pm s_i(M)$. The second assertion is verified by direct calculation. □

The next lemma shows how the concept of majorization relates to eigenvalues/singular values of a matrix.

LEMMA 1.14 (Spectrum majorizes the diagonal). *Let $M \in \mathsf{M}_n$ be a self-adjoint matrix, let $d(M) = (m_{ii}) \in \mathbb{R}^m$ be the vector of diagonal entries of M, and let $\operatorname{spec}(M) = (\lambda_i) \in \mathbb{R}^m$ be the vector of eigenvalues of M, arranged in non-increasing order. Then $d(M) \prec \operatorname{spec}(M)$.*

PROOF. First, it is known from linear algebra that $\sum_i m_{ii} = \sum_i \lambda_i$, so majorization is in principle possible. Write M as $M = U\Lambda U^\dagger$, where Λ is a diagonal matrix whose entries are the eigenvalues of M, and U is a unitary matrix. We then have

$$m_{ii} = \sum_j u_{ij}\lambda_j \overline{u_{ji}} = \sum_j |u_{ij}|^2 \lambda_j.$$

Since the matrix with entries $|u_{ij}|^2$ is bistochastic, the assertion follows from Proposition 1.12 (iii). □

We now state the Davis convexity theorem, which gives a characterization of all convex functions f on $\mathsf{M}_m^{\mathrm{sa}}$ that are unitarily invariant.

PROPOSITION 1.15 (Davis convexity theorem). *Let $f : \mathsf{M}_m^{\mathrm{sa}} \to \mathbb{R}$ a function which is unitarily invariant, i.e., such that $f(UAU^\dagger) = f(A)$ for any self-adjoint matrix A and any unitary matrix U. Then f is convex if and only if the restriction of f to the subspace of diagonal matrices is convex.*

PROOF. Assume that the restriction of f to diagonal matrices is convex (the converse implication being obvious). This restriction, when considered as a function on $\mathbb{R}^m$, is permutationally invariant, as can be checked by choosing for U a permutation matrix. Given $0 < \lambda < 1$ and $A, B \in \mathsf{M}_m^{\mathrm{sa}}$, we need to show that

$$f(\lambda A + (1-\lambda)B) \leqslant \lambda f(A) + (1-\lambda) f(B). \tag{1.30}$$

Since f is unitarily invariant, we may assume that the matrix $\lambda A + (1-\lambda)B$ is diagonal. Denoting by $\operatorname{diag} A$ the matrix obtained from a matrix A by changing all its off-diagonal elements to 0, the hypothesis on f implies

$$f(\lambda A + (1-\lambda)B) \leqslant \lambda f(\operatorname{diag} A) + (1-\lambda) f(\operatorname{diag} B).$$

Using Lemma 1.14 and Proposition 1.12(iv), it follows that $f(\operatorname{diag} A) \leqslant f(A)$ and $f(\operatorname{diag} B) \leqslant f(B)$, showing (1.30). □

An immediate consequence of the Davis convexity theorem is that the Schatten p-norms satisfy the triangle inequality.

PROPOSITION 1.16. *For $1 \leqslant p \leqslant \infty$, if $M, N \in \mathsf{M}_{m,n}$, we have*

$$\|M+N\|_p \leqslant \|M\|_p + \|N\|_p.$$

PROOF. By the first assertion of Lemma 1.13, it is enough to consider the case of $m = n$ and self-adjoint M, N. We now use Proposition 1.15 for the unitarily invariant function $f(\cdot) = \|\cdot\|_p$. The restriction of $\|\cdot\|_p$ to the subspace of diagonal matrices identifies with the usual (commutative) ℓ_p-norm on $\mathbb{R}^n$, and hence, by Proposition 1.15, the function $\|\cdot\|_p$ is convex on $\mathsf{M}_m^{\mathrm{sa}}$. Since it is also positively homogeneous, the triangle inequality follows. □

Obviously, the Schatten p-norms of a given matrix satisfy the same inequalities as the ℓ_p-norms: if $1 \leqslant p \leqslant q \leqslant \infty$, and M is an $m \times n$ matrix (with $m \leqslant n$; what is important is that the rank of M is at most m), then

$$\|M\|_q \leqslant \|M\|_p \leqslant m^{1/p-1/q} \|M\|_p. \tag{1.31}$$

Duality between Schatten p-norms holds as in the commutative case.

PROPOSITION 1.17 (The non-commutative Hölder inequality). *Let $1 \leqslant p, q \leqslant \infty$ such that $1/p + 1/q = 1$, and $M \in \mathsf{M}_{m,n}$, $N \in \mathsf{M}_{n,m}$. We have*

$$|\operatorname{Tr} MN| \leqslant \|M\|_p \|N\|_q. \tag{1.32}$$

As a consequence, the Schatten p-norm and q-norm are dual to each other. This holds in all settings: for rectangular matrices (real or complex), for Hermitian matrices, and for real symmetric matrices.

As in the case of ℓ_p^n-spaces, the above duality relation can be equivalently expressed in terms of polars. Denote by $S_p^{m,n}$ the unit ball associated to the Schatten norm $\|\cdot\|_p$ on $\mathsf{M}_{m,n}$ and $S_p^{m,\mathrm{sa}} := S_p^{m,m} \cap \mathsf{M}_m^{\mathrm{sa}}$. (Again, there are two settings, real and complex, and some care needs to be exercised as minor subtleties occasionally arise.) We then have

COROLLARY 1.18. *If* $1 \leqslant p, q \leqslant \infty$ *with* $1/p + 1/q = 1$, *then*

$$S_q^{m,n} = \{A \in \mathsf{M}_{m,n} : |\langle X, A\rangle| \leqslant 1 \text{ for all } X \in S_p^{m,n}\} \tag{1.33}$$

$$= \{A \in \mathsf{M}_{m,n} : \mathrm{Re}\langle X, A\rangle \leqslant 1 \text{ for all } X \in S_p^{m,n}\} \tag{1.34}$$

$$S_q^{m,\mathrm{sa}} = \left(S_p^{m,\mathrm{sa}}\right)^\circ, \tag{1.35}$$

where $\langle\cdot,\cdot\rangle$ *and* $^\circ$ *are meant in the sense of trace duality* (0.4).

While (1.33) and (1.35) are simply straightforward reformulations of duality relations from Proposition 1.17, the equality in (1.34) needs to be justified (only the inclusion "$\subset$" is immediate). Given $A \in \mathsf{M}_{m,n}$ and $X \in S_p^{m,n}$ such that $|\langle X, A\rangle| > 1$, let $\xi = \frac{\langle X,A\rangle}{|\langle X,A\rangle|}$. Then, setting $X' = \bar{\xi}X$, we see that $X' \in S_p^{m,n}$, while $\mathrm{Re}\langle X', A\rangle = |\langle X, A\rangle| > 1$, which yields the other inclusion "$\supset$" in (1.34). The expression in (1.34) can be thought of as a definition of the polar $\left(S_p^{m,n}\right)^\circ$ by "dropping the complex structure"; see Exercise 1.48 for the general principle. Another potential complication is that, in the complex setting, the identification with the dual space is anti-linear, see Section 0.2. Note that no issues of such nature arise in defining the polar of $S_p^{m,\mathrm{sa}}$, as that set "lives" in a real inner product space irrespectively of the setting.

PROOF OF PROPOSITION 1.17. Consider first the Hermitian case. By unitary invariance, we may assume that M is diagonal. We then have

$$|\operatorname{Tr}(MN)| = \Big|\sum_i m_{ii}n_{ii}\Big| \leqslant \|(m_{ii})\|_p \|(n_{ii})\|_q \leqslant \|M\|_p \|N\|_q,$$

where we used the commutative Hölder inequality, Lemma 1.14, and Proposition 1.12 (iv).

In the general case, Lemma 1.13 and the Hermitian case of (1.32) shown above imply that, for all $M, N \in \mathsf{M}_{n,m}$,

$$\mathrm{Re}\,\mathrm{Tr}\, M^\dagger N \leqslant \|M\|_p \|N\|_q,$$

and the same bound for $|\operatorname{Tr}(MN)|$ (or $|\operatorname{Tr}(M^\dagger N)|$) follows by the same trick as the one used to establish equality in (1.34) (see the paragraph following Corollary 1.18).

As in the commutative case, Hölder's inequality constitutes "the hard part" of the duality assertion, such as the inclusion $S_q^{m,\mathrm{sa}} \subset \left(S_p^{m,\mathrm{sa}}\right)^\circ$ in (1.35). "The easy part" involves establishing that for every M, there is $N \neq 0$ such that we have equality in (1.32). In the Hermitian case, this follows readily by restricting attention to matrices that diagonalize in the same orthonormal basis as M and by appealing to the analogous statement for the usual ℓ_p-norm. In the general case one considers similarly the singular value decomposition (SVD) of M. □

EXERCISE 1.39 (Davis convexity theorem, the real case). State and prove a real version of Proposition 1.15, i.e., for functions defined on the set of real symmetric matrices.

EXERCISE 1.40 (Klein's lemma). Show that if the function $\phi : \mathbb{R} \to \mathbb{R}$ is convex, then $X \mapsto \operatorname{Tr} \phi(X)$ is convex on the set of self-adjoint matrices, and similarly for $\phi : I \to \mathbb{R}$ and the set of self-adjoint matrices with spectrum in I, where $I \subset \mathbb{R}$ is an interval.

EXERCISE 1.41. Show that the function $X \mapsto \log \operatorname{Tr} \exp(X)$ is convex on the set of self-adjoint matrices.

EXERCISE 1.42 (Log-concavity of the determinant). Show that the function $\log\det$ is strictly concave on the interior of $\mathcal{PSD}$.

EXERCISE 1.43. Show that if a function $X \mapsto \Phi(X)$ is convex on M_n and unitarily invariant, then $\Phi(\operatorname{diag} X) \leqslant \Phi(X)$ for any $X \in \mathsf{M}_n$ (and similarly for $\mathsf{M}_n^{\mathrm{sa}}$ in place of $X \in \mathsf{M}_n$). If Φ is strictly convex and X is not diagonal, then the inequality is strict.

EXERCISE 1.44 (Extreme points of Schatten unit balls). What are the extreme points of $S_1^{m,n}$? Of $S_1^{m,\mathrm{sa}}$? Of $S_\infty^{m,n}$? $S_\infty^{m,\mathrm{sa}}$? For the latter, how many connected components does the set of extreme points have?

EXERCISE 1.45 (Spectral theorem and SVD vs. Carathéodory's theorem). Let K be one of $S_\infty^n, S_1^n, S_\infty^{n,\mathrm{sa}}, S_1^{n,\mathrm{sa}}$. Show that every element of K can be written as a convex combination of $n+1$ extreme points of K. Compare this fact with what one obtains by a direct application of the Carathéodory's Theorem 1.2 in the respective matrix space.

EXERCISE 1.46 (The real Schatten balls). In the real case, the space $\mathsf{M}_2^{\mathrm{sa}}$ is 3-dimensional. Which familiar solids are $S_1^{2,\mathrm{sa}}$ and $S_\infty^{2,\mathrm{sa}}$?

EXERCISE 1.47 (Characterization of unitarily invariant norms). Let $m \leqslant n$, and $\|\cdot\|$ be a norm on $\mathbb{R}^m$ such that

$$\|(\varepsilon_1 x_{\sigma(1)}, \ldots, \varepsilon_m x_{\sigma(m)})\| = \|(x_1, \ldots, x_m)\|$$

for any $x \in \mathbb{R}^m$, $\varepsilon \in \{-1,1\}^m$ and $\sigma \in \mathfrak{S}_m$. (We call such norms *permutationally symmetric*.) Show that $M \mapsto \|s(M)\|$ is a norm on $\mathsf{M}_{m,n}$ and that every norm which is bi-unitarily invariant (i.e., verifying $\|UMV\| = \|M\|$ for $U \in \mathsf{U}(m)$ and $V \in \mathsf{U}(n)$) can be defined in this way.

EXERCISE 1.48 (Polarity in the complex setting). If $\mathcal{H}$ is a complex Hilbert space and K a closed convex subset, the polar of K can be defined via $K^\circ := \{y \in \mathcal{H} : \operatorname{Re}\langle x, y\rangle \leqslant 1 \text{ for all } x \in K\}$, i.e., by dropping the complex structure, as described in Section 0.5. Show that $K^\circ := \{y \in \mathcal{H} : |\langle x, y\rangle| \leqslant 1 \text{ for all } x \in K\}$ if and only if K is circled.

1.3.3. Von Neumann and Rényi entropies. Let $\mathrm{D}(\mathbb{C}^d)$ be the set of quantum states on $\mathbb{C}^d$ (see Section 0.10) and $\sigma \in \mathrm{D}(\mathbb{C}^d)$. The *von Neumann entropy* of σ is defined as

$$S(\sigma) = -\operatorname{Tr}(\sigma \log \sigma), \tag{1.36}$$

where log is the natural logarithm. (Note that many texts use base 2 logarithm to define entropy, see Notes and Remarks.)

PROPOSITION 1.19. *The von Neumann entropy S satisfies the following properties:*
(i) it is a concave function from $\mathrm{D}(\mathbb{C}^d)$ onto $[0, \log d]$,
(ii) for $\sigma \in \mathrm{D}(\mathbb{C}^d)$, we have $S(\sigma) = 0$ if and only if σ is pure (i.e., has rank 1),
(iii) for $\sigma \in \mathrm{D}(\mathbb{C}^d)$, we have $S(\sigma) = \log d$ if and only if $\sigma = \mathrm{I}/d$,
(iv) if $\sigma \in \mathrm{D}(\mathbb{C}^d)$ and $U \in \mathsf{U}(d)$, then $S(\sigma) = S(U\sigma U^\dagger)$,
(v) if $\sigma \in \mathrm{D}(\mathbb{C}^d)$ and $\tau \in \mathrm{D}(\mathbb{C}^n)$, then $S(\sigma \otimes \tau) = S(\sigma) + S(\tau)$.

PROOF. All these properties are straightforward to show, except perhaps the concavity which follows from the concavity of $x \mapsto -x \log x$, together with Klein's lemma (Exercise 1.40). □

The following lemma quantifies the fact that very mixed states have large entropy.

LEMMA 1.20. *Let $\rho \in \mathrm{D}(\mathbb{C}^d)$ be a state with spectrum in the interval $\left[\frac{1-\varepsilon}{d}, \frac{1+\varepsilon}{d}\right]$ for some $\varepsilon \in [0,1]$. Then $S(\rho) \geqslant \log d - h(\varepsilon)$, where*

$$h(\varepsilon) = \frac{1+\varepsilon}{2}\log(1+\varepsilon) + \frac{1-\varepsilon}{2}\log(1-\varepsilon).$$

Note that $h(\varepsilon) \sim \varepsilon^2/2$ as ε goes to 0.

PROOF. Assume that d is even and consider a state $\sigma \in \mathrm{D}(\mathbb{C}^d)$ with $d/2$ eigenvalues equal to $(1+\varepsilon)/d$ and $d/2$ eigenvalues equal to $(1-\varepsilon)/d$. One finds directly from the definition of majorization that $\mathrm{spec}(\rho) \prec \mathrm{spec}(\sigma)$. It follows then from Proposition 1.12 (iv) that

$$S(\rho) \geqslant S(\sigma) = \log(d) - h(\varepsilon).$$

If d is odd, a similar argument applies where σ has $(d-1)/2$ eigenvalues equal to $(1 \pm \varepsilon)/d$ and one eigenvalue equal to $1/d$. One checks by direct computation that $S(\sigma) > \log(d) - h(\varepsilon)$. □

REMARK 1.21. Note that while the entropy of (normalized) quantum states (i.e., $\rho \in \mathrm{D}$) is of primary physical interest, the definition makes sense for, and most properties generalize to, $\rho \in \mathcal{PSD}$.

Let σ be a state on $\mathbb{C}^d$, and $p \in (0, \infty)$. The *p-Rényi entropy* of σ is

$$S_p(\sigma) = \frac{1}{1-p}\log \mathrm{Tr}(\sigma^p). \tag{1.37}$$

The definition for $p = 1$ should be understood as the limit as $p \to 1$. We then recover the von Neumann entropy, so that $S_1 = S$. Other limit cases are $p \to 0$, which gives $S_0(\sigma) = \log \mathrm{rank}\, \sigma$, and $p \to \infty$, which gives $S_\infty(\sigma) = -\log \|\sigma\|_\infty$. When $p > 1$, the Rényi entropy is connected to the Schatten p-norm by the formula $S_p(\sigma) = \frac{p}{1-p}\log \|\sigma\|_p$. Just like the von Neumann entropy is a generalization of *Shannon entropy*, defined for classical states (probability mass functions) $\mathbf{p} = (p_k) \in \Delta_n$ by

$$H(\mathbf{p}) := -\sum_k p_k \log p_k, \tag{1.38}$$

the Rényi entropy may be thought of as a generalization of the ℓ_p-norm (up to logarithmic change of variables and rescaling; it also has a classical variant defined via $H_p(\mathbf{q}) := \frac{p}{1-p}\log \|\mathbf{q}\|_p$).

EXERCISE 1.49 (Properties of Rényi entropies). Verify that, for $p \in (0, \infty]$, S_p satisfies properties (i)–(v) from Proposition 1.19. Note that (iii) fails for $p = 0$.

EXERCISE 1.50 (Entropy of the state vs. entropy of the diagonal). Show that, for any $\rho \in \mathrm{D}$, $S(\operatorname{diag} \rho) \geqslant S(\rho)$, with equality only if ρ is diagonal.

EXERCISE 1.51 (Monotonicity of Rényi entropies). Show that $S_p(\sigma)$ and $H_p(\mathbf{q})$ are non-increasing in p for fixed $\sigma, \mathbf{q}$.

Notes and Remarks

A presentation of convex analysis oriented towards applications (notably to computer science) can be found in [**Bar02**]. An older but still valuable reference is the book [**Roc70**].

Section 1.1. Following the customary usage in functional analysis, we name Theorem 1.3 after Krein–Milman. However, it should be pointed out that the main contribution by Krein–Milman is an extension to infinite-dimensional locally convex spaces; the finite-dimensional case, which is presented here, is due to Minkowski [**Min11**].

The inequality (1.5) proved in Exercise 1.11 is due to Hanner [**Han56**]; it belongs to the family of inequalities (including the earlier Clarkson inequalities [**Cla36**]) that degenerate into the parallelogram identity when $p = 2$. The inequality (1.6) is the so-called "2-uniform convexity" of the p-norm for $p \in (1, 2]$. For $p \geqslant 2$, the inequality is reversed (2-uniform smoothness); for $p = 1$, it degenerates into the triangle inequality. One establishes similarly p-uniform convexity for $p \in [2, \infty)$ and p-uniform smoothness for $p \in (1, 2]$.

It is natural to ask whether these inequalities remain valid for the Schatten p-norm, i.e., when x, y are matrices. This is known to be true for inequality (1.6) when $1 \leqslant p \leqslant 2$ (and for its reversed form when $p \geqslant 2$). However, the stronger Hanner inequality (1.5) for matrices has been proved only in the range $1 \leqslant p \leqslant 4/3$ (or, for the reversed inequality, in the range $p \geqslant 4$). For proofs and references, see [**BCL94, CL06**].

Section 1.2. Lemma 1.6 seems to be a folklore result, but does not appear in standard references for convexity (the best source we were pointed to after consulting specialists was Exercise 6, §3.4 of [**Grü03**]). However, once stated, the lemma is straightforward to prove.

Convex cones play a fundamental role in the theory of convex optimization and in linear and semi-definite programming, all of which have their own links to quantum information. We do not develop any of these areas or connections here. We refer the interested reader to the books [**BV04**] and [**BTN01a**], the survey [**Nem07**], and, for sample links, to [**Rei08, KL09, BH13, HNW15**].

Section 1.3. A comprehensive reference for majorization and for connections to matrix inequalities is the book [**Bha97**]. Klein's lemma originates from [**Kle32**]. Davis convexity theorem appears in [**Dav57**]. Early references for Schatten norms include [**Sch50, Sch70**].

The concept of von Neumann entropy is crucial in quantum information theory and quantum Shannon theory. A reason for this is that von Neumann entropy and its variants (quantum relative entropy, quantum mutual information) have several *operational* interpretations, i.e., quantify the rate at which basic information processing tasks (transmission, encoding, decoding) can be performed. This point of view is hardly mentioned in this book. For an accessible introduction to quantum Shannon theory we refer to [**Wil17**]. Interestingly, the concept of von Neumann entropy appears already in [**von27**, **von32**] (see [**Pet01**] for historical background) and predates the development of its classical counterpart, the Shannon entropy which—like much of modern information theory—has its roots in the 1948 two-part article by Claude Shannon [**Sha48**].

Many texts use base 2 logarithm to define entropy. While using the natural logarithm simplifies some calculations, the choice of the base is immaterial in our context; as a rule, the stated identities and estimates typically hold for any base, as long as one is consistent. The few exceptions to this principle are clearly marked.

CHAPTER 2

The mathematics of quantum information theory

This chapter puts into mathematical perspective some basic concepts of quantum information theory. (For a physically motivated approach, see Chapter 3.) We discuss the geometry of the set of quantum states, the entanglement vs. separability dichotomy, and introduce completely positive maps and quantum channels. All these concepts will be extensively used in Chapters 8–12.

2.1. On the geometry of the set of quantum states

2.1.1. Pure and mixed states. In this section we take a closer look at the set $\mathrm{D}(\mathcal{H})$ (or simply D) of quantum states on a finite-dimensional complex Hilbert space $\mathcal{H}$. By definition (see Section 0.10), we have

$$\mathrm{D}(\mathcal{H}) = \{\rho \in B_{\mathrm{sa}}(\mathcal{H}) \;:\; \rho \geqslant 0, \operatorname{Tr}\rho = 1\}. \tag{2.1}$$

If $\mathcal{H} = \mathbb{C}^d$, the definition (2.1) simply says that $\mathrm{D}(\mathbb{C}^d)$ is the base of the positive semi-definite cone $\mathcal{PSD}(\mathbb{C}^d)$ defined by the hyperplane $H_1 \subset \mathsf{M}_d^{\mathrm{sa}}$ of trace one Hermitian matrices (cf. (1.22)). The (real) dimension of the set $\mathrm{D}(\mathbb{C}^d)$ equals $d^2 - 1$: it has non-empty interior inside H_1. (This follows from $\mathcal{PSD}(\mathbb{C}^d)$ being a full cone.)

A state $\rho \in \mathrm{D}(\mathcal{H})$ is called *pure* if it has rank 1, i.e., if there is a unit vector $\psi \in \mathcal{H}$ such that

$$\rho = |\psi\rangle\langle\psi|.$$

Note that $|\psi\rangle\langle\psi|$ is the orthogonal projection onto the (complex) line spanned by ψ. We sometimes use the terminology "consider a pure state ψ" (such language is prevalent in physics literature). What we mean is that ψ is a unit vector and we consider the corresponding pure state $|\psi\rangle\langle\psi|$. We use the terminology of *mixed* states when we want to emphasize that we consider the set of all states, not necessarily pure.

Let ψ, χ be unit vectors in $\mathcal{H}$. Then the pure states $|\psi\rangle\langle\psi|$ and $|\chi\rangle\langle\chi|$ coincide if and only if there is a complex number λ with $|\lambda| = 1$ such that $\chi = \lambda\psi$. Therefore the set of pure states identifies with $\mathsf{P}(\mathcal{H})$, the projective space on $\mathcal{H}$. (See Appendix B.2; note that the space $\mathsf{P}(\mathbb{C}^d)$ is more commonly denoted by $\mathbb{CP}^{d-1}$.)

The set $\mathrm{D}(\mathcal{H})$ is a compact convex set, and it is easily checked that the extreme points of $\mathrm{D}(\mathcal{H})$ are exactly the pure states (cf. Proposition 1.9 and Corollary 1.10).

It follows from general convexity theory (Krein–Milman and Carathéodory's theorems) that any state is a convex combination of at most $(\dim\mathcal{H})^2$ pure states. However, using the spectral theorem instead tells us more: any state is a convex combination of at most $\dim\mathcal{H}$ pure states $|\psi_i\rangle\langle\psi_i|$, where (ψ_i) are pairwise orthogonal unit vectors (cf. Exercise 1.45). A fundamental consequence is that whenever we want to maximize a convex function (or minimize a concave function) over the set $\mathrm{D}(\mathcal{H})$, the extremum is achieved on a pure state, which significantly reduces the dimension of the problem.

As opposed to pure states, which are extremal, the "most central" element in $\mathrm{D}(\mathcal{H})$ is the state $\mathrm{I}/\dim\mathcal{H}$, which is called the *maximally mixed state* and denoted by ρ_* when there is no ambiguity. We also note that the set of states on $\mathcal{H}$ which are diagonal with respect to a given orthonormal basis $(e_i)_{i\in I}$ naturally identifies with the set of classical states on I.

EXERCISE 2.1. Describe states which belong to the boundary of $\mathrm{D}(\mathcal{H})$.

EXERCISE 2.2 (Every state is an average of pure states). Show that every state $\rho \in \mathrm{D}(\mathbb{C}^d)$ can be written as $\frac{1}{d}(|\psi_1\rangle\langle\psi_1| + \cdots + |\psi_d\rangle\langle\psi_d|)$ for some unit vectors $\psi_1, \dots, \psi_d$ in $\mathbb{C}^d$.

2.1.2. The Bloch ball $\mathrm{D}(\mathbb{C}^2)$. The situation for $d = 2$ is very special. Let $\rho \in \mathsf{M}_2^{\mathrm{sa}}$, with $\operatorname{Tr}\rho = 1$. Then ρ has two eigenvalues, which can be written as $1/2-\lambda$ and $1/2+\lambda$ for some $\lambda \in \mathbb{R}$. Moreover, $\rho \geqslant 0$ if and only if $|\lambda| \leqslant 1/2$. On the other hand, we have

$$\|\rho - \rho_*\|_{\mathrm{HS}} = \sqrt{2}|\lambda|.$$

Therefore, ρ is a state if and only if $\|\rho - \rho_*\|_{\mathrm{HS}} \leqslant 1/\sqrt{2}$. What we have proved is that, inside the space of trace one self-adjoint operators, the set of states is a Euclidean ball centered at ρ_* and with radius $1/\sqrt{2}$. This ball is called the *Bloch ball* and its boundary is called the *Bloch sphere*. Once we introduce the *Pauli matrices*

$$\sigma_x = \begin{bmatrix} 0 & 1 \\ 1 & 0 \end{bmatrix}, \qquad \sigma_y = \begin{bmatrix} 0 & -i \\ i & 0 \end{bmatrix}, \qquad \sigma_z = \begin{bmatrix} 1 & 0 \\ 0 & -1 \end{bmatrix}, \tag{2.2}$$

a convenient orthonormal basis (with respect to the Hilbert–Schmidt inner product) in $\mathsf{M}_2^{\mathrm{sa}}$ is

$$\Big(\frac{1}{\sqrt{2}}\mathrm{I}, \frac{1}{\sqrt{2}}\sigma_x, \frac{1}{\sqrt{2}}\sigma_y, \frac{1}{\sqrt{2}}\sigma_z\Big). \tag{2.3}$$

A very useful consequence of $\mathrm{D}(\mathbb{C}^2)$ being a ball is the fact—mentioned already in Section 1.2.1—that the cone $\mathcal{PSD}(\mathbb{C}^2)$ is isomorphic (or even isometric in the appropriate sense) to the Lorentz cone $\mathcal{L}_4$. A popular explicit isomorphism, inducing the so-called *spinor map* (see Appendix C), is given by

$$\mathbb{R}^4 \ni \mathbf{x} = (t, x, y, z) \mapsto \begin{bmatrix} t+z & x-iy \\ x+iy & t-z \end{bmatrix} = X \in \mathsf{M}_2^{\mathrm{sa}}. \tag{2.4}$$

The formula for X can be rewritten in terms of the Pauli matrices (2.2) as

$$X = t\,\mathrm{I} + x\sigma_x + y\sigma_y + z\sigma_z, \tag{2.5}$$

and so a convenient expression for it is $X = \mathbf{x}\cdot\sigma$, where σ is a shorthand for $(\mathrm{I}, \sigma_x, \sigma_y, \sigma_z)$, and "$\cdot$" is a "formal dot product". Since $\{\mathrm{I}, \sigma_x, \sigma_y, \sigma_z\}$ is a multiple of the orthonormal basis (2.3) of $\mathsf{M}_2^{\mathrm{sa}}$, it follows that the map given by (2.4) is likewise a multiple of isometry (with respect to the Euclidean metric in the domain and the Hilbert–Schmidt metric in the range). Next, it is readily verified that

$$\frac{1}{2}\operatorname{Tr} X = t, \qquad \det X = t^2 - x^2 - y^2 - z^2 =: q(\mathbf{x}), \tag{2.6}$$

where q is the quadratic form of the Minkowski spacetime, which confirms that $X \in \mathcal{PSD}(\mathbb{C}^2)$ iff $\mathbf{x} \in \mathcal{L}_4$. The isomorphism $\mathbf{x} \mapsto \mathbf{x}\cdot\sigma$ will be useful in understanding automorphisms of the cones $\mathcal{L}_4$ and $\mathcal{PSD}(\mathbb{C}^2)$, and when proving Størmer's theorem in Section 2.4.5.

When $d > 2$, the set $\mathrm{D}(\mathbb{C}^d)$ is no longer a ball, but rather the non-commutative analogue of a simplex. Its symmetrization (see Section 4.1.2)

$$\mathrm{D}(\mathbb{C}^d)_{\oslash} = \mathrm{conv}\left(\mathrm{D}(\mathbb{C}^d) \cup -\mathrm{D}(\mathbb{C}^d)\right) = \{A \in \mathsf{M}_d^{\mathrm{sa}} \ : \ \|A\|_1 \leqslant 1\}$$

is $S_1^{d,\mathrm{sa}}$, the unit ball of the self-adjoint part of the 1-Schatten space (see Section 1.3.2).

One way to quantify the fact that the set $\mathrm{D}(\mathbb{C}^d)$ is different from a ball when $d > 2$ is to compute the radius of its inscribed and circumscribed Hilbert–Schmidt balls. The former equals $1/\sqrt{d(d-1)}$ while the latter is $\sqrt{(d-1)/d}$ (the same values as for the set Δ_{d-1} of classical states on $\{1, \dots, d\}$, and for the same reasons). In other words, if we denote by $B(\rho_*, r)$ the ball centered at ρ_* and with Hilbert–Schmidt radius r inside the hyperplane $H_1 = \{\mathrm{Tr}(\cdot) = 1\} \subset \mathsf{M}_d^{\mathrm{sa}}$, we have

$$B\left(\rho_*, \frac{1}{\sqrt{d(d-1)}}\right) \subset \mathrm{D}(\mathbb{C}^d) \subset B\left(\rho_*, \sqrt{\frac{d-1}{d}}\right), \tag{2.7}$$

and these values—differing by the factor of $d-1$—are the best possible.

EXERCISE 2.3 (The Bloch sphere is a sphere). Show that the matrix X given by (2.5) has eigenvalues 1 and -1 if and only if $t = 0$ and $x^2 + y^2 + z^2 = 1$.

EXERCISE 2.4 (Composition rules for Pauli matrices). Verify the composition rules for Pauli matrices. (i) $\sigma_a^2 = \mathrm{I}$. (ii) If a, b, c are all different, then $\sigma_a \sigma_b = i\varepsilon\sigma_c$, where $\varepsilon = \pm 1$ is the sign of the permutation $(x, y, z) \mapsto (a, b, c)$; in particular, if $a \neq b$, then $\sigma_a\sigma_b = -\sigma_b\sigma_a$.

2.1.3. Facial structure.

PROPOSITION 2.1 (Characterization of faces of D). *There is a one-to-one correspondence between nontrivial subspaces of $\mathbb{C}^d$ and proper faces of $\mathrm{D}(\mathbb{C}^d)$. Given a subspace $\{0\} \subsetneq E \subsetneq \mathbb{C}^d$, the corresponding face $\mathrm{D}(E)$ is the set of states whose range is contained in E:*

$$\mathrm{D}(E) = \{\rho \in \mathrm{D}(\mathbb{C}^d) \ : \ \rho(\mathbb{C}^d) \subset E\}.$$

In particular, pure states (extreme points, i.e., minimal, 0-dimensional faces) correspond to the case $\dim E = 1$. In the direction opposed to a pure state $|x\rangle\langle x|$ lies a face which corresponds to all states with a range orthogonal to x; these are maximal proper faces.

REMARK 2.2. All faces of $\mathrm{D}(\mathbb{C}^d)$ are exposed (as defined in Exercise 1.5) since $\mathrm{D}(E)$ is the intersection of $\mathrm{D}(\mathbb{C}^d)$ with the hyperplane $\{X \ : \ \mathrm{Tr}(XP_E) = 1\}$.

PROOF OF PROPOSITION 2.1. Denote by $\mathrm{range}(\rho) = \rho(\mathbb{C}^d)$ the range of a state $\rho \in \mathrm{D}(\mathbb{C}^d)$. We use the following observation: if $\rho, \sigma \in \mathrm{D}(\mathbb{C}^d)$ and $\lambda \in (0,1)$, then

$$\mathrm{range}(\lambda\rho + (1-\lambda)\sigma) = \mathrm{range}(\rho) + \mathrm{range}(\sigma). \tag{2.8}$$

We first check that, for any nontrivial subspace $E \subset \mathbb{C}^d$, $\mathrm{D}(E)$ is a face of $\mathrm{D}(\mathbb{C}^d)$. For indeed, if $\rho \in \mathrm{D}(E)$ can be written as $\lambda\rho_1 + (1-\lambda)\rho_2$ for $\rho_1, \rho_2 \in \mathrm{D}(\mathbb{C}^d)$ and $\lambda \in (0,1)$, then (2.8) implies that $\mathrm{range}(\rho_1) \subset E$ and $\mathrm{range}(\rho_2) \subset E$.

Conversely, let $F \subset \mathrm{D}(\mathbb{C}^d)$ be a proper face. Define $E = \bigcup\{\mathrm{range}(\rho) \ : \ \rho \in F\}$. It follows—from (2.8) and from the fact that F is convex—that E is actually a subspace and that F contains an element ρ such that range$(\rho) = E$. We now claim that $F = \mathrm{D}(E)$. The direct inclusion is obvious. Conversely, consider $\sigma \in \mathrm{D}(E)$. For $\lambda > 0$ small enough, the operator $\tau = \frac{1}{1-\lambda}(\rho - \lambda\sigma)$ is a state. Since $\rho = \lambda\sigma + (1-\lambda)\tau$, we conclude that the segment joining σ and τ is contained in F; in particular $\sigma \in F$. □

EXERCISE 2.5. Show directly (i.e., without appealing to Proposition 2.1) that any *exposed* face of $\mathrm{D}(\mathbb{C}^d)$ has the form $\mathrm{D}(E)$ for some subspace $E \subset \mathbb{C}^d$.

2.1.4. Symmetries. We now describe the symmetries of $\mathrm{D}(\mathbb{C}^d)$. This is closely related to the famous theorem of Wigner that characterizes the isometries of complex projective space as a metric space. Recall (see Appendix B.2) that $[\psi]$ denotes the equivalence class in $\mathsf{P}(\mathbb{C}^d)$ of a unit vector $\psi \in S_{\mathbb{C}^d}$.

THEOREM 2.3 (Wigner's theorem). *Denote by $\mathsf{P}(\mathbb{C}^d)$ the projective space over $\mathbb{C}^d$, equipped with the Fubini–Study metric* (B.5). *A map $f : \mathsf{P}(\mathbb{C}^d) \to \mathsf{P}(\mathbb{C}^d)$ is an isometry if and only if there is a map U on $\mathbb{C}^d$ which is either unitary or anti-unitary such that, for any unit vector ψ,*

$$f([\psi]) = [U(\psi)]. \tag{2.9}$$

A map $U : \mathbb{C}^d \to \mathbb{C}^d$ is anti-unitary if it is the composition of a unitary map with complex conjugation.

PROOF. We outline the proof of Wigner's theorem for $d = 2$. Since the projective space over $\mathbb{C}^2$ identifies with the Bloch sphere, its group of isometries is given by the orthogonal group $\mathsf{O}(3)$, and splits into direct isometries (rotations, or $\mathsf{SO}(3)$) and indirect isometries.

Let f be a direct isometry of the Bloch ball. It has two opposite fixed points $[\varphi_1]$ and $[\varphi_2]$, with $\varphi_1 \perp \varphi_2$, and is a rotation of angle θ in the plane $\{[\frac{1}{\sqrt{2}}(\varphi_1 + e^{i\alpha}\varphi_2)] : \alpha \in \mathbb{R}\}$. One checks that (2.9) is satisfied when U is given by $U(\varphi_1) = \varphi_1$ and $U(\varphi_2) = e^{i\theta}\varphi_2$. Note that U is determined up to a global phase. In particular, if we insist on having $U \in \mathsf{SU}(2)$, we are led to the choice $U(\varphi_1) = e^{-i\theta/2}\varphi_1$ and $U(\varphi_2) = e^{i\theta/2}\varphi_2$ involving the half-angle. (We point out the isomorphism $\mathsf{PSU}(2) \leftrightarrow \mathsf{SO}(3)$, see Exercise B.4.)

The complex conjugation with respect to an orthonormal basis (ψ_1, ψ_2) in $\mathbb{C}^2$ induces on the Bloch ball the reflection R in the plane $\{[\cos\theta\psi_1 + \sin\theta\psi_2] : \theta \in \mathbb{R}\}$. Since any indirect isometry of the Bloch ball is the composition of R with a direct isometry, the result follows.

The case $d > 2$ can be deduced from the $d = 2$ case; we do not include the argument here (see Notes and Remarks). □

When $\mathsf{P}(\mathbb{C}^d)$ is identified with the set of pure states on $\mathbb{C}^d$, the isometries from Theorem 2.3 act as $\rho \mapsto U\rho U^\dagger$ or $\rho \mapsto U\rho^T U^\dagger$ for $U \in \mathsf{U}(d)$. Here ρ^T denotes the transposition of a state ρ with respect to a distinguished basis (since $\rho = \rho^\dagger$, ρ^T is also the complex conjugate of ρ with respect to that basis).

THEOREM 2.4 (Kadison's theorem). *Affine maps preserving globally $\mathrm{D}(\mathbb{C}^d)$ are of the form $\rho \mapsto U\rho U^\dagger$ or $\rho \mapsto U\rho^T U^\dagger$ for $U \in \mathsf{U}(d)$. In particular, they are isometries with respect to the Hilbert–Schmidt distance.*

PROOF. Let Φ be an affine map on $\mathsf{M}_d^{\mathrm{sa}}$ such that $\Phi(\mathrm{D}(\mathbb{C}^d)) = \mathrm{D}(\mathbb{C}^d)$. Then Φ preserves the set of faces of $\mathrm{D}(\mathbb{C}^d)$, which are described in Proposition 2.1. In particular, Φ preserves the set of minimal faces which identify with pure states. Therefore Φ induces a bijection on $\mathsf{P}(\mathbb{C}^d)$. We claim that Φ is an isometry with respect to the Fubini–Study distance (B.5), which is equivalent to

$$\operatorname{Tr}\left(\Phi(|\psi\rangle\langle\psi|)\cdot\Phi(|\varphi\rangle\langle\varphi|)\right) = |\langle\psi,\varphi\rangle|^2$$

for $\psi,\varphi \in \mathbb{C}^d$. If $[\psi] = [\varphi]$, this is clear. Otherwise, let $M \subset \mathbb{C}^d$ be the 2-dimensional subspace generated by ψ and φ. By Proposition 2.1, the set $\mathrm{D}(M)$ canonically identifies with a (3-dimensional) face of $\mathrm{D}(\mathbb{C}^d)$. Consequently, $\Phi(\mathrm{D}(M))$ is also a face, which identifies with $\mathrm{D}(M')$ for some 2-dimensional subspace $M' \subset \mathbb{C}^d$. Since $\mathrm{D}(M)$ and $\mathrm{D}(M')$ are Bloch balls, the map Φ restricted to $\mathrm{D}(M)$ must be an isometry (affine maps preserving S^2 are isometries). We may now apply Wigner's theorem: there is $U \in \mathsf{U}(d)$ such that either $\Phi(\rho) = U\rho U^\dagger$ whenever ρ is a pure state, or $\Phi(\rho) = U\rho^T U^\dagger$ for all pure states ρ. Since Φ is affine, one of the two formulas is valid for all $\rho \in \mathrm{D}(\mathbb{C}^d)$. □

Although for $d > 2$ the set $\mathrm{D}(\mathbb{C}^d)$ is not centrally symmetric, we may argue that the maximally mixed state ρ_* plays the role of a center. In particular, we have

PROPOSITION 2.5. *Let $\rho \in \mathrm{D}(\mathbb{C}^d)$ be a state which is fixed by all the isometries of $\mathrm{D}(\mathbb{C}^d)$ (with respect to the Hilbert–Schmidt distance). Then $\rho = \rho_*$.*

PROOF. We have $U\rho U^\dagger = \rho$ for every unitary matrix U. Since $\mathsf{U}(d)$ spans M_d as a vector space, ρ commutes with any matrix, therefore it equals $\alpha\mathrm{I}$ for some $\alpha \in \mathbb{C}$, and the trace constraint forces $\alpha = 1/d$. □

One consequence of Proposition 2.5 is that ρ_* is the centroid of $\mathrm{D}(\mathbb{C}^d)$. Kadison's theorem also implies that D has enough symmetries in the sense of Section 4.2.2 (see Exercise 4.25). Another consequence of Kadison's Theorem 2.4 is a characterization of affine automorphisms of the cone of positive semi-definite matrices, which will be presented in Proposition 2.29.

EXERCISE 2.6. Show that the affine automorphisms of $\mathrm{D}(\mathbb{C}^2)$ form a group which is isomorphic to $\mathsf{O}(3)$.

EXERCISE 2.7. Show that the affine automorphisms of $\mathrm{D}(\mathbb{C}^d)$ form a group which is isomorphic to the semidirect product of $\mathsf{PSU}(d)$ and $\mathbb{Z}_2$ with respect to the action of $\mathbb{Z}_2$ on $\mathsf{PSU}(d)$ induced by the complex conjugation.

EXERCISE 2.8. State and prove the real version of Wigner's theorem.

EXERCISE 2.9. Let ρ be a state which is invariant under transposition with respect to any basis. Show that $\rho = \rho_*$.

2.2. States on multipartite Hilbert spaces

2.2.1. Partial trace. A fundamental concept in quantum information theory is the *partial trace* (for a physically motivated approach, see Section 3.4). Let $\mathcal{H} = \mathcal{H}_1 \otimes \mathcal{H}_2$ be a bipartite Hilbert space. The partial trace over $\mathcal{H}_2$ is the map (or the superoperator, see Section 0.9) $\operatorname{Tr}_{\mathcal{H}_2} : B(\mathcal{H}_1) \otimes B(\mathcal{H}_2) \to B(\mathcal{H}_1)$ defined as $\mathrm{Id}_{B(\mathcal{H}_1)} \otimes \operatorname{Tr}$. Its action on product operators is given by

$$\operatorname{Tr}_{\mathcal{H}_2}(A \otimes B) = (\operatorname{Tr} B)A$$

for $A \in B(\mathcal{H}_1)$, $B \in B(\mathcal{H}_2)$. Similarly, the partial trace with respect to $\mathcal{H}_1$ is defined as $\operatorname{Tr}_{\mathcal{H}_1} = \operatorname{Tr} \otimes \mathrm{Id}_{B(\mathcal{H}_2)}$

In particular, if ρ is a state on $\mathcal{H}_1 \otimes \mathcal{H}_2$, then $\operatorname{Tr}_{\mathcal{H}_1} \rho$ is a state on $\mathcal{H}_2$, and $\operatorname{Tr}_{\mathcal{H}_2}$ is a state on $\mathcal{H}_1$. Note also the formulas $\operatorname{Tr}_{\mathcal{H}_1}(\rho_1 \otimes \rho_2) = \rho_2$ and $\operatorname{Tr}_{\mathcal{H}_2}(\rho_1 \otimes \rho_2) = \rho_1$ for states $\rho_1 \in \mathrm{D}(\mathcal{H}_1)$, $\rho_2 \in \mathrm{D}(\mathcal{H}_2)$.

We sometimes write Tr_1 for $\operatorname{Tr}_{\mathcal{H}_1}$ and Tr_2 for $\operatorname{Tr}_{\mathcal{H}_2}$. The definition of partial trace extends naturally to the multipartite setting: if $\mathcal{H} = \mathcal{H}_1 \otimes \cdots \otimes \mathcal{H}_k$, then for $1 \leqslant i \leqslant k$ we denote by $\operatorname{Tr}_{\mathcal{H}_i}$ or Tr_i the operation

$$\mathrm{Id}_{B(\mathcal{H}_1)} \otimes \cdots \otimes \mathrm{Id}_{B(\mathcal{H}_{i-1})} \otimes \operatorname{Tr} \otimes \mathrm{Id}_{B(\mathcal{H}_{i+1})} \otimes \cdots \otimes \mathrm{Id}_{B(\mathcal{H}_k)} .$$

2.2.2. Schmidt decomposition. We recall the *singular value decomposition* (SVD) for matrices: any real or complex matrix $A \in \mathsf{M}_{k,d}$ can be decomposed as $A = U\Sigma V^\dagger$, when U and V are unitary matrices of sizes k and d respectively, and $\Sigma = (\Sigma_{ij}) \in \mathsf{M}_{k,d}$ is a "rectangular diagonal" (i.e., such that $\Sigma_{ij} = 0$ whenever $i \neq j$) nonnegative matrix. Moreover, up to permutation, the "diagonal" elements of Σ are uniquely determined by A and are called the *singular values* of A. We often denote the singular values of A by $s_1(A) \geqslant \cdots \geqslant s_{\min(k,d)}(A)$. The singular values of A coincide with the eigenvalues of $(AA^\dagger)^{1/2}$ when $k \leqslant d$, and with the eigenvalues of $(A^\dagger A)^{1/2}$ when $k \geqslant d$. Note that, in any case, $AA^\dagger$ and $A^\dagger A$ share the same nonzero eigenvalues.

An equivalent presentation of the SVD is as follows: there exist orthonormal sequences (u_i) (in $\mathbb{R}^k$ or $\mathbb{C}^k$, depending on the context) and (v_i) (in $\mathbb{R}^d$ or $\mathbb{C}^d$), and a non-increasing sequence of nonnegative scalars (s_i) such that

$$A = \sum_i s_i |u_i\rangle\langle v_i|. \tag{2.10}$$

When translated into the language of tensors (see Section 0.4), the singular value decomposition becomes the *Schmidt decomposition*, which is widely used in quantum information. We note that, besides the bipartite situation, there is no analogue of the Schmidt decomposition in multipartite Hilbert spaces.

PROPOSITION 2.6 (Easy). *Let ψ be a vector in a (real or complex) bipartite Hilbert space $\mathcal{H}_1 \otimes \mathcal{H}_2$, with $d_1 = \dim \mathcal{H}_1$ and $d_2 = \dim \mathcal{H}_2$. Set $d := \min(d_1, d_2)$. Then there exist nonnegative scalars $(\lambda_i)_{1 \leqslant i \leqslant d}$, and orthonormal vectors $(\chi_i)_{1 \leqslant i \leqslant d}$ in $\mathcal{H}_1$ and $(\varphi_i)_{1 \leqslant i \leqslant d}$ in $\mathcal{H}_2$, such that*

$$\psi = \sum_{i=1}^d \lambda_i \chi_i \otimes \varphi_i. \tag{2.11}$$

The numbers $(\lambda_1, \ldots, \lambda_d)$ are uniquely determined if we require that $\lambda_1 \geqslant \cdots \geqslant \lambda_d$ and are called the Schmidt coefficients of ψ.

Note that $\lambda_1^2 + \cdots + \lambda_d^2 = |\psi|^2$. We may write $\lambda_i(\psi)$ instead of λ_i to emphasize the dependence on ψ. The largest r such that $\lambda_r(\psi) > 0$ is called the *Schmidt rank* of ψ. If $\psi \in \mathbb{C}^k \otimes \mathbb{C}^d$ is identified with a matrix $M \in \mathsf{M}_{k,d}$ as in Section 0.8, then

$$\operatorname{Tr}_{\mathbb{C}^d} |\psi\rangle\langle\psi| = MM^\dagger. \tag{2.12}$$

Via this identification, Schmidt coefficients of ψ coincide with singular values of M, and the Schmidt rank of ψ coincides with the rank of M. States of Schmidt rank 1 are exactly product vectors. The largest and the smallest Schmidt coefficients of $\psi \in \mathcal{H}_1 \otimes \mathcal{H}_2$ are also given by the variational formulas

$$\lambda_1(\psi) = \max\{|\langle\psi, \chi \otimes \varphi\rangle| \ : \ \chi \in \mathcal{H}_1, \varphi \in \mathcal{H}_2, |\chi| = |\varphi| = 1\}, \tag{2.13}$$

often referred to as the maximal *overlap* with a product vector, and

$$\lambda_d(\psi) = \min_{\chi \in \mathcal{H}_1, |\chi|=1} \max_{\varphi \in \mathcal{H}_2, |\varphi|=1} |\langle \psi, \chi \otimes \varphi \rangle|. \tag{2.14}$$

The above are fully analogous to the (special cases of) Courant–Fischer variational formulas for singular values of a matrix.

2.2.3. A fundamental dichotomy: Separability vs. entanglement. We now introduce a fundamental concept: the dichotomy between separability and entanglement for quantum states. Let $\mathcal{H}$ be a **complex** Hilbert space admitting a tensor decomposition

$$\mathcal{H} = \mathcal{H}_1 \otimes \cdots \otimes \mathcal{H}_k. \tag{2.15}$$

Recall that since 1-dimensional factors may be dropped, we may—and usually will—assume that all the factors are of dimension at least 2.

DEFINITION 2.7. A pure state $\rho = |\chi\rangle\langle\chi|$ on $\mathcal{H}$ is said to be *pure separable* if the unit vector χ is a product vector, i.e., if there exist unit vectors $\chi_1, \ldots, \chi_k$ such that $\chi = \chi_1 \otimes \cdots \otimes \chi_k$. In that case,

$$\rho = |\chi_1\rangle\langle\chi_1| \otimes \cdots \otimes |\chi_k\rangle\langle\chi_k|. \tag{2.16}$$

Extending the definition of separability to mixed states requires us to consider convex combinations (we study in detail the convex hull operation $A \mapsto \operatorname{conv}(A)$ in Section 1.1.2).

DEFINITION 2.8. A mixed state $\rho = |\chi\rangle\langle\chi|$ on $\mathcal{H}$ is said to be *separable* if it can be written as a convex combination of pure separable states. We denote by $\operatorname{Sep}(\mathcal{H})$ (or simply by Sep) the set of separable states on $\mathcal{H}$. We have

$$\operatorname{Sep}(\mathcal{H}) = \operatorname{conv}\{|\chi_1 \otimes \cdots \otimes \chi_k\rangle\langle\chi_1 \otimes \cdots \otimes \chi_k| \ : \ \chi_1 \in \mathcal{H}_1, \ldots, \chi_k \in \mathcal{H}_k\}. \tag{2.17}$$

States which are not separable are called *entangled.* Since pure states are the extreme points even of the larger set $\mathrm{D}(\mathcal{H})$ (Proposition 2.1), it follows that the pure separable states (i.e., those given by (2.16)) are exactly the extreme points of $\operatorname{Sep}(\mathcal{H})$. Since there are vectors that are not product vectors, the set $\operatorname{Sep}(\mathcal{H})$ is a proper subset of $\mathrm{D}(\mathcal{H})$. A schematic representation of the inclusion $\mathrm{Sep} \subset \mathrm{D}$ and of the corresponding extreme points can be found in Figure 2.1.

An alternative description of the set $\operatorname{Sep}(\mathcal{H})$ is the following: it is the convex hull of product states.

$$\operatorname{Sep}(\mathcal{H}) = \operatorname{conv}\{\rho_1 \otimes \cdots \otimes \rho_k \ : \ \rho_1 \in \mathrm{D}(\mathcal{H}_1), \ldots, \rho_k \in \mathrm{D}(\mathcal{H}_k)\}. \tag{2.18}$$

It is noteworthy that $\operatorname{Sep}(\mathcal{H})$ and $\mathrm{D}(\mathcal{H})$ have the same dimension. This can be seen from the following observation. Let $V_1, \ldots, V_k$ be real or complex vector spaces and, for each i, let $\mathcal{F}_i$ be a family of linear independent vectors in V_i. Then the family

$$\bigotimes \mathcal{F}_i = \{f_1 \otimes \cdots \otimes f_k \ : \ f_i \in \mathcal{F}_i\}$$

is linearly independent in $\bigotimes V_i$. We apply the observation with $V_i = B^{\mathrm{sa}}(\mathcal{H}_i)$ and with $\mathcal{F}_i$ being a basis of $B^{\mathrm{sa}}(\mathcal{H}_i)$ consisting of states. This way, we obtain a family of $(\dim \mathcal{H})^2$ linearly independent product states which are elements of $\operatorname{Sep}(\mathcal{H})$. This shows that $\operatorname{Sep}(\mathcal{H})$ has dimension $(\dim \mathcal{H})^2 - 1$. Note that this argument uses the fact that the field is $\mathbb{C}$: in real quantum mechanics, the set of separable states has empty interior (cf. Section 0.4).

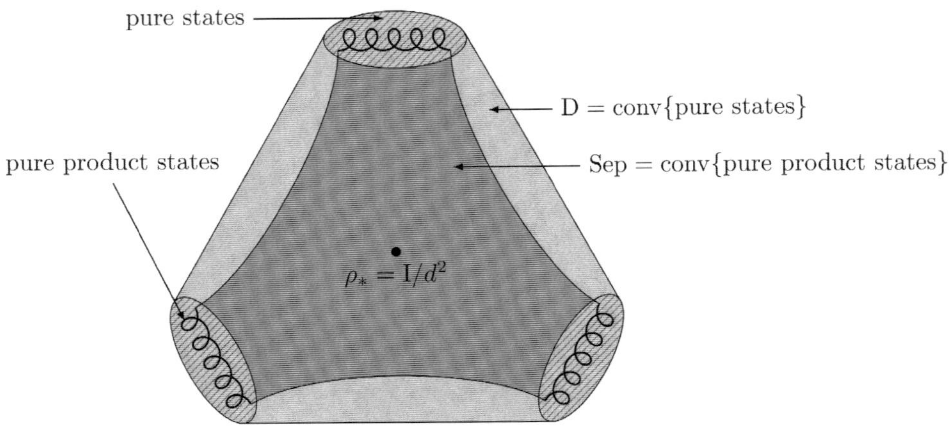

FIGURE 2.1. The sets of states (D) and of separable states (Sep) on $\mathbb{C}^d \otimes \mathbb{C}^d$. Pure product states have measure zero inside the set of pure states; however both convex hulls have the same dimension. The picture does not respect convexity of Sep, but it is supposed to reflect the relative rarity of separability.

A deeper result asserts that, in the bipartite case, not only do Sep and D have the same dimension, they also have the same inradius. This may look surprising since Sep is defined as the convex hull of a very small subset of the set of extreme points of D. This remarkable fact was discovered by Gurvits and Barnum and will be proved later (see Theorem 9.15).

It is often useful to consider the cone

$$\mathcal{SEP}(\mathcal{H}) = \{\lambda\rho \,:\, \lambda \geqslant 0,\ \rho \in \mathrm{Sep}(\mathcal{H})\}$$

of separable operators; we will return to this in Section 2.4.

We emphasize that the notion of separability depends crucially on the tensor decomposition (2.15) of $\mathcal{H}$. As a concrete example, consider a tripartite space $\mathcal{H} = \mathcal{H}_1 \otimes \mathcal{H}_2 \otimes \mathcal{H}_3$. There are several different notions of separability on $\mathcal{H}$: separability with respect to the tripartition $\mathcal{H}_1 : \mathcal{H}_2 : \mathcal{H}_3$, and separability with respect to each of the three bipartitions $\mathcal{H}_1 : \mathcal{H}_2 \otimes \mathcal{H}_3$, $\mathcal{H}_2 : \mathcal{H}_1 \otimes \mathcal{H}_3$, and $\mathcal{H}_3 : \mathcal{H}_1 \otimes \mathcal{H}_2$ or combinations thereof. Moreover, some authors introduce the concept of "absolute" properties. For example, a state $\rho \in \mathrm{D}(\mathcal{H}_1 \otimes \cdots \otimes \mathcal{H}_k)$ is *absolutely separable* if $U\rho U^\dagger$ is separable for any unitary operator U on $\mathcal{H}_1 \otimes \cdots \otimes \mathcal{H}_k$. However, in this book we will focus primarily on the setting in which all partitions are fixed.

Although the extreme points of Sep are very easy to describe (as noted earlier, they are precisely the pure product states), there is no simple description of the facial structure of Sep available (compare with Proposition 2.1, which describes all the faces of D). The complexity of the facial structure of Sep can be related to the fact that deciding whether a state is separable is known to be, in the general setting, NP-hard. This makes calculating some parameters of Sep highly nontrivial; we will run into this problem in Chapter 9 (see, e.g., Theorem 9.6). Finally, in view of the dual formulation of the problem of describing faces of a convex body (see Section 1.1.5, and particularly Proposition 1.5), characterizing maximal faces of Sep is essentially equivalent to describing extreme points of the object dual to Sep (see (2.47)), which are well understood only for very small dimensions. (Appendix C discusses closely related issues.)

EXERCISE 2.10 (The length of separable representations). (i) Using Carathéodory's theorem (see Section 1.1.2), show that any separable state on $\mathbb{C}^d \otimes \mathbb{C}^d$ can be written as the convex combination of at most d^4 pure product states. (ii) Using a dimension-counting argument, prove that there exist separable states on $\mathbb{C}^d \otimes \mathbb{C}^d$ which cannot be written as a convex combination of less than cd^3 pure product states, for some constant $c > 0$.

EXERCISE 2.11 (Edges of Sep). Let $d_1, d_2 \geqslant 2$. Show that $\mathrm{Sep}(\mathbb{C}^{d_1} \otimes \mathbb{C}^{d_2})$ has a face (as defined in Section 1.1.3) which is 1-dimensional.

2.2.4. Some examples of bipartite states. We now present some examples of states on $\mathbb{C}^d \otimes \mathbb{C}^d$ that are widely used in quantum information theory.

2.2.4.1. *Maximally entangled states.* A pure state on $\mathbb{C}^d \otimes \mathbb{C}^d$ is called *maximally entangled* if it has the form $\rho = |\psi\rangle\langle\psi|$ with

$$\psi = \frac{1}{\sqrt{d}} \sum_{i=1}^{d} e_i \otimes f_i, \tag{2.19}$$

where $(e_i)_{1 \leqslant i \leqslant d}$ and $(f_i)_{1 \leqslant i \leqslant d}$ are two orthonormal bases in $\mathbb{C}^d$. Such a vector ψ is called a *maximally entangled vector.*

In the special case of $d = 2$, i.e., for systems formed of 2 qubits, the maximally entangled states are called *Bell states*. Many quantum information protocols, such as quantum teleportation, use Bell states as a fundamental resource.

If we identify vectors and matrices as explained in Section 0.8, the set of all maximally entangled vectors on $\mathbb{C}^d \otimes \mathbb{C}^d$ (or, more precisely, on $\overline{\mathbb{C}^d} \otimes \mathbb{C}^d$) identifies with the unitary group $\mathsf{U}(d) \subset \mathsf{M}_d$.

EXERCISE 2.12 (Maximally entangled states and trace duality). Let ψ be the maximally entangled state given by (2.19), with (e_i) and (f_i) both equal to the canonical basis $(|i\rangle)_{1 \leqslant i \leqslant d}$, and let $\rho = |\psi\rangle\langle\psi|$. Show that $\mathrm{Tr}\,\big(\rho(X \otimes Y)\big) = \frac{1}{d}\,\mathrm{Tr}(XY^T)$ for any $X, Y \in B(\mathbb{C}^d)$.

EXERCISE 2.13 (Maximal entanglement and the distance to Seg). Let ψ be a unit vector in $\mathbb{C}^d \otimes \mathbb{C}^d$ and $\mathrm{Seg} \subset S_{\mathbb{C}^d \otimes \mathbb{C}^d}$ the set of unit product vectors (see (B.6)). Show that $|\psi\rangle\langle\psi|$ is maximally entangled if and only if $\mathrm{dist}(\psi, \mathrm{Seg})$ is maximal. For extensions to the multipartite case, see Section 8.5.

2.2.4.2. *Isotropic states.* Isotropic states are states which are a convex (or affine) combination of the maximally mixed state and a maximally entangled state. They have the form

$$\rho_\beta = \beta|\psi\rangle\langle\psi| + (1-\beta)\frac{\mathrm{I}}{d^2}, \tag{2.20}$$

where ψ is as in (2.19) and $-\frac{1}{d^2-1} \leqslant \beta \leqslant 1$.

2.2.4.3. *Werner states.* Consider the *flip operator* $F \in B^{\mathrm{sa}}(\mathbb{C}^d \otimes \mathbb{C}^d)$ defined on pure tensors by $F(x \otimes y) = y \otimes x$ and extended by linearity. Its eigenspaces are the *symmetric subspace*

$$\mathrm{Sym}_d = \{\psi \in \mathbb{C}^d \otimes \mathbb{C}^d \ : \ F(\psi) = \psi\}$$

and the *antisymmetric subspace*

$$\mathrm{Asym}_d = \{\psi \in \mathbb{C}^d \otimes \mathbb{C}^d \ : \ F(\psi) = -\psi\}.$$

The corresponding projectors are $P_{\mathrm{Sym}_d} = \frac{1}{2}(\mathrm{I}+F)$ and $P_{\mathrm{Asym}_d} = \frac{1}{2}(\mathrm{I}-F)$. We need to know that the symmetric and antisymmetric subspaces are irreducible for the action $U \mapsto U \otimes U$ of the unitary group.

PROPOSITION 2.9 (see Exercise 2.15). *Let $E \subsetneq \mathbb{C}^d \otimes \mathbb{C}^d$ be a nonzero subspace such that for every $U \in \mathsf{U}(d)$ and $\psi \in E$, we have $(U \otimes U)\psi \in E$. Then either $E = \mathrm{Sym}_d$ or $E = \mathrm{Asym}_d$.*

Note that $\dim \mathrm{Sym}_d = d(d+1)/2$ while $\dim \mathrm{Asym}_d = d(d-1)/2$. The *symmetric* and *antisymmetric states* are defined respectively as

$$\pi_s = \frac{2}{d(d+1)} P_{\mathrm{Sym}_d} \quad \text{and} \quad \pi_a = \frac{2}{d(d-1)} P_{\mathrm{Asym}_d}.$$

For $\lambda \in [0,1]$, consider the state w_λ (called the *Werner state*) obtained as a convex combination of these two projectors

$$w_\lambda = \lambda \pi_s + (1-\lambda)\pi_a. \tag{2.21}$$

Another equivalent expression is

$$w_\lambda = \frac{1}{d^2 - d\alpha}(\mathrm{I} - \alpha F), \tag{2.22}$$

where

$$\alpha = \frac{1 + d(1-2\lambda)}{1 + d - 2\lambda} \in [-1, 1]. \tag{2.23}$$

When $d = 2$, the space Asym_2 has dimension one, and Werner states are then a special case of isotropic states.

EXERCISE 2.14 (Polarization formulas in Sym_d and Asym_d). Prove that $\mathrm{Sym}_d = \mathrm{span}\{x \otimes x \ : \ x \in \mathbb{C}^d\}$ and $\mathrm{Asym}_d = \mathrm{span}\{x \otimes y - y \otimes x \ : \ x, y \in \mathbb{C}^d\}$.

EXERCISE 2.15 (Irreducibility of Sym_d and Asym_d).
Denote by $\mathscr{A} = \mathrm{span}\{U \otimes U \ : \ U \in \mathsf{U}(d)\}$.
(i) Prove that for every subspace $E \subset \mathbb{C}^d$, $P_E \otimes P_E \in \mathscr{A}$.
(ii) Show that for every nonzero vectors $\varphi, \psi \in \mathrm{Sym}_d$, there is $V \in \mathscr{A}$ such that $\langle \varphi | V | \psi \rangle \neq 0$.
(iii) Show that for every nonzero vectors $\varphi, \psi \in \mathrm{Asym}_d$, there is $V \in \mathscr{A}$ such that $\langle \varphi | V | \psi \rangle \neq 0$.
(iv) Deduce Proposition 2.9.

EXERCISE 2.16 (The twirling channel and Werner states).
(i) Show that a state $\rho \in \mathrm{D}(\mathbb{C}^d \otimes \mathbb{C}^d)$ satisfies $(V \otimes V)\rho(V \otimes V)^\dagger = \rho$ for all $V \in \mathsf{U}(d)$ if and only if it is a Werner state.
(ii) Show that if U is chosen at random with respect to the Haar measure on $\mathsf{U}(\mathbb{C}^d)$, then for any $\rho \in \mathrm{D}(\mathbb{C}^d \otimes \mathbb{C}^d)$, $\mathbf{E}(U \otimes U)\rho(U \otimes U)^\dagger = w_\lambda$ with $\lambda = \mathrm{Tr}(\rho P_{\mathrm{Sym}_d})$. (The map $\rho \mapsto \mathbf{E}(U \otimes U)\rho(U \otimes U)^\dagger$ is called the *twirling channel.*)
(iii) Show that if $\psi \in S_{\mathbb{C}^d}$ is chosen uniformly at random, then $\mathbf{E}\, |\psi \otimes \psi\rangle\langle \psi \otimes \psi| = \pi_s$.

2.2.5. Entanglement hierarchies.

2.2.5.1. *k-extendible states.* Consider a bipartite Hilbert space $\mathcal{H}_1 \otimes \mathcal{H}_2$ and $k \geqslant 2$. For $i \in \{1, \dots, k\}$, we denote by

$$\mathrm{Tr}_{\text{all but } i} : B(\mathcal{H}_1 \otimes \mathcal{H}_2^{\otimes k}) \to B(\mathcal{H}_1 \otimes \mathcal{H}_2)$$

the partial trace with respect to all copies of $\mathcal{H}_2$, except for the ith. A state $\rho \in \mathrm{D}(\mathcal{H}_1 \otimes \mathcal{H}_2)$ is said to be *k-extendible* (with respect to $\mathcal{H}_2$) if there exists a state $\rho_k \in \mathrm{D}(\mathcal{H}_1 \otimes \mathcal{H}_2^{\otimes k})$ with the property that

$$\mathrm{Tr}_{\text{all but } i}\, \rho_k = \rho$$

for every $i \in \{1, \dots, k\}$. The state ρ_k is called a k-extension of ρ. The main result regarding k-extendible states is the following theorem.

THEOREM 2.10 (Not proved here). *A quantum state on $\mathcal{H}_1 \otimes \mathcal{H}_2$ is separable if and only if it is k-extendible for every $k \geqslant 2$.*

The "only if" direction is easy (see Exercise 2.17), while the "if" direction relies on the quantum de Finetti theorem and is beyond the scope of this book.

EXERCISE 2.17. For $k \geqslant 2$, denote by k-Ext the set of k-extendible states on $\mathcal{H}_1 \otimes \mathcal{H}_2$. Show that k-Ext is convex and check the inclusions $\mathrm{Sep} \subset l\text{-Ext} \subset k\text{-Ext}$ for $k \leqslant l$.

EXERCISE 2.18 (2-extendibility of pure states). (i) Let $\rho \in \mathrm{D}(\mathcal{H}_1 \otimes \mathcal{H}_2)$ be a state such that $\mathrm{Tr}_{\mathcal{H}_2} \rho = |\psi\rangle\langle\psi|$ for some $\psi \in \mathcal{H}_1$. Show that $\rho = |\psi\rangle\langle\psi| \otimes \sigma$ for some $\sigma \in \mathrm{D}(\mathcal{H}_2)$. (ii) Let $\chi \in \mathcal{H}_1 \otimes \mathcal{H}_2$ be a unit vector. Show that $|\chi\rangle\langle\chi|$ is 2-extendible if and only if χ is a product vector.

2.2.5.2. *k-entangled states.* A quantum state on $\mathcal{H} = \mathcal{H}_1 \otimes \mathcal{H}_2$ is said to be *k-entangled* if it can be written as a convex combination

$$\sum_i \lambda_i |\psi_i\rangle\langle\psi_i|$$

where each unit vector $\psi_i \in \mathcal{H}_1 \otimes \mathcal{H}_2$ has Schmidt rank at most k. Note that separable states are exactly 1-entangled states.

2.2.6. Partial transposition.

Let $\mathcal{H}$ be a complex Hilbert space, and let (e_j) be an orthonormal basis in $\mathcal{H}$. We can identify $B(\mathcal{H})$ with the set of $n \times n$ matrices by associating a matrix (a_{ij}) with the operator

$$\sum_{i,j} a_{ij} |e_i\rangle\langle e_j|.$$

Once the basis is fixed, it makes sense to consider the transposition $T : B(\mathcal{H}) \to B(\mathcal{H})$ with respect to that basis, defined as

$$T\Big(\sum_{i,j} a_{ij} |e_i\rangle\langle e_j|\Big) = \sum_{i,j} a_{ij} |e_j\rangle\langle e_i|.$$

We will sometimes use the alternative notation $A^T = T(A)$. Note that T is *not* canonical and depends on the choice of the basis in $\mathcal{H}$. The standard usage in linear algebra refers to the transposition with respect to the standard basis $(|j\rangle)_{j=1}^{\dim \mathcal{H}}$.

We now define the *partial transposition*: if $\mathcal{H} = \mathcal{H}_1 \otimes \mathcal{H}_2$ is a bipartite Hilbert space, and if T denotes the transposition on $B(\mathcal{H}_1)$ (with respect to a specified basis) and Id is the identity operation of $B(\mathcal{H}_2)$, then the partial transposition (or partial transpose) is the operation

$$\Gamma = T \otimes \mathrm{Id} : B(\mathcal{H}_1 \otimes \mathcal{H}_2) \to B(\mathcal{H}_1 \otimes \mathcal{H}_2).$$

The partial transposition of a state $\rho \in \mathrm{D}(\mathcal{H}_1 \otimes \mathcal{H}_2)$ is denoted by $\rho^\Gamma = \Gamma(\rho)$. What we have defined is actually the partial transposition with respect to the first factor. The partial transposition with respect to the second factor is defined by switching the roles of $\mathcal{H}_1$ and $\mathcal{H}_2$.

Partial transposition applies nicely to states represented as block matrices (see Section 0.7): if $\rho \in \mathrm{D}(\mathcal{H}_1 \otimes \mathcal{H}_2)$ corresponds to the block operator (A_{ij}), with $A_{ij} \in B(\mathcal{H}_2)$, then ρ^Γ corresponds to the block operator (A_{ji}). Similarly, partial transposition of ρ with respect to the second factor corresponds to the block operator (A_{ij}^T). We illustrate this by computing the partial transposition of the (maximally entangled) Bell state: if $\psi = \frac{1}{\sqrt{2}}(|00\rangle + |11\rangle)$, then (assuming transposition is taken with respect to the canonical basis of $\mathbb{C}^2$)

$$|\psi\rangle\langle\psi| = \frac{1}{2}\left[\begin{array}{cc|cc} 1 & 0 & 0 & 1 \\ 0 & 0 & 0 & 0 \\ \hline 0 & 0 & 0 & 0 \\ 1 & 0 & 0 & 1 \end{array}\right], \quad |\psi\rangle\langle\psi|^\Gamma = \frac{1}{2}\left[\begin{array}{cc|cc} 1 & 0 & 0 & 0 \\ 0 & 0 & 1 & 0 \\ \hline 0 & 1 & 0 & 0 \\ 0 & 0 & 0 & 1 \end{array}\right]. \tag{2.24}$$

As for the usual transposition, the partial transposition depends on a choice of basis. However, we have the following result.

PROPOSITION 2.11. *The eigenvalues of the partial transposition of an operator do not depend on a choice of basis.*

PROOF. Let (e_i) and (e'_i) be two orthonormal bases in $\mathcal{H}_1$, and T and T' denote the transpositions with respect to each basis. Let U be the unitary transformation such that $e'_j = U(e_j)$. We claim that, for every operator $X \in B(\mathcal{H}_1)$,

$$T'(X) = V^\dagger T(X) V, \tag{2.25}$$

where $V = UT(U)$. By linearity, it is enough to check (2.25) when $X = |e'_i\rangle\langle e'_j|$, in which case $T'(X) = |e'_j\rangle\langle e'_i|$. On the other hand, since $X = U|e_i\rangle\langle e_j|U^\dagger$, we then have

$$T(X) = T(U^\dagger)|e_j\rangle\langle e_i|T(U) = T(U^\dagger)U^\dagger|e'_j\rangle\langle e'_i|UT(U) = T(U)^\dagger U^\dagger|e'_j\rangle\langle e'_i|UT(U),$$

as claimed. This shows that the partial transpositions with respect to the two bases are conjugated via the unitary transformation $V \otimes \mathrm{I}$, and the claim follows since unitary conjugation preserves the spectrum. □

Partial transposition naturally extends to the multipartite setting: if $\mathcal{H} = \mathcal{H}_1 \otimes \cdots \otimes \mathcal{H}_k$, then for any $i \in \{1, \ldots, k\}$ we may define the partial transposition with respect to the ith factor as

$$\Gamma_i := \mathrm{Id}_{B(\mathcal{H}_1)} \otimes \cdots \otimes \mathrm{Id}_{B(\mathcal{H}_{i-1})} \otimes T \otimes \mathrm{Id}_{B(\mathcal{H}_{i+1})} \otimes \cdots \otimes \mathrm{Id}_{B(\mathcal{H}_k)}\,.$$

EXERCISE 2.19 (Eigenvalues of the partial transpose of a pure state). Find all eigenvalues of the partial transpose of a pure state in terms of the Schmidt coefficients of that state.

EXERCISE 2.20 (Partial transpose and the flip operator). Let $\psi = \frac{1}{\sqrt{d}} \sum_{i=1}^{d} e_i \otimes e_i$ be a maximally entangled state on $\mathbb{C}^d \otimes \mathbb{C}^d$ and assume that partial transposition is computed with respect to the basis (e_i). Show that $|\psi\rangle\langle\psi|^\Gamma = \frac{1}{d} F$ where $F : x \otimes y \mapsto y \otimes x$ is the flip operator.

EXERCISE 2.21. Find an error in the following argument that purports to mimic the proof of Proposition 2.11 to show that the partial transpose of any state is positive.
If $X \in B^{\mathrm{sa}}(\mathcal{H}_1)$, then $T(X)$ (with respect to some fixed basis) has the same spectrum as X and so there is a unitary operator V such that $T(X) = V^\dagger X V$. This shows that the partial transpose with respect to the same basis is given by conjugation by the unitary transformation $V \otimes \mathrm{I}$. Since such conjugation preserves spectra, it follows that the partial transpose of any state is positive.

2.2.7. PPT states.

DEFINITION 2.12. A state $\rho \in \mathrm{D}(\mathcal{H}_1 \otimes \mathcal{H}_2)$ is said to have a *positive partial transpose* (or to be PPT) if the operator ρ^Γ is positive. We denote by $\mathrm{PPT}(\mathcal{H}_1 \otimes \mathcal{H}_2)$, or simply PPT, the set of PPT states (note that this set is convex).

Proposition 2.11 implies that the definition of PPT states is basis-independent. Similarly, we do not need to specify whether we apply the partial transposition to the first or the second factor; one passes from one to the other by applying the full transposition, which is a spectrum-preserving operation.

Let ρ be a state on $\mathcal{H}_1 \otimes \mathcal{H}_2$. Since the partial transposition preserves the trace, we have $\operatorname{Tr} \rho^\Gamma = 1$, and therefore ρ is PPT if and only if ρ^Γ is a state. Geometrically, the set of PPT states can therefore be described as the intersection

$$\mathrm{PPT} = \mathrm{D} \cap \Gamma(\mathrm{D}). \tag{2.26}$$

For a schematic illustration of (2.26), see Figure 2.2. The map Γ is a linear map which preserves the Hilbert–Schmidt norm, and therefore behaves as an isometry (see Exercise 2.22). This map is not a canonical object and depends on the choice of a basis. However, the intersection $\mathrm{D} \cap \Gamma(\mathrm{D})$ does not depend on the particular basis used.

The next proposition lies at the root of the relevance of the concept of PPT states to quantum information theory. (For a schematic illustration of the inclusion (2.27), see again Figure 2.2.)

PROPOSITION 2.13 (Peres–Horodecki criterion). *Let ρ be a state on $\mathcal{H}_1 \otimes \mathcal{H}_2$. If ρ is separable, then ρ is PPT. In other words, we have the inclusion*

$$\mathrm{Sep}(\mathcal{H}_1 \otimes \mathcal{H}_2) \subset \mathrm{PPT}(\mathcal{H}_1 \otimes \mathcal{H}_2). \tag{2.27}$$

PROOF. Since the set PPT is convex, it suffices to show that the extreme points of $\mathrm{Sep}(\mathcal{H}_1 \otimes \mathcal{H}_2)$ are PPT. The extreme points of $\mathrm{Sep}(\mathcal{H}_1 \otimes \mathcal{H}_2)$ are pure product states, i.e., states of the form

$$\rho = |\psi_1 \otimes \psi_2\rangle\langle\psi_1 \otimes \psi_2| = |\psi_1\rangle\langle\psi_1| \otimes |\psi_2\rangle\langle\psi_2|$$

for unit vectors $\psi_1 \in \mathcal{H}_1, \psi_2 \in \mathcal{H}_2$. The partial transpose of such a state is

$$\rho^\Gamma = |\psi_1\rangle\langle\psi_1|^T \otimes |\psi_2\rangle\langle\psi_2| = |\overline{\psi_1}\rangle\langle\overline{\psi_1}| \otimes |\psi_2\rangle\langle\psi_2|,$$

where $\overline{\psi_1}$ is the vector obtained by applying the complex conjugation to each coordinate of ψ_1. It follows that ρ^Γ is positive, hence ρ is PPT. □

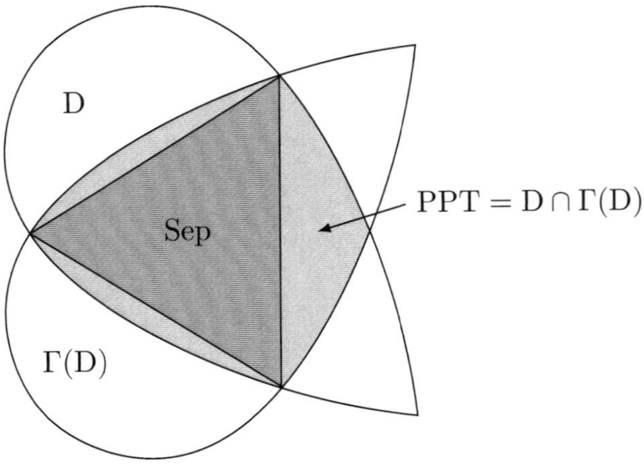

FIGURE 2.2. An illustration of the inclusion Sep $\subset$ PPT $=$ D $\cap$ Γ(D). The inclusion is strict if and only if $\dim \mathcal{H}_1 \dim \mathcal{H}_2 > 6$, see Theorem 2.15. The set Sep is not a polytope, but the set of its extreme points is much "thinner" than those of D and of PPT if the dimension is large.

The Peres–Horodecki criterion (or the PPT criterion) is shown in action in (2.24), where it certifies nonseparability of the Bell state: the partial transpose $|\psi\rangle\langle\psi|^\Gamma$ is clearly non-positive. However, positivity of ρ^Γ is, in general, only a necessary condition for separability of ρ as, without additional assumptions, the inclusion (2.27) is strict. Still, there are two important cases where PPT states are guaranteed to be separable: pure states and states in low dimensions, specifically in $\mathbb{C}^2 \otimes \mathbb{C}^2$ and $\mathbb{C}^2 \otimes \mathbb{C}^3$.

LEMMA 2.14. *A pure state is PPT if and only if it is separable.*

PROOF. Let $\rho = |\psi\rangle\langle\psi|$ be a pure state, and let $\psi = \sum \lambda_i \chi_i \otimes \psi_i$ be a Schmidt decomposition. If we compute the partial transposition with respect to a basis including (χ_i), we obtain

$$\rho^\Gamma = \sum_{i,j} \lambda_i \lambda_j |\chi_i \otimes \psi_j\rangle\langle\chi_j \otimes \psi_i|. \tag{2.28}$$

Suppose there exist two nonzero Schmidt coefficients (say, λ_i and λ_j with $i \neq j$). Then one checks from (2.28) that the restriction of ρ^Γ to $\operatorname{span}\{\chi_i \otimes \psi_j, \chi_j \otimes \psi_i\}$ is not positive. It follows that ρ is PPT if and only if only one Schmidt coefficient of ψ is nonzero, which means that ψ is a product vector and, consequently, ρ is separable. (See Exercise 2.19 for a complete description of the spectrum of ρ^Γ.) □

THEOREM 2.15 (Størmer–Woronowicz theorem; see Section 2.4.5 for the $2 \otimes 2$ case, the $2 \otimes 3$ case is not proved here). *If $\mathcal{H} = \mathbb{C}^2 \otimes \mathbb{C}^2$ or $\mathcal{H} = \mathbb{C}^3 \otimes \mathbb{C}^2$ or $\mathcal{H} = \mathbb{C}^2 \otimes \mathbb{C}^3$, then every PPT state on $\mathcal{H}$ is separable.*

Examples of entangled PPT states are known for any other (nontrivial) pairs of dimensions.

Besides pure and low-dimensional states, another family of states for which separability and the PPT property are equivalent are the Werner states. We have

PROPOSITION 2.16 (Separability of Werner states). *For $\lambda \in [0,1]$, let w_λ be the Werner state on $\mathcal{H} = \mathbb{C}^d \otimes \mathbb{C}^d$ as defined in* (2.21). *The following are equivalent:*
(i) w_λ *is separable,*
(ii) w_λ *is PPT,*
(iii) $\operatorname{Tr} w_\lambda F \geqslant 0$,
(iv) $\lambda \geqslant 1/2$.

PROOF. The equivalence (iii) $\iff$ (iv) is a straightforward calculation (we have $\operatorname{Tr} w_\lambda F = 2\lambda - 1$). To show that (ii) $\iff$ (iv), we compute the partial transpose of Werner states in the form (2.22) to obtain (see also Exercise 2.20)

$$w_\lambda^\Gamma = \frac{1}{d^2 - d\alpha}\big(\mathrm{I} - \alpha d|x\rangle\langle x|\big),$$

where x is the maximally entangled vector in the canonical basis $(|i\rangle)_{1\leqslant i\leqslant d}$. It follows that $w_\lambda^\Gamma \geqslant 0 \iff \alpha \leqslant 1/d \iff \lambda \geqslant 1/2$ (see (2.23) for the second equivalence). It remains to prove that (iv) implies (i); since Sep is convex, it is enough to establish that w_1 and $w_{1/2}$ are separable. The separability of $w_1 = \pi_s$ is clear from part (iii) of Exercise 2.16. To show that $w_{1/2}$ is separable, we proceed as follows. For $j \neq k$ and a complex number ξ with modulus one, denote $v^\pm = |j\rangle \pm \xi|k\rangle$. Next, think of ξ as a random variable uniformly distributed on the unit circle. The operator $\mathbf{E}\,|v^+\rangle\langle v^+| \otimes |v^-\rangle\langle v^-|$ belongs to the separable cone $\mathcal{SEP}$. We compute

$$\mathbf{E}\,|v^+v^-\rangle\langle v^+v^-| = |jj\rangle\langle jj| + |kk\rangle\langle kk| + |jk\rangle\langle jk| + |kj\rangle\langle kj| - |jk\rangle\langle kj| - |kj\rangle\langle jk|,$$

where we omitted the symbols $\otimes$ to reduce the clutter. Summing over $j \neq k$, we obtain that

$$A := 2d\sum_j |j\rangle\langle j| \otimes |j\rangle\langle j| + 2\sum_{j\neq k} |j\rangle\langle j| \otimes |k\rangle\langle k| - 2F \in \mathcal{SEP}.$$

The separability of $w_{1/2}$ follows now from the identity

$$w_{1/2} = \frac{1}{d(d^2-1)}(d\mathrm{I} - F) = \frac{1}{d(d^2-1)}\Big(\frac{A}{2} + (d-1)\sum_{j\neq k} |j\rangle\langle j| \otimes |k\rangle\langle k|\Big),$$

where the first equality is just (2.22) (note that $\lambda = 1/2$ implies $\alpha = 1/d$ by (2.23)). □

EXERCISE 2.22 (Partial transposition as a reflection). Find a subspace $E \subset B^{\mathrm{sa}}(\mathcal{H}_1 \otimes \mathcal{H}_2)$ such that $\Gamma = 2P_E - \mathrm{Id}$, where P_E denotes the orthogonal projection onto E. Geometrically, Γ identifies with the reflection with respect to E.

EXERCISE 2.23 (Separability of isotropic states). For $-\frac{1}{d^2-1} \leqslant \beta \leqslant 1$, let $\rho_\beta \in \mathrm{D}(\mathbb{C}^d \otimes \mathbb{C}^d)$ be the isotropic state as defined in (2.20). Show that ρ_β is separable if and only if $\beta \leqslant \frac{1}{d+1}$.

EXERCISE 2.24 (The realignment criterion). The *realignment* $A^R \in B(\mathbb{C}^{d_2} \otimes \mathbb{C}^{d_2}, \mathbb{C}^{d_1} \otimes \mathbb{C}^{d_1})$ of an operator $A \in B(\mathbb{C}^{d_1} \otimes \mathbb{C}^{d_2})$ is defined as follows: the map $A \mapsto A^R$ is $\mathbb{C}$-linear, and $|ij\rangle\langle kl|^R = |ik\rangle\langle jl|$.
(i) Let $\rho \in \mathrm{D}(\mathbb{C}^{d_1} \otimes \mathbb{C}^{d_2})$ be a separable state. Show that $\|\rho^R\|_1 \leqslant 1$. (The trace norm $\|\cdot\|_1$ is defined in Section 1.3.2).
(ii) Let $\rho \in \mathrm{D}(\mathbb{C}^{d_1} \otimes \mathbb{C}^{d_2})$ be a pure entangled state. Show that $\|\rho^R\|_1 > 1$.
The condition $\|\rho^R\|_1 \leqslant 1$ is usually called the *realignment criterion*. Just as for

the PPT criterion, this is a necessary (but generally not sufficient) condition for separability.

2.2.8. Local unitaries and symmetries of Sep. Let us state an analogue of Kadison's theorem (Theorem 2.4), which characterizes affine maps preserving the set Sep. This can be seen as a motivation for the study of partial transposition.

THEOREM 2.17 (Not proved here). *Let $\mathcal{H} = \mathbb{C}^{d_1} \otimes \cdots \otimes \mathbb{C}^{d_k}$ be a multipartite Hilbert space. An affine map $\Phi : B^{\mathrm{sa}}(\mathcal{H}) \to B^{\mathrm{sa}}(\mathcal{H})$ satisfies $\Phi(\mathrm{Sep}) = \mathrm{Sep}$ if and only if it can be written as the composition of maps of the following forms:*

(i) local unitaries

$$\rho \mapsto (U_1 \otimes \cdots \otimes U_k)\rho(U_1 \otimes \cdots \otimes U_k)^\dagger$$

for $U_i \in \mathsf{U}(d_i)$,

(ii) partial transpositions

$$\rho_1 \otimes \cdots \otimes \rho_i \otimes \cdots \otimes \rho_k \mapsto \rho_1 \otimes \cdots \otimes \rho_i^T \otimes \cdots \otimes \rho_k$$

for some $i \in \{1, \ldots, d\}$,

(iii) swaps

$$\rho_1 \otimes \cdots \otimes \rho_i \otimes \cdots \otimes \rho_j \otimes \cdots \otimes \rho_k \mapsto \rho_1 \otimes \cdots \otimes \rho_j \otimes \cdots \otimes \rho_i \otimes \cdots \otimes \rho_k$$

for some $i < j$ such that $d_i = d_j$.

All these maps are also isometries with respect to the Hilbert–Schmidt distance.

Although $\mathrm{Sep}(\mathcal{H})$ has a much smaller group of isometries than $\mathrm{D}(\mathcal{H})$, the conclusion of Proposition 2.5 still holds for Sep: the only fixed point is ρ_*. This implies for example that ρ_* is the centroid of Sep.

PROPOSITION 2.18. *Consider $\mathcal{H} = \mathcal{H}_1 \otimes \cdots \otimes \mathcal{H}_k$, and let $A \in B^{\mathrm{sa}}(\mathcal{H})$ be an operator which is invariant under local unitaries, i.e., such that*

$$A = (U_1 \otimes \cdots \otimes U_k)A(U_1 \otimes \cdots \otimes U_k)^\dagger$$

for any unitary matrices U_i on $\mathcal{H}_i$. Then A is a multiple of identity. In particular, if A is a state, then $A = \rho_*$.

PROOF. We use the following elementary fact: an operator $A_j \in B(\mathcal{H}_j)$ which commutes with any unitary operator actually commutes with any operator and is therefore a multiple of identity. We can write A as a linear combination of product operators

$$A = \sum_i c_i A_1^{(i)} \otimes \cdots \otimes A_k^{(i)},$$

where $A_j^{(i)} \in B^{\mathrm{sa}}(\mathcal{H}_j)$. Let $U = U_1 \otimes \cdots \otimes U_k$, where (U_j) are random unitary matrices, independent and Haar-distributed on the corresponding unitary groups. By the translation-invariance of the Haar measure (see Appendix B.3), the operator $\mathbf{E}\, U_j A_j^{(i)} U_j^\dagger$ commutes with any unitary operator on $\mathcal{H}_j$ and therefore (by the preceding fact) equals $\alpha_{i,j}\,\mathrm{I}_{\mathcal{H}_j}$ for some $\alpha_{i,j} \in \mathbb{R}$. By independence, it follows that

$$\begin{aligned}\mathbf{E}\,UAU^\dagger &= \sum_i c_i\,\mathbf{E}\left(U_1A_1^{(i)}U_1^\dagger \otimes \cdots \otimes U_kA_k^{(i)}U_k^\dagger\right) \\ &= \sum_i c_i(\mathbf{E}\,U_1A_1^{(i)}U_1^\dagger) \otimes \cdots \otimes (\mathbf{E}\,U_kA_k^{(i)}U_k^\dagger) \\ &= \left(\sum_i c_i \prod_{j=1}^k \alpha_{i,j}\right)\mathrm{I}_{\mathcal{H}}\,.\end{aligned}$$

Since $UAU^\dagger = A$, the conclusion follows. □

However, the group of local unitaries does not act irreducibly: there are non-trivial invariant subspaces which are described by the following lemma.

LEMMA 2.19 (Not proved here). *Let $\mathcal{H} = \mathbb{C}^{d_1} \otimes \cdots \otimes \mathbb{C}^{d_k}$ be a multipartite Hilbert space, and*

$$\mathsf{G} = \{U_1 \otimes \cdots \otimes U_k \ :\ U_i \in \mathsf{U}(d_i)\}$$

be the group of local unitaries. For $1 \leqslant i \leqslant k$, write $\mathsf{M}_{d_i}^{\mathrm{sa}} = V_i^1 \oplus V_i^2$, where V_i^1 denotes the hyperplane of trace zero Hermitian matrices, and $V_i^2 = \mathbb{R}\,\mathrm{I}$.

A subspace $E \subset B^{\mathrm{sa}}(\mathcal{H})$ is invariant under G if and only if it can be decomposed as a direct sum of subspaces of the form

$$V_{i_1}^{\alpha_1} \otimes \cdots \otimes V_{i_k}^{\alpha_k}$$

for some choice $(\alpha_1, \ldots, \alpha_k) \in \{1,2\}^k$.

2.3. Superoperators and quantum channels

We now turn our attention to maps acting between spaces of operators, hence the name *superoperators*. Other terms that will be used to describe these objects are *quantum maps* and *quantum operations*. The crucial observation is that with any such map one can naturally associate usual operators acting on larger Hilbert spaces.

2.3.1. The Choi and Jamiołkowski isomorphisms.

As usual, let $\mathcal{H}_1$ and $\mathcal{H}_2$ denote complex (finite-dimensional) Hilbert spaces. Recall (see Sections 0.4 and 0.8) the canonical isomorphisms $(\mathcal{H}_1 \otimes \mathcal{H}_2)^* \leftrightarrow \mathcal{H}_1^* \otimes \mathcal{H}_2^*$ and

$$\mathcal{H}_1^* \otimes \mathcal{H}_2 \leftrightarrow B(\mathcal{H}_1, \mathcal{H}_2). \tag{2.29}$$

It follows that there is a canonical isomorphism

$$B(\mathcal{H}_1, \mathcal{H}_2)^* \leftrightarrow B(\mathcal{H}_2, \mathcal{H}_1).$$

This isomorphism can be seen more concretely via trace duality: a map $S \in B(\mathcal{H}_2, \mathcal{H}_1)$ is identified with the linear form on $B(\mathcal{H}_1, \mathcal{H}_2)$ defined by $T \mapsto \operatorname{Tr} ST$.

By iterating (2.29), we deduce that there is a canonical isomorphism

$$J : B(B(\mathcal{H}_1), B(\mathcal{H}_2)) \longrightarrow B(\mathcal{H}_2 \otimes \mathcal{H}_1)$$

(both spaces being canonically isomorphic to $\mathcal{H}_1 \otimes \mathcal{H}_1^* \otimes \mathcal{H}_2 \otimes \mathcal{H}_2^*$), which is called the *Jamiołkowski isomorphism*. A concrete representation of the Jamiołkowski isomorphism is as follows: fix any basis (e_i) in $\mathcal{H}_1$ and denote by E_{ij} the operator

$|e_i\rangle\langle e_j| \in B(\mathcal{H}_1)$. Then J is described as

$$J : B(B(\mathcal{H}_1), B(\mathcal{H}_2)) \longrightarrow B(\mathcal{H}_2 \otimes \mathcal{H}_1) \qquad \Phi \longmapsto \sum_{i,j} \Phi(E_{ij}) \otimes E_{ji}. \tag{2.30}$$

It turns out that there is another related isomorphism, called the *Choi isomorphism*, which is often more useful. Once a basis in $\mathcal{H}_1$ is fixed, the Choi isomorphism is the $\mathbb{C}$-linear bijective map

$$C : B(B(\mathcal{H}_1), B(\mathcal{H}_2)) \longrightarrow B(\mathcal{H}_2 \otimes \mathcal{H}_1) \qquad \Phi \longmapsto \sum_{i,j} \Phi(E_{ij}) \otimes E_{ij}. \tag{2.31}$$

We call $C(\Phi)$ the *Choi matrix* of Φ. Note that the Choi isomorphism is basis-dependent, whereas the Jamiołkowski isomorphism is not. The relation between the isomorphisms J and C is given by the partial transposition: if Γ denotes the partial transposition on $\mathcal{H}_2 \otimes \mathcal{H}_1$ with respect to $\mathcal{H}_1$, then $C = \Gamma \circ J$.

Here is a simple lemma which identifies the elements in $B(B(\mathcal{H}_1), B(\mathcal{H}_2))$ that correspond to rank 1 operators under the Choi isomorphism.

LEMMA 2.20. *Given $A, B \in B(\mathcal{H}_1, \mathcal{H}_2)$, consider the map $\Phi : B(\mathcal{H}_1) \to B(\mathcal{H}_2)$ defined by*

$$\Phi(X) = AXB^\dagger$$

for $X \in B(\mathcal{H}_1)$. Then $C(\Phi) = |a\rangle\langle b|$, where $a = \mathrm{vec}(A)$ and $b = \mathrm{vec}(B)$ are the vectors in $\mathcal{H}_2 \otimes \mathcal{H}_1$ associated to the operators A and B (see Section 0.8). Note also that A has rank 1 if and only if a is a product vector.

PROOF. By $\mathbb{C}$-linearity it is enough to consider $A = |\psi\rangle\langle e_j|$ and $B = |\chi\rangle\langle e_i|$ for some $\psi, \chi \in \mathcal{H}_2$ and some basis vectors $e_i, e_j \in \mathcal{H}_1$. A simple computation shows that then $C(\Phi) = |\psi\rangle\langle\chi| \otimes E_{ij}$, while $a = \psi \otimes e_j$ and $b = \chi \otimes e_i$, and the Lemma follows. □

Finally, let us mention a connection with the notion of realignment defined in Exercise 2.24. If $\Phi : B(\mathbb{C}^{d_1}) \to B(\mathbb{C}^{d_2})$ is a superoperator, the matrix of Φ with respect to the bases $(E_{ij})_{1 \leqslant i,j \leqslant d_1}$ and $(E_{kl})_{1 \leqslant k,l \leqslant d_2}$ is given by the realigned Choi matrix $C(\Phi)^R$.

2.3.2. Positive and completely positive maps. A map $\Phi : B(\mathcal{H}_1) \to B(\mathcal{H}_2)$ is called *self-adjointness-preserving* if $\Phi(B^{\mathrm{sa}}(\mathcal{H}_1)) \subset B^{\mathrm{sa}}(\mathcal{H}_2)$. It is easily checked that the following are equivalent:

(1) Φ is self-adjointness-preserving,
(2) $\Phi(X^\dagger) = (\Phi(X))^\dagger$ for any $X \in B(\mathcal{H}_1)$,
(3) $J(\Phi) \in B^{\mathrm{sa}}(\mathcal{H}_2 \otimes \mathcal{H}_1)$,
(4) $C(\Phi) \in B^{\mathrm{sa}}(\mathcal{H}_2 \otimes \mathcal{H}_1)$.

An elegant way to rewrite the definition (2.31) of Choi's matrix is as follows:

$$C(\Phi) = \left(\Phi \otimes \mathrm{Id}_{B(\mathcal{H}_1)}\right)(|\chi\rangle\langle\chi|), \tag{2.32}$$

where $\chi = \sum_i e_i \otimes e_i \in \mathcal{H}_1 \otimes \mathcal{H}_1$ is (a multiple of) a maximally entangled vector.

(Recall that we fixed a basis (e_i) in $\mathcal{H}_1$ when defining the Choi isomorphism.) We also note that there is a one-to-one correspondence between

(a) self-adjointness-preserving $\mathbb{C}$-linear maps $\Phi : B(\mathcal{H}_1) \to B(\mathcal{H}_2)$ and
(b) $\mathbb{R}$-linear maps $\Psi : B^{\mathrm{sa}}(\mathcal{H}_1) \to B^{\mathrm{sa}}(\mathcal{H}_2)$.

The correspondence is straightforward: Ψ is obtained from Φ by restriction, whereas Φ is obtained from Ψ by complexification (see Section 0.5).

In what follows, we will occasionally refer to maps of the form $\Phi \otimes \mathrm{Id}_{B(\mathcal{H}_1)}$ as *extensions* of Φ (not to be confused with k-extensions of *states* defined in Section 2.2.5.1). As an example, the partial transposition Γ is an extension of the transposition T.

Throughout this section, we consider a self-adjointness-preserving linear map $\Phi : B(\mathcal{H}_1) \to B(\mathcal{H}_2)$. The *adjoint* of Φ is the unique map $\Phi^* : B(\mathcal{H}_2) \to B(\mathcal{H}_1)$ such that

$$\mathrm{Tr}(X\Phi(Y)) = \mathrm{Tr}(\Phi^*(X)Y)$$

for any $X \in B(\mathcal{H}_2)$ and $Y \in B(\mathcal{H}_1)$. Note that Φ^* is automatically self-adjointness-preserving if Φ is.

The map Φ is said to be *positivity preserving*—shortened to *positive* when this does not lead to ambiguity—if the image of every positive operator is a positive operator. The map Φ is said to be *n-positive* if $\Phi \otimes \mathrm{Id} : B^{\mathrm{sa}}(\mathcal{H}_1 \otimes \mathbb{C}^n) \to B^{\mathrm{sa}}(\mathcal{H}_2 \otimes \mathbb{C}^n)$ is positive. (Note that n-positivity formally implies k-positivity for any $k < n$.) Finally, the map Φ is said to be *completely positive* if it is n-positive for every integer n. (However, only $n = \min(\dim \mathcal{H}_1, \dim \mathcal{H}_2)$ needs to be checked, see Exercise 2.28.) We denote by $\boldsymbol{CP}(\mathcal{H}_1, \mathcal{H}_2)$ the set of completely positive maps from $B(\mathcal{H}_1)$ to $B(\mathcal{H}_2)$. It is immediate from the definition that $\boldsymbol{CP}(\mathcal{H}_1, \mathcal{H}_2)$ is a convex cone; more about this aspect of the theory in Section 2.4.

The transposition is an example of a map which is positive but not 2-positive; this can be seen, e.g., from (2.24) in Section 2.2.6 or from Exercise 2.32. Here is an important structure theorem concerning completely positive maps.

THEOREM 2.21 (Choi's theorem). *Let $\Phi : B(\mathcal{H}_1) \to B(\mathcal{H}_2)$ be self-adjointness-preserving. The following are equivalent:*

(1) the map Φ is completely positive,
(2) the Choi matrix $C(\Phi)$ is positive semi-definite,
(3) there exist finitely many operators $A_1, \ldots, A_N \in B(\mathcal{H}_1, \mathcal{H}_2)$ such that, for any $X \in B(\mathcal{H}_1)$,

$$\Phi(X) = \sum_{i=1}^{N} A_i X A_i^\dagger. \tag{2.33}$$

A decomposition of Φ in the form (2.33) is called a *Kraus decomposition* of Φ. The smallest integer N such that a Kraus decomposition is possible is called the *Kraus rank* of Φ. As will be clear from the proof, the Kraus rank of Φ is the same as the rank of $C(\Phi)$ in the usual (linear algebra) sense. In particular, it will follow that the Kraus rank of $\Phi : B(\mathcal{H}_1) \to B(\mathcal{H}_2)$ is at most $\dim \mathcal{H}_1 \dim \mathcal{H}_2$.

PROOF. It is easily checked that (3) implies (1). The implication (1) $\Rightarrow$ (2) follows from the representation (2.32) of the Choi matrix. We now prove (2) $\Rightarrow$ (3). By the spectral theorem, there exist vectors $a_i \in \mathcal{H}_1 \otimes \mathcal{H}_2$ such that

$$C(\Phi) = \sum_i |a_i\rangle\langle a_i|. \tag{2.34}$$

By Lemma 2.20, $|a_i\rangle\langle a_i|$ is the Choi matrix of the map $X \mapsto A_i X A_i^\dagger$, where $A_i \in B(\mathcal{H}_1, \mathcal{H}_2)$ is associated to a_i via the relation $a_i = \mathrm{vec}(A_i)$. A representation of type (3) follows now from the linearity of the Choi isomorphism. □

There is a simple relation between Kraus decompositions of a completely positive map and its adjoint: if Φ is given by (2.33), then for any $Y \in B(\mathcal{H}_2)$,

$$\Phi^*(Y) = \sum_{i=1}^{N} A_i^\dagger Y A_i. \tag{2.35}$$

It is clear from the above analysis that Φ^* is completely positive if and only if Φ is. It is also readily checked that Φ^* is positivity-preserving if and only if Φ is; this and related properties are explored in Exercises 2.25–2.33, and discussed in a more general setting in Section 2.4.

EXERCISE 2.25. Let $\Phi : B(\mathcal{H}_1) \to B(\mathcal{H}_2)$ be self-adjointness-preserving. Show that Φ^* is positive if and only if Φ is positive, and that for any n, Φ^* is n-positive if and only if Φ is n-positive.

EXERCISE 2.26. Show that if Φ and Ψ are completely positive, so are $\Phi \otimes \Psi$ and $\Phi \circ \Psi$ (the composition, assuming it is defined).

EXERCISE 2.27. Show that any self-adjointness-preserving map $\Phi : B(\mathcal{H}_1) \to B(\mathcal{H}_2)$ is the difference of two completely positive maps.

EXERCISE 2.28. Show that the assertions of Theorem 2.21 are also equivalent to the fact that Φ is n-positive, with $n = \min(\dim \mathcal{H}_1, \dim \mathcal{H}_2)$.

EXERCISE 2.29. Let $k < n$ be integers. Show that the map $\Phi : \mathsf{M}_n \to \mathsf{M}_n$ defined by $\Phi(X) = k\,\mathrm{Tr}(X)\,\mathrm{I} - X$ is k-positive but not $(k+1)$-positive.

2.3.3. Quantum channels and Stinespring representation. Consider a self-adjointness-preserving map $\Phi : B(\mathcal{H}_1) \to B(\mathcal{H}_2)$. We say that Φ is *unital* if $\Phi(\mathrm{I}_{\mathcal{H}_1}) = \mathrm{I}_{\mathcal{H}_2}$. We say that Φ is *trace-preserving* if $\mathrm{Tr}\,\Phi(X) = \mathrm{Tr}\,X$ for any $X \in B(\mathcal{H}_1)$. It is easily checked that these properties are dual to each other:

$$\Phi \text{ is unital} \iff \Phi^* \text{ is trace-preserving.} \tag{2.36}$$

We now introduce a fundamental concept in quantum information theory.

DEFINITION 2.22. A *quantum channel* $\Phi : B(\mathcal{H}_1) \to B(\mathcal{H}_2)$ is a completely positive and trace-preserving map.

The reasons why we require quantum channels to be positivity- and trace-preserving are clear: since Φ is supposed to represent some physically possible process, we want states to be mapped to states. (The motivation behind the *complete* positivity condition is more subtle; we attempt to explain it in Section 3.5.) A channel that is additionally unital (i.e., if both Φ and Φ^* are channels) is called *doubly stochastic* or *bistochastic*. Clearly, such channels exist only if $\dim \mathcal{H}_1 = \dim \mathcal{H}_2$. (However, see Proposition 2.32 for a notion that makes sense also when $\dim \mathcal{H}_1 \neq \dim \mathcal{H}_2$.)

REMARK 2.23. It follows immediately from the relation (2.33) that the condition $\sum_{i=1}^{N} A_i A_i^\dagger = \mathrm{I}_{\mathcal{H}_2}$ is equivalent to $\Phi\big(\mathrm{I}_{\mathcal{H}_1}\big) = \mathrm{I}_{\mathcal{H}_2}$, i.e., to Φ being unital. It is less obvious, but easily checked, that $\sum_{i=1}^N A_i^\dagger A_i = \mathrm{I}_{\mathcal{H}_1}$ is equivalent to Φ being trace-preserving. Indeed, if the condition holds, then for any $\xi \in \mathcal{H}_1$,

$$\operatorname{Tr}(|\xi\rangle\langle\xi|) = \operatorname{Tr}\Big(\sum_{i=1}^N A_i^\dagger A_i|\xi\rangle\langle\xi|\Big) = \operatorname{Tr}\Big(\sum_{i=1}^N A_i|\xi\rangle\langle\xi|A_i^\dagger\Big).$$

In other words, $\operatorname{Tr}\Phi(X) = \operatorname{Tr} X$ if $X = |\xi\rangle\langle\xi|$ and hence, by linearity, for any $X \in B^{\mathrm{sa}}(\mathcal{H}_1)$. Furthermore, the argument is clearly reversible, so we have equivalence.

We now state the Stinespring representation theorem, which plays a fundamental role in understanding the structure of quantum maps.

THEOREM 2.24 (Stinespring theorem). *Let $\Phi : B(\mathcal{H}_1) \to B(\mathcal{H}_2)$ be a completely positive map. Then there exist a finite-dimensional Hilbert space $\mathcal{H}_3$ and an embedding $V : \mathcal{H}_1 \to \mathcal{H}_2 \otimes \mathcal{H}_3$ such that, for any $X \in B(\mathcal{H}_1)$,*

$$\Phi(X) = \operatorname{Tr}_{\mathcal{H}_3} VXV^\dagger. \tag{2.37}$$

Moreover, Φ is a quantum channel if and only if V is an isometry. Conversely, for any isometric embedding V, the map Φ defined via (2.37) is a quantum channel.

The proof shows that the smallest possible dimension for $\mathcal{H}_3$ equals the Kraus rank of Φ; in particular we can require that $\dim(\mathcal{H}_3) \leqslant \dim(\mathcal{H}_1)\dim(\mathcal{H}_2)$.

PROOF. Start from a Kraus decomposition (2.33) for Φ. Set $\mathcal{H}_3 := \mathbb{C}^N$, and let $(|i\rangle)_{1\leqslant i\leqslant N}$ be its canonical basis. Define V by the formula

$$V|\psi\rangle = \sum_{i=1}^N A_i|\psi\rangle \otimes |i\rangle \quad \text{for} \quad \psi \in \mathcal{H}_1. \tag{2.38}$$

We claim that, for any $X \in B(\mathcal{H}_1)$,

$$VXV^\dagger = \sum_{i,j=1}^N A_i X A_j^\dagger \otimes |i\rangle\langle j|.$$

As in Remark 2.23, this follows by linearity from the special case $X = |\psi\rangle\langle\psi|$. This implies the identity (2.37). We also see from (2.38) that $V^\dagger V = \sum_{i=1}^N A_i^\dagger A_i$. By Remark 2.23 it follows that Φ is a quantum channel if and only if $V^\dagger V = \mathrm{I}_{\mathcal{H}_1}$, which is equivalent to V being an isometry. Finally, the last assertion is straightforward: complete positivity follows from (the easy direction of) Choi's Theorem 2.21 and the trace preserving property is immediate. □

When $\mathcal{H}_1 = \mathcal{H}_2$, the Stinespring theorem can be reformulated as follows: any quantum channel can be lifted to a unitary transformation using some ancillary Hilbert space.

THEOREM 2.25. *Let $\Phi : B(\mathcal{H}) \to B(\mathcal{H})$ be a quantum channel. Then there exist a finite-dimensional Hilbert space $\mathcal{H}'$, a unit vector $\psi \in \mathcal{H}'$, and a unitary transformation U on $\mathcal{H} \otimes \mathcal{H}'$ such that, for any X in $B(\mathcal{H})$,*

$$\Phi(X) = \operatorname{Tr}_{\mathcal{H}'} U(X \otimes |\psi\rangle\langle\psi|)U^\dagger. \tag{2.39}$$

PROOF. Let $V : \mathcal{H} \to \mathcal{H} \otimes \mathcal{H}'$ be given by Theorem 2.24 (with $\mathcal{H}' = \mathcal{H}_3$). Choose any vector $\psi \in \mathcal{H}'$. The map $\varphi \otimes \psi \mapsto V(\varphi)$ (defined on the subspace $\mathcal{H} \otimes \psi \subset \mathcal{H} \otimes \mathcal{H}'$) is an isometry, and therefore can be extended to a unitary U on $\mathcal{H} \otimes \mathcal{H}'$. One checks easily that (2.39) holds. □

We mention in passing that a popular way to quantify how different two quantum channels are is the *diamond norm.* For a self-adjointness-preserving map $\Phi : B(\mathcal{H}_1) \to B(\mathcal{H}_2)$, define

$$\|\Phi\|_\diamond = \sup_{k\in\mathbb{N}} \sup_{\rho\in \mathrm{D}(\mathbb{C}^k)} \|(\Phi \otimes \mathrm{I}_{B(\mathbb{C}^k)})(\rho)\|_1.$$

EXERCISE 2.30. Show that any positive unital map $\Phi : \mathsf{M}_m^{\mathrm{sa}} \to \mathsf{M}_n^{\mathrm{sa}}$ is a contraction with respect to the operator norm $\|\cdot\|_\infty$.

EXERCISE 2.31. Show that any positive trace-preserving map $\Phi : \mathsf{M}_m^{\mathrm{sa}} \to \mathsf{M}_n^{\mathrm{sa}}$ is a contraction with respect to the trace norm $\|\cdot\|_1$ (cf. Proposition 8.4).

EXERCISE 2.32. (i) Let $\Phi : \mathsf{M}_m^{\mathrm{sa}} \to \mathsf{M}_n^{\mathrm{sa}}$ be a trace preserving map. Show that Φ is k-positive if and only if $\Phi \otimes \mathrm{Id} : B^{\mathrm{sa}}(\mathbb{C}^m \otimes \mathbb{C}^k) \to B^{\mathrm{sa}}(\mathbb{C}^n \otimes \mathbb{C}^k)$ is a contraction with respect to the trace norm $\|\cdot\|_1$. (ii) Let $T : \mathsf{M}_n \to \mathsf{M}_n$ be the transposition map. Calculate the norm of $T \otimes \mathrm{Id}$ considered as a map on $\big(B^{\mathrm{sa}}(\mathbb{C}^m \otimes \mathbb{C}^2), \|\cdot\|_1\big)$ and give an example of an operator on which that norm is attained. (iii) Repeat (ii) for the operator norm $\|\cdot\|_\infty$.

EXERCISE 2.33. Show that any positive, unital, and trace-preserving map $\Phi : \mathsf{M}_n^{\mathrm{sa}} \to \mathsf{M}_n^{\mathrm{sa}}$ is rank nondecreasing, i.e., $\operatorname{rank} \Phi(\rho) \geqslant \operatorname{rank} \rho$ for any $\rho \in \mathrm{D}(\mathbb{C}^n)$.

2.3.4. Some examples of channels. In this section we list some important classes and examples of quantum channels or, more generally, of superoperators. (Sometimes it is convenient to drop the trace-preserving constraint.)

2.3.4.1. *Unitary channels. Unitary channels* are the completely positive isometries of the set of states identified in Theorem 2.4, i.e., the maps that are of the form $\rho \mapsto U\rho U^\dagger$ for some $U \in \mathsf{U}(d)$.

2.3.4.2. *Mixed-unitary channels.* A *mixed-unitary channel* $\Phi : B(\mathbb{C}^d) \to B(\mathbb{C}^d)$ is a channel which is a convex combination of unitary channels, i.e., is of the form

$$\Phi(\rho) = \sum_{i=1}^N \lambda_i U_i \rho U_i^\dagger, \tag{2.40}$$

where (λ_i) is a convex combination and $U_i \in \mathsf{U}(\mathbb{C}^d)$. Such channels are automatically unital. A remarkable fact is that the converse is true when $d = 2$.

PROPOSITION 2.26 (See Exercise 2.34). *Let $\Phi : B(\mathbb{C}^2) \to B(\mathbb{C}^2)$ be a unital quantum channel. Then Φ is mixed-unitary.*

EXERCISE 2.34 (Proof of Proposition 2.26).
(i) Argue that it is enough to prove Proposition 2.26 for channels which are diagonal with respect to the basis of Pauli matrices (2.2).
(ii) Given real numbers a, b, c, check that the superoperator

$$\frac{1}{2}\big(|\mathrm{I}\rangle\langle\mathrm{I}| + a|\sigma_x\rangle\langle\sigma_x| + b|\sigma_y\rangle\langle\sigma_y| + c|\sigma_z\rangle\langle\sigma_z|\big)$$

is completely positive if and only if $(a+b)^2 \leqslant (1+c)^2$ and $(a-b)^2 \leqslant (1-c)^2$.
(iii) Rewrite the conditions from part (ii) as a system of four linear inequalities and conclude the proof.

EXERCISE 2.35. Show that any mixed-unitary channel $\Phi : B(\mathbb{C}^d) \to B(\mathbb{C}^d)$ can be expressed as in (2.40) with $N \leqslant d^4 - 2d^2 + 2$. Note that the argument from Exercise 2.34 gives $N \leqslant 4$ (which is optimal) for $d = 2$.

2.3.4.3. *Depolarizing and dephasing channels.* The *completely depolarizing* (or *completely randomizing*) *channel* is the channel $R : B(\mathbb{C}^d) \to B(\mathbb{C}^d)$ defined as $R(X) = \operatorname{Tr} X \frac{\mathrm{I}}{d}$. It maps every state to the maximally mixed state. The completely dephasing channel is the channel $D : B(\mathbb{C}^d) \to B(\mathbb{C}^d)$ that maps any operator to its diagonal part (with respect to a fixed basis).

EXERCISE 2.36 (Depolarizing channels and isotropic states). The family of depolarizing channels is defined as $R_\lambda = \lambda \mathrm{I} + (1-\lambda)R$ for $-\frac{1}{d^2-1} \leqslant \lambda \leqslant 1$. Check that the Choi matrix of Φ_λ is $d\rho_\lambda$, where ρ_λ is the isotropic state defined in (2.20).

EXERCISE 2.37. Show that the completely depolarizing and completely dephasing channels are mixed-unitaries (see also Exercise 8.6).

2.3.4.4. *POVMs, quantum-classical channels.* A *POVM* (Positive Operator-Valued Measure) on $\mathcal{H}$ is a finite family of positive operators $(M_i)_{1 \leqslant i \leqslant N}$ with the property that $\sum M_i = \mathrm{I}$. Given a POVM, we can associate to it a quantum channel (called sometimes a quantum-classical or q-c channel) $\Phi : B(\mathcal{H}) \to B(\mathbb{C}^N)$ defined as

$$\Phi(\rho) = \sum_{i=1}^{N} |i\rangle\langle i| \operatorname{Tr}(M_i \rho). \tag{2.41}$$

The dual concept is the notion of a classical-quantum or c-q channel $\Psi : B(\mathbb{C}^N) \to B(\mathcal{H})$. This is a channel of the form

$$\Psi(\rho) = \sum_{i=1}^{N} \rho_i \langle i|\rho|i\rangle,$$

where (ρ_i) are states on $\mathcal{H}$.

EXERCISE 2.38 (Duality between c-q and q-c channels). Let Φ be a q-c channel of the form (2.41). Under what condition on (M_i) is Φ unital? When this condition is satisfied, show that the dual map Φ^* is a c-q channel.

2.3.4.5. *Entanglement-breaking maps.* A map $\Phi \in \boldsymbol{CP}(\mathcal{H}^{in}, \mathcal{H}^{out})$ is said to be *entanglement-breaking* if, for any integer d and for any positive operator $X \in B^{\mathrm{sa}}(\mathcal{H}^{in} \otimes \mathbb{C}^d)$, the operator $(\Phi \otimes \mathrm{Id}_{\mathsf{M}_d})(X)$ belongs to the cone $\mathcal{SEP}(\mathcal{H}^{out} \otimes \mathbb{C}^d)$ of separable operators. Here are equivalent descriptions of entanglement-breaking maps (for a description based on duality see Table 2.2 and the comments following it):

LEMMA 2.27 (Characterization of entanglement-breaking maps, see Exercise 2.39). *Let $\Phi : B(\mathcal{H}^{in}) \to B(\mathcal{H}^{out})$ be completely positive. The following are equivalent:*
(i) Φ is entanglement-breaking,
(ii) the Choi matrix $C(\Phi)$ lies in the separable cone $\mathcal{SEP}(\mathcal{H}^{out} \otimes \mathcal{H}^{in})$,
(iii) there is a Kraus decomposition of Φ (2.33) where all the Kraus operators A_i have rank 1.

Entanglement-breaking quantum channels are sometimes called *q-c-q channels*. This reflects the fact that a quantum channel Φ is entanglement-breaking if and only if it can be written as the composition of a q-c channel with a c-q channel. Still another alternative terminology for entanglement-breaking channels is that of *super-positive maps*.

EXERCISE 2.39. Prove Lemma 2.27.

EXERCISE 2.40 (Once broken, always broken). Let Φ, Ψ be two completely positive maps, with one of them being entanglement-breaking. Show that $(\Phi \otimes \Psi)(X) \in \mathcal{SEP}$ for any positive operator X.

2.3.4.6. *PPT-inducing maps.* A map $\Phi \in \boldsymbol{CP}(\mathcal{H}^{in}, \mathcal{H}^{out})$ is said to be *PPT-inducing* if for any integer d and any positive operator $X \in B^{\mathrm{sa}}(\mathcal{H}^{in} \otimes \mathbb{C}^d)$, the operator $(\Phi \otimes \mathrm{Id}_{\mathsf{M}_d})(X)$ has positive partial transpose.

LEMMA 2.28 (Characterization of PPT-inducing maps, see Exercise 2.41). *A completely positive map Φ is PPT-inducing if and only if $J(\Phi) = C(\Phi)^\Gamma$ is positive semi-definite.*

EXERCISE 2.41. Prove Lemma 2.28.

2.3.4.7. *Schur channels.* Given matrices $A, B \in \mathsf{M}_d$, their Schur product $A \odot B$ is defined as the entrywise product: $(A \odot B)_{ij} = A_{ij}B_{ij}$. Given $A \in \mathsf{M}_d$, the map $\Theta_A : \mathsf{M}_d \to \mathsf{M}_d$ defined as $\Theta_A(X) = A \odot X$ is called a Schur multiplier. When A is positive with $A_{ii} = 1$ for all i, the map Θ_A is a quantum channel called a *Schur channel.*

EXERCISE 2.42 (Positivity of Schur multipliers). Let $A \in \mathsf{M}_d$. Show that the following are equivalent:
(i) A is positive semi-definite,
(ii) Θ_A is positive,
(iii) Θ_A is completely positive.

EXERCISE 2.43 (Kraus decompositions of Schur channels). Let $\Phi : \mathsf{M}_d \to \mathsf{M}_d$ be a quantum channel. Show that Φ is a Schur channel if and only if it admits a Kraus decomposition (2.33) where A_i are diagonal operators.

2.3.4.8. *Separable and LOCC superoperators.* We now assume that $\mathcal{H}^{in}$ and $\mathcal{H}^{out}$ are bipartite spaces, say $\mathcal{H}^{in} = \mathcal{H}_1^{in} \otimes \mathcal{H}_2^{in}$ and $\mathcal{H}^{out} = \mathcal{H}_1^{out} \otimes \mathcal{H}_2^{out}$. A map $\Phi \in \boldsymbol{CP}(\mathcal{H}^{in}, \mathcal{H}^{out})$ is called *separable* if it admits a Kraus decomposition involving product operators, i.e., if there exist operators $A_i^{(1)} : \mathcal{H}_1^{in} \to \mathcal{H}_1^{out}$ and $A_i^{(2)} : \mathcal{H}_2^{in} \to \mathcal{H}_2^{out}$ such that for any $X \in B(\mathcal{H}^{in})$,

$$\Phi(X) = \sum_{i=1}^{N} (A_i^{(1)} \otimes A_i^{(2)}) X (A_i^{(1)} \otimes A_i^{(2)})^\dagger.$$

A widely used class is the class of LOCC channels (LOCC standing for "Local Operations and Classical Communication"). Without defining this class, we simply note that any LOCC channel is separable, and that any convex combination of product channels (of the form $\Phi_1 \otimes \Phi_2$) is an LOCC channel. (Note that these notions *are not* all equivalent; see Exercise 2.44.) More properties of this class will be presented in Section 12.2.

EXERCISE 2.44. Consider the following operators on $\mathbb{C}^2 \otimes \mathbb{C}^2$

$$A_1 = |0\rangle\langle 0| \otimes |0\rangle\langle 0|, \quad A_2 = |0\rangle\langle 0| \otimes |0\rangle\langle 1|,$$
$$A_3 = |1\rangle\langle 1| \otimes |1\rangle\langle 1|, \quad A_4 = |1\rangle\langle 1| \otimes |1\rangle\langle 0|.$$

Show that the channel on $B(\mathbb{C}^2 \otimes \mathbb{C}^2)$ defined as $\Phi(X) = \sum_{i=1}^4 A_i X A_i^\dagger$ is a separable channel which cannot be written as a convex combination of product channels.

2.3.4.9. *Direct sums.* Let $\Phi_1 : B(\mathcal{H}_1^{in}) \to B(\mathcal{H}_1^{out})$ and $\Phi_2 : B(\mathcal{H}_2^{in}) \to B(\mathcal{H}_2^{out})$ be two quantum channels. Their *direct sum*

$$\Phi_1 \oplus \Phi_2 : B(\mathcal{H}_1^{in} \oplus \mathcal{H}_2^{in}) \to B(\mathcal{H}_1^{out} \oplus \mathcal{H}_2^{out})$$

is the quantum channel defined by its action on block operators as

$$(\Phi_1 \oplus \Phi_2)\left(\begin{bmatrix} X_{11} & X_{12} \\ X_{21} & X_{22} \end{bmatrix}\right) = \begin{bmatrix} \Phi_1(X_{11}) & 0 \\ 0 & \Phi_2(X_{22}) \end{bmatrix}. \tag{2.42}$$

EXERCISE 2.45. Describe the Kraus operators of $\Phi_1 \oplus \Phi_2$ in terms of the Kraus operators of Φ_1 and Φ_2.

2.4. Cones of QIT

In this section we will review some of the cones used commonly in quantum information theory. We will distinguish between cones of operators and cones of superoperators, and emphasize the distinction by using two different fonts: $\mathcal{C}$ denotes a generic cone of operators and $\boldsymbol{C}$ a generic cone of superoperators.

2.4.1. Cones of operators. We start by describing some cones of operators and by identifying their bases and their dual cones (Table 2.1). We work in a Hilbert space $\mathcal{H}$ and the corresponding space $B^{\mathrm{sa}}(\mathcal{H})$ of self-adjoint operators. The vector e chosen to define the base in (1.22) is the maximally mixed state. Here and in what follows, we assume that separability and the PPT property are defined with respect to a fixed bipartition $\mathcal{H} = \mathcal{H}_1 \otimes \mathcal{H}_2$. However, most considerations extend to multipartite variants and settings allowing flexibility in the choice of the partition. In order to lighten the notation, we often write $\mathcal{PSD}$ and $\mathcal{SEP}$ instead of $\mathcal{PSD}(\mathcal{H})$ and $\mathcal{SEP}(\mathcal{H}_1 \otimes \mathcal{H}_2)$ unless this may cause ambiguity.

TABLE 2.1. List of cones of operators. All cones live in $B^{\mathrm{sa}}(\mathcal{H})$, the space of self-adjoint operators on a bipartite Hilbert space $\mathcal{H} = \mathcal{H}_1 \otimes \mathcal{H}_2$ with dimension $n = \dim \mathcal{H}$. The base is taken with respect to the distinguished vector $e = \mathrm{I}/n$. The cones $\mathcal{C}$ are listed in the decreasing order (with respect to inclusion) from top to bottom and, consequently, the dual cones $\mathcal{C}^*$ are in the increasing order from top to bottom. Most inclusions/duality relations are straightforward and/or were pointed out earlier in this chapter; the remaining few are clarified in this subsection.

Cone of operators $\mathcal{C}$		base $\mathcal{C}^{\mathrm{b}}$	dual cone $\mathcal{C}^*$
Block-positive	$\mathcal{BP}$	BP	$\mathcal{SEP}$
Decomposable	co-$\mathcal{PSD} + \mathcal{PSD}$	$\mathrm{conv}(\mathrm{D} \cup \Gamma(\mathrm{D}))$	$\mathcal{PPT}$
Positive	$\mathcal{PSD}$	D	$\mathcal{PSD}$
Pos. partial transpose	$\mathcal{PPT}$	PPT	co-$\mathcal{PSD} + \mathcal{PSD}$
Separable	$\mathcal{SEP}$	Sep	$\mathcal{BP}$

In the same way that $\mathcal{PSD}$ is associated with its base D, the set of separable states Sep gives rise to the separable cone $\mathcal{SEP}$, and the set PPT of states with positive partial transpose leads to the $\mathcal{PPT}$ cone. Another example is the cone of k-entangled matrices (cf. Section 2.2.5). In general, whenever a definition of a set of matrices involves linear matrix inequalities and a trace constraint, dropping that constraint gives us a cone. When the original set of matrices is compact, the resulting cone is pointed, with the hyperplane of trace zero matrices isolating 0 as an exposed point (cf. Corollary 1.8). All the cones cataloged in this section have this property and are in fact nondegenerate.

One more convenient concept is that of co-$\mathcal{PSD}$ matrices

$$\text{co-}\mathcal{PSD} := \Gamma(\mathcal{PSD}) = \{\rho \in \mathsf{M}_n^{\text{sa}} : \rho^\Gamma \in \mathcal{PSD}\}, \tag{2.43}$$

where Γ is the partial transpose defined in Section 2.2.6. It allows a compact description of the cone dual to $\mathcal{PPT}$: since $\mathcal{PPT} = \text{co-}\mathcal{PSD} \cap \mathcal{PSD}$, it follows from (1.20) (see also Exercise (1.36)) that

$$\mathcal{PPT}^* = \text{co-}\mathcal{PSD} + \mathcal{PSD}, \tag{2.44}$$

the cone of *decomposable* matrices. Note that, except in trivial cases, this cone is strictly larger than $\mathcal{PSD}$ and so its base contains matrices that are not states.

To conclude the review of the standard cones, we will identify the cone $\mathcal{SEP}^*$. To that end, it is convenient to think of operators on a composite Hilbert space $\mathbb{C}^m \otimes \mathbb{C}^n$ as *block matrices* $M = (M_{jk})_{j,k=1}^m$, where $M_{jk} \in \mathsf{M}_n$ (see Section 0.7). Since the extreme rays of $\mathcal{SEP}$ are generated by pure separable states $|\xi\otimes\eta\rangle\langle\xi\otimes\eta|$ (see Section 2.2.3), we have

$$M \in \mathcal{SEP}^* \iff \forall\xi \in \mathbb{C}^m, \quad \forall\eta \in \mathbb{C}^n, \quad \operatorname{Tr}\big(M|\xi\otimes\eta\rangle\langle\xi\otimes\eta|\big) \geqslant 0 \tag{2.45}$$

$$\iff \forall\xi \in \mathbb{C}^m, \quad \sum_{j,k=1}^m \xi_j\overline{\xi_k}M_{jk} \in \mathcal{PSD}(\mathbb{C}^n). \tag{2.46}$$

The condition in (2.46) is usually referred to as $M = (M_{jk})$ being *block-positive*. (We note that the definition treats m and n symmetrically, even though this is not apparent in (2.46).) In other words, the dual to the cone of separable matrices is that of block-positive matrices, denoted by $\mathcal{BP}$. As a consequence, the polar of Sep can be identified: we obtain from Lemma 1.6 that

$$\text{Sep}^\circ = -d^2\text{BP}, \tag{2.47}$$

where BP denotes the set of block-positive matrices with unit trace and the minus sign stands for the point reflection with respect to the appropriately normalized identity matrix.

2.4.2. Cones of superoperators. We next turn our attention to the classes of superoperators considered in Section 2.3.2. We consider superoperators acting from $B^{\text{sa}}(\mathcal{H})$ to $B^{\text{sa}}(\mathcal{K})$ and denote the corresponding cones as $\boldsymbol{C}(\mathcal{H},\mathcal{K})$, or as $\boldsymbol{C}(\mathcal{H})$ when $\mathcal{H} = \mathcal{K}$, or simply as $\boldsymbol{C}$ when there is no ambiguity. The cones we consider most frequently are gathered in Table 2.2. (See Exercise 2.48 for a discussion of identification and duality relations for k-positive superoperators and k-entangled states.)

In the language of cones, a positivity-preserving superoperator $\Phi : B^{\text{sa}}(\mathcal{H}) \to B^{\text{sa}}(\mathcal{K})$ may be defined via the condition $\Phi\big(\mathcal{PSD}(\mathcal{H})\big) \subset \mathcal{PSD}(\mathcal{K})$. It is readily seen that the set of positivity-preserving maps is itself a cone (which we will denote by $\boldsymbol{P}(\mathcal{H},\mathcal{K})$) in the space $B\big(B^{\text{sa}}(\mathcal{H}), B^{\text{sa}}(\mathcal{K})\big)$.

As was noted in Section 2.3.2, $\Phi \in \boldsymbol{P}(\mathcal{H},\mathcal{K})$ iff $\Phi^* \in \boldsymbol{P}(\mathcal{K},\mathcal{H})$. As we shall see, it would be erroneous to take this to mean that $\boldsymbol{P}$ is self-dual. Instead, this is a special case of a very general elementary fact: *If V_1, V_2 are vector spaces, if*

TABLE 2.2. Cones of superoperators. To each cone $\boldsymbol{C}$ from the first (double) column we associate a cone $\mathcal{C}$ which consists of Choi matrices of elements from $\boldsymbol{C}$. They are connected by the relation $\Phi \in \boldsymbol{C} \iff C(\Phi) \in \mathcal{C}$. We note that $\boldsymbol{C}$ is a subset of $B(B^{\mathrm{sa}}(\mathcal{H}), B^{\mathrm{sa}}(\mathcal{K}))$ while $\mathcal{C}$ is a subset of $B^{\mathrm{sa}}(\mathcal{K} \otimes \mathcal{H})$. The cones $\boldsymbol{C}$ and $\mathcal{C}$ are in decreasing order from top to bottom and the dual cones $\mathcal{C}^*$ and $\boldsymbol{C}^*$ are in increasing order from top to bottom.

Cone of superoperators $\boldsymbol{C}$		$\mathcal{C}$	$\mathcal{C}^*$	$\boldsymbol{C}^*$
Positivity-preserving	$\boldsymbol{P}$	$\mathcal{BP}$	$\mathcal{SEP}$	$\boldsymbol{EB}$
Decomposable	$\boldsymbol{DEC}$	co-$\mathcal{PSD}+\mathcal{PSD}$	$\mathcal{PPT}$	$\boldsymbol{PPT}$
Completely positive	$\boldsymbol{CP}$	$\mathcal{PSD}$	$\mathcal{PSD}$	$\boldsymbol{CP}$
PPT-inducing	$\boldsymbol{PPT}$	$\mathcal{PPT}$	co-$\mathcal{PSD}+\mathcal{PSD}$	$\boldsymbol{DEC}$
Entanglement-breaking	$\boldsymbol{EB}$	$\mathcal{SEP}$	$\mathcal{BP}$	$\boldsymbol{P}$

$\mathcal{C}_1 \subset V_1$, $\mathcal{C}_2 \subset V_2$ are closed convex cones, and if $\Phi : V_1 \to V_2$ is linear, then $\Phi(\mathcal{C}_1) \subset \mathcal{C}_2$ iff $\Phi^*(\mathcal{C}_2^*) \subset \mathcal{C}_1^*$.

The most important cone of superoperators is arguably that of completely positive maps, denoted by $\boldsymbol{CP}$. By Choi's Theorem 2.21, $\Phi \in \boldsymbol{CP}$ iff the Choi matrix $C(\Phi)$ is positive semi-definite. In other words, $\boldsymbol{CP}(\mathbb{C}^m, \mathbb{C}^n)$ is isomorphic to $\mathcal{PSD}(\mathbb{C}^n \otimes \mathbb{C}^m)$. This means that—with proper identifications, see Exercise 2.47—the cone $\boldsymbol{CP}$ is self-dual. Choi's correspondence $\Phi \mapsto C(\Phi)$ relates similarly the cone $\boldsymbol{EB}(\mathbb{C}^m, \mathbb{C}^n)$ of entanglement-breaking maps from $\mathsf{M}_m^{\mathrm{sa}}$ to $\mathsf{M}_n^{\mathrm{sa}}$ to $\mathcal{SEP}(\mathbb{C}^n \otimes \mathbb{C}^m)$, as well as the cone $\boldsymbol{PPT}(\mathbb{C}^m, \mathbb{C}^n)$ of PPT-inducing maps to $\mathcal{PPT}(\mathbb{C}^n \otimes \mathbb{C}^m)$.

A map $\Phi : \mathsf{M}_m^{\mathrm{sa}} \to \mathsf{M}_n^{\mathrm{sa}}$ is said to be *co-completely positive* if $C(\Phi) \in$ co-$\mathcal{PSD}$. Similarly, one says that Φ is *decomposable* if it can be represented as a sum of a completely positive map and a co-completely positive map. It follows that the correspondence $\Phi \mapsto C(\Phi)$ relates the cone $\boldsymbol{DEC}(\mathbb{C}^n, \mathbb{C}^m)$ of decomposable maps to the cone of decomposable matrices.

Interestingly, $\mathcal{SEP}(\mathbb{C}^n \otimes \mathbb{C}^m)^*$ identifies with $\boldsymbol{P}(\mathbb{C}^m, \mathbb{C}^n)$. This last identification is in fact easy to see directly from (2.45)–(2.46). Indeed, $C(\Phi) = (M_{jk})$ means that $M_{jk} = \Phi(|e_j\rangle\langle e_k|)$ and hence if $\xi = (\xi_j)_{j=1}^m \in \mathbb{C}^m$, then $\Phi(|\xi\rangle\langle\xi|) = \sum_{j,k=1}^m \xi_j \overline{\xi_k} M_{jk}$. Consequently,

$$\begin{aligned} C(\Phi) \in \mathcal{SEP}(\mathbb{C}^n \otimes \mathbb{C}^m)^* \quad &\iff \quad \Phi(|\xi\rangle\langle\xi|) \in \mathcal{PSD}(\mathbb{C}^n) \text{ for } \xi \in \mathbb{C}^m \\ &\iff \quad \Phi \in \boldsymbol{P}, \end{aligned}$$

which is the claimed identification. The first equivalence is simply (2.45)–(2.46) for the choice $M = C(\Phi)$, whereas the second one reflects the fact that the property of "preserving positivity" needs to be checked only on the extreme rays of the $\mathcal{PSD}$ cone, i.e., on operators of the form $|\xi\rangle\langle\xi|$. (See Section 1.2.2 and particularly Corollary 1.10.)

EXERCISE 2.46 (Composition rules for maps). Show that a composition of two co-completely positive maps is completely positive. Similarly, show that a composition of a co-completely positive map and a completely positive map is co-completely positive.

EXERCISE 2.47 (The completely positive cone is self-dual). Show that

$$\boldsymbol{CP}(\mathbb{C}^n,\mathbb{C}^m) = \{\Psi \in B(\mathsf{M}_n^{\mathrm{sa}},\mathsf{M}_m^{\mathrm{sa}}) \ : \ \mathbf{Tr}(\Psi \circ \Phi) \geqslant 0 \ \ \forall \Phi \in \boldsymbol{CP}(\mathbb{C}^m,\mathbb{C}^n)\},$$

where $\mathbf{Tr}$ denotes the trace on $B(\mathsf{M}_n^{\mathrm{sa}})$.

EXERCISE 2.48 (k-positive superoperators and k-entangled states). Let $1 \leqslant k \leqslant \min(m,n)$ and $\Phi : \mathsf{M}_n \to \mathsf{M}_m$ be self-adjointness-preserving. Show that the following are equivalent:
(1) Φ is k-positive,
(2) for every $x \in \mathbb{C}^m \otimes \mathbb{C}^n$ with Schmidt rank at most k, we have $\langle x|C(\Phi)|x\rangle \geqslant 0$,
(3) for every $A \in \mathsf{M}_{k,m}$ and $B \in \mathsf{M}_{k,n}$, the operator $(A\otimes B)^\dagger C(\Phi)(A\otimes B)$ is positive.
In other words, the cone of Choi matrices of k-positive superoperators is dual to the cone generated by the set of k-entangled states (as defined in Section 2.2.5).

2.4.3. Symmetries of the $\mathcal{PSD}$ cone. The results of Sections 2.1.4 allow us to deduce a description of the groups of affine automorphisms of some of the cones cataloged in the present section. The argument is based on the following two simple observations: first, since affine automorphisms preserve facial structure, and since 0 is the only extreme point of all the cones considered above, any affine automorphism must be linear. Next, if $\Phi : \mathsf{M}_m^{\mathrm{sa}} \to \mathsf{M}_n^{\mathrm{sa}}$ is such that $A = \Phi(\mathrm{I})$ is positive definite, then Ψ defined by $\Psi(\rho) = A^{-1/2}\Phi(\rho)A^{-1/2}$ is unital, and its adjoint, Ψ^*, is trace-preserving (see (2.36)). This often allows to reduce the analysis of general maps to that of unital or trace-preserving maps. As an example of such reduction we will prove the following statement.

PROPOSITION 2.29 (Characterization of automorphisms of the $\mathcal{PSD}$ cone). *Let $\Phi : \mathsf{M}_n^{\mathrm{sa}} \to \mathsf{M}_n^{\mathrm{sa}}$ be an affine map which satisfies $\Phi(\mathcal{PSD}(\mathbb{C}^n)) = \mathcal{PSD}(\mathbb{C}^n)$. Then Φ is a linear automorphism of $\mathcal{PSD}(\mathbb{C}^n)$ and is of one of two possible forms: $\Phi(\rho) = V\rho V^\dagger$ or $\Phi(\rho) = V\rho^T V^\dagger$, for some $V \in \mathsf{GL}(n,\mathbb{C})$. In the first case Φ is completely positive, whereas in the second case Φ is co-completely positive.*

PROOF. Since $\operatorname{rank}\Phi \geqslant \dim \mathcal{PSD}(\mathbb{C}^n) = \dim \mathsf{M}_n^{\mathrm{sa}}$, it follows that Φ is surjective and hence injective, so it is indeed an automorphism of $\mathcal{PSD}(\mathbb{C}^n)$ (and, consequently, so is Φ^{-1}). By the earlier remark, Φ must be linear. Since the adjoint of a positive map is positive (see Section 2.3.2), it follows that Φ^* and $(\Phi^*)^{-1} = (\Phi^{-1})^*$ are positive. Hence they are both automorphisms of $\mathcal{PSD}(\mathbb{C}^n)$. Let $A = \Phi^*(\mathrm{I}) \in \mathcal{PSD}(\mathbb{C}^n)$. We claim that A belongs to the interior of $\mathcal{PSD}(\mathbb{C}^n)$ and, consequently, is positive definite (and invertible). This follows from topological considerations, but can also be deduced from Proposition 1.4: if $A = \Phi^*(\mathrm{I})$ lay on the boundary of $\mathcal{PSD}(\mathbb{C}^n)$, we would have $A \in F$ for some face of $\mathcal{PSD}(\mathbb{C}^n)$, which would imply $\Phi^*(\mathcal{PSD}(\mathbb{C}^n)) \subset F$, contradicting injectivity of Φ^*. Having established the claim, we set $\Psi(\sigma) = A^{-1/2}\Phi^*(\sigma)A^{-1/2}$, so that Ψ is a unital automorphism of $\mathcal{PSD}(\mathbb{C}^n)$. Consequently, Ψ^* is a trace-preserving automorphism of $\mathcal{PSD}(\mathbb{C}^n)$, which is only possible if $\Psi^*(\mathrm{D}) = \mathrm{D}$. It now follows from Kadison's Theorem 2.4 that, for some $U \in \mathsf{U}(n)$, either (i) $\Psi^*(\tau) = U\tau U^\dagger$ or (ii) $\Psi^*(\tau) = U\tau^T U^\dagger$ (for all $\tau \in \mathsf{M}_n^{\mathrm{sa}}$). The rest of the argument is just bookkeeping. First, the definition of Ψ—and that of an adjoint map—imply that Ψ^* is given by the formula $\Psi^*(\tau) = \Phi(A^{-1/2}\tau A^{-1/2})$. In case (i), this shows that $\Phi(A^{-1/2}\tau A^{-1/2}) = U\tau U^\dagger$ or, substituting $\rho = A^{-1/2}\tau A^{-1/2}$, $\Phi(\rho) = UA^{1/2}\rho A^{1/2}U^\dagger = V\rho V^\dagger$, where $V = UA^{1/2}$, as needed. The fact that Φ is then completely positive is the easy implication of Choi's Theorem 2.21. Case (ii) is handled in the same way. □

We have an immediate corollary.

COROLLARY 2.30. *Completely positive automorphisms of the cone $\mathcal{PSD}(\mathbb{C}^n)$, all of which are of the form $\Phi_V(\rho) = V\rho V^\dagger$ for some $V \in \mathsf{GL}(n, \mathbb{C})$, act transitively on the interior of that cone.*

For future reference, we state here a slightly more general form of the principle that is implicit in the proof of Proposition 2.29.

LEMMA 2.31. *If $\Phi : \mathsf{M}_m^{\mathrm{sa}} \to \mathsf{M}_n^{\mathrm{sa}}$ is a positivity-preserving linear map such that $A = \Phi(\mathrm{I})$ is positive definite, then $\tilde{\Phi}$ defined by $\tilde{\Phi}(\rho) = A^{-1/2}\Phi(\rho)A^{-1/2}$ is unital and positivity-preserving. Similarly, if Ψ is a positivity-preserving linear map such that $\Psi(\rho) \neq 0$ for $\rho \in \mathcal{PSD}(\mathbb{C}^m)\backslash\{0\}$, then $\tilde{\Psi}(\rho) = \Psi(B^{-1/2}\rho B^{-1/2})$ is trace-preserving and positivity-preserving, where $B = \Psi^*(\mathrm{I})$ (necessarily positive definite).*

We emphasize that the map Φ in Lemma 2.31 is not assumed to be an automorphism of the $\mathcal{PSD}$ cone (as was the case in Proposition 2.29), only positivity-preserving. Moreover, we also allow the dimensions in the domain and in the range to be different. Finally, recall that, by Lemma 1.7, the properties "$\Phi(\mathrm{I})$ is positive definite" and "$\Psi(\rho) \neq 0$ for $\rho \in \mathcal{PSD}(\mathbb{C}^m)\backslash\{0\}$" are dual to each other.

In view of the above result, it is natural to wonder when a positivity-preserving map is equivalent, in the sense of Lemma 2.31, to a map which is *both* unital and trace-preserving. (Of course if the dimensions in the domain and in the range are different, this is only possible if we use the normalized trace or, alternatively, if we ask that the maximally mixed state be mapped to the maximally mixed state.) It turns out that this can be ensured if just a little more regularity is assumed. (See Exercise 2.52 for examples exploring the necessity of the stronger hypothesis.) We have:

PROPOSITION 2.32 (Sinkhorn's normal form for positive maps). *Let $\Phi : \mathsf{M}_m^{\mathrm{sa}} \to \mathsf{M}_n^{\mathrm{sa}}$ be a linear map which belongs to the interior of $\boldsymbol{P}$, the cone of positivity-preserving maps. Then there exist positive operators $A \in \mathcal{PSD}(\mathbb{C}^n)$ and $B \in \mathcal{PSD}(\mathbb{C}^m)$ such that the map $\tilde{\Phi}(\rho) = A\Phi(B\rho B)A$ is trace-preserving and maps the maximally mixed state to the maximally mixed state (and is necessarily positivity-preserving).*

PROOF. Let us first focus on the case $m = n$. Given positive definite A, B, let $\tilde{\Phi}$ be given by the formula from the Proposition. Then

$$\tilde{\Phi} \text{ is unital} \Leftrightarrow A\Phi(B^2)A = \mathrm{I} \Leftrightarrow \Phi(B^2) = A^{-2} \Leftrightarrow \Phi(B^2)^{-1} = A^2. \tag{2.48}$$

We next note that, in the notation of Corollary 2.30, $\tilde{\Phi} = \Phi_A \circ \Phi \circ \Phi_B$ and so $\tilde{\Phi}^* = \Phi_B \circ \Phi^* \circ \Phi_A$ (this uses the identity $\Phi_M^* = \Phi_M$, valid when M is self-adjoint). Accordingly, by (2.36),

$$\tilde{\Phi} \text{ is trace-preserving} \Leftrightarrow \tilde{\Phi}^* \text{ is unital} \Leftrightarrow B\Phi^*(A^2)B = \mathrm{I} \Leftrightarrow \Phi^*(A^2) = B^{-2}. \tag{2.49}$$

Solving the last equation in (2.49) for B^2 and substituting it in (2.48) we are led to a system of equations

$$B^2 = \Phi^*(A^2)^{-1} \quad \text{and} \quad \Phi\big(\Phi^*(A^2)^{-1}\big)^{-1} = A^2. \tag{2.50}$$

The second equation in (2.50) says that $S = A^2$ is a fixed point of the function

$$S \mapsto f(S) := \Phi\big(\Phi^*(S)^{-1}\big)^{-1}. \tag{2.51}$$

Conversely, if S is a positive definite fixed point of f, then $A = S^{1/2}$ and $B = \Phi^*(A^2)^{-1/2}$ (i.e., B defined so that the first equation in (2.50) holds) satisfy (2.48) and (2.49) and yield $\tilde{\Phi}$ that is unital and trace-preserving. (The hypothesis "Φ belongs to the interior of $\boldsymbol{P}$" guarantees that all the inverses and negative powers above make sense, and that f is well-defined and continuous on $\mathcal{PSD}\backslash\{0\}$; see Exercises 2.50 and 2.51;)

To find a fixed point of f we want to use Brouwer's fixed-point theorem, which requires a (continuous) function that is a self-map of a compact convex set. One way to arrive at such setting is to consider $f_1 : \mathrm{D}(\mathbb{C}^n) \to \mathrm{D}(\mathbb{C}^n)$ defined by

$$f_1(\sigma) = \frac{f(\sigma)}{\operatorname{Tr} f(\sigma)}. \tag{2.52}$$

It then follows that there is $\sigma_0 \in D(\mathbb{C}^n)$ such that $f_1(\sigma_0) = \sigma_0$ and hence $f(\sigma_0) = t\sigma_0$, where $t = \operatorname{Tr} f(\sigma_0) > 0$. The final step is to note that if we choose, as before, $A = \sigma_0^{1/2}$ and $B = \Phi^*(A^2)^{-1/2}$, then the corresponding $\tilde{\Phi}$ is trace-preserving and satisfies $\tilde{\Phi}(\mathrm{I}) = t^{-1}\,\mathrm{I}$. If $m = n$, this is only possible if $t = 1$. In other words, σ_0 is a fixed point of f that we needed in order to conclude the argument. In the general case, the same argument yields $t = n/m$, which translates to $\tilde{\Phi}(\mathrm{I}/m) = \mathrm{I}/n$, again as needed. □

EXERCISE 2.49. Show that $\Phi \in \boldsymbol{P}(\mathbb{C}^n)$ is an automorphism of $\mathcal{PSD}(\mathbb{C}^n)$ if and only if it is rank-preserving.

EXERCISE 2.50 (Descriptions of the interior of the positive cone). Show that Φ belongs to the interior of $\boldsymbol{P}(\mathbb{C}^n)$ iff Φ maps $\mathcal{PSD}(\mathbb{C}^n)\backslash\{0\}$ to the interior of $\mathcal{PSD}(\mathbb{C}^n)$ iff there exists $\delta > 0$ such that $\Phi(\rho) \geqslant \delta(\operatorname{Tr}\rho)\,\mathrm{I}$ for all $\rho \in \mathcal{PSD}$.

EXERCISE 2.51 (Interior of the positive cone is self-dual). Show that Φ verifies $\Phi(\rho) \geqslant \delta(\operatorname{Tr}\rho)\,\mathrm{I}$ (for all $\rho \in \mathcal{PSD}$) iff Φ^* does.

EXERCISE 2.52 (Discussion of the necessity of the hypothesis of Proposition 2.32). Give examples of $\Phi, \Psi \in \boldsymbol{P}(\mathbb{C}^2)$ such that (a) $\Phi(\mathrm{I})$ and $\Phi^*(\mathrm{I})$ are positive definite, but Φ is not equivalent (in the sense of Proposition 2.32) to a unital, trace-preserving map, and (b) Ψ is unital and trace-preserving, but $\Psi \in \partial\boldsymbol{P}$.

EXERCISE 2.53 (Rank nondecreasing and Sinkhorn's normal form). Give an example of map $\Phi \in \boldsymbol{P}(\mathbb{C}^2, \mathbb{C}^2)$ which is rank nondecreasing (i.e., verifies $\operatorname{rank}\Phi(\rho) \geqslant \operatorname{rank}\rho$ for any $\rho \in \mathrm{D}(\mathbb{C}^2)$), but which does not satisfy the conclusion of Proposition 2.32.

2.4.4. Entanglement witnesses. The formalism of cones and their duality allows us to conveniently discuss the concept of *entanglement witnesses*. We start with the following simple observation, which is a direct consequence of the identifications of the dual cone $\mathcal{SEP}^*$ as $\mathcal{BP}$ (see Table 2.1 in Section 2.4), and of the corresponding cone of superoperators as $\boldsymbol{P}$ (Table 2.2).

PROPOSITION 2.33 (Entanglement witnesses, take #1). *Let $\mathcal{H} = \mathbb{C}^m \otimes \mathbb{C}^n$ and let ρ be a state on $\mathcal{H}$. Then the following conditions are equivalent:*
(i) ρ is entangled,
(ii) there exists $\sigma \in \mathcal{SEP}(\mathcal{H})^* = \mathcal{BP}$ such that $\langle\sigma, \rho\rangle_{\mathrm{HS}} = \operatorname{Tr}(\sigma\rho) < 0$,
(iii) there exists a positivity-preserving linear map $\Psi : \mathsf{M}_n^{\mathrm{sa}} \to \mathsf{M}_m^{\mathrm{sa}}$ such that $\operatorname{Tr}(C(\Psi)\rho) < 0$.

The next result is a simple corollary of the above observation, but it goes well beyond a straightforward reformulation.

THEOREM 2.34 (Horodecki's entanglement witness theorem). *Let* $\mathcal{H} = \mathbb{C}^m \otimes \mathbb{C}^n$ *and let* ρ *be a state on* $\mathcal{H}$. *Then* ρ *is entangled iff there exists a positivity-preserving map* $\Phi : \mathsf{M}_m^{\mathrm{sa}} \to \mathsf{M}_n^{\mathrm{sa}}$ *such that the operator* $(\Phi \otimes \mathrm{Id}_{\mathsf{M}_n^{\mathrm{sa}}})\rho$ *is not positive semi-definite.*

In the setting of Proposition 2.33 and Theorem 2.34, the operator σ or the map Φ are said to witness the entanglement present in ρ, hence the term "entanglement witnesses".

PROOF OF THEOREM 2.34. The sufficiency is obvious: if $\rho = \tau \otimes \tau'$ is a product state and Φ is positivity-preserving, then $(\Phi \otimes \mathrm{Id})\rho = \Phi(\tau) \otimes \tau'$, which is clearly positive; the case of convex combinations of product states easily follows. To show necessity, let $\Psi : \mathsf{M}_n^{\mathrm{sa}} \to \mathsf{M}_m^{\mathrm{sa}}$ be the positivity-preserving map given by Proposition 2.33. If $\chi \in \mathbb{C}^n \otimes \mathbb{C}^n$ is the maximally entangled vector as in (2.32), then

$$\begin{aligned} 0 &> \mathrm{Tr}(C(\Psi)\rho) = \langle C(\Psi), \rho \rangle_{\mathrm{HS}} = \langle (\Psi \otimes \mathrm{Id}_{\mathsf{M}_n^{\mathrm{sa}}})|\chi\rangle\langle\chi|, \rho \rangle_{\mathrm{HS}} \\ &= \langle |\chi\rangle\langle\chi|, (\Psi^* \otimes \mathrm{Id}_{\mathsf{M}_n^{\mathrm{sa}}})\rho \rangle_{\mathrm{HS}} = \langle \chi | (\Psi^* \otimes \mathrm{Id}_{\mathsf{M}_n^{\mathrm{sa}}})\rho | \chi \rangle, \end{aligned}$$

which implies that $(\Psi^* \otimes \mathrm{Id}_{\mathsf{M}_n^{\mathrm{sa}}})\rho$ is not positive. Given that Ψ^* is positivity-preserving if and only if Ψ is (see Section 2.3.2), the choice of $\Phi = \Psi^*$ works as needed. □

REMARK 2.35. It follows from general considerations that the entanglement witnesses σ, Φ may be required to satisfy various additional properties. First, one may include a normalizing condition such as $\mathrm{Tr}\,\sigma = 1$ or $\mathrm{Tr}\,\Phi(\mathrm{I}) = 1$, which reduces the search for a witness to a convex compact set. Next, since linear functions (restricted to compact sets) attain extreme values on extreme points, one may insist that σ or Φ belong to an extreme ray of the respective cone (or even, by a density argument, to an exposed ray; cf. Exercise 1.5). Finally, another acceptable normalizing condition is to require that Φ be unital or trace-preserving. To see that Φ can be assumed unital, we note first that by a density argument the operator $\Phi(\mathrm{I})$ may be assumed to be positive definite, in which case Lemma 2.31 applies. The case of the trace-preserving restriction is slightly more involved and requires increasing the dimension of the range of Φ. We relegate the details of the arguments to Exercises 2.54 and 2.55.

EXERCISE 2.54 (Unital witnesses suffice). Show that in Theorem 2.34 one can require that Φ be unital

EXERCISE 2.55 (Trace-preserving witnesses suffice). Show that in Theorem 2.34 one can require that Φ be trace-preserving, at the cost of allowing the range of Φ to be $\mathsf{M}_{m+n}^{\mathrm{sa}}$.

EXERCISE 2.56 (Optimal entanglement witnesses). We work in the Hilbert space $\mathcal{H} = \mathbb{C}^m \otimes \mathbb{C}^n$. For $\sigma \in \mathcal{BP}$, we denote by $\mathrm{E}(\sigma) = \{\rho \in \mathrm{D} \;:\; \mathrm{Tr}(\rho\sigma) < 0\}$ the set of states detected to be entangled by σ. We say that σ is an optimal entanglement witness if $\mathrm{E}(\sigma)$ is maximal (i.e., whenever $\mathrm{E}(\sigma) \subset \mathrm{E}(\tau)$ for $\tau \in \mathcal{BP}$, then $\mathrm{E}(\sigma) = \mathrm{E}(\tau)$). Use the S-lemma (Lemma C.4) to show that if σ lies on an extreme ray of $\mathcal{BP}$ and $\sigma \notin \mathcal{PSD}$, then σ is an optimal entanglement witness.

2.4.5. Proofs of Størmer's theorem. In this section we will present two rather different proofs of the $\mathbb{C}^2 \otimes \mathbb{C}^2$ case of Theorem 2.15, which we state here in a slightly more general form. (See Notes and Remarks for comments regarding the $\mathbb{C}^2 \otimes \mathbb{C}^3$ case.)

THEOREM 2.36 (Størmer's theorem). *If $\mathcal{H} = \mathbb{C}^2 \otimes \mathbb{C}^2$, then the separable cone $\mathcal{SEP}(\mathcal{H})$ and the cone $\mathcal{PPT}(\mathcal{H})$ coincide. Equivalently, $\boldsymbol{P}(\mathbb{C}^2) = \boldsymbol{DEC}(\mathbb{C}^2)$.*

The equivalence of the two assertions of the Theorem follows from Choi's correspondence and duality (see Section 2.4 and particularly Table 2.2). We will focus on the second assertion. Since the inclusion $\boldsymbol{DEC}(\mathcal{H}) \subset \boldsymbol{P}(\mathcal{H})$ always holds, we only need to establish that every positivity-preserving map on $\mathsf{M}_2^{\mathrm{sa}}$ is decomposable.

In a nutshell, the first proof depends on noticing that Proposition 2.32 effectively reduces the general case to that of unital, trace-preserving maps, which in turn follows easily from *very* classical facts. The second proof handles first the maps generating extreme rays of $\boldsymbol{P}(\mathbb{C}^2)$, and concludes via the Krein–Milman theorem. Here are the details.

PROOF # 1 OF THEOREM 2.36. The crucial observation is that it suffices to show that the interior of $\boldsymbol{P}(\mathbb{C}^2)$ is contained in $\boldsymbol{DEC}(\mathbb{C}^2)$. The needed inclusion $\boldsymbol{P}(\mathbb{C}^2) \subset \boldsymbol{DEC}(\mathbb{C}^2)$ follows then from both cones being closed, and being the closures of their interiors.

To that end, suppose that Φ belongs to the interior of $\boldsymbol{P}(\mathbb{C}^2)$. Proposition 2.32 implies then that there exist positive operators $A, B \in \mathsf{M}_2^{\mathrm{sa}}$ and a positivity-preserving, unital and trace-preserving map $\tilde{\Phi} : \mathsf{M}_2^{\mathrm{sa}} \to \mathsf{M}_2^{\mathrm{sa}}$ such that $\Phi(\rho) = A^{-1}\tilde{\Phi}\big(B^{-1}\rho B^{-1}\big)A^{-1}$ for all $\rho \in \mathsf{M}_2^{\mathrm{sa}}$. In other words, $\Phi = \Phi_{A^{-1}} \circ \tilde{\Phi} \circ \Phi_{B^{-1}}$, where $\Phi_M(\rho) := M\rho M^\dagger$. Since every Φ_M is completely positive, the composition rules for completely positive and co-completely positive maps (see Exercises 2.26 and 2.46) show that the problem reduces to establishing decomposability of $\tilde{\Phi}$.

Up to now, the argument worked in any dimension; presently, we will exploit the special features of dimension 2. Since $\tilde{\Phi}$ is an affine self-map of the Bloch ball that preserves the center, it may be thought of as a linear map $R \in B(\mathbb{R}^3)$ with $\|R\|_\infty \leqslant 1$. Such maps are convex combinations of elements of $\mathsf{O}(3)$ (cf. Exercises 1.44 and 1.45), which in turn correspond to maps of the form (i) $\rho \mapsto U\rho U^\dagger$ or (ii) $\rho \mapsto U\rho^T U^\dagger$ for some $U \in \mathsf{U}(2)$ (depending on whether the said element of $\mathsf{O}(3)$ belongs to $\mathsf{SO}(3)$ or not). This is a very special and elementary case of Kadison's Theorem 2.4, and was explained in the proof of Wigner's Theorem 2.3 (see also Exercise B.4 for the isomorphism $\mathsf{PSU}(2) \leftrightarrow \mathsf{SO}(3)$). It remains to recall that the maps of form (i) are completely positive and those of form (ii) are co-completely positive. □

REMARK 2.37. The above argument, when combined with the result from Exercise 1.45, shows that every $\Phi \in \boldsymbol{P}(\mathbb{C}^2)$ can be represented as $\Phi = \sum_j \Phi_{A_j} + \sum_k \Phi_{B_k} \circ T$ so that the total number of terms does not exceed four.

PROOF # 2 OF THEOREM 2.36. Again, we will prove the inclusion $\boldsymbol{P}(\mathbb{C}^2) \subset \boldsymbol{DEC}(\mathbb{C}^2)$. Since $\boldsymbol{P}(\mathbb{C}^2)$ is convex and nondegenerate, it is enough to verify that its extreme rays consist of decomposable maps (see the comment following Proposition 1.9). The following characterization of such extreme rays comes in handy.

PROPOSITION 2.38 (see Appendix C). *Let $\Phi : \mathsf{M}_2^{\mathrm{sa}} \to \mathsf{M}_2^{\mathrm{sa}}$ be a map which generates an extreme ray of $\boldsymbol{P}(\mathbb{C}^2)$. Then either Φ is an automorphism of $\mathcal{PSD}(\mathbb{C}^2)$,* *in which case it is described by Proposition* 2.29, *or* Φ *is of rank one, in which case it is of the form* $\Phi(\rho) = \mathrm{Tr}(\rho|\varphi\rangle\langle\varphi|)|\psi\rangle\langle\psi| = |\psi\rangle\langle\varphi|\rho|\varphi\rangle\langle\psi|$ *for some* $\varphi, \psi \in \mathbb{C}^2\backslash\{0\}$.

Proposition 2.38 is a special case of the characterization of the extreme rays of the maps preserving the Lorentz cone $\mathcal{L}_n$ (remember that the cone $\mathcal{PSD}(\mathbb{C}^2)$ is isomorphic to the Lorentz cone $\mathcal{L}_4$) that will be proved in Appendix C. The proof is based on the so-called S-lemma, a well-known fact from control theory and quadratic/semi-definite programming.

Once we assume the above Proposition, concluding the proof is easy. Indeed, if Φ is an automorphism of $\mathcal{PSD}(\mathbb{C}^2)$, then, by Proposition 2.29, it is either completely positive or co-completely positive, so *a fortiori* decomposable. On the other hand, if Φ is of rank one and $\Phi(\rho) = |\psi\rangle\langle\varphi|\rho|\varphi\rangle\langle\psi|$, then Φ is clearly completely positive with Kraus rank one and the single Kraus operator $A = |\psi\rangle\langle\varphi|$ (see Choi's Theorem 2.21; actually, since A is itself of rank one, it follows that $C(\Phi)$ is in fact separable and hence that Φ is entanglement-breaking; see Lemmas 2.20 and 2.27). □

Notes and Remarks

Classical references for the mathematical aspects of quantum information theory are [**NC00, Hol12, Wil17**]. We also recommend [**Wat**].

Section 2.1. A general reference for the geometry of quantum states is the book [**BŻ06**]. Wigner's theorem appears in [**Wig59**] and Kadison's theorem in [**Kad65**] in a broader context. Elementary proofs can be found in [**Hun72, Sim76**] and recent generalizations in [**SCM16, Stø16**].

Section 2.2. The definition of separability for mixed states was introduced in [**Wer89**]. The NP-hardness of deciding whether a state is separable was shown in [**Gur03**]. The argument sketched in Exercise 2.10 about the number of product vectors needed to represent any separable state is from [**CĐ13**].

Werner states were introduced in [**VW01**], where the question of their separability (Proposition 2.16) is also discussed.

Theorem 2.10 was proved in [**DPS04**]. For more information about k-extendibility and the symmetric subspace (also in the multipartite setting) we refer to the survey [**Har13**]. An early reference for k-entangled states is [**TH00**]. See Notes and Remarks on Chapter 9 for quantitative results about the hierarchies defined in Section 2.2.5.

The observation that non-PPT states are entangled (Peres–Horodecki criterion, Proposition 2.13) goes back to [**Per96**], see also [**HHH96**].

It was observed in [**HHH96**] that Theorem 2.15 is a consequence of results by Størmer [**Stø63**] and Woronowicz [**Wor76**]. See Notes and Remarks on Section 2.4 for more information.

For examples of PPT entangled states in $\mathbb{C}^3 \otimes \mathbb{C}^3$ or $\mathbb{C}^2 \otimes \mathbb{C}^4$, see [**Hor97**]; an early result going in the same direction can be found in [**Cho75b**]. Less *ad hoc* examples (in higher dimensions) are presented, e.g., in [**BDM$^+$99**]. A geometric (non-constructive) argument is given in Chapter 9 (see Propositions 9.18 and 9.20; this approach works if the dimension is sufficiently large).

The realignment criterion to detect entanglement (also called cross-norm criterion) presented in Exercise 2.24 is from [**CW03, Rud05**]. It is neither weaker nor stronger than the PPT criterion. For more separability criteria, see the survey [**HHHH09**].

Theorem 2.17 was proved in [**AS10**] in the bipartite case and in [**FLPS11**] in the general case.

The geometry of the set of absolutely separable states is poorly understood. By definition, whether a state ρ is absolutely separable depends only on its spectrum. An explicit description is known for $\mathbb{C}^2 \otimes \mathbb{C}^2$: a state ρ with eigenvalues $\lambda_1 \geqslant \lambda_2 \geqslant \lambda_3 \geqslant \lambda_4$ is absolutely separable if and only if $\lambda_1 \leqslant \lambda_3 + 2\sqrt{\lambda_2\lambda_4}$ [**VADM01**].

Similarly to absolute separability, one may say that a state $\rho \in \mathcal{H}_1 \otimes \mathcal{H}_2$ is absolutely PPT if $U\rho U^\dagger$ is PPT for any unitary U on $\mathcal{H}_1 \otimes \mathcal{H}_2$. An intriguing open problem is whether every absolutely PPT state is absolutely separable; see [**AJR15**].

Lemma 2.19 can be proved via elementary representation theory; see, e.g., Appendix C in [**ASY14**].

Section 2.3. The Jamiołkowski isomorphism can be traced to [**Jam72**]. Choi's and Jamiołkowski's isomorphisms are seldom distinguished in the literature; a discussion of the difference between the two appears in [**LS13**].

Choi's Theorem 2.21 as stated was proved in [**Cho75a**], which also contains a description of extreme completely positive unital maps. Closely related statements (including variants of Stinespring's Theorem 2.24) varying by the level of abstractness were arrived at (largely) independently by various authors, see, e.g., [**Sti55, Kra71, Kra83**].

Proposition 2.26 is from [**LS93**] and the argument from Exercise 2.34 is based on more general results from [**RSW02**] which give various descriptions of all quantum channels between qubits and of extreme points of the set of such channels.

For elementary properties of the diamond norm, see Section 3.3.4 in [**Wat**] (where it is studied under the name *completely bounded trace norm*). Entanglement-breaking channels were studied in detail in [**HSR03**].

The example from Exercise 2.29 is from [**Tom85**]. Exercise 2.44 is from [**Wat**], to which we also refer for a discussion of the class of LOCC channels.

Section 2.4. Proposition 2.29 is a folklore result which appears explicitly in [**Sch65**]. Many similar results involve classification of "linear preservers", i.e., linear maps on M_d which preserve some property of matrices. Here is a typical statement due to Frobenius: a linear map $\Phi : \mathsf{M}_d \to \mathsf{M}_d$ satisfies the equation $\det \Phi(X) = \det X$ if and only if it has the from $X \mapsto AXB$ or $A \mapsto AX^TB$ for $A, B \in \mathsf{M}_d$ with $\det(AB) = 1$. For a survey on linear preserver problems, see [**LT92**].

The result from Proposition 2.32 and its derivation from Brouwer's fixed-point theorem appear in [**Ide13, Ide16, AS15**]. A similar statement (proved via an iterative construction) appeared in [**Gur03**] for positive maps Φ which are "rank non-decreasing" (however, not all such maps satisfy the conclusion of Proposition 2.32, see Exercise 2.53). The validity of Proposition 2.32 for completely positive maps is simpler and well known, see for example [**GGHE08**] and its references. The original Sinkhorn's theorem (for matrices, or for maps preserving the positive orthant in $\mathbb{R}^n$) goes back to [**Sin64**]; see [**Ide16**] for an extensive survey of related topics.

Theorem 2.34 is from [**HHH96**]. The concept of optimal entanglement witness which appears in Exercise 2.56 was investigated in [**LKCH00**].

Størmer's Theorem 2.36 was initially proved in [**Stø63**]; the original formulation involved the second of the two statements. The first proof presented here seems to be new and was a byproduct of the work on this book [**AS15**]. The scheme behind the second proof was apparently folklore for some time; it was documented in [**MO15**]. The novelty of its current presentation, if any, consists in streamlining the proof of Proposition 2.38. (For more background information on Proposition 2.38, see Appendix C.) Other proofs (of either of the two versions given in Theorem 2.36) appeared in [**KCKL00, VDD01, LMO06, KVSW09, Stø13**]. A recent study of positivity-preserving maps on M_3 can be found in [**MO16**]. While [**MO16**] is focused on the unital trace-preserving case, it is likely that (particularly when combined with our Proposition 2.32) it may provide a clear picture of the more general setting. In particular, it may lead to a simple and transparent proof of the $\mathbb{C}^2 \otimes \mathbb{C}^3$ case of Theorem 2.15 (Woronowicz's theorem).

CHAPTER 3

Quantum mechanics for mathematicians

This section is addressed primarily to mathematicians who are new to quantum information theory. Its purpose is to indicate why various mathematical concepts enter the theory, and to give an idea of their physical meaning or interpretation. We make no attempt at being comprehensive; our attention is restricted to the constructs that play a central role in this book and that we ourselves have found (and still find) puzzling, such as mixed states and completely positive maps. In any case, neither of the authors being a physicist, the scope (and the depth) of the presentation will necessarily be limited.

This section is designed to be essentially independent of the rest of the book. The only "non-mainstream" technical device that is indispensable for following it is the Dirac bra-ket notation (see Section 0.3). The discussion will be occasionally informal in order for readers to acquaint themselves with concepts that are presented more rigorously elsewhere in the book.

3.1. Simple-minded quantum mechanics

The state of a physical system (say, a particle) is described by a *wave function* $\psi \in L_2(\mathbb{R}^3)$, which is generally time-dependent and complex-valued. Its dependence on time is governed by some evolution equation (for example, the Schrödinger equation) and is necessarily unitary: given $t > 0$, there is a unitary operator U_t such that if the state of the particle at time 0 is described by ψ_0 (a priori unknown), its state at time t will be $U_t\psi_0$. The probability of finding the particle at $x \in \mathbb{R}^3$ (assuming the appropriate measurement is performed) is given by the probability density function $|\psi(x)|^2$, according to the Copenhagen interpretation. This forces wave functions to be normalized in L_2 and justifies the postulate of unitary evolution. Other physical properties of the particle are exhibited similarly. In particular, if a given physical quantity is discrete, then there is an orthonormal sequence (or basis) (u_j), indexed by possible values of the quantity in question, such that the probability of obtaining the jth value during measurement is $|\langle\psi, u_j\rangle|^2$. This is the simplest case of the so-called Born rule. In a way, the actual values of the physical quantity are of secondary importance and one simply says that "a measurement was performed in the basis (u_j)" or that "(u_j) is the *computational basis*" for this particular measuring/experimental setup. (We will briefly discuss other, more general measurement schemes in Section 3.6.)

It should be emphasized that it is possible for measurement results to be deterministic. If the basis (u_j) is such that $\psi = u_{j_0}$ for some j_0, then measuring ψ in the basis (u_j) will yield the j_0th outcome with probability 1. For the same reason, two states ψ and φ are *in principle* perfectly distinguishable if (and only if) they are orthogonal; one then "merely" needs to arrange a measurement in a basis that contains both ψ and φ.

3.2. Finite vs. infinite dimension, projective spaces, and matrices

In the previous section the "state space" is the infinite-dimensional Hilbert space $\mathcal{H} = L_2(\mathbb{R}^3)$. However, if the number of possible values of a physical quantity is finite (and it may be argued that this is always the case, the "infinite" being just a useful abstraction of "large"), the interesting part of the Hilbert space is finite-dimensional and, consequently, may be identified with $\mathbb{C}^d$ for some $d \in \mathbb{N}$ (a d-level system). A state is then simply a unit vector $\psi \in \mathbb{C}^d$. *A priori* d may be very large, but even the simple case of $d = 2$ (a qubit) is of interest: it may describe for example the spin of an electron or the polarization of a photon.

Next, it is apparent from the discussion in Section 3.1 that no measurement can distinguish between the wave functions ψ and $\omega\psi$, where $\omega \in \mathbb{C}$ with $|\omega| = 1$, and so the "true" state space is the complex projective space $\mathsf{P}(\mathbb{C}^d)$ (or $\mathbb{CP}^{d-1}$) for d-level systems. Another mathematical scheme that conveniently disregards scalar factors is to consider not a unit vector $\psi \in \mathbb{C}^d$, but the orthogonal projection onto $\mathbb{C}\psi$ or, in the language of matrices, the outer product $\rho = |\psi\rangle\langle\psi| \in \mathsf{M}_d$. In that language, when a measurement is performed in some basis (u_j), the probability of the jth outcome is

$$|\langle\psi, u_j\rangle|^2 = \langle u_j, \psi\rangle\langle\psi, u_j\rangle = \langle u_j|\rho|u_j\rangle = \operatorname{Tr}\left(\rho|u_j\rangle\langle u_j|\right). \tag{3.1}$$

3.3. Composite systems and quantum marginals: Mixed states

This section gives motivation to the definition of (mixed) quantum states which appeared in Section 2.1.1.

For classical systems, the state space of a system consisting of components is the Cartesian product of the corresponding state spaces. In the quantum setting, if the state spaces of the components (subsystems, particles, ...) are Hilbert spaces $\mathcal{H}_1, \dots, \mathcal{H}_m$, the state space of the composite system is the tensor product $\mathcal{K} = \mathcal{H}_1 \otimes \cdots \otimes \mathcal{H}_m$. However, the Cartesian product of orthonormal bases of $\mathcal{H}_k$'s is an orthonormal basis of $\mathcal{K}$. This is as far as the similarities to the classical case go.

Consider now a bipartite system $\mathcal{K} = \mathcal{H} \otimes \mathcal{E}$ and assume that we have access only to the $\mathcal{H}$ part. (This may be the case when $\mathcal{H}$ describes the state inside an apparatus in a laboratory and $\mathcal{E}$ the environment, or if we decide to focus only on the first subsystem.) Suppose that our system is in the state described by $\psi \in \mathcal{K}$ and let us try to figure out the $\mathcal{H}$-marginal of ψ, i.e., the state on $\mathcal{H}$, measurements of which "within $\mathcal{H}$" are consistent with hypothetical measurements of the complete state ψ.

If $\psi = \xi \otimes \eta$ (a product vector), the result is as expected: the $\mathcal{H}$-marginal of ψ is ξ. To check this, we note that if we measure ξ in some basis (u_j) of $\mathcal{H}$, we obtain the jth outcome with probability $p_j = |\langle\psi, u_j\rangle|^2$. For a different point of view, suppose that we have access to the entire system and that we perform a measurement in the basis $(u_j \otimes v_k)_{j,k}$, where (v_k) is some basis of $\mathcal{E}$. The probability of obtaining the (j,k)th outcome is then $q_{jk} = |\langle\xi \otimes \eta, u_j \otimes v_k\rangle|^2 = |\langle\xi, u_j\rangle|^2 \cdot |\langle\eta, v_k\rangle|^2$. Summing over k, we again find that the probability of the jth outcome on the first component is $|\langle\psi, u_j\rangle|^2 = p_j$. This is simply a verification that the probability distribution (p_j) is the (first) marginal of (q_{jk}) and that, moreover, product vectors lead to product distributions, or to independent random variables. Another way to express this marginal probability is $p_j = \operatorname{Tr} \rho P_{u_j}$, where $\rho = |\psi\rangle\langle\psi|$, and where $P_u = |u\rangle\langle u| \otimes \mathrm{I}_{\mathcal{E}}$ is the orthogonal projection onto the subspace $u \otimes \mathcal{E}$ of $\mathcal{H} \otimes \mathcal{E}$. This calculation perfectly makes sense even if ξ is not a product vector, and it makes clear that p_j does not depend on, say, the choice of the basis of $\mathcal{E}$.

Consider now $\psi \in \mathcal{H} \otimes \mathcal{E}$, which is *not* a product vector. Let

$$\psi = \sum_{i=1}^{r} a_i\, \xi_i \otimes \eta_i \tag{3.2}$$

be its Schmidt decomposition (see Section 2.2.2), necessarily with $r \geqslant 2$. Since the $\mathcal{H}$-marginal of $\xi_i \otimes \eta_i$ is ξ_i, it is tempting to guess that the $\mathcal{H}$-marginal of ψ is $\sum_{i=1}^{r} a_i\, \xi_i$. However, one should immediately become suspicious: for any choice of (complex) signs ω_i, the vector $\sum_{i=1}^{r} a_i\omega_i\xi_i$ is an equally valid candidate, and while the state remains unchanged if you multiply a vector by a complex number ω with $|\omega| = 1$, it may change radically if you multiply different (non-zero) components by different numbers. A more careful analysis is needed, and it turns out that the proper language to describe marginals is that of matrices. In the notation of the preceding paragraph we have

$$\begin{aligned}
p_j = \operatorname{Tr}\left(|\psi\rangle\langle\psi| P_{u_j}\right) &= \operatorname{Tr}\Big[\Big(\sum_{i,l=1}^{r} a_i\bar{a}_l\, |\xi_i\rangle\langle\xi_l| \otimes |\eta_i\rangle\langle\eta_l|\Big)\big(|u_j\rangle\langle u_j| \otimes \mathrm{I}_{\mathcal{E}}\,\big)\Big] \\
&= \sum_{i,l=1}^{r} a_i\bar{a}_l\, \operatorname{Tr}\left[\left(|\xi_i\rangle\langle\xi_l|\right)\left(|u_j\rangle\langle u_j|\right)\right] \operatorname{Tr}\left(|\eta_i\rangle\langle\eta_l|\right) \\
&= \operatorname{Tr}\Big[\Big(\sum_{i=1}^{r} |a_i|^2 |\xi_i\rangle\langle\xi_i|\Big)|u_j\rangle\langle u_j|\Big] \\
&= \langle u_j|\Big(\sum_{i=1}^{r} |a_i|^2 |\xi_i\rangle\langle\xi_i|\Big)|u_j\rangle.
\end{aligned} \tag{3.3}$$

In other words, the probability that a measurement performed in a basis (u_j) yields the jth outcome is $\langle u_j|\rho_{\mathcal{H}}|u_j\rangle = \operatorname{Tr}\left(\rho_{\mathcal{H}}|u_j\rangle\langle u_j|\right)$, where

$$\rho_{\mathcal{H}} = \sum_{i=1}^{r} |a_i|^2 |\xi_i\rangle\langle\xi_i|. \tag{3.4}$$

So the *mixed state* $\rho_{\mathcal{H}}$ fits the role of the $\mathcal{H}$-marginal of the "global" state $\rho = \rho_{\mathcal{H}\mathcal{E}} = |\psi\rangle\langle\psi|$. Therefore, while *in principle* the state of a quantum system is described by a vector (or a rank one projection, or an element of a projective space, or a wave function), i.e., by a *pure* state, we seldom, if ever, will be able to perform a measurement in a global basis, and we therefore have to rely on mixed states for modeling such systems. To use the Platonic analogy, a mixed state is "the shadow on the wall" of our cave, comprising all the features of the "idea" (or "form") ψ that are accessible to our perception.

A more heuristic explanation of the formula (3.4) for the marginal is that, from the perspective of $\mathcal{H}$, the state of our system is ξ_i with probability $p_i = |a_i|^2$, and so we need to compute the weighted average of probabilities corresponding to $\rho = |\xi_i\rangle\langle\xi_i|$. Since the expression $\operatorname{Tr}\rho\big(|u\rangle\langle u|\big)$ is linear in ρ, the average can be performed inside the trace, implying the formula for $\rho_{\mathcal{H}}$. We encourage readers who are not used to the bra-ket formalism to work out the details of several variants of this calculation outlined in Exercise 3.1.

The key features of the marginal $\rho_{\mathcal{H}}$ are that it is canonical (for example, it does not depend on the basis (u_j) of $\mathcal{H}$ in which the measurement is performed) and that it encodes all the information that can be obtained about the global state by measurements inside $\mathcal{H}$. In particular, if $\rho_{\mathcal{H}}$ is truly mixed (i.e., not pure, with $r \geqslant 2$ in (3.2) or in (3.4)), then there are no measurements inside $\mathcal{H}$ that are deterministic.

A simple but spectacular demonstration of this phenomenon are the Bell states on $\mathbb{C}^2 \otimes \mathbb{C}^2$: $\rho = |\psi\rangle\langle\psi|$ with ψ being (for example) one of the four *Bell vectors*

$$\varphi^{\pm} = \frac{1}{\sqrt{2}}(|00\rangle \pm |11\rangle), \quad \psi^{\pm} = \frac{1}{\sqrt{2}}(|01\rangle \pm |10\rangle),$$

where $|0\rangle, |1\rangle$ is the canonical basis of $\mathbb{C}^2$ (recall that $|00\rangle$ stands for $|0\rangle \otimes |0\rangle$). It is easily seen that in each case the marginal of ρ on either $\mathbb{C}^2$ factor is $(|0\rangle\langle 0| + |1\rangle\langle 1|)/2 = \mathrm{I}/2$. Consequently, when measuring in any basis (u_1, u_2) (of, say, the first factor), each of the two outcomes occurs with probability $1/2$, and so the results of such measurements, in and of themselves, tell us nothing. In particular, they cannot help us distinguish between $\varphi^+, \varphi^-, \psi^+, \psi^-$, even though a global measurement performed in the basis consisting of these four vectors would tell them apart perfectly.

EXERCISE 3.1. Perform alternative calculations of the probabilities from (3.3) according to the following outline. Consider a product basis $(u_j \otimes v_k)_{j,k}$ of $\mathcal{H} \otimes \mathcal{E}$. If $\rho = |\psi\rangle\langle\psi|$ with $\psi = \sum_{i=1}^r a_i\, \xi_i \otimes \eta_i$, the probability of the (j,k)th outcome will be, by (3.1),

$$q_{jk} = \Big|\Big\langle \sum_{i=1}^{r} a_i\, \xi_i \otimes \eta_i, u_j \otimes v_k \Big\rangle\Big|^2 = \operatorname{Tr} \rho(|u_j\rangle\langle u_j| \otimes |v_k\rangle\langle v_k|).$$

Finally, retrieve $p_j = \sum_k q_{jk}$ by expanding either the second or the third expression in the above.

3.4. The partial trace: Purification of mixed states

The discussion in the previous section shows that, in some cases, a natural way of modeling the state of a subsystem of a quantum system is to consider operators rather than unit vectors. An elegant way to describe *quantum marginals* is via the concept of *partial trace*, which is defined as follows (see also Section 2.2.1). First, for any operator (self-adjoint or not) on a composite Hilbert space $\mathcal{H} \otimes \mathcal{E}$ which is a tensor product of operators, we define its partial trace with respect to $\mathcal{E}$ as

$$\operatorname{Tr}_{\mathcal{E}}(\sigma \otimes \tau) = \operatorname{Tr}(\tau)\sigma.$$

Next, we extend this operation to all operators by linearity (which is possible because of the universal property of the tensor product). Clearly, if $\xi \in \mathcal{H}, \eta \in \mathcal{E}$ are unit vectors, then

$$\operatorname{Tr}_{\mathcal{E}}(|\xi \otimes \eta\rangle\langle \xi \otimes \eta|) = \operatorname{Tr}_{\mathcal{E}}(|\xi\rangle\langle\xi| \otimes |\eta\rangle\langle\eta|) = |\xi\rangle\langle\xi|.$$

Similarly, if $\psi = \sum_{i=1}^r a_i\, \xi_i \otimes \eta_i$ is a Schmidt decomposition, then

$$\operatorname{Tr}_{\mathcal{E}}(|\psi\rangle\langle\psi|) = \sum_{i=1}^{r} |a_i|^2 |\xi_i\rangle\langle\xi_i|.$$

In other words, $\operatorname{Tr}_{\mathcal{E}}(\rho) = \rho_{\mathcal{H}}$, the $\mathcal{H}$-marginal of ρ defined by (3.4). The notation may be a little confusing since in order to find the $\mathcal{H}$-marginal we need to calculate the partial trace with respect to $\mathcal{E}$, but it is generally accepted. It simply corresponds to the following fact from elementary probability: given two random variables X, Y with joint density $f(x, y)$, the marginal density of X is obtained by integrating f with respect to y.

Another point which needs to be clarified is that the set of mixed states on $\mathcal{H}$ that may be obtained as $\mathcal{H}$-marginals of pure states on composite systems $\mathcal{H} \otimes \mathcal{E}$ (for some auxiliary space $\mathcal{E}$) is exactly the set $\mathrm{D}(\mathcal{H})$ of positive semi-definite trace one operators (usually referred to as *density matrices*, particularly if $\mathcal{H} = \mathbb{C}^d$). This is the consequence of the following computation: if $\rho \in \mathrm{D}(\mathcal{H})$, and $\rho = \sum_i \lambda_i |\xi_i\rangle\langle\xi_i|$ is its spectral decomposition, then choosing $\mathcal{E} = \mathcal{H}$ and $\psi = \sum_i \sqrt{\lambda_i}\, \xi_i \otimes \xi_i$ ensures that $\mathrm{Tr}_{\mathcal{E}}(|\psi\rangle\langle\psi|) = \rho$. We say that $|\psi\rangle\langle\psi|$ (or simply ψ) is a *purification* of ρ. Clearly, the Schmidt rank of ψ (always) equals $\operatorname{rank} \rho =: r$. Moreover, the minimal dimension of $\mathcal{E}$ for which a purification of ρ exists in $\mathcal{H} \otimes \mathcal{E}$ is also equal to r. Even though this construction is abstract, it is canonical in the following sense: if ρ is a *physical* state on $\mathcal{H}$ that is the $\mathcal{H}$-marginal of a physical pure state $\psi \in \mathcal{H} \otimes \mathcal{E}$ (where $\mathcal{E}$ is the environment relative to $\mathcal{H}$), then we must have $\psi = \sum_{i=1}^r \sqrt{\lambda_i}\, \xi_i \otimes \eta_i$ for some basis (η_i) of $\mathcal{E}$. (The only catch is that (η_i) may not be the most natural basis of $\mathcal{E}$.)

3.5. Unitary evolution and quantum operations: The completely positive maps

As mentioned earlier, the evolution of a quantum system is unitary, i.e., if $t_0 < t_1$, then there is a unitary operator U such that if the state of the system at time t_0 (the initial state) is described by a vector ψ (which is a priori general and/or unknown), then its state at time t_1 (the terminal state) will be $U\psi$. (U depends on the physical laws governing the evolution, and we may be able to control some of its parameters, but it is independent of ψ.) If we switch to the language of density matrices, the formula $\psi \mapsto U\psi$ becomes $|\psi\rangle\langle\psi| = \rho \mapsto U\rho U^\dagger$. (These are the unitary channels defined in Section 2.3.4.)

We now want to understand how the formalism needs to be adapted to describe subsystems, i.e., when we pass to the more general context of mixed states. Assume that our evolution operator U acts on a composite space $\mathcal{H} \otimes \mathcal{E}$ and—to begin with—takes the form $V \otimes W$, where V and W are unitary operators on $\mathcal{H}$ and $\mathcal{E}$ respectively. If $\psi = \xi \otimes \eta$ is also a product vector, then the evolution of the subsystem $\mathcal{H}$ is clearly given by $\xi \mapsto V\xi$, or by $\sigma \mapsto V\sigma V^\dagger$ in the language of density matrices. The latter formula remains valid if $\psi \in \mathcal{H} \otimes \mathcal{E}$ is an arbitrary (unit) vector, and $\sigma = \mathrm{Tr}_{\mathcal{E}}(|\psi\rangle\langle\psi|)$ is the corresponding $\mathcal{H}$-marginal. (This follows from the identity $V \, \mathrm{Tr}_{\mathcal{E}}(\rho) V^\dagger = \mathrm{Tr}_{\mathcal{E}}(U\rho U^\dagger)$, valid for $U = V \otimes W$ and for any matrix ρ.)

The situation becomes more complicated in a case where the evolution of the subsystem $\mathcal{H}$ and the environment $\mathcal{E}$ are not decoupled, i.e., where U is not a product of two unitaries. Even if the initial state of the system is a product vector $\psi = \xi \otimes \eta$, there is no reason why the terminal state $U\psi$, which can *a priori* be arbitrary, should be of that form. In other words, even if the initial $\mathcal{H}$-marginal $\sigma = |\xi\rangle\langle\xi|$ is pure, the terminal marginal may be mixed. In particular, the evolution of the marginal is not necessarily unitary. Moreover, for fixed ξ, different values of the initial $\mathcal{E}$-marginal η may result in radically different values of the terminal $\mathcal{H}$-marginal.

However, this is neither surprising nor fatal. First, if *there is* interaction between our subsystem $\mathcal{H}$ and the environment $\mathcal{E}$, it is to be expected that the terminal state of $\mathcal{H}$ possibly depends on the state of $\mathcal{E}$. Second, while we may not know what the initial state of $\mathcal{E}$ is, we can simply think of it as an external parameter affecting the evolution of our subsystem $\mathcal{H}$, which is the only one we can manipulate, control and measure.

We now want to come up with a formula that generalizes the unitary evolution $\rho \mapsto U\rho U^\dagger$ or, more precisely, that is the "shadow on the wall of our cave" of the unitary evolution. Let us start again with the global initial state being a product vector $\psi = \xi \otimes \eta$; the terminal state is then represented by the vector $U(\xi \otimes \eta)$. Since η is assumed to be fixed, we can omit the dependence on η in the description and simply talk about an (*a priori* arbitrary) *isometry* $\xi \mapsto V\xi \in \mathcal{H} \otimes \mathcal{E}$. (Of course, since by definition $V\xi = U(\xi \otimes \eta)$, V does implicitly depend on η.) In the language of density matrices, the evolution of the $\mathcal{H}$-marginal is then given by

$$\sigma \mapsto \operatorname{Tr}_{\mathcal{E}} V\sigma V^\dagger, \tag{3.5}$$

where $\sigma = |\xi\rangle\langle\xi|$ is the initial marginal (cf. Theorem 2.24).

If we want to give a description of the evolution that is intrinsic to $\mathcal{H}$, we may proceed as follows. Let (v_i) be an orthonormal basis of $\mathcal{E}$. The isometry V can be represented as $V\xi = \sum_i (A_i\xi) \otimes v_i$ for some operators $A_i \in B(\mathcal{H})$. Consequently,

$$V\sigma V^\dagger = \sum_{i,j} |A_i\xi\rangle\langle A_j\xi| \otimes |v_i\rangle\langle v_j| = \sum_{i,j} \left(A_i|\xi\rangle\langle\xi|A_j^\dagger\right) \otimes |v_i\rangle\langle v_j|$$

and further,

$$\operatorname{Tr}_{\mathcal{E}} V\sigma V^\dagger = \sum_{i,j} \left(A_i|\xi\rangle\langle\xi|A_j^\dagger\right) \operatorname{Tr} |v_i\rangle\langle v_j| = \sum_i A_i|\xi\rangle\langle\xi|A_i^\dagger.$$

Accordingly, an alternative description of the evolution is

$$\sigma \mapsto \sum_i A_i \sigma A_i^\dagger\,. \tag{3.6}$$

This is a description intrinsic to $\mathcal{H}$, since $A_i \in B(\mathcal{H})$. Moreover, according to Choi's theorem (Theorem 2.21), the evolution described by (3.5)–(3.6) is given by a completely positive map on $B(\mathcal{H})$. The operators A_i aren't completely arbitrary, since the resulting map $\xi \mapsto V\xi = \sum_i (A_i\xi) \otimes v_i$ needs to be an isometry. For this to happen we must have, for every $\xi \in \mathcal{H}$,

$$\langle \xi, \xi\rangle = \langle V\xi, V\xi\rangle = \sum_{i,j} \langle A_i\xi, A_j\xi\rangle\langle v_i, v_j\rangle = \sum_i \langle A_i\xi, A_i\xi\rangle = \langle\xi| \sum_i A_i^\dagger A_i |\xi\rangle.$$

Given that for self-adjoint operators $A, B \in B(\mathcal{H})$ the condition $\langle\xi|A|\xi\rangle = \langle\xi|B|\xi\rangle$ for all $\xi \in \mathcal{H}$ implies $A = B$, it follows that V being an isometry is equivalent to

$$\sum_i A_i^\dagger A_i = \mathrm{I}_{\mathcal{H}}, \tag{3.7}$$

which in turn (see Remark 2.23) is equivalent to the map given by (3.6) being trace-preserving. This should not come as a surprise, since we want the evolution equation to map density matrices to density matrices, which for linear evolutions is equivalent to preserving the trace.

To summarize, under the hypothesis of unitary evolution of the global system $\mathcal{H} \otimes \mathcal{E}$, the relationship $\sigma \mapsto \Phi(\sigma)$ between the initial state σ of subsystem $\mathcal{H}$ (the initial $\mathcal{H}$-marginal) and its terminal state $\Phi(\sigma)$ is described by a *completely positive trace-preserving map* (CPTP) Φ acting on $B(\mathcal{H})$. CPTP maps are also called *quantum channels*.

We derived the above characterization of quantum evolution maps Φ under the assumption that the initial global state was given by a product vector $\psi = \xi \otimes \eta$, with $\Phi \in B(\mathcal{H})$ depending on the (*a priori* unknown, but specific) $\mathcal{E}$-marginal described by η. One could ask whether a similar (or some other) characterization can be derived in a more general case where the initial state, while still a vector, is no longer separable. However, there appears to be no straightforward way to produce a canonical map in that setting. One natural approach would be to try to associate an evolution map $\Phi : B(\mathcal{H}) \to B(\mathcal{H})$, acting in a consistent manner on $\mathcal{H}$-marginals, to a given global unitary evolution induced by U and a given $\mathcal{E}$-marginal $\tau \in B(\mathcal{E})$. However, while knowing $\mathcal{H}$- and $\mathcal{E}$-marginals of a pure state tells us a lot about the structure of that state, it still leaves a lot of uncertainty. For example, $\mathcal{H}$- and $\mathcal{E}$-marginals of all four Bell states $\varphi^+, \varphi^-, \psi^+, \psi^-$ on $\mathbb{C}^2 \otimes \mathbb{C}^2$ are identical: they are maximally mixed states $\frac{1}{2}\,\mathrm{I}_{\mathbb{C}^2}$. On the other hand, in the absence of some strong restrictions on the form of the global unitary evolution U, there is no reason to expect the $\mathcal{H}$-marginals of $U\varphi^+, U\varphi^-, U\psi^+, U\psi^-$ to be the same. (In fact, various quantum algorithms exploit the fact that those marginals may be quite different.) In other words, such a map Φ cannot be consistently defined.

In physics texts this characterization, and specifically the postulate of complete positivity, is usually arrived at in a somewhat different way. First, it is noted that a quantum evolution map (or a quantum operation) $\Phi : B(\mathcal{H}) \to B(\mathcal{H})$ should map density matrices to density matrices. Under the assumption of linearity, this is equivalent to Φ being positive and trace-preserving (see Section 2.3.2). Second, when Φ is coupled with an identity map on the environment $\mathcal{E}$, then the resulting map $\Phi \otimes \mathrm{Id}_{B(\mathcal{E})}$ should also be an allowed quantum operation and in particular, it should be positive. If $\dim \mathcal{E}$ is at least as large as $\dim \mathcal{H}$, this is equivalent to complete positivity of Φ. The argument presented earlier in this section is substantially more involved, but seems to us more physically natural (and less formal).

3.6. Other measurement schemes

Throughout our discussion we assumed that a measurement is performed in some basis (u_j) of the entire space, or of the space corresponding to the accessible subsystem, with the probability of the jth outcome being either $|\langle \psi, u_j \rangle|^2$ or $\langle u_j|\rho|u_j\rangle = \mathrm{Tr}\,(\rho|u_j\rangle\langle u_j|)$ (depending on whether the state of the system is pure or mixed). A slightly more general scheme is that of a *projective measurement*, where the measuring apparatus is modeled by a sequence of mutually orthogonal projections (P_i) and the probability of the ith outcome is

$$|P_i\psi|^2 = \langle \psi|P_i|\psi\rangle = \mathrm{Tr}\,\rho P_i. \tag{3.8}$$

However, this is barely more general: we can think of the instrument as being related to a basis (u_j), but as providing only a coarse-grained view, where some of the basis elements u_j are merged into one projection P_i.

A more substantive generalization is derived from basis/projective measurements in a similar way that CPTP maps were derived from unitary operations. Suppose that a projective measurement (P_i) on $\mathcal{H} \otimes \mathcal{E}$ (rank one or not) is performed and consider the effects of applying it to a product state $\psi = \xi \otimes \eta$. The probability of the ith outcome is then

$$p_i = \langle\psi|P_i|\psi\rangle = \operatorname{Tr}\big(|\psi\rangle\langle\psi|P_i\big) = \operatorname{Tr}\big((|\xi\rangle\langle\xi| \otimes |\eta\rangle\langle\eta|)P_i\big) \\ = \operatorname{Tr}\big(|\xi\rangle\langle\xi| \operatorname{Tr}_{\mathcal{E}}(\mathrm{I}\otimes|\eta\rangle\langle\eta|)P_i\big). \tag{3.9}$$

In the last equality we used the identity

$$\operatorname{Tr}\big((\tau\otimes \mathrm{I})X\big) = \operatorname{Tr}\big(\tau \operatorname{Tr}_{\mathcal{E}} X\big), \tag{3.10}$$

which is easily verified if X is a product operator and follows by linearity for arbitrary X. In other words, there are operators (M_i) on $\mathcal{H}$ such that

$$p_i = \operatorname{Tr}(|\xi\rangle\langle\xi|M_i). \tag{3.11}$$

Varying ξ and using the fact that $\sum_i P_i = \mathrm{I}_{\mathcal{H}\otimes\mathcal{E}}$ we deduce that

$$\sum_i M_i = \mathrm{I}_{\mathcal{H}} \tag{3.12}$$

and that M_i is positive for each i. Even though Born's rule (3.11) was derived for a pure state $\rho = |\xi\rangle\langle\xi|$, it extends by linearity to a general (possibly mixed) mixed state ρ on $\mathcal{H}$ via the formula

$$p_i = \operatorname{Tr}(\rho M_i). \tag{3.13}$$

A system (M_i) verifying the condition (3.12) is called a *positive operator-valued measure* (POVM) and the associated measurement scheme a *POVM measurement.* The reason for invoking the term "measure" is that there are also continuous variants, namely operator-valued measures integrating to identity.

3.7. Local operations

This short section aims at explaining the meaning of the word "local", which is often used in quantum information theory. Up to now we have focused on a Hilbert space denoted $\mathcal{H}$. Moreover, the standard framework of quantum information theory assumes that $\mathcal{H}$ is endowed with a tensor decomposition $\mathcal{H} = \mathcal{H}_A \otimes \mathcal{H}_B$ (or a multipartite variant), where $\mathcal{H}_A$ is the Hilbert space of Alice's system and $\mathcal{H}_B$ is the Hilbert space of Bob's system. The usual assumption is that Alice and Bob are surrogates for two distant experimentalists who share a quantum system $\mathcal{H}$.

In this context, operations that can be performed "privately" by Alice and Bob are called *local operations*. For example, *local unitaries* on $\mathcal{H}$ are unitary operators of the form $U = U_A \otimes U_B$, where U_A (resp., U_B) is a unitary operator on $\mathcal{H}_A$ (resp., on $\mathcal{H}_B$). Similarly, local POVMs on $\mathcal{H}$ are of the form $(M_i \otimes N_j)$, where (M_i) is a POVM on $\mathcal{H}_A$ and (N_j) is a POVM on $\mathcal{H}_B$. A local channel $\Phi : B(\mathcal{H}) \to B(\mathcal{H})$ is of form $\Phi_A \otimes \Phi_B$, where $\Phi_A : B(\mathcal{H}_A) \to B(\mathcal{H}_A)$ and $\Phi_B : B(\mathcal{H}_B) \to B(\mathcal{H}_B)$ are quantum channels.

A related concept is that of *Local Operations with Classical Communication* (LOCC). In a nutshell, the LOCC class is obtained by combining the local operations described above with classical communication between the parties. However, its precise mathematical definition is actually quite intricate (see Section XI in [**HHHH09**]). We consider some aspects of LOCC operations in Section 12.2.1.

3.8. Spooky action at a distance

We conclude this chapter by presenting a baby version of Einstein's "spooky action at a distance" consequence of a quantum description of the physical reality. Suppose that each of two distant experimentalists, Alice and Bob, has in their lab a particle that they can *locally* measure, and that each particle can be in one of two possible states $|0\rangle$ or $|1\rangle$. Suppose further that, as a system, the two particles are in a Bell quantum state $\psi^+ = \frac{1}{\sqrt{2}}(|01\rangle + |10\rangle)$ (on the Hilbert space $\mathcal{H} = \mathcal{H}_A \otimes \mathcal{H}_B = \mathbb{C}^2 \otimes \mathbb{C}^2$). As described in Section 3.3, independently of the choice of measurement bases in $\mathcal{H}_A$ and $\mathcal{H}_B$, both outcomes of Alice's (resp., Bob's) measurement will be equally likely. However, some combinations of the outcomes are more likely than others. For example, suppose that each of them performs the measurement in their computational basis $(|0\rangle, |1\rangle)$, which, in the terminology of Section 3.7, corresponds to a local POVM with $(M_i) = (N_j) = (|0\rangle\langle 0|, |1\rangle\langle 1|)$. Table 3.1 shows the resulting joint probability distribution. Note that Alice's and Bob's outcomes are always different. This is not immediately fatal as it may just be the case that—perhaps because of some conservation law in their interaction in the past—the two particles are in opposite states, we just don't know which. However, on further reflection, this indicates that either the description of the reality given by ψ^+ is incomplete, with some other *hidden variable* controlling the outcomes of measurements, or that the fact of Alice's performing her experiment *instantaneously* affects the particle that is in Bob's possession.

TABLE 3.1. Joint probability distribution of Alice's and Bob's measurement outcomes.

Alice \ Bob	$\|0\rangle$	$\|1\rangle$
$\|0\rangle$	0	$\frac{1}{2}$
$\|1\rangle$	$\frac{1}{2}$	0

Moreover, this phenomenon is just a harbinger of more involved schemes, but based on very similar principles, which lead to effects that cannot be explained by a hidden variable model, and to phenomena such as pseudotelepathy or quantum teleportation. We will briefly explore some of these examples later on, mostly in Chapter 11.

EXERCISE 3.2. Verify the details of the calculation of probabilities in Table 3.1.

Notes and Remarks

There are many books which present quantum mechanics for specific audiences. In addition to the references given at the end of Chapter 2, we point out [**Mer07**] (mostly directed at computer scientists) and [**RP11**]. Other references targeting mathematicians are [**Tak08**] and [**Sha08**].

Part 2

Banach and His Spaces

Asymptotic Geometric Analysis Miscellany

CHAPTER 4

More convexity

The focus of this chapter is concepts, invariants, and operations related to finite-dimensional convex bodies. The primary objectives are to be able to describe, tell apart, and measure the size of such bodies. While some of the results are relatively new, they all have roots in classical convex geometry and, most notably, in the work of Hermann Minkowski in the late 19th and early 20th century. Other, more modern aspects of the theory of convex bodies will be addressed in Chapters 5 and 7.

4.1. Basic notions and operations

4.1.1. Distances between convex sets. A natural way to quantify how different two subsets of a metric space are is the Hausdorff distance. When we consider convex bodies $K, L \subset \mathbb{R}^n$ containing the origin in their interiors, and identified when related by a homothetic transformations, a more relevant notion is often their *geometric distance*, defined as

$$d_g(K, L) = \inf\{\alpha\beta \ : \ \alpha, \beta > 0, \ K \subset \alpha L, \ L \subset \beta K\}. \tag{4.1}$$

Equivalently,

$$d_g(K, L) = \sup_{x \in \mathbb{R}^n, x \neq 0} \frac{\|x\|_K}{\|x\|_L} \times \sup_{x \in \mathbb{R}^n, x \neq 0} \frac{\|x\|_L}{\|x\|_K}.$$

This "distance" satisfies the multiplicative version of the triangle inequality

$$d_g(K, M) \leqslant d_g(K, L) d_g(L, M).$$

If we want to consider the family of n-dimensional convex bodies up to affine transformations, the proper tool is the *Banach–Mazur distance*

$$d_{BM}(K, L) = \inf\{d_g(K + a, TL + b) \ : \ T \in \mathsf{GL}(n, \mathbb{R}), \ a, b \in \mathbb{R}^n\}. \tag{4.2}$$

In the case where K and L are symmetric (i.e., 0-symmetric), which is the setting most frequently encountered in the literature, we can restrict the infimum in (4.2) to $a = b = 0$. In either case, we are led to a compact set (see Exercise 4.3), usually called the *Banach–Mazur compactum* (or *Minkowski compactum*). As a consequence of the compactness, whenever a (reasonable) functional $f(K)$ defined on convex bodies in $\mathbb{R}^n$ (or on symmetric convex bodies) has the property that it is affine-invariant, it attains its extreme values on specific equivalence classes of convex bodies. It is sometimes challenging to identify those extremal bodies.

EXERCISE 4.1 (Two hyperplane sections are close). Let $H \subset \mathbb{R}^n$ be a hyperplane, and K, L two symmetric convex bodies such that $K \cap H = L \cap H$. Show that $d_{BM}(K, L) \leqslant C$ for some absolute constant C. Deduce that if H_1, H_2 are two linear hyperplanes, then $d_{BM}(K \cap H_1, K \cap H_2) \leqslant C$. (We tacitly identify H_1 and H_2 with $\mathbb{R}^{n-1}$.)

EXERCISE 4.2 (Boundedness of the space of convex bodies). Let $K \subset \mathbb{R}^n$ be a convex body and let Δ be the simplex of largest volume contained in K. Show that if 0 is the centroid of Δ, then $K \subset -n\Delta \subset n^2\Delta$ and $K \subset (n+1)\Delta$. In particular, if Δ_n is the regular n-dimensional simplex, then $d_{BM}(K, \Delta_n) \leqslant n+1$.

EXERCISE 4.3 (Compactness of the space of convex bodies). Deduce from the previous exercise that the set of convex bodies in $\mathbb{R}^n$ (up to identification via invertible affine transformation), equipped with the distance $\log d_{BM}$, is a compact metric space.

4.1.2. Symmetrization. If $K \subset \mathbb{R}^n$ is a non-symmetric convex body containing 0, there are several symmetric convex bodies that can be associated with K (see Figure 4.1). Such symmetrization operations are useful because symmetric convex bodies are often easier to deal with, whereas the symmetrized set still "remembers" many features of K.

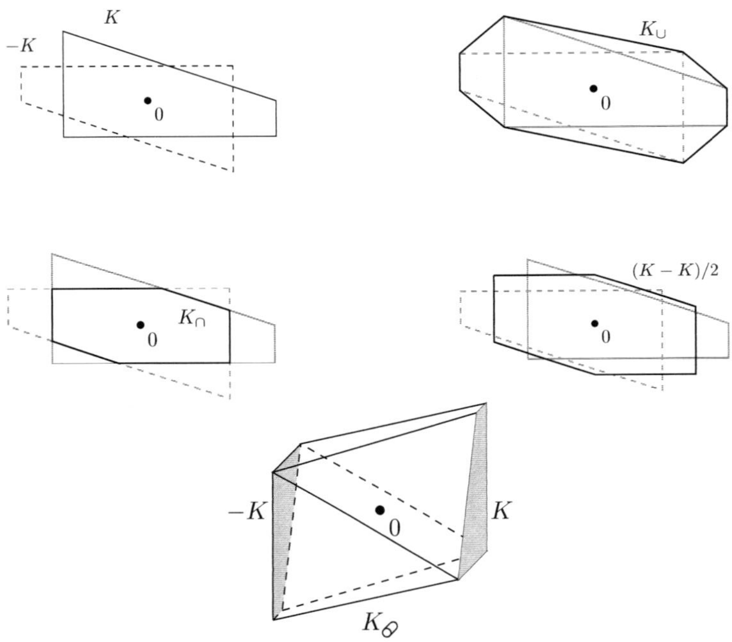

FIGURE 4.1. A convex body $K \subset \mathbb{R}^2$ (top left) and its four kinds of symmetrizations $K_\cup$ (top right), $K_\cap$ (middle left), $(K-K)/2$ (middle right) and $K_\oslash$ (bottom).

We may define the following convex bodies

$$K_\cup = \operatorname{conv}(K \cup (-K)). \tag{4.3}$$

If K also contains 0 in its interior, we may also consider

$$K_\cap = K \cap (-K). \tag{4.4}$$

These operations are dual to each other since we have, by the bipolar theorem (1.11),

$$K \cap (-K) = \operatorname{conv}(K^\circ, -K^\circ)^\circ. \tag{4.5}$$

Still another possible symmetrization is $(K-K)/2 := \{(x-y)/2 : x, y \in K\}$ (cf. the definitions (4.7) below). This choice is appealing since it is invariant under translations of K and makes sense even if $0 \notin K$. However, the description of the polar of $(K-K)/2$ is somewhat awkward. The set $K-K$ is often called in the literature the *difference body*. Obviously if K is already 0-symmetric then $K_\cap = K_\cup = (K-K)/2 = K$.

Several examples of n-dimensional convex bodies naturally lie inside an affine hyperplane in $\mathbb{R}^{n+1}$. This is the case for the regular simplex (the set of classical states) and for the set of quantum states (see Section 0.10). In this situation still another symmetrization is useful. If $H \subset \mathbb{R}^{n+1}$ is an affine hyperplane not containing 0, and K is a convex body in H (so that K is n-dimensional), one may consider

$$K_{\oslash} = \operatorname{conv}(K \cup (-K)). \tag{4.6}$$

The symbol $\oslash$ depicts a cylinder. This is motivated by the observation that when K is a Euclidean disk, the resulting body $K_{\oslash}$ is a cylinder. It coincides with what is commonly called a *generalized cylinder* if K is centrally symmetric.

The set $K_{\oslash}$ is an $(n+1)$-dimensional convex body, so while Equation (4.6) is identical to (4.3), we distinguish the two operations since they will be applied in different contexts (for a description of $(K_{\oslash})^\circ$, see Exercise 4.5). For example, if $K = \Delta_n$ is the regular simplex defined as the convex hull of the canonical basis in $\mathbb{R}^{n+1}$, the convex body obtained after symmetrization is $(\Delta_n)_{\oslash} = B_1^{n+1}$.

All these symmetrizations turn a non-symmetric convex body into a *centrally* symmetric convex body. The word "symmetrization" is also used to describe operations for which the output has some other symmetry properties. One example of such an operation is the Steiner symmetrization as described in Exercise 4.31. One of its important features is that for any convex body there is a sequence of successive Steiner symmetrizations converging to a Euclidean ball, which is very handy for proving geometric inequalities. For other examples of similar nature, see the Notes and Remarks on Section 5.2.

EXERCISE 4.4 (Origin shifting and symmetrization). Show that for any convex body $K \subset \mathbb{R}^n$ and $a, b \in K$,

$$d_{BM}((K-a)_\cup, \quad (K-b)_\cup) \leqslant 4.$$

EXERCISE 4.5 (The polar of cylindrical symmetrization). Let H_e be defined as (1.21), K be a convex body inside H_e, and $K_{\oslash}$ its symmetrization defined as in (4.6). Denote by $\mathcal{C} = \mathbb{R}_+ K$ the cone generated by K, and show that

$$(K_{\oslash})^\circ = \left(\frac{e}{|e|^2} - \mathcal{C}^*\right) \cap \left(-\frac{e}{|e|^2} + \mathcal{C}^*\right).$$

If we write $x \leqslant y$ when $y - x \in \mathcal{C}^*$, this is the "interval" $\{x \in \mathbb{R}^n \ : \ -e/|e|^2 \leqslant x \leqslant e/|e|^2\}$ in the order induced by $\mathcal{C}^*$.

4.1.3. Zonotopes and zonoids. A crucial notion in convex geometry is that of *Minkowski operations* on sets. If $A, B \subset \mathbb{R}^n$ and $t \in \mathbb{R}$, we set

$$A + B := \{x + y \ : \ x \in A,\, y \in B\}, \quad tA := \{tx \ : \ t \in \mathbb{R},\, x \in A\}. \tag{4.7}$$

The definition of the Minkowski sum extends to the case of finitely many convex bodies.

A convex body $K \subset \mathbb{R}^n$ is called a *zonotope* if it is the sum of finitely many segments. For example the cube $[-1,1]^n$ is a zonotope since

$$[-1,1]^n = [-e_1, e_1] + \cdots + [-e_n, e_n],$$

where $[-e_i, e_i]$ denotes the segment joining the ith canonical basis vector and its opposite.

A convex body $K \subset \mathbb{R}^n$ is called a *zonoid* if it can be written as a limit of zonotopes (in the Hausdorff distance). Note that the class of zonotopes (or zonoids) is invariant under affine transformations, so we could alternatively use the Banach–Mazur distance instead of the Hausdorff distance.

Observe that zonotopes and zonoids are automatically centrally symmetric. We will usually assume that the center of symmetry is at the origin. Here is a useful characterization of zonoids as polars of unit balls of subspaces of L_1.

PROPOSITION 4.1 (Not proved here). *Let $K \subset \mathbb{R}^n$ be a symmetric convex body. The following are equivalent:*
(i) K is a zonoid.
(ii) There is a positive Borel measure μ_K on S^{n-1} such that, for any $x \in \mathbb{R}^n$,

$$\|x\|_{K^\circ} = \int_{S^{n-1}} |\langle x, \theta\rangle| \, \mathrm{d}\mu_K(\theta). \tag{4.8}$$

We emphasize that μ_K is not *assumed to be a probability measure.*

It follows in particular that every ellipsoid is a zonoid (use $\mu_K = \sigma$ in (4.8), then affine equivalence). Note also that, for a given zonoid $K \subset \mathbb{R}^n$, the Borel measure μ_K on S^{n-1} satisfying (4.8) is unique if we additionally require it to be *even* (i.e., to verify $\mu_K(-B) = \mu_K(B)$ for every Borel set $B \subset S^{n-1}$).

EXERCISE 4.6 (A formula for μ_K). Let $K = [-u_1, u_1] + \cdots + [-u_p, u_p]$ be a zonotope, where $u_1, \ldots, u_p$ are vectors in $\mathbb{R}^n$. What is the measure μ_K appearing in (4.8)?

EXERCISE 4.7 (Planar zonotopes and zonoids). Show that every centrally symmetric polygon is a zonotope, and that any centrally symmetric convex body $K \subset \mathbb{R}^2$ is a zonoid.

EXERCISE 4.8 (Octahedron is not a zonotope). Show that B_1^3 is not a zonotope.

EXERCISE 4.9. Let K_1, K_2 be convex bodies in $\mathbb{R}^n$ such that $K_1 + K_2 = B_2^n$. Does it follow that K_1, K_2 are Euclidean balls?

4.1.4. Projective tensor product. If K and K' are closed convex sets in $\mathbb{R}^n$ and $\mathbb{R}^{n'}$ respectively, their *projective tensor product* is the closed convex set $K \mathbin{\widehat{\otimes}} K'$ in $\mathbb{R}^n \otimes \mathbb{R}^{n'} \leftrightarrow \mathbb{R}^{nn'}$ defined as follows:

$$K \mathbin{\widehat{\otimes}} K' = \overline{\mathrm{conv}}\{x \otimes x' \ : \ x \in K, x' \in K'\}. \tag{4.9}$$

This terminology is motivated by the fact that when K and K' are unit balls with respect to some norms, the set $K \mathbin{\widehat{\otimes}} K'$ is the unit ball of the corresponding projective tensor product norm on $\mathbb{R}^n \otimes \mathbb{R}^{n'}$. Recall that, given two finite-dimensional normed spaces $(V, \|\cdot\|)$ and $(V', \|\cdot\|)$, their projective tensor product (denoted by $V \mathbin{\widehat{\otimes}} V'$) is the space $V \otimes V'$ equipped with the norm

$$\|z\|_\wedge = \inf\left\{\sum \|x_i\| \, \|y_i\| \ : \ z = \sum x_i \otimes y_i\right\}.$$

It is easily checked that $B_1^m \hat{\otimes} B_1^n$ identifies with B_1^{mn} when the space $\mathbb{R}^m \otimes \mathbb{R}^n$ is identified with $\mathbb{R}^{mn}$ (see also Exercise 4.16), and that $B_2^m \hat{\otimes} B_2^n$ identifies with $S_1^{m,n}$ when $\mathbb{R}^m \otimes \mathbb{R}^n$ is identified with $\mathsf{M}_{m,n}$.

There is a dual notion to the projective tensor product, which is called the *injective tensor product.* It can be defined via polarity: if $K \subset \mathbb{R}^n$ and $K' \subset \mathbb{R}^{n'}$ are convex bodies containing 0 in the interior, their injective tensor product is the convex body $K \check{\otimes} K'$ in $\mathbb{R}^n \otimes \mathbb{R}^{n'} \leftrightarrow \mathbb{R}^{nn'}$ defined as follows:

$$K \check{\otimes} K' = \big(K^\circ \hat{\otimes} (K')^\circ\big)^\circ . \tag{4.10}$$

This definition does not depend on the particular choice of Euclidean structures on $\mathbb{R}^n$ and $\mathbb{R}^{n'}$, provided one considers the Euclidean structure on $\mathbb{R}^n \otimes \mathbb{R}^{n'}$ obtained as their Hilbertian tensor product.

The relevance of the above notions to information theoretical context—quantum or classical—is evident. The set of separable states is the projective tensor product of the sets of states on factor spaces. More precisely, if $\mathcal{H} = \mathcal{H}_1 \otimes \mathcal{H}_2$, then

$$\mathrm{Sep}(\mathcal{H}) = \mathrm{D}(\mathcal{H}_1) \hat{\otimes} \mathrm{D}(\mathcal{H}_2). \tag{4.11}$$

(These objects were defined in Section 2.2.) Similarly, for classical states, the projective tensor product $\Delta_{m-1} \hat{\otimes} \Delta_{n-1}$ identifies with Δ_{mn-1}.

The definition of $K \hat{\otimes} K'$ (similarly to other definitions and comments of this section) immediately generalizes to tensor products of any finite number of factors. However, for the sake of transparency we shall concentrate in this section on the case of two convex bodies. We also point out that the definition (4.9) makes sense when K, K' are subsets of complex spaces.

It is easy to see that the operation $\hat{\otimes}$ commutes with some of the symmetrizations we introduced earlier, e.g.,

$$K_\cup \hat{\otimes} K'_\cup = (K \hat{\otimes} K')_\cup \tag{4.12}$$

and

$$K_\wp \hat{\otimes} K'_\wp = (K \hat{\otimes} K')_\wp. \tag{4.13}$$

To check that (4.13) makes sense, we note that if K (resp., K') is a convex body in the affine hyperplane $H_e \subset \mathbb{R}^n$ (resp., $H_{e'} \subset \mathbb{R}^{n'}$) defined as in (1.21), then $K \hat{\otimes} K'$ is a convex body in the affine hyperplane $H_{e\otimes e'} \subset \mathbb{R}^n \otimes \mathbb{R}^{n'}$ (cf. Exercises 4.13 and 4.15).

A specific situation where (4.13) holds, which will be fundamental in Chapter 9, is when K is the set of quantum states on a Hilbert space. Since $\mathrm{D}(\mathbb{C}^d)_\wp = S_1^{d,\mathrm{sa}}$, it follows that

$$\mathrm{Sep}(\mathbb{C}^d \otimes \mathbb{C}^{d'})_\wp = S_1^{d,\mathrm{sa}} \hat{\otimes} S_1^{d',\mathrm{sa}}. \tag{4.14}$$

To put it in words, the symmetrization of the set of separable states is canonically identified with the projective tensor product of two copies of the self-adjoint part of the unit ball for the trace norm and, consequently, is the unit ball in the projective tensor product norm of (the self-adjoint parts of) two 1-Schatten spaces.

EXERCISE 4.10 (Projective tensor product and compactness). Show that if K, K' are compact convex sets, then $\mathrm{conv}\{x \otimes x' \ : \ x \in K, x' \in K'\}$ is compact and hence equal to $K \hat{\otimes} K'$. Give an example of closed convex sets K, K' such that the set $\mathrm{conv}\{x \otimes x' \ : \ x \in K, x' \in K'\}$ is not closed.

EXERCISE 4.11 (Linear invariance of the projective tensor product). Let $K_i \subset \mathbb{R}^{n_i}$ and let $T_i : \mathbb{R}^{n_i} \to \mathbb{R}^{m_i}$ be linear maps, $i = 1, 2$. Show that $(T_1 \otimes T_2)(K_1 \hat{\otimes} K_2) = (T_1 K_1) \hat{\otimes} (T_2 K_2)$.

EXERCISE 4.12 (Projective tensor product with a linear subspace). Let $K \subset \mathbb{R}^n$ be a closed convex set, let $V = \operatorname{span} K$, and let $V' \subset \mathbb{R}^{n'}$ be a vector subspace. Show that $K \hat{\otimes} V' = V \hat{\otimes} V' = V \otimes V'$.

EXERCISE 4.13 (Projective tensor product of affine subspaces). Let $V_i \subset \mathbb{R}^{n_i}$ be affine subspaces for $i = 1, 2$. Show that $V_1 \hat{\otimes} V_2$ is an affine subspace of $\mathbb{R}^{n_1} \otimes \mathbb{R}^{n_2}$ and find its dimension.

EXERCISE 4.14 (Projective tensor product of cones). Show that if $\mathcal{C}$ and $\mathcal{C}'$ are closed convex cones, then the set $\operatorname{conv}\{x \otimes x' \ : \ x \in \mathcal{C}, x' \in \mathcal{C}'\}$ is a closed convex cone and in particular equals $\mathcal{C} \hat{\otimes} \mathcal{C}'$.

EXERCISE 4.15 (Projective tensor product of bodies are bodies). Show that if $K_i \subset \mathbb{R}^{n_i}$ are convex bodies, then $K_1 \hat{\otimes} K_2$ is a convex body in $\mathbb{R}^{n_1} \otimes \mathbb{R}^{n_2}$. Similarly, if each K_i is a convex body in an affine subspace $V_i \subset \mathbb{R}^{n_i}$, then $K_1 \hat{\otimes} K_2$ is a convex body in $V_1 \hat{\otimes} V_2$.

EXERCISE 4.16 (Projective tensor product with B_1^n). Let K be a symmetric convex body in $\mathbb{R}^m$. (i) What is then $B_1^k \hat{\otimes} K$? (ii) Show that

$$\operatorname{vol}(B_1^k \hat{\otimes} K) = \frac{(m!)^k}{(km)!} \operatorname{vol}(K)^k.$$

EXERCISE 4.17 (Extreme points of projective tensor products). If K and K' are symmetric convex bodies, show that the set of extreme points of $K \hat{\otimes} K'$ is exactly the set of elements $x \otimes x'$, where x is an extreme point of K and x' is an extreme point of K'. Show that this may be false if either K or K' is not symmetric.

EXERCISE 4.18 (Injective tensor products and bilinear forms). If $K = B_X$ and $K' = B_{X'}$, show that $K \check{\otimes} K'$ identifies with the set of bilinear maps $F : X \times X' \to \mathbb{R}$ such that $|F(x, x')| \leqslant \|x\| \cdot \|x'\|$ for all x, x' (i.e., with the unit ball in the space of bilinear maps).

4.2. John and Löwner ellipsoids

4.2.1. Definition and characterization. We start with the following proposition.

PROPOSITION 4.2. *For every convex body $K \subset \mathbb{R}^n$*
(i) there is a unique ellipsoid $\mathscr{E} \subset \mathbb{R}^n$ with maximal volume under the constraint $\mathscr{E} \subset K$ and
(ii) there is a unique ellipsoid $\mathscr{F} \subset \mathbb{R}^n$ with minimal volume under the constraint $\mathscr{F} \supset K$.

The ellipsoid $\mathscr{E}$ appearing in (i) is called the *John ellipsoid* of K and denoted by $\operatorname{John}(K)$. The ellipsoid $\mathscr{F}$ appearing in (ii) is called the *Löwner ellipsoid* of K and denoted by $\operatorname{Löw}(K)$. By a compactness argument, the existence of an ellipsoid of maximal/minimal volume is clear in (i) and (ii). Note also that these ellipsoids are affine invariants: for any affine map T, we have $\operatorname{John}(TK) = T \operatorname{John}(K)$ and $\operatorname{Löw}(TK) = T \operatorname{Löw}(K)$. We say that K is in *John position* if $\operatorname{John}(K) = B_2^n$, and that K is in *Löwner position* if $\operatorname{Löw}(K) = B_2^n$ (see Figure 4.2).

Uniqueness deserves an argument (the proof will be elementary, but to show part (ii) in full generality we will need a trick implicit in Proposition 4.4). For (i) this is fairly straightforward: assume that $\mathscr{E} \neq \mathscr{E}'$ are two distinct ellipsoids of maximal volume contained in K, then write $\mathscr{E} = S(B_2^n) + x$ and $\mathscr{E}' = S'(B_2^n) + x'$ for $S, S' \in \mathcal{PSD}$ and $x, x' \in \mathbb{R}^n$. Since $\mathscr{E} \neq \mathscr{E}'$, we necessarily have $(S, x) \neq (S', x')$. By linear invariance, we may assume that $S = \mathrm{I}$, which implies that $\det(S') = 1$. If $S' = \mathrm{I}$, then $\mathscr{E}$ and $\mathscr{E}'$ are two distinct balls of radius 1, and it is easy to see that $\operatorname{conv}(\mathscr{E}, \mathscr{E}')$ (and hence K) contains an ellipsoid centered at $\frac{x+x'}{2}$ of volume larger than $\operatorname{vol}(\mathscr{E})$, a contradiction. If $S' \neq \mathrm{I}$, then K contains the ellipsoid $T(B_2^n) + y$ with $T = \frac{\mathrm{I}+S}{2}$ and $y = \frac{x+x'}{2}$. Since $\det T > 1$ (see Exercise 1.42), this ellipsoid is of a volume greater than $\operatorname{vol}(\mathscr{E})$, also a contradiction.

The uniqueness in (ii) follows by duality when K is centrally symmetric. Indeed, the minimization problem in (ii) can be restricted in that case to 0-symmetric ellipsoids (by essentially the same argument as in the case of $S' = \mathrm{I}$ above). Since for a 0-symmetric ellipsoid $\mathscr{F}$ we have $\operatorname{vol}(\mathscr{F}) \operatorname{vol}(\mathscr{F}^\circ) = \operatorname{vol}(B_2^n)^2$ by (1.10), and since $K \subset \mathscr{F} \iff \mathscr{F}^\circ \subset K^\circ$, the uniqueness follows, together with the relation $\text{Löw}(K) = \text{John}(K^\circ)^\circ$.

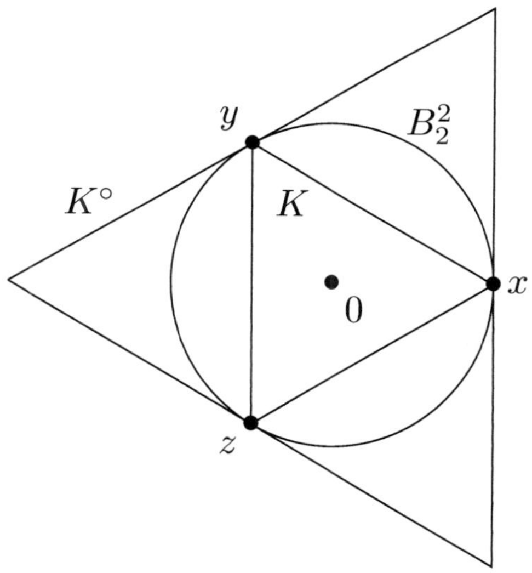

FIGURE 4.2. An equilateral triangle K in Löwner position. The polar body K° is in John position. The contact points x, y, z satisfy the relations $x + y + z = 0$ and $\frac{2}{3}|x\rangle\langle x| + \frac{2}{3}|y\rangle\langle y| + \frac{2}{3}|z\rangle\langle z| = \mathrm{I}$ as in Definition 4.5.

The uniqueness in (ii) in the general case is not obvious at this point; we postpone its justification until after Proposition 4.4.

We will now present a general trick that makes it possible to reduce the search for the Löwner ellipsoid of the not-necessarily-symmetric bodies to the symmetric case. To that end, fix $h > 0$ and consider the affine hyperplane

$$H := \{(h, x) \; : \; x \in \mathbb{R}^n\} \subset \mathbb{R}^{n+1}.$$

To each ellipsoid $\mathscr{E} \subset H$ we associate the symmetrization $\mathscr{E}_{\oslash} = \operatorname{conv}(-\mathscr{E} \cup \mathscr{E})$, which is an ellipsoidal cylinder in $\mathbb{R}^{n+1}$. The following lemma describes the Löwner ellipsoid of $\mathscr{E}_{\oslash}$.

LEMMA 4.3. *Let $S \in \mathsf{GL}(n, \mathbb{R})$ and $a \in \mathbb{R}^n$, and consider the ellipsoid*

$$\mathscr{E} = \{(h, Sx + a) \ : \ x \in B_2^n\} \subset H.$$

Then $\text{Löw}(\mathscr{E}_{\oslash}) = T(B_2^{n+1})$, *where*

$$T = \begin{bmatrix} \sqrt{n+1}\, h & 0 \\ \sqrt{n+1}\, ha & \sqrt{1 + \frac{1}{n}}\, S \end{bmatrix}.$$

In particular,

$$\operatorname{vol}(\text{Löw}(\mathscr{E}_{\oslash})) = c_n h \operatorname{vol}(\mathscr{E}) \tag{4.15}$$

for some constant c_n depending only on n.

PROOF. Consider first the special case (denoted by $\mathscr{E}_0$) where $S = \mathrm{I}$, $h = 1$, and $a = 0$. It follows from the uniqueness—which has already been fully proved in the symmetric case—that $\text{Löw}((\mathscr{E}_0)_{\oslash})$ inherits all the symmetries of $(\mathscr{E}_0)_{\oslash}$ and therefore has the form $T_0(B_2^{n+1})$, where T_0 is a diagonal matrix with coefficients $(\alpha, \beta, \ldots, \beta)$, with $\alpha, \beta > 0$ to be determined. Since $(\mathscr{E}_0)_{\oslash} \subset T_0(B_2^{n+1})$ if and only if $\frac{1}{\alpha^2} + \frac{1}{\beta^2} \leqslant 1$ and $\operatorname{vol}(T_0(B_2^{n+1})) = \alpha\beta^n \operatorname{vol}(B_2^{n+1})$, the minimization problem yields the values $\alpha = \sqrt{n+1}$, $\beta = \sqrt{1 + 1/n}$, as needed.

For the general case, note that $\mathscr{E} = A(\mathscr{E}_0)$, where

$$A = \begin{bmatrix} h & 0 \\ a & S \end{bmatrix} \in \mathsf{M}_{n+1}.$$

Since $\text{Löw}(\mathscr{E}_{\oslash}) = \text{Löw}(A(\mathscr{E}_0)_{\oslash}) = A\,\text{Löw}((\mathscr{E}_0)_{\oslash})$ by invariance, it follows that $T = AT_0$ as claimed. The relation (4.15) follows by expressing $\det T$ in terms of $\det S$. □

PROPOSITION 4.4. *Let $K \subset H$ be a convex body and $\mathscr{E} \subset H$ an ellipsoid. The following are equivalent:*
(i) $\mathscr{E}$ is a minimal volume ellipsoid containing K.
(ii) $\text{Löw}(\mathscr{E}_{\oslash}) = \text{Löw}(K_{\oslash})$.

Since $\mathscr{E} = \text{Löw}(\mathscr{E}_{\oslash}) \cap H$, Proposition 4.4 implies in particular uniqueness of the Löwner ellipsoid for not-necessarily-symmetric convex bodies, completing the proof of Proposition 4.2.

PROOF OF PROPOSITION 4.4. Assuming (i), let $\mathscr{F} = \text{Löw}(K_{\oslash}) \cap H$. Since $\mathscr{F}$ is an ellipsoid containing K, we have $\operatorname{vol}(\mathscr{F}) \geqslant \operatorname{vol}(\mathscr{E})$, which by (4.15) implies $\operatorname{vol}(\text{Löw}(\mathscr{F}_{\oslash})) \geqslant \operatorname{vol}(\text{Löw}(\mathscr{E}_{\oslash}))$. Next, since $K_{\oslash} \subset \mathscr{F}_{\oslash} \subset \text{Löw}(K_{\oslash})$, it follows that $\text{Löw}(K_{\oslash}) = \text{Löw}(\mathscr{F}_{\oslash})$. Given that $\text{Löw}(\mathscr{E}_{\oslash})$ is an ellipsoid containing $K_{\oslash}$ with volume not exceeding the minimum possible, it must coincide with $\text{Löw}(K_{\oslash})$.

Assume now (ii), and let $\mathscr{F}$ be an ellipsoid containing K. Since $\text{Löw}(\mathscr{F}_{\oslash})$ contains $K_{\oslash}$, it follows that $\operatorname{vol}(\text{Löw}(\mathscr{F}_{\oslash})) \geqslant \operatorname{vol}(\text{Löw}(K_{\oslash})) = \operatorname{vol}(\text{Löw}(\mathscr{E}_{\oslash}))$. By (4.15), this means that $\operatorname{vol}(\mathscr{F}) \geqslant \operatorname{vol}(\mathscr{E})$, as needed. □

The following concept will be useful for our purposes.

DEFINITION 4.5. A *resolution of identity* in $\mathbb{R}^n$ is a finite family $(x_i, c_i)_{i\in I}$, where $(x_i)_{i\in I}$ belong to S^{n-1} and $(c_i)_{i\in I}$ are positive numbers, such that

$$\sum_i c_i \, |x_i\rangle\langle x_i| = \mathrm{I}_n \,. \tag{4.16}$$

A resolution is called *unbiased* if, additionally,

$$\sum_i c_i x_i = 0. \tag{4.17}$$

If K is a convex body in $\mathbb{R}^n$ and all points x_i belong to ∂K, we will say that $(x_i, c_i)_{i\in I}$ is *associated to* K. Note that if, additionally, $K \subset B_2^n$ or $B_2^n \subset K$ (which will be usually the case), then all points x_i are contact points of K and the unit sphere, i.e., such that $\|x_i\|_K = \|x_i\|_{K^\circ} = |x_i|$.

Taking the trace of both sides in condition (4.16), we see that necessarily $\sum c_i = n$. More generally, if $T \in B(\mathbb{R}^n)$, then

$$\operatorname{Tr} T = \sum_i c_i \, \langle T x_i, x_i \rangle \tag{4.18}$$

(see Exercise 4.19). Note also that condition (4.17) is redundant for symmetric convex bodies, since one can always enforce it by replacing every couple (c_i, x_i) in the decomposition by two couples $(\frac{1}{2}c_i, x_i)$ and $(\frac{1}{2}c_i, -x_i)$.

The following pair of propositions characterizes John and Löwner positions via resolutions of identity. The presentations of these results that are easily available in the literature focus on the class of symmetric bodies, and we will assume henceforth that they are both known to be true in that setting (for a reference, see Theorem 2.1.15 in [**AAGM15**] or Theorem 3.1 in [**Bal97**]). It is also easy to see that in the symmetric case the two statements are formally equivalent by duality (i.e., by passing to polars).

PROPOSITION 4.6. *Let K be a convex body in $\mathbb{R}^n$. The following are equivalent.*
(i) K is in Löwner position.
(ii) $K \subset B_2^n$ and there exists an unbiased resolution of identity associated to K.

PROPOSITION 4.7. *Let K be a convex body in $\mathbb{R}^n$. The following are equivalent.*
(i) K is in John position.
(ii) $K \supset B_2^n$ and there exists an unbiased resolution of identity associated to K.

PROOF OF PROPOSITION 4.6 (ASSUMING THE SYMMETRIC CASE). To a convex body K we associate

$$\widetilde{K} = \left\{ \left(\frac{1}{\sqrt{n+1}}, \sqrt{\frac{n}{n+1}}\, x \right) \; : \; x \in K \right\} \subset \mathbb{R}^{n+1}.$$

It follows from Lemma 4.3 that $B_2^{n+1} = \text{Löw}\left((\widetilde{B_2^n})_{\diamond}\right)$. In view of Proposition 4.4, we have the equivalence

$$K \text{ is in Löwner position} \iff \widetilde{K}_{\diamond} \text{ is in Löwner position.}$$

Consequently, our task is reduced to showing that K has an unbiased resolution of identity (in $\mathbb{R}^n$) if and only if $\widetilde{K}_{\diamond}$ has a resolution of identity (in $\mathbb{R}^{n+1}$). To that end, let $e_0 = (1, 0, \ldots, 0) \in \mathbb{R}^{n+1}$ and let (x_i, c_i) be a resolution of identity for $\widetilde{K}_{\diamond}$. The points x_i are extreme points of $\widetilde{K}_{\diamond}$, and since we have freedom to replace x_i

by $-x_i$, we may assume that each x_i has the form $x_i = \left(\frac{1}{\sqrt{n+1}}, \sqrt{\frac{n}{n+1}}\, y_i\right)$ with $y_i \in K \cap S^{n-1}$. Setting $z = \sum c_i y_i$, we have

$$\begin{aligned}
\mathrm{I}_{n+1} &= \sum_i c_i\, |x_i\rangle\langle x_i| \\
&= \sum_i c_i \Big| \frac{1}{\sqrt{n+1}}\, e_0 + \sqrt{\frac{n}{n+1}}\,(0, y_i) \Big\rangle\Big\langle \frac{1}{\sqrt{n+1}}\, e_0 + \sqrt{\frac{n}{n+1}}\,(0, y_i) \Big| \\
&= |e_0\rangle\langle e_0| + \frac{\sqrt{n}}{n+1}\big(|e_0\rangle\langle (0,z)| + |(0,z)\rangle\langle e_0|\big) + \frac{n}{n+1}\sum_i c_i |(0,y_i)\rangle\langle (0,y_i)|,
\end{aligned}$$

where in the last equality we used the fact that $\sum_i c_i = n+1$. By applying this operator equality to the vector e_0, we obtain $z = 0$. Thus the middle term in the last line above vanishes, which easily implies that $\left(y_i, \frac{n}{n+1} c_i\right)$ is an unbiased resolution of identity for K. The reverse argument simply retraces the above calculation backwards; the reader is encouraged to verify the details. (Note that $z = 0$ then follows from the hypothesis.) □

Proof of Proposition 4.7. Assume that K is in John position. We claim that K° is in Löwner position. To check this, let $\mathscr{E}$ be an ellipsoid containing K°. We then have $\mathscr{E}^\circ \subset K$. We know from Exercise 1.26 (or from Exercise D.3, which outlines a simpler but less elementary proof) that $\mathscr{E}^\circ$ is an ellipsoid and that $\mathrm{vol}(\mathscr{E})\,\mathrm{vol}(\mathscr{E}^\circ) \geqslant \mathrm{vol}(B_2^n)^2$, with equality iff $\mathscr{E}$ is 0-symmetric. Since $\mathrm{vol}(\mathscr{E}^\circ) \leqslant \mathrm{vol}(B_2^n)$ by definition of the John ellipsoid, it follows that $\mathrm{vol}(\mathscr{E}) \geqslant \mathrm{vol}(B_2^n)$, showing that K° is in Löwner position. By Proposition 4.6, K° admits an unbiased resolution of identity, and so does K.

Conversely, suppose that $B_2^n \subset K$ and that (x_i, c_i) is an unbiased resolution of identity for K. We note that K is contained in $\bigcap_i \{\langle \cdot\,, x_i\rangle \leqslant 1\}$ (indeed, since $x_i \in \partial K \cap S^{n-1}$, the support hyperplane for K at x_i is necessarily orthogonal to x_i). Let $\mathscr{E} \subset K$ be an ellipsoid. Write $\mathscr{E} = S(B_2^n) + a$ for $S \in \mathcal{PSD}$ and $a \in \mathbb{R}^n$. Since $Sx_i + a \in \mathscr{E} \subset K$, we have $\langle Sx_i + a, x_i\rangle \leqslant 1$ for all i. Since $\sum c_i x_i = 0$, this shows that

$$n = \sum_i c_i \geqslant \sum_i c_i \langle Sx_i + a, x_i\rangle = \sum_i c_i \langle Sx_i, x_i\rangle = \operatorname{Tr} S,$$

the last equality following from (4.18). The arithmetic-geometric mean inequality now implies that $\det S \leqslant 1$, and hence that $\mathrm{vol}(\mathscr{E}) \leqslant \mathrm{vol}(B_2^n)$. Since $\mathscr{E} \subset K$ was arbitrary, this shows that K is in John position. □

John's theorem implies estimates on the diameter of the Banach–Mazur compactum which are essentially sharp in the symmetric case only (see Exercises 4.20–4.21, and Notes and Remarks for further comments).

Exercise 4.19. Prove identity (4.18).

Exercise 4.20 (The diameter of the Banach–Mazur compactum). Let $K \subset B_2^n$ (resp., $K \supset B_2^n$) be a *symmetric* convex body and assume that there exists a resolution of identity associated to K. Show that $K \supset \frac{1}{\sqrt{n}} B_2^n$ (resp., $K \subset \sqrt{n} B_2^n$) and so, in particular, $d_g(B_2^n, K) \leqslant \sqrt{n}$. Conclude that any pair K, L of symmetric convex bodies in $\mathbb{R}^n$ satisfies $d_{BM}(K, L) \leqslant n$.

EXERCISE 4.21 (Bounds on the diameter of Banach–Mazur compactum, the non-symmetric case). Let $K \subset B_2^n$ (resp., $K \supset B_2^n$) be a convex body and assume that there exists an unbiased resolution of identity associated to K. Show that $K \supset \frac{1}{n} B_2^n$ (resp., $K \subset n B_2^n$). Conclude that any pair K, L of convex bodies in $\mathbb{R}^n$ satisfies $d_{BM}(K, L) \leqslant n^2$.

EXERCISE 4.22 (The length of resolutions of identity). Show that in Propositions 4.6 and 4.7, the length of the resolution of identity associated to K can be assumed to be at most $\frac{n(n+3)}{2}$ in the general case, and at most $\frac{n(n+1)}{2}$ if K is symmetric.

EXERCISE 4.23 (The radius of the Banach–Mazur compactum). Show that the first estimate from Exercise 4.20 is optimal by verifying that $d_{BM}(B_2^n, B_\infty^n) = \sqrt{n}$.

4.2.2. Convex bodies with enough symmetries. In this section we describe a class of convex bodies "with enough symmetries", which in particular admit a unique Euclidean structure compatible with those symmetries. These properties force the John and Löwner ellipsoids (or any other ellipsoids "functorially associated" with such bodies) to be balls with respect to that Euclidean structure.

Let $K \subset \mathbb{R}^n$ be a convex body. We consider *symmetries* of K, i.e., invertible affine maps $T : \mathbb{R}^n \to \mathbb{R}^n$ such that $T(K) = K$. We start by making two observations. First, such maps necessarily fix the centroid of K. If the centroid is at the origin (which may be assumed by translating K), the set of symmetries becomes a subgroup of $\mathsf{GL}(n, \mathbb{R})$. Second, since this subgroup is compact, it must preserve a scalar product (consider any scalar product and average it with respect to the Haar measure on the group of symmetries). Equivalently, by replacing K with a linear image we may ensure that all symmetries of K are (Euclidean) isometries; in virtually all applications this property will be automatically satisfied. This is tacitly assumed in what follows, although the definitions and the proposition can be easily rephrased to make sense and/or hold without that assumption.

We therefore consider $K \subset \mathbb{R}^n$ a convex body with the centroid at the origin. An *isometry* of K is an orthogonal transformation $O \in \mathsf{O}(n)$ such that $O(K) = K$. The isometries of K form a subgroup of $\mathsf{O}(n)$, which will be called the *isometry group* of K and denoted by $\mathrm{Iso}(K)$. This definition extends *mutatis mutandis* to convex bodies $K \subset \mathbb{C}^n$; in that case $\mathrm{Iso}(K)$ is a subgroup of $\mathsf{U}(n)$.

We say that K has *enough symmetries* if $\mathrm{Iso}(K)' = \mathbb{R}\,\mathrm{I}$ (or $\mathbb{C}\,\mathrm{I}$ in the complex case). Here G' denotes the commutant of G, i.e., the set of linear maps S such that $SO = OS$ for every $O \in \mathsf{G}$.

There is a closely related notion (and possibly a source of confusion): one says that $\mathrm{Iso}(K)$ acts *irreducibly* if any $\mathrm{Iso}(K)$-invariant subspace is either $\{0\}$ or $\mathbb{R}^n$ (or $\mathbb{C}^n$ in the complex case; a subspace E is G-invariant if $O(E) = E$ for any $O \in \mathsf{G}$). One checks that $\mathrm{Iso}(K)$ acts irreducibly if and only if $\mathrm{Iso}(K)'$ contains no nontrivial orthogonal projection, and also if and only if $\mathrm{Iso}(K)' \cap \mathsf{M}_n^{\mathrm{sa}} = \mathbb{R}\,\mathrm{I}$; this idea is also used in Proposition 4.8.

It is immediate that when K has enough symmetries, $\mathrm{Iso}(K)$ acts irreducibly. In the complex case, the reverse implication also holds (this is the content of Schur's lemma) and both notions are equivalent. In the real case, the notions are different (see Exercise 4.26).

The following proposition shows that ellipsoids associated to a convex body in a "functorial" way (such as the John and Löwner ellipsoids, or the ℓ-ellipsoid introduced in Section 7.1) inherit its symmetries.

PROPOSITION 4.8. *Let $K \subset \mathbb{R}^n$ be a convex body and let $\mathscr{E}$ be an ellipsoid such that $O(\mathscr{E}) = \mathscr{E}$ for any $O \in \mathrm{Iso}(K)$. Then there exist pairwise orthogonal subspaces $E_1, \ldots, E_k$, which are invariant under $\mathrm{Iso}(K)$, and positive numbers $\lambda_1, \ldots, \lambda_k$ such that*

$$\mathscr{E} = TB_2^n, \quad \text{where } T = \lambda_1 P_{E_1} + \cdots + \lambda_k P_{E_k}.$$

In particular, when $\mathrm{Iso}(K)$ acts irreducibly, $\mathscr{E}$ is a Euclidean ball.

PROOF. Let T be the unique positive matrix such that $\mathscr{E} = T(B_2^n)$. For every $O \in \mathrm{Iso}(K)$, we have $\mathscr{E} = O(\mathscr{E}) = OTO^\dagger(B_2^n)$, thus $OTO^\dagger = T$. Write $T = \sum_i \lambda_i P_i$, where $\lambda_i > 0$ are distinct positive numbers and P_i pairwise orthogonal projectors. From the relation $OTO^\dagger = T$ we deduce that, for every i, we have $OP_iO^\dagger = P_i$ for all $O \in \mathrm{Iso}(K)$, and therefore that the range of P_i is invariant under $\mathrm{Iso}(K)$. □

We conclude this section with two examples of groups of symmetries of $\mathbb{R}^n$ (or $\mathbb{C}^n$) which play an important role in geometric functional analysis

$$\mathsf{G}_{\mathrm{unc}} := \{(x_1, \ldots, x_n) \mapsto (\varepsilon_1 x_1, \ldots, \varepsilon_n x_n) : |\varepsilon_j| = 1\} \tag{4.19}$$

$$\mathsf{G}_{\mathrm{sym}} := \{(x_1, \ldots, x_n) \mapsto (\varepsilon_1 x_{\pi(1)}, \ldots, \varepsilon_n x_{\pi(n)}) : |\varepsilon_j| = 1\}, \tag{4.20}$$

where $\varepsilon_1, \ldots, \varepsilon_n$ are scalars and $\pi \in \mathfrak{S}_n$ is the group of permutations. A convex body K (resp., the norm or the space, for which K is the unit ball) is called *unconditional* (with respect to the standard basis) if $\mathrm{Iso}(K) \supset \mathsf{G}_{\mathrm{unc}}$ and, similarly, *permutationally symmetric* if $\mathrm{Iso}(K) \supset \mathsf{G}_{\mathrm{sym}}$. Bodies of the second kind have enough symmetries, but bodies of the first kind not necessarily; see Exercise 4.24. (In functional analysis, the standard terminology for the latter is "symmetric", but we prefer to avoid the confusion with the notion of being *centrally* symmetric.) More generally, one may consider bodies (or norms) that are unconditional (resp., permutationally symmetric) with respect to some other basis (u_j), i.e., invariant under maps of the form $u_j \mapsto \varepsilon_j u_j$, $j = 1, \ldots, n$ (resp., $u_j \mapsto \varepsilon_j u_{\pi(j)}$, $j = 1, \ldots, n$). The basis (u_j) is then called unconditional (resp., permutationally symmetric), and the property of having a basis of either kind is a linear invariant.

EXERCISE 4.24 (Permutationally symmetric or unconditional vs. enough symmetries). Show that every permutationally symmetric convex body has enough symmetries. Give an example of an unconditional body which does not have enough symmetries.

EXERCISE 4.25 (Examples of bodies with enough symmetries). Let $1 \leqslant p \leqslant \infty$, $p \neq 2$. For each convex body in the following list, determine if it has enough symmetries.
(i) The ℓ_p ball B_p^n,
(ii) its non-commutative analogues $S_p^{n,m}$,
(iii) the self-adjoint version $S_p^{n,\mathrm{sa}}$, and its intersection with the hyperplane of trace 0 matrices,
(iv) the regular simplex,
(v) the set $\mathrm{D}(\mathbb{C}^n)$ of quantum states.

EXERCISE 4.26 (Enough symmetries vs. irreducible action).
(i) Let $R \in \mathsf{SO}(2)$ be the rotation of angle $2\pi/p$ for an integer $p \geqslant 3$. Construct a convex body $K \subset \mathbb{R}^2$ whose isometry group is exactly $\{R^k : 0 \leqslant k \leqslant p-1\}$. Show that K does not have enough symmetries although $\mathrm{Iso}(K)$ acts irreducibly.
(ii) For any n, give an example of a convex body $L \subset \mathbb{R}^{2n}$ without enough symmetries although $\mathrm{Iso}(L)$ acts irreducibly.

EXERCISE 4.27 (Projective tensor product and enough symmetries). Let $K \subset \mathbb{R}^m$ and $L \subset \mathbb{R}^n$ be convex bodies with enough symmetries. Show that $K \mathbin{\widehat{\otimes}} L$ has enough symmetries.

4.2.3. Ellipsoids and tensor products. It turns out that Löwner ellipsoids behave well with respect to the projective tensor product, as the following lemma shows. Note that the analogous statement *does not* hold for the John ellipsoid (see Exercise 4.28).

LEMMA 4.9. *Let $K \subset \mathbb{R}^n$ and $K' \subset \mathbb{R}^{n'}$ be two convex bodies and assume that the ellipsoids* $\mathrm{Löw}(K)$ *and* $\mathrm{Löw}(K')$ *are 0-symmetric. Then the Löwner ellipsoid of their projective tensor product is the Hilbertian tensor product of the respective Löwner ellipsoids.*

In terms of scalar products, for every x, y in $\mathbb{R}^n$ and x', y' in $\mathbb{R}^{n'}$, we have

$$\langle x \otimes x', y \otimes y' \rangle_{\mathrm{Löw}(K \widehat{\otimes} K')} = \langle x, y \rangle_{\mathrm{Löw}(K)} \langle x', y' \rangle_{\mathrm{Löw}(K')}$$

PROOF. First suppose that $\mathrm{Löw}(K) = B_2^n$ and $\mathrm{Löw}(K') = B_2^{n'}$. By Proposition 4.6, there exist unbiased resolutions of identity for K and K', respectively (x_i, c_i) and (x'_j, c'_j). We easily check that $K \mathbin{\widehat{\otimes}} K' \subset B_2^{nn'} = B_2^n \otimes_2 B_2^{n'}$. We may verify that $(x_i \otimes x'_j, c_i c'_j)$ is an unbiased resolution of identity for $K \mathbin{\widehat{\otimes}} K'$ by writing

$$\sum_i \sum_j c_i c'_j x_i \otimes x'_j = \Big(\sum_i c_i x_i\Big) \otimes \Big(\sum_j c'_j x'_j\Big) = 0,$$

$$\sum_i \sum_j c_i c'_j |x_i \otimes x'_j\rangle\langle x_i \otimes x'_j| = \Big(\sum_i c_i |x_i\rangle\langle x_i|\Big) \otimes \Big(\sum_j c'_j |x'_j\rangle\langle x'_j|\Big) = \mathrm{I}.$$

It follows from Proposition 4.6 that $\mathrm{Löw}(K \mathbin{\widehat{\otimes}} K') = B_2^{nn'}$. For the general case, let T and T' be linear maps such that $T\,\mathrm{Löw}(K) = B_2^n$ and $T'\,\mathrm{Löw}(K') = B_2^{n'}$. Using the elementary identities $\mathrm{Löw}(TK) = T\,\mathrm{Löw}(K)$ and $(T \otimes T')(K \mathbin{\widehat{\otimes}} K') = (TK) \mathbin{\widehat{\otimes}} (T'K')$, the result follows from the previous special case. □

EXERCISE 4.28 (Projective tensor product and the John ellipsoid). Compare $\mathrm{John}(K \widehat{\otimes} L)$ and $\mathrm{John}(K) \widehat{\otimes} \mathrm{John}(L)$ when $K = L = B_2^n$ and when $K = L = \sqrt{n} B_1^n$.

4.3. Classical inequalities for convex bodies

In this section we review classical inequalities involving various geometric invariants of convex bodies, most notably the volume and the mean width. We use the Minkowski operations defined in (4.7).

4.3.1. The Brunn–Minkowski inequality. The Brunn–Minkowski inequality is a fundamental inequality which governs the behavior of the volume of sets under operations related to convexity. It asserts that the volume (the Lebesgue measure on $\mathbb{R}^n$) is log-concave with respect to Minkowski operations, in the following sense.

THEOREM 4.10 (Brunn–Minkowski, not proved here). *Let $K, L \subset \mathbb{R}^n$ be Borel sets and $\lambda \in [0,1]$. Then*

$$\mathrm{vol}(\lambda K + (1-\lambda)L) \geqslant \mathrm{vol}(K)^\lambda \, \mathrm{vol}(L)^{1-\lambda}. \tag{4.21}$$

Another formulation of the Brunn–Minkowski inequality can be given (see Exercise 4.30) as follows: under the same assumptions,

$$\mathrm{vol}(K+L)^{1/n} \geqslant \mathrm{vol}(K)^{1/n} + \mathrm{vol}(L)^{1/n}. \tag{4.22}$$

The Brunn–Minkowski inequality implies the famous isoperimetric inequality in $\mathbb{R}^n$: among sets of given volume, the balls have the smallest surface area. If $K \subset \mathbb{R}^n$ is sufficiently regular, the surface area can be defined as the first-order variation of the volume of the "enlarged" set $K + \varepsilon B_2^n$ when ε goes to 0,

$$\mathrm{area}(K) := \lim_{\varepsilon \to 0} \frac{\mathrm{vol}(K + \varepsilon B_2^n) - \mathrm{vol}(K)}{\varepsilon}. \tag{4.23}$$

Note that for a general subset $K \subset \mathbb{R}^n$, some care is needed in defining area since the limit in (4.23) may not exist or may not coincide with other notions of surface area. However, such problems do not arise for convex sets.

A convenient formulation of the isoperimetric inequality uses the concept of *volume radius.* Given a bounded measurable $K \subset \mathbb{R}^n$, its volume radius $\mathrm{vrad}(K)$ is defined as

$$\mathrm{vrad}(K) := \left(\frac{\mathrm{vol}(K)}{\mathrm{vol}(B_n^2)} \right)^{\frac{1}{n}}. \tag{4.24}$$

In words, the volume radius of K is the radius of the Euclidean ball which has the same volume of K. A standard computation shows that

$$\mathrm{vol}\left(B_2^n\right) = \frac{\pi^{n/2}}{\Gamma\left(\frac{n}{2}+1\right)}. \tag{4.25}$$

Notice that, as a function of n, $\mathrm{vol}(B_2^n)$ decreases super-exponentially fast to 0 as $n \to \infty$. In particular, $\mathrm{vol}(B_2^n)^{1/n}$ is equivalent to $\sqrt{2\pi e/n}$ as n tends to infinity. When $K \subset \mathbb{R}^n$ is a convex body containing 0 in the interior, another useful formula for the volume radius of K (proved via integrating in spherical coordinates) is

$$\mathrm{vrad}(K) = \left(\int_{S^{n-1}} \|\theta\|_K^{-n} \mathrm{d}\sigma(\theta) \right)^{1/n}. \tag{4.26}$$

Here is the statement of the isoperimetric inequality in $\mathbb{R}^n$ employing the notion of volume radius.

PROPOSITION 4.11 (Isoperimetric inequality). *Let $K \subset \mathbb{R}^n$ be bounded and denote $r = \mathrm{vrad}(K)$. Then, for every $\varepsilon > 0$,*

$$\mathrm{vol}(K + \varepsilon B_2^n) \geqslant \mathrm{vol}(rB_2^n + \varepsilon B_2^n) \tag{4.27}$$

or, equivalently, $\mathrm{vrad}(K + \varepsilon B_2^n) \geqslant \mathrm{vrad}(K) + \varepsilon$. *Consequently, whenever the limit in* (4.23) *exists, we have* $\mathrm{area}(K) \geqslant \mathrm{area}(rB_2^n)$.

PROOF. It follows from the Brunn–Minkowski inequality (4.22) that

$$\begin{aligned} \mathrm{vol}(K + \varepsilon B_2^n)^{1/n} &\geqslant \mathrm{vol}(K)^{1/n} + \mathrm{vol}(\varepsilon B_2^n)^{1/n} \\ &= (r+\varepsilon)\, \mathrm{vol}(B_2^n)^{1/n} \\ &= \mathrm{vol}(rB_2^n + \varepsilon B_2^n)^{1/n}. \end{aligned}$$

□

EXERCISE 4.29 (Superadditivity of the volume radius). Show that the Brunn–Minkowski inequality can be restated as $\operatorname{vrad}(K+L) \geqslant \operatorname{vrad}(K) + \operatorname{vrad}(K)$.

EXERCISE 4.30 (Superadditivity and log-concavity). Show that the inequalities (4.21) and (4.22) are formally globally equivalent.

EXERCISE 4.31 (Steiner-like symmetrizations). Show that the following statement is equivalent to the Brunn–Minkowski inequality for convex bodies. Let $K \subset \mathbb{R}^n$ a convex body and $E \subset \mathbb{R}^n$ a k-dimensional subspace with $0 < k < n$. Define a set $L \subset E \times E^\perp$ by the following (where $x \in E, y \in E^\perp$):

$$(x, y) \in L \iff |x| \leqslant \operatorname{vrad}(K \cap (E + y)),$$

where the volume radius is measured in $E + y$. Then L is convex. (When E is a hyperplane, the map $K \mapsto L$ defined above is called *Steiner symmetrization*.)

EXERCISE 4.32. Let $\mathscr{E} \subset \mathbb{R}^m$ and $\mathscr{F} \subset \mathbb{R}^n$ be two 0-symmetric ellipsoids. Show the formula $\operatorname{vrad}(\mathscr{E} \otimes_2 \mathscr{F}) = \operatorname{vrad}(\mathscr{E}) \operatorname{vrad}(\mathscr{F})$.

4.3.2. log-concave measures. Closely related to the Brunn–Minkowski inequality is the concept of a log-concave measure. In our setting, log-concave measures appear as (limits of) marginals of uniform measures on convex sets.

Let μ be a measure on $\mathbb{R}^n$ with density f with respect to the Lebesgue measure. We say that μ is log-concave if $\log f$ is a concave function. Similarly, given $\alpha > 0$, we say that μ is α-concave if the function f^α is concave when restricted to the support of μ. We now state basic facts about log- and α-concave measures and relegate the proofs to exercises.

LEMMA 4.12 (See Exercise 4.34). *Let μ be a finite* log-*concave measure on $\mathbb{R}^n$. Then there is a sequence $(\mu_s)_{s\in\mathbb{N}}$ of measures on $\mathbb{R}^n$ converging weakly to μ such that μ_s is $1/s$-concave.*

LEMMA 4.13 (See Exercise 4.35). *Let μ be a measure on $\mathbb{R}^n$, and $s \in \mathbb{N}$. The following are equivalent:*

(1) The measure μ is $1/s$-concave.
(2) There is a closed convex set $K \subset \mathbb{R}^n \times \mathbb{R}^s$ such that μ is the marginal over $\mathbb{R}^s$ of the Lebesgue measure restricted to K, i.e., such that, for any Borel set $B \subset \mathbb{R}^n$,

$$\mu(B) = \operatorname{vol}_{n+s}\left((B \times \mathbb{R}^s) \cap K\right).$$

As a corollary to Lemmas 4.12 and 4.13, we obtain the following characterization of log-concave measures.

PROPOSITION 4.14 (Characterization of log-concave measures; see Exercise 4.36). *Let μ be a finite and absolutely continuous measure on $\mathbb{R}^n$. The following are equivalent:*

(1) The measure μ is log-*concave.*
(2) The measure μ satisfies the following analogue of (4.21): for any Borel sets $K, L \in \mathbb{R}^n$ and $\lambda \in [0,1]$,

$$\mu(\lambda K + (1-\lambda)L) \geqslant \mu(K)^\lambda \mu(L)^{1-\lambda}. \tag{4.28}$$

To summarize, log-concave measures on $\mathbb{R}^n$ are uniform measures on convex bodies, marginals of uniform measures on convex bodies in $\mathbb{R}^N$ for $N > n$ (see

Exercise 4.33), and their limits. Archetypical examples of log-concave measures include the *standard Gaussian measure* γ_n or *any* Gaussian measure (see Appendix A.2 and Notes and Remarks on Section 4.3).

EXERCISE 4.33 (α-concavity and log-concavity). Check that an α-concave measure is log-concave, and also β-concave for any $\beta \in (0, \alpha]$.

EXERCISE 4.34 (More on α-concavity vs. log-concavity). Prove Lemma 4.12.

EXERCISE 4.35 (α-concavity and marginals). Deduce Lemma 4.13 from the Brunn–Minkowski inequality (4.22) applied in $\mathbb{R}^s$.

EXERCISE 4.36 (Characterization of log-concave measures). Deduce Proposition 4.14 from Lemmas 4.12 and 4.13.

4.3.3. Mean width and the Urysohn inequality. Given a nonempty and bounded set $K \subset \mathbb{R}^n$ and a vector $u \in \mathbb{R}^n$, we define the quantity

$$w(K, u) := \sup_{x \in K} \langle u, x \rangle. \tag{4.29}$$

In the particular case when K is a convex body containing 0 in the interior, we have $w(K, u) = \|u\|_{K^\circ}$ (see (1.8)). If $|u| = 1$, then $w(K, u)$ is called the *support function* of K in direction u. (An alternative notation for the support function, widely used in convex geometry, is $h_K(u)$.) Geometrically, $w(K, u)$ is then the distance from the origin to the hyperplane tangent to K in the direction u (that is, with u being normal to the hyperplane, and outer to K). In particular, $w(K, u) + w(K, -u)$ is the width of the smallest strip in the direction orthogonal to u which contains K (see Figure 4.3).

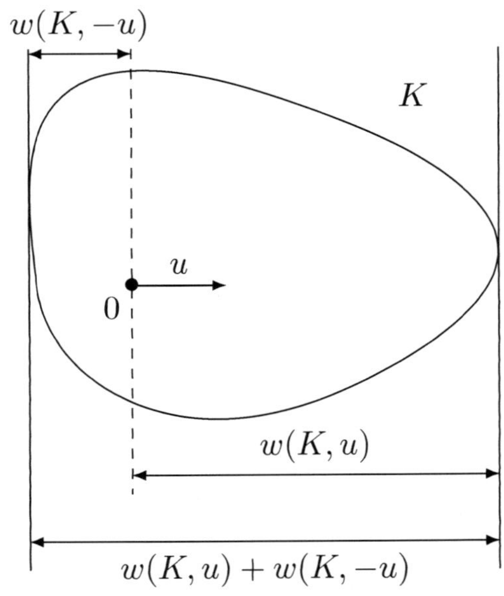

FIGURE 4.3. If $|u| = 1$, then $w(K, u) + w(K, -u)$ is the width of K in the direction of u.

For a nonempty bounded subset $K \subset \mathbb{R}^n$, we may define the *mean width* of K as the average of $w(K, \cdot)$ over the unit sphere

$$w(K) := \int_{S^{n-1}} w(K, u)\, \mathrm{d}\sigma(u), \tag{4.30}$$

where σ is the Lebesgue measure on the sphere, normalized so that $\sigma(S^{n-1}) = 1$. Although the definition makes sense for every bounded set K, we mostly consider the case where K is also closed and convex. This is not really a restriction since $w(K, \cdot) = w(\overline{\operatorname{conv} K}, \cdot)$.

From the geometric point of view, it might have been more accurate to call $w(K)$ the mean *half-width* (or, as some authors do, to include an additional factor 2 in the definition; observe that $w(K)$ is *half* of the average of $w(K, u) + w(K, -u)$). However, we opted for simplicity. Note that, under our convention, one has $w(B_2^n) = 1$, and that if K is a convex body which contains the origin in the interior, then

$$w(K) = \int_{S^{n-1}} \|u\|_{K^\circ}\, \mathrm{d}\sigma(u).$$

It is often convenient to consider the Gaussian variant of the mean width. Let G be a *standard Gaussian vector* in $\mathbb{R}^n$, i.e., a $\mathbb{R}^n$-valued random variable whose coordinates in any orthonormal basis are independent and follow the $N(0, 1)$ distribution (see Appendix A). For any nonempty bounded set $K \subset \mathbb{R}^n$, we define the *Gaussian mean width* of K as

$$w_G(K) := \mathbf{E}\, w(K, G) = \frac{1}{(2\pi)^{n/2}} \int_{\mathbb{R}^n} \sup_{x \in K} \langle u, x\rangle \exp(-|u|^2/2)\, \mathrm{d}u. \tag{4.31}$$

Using (A.7), one checks that

$$w_G(K) = \kappa_n\, w(K), \tag{4.32}$$

where κ_n depends only on n and is of order $\sqrt{n}$ (more precise estimates appear in Proposition A.1). We take the convention that whenever we write $w(K)$ or $w_G(K)$ for a set $K \subset \mathbb{R}^n$, it is tacitly assumed that K is nonempty.

Given bounded subsets K, L in $\mathbb{R}^n$ and a vector u, one checks that $w(K+L, u) = w(K, u) + w(L, u)$. Integration yields

$$w(K + L) = w(K) + w(L), \tag{4.33}$$

and similarly for w_G. In the special case when L is a singleton, this shows that the mean width (Gaussian or not) is translation-invariant.

An advantage of the Gaussian mean width is that it does not depend on the ambient dimension. Indeed, suppose that K is a bounded subset in a subspace $E \subset \mathbb{R}^n$. Then the value of $w_G(K)$ does not depend on whether it is computed in E or in $\mathbb{R}^n$, while the value of $w(K)$ *does*.

The following result, known as the Urysohn inequality, asserts that among sets of given volume, the mean width is minimized for Euclidean balls.

PROPOSITION 4.15 (Urysohn's inequality; see Exercise 4.49). *Let $K \subset \mathbb{R}^n$ be a bounded Borel set. Then*

$$\operatorname{vrad}(K) \leqslant w(K). \tag{4.34}$$

The Urysohn inequality can be seen as a consequence of the Brunn–Minkowski inequality, see Exercise 4.49. Among closed sets, the Urysohn inequality is an equality if and only if K is a Euclidean ball.

Define the *outradius* of a bounded set $K \subset \mathbb{R}^n$ as the smallest radius (denoted $\text{outrad}(K)$) of a Euclidean ball that contains K (such a ball is unique, see Exercise 4.41), and the *inradius* of a convex body $K \subset \mathbb{R}^n$ as the largest radius (denoted $\text{inrad}(K)$) of a Euclidean ball contained in K. (Such a ball is not necessarily unique; however, when K is symmetric, the inradius is witnessed by Euclidean balls centered at the origin.) We have the chain of inequalities

$$\text{inrad}(K) \leqslant \text{vrad}(K) \leqslant w(K) \leqslant \text{outrad}(K). \tag{4.35}$$

For a longer chain of inequalities which also includes dual quantities, see Exercise 4.51. It is instructive to compare in Table 4.1 the values of these quantities for the most standard examples of convex bodies. For a derivation, see Exercises 4.38 and 6.6 (we postpone the nontrivial mean width computations to Chapter 6, where they fit more naturally).

TABLE 4.1. Radii for standard convex bodies in $\mathbb{R}^n$. Quantities in each row are non-decreasing from left to right, see (4.35) and Exercise 4.51. The simplex K is normalized to be a regular simplex inscribed in the Euclidean ball of radius $\sqrt{n}$ centered at the origin. This normalization is appealing since it has the property that $K^\circ = -K$. When compared to the simplex Δ_n as defined in Section 1.1.2, K is congruent to $\sqrt{n+1}\,\Delta_n$.

K	$\text{inrad}(K)$	$w(K^\circ)^{-1}$	$\text{vrad}(K)$	$w(K)$	$\text{outrad}(K)$
B_2^n	1	1	1	1	1
B_1^n	$1/\sqrt{n}$	$\sim \sqrt{\pi/2n}$	$\sim \sqrt{2e/\pi n}$	$\sim \sqrt{2\log n}/\sqrt{n}$	1
B_∞^n	1	$\sim \sqrt{n}/\sqrt{2\log n}$	$\sim \sqrt{2n/\pi e}$	$\sim \sqrt{2n/\pi}$	$\sqrt{n}$
simplex	$1/\sqrt{n}$	$\sim 1/\sqrt{2\log n}$	$\sim \sqrt{e/2\pi}$	$\sim \sqrt{2\log n}$	$\sqrt{n}$

We check in Table 4.1 that for all these basic examples of convex bodies, the volume radius and the mean width are of comparable order of magnitude, at least up to a logarithmic factor. This cannot be true for general convex bodies (see Exercise 4.42), but a convex body such that $\text{vrad}(K)$ is much smaller than $w(K)$ has to be strongly "non-isotropic", cf. Corollary 7.11.

The Urysohn inequality has a "dual" version, which is actually easier to prove since it depends only on the Hölder inequality.

PROPOSITION 4.16. *For every convex body $K \subset \mathbb{R}^n$ containing the origin in its interior, we have*

$$\text{vrad}(K) \geqslant w(K^\circ)^{-1}. \tag{4.36}$$

PROOF. This follows from Hölder's inequality

$$\begin{aligned} 1 &= \int_{S^{n-1}} \|\theta\|_K^{\frac{n}{n+1}} \|\theta\|_K^{-\frac{n}{n+1}} \,\mathrm{d}\sigma(\theta) \\ &\leqslant \left(\int_{S^{n-1}} \|\theta\|_K \,\mathrm{d}\sigma(\theta) \right)^{\frac{n}{n+1}} \cdot \left(\int_{S^{n-1}} \|\theta\|_K^{-n} \,\mathrm{d}\sigma(\theta) \right)^{\frac{1}{n+1}} \\ &= \left(w(K^\circ)\, \text{vrad}(K) \right)^{\frac{n}{n+1}}, \end{aligned}$$

where we used (4.26) to compute the volume radius. □

EXERCISE 4.37 (The mean width of the polar). Let $K \subset \mathbb{R}^n$ be a convex body. Show that $w(K)w(K^\circ) \geqslant 1$.

EXERCISE 4.38. Derive the estimates about inradius, volume radius and outradius in Table 4.1. For the mean width, see Exercise 6.6.

EXERCISE 4.39 (Rough bounds on volume radius of B_p^n). Use the inequalities (1.4) between ℓ_p-norms and the information on volume radii from Table 4.1 (or direct calculations) to conclude that $\operatorname{vrad}(B_p^n) \simeq n^{1/2-1/p}$ for $1 \leqslant p \leqslant \infty$.

EXERCISE 4.40 (Volume of B_p^n). Let $1 \leqslant p \leqslant \infty$. By calculating $\int_{\mathbb{R}^n} e^{-\|x\|_p^p}\,\mathrm{d}x$ in two different ways, show that $\operatorname{vol}(B_p^n) = \big(2\Gamma(1+\frac{1}{p})\big)^n/\Gamma(1+\frac{n}{p})$. Deduce that, for large n, $\operatorname{vrad}(B_p^n) \sim 2\Gamma(1+\frac{1}{p})(pe)^{1/p}n^{1/2-1/p}$.

EXERCISE 4.41 (Uniqueness of outradius witness). Show that there is a unique Euclidean ball of minimal radius containing a given set $K \subset \mathbb{R}^n$.

EXERCISE 4.42 (The gap in Urysohn's inequality). Give examples of convex bodies $K \subset \mathbb{R}^2$ such that the ratio $w(K)/\operatorname{vrad}(K)$ is arbitrarily large.

EXERCISE 4.43 (The mean width and the diameter). Show that for a convex body $K \subset \mathbb{R}^n$, $w(K) \geqslant \frac{1}{2}\frac{\kappa_1}{\kappa_n}\operatorname{diam} K$.

EXERCISE 4.44 (The mean width and the perimeter). For a convex body $K \subset \mathbb{R}^2$, show that $w(K)$ is equal to $(2\pi)^{-1}$ times the perimeter of K. For convex planar sets, the Urysohn inequality is therefore equivalent to the isoperimetric inequality.

EXERCISE 4.45 (The mean width of a projection). Let $K \subset \mathbb{R}^n$ be bounded and P_E be the orthogonal projection onto a subspace $E \subset \mathbb{R}^n$. Show that $w_G(P_E K) \leqslant w_G(K)$.

EXERCISE 4.46 (The mean width of an affine contraction). Let $A : \mathbb{R}^n \to \mathbb{R}^n$ be an affine contraction (i.e., such that $|Ax - Ay| \leqslant |x-y|$ for every $x, y \in \mathbb{R}^n$). Show that for every bounded set $K \subset \mathbb{R}^n$, we have $w_G(AK) \leqslant w_G(K)$.

EXERCISE 4.47 (The mean width of a union). If K, L are convex bodies in $\mathbb{R}^n$ with $K \cap L \neq \varnothing$, then $w(K \cup L) \leqslant w(K) + w(L)$. For an improvement on this, see Exercise 5.28.

EXERCISE 4.48 (Geometric mean width). Prove the following strengthening of the inequality from Proposition 4.16: $\exp\big(\int_{S^{n-1}} \log\|\theta\|_K \,\mathrm{d}\sigma(\theta)\big) \geqslant \operatorname{vrad}(K)^{-1}$. In other words, the "geometric mean" of $\|\cdot\|_K$ is at least as large as $\operatorname{vrad}(K)^{-1}$, while inequality (4.36) asserts the same only about the "arithmetic mean" $w(K^\circ) = \int_{S^{n-1}} \|\theta\|_K \,\mathrm{d}\sigma(\theta)$.

EXERCISE 4.49 (A proof of Urysohn's inequality).
(i) Explain in which sense the following generalization of the Brunn–Minkowski holds and prove it: if $(\Omega, \mathcal{F}, \mu)$ is a measure space and $K_t \subset \mathbb{R}^n$ a convex body depending in a measurable way on a parameter $t \in \Omega$, then

$$\int_\Omega \operatorname{vol}(K_t)^{1/n}\,\mathrm{d}\mu(t) \leqslant \left(\operatorname{vol}\left(\int_\Omega K_t\,\mathrm{d}\mu_t\right)\right)^{1/n}. \tag{4.37}$$

(ii) Fix a convex body $K \subset \mathbb{R}^n$. By choosing (Ω, μ) to be the orthogonal group $\mathsf{O}(n)$ equipped with the Haar measure, and $K_t = t(K)$ for $t \in \mathsf{O}(n)$, prove (4.34).

4.3.4. The Santaló and the reverse Santaló inequalities. When dealing with convex bodies, it is often convenient to consider the dual picture, involving the polar bodies. It turns out that the volume is especially well behaved with respect to the polar operation. This is the content of the Santaló and reverse Santaló inequalities.

THEOREM 4.17 (Santaló and reverse Santaló inequalities, not proved here; see Exercise 7.33). *There is a constant $c > 0$ such that the following holds: for any $n \in \mathbb{N}$ and for any symmetric convex body $K \subset \mathbb{R}^n$, we have*

$$c \leqslant \operatorname{vrad}(K) \operatorname{vrad}(K^\circ) \leqslant 1. \tag{4.38}$$

For a non-symmetric convex body $K \subset \mathbb{R}^n$, the product $\operatorname{vrad}(K)\operatorname{vrad}(K^\circ)$ may be arbitrary large (and even infinite, if 0 belongs to the boundary of K). The correct version of the Theorem in that context is as follows: *any convex body $K \subset \mathbb{R}^n$ can be translated so that* (4.38) *holds.* Moreover, it is known (see Proposition D.2 in Appendix D) that among the translates of K, the minimum of the volume of the polar (and hence of the product of the volume radii) occurs when the polar has the centroid at 0. Such a point is unique and called the Santaló point of K.

The upper bound in (4.38) is also known as the Blaschke–Santaló inequality and can be proved through a symmetrization procedure. Note that a 0-symmetric ellipsoid $\mathscr{E} \subset \mathbb{R}^n$ satisfies $\operatorname{vrad}(\mathscr{E})\operatorname{vrad}(\mathscr{E}^\circ) = 1$ and no other bodies saturate the upper bound. Concerning the lower bound, the best constants to date are $c = 1/2$ in the symmetric case and $c = 1/4$ in the general case (cf. Exercise 4.57).

EXERCISE 4.50 (Santaló implies Urysohn). Using the Santaló inequality, deduce the Urysohn inequality (4.34) from its dual version (Proposition 4.16).

EXERCISE 4.51 (Inequalities between various radii). Show that if $K \subset \mathbb{R}^n$ is a symmetric convex body, then

$$\operatorname{inrad}(K) \leqslant w(K^\circ)^{-1} \leqslant \operatorname{vrad}(K) \leqslant \operatorname{vrad}(K^\circ)^{-1} \leqslant w(K) \leqslant \operatorname{outrad}(K).$$

Show that these inequalities also hold if K is a convex body such that the only fixed point of $\operatorname{Iso}(K)$ is 0.

EXERCISE 4.52 (Minimizers in the reverse Santaló inequality). Show that we have $\operatorname{vrad}(K)\operatorname{vrad}(K^\circ) = \operatorname{vrad}(B_1^6)\operatorname{vrad}(B_\infty^6)$ when $K = B_1^3 \times B_1^3 \subset \mathbb{R}^6$. This exemplifies nonuniqueness of the conjectured extremal case in reverse Santaló inequality, or (the symmetric version of) the Mahler conjecture (see Notes and Remarks).

4.3.5. Symmetrization inequalities. We described in Section 4.1.2 several natural ways to construct a symmetric convex body associated to a given (non-symmetric) convex body. In each case, it is possible to control the volume of the symmetric body in terms of the volume of the initial body.

4.3.5.1. *Milman–Pajor inequality.*

PROPOSITION 4.18. *Let K, L be two convex bodies in $\mathbb{R}^n$ with the same centroid. We have*

$$\operatorname{vol}(K)\operatorname{vol}(L) \leqslant \operatorname{vol}(K \cap L)\operatorname{vol}(K - L).$$

In particular, if $K \subset \mathbb{R}^n$ is a convex body with centroid at the origin, then

$$\operatorname{vol}(K_\cap) \geqslant 2^{-n}\operatorname{vol}(K) \tag{4.39}$$

Recall that $K_\cap = (-K) \cap K$. The factor 2^{-n} may appear small, but remember that it is the n-th root of the volume that is the relevant quantity. In particular, in terms of volume radii, the conclusion of the second part of Proposition 4.18 simply becomes $\operatorname{vrad}(K_\cap) \geqslant \frac{1}{2} \operatorname{vrad}(K)$. In is natural to conjecture that among convex bodies of fixed volume with centroid at the origin, the volume of $K_\cap$ is minimized when K is a simplex. This would lead to a constant $((2/e + o(1))^n$ instead of 2^{-n} in (4.39).

To prove Proposition 4.18, we use the following lemma (which is much simpler to prove for symmetric convex bodies, see Exercise 4.53).

LEMMA 4.19 (Spingarn inequality). *Let $K \subset \mathbb{R}^n$ be a convex body with the centroid at the origin. If $E \subset \mathbb{R}^n$ is a (vector) subspace and $F = E^\perp$, we have the inequality*

$$\operatorname{vol}(K) \leqslant \operatorname{vol}_E(K \cap E) \operatorname{vol}_F(P_F K).$$

Recall that vol_H refers to the Lebesgue measure on an affine subspace $H \subset \mathbb{R}^n$.

PROOF OF LEMMA 4.19. Define a function $\Phi : P_F K \to \mathbb{R}_+$ by

$$\Phi(x) = \operatorname{vol}_{E+x}(K \cap (E + x))^{1/k},$$

where $k = \dim E$. The Brunn–Minkowski inequality (4.22) implies that the function Φ is concave (see Exercise 4.31). Since concave functions can be realized as minima of affine functions, there exists a $y \in F$ such that for any $x \in P_F K$,

$$\Phi(x) \leqslant \langle x, y\rangle + \Phi(0). \tag{4.40}$$

By the Fubini–Tonelli theorem and the Hölder inequality, we have

$$\operatorname{vol}(K) = \int_{P_F K} \Phi(x)^k \, dx \leqslant \operatorname{vol}_F(P_F K)^{\frac{1}{k+1}} \left(\int_{P_F K} \Phi(x)^{k+1} \, dx \right)^{k/(k+1)}. \tag{4.41}$$

Next, by (4.40),

$$\int_{P_F K} \Phi(x)^{k+1} \, dx \leqslant \int_{P_F K} \Phi(x)^k (\langle x, y\rangle + \Phi(0)) \, dx. \tag{4.42}$$

Since 0 is the centroid of K, we have $\int_{P_F K} \Phi(x)^k \langle x, y \rangle \, dx = 0$. Consequently, combining (4.41) and (4.42), we are led to

$$\operatorname{vol}(K) \leqslant \operatorname{vol}_F(P_F K)^{\frac{1}{k+1}} \Phi(0)^{\frac{k}{k+1}} \operatorname{vol}(K)^{\frac{k}{k+1}}.$$

Since $\Phi(0)^k = \operatorname{vol}_E(K \cap E)$, the inequality follows. □

PROOF OF PROPOSITION 4.18. We may assume, by translating them if necessary, that K and L have centroids at the origin. We apply Lemma 4.19 to the convex body $K \times L \subset \mathbb{R}^n \times \mathbb{R}^n \sim \mathbb{R}^{2n}$ and to the subspaces $E = \{(x, x) \,:\, x \in \mathbb{R}^n\}$ and $F = \{(x, -x) \,:\, x \in \mathbb{R}^n\}$. We note that $\operatorname{vol}_{2n}(K \times L) = \operatorname{vol}_n(K) \operatorname{vol}_n(L)$, $\operatorname{vol}_n(K \cap E) = 2^{n/2} \operatorname{vol}_n(K \cap L)$ and $\operatorname{vol}_n(P_F K) = 2^{-n/2} \operatorname{vol}_n(K - L)$. The conclusion follows. □

EXERCISE 4.53 (Spingarn inequality for symmetric bodies). Why is Lemma 4.19 very simple to prove when K is centrally symmetric?

4.3.5.2. *Rogers–Shephard inequalities.* There is a converse to Lemma 4.19 which is simpler since it does not require any hypothesis on the centroid.

LEMMA 4.20. *Let $K \subset \mathbb{R}^n$ be a convex body. If $E \subset \mathbb{R}^n$ is an affine subspace of dimension k and $F = E^\perp$, we have the inequality*

$$\mathrm{vol}(K) \geqslant \binom{n}{k}^{-1} \mathrm{vol}_E(K \cap E)\,\mathrm{vol}_F(P_F K).$$

PROOF. Let $\Phi : P_F K \to \mathbb{R}_+$ as in the proof of Lemma 4.19. The function Φ is concave and vanishes on the boundary of $P_F K$, therefore, for any $x \in P_F K$,

$$\Phi(x) \geqslant \Phi(0)(1 - \|x\|_{P_F K}).$$

It follows that

$$\mathrm{vol}(K) = \int_{P_F K} \Phi(x)^k \, dx \geqslant \mathrm{vol}_E(K \cap E) \int_{P_F K} (1 - \|x\|_{P_F K})^k \, dx$$

and the last integral reduces to a Beta integral and equals $\mathrm{vol}_F(P_F K)\binom{n}{k}^{-1}$. □

Lemma 4.20 implies a series of inequalities, all due to Rogers and Shephard, stating that the simplex is the convex body for which the volume increase is the largest after symmetrization. Their proofs are relegated to exercises.

THEOREM 4.21 (See Exercise 4.54). *If $K \subset \mathbb{R}^n$ is a convex body,*

$$\mathrm{vol}(K) \leqslant \mathrm{vol}((K - K)/2) \leqslant 2^{-n}\binom{2n}{n} \mathrm{vol}(K). \tag{4.43}$$

As a consequence

$$\mathrm{vrad}(K) \leqslant \mathrm{vrad}((K - K)/2) \leqslant 2\,\mathrm{vrad}(K). \tag{4.44}$$

THEOREM 4.22 (See Exercise 4.55). *Let H be an affine hyperplane in $\mathbb{R}^{n+1}$, not containing the origin, and $h > 0$ be the distance between H and the origin. Let K be a convex body in H. We have the following inequalities*

$$2h\,\mathrm{vol}_H(K) \leqslant \mathrm{vol}_{n+1}(K_{\diamond}) \leqslant 2h\,\frac{2^n}{n+1}\,\mathrm{vol}_H(K). \tag{4.45}$$

If $0 \in K$, then $K_\cup \subset K - K$ and so, by (4.43), $\mathrm{vol}(K_\cup) \leqslant \binom{2n}{n} \mathrm{vol}(K) \leqslant 4^n \mathrm{vol}(K)$. However, the constant 4 can be improved to the optimal value of 2.

THEOREM 4.23 (See Exercise 4.56). *If $K \subset \mathbb{R}^n$ is a convex body with $0 \in K$, then*

$$\mathrm{vol}(K_\cup) \leqslant 2^n \,\mathrm{vol}(K).$$

EXERCISE 4.54. Deduce Theorem 4.21 from Lemma 4.20.

EXERCISE 4.55. Deduce Theorem 4.22 from Lemma 4.20.

EXERCISE 4.56. Deduce Theorem 4.23 from Theorem 4.22 and Lemma 4.20.

EXERCISE 4.57 (Symmetric vs. non-symmetric reverse Santaló inequality). Show that whenever the reverse Santaló inequality (the lower bound in Theorem 4.17) holds with a constant $c > 0$ for symmetric convex bodies, it holds with constant $c/2$ for all convex bodies.

4.3.6. Functional inequalities. Most classical inequalities for convex bodies described in this section admit functional variants. As an example, we will state the *Prékopa–Leindler inequality*, which is a generalization of the Brunn–Minkowski inequality.

THEOREM 4.24 (Prékopa–Leindler inequality, not proved here; see Exercise 4.58). *Let $\lambda \in (0,1)$ and let f, g, h be nonnegative integrable functions on $\mathbb{R}^n$ such that*

$$h(\lambda x + (1-\lambda)y) \geqslant f(x)^\lambda g(y)^{1-\lambda} \tag{4.46}$$

for all $x, y \in \mathbb{R}^n$. Then

$$\int_{\mathbb{R}^n} h(x)\,\mathrm{d}x \geqslant \left(\int_{\mathbb{R}^n} f(x)\,\mathrm{d}x\right)^\lambda \left(\int_{\mathbb{R}^n} g(x)\,\mathrm{d}x\right)^{1-\lambda}. \tag{4.47}$$

The Brunn–Minkowski inequality in the form (4.21) follows immediately from Theorem 4.24 applied with $f = \mathbf{1}_K$, $g = \mathbf{1}_L$, and $h = \mathbf{1}_{\lambda K+(1-\lambda)L}$ (the indicator functions of K, L, and $\lambda K + (1-\lambda)L$). See Notes and Remarks for pointers to other functional inequalities.

EXERCISE 4.58. Using induction on the dimension, derive the general Prékopa–Leindler inequality from the case $n = 1$.

4.4. Volume of central sections and the isotropic position

Let $K \subset \mathbb{R}^n$ be a convex body with centroid at the origin. The *inertia matrix* of K is defined as

$$I_K = \frac{1}{\operatorname{vol} K}\int_K |x\rangle\langle x|\,\mathrm{d}x.$$

Note that I_K is invertible (because it is positive definite). One says that K is *isotropic* (or is in the isotropic position) if I_K is a multiple of identity.

If $T \in \mathsf{GL}(n, \mathbb{R})$, one checks that $I_{TK} = TI_KT^\dagger$. It follows that any convex body with the centroid at the origin has a linear image which is isotropic. Moreover, this position is unique in the following sense: if both K and TK are isotropic for some $T \in \mathsf{GL}(n, \mathbb{R})$, then T is a multiple of an orthogonal matrix. In particular, we have the following.

PROPOSITION 4.25 (Easy). *Convex bodies with enough symmetry are isotropic.*

Isotropic convex bodies have the remarkable property that all their central hyperplane sections have comparable volumes.

PROPOSITION 4.26 (See Exercise 4.59). *Let $K \subset \mathbb{R}^n$ be a convex body with the centroid at the origin, and assume that $I_K = \lambda^2 \mathrm{I}$ for some $\lambda > 0$. Then, for any linear hyperplane $H \subset \mathbb{R}^n$,*

$$c\frac{\operatorname{vol}_n(K)}{\lambda} \leqslant \operatorname{vol}_{n-1}(K \cap H) \leqslant C\frac{\operatorname{vol}_n(K)}{\lambda}, \tag{4.48}$$

where $c = \frac{1}{2\sqrt{3}}$ and $C = \frac{1}{\sqrt{2}}$.

A very important open problem is how the two parameters λ and $\operatorname{vol}_n(K)$ appearing in (4.48) are related. The *hyperplane conjecture* postulates that, for every convex body K with $\operatorname{vol}_n(K) = 1$ and $I_K = \lambda^2 \mathrm{I}$, we have $\lambda \leqslant C_0$ for an absolute constant C_0; see Notes and Remarks for more background on this conjecture.

For some special bodies much more precise estimates are available.

PROPOSITION 4.27 (Sections of the cube, not proved here). *Let H be a k-codimensional vector subspace of $\mathbb{R}^n$. Then*

$$1 \leqslant \mathrm{vol}_{n-k}\left(\tfrac{1}{2}B_\infty^n \cap H\right) \leqslant 2^{k/2}. \tag{4.49}$$

We conclude the section by presenting a statement in the spirit of Proposition 4.26 for the volume radius. Since the volume radius is a more robust parameter than the volume itself, it allows us to infer in many situations (including non-isotropic convex bodies) that the volume radius of a convex set is comparable to the volume radius of sections through its centroid. (The reader who wonders why such relationships may be relevant in the context of this book may check Section 9.3.)

PROPOSITION 4.28. *Let K be an n-dimensional convex body with centroid at a, and let H be a k-codimensional affine subspace passing through a. Denote $\theta = k/n$ and let r and R be the inradius and outradius of K with respect to a. Then*

$$R^{-\theta}\, b(n,k) \leqslant \frac{\mathrm{vrad}(K\cap H)^{1-\theta}}{\mathrm{vrad}(K)} \leqslant r^{-\theta}\, b(n,k)\binom{n}{k}^{\frac{1}{n}}, \tag{4.50}$$

where

$$b(n,k) := \left(\frac{\mathrm{vol}_n(B_2^n)}{\mathrm{vol}_k(B_2^k)\,\mathrm{vol}_{n-k}(B_2^{n-k})}\right)^{\frac{1}{n}} = \left(\frac{\Gamma(\frac{k}{2}+1)\Gamma(\frac{n-k}{2}+1)}{\Gamma(\frac{n}{2}+1)}\right)^{1/n}. \tag{4.51}$$

PROOF. We may assume that $a=0$ (otherwise consider $K-a$). By hypothesis, we have then

$$rB_2^n \subset K \subset RB_2^n, \tag{4.52}$$

where B_2^n is the n-dimensional unit Euclidean ball. For a subspace E, denote by P_E the orthogonal projection onto E. Then, by Lemma 4.19,

$$\mathrm{vol}_n(K) \leqslant \mathrm{vol}_s(K\cap H)\ \ \mathrm{vol}_k(P_{H^\perp}K), \tag{4.53}$$

where $H^\perp$ is the k-dimensional space orthogonal to H and $s=n-k$. Therefore

$$\frac{\mathrm{vol}_n(K)}{\mathrm{vol}_n(B_2^n)} \leqslant \frac{\mathrm{vol}_s(K\cap H)}{\mathrm{vol}_s(B_2^s)}\ \frac{\mathrm{vol}_k(P_{H^\perp}K)}{\mathrm{vol}_k(B_2^k)}\ \frac{\mathrm{vol}_s(B_2^s)\,\mathrm{vol}_k(B_2^k)}{\mathrm{vol}_n(B_2^n)}.$$

Hence, using (4.52),

$$\mathrm{vrad}(K)^n \ \leqslant\ \mathrm{vrad}(K\cap H)^s\ R^k\ \frac{\mathrm{vol}_s(B_2^s)\ \mathrm{vol}_k(B_2^k)}{\mathrm{vol}_n(B_2^n)},$$

which is the first inequality in (4.50). For the second inequality, we note that by Lemma 4.20, which does not even require that H passes through the centroid of K,

$$\mathrm{vol}_n(K) \geqslant \binom{n}{k}^{-1} \mathrm{vol}_s(K\cap H)\ \mathrm{vol}_k(P_{H^\perp}K). \tag{4.54}$$

As earlier, this can be rewritten in terms of volume radii as

$$\binom{n}{k}\mathrm{vrad}(K)^n \ \geqslant\ \mathrm{vrad}(K\cap H)^s\ r^k\ \frac{\mathrm{vol}_s(B_2^s)\ \mathrm{vol}_k(B_2^k)}{\mathrm{vol}_n(B_2^n)},$$

which is the second inequality in (4.50). □

REMARK 4.29. Although the argument that led to bounds (4.50) looks rough, we note that we always have (see Exercise 4.60)

$$\frac{1}{\sqrt{2}} < b(n,k) < 1 < b(n,k)\binom{n}{k}^{\frac{1}{n}} < \sqrt{2}. \tag{4.55}$$

EXERCISE 4.59 (Isotropic position and central sections).
(i) Let $f : \mathbb{R} \to \mathbb{R}_+$ be an even function such that $\log f$ is concave and $\int f(x)\,dx = 1$. Show that $\frac{1}{12f(0)^2} \leqslant \int x^2 f(x)\,dx \leqslant \frac{1}{2f(0)^2}$. (This conclusion also holds if the assumption "f is even" is replaced by "$\int xf(x)\,dx = 0$," but the proof is more involved, see [**Fra99**].)
(ii) Use (i) to prove Proposition 4.26.

EXERCISE 4.60. Prove the bounds (4.55).

Notes and Remarks

A comprehensive reference for geometry and for convex bodies focusing on the issues related to the Brunn–Minkowski inequality is the book [**Sch14**].

Section 4.1. The Banach–Mazur distance is most frequently defined in the category of normed spaces with

$$d(X,Y) := \inf\{\|T\| \cdot \|T^{-1}\| \ : \ T : X \to Y \text{ an isomorphism}\}.$$

This corresponds to definition (4.2) with K, L being 0-symmetric (and, consequently, $a = b = 0$).

It is shown in [**GLMP04**] that $d_{BM}(K, \Delta_n) \leqslant n$ for every convex body $K \subset \mathbb{R}^n$. This was known to Grünbaum for $n = 2$. It would be nice to have a simple proof for $n > 2$ (cf. Exercise 4.2).

The question of computing the diameter of (various versions of) the Banach–Mazur compactum has attracted a lot of attention. It follows from Exercise 4.20 that the diameter is at most n. In an important and short paper [**Glu81**], Gluskin showed that this estimate is asymptotically sharp via the probabilistic method. A variant of his argument shows that if we denote by K_n, K_n' two randomly and independently chosen n-dimensional sections of the $3n$-dimensional cube, then with large probability $d_{BM}(K_n, K_n') \gtrsim n$. Remarkably, no explicit example of a pair of convex bodies more than $C\sqrt{n}$ apart is known. It is proved in [**Sza90**] that $d_{BM}(K_n, B_\infty^n) \gtrsim \sqrt{n}\log n$ for some randomly constructed K_n.

In the non-symmetric case, the order of growth of the diameter of the Banach–Mazur compactum is not known, and determining it is an important open problem. It is clearly $\Omega(n)$, and we do not know whether this inequality is strict. Conversely, an upper bound of $Cn^{4/3}\log^C n$ was shown in [**Rud00**], which improves on the trivial bound $O(n^2)$ (see also [**BLPS99**]).

For more information and references on the Banach–Mazur distance and the Banach–Mazur compactum see the website [**@3**].

For more information on zonotopes and zonoids, we refer to the surveys [**SW83**, **GW93**].

We also point out that while the definition of the projective tensor product appears to be well-adapted to 0-symmetric sets and cones, with linear maps as morphisms, the projective tensor product *is not* invariant under affine maps. We refer to [**Sve81**], Chapter 2, for a discussion of related categorical issues and to [**DF93**] for exhaustive treatment of tensor products of normed spaces.

The result from Exercise 4.17 appears in [**Ceĭ76**].

Note that, in general, the Minkowski sum of Borel sets does not need to be Borel [**ES70**]. However, it is always measurable [**Kec95**]. The Minkowski sum also behaves strangely with respect to smoothness: for example the Minkowski sum of two planar convex bodies with real-analytic boundary is always of class C^6 but possibly not of class C^7 [**Kis87**]. (See also [**Bom90b, Bom90a**].)

Section 4.2. John's theorem was first proved (in a slightly different form) in [**Joh48**]. We refer to [**Bal97**] for a modern proof (arguments already appeared in [**Bal92a**]) and to [**Hen12**] for historical aspects. The reduction of the general setting to the symmetric case presented here (Proposition 4.4, and the proofs of Propositions 4.6 and 4.7) appears to be new.

The concept of convex bodies with "enough symmetries" was defined in [**GG71**]; see also Chapter 16 in [**TJ89**].

The affinity between projective tensor products and Löwner ellipsoids (Lemma 4.9) was noted in [**Sza05, AS06**].

Section 4.3. The Brunn–Minkowski inequality (4.22) was first proved in dimensions 2 and 3 by Brunn and extended by Minkowski to higher dimensions. The equality case is known: when K, L are convex bodies and $0 < \lambda < 1$, the inequality (4.21) is an equality if and only if K and L are homothetic. The equality case was extended by Lusternik to the general case and is essentially the same up to null sets; for precise statements, and for a panorama of inequalities connected to the isoperimetric inequalities, we refer to the survey [**Gar02**]. Far-reaching generalizations of the Brunn–Minkowski inequality are the Alexandrov–Fenchel inequalities, for which we refer to [**Sch14**].

The two sides of the inequality (4.22) can be very different; for example, if K and L are perpendicular segments in $\mathbb{R}^2$ (hence of volume 0), $K + L$ is a rectangle, and this behavior can be approximated in the category of convex bodies by replacing segments with narrow rectangles. It is therefore surprising that the Brunn–Minkowski inequality admits—after some tweaking—a reverse: *any two n-dimensional convex bodies have affine images (of the same dimension), for which* (4.22) *can be reversed, up to a universal constant* (see (7.32) in Notes and Remarks on Section 7.2). A vaguely similar reverse of Urysohn inequality (4.34) can be found in Chapter 7 (Corollary 7.11).

Another variant of (4.22) that has information-theoretic links is the *restricted Brunn–Minkowski inequality* [**SV96, SV00**]. It asserts that when $K, L \subset \mathbb{R}^n$ satisfy some minimal non-degeneracy assumptions and $\Theta \subset K \times L \subset \mathbb{R}^{2n}$ is not too small (e.g., $\mathrm{vol}_{2n}(\Theta) \geqslant c \,\mathrm{vol}_n(K)\,\mathrm{vol}_n(L)$ for appropriate universal constant $c \in (0,1)$), then $\mathrm{vol}(K +_\Theta L)^{2/n} \geqslant \mathrm{vol}(K)^{2/n} + \mathrm{vol}(L)^{2/n}$, where $K +_\Theta L := \{x + y \;:\; (x,y) \in \Theta\}$ is the restricted (to Θ) Minkowski sum.

The characterization of log-concave measures (Proposition 4.14) holds without the absolute continuity assumption: by a result of Borell [**Bor75a**], any Radon measure on $\mathbb{R}^n$ which satisfies part (2) of Proposition 4.14 necessarily has a density with respect to the Lebesgue measure on some affine subspace, and this density is a log-concave function.

The upper bound (known as the Blaschke–Santaló inequality) in Theorem 4.17 was proved by Blaschke in dimensions 2 and 3 and by Santaló in any dimension. The first proof of the lower bound is due to Bourgain and Milman [**BM87**]. Other—quite different—proofs were given later by Kuperberg [**Kup08**] (which gives the values of c quoted in the text) and Nazarov [**Naz12**] (we recommend the notes [**RZ14**] for a detailed presentation of Nazarov's argument). However, no elementary proof is known (a simple argument giving a lower bound $\mathrm{vrad}(K)\,\mathrm{vrad}(K^\circ) \gtrsim 1/\log n$ appears in [**Kup92**]).

It is conjectured that the product $\mathrm{vrad}(K)\,\mathrm{vrad}(K^\circ)$ in (4.38) is minimized for the pair (B_1^n, B_∞^n) (and for the family of Hanner polytopes, defined as the smallest class of polytopes containing $[-1,1]$ and stable under the operations $K \mapsto K^\circ$ and $(K,L) \mapsto K \times L$; cf. Exercise 4.52) and, in the non-symmetric case, for $K = \Delta_n$ (the minimum being then conjectured to be unique). This is the content of the so-called *Mahler conjecture.*

Several inequalities, for which the Euclidean ball is the extremal case, such that the isoperimetric inequality, the Urysohn inequality and the Santaló inequality (the upper bound in (4.38)), can be proved using symmetrizations. For example one may consider the Steiner symmetrizations as defined in Exercise 4.31. A useful result is then the fact that, given any convex body $K \subset \mathbb{R}^n$, there is choice of successive Steiner symmetrizations that converge to a Euclidean ball of radius $\mathrm{vrad}(K)$ (see, e.g., Theorem 1.1.16 in [**AAGM15**] for a sketch of proof).

Proposition 4.18 appears in [**MP00**] and Lemma 4.19 in [**Spi93**]. Lemma 4.20 is from [**RS58**]; a simpler proof can be found in [**Cha67**].

Theorem 4.24 was shown in [**Lei72**] and [**Pré71, Pré73**]; see also [**BL75, BL76**]. A complete compact proof can be found in [**AAGM15**] or [**Gar02**], the latter of which also sketches historical background and contains many further references.

Other functional versions of inequalities presented in this section include analogues of the Santaló inequality that can be traced to K. Ball's Ph.D. thesis [**Bal86**] (see also [**AAKM04**]), and of its reverse [**KM05**]; see also [**AAS15**] and [**CFG+16**] for more recent contributions and references.

Functional versions of Rogers–Shephard inequalities were considered starting from [**Col06**]; see also [**AGMJV16**].

Section 4.4. A very complete reference about the geometry of convex bodies in isotropic position (including the most recent developments) is the book [**BGVV14**]. Proposition 4.26 was proved by Hensley [**Hen80**] for symmetric convex bodies and the symmetry assumption was removed in [**Fra99**].

The hyperplane conjecture (also known as the "slicing problem") asserts that any convex body of volume 1 in $\mathbb{R}^n$ admits a hyperplane section of volume larger than c_0, for some absolute constant $c_0 > 0$. This is equivalent to the statement mentioned in the text: if an isotropic convex body K satisfies $\mathrm{vol}(K) = 1$ and $I_K = \lambda^2\,\mathrm{I}$, does $\lambda \leqslant C_0$ for some absolute constant C_0? (It is even conceivable that the above are true with $c_0 = C_0 = 1$.) The answer is known to be positive for many natural classes of bodies; of those that are particularly relevant to the subject of this book we mention unit balls in Schatten p-norms, see [**KMP98**]. However, the best known estimate in the general case is only $\lambda = O(n^{1/4})$ [**Kla06**]; we refer to [**BGVV14**] for more references and an extensive discussion of related questions.

The hyperplane conjecture can be seen as an isomorphic version of the classical (now fully solved) Busemann–Petty problem, which asks the following: *if two symmetric convex bodies* $K, L \subset \mathbb{R}^n$ *satisfy* $\mathrm{vol}_{n-1}(K \cap H) \leqslant \mathrm{vol}_{n-1}(L \cap H)$ *for every hyperplane* H *containing the origin, can we conclude that* $\mathrm{vol}_n(K) \leqslant \mathrm{vol}_n(L)$*?* It is known that the answer is affirmative when $n \leqslant 4$ and negative when $n \geqslant 5$ (see [**Kol05**] for references).

Proposition 4.27 is due to Vaaler ([**Vaa79**], the lower bound) and Ball ([**Bal89**], the upper bound).

Proposition 4.28 is from [**SWŻ08**]. It is instructive to compare Propositions 4.26 and 4.28. The first one gives very precise estimates for volumes of hyperplane sections in the isotropic position, while the second one deals with sections of proportional (or subproportional) codimension, but only at the level of the volume radius, that is, after raising the volumes to the power of 1 over the dimension.

CHAPTER 5

Metric entropy and concentration of measure in classical spaces

This chapter presents two fundamental concepts which will be applied in later chapters: the metric entropy (a.k.a. packing and covering) and the concentration of measure. Their conjunction leads to the Dvoretzky theorem, which will be presented in Chapter 7.

5.1. Nets and packings

We will introduce now the complementary concepts of covering numbers (also called metric entropy) and packing numbers, which quantify the complexity of a given compact metric set. It will turn out that these parameters are closely related to the volume and the mean width considered in the preceding chapter.

We first analyze the special but fundamental cases of the sphere and the discrete cube. We subsequently discuss classical groups and manifolds, and general convex bodies.

5.1.1. Definitions. If K is a compact subset of a metric space (M, d), a finite subset $\mathcal{N} \subset K$ is called an *ε-net* of K if, for every $x \in K$, $\operatorname{dist}(x, \mathcal{N}) \leqslant \varepsilon$. Since this is equivalent to the union of the corresponding balls containing K, an alternative terminology is that of a *covering*, see Figure 5.1. We denote by $N(K, \varepsilon)$ (or by $N(K, d, \varepsilon)$, if there is an ambiguity as to the choice of the metric) the minimal cardinality of an ε-net in K.

A subset $\mathcal{P} \subset K$ is called *ε-separated* if any pair (x, y) of distinct elements from $\mathcal{P}$ satisfies $d(x, y) > \varepsilon$. This property implies that the balls of radius $\varepsilon/2$ centered at elements of $\mathcal{P}$ are disjoint (a configuration usually referred to as *packing*, whence the usage of the letter P; see Figure 5.1), and in most contexts the two properties are essentially equivalent. We denote by $P(K, \varepsilon)$ or $P(K, d, \varepsilon)$ the largest cardinality of an ε-separated set in K. The quantities $N(K, \varepsilon)$ and $P(K, \varepsilon)$ are called, respectively, *covering numbers* and *packing numbers*. The function $\varepsilon \mapsto N(K, d, \varepsilon)$, and its various generalizations, is also often referred to as the *metric entropy* of (K, d).

For any compact metric space K, the following two relations between nets and packings are fundamental. First, if $\mathcal{P}$ is a 2ε-separated set and $\mathcal{N}$ is an ε-net, then the open balls of radius ε centered at elements from $\mathcal{N}$ cover K, and each ball contains at most one element of $\mathcal{P}$. Second, an ε-separated set which is maximal (with respect to inclusion) is an ε-net (the reader not familiar with this circle of ideas is encouraged to check these elementary facts). It follows that we have the inequalities

$$P(K, 2\varepsilon) \leqslant N(K, \varepsilon) \leqslant P(K, \varepsilon). \tag{5.1}$$

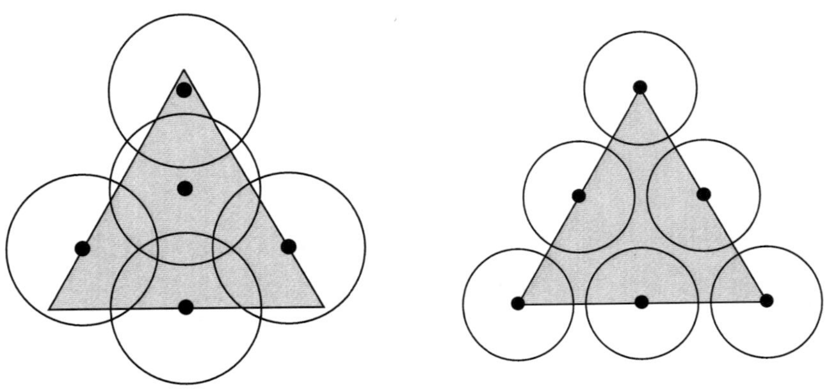

FIGURE 5.1. A net (left) and a packing (right) for an equilateral triangle (with the Euclidean metric in $\mathbb{R}^2$). For optimal packings or coverings with few "classical" convex bodies in the plane (squares, circles or triangles), see the website [**@1**].

Packings and coverings have been extensively studied, particularly for "standard" metric spaces. In various applications it is useful to know that there exist "large" packings and/or "small" nets, and often to be able to exhibit them in a constructive manner. By (5.1), both notions are equivalent whenever the resolution parameter ε is specified only up to a multiplicative constant. On the other hand, for some applications, such as coding theory, very precise results are in high demand.

In many situations the isometry group of K acts transitively and preserves a natural probability measure μ. In particular, all balls of radius ε have then the same measure, denoted by $V(\varepsilon)$, and we have the simple inequalities

$$\frac{1}{V(\varepsilon)} \leqslant N(K,\varepsilon) \leqslant P(K,\varepsilon) \leqslant \frac{1}{V(\varepsilon/2)}. \tag{5.2}$$

EXERCISE 5.1. Here, we introduce variations on the definitions and check their equivalence. Let M be a metric space and K a compact subset. Denote by $N'(K,\varepsilon)$ the smallest cardinality of a family of closed balls of radius ε in M whose union contains K (the difference with the definition of $N(K,\varepsilon)$ is that the centers are not required to be in K). It is sometimes more convenient to allow sets of *diameter* $\leqslant 2\varepsilon$ in place of balls of *radius* ε; call the resulting the quantity $N''(K,\varepsilon)$. Let also $P'(K,\varepsilon)$ be the largest cardinality of a family of disjoint open balls of radius $\varepsilon/2$ with centers in K. Check the inequalities

$$N''(K,\varepsilon) \leqslant N'(K,\varepsilon) \leqslant N(K,\varepsilon) \leqslant P(K,\varepsilon) \leqslant N''(K,\varepsilon/2)$$

and

$$P(K,\varepsilon) \leqslant P'(K,\varepsilon) \leqslant N(K,\varepsilon/2).$$

Give examples showing that the above inequalities may be strict (see also Exercise 5.16).

5.1.2. Nets and packings on the Euclidean sphere. We first consider the specific case of the sphere S^{n-1} for $n \geqslant 2$; denote by g the geodesic distance and by σ the normalized Haar measure. In some cases, it is more appropriate to consider the extrinsic distance inherited from $\mathbb{R}^n$. However, any result about one distance transfers automatically to the other distance (see Appendix B.1 for details). We give a brief overview of known estimates for packing and covering numbers for the sphere. The first point of business will be a discussion of volumes of spherical caps, which enter the subject via (5.2).

5.1.2.1. *Estimates on volumes of spherical caps.* Given $x_0 \in S^{n-1}$, let $C(x_0, \varepsilon)$ be the cap of center x_0 and geodesic radius ε, and denote $V(\varepsilon) = \sigma(C(x_0, \varepsilon))$ ($\varepsilon \in [0, \pi]$ is tacitly assumed). We have

$$V(\varepsilon) = \frac{\int_0^\varepsilon \sin^{n-2}\theta \, d\theta}{\int_0^\pi \sin^{n-2}\theta \, d\theta}. \tag{5.3}$$

The denominator at the right-hand side of (5.3) (the Wallis integral) equals $\sqrt{2\pi}/\kappa_{n-1}$. Note that $V(\pi - \varepsilon) = 1 - V(\varepsilon)$, in particular $V(\pi/2) = 1/2$. For fixed $0 < \varepsilon < \pi/2$, $V(\varepsilon)$ tends to 0 exponentially fast in the dimension: one has $V(\varepsilon)^{1/n} \sim \sin(\varepsilon)$. The following proposition gives elementary but reasonably precise bounds. The first one is sharp when the radius is small, and the second one is sharp for a radius slightly smaller than $\pi/2$.

Proposition 5.1. *If $0 \leqslant t \leqslant \pi/2$, then $V(t) \leqslant \frac{1}{2}\sin^{n-1}(t)$. More precisely*

$$(\sqrt{2\pi}\kappa_n)^{-1}(\sin t)^{n-1} \leqslant V(t) \leqslant (\sqrt{2\pi}\kappa_n \cos t)^{-1}(\sin t)^{n-1}, \tag{5.4}$$

where $\kappa_n \sim \sqrt{n}$ is given by (A.8). *Moreover, if $n > 2$, then*

$$V(\pi/2 - t) \leqslant \frac{1}{2}\exp(-nt^2/2). \tag{5.5}$$

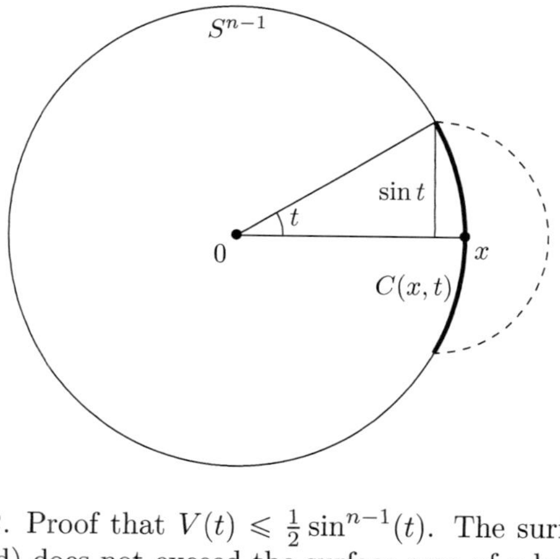

Figure 5.2. Proof that $V(t) \leqslant \frac{1}{2}\sin^{n-1}(t)$. The surface area of $C(x,t)$ (bold) does not exceed the surface area of a half-sphere of radius $\sin t$ (dashed).

A proof of (5.4) is sketched in Exercise 5.4. It is based on the fact that, for convex sets, surface area is monotone with respect to inclusion (Exercise 5.2). The inequality (5.5) is from [**Jen13**] (see also [**JS**]); a version with $n - 1$ instead of n in the exponent is proved in Exercise 5.3.

The following fact is only marginally used in what follows, but we include it since we did not encounter it in the convexity/functional analysis literature.

PROPOSITION 5.2 (Concavity properties of $V(\cdot)$, see Exercise 5.5). *If $V(r)$ is the measure of a spherical cap of radius r, then the function $t \mapsto \log V(e^t)$ is concave. A fortiori, the function $r \mapsto \log V(r)$ is strictly concave on $[0, \pi]$.*

A consequence of Proposition 5.2 is that, for $0 \leqslant s \leqslant t \leqslant \pi$,

$$V(t) \leqslant \left(\frac{t}{s}\right)^{n-1} V(s). \tag{5.6}$$

Inequality (5.6) is a well-known fact in differential geometry; for example, it constitutes the trivial case of the Gromov–Bishop comparison theorem. It is very likely that Proposition 5.2 also follows from similar general results.

EXERCISE 5.2 (Surface area is monotone with respect to inclusion). Show that if $K \subset L$ are convex bodies, then $\operatorname{area}(K) \leqslant \operatorname{area}(L)$.

EXERCISE 5.3. Using Exercise 5.2, show that for $t \in [0, \pi/2]$, we have $V(t) \leqslant \frac{1}{2}\sin^{n-1}(t)$. Conclude that

$$V(\pi/2 - t) \leqslant \frac{1}{2}(\cos t)^{n-1} \leqslant \frac{1}{2}\exp(-(n-1)t^2/2).$$

This is only slightly weaker than the bound (5.5) and sharper than the estimates typically cited in the literature.

EXERCISE 5.4 (Sharp bounds for volumes of caps). Using Exercise 5.2, show the inequalities (5.4). Then strengthen the lower bound to $(\sqrt{2\pi}\,\kappa_n \cos(t/2))^{-1}\sin^{n-1} t$.

EXERCISE 5.5 (Concavity properties of $V(\cdot)$). Prove Proposition 5.2 and derive the inequality (5.6).

5.1.2.2. *Nets in the sphere.* If $\varepsilon \in [\pi/2, \pi)$, we clearly have $N(S^{n-1}, g, \varepsilon) = 2$. The interesting case is when $\varepsilon \in (0, \pi/2)$. In that range, the proportion $V(\varepsilon)$ of the sphere covered by a cap of geodesic radius ε decays exponentially with n. It follows that the cardinality of ε-nets grows also exponentially fast. For example, the first estimate from Proposition 5.1 implies that, for $\varepsilon \in (0, \pi/2)$,

$$N(S^{n-1}, g, \varepsilon) \geqslant V(\varepsilon)^{-1} \geqslant \frac{2}{\sin^{n-1}\varepsilon}. \tag{5.7}$$

A basic and extremely useful bound for ε-nets (formulated in the extrinsic distance) is the following

LEMMA 5.3. *For every dimension n and every $\varepsilon \leqslant 1$, there is an ε-net in $(S^{n-1}, |\cdot|)$ with less than $(2/\varepsilon)^n$ elements. In other words, $N(S^{n-1}, |\cdot|, \varepsilon) \leqslant (2/\varepsilon)^n$.*

The standard and often quoted volumetric argument (which is a special case of Lemma 5.8 below) gives a slightly worse bound $(1 + 2/\varepsilon)^n$. The improved bound $(2/\varepsilon)^n$ can be achieved by a finer analysis combining a version (based on [**Dum07**]) of Proposition 5.4 below with the use of explicit nets in lower dimensions, see [**Swe**]. We also note that there exist simple *explicit* ε-nets in S^{n-1} with cardinality at most $(C/\varepsilon)^n$ (see Exercise 5.22).

To discuss finer results it is more convenient to switch to the geodesic distance. We know from the volume argument (5.2) that $N(S^{n-1}, g, \varepsilon) \geqslant V(\varepsilon)^{-1}$. It turns out that this trivial estimate is remarkably sharp: an almost-matching upper estimate is provided by an elegant random covering argument due to Rogers.

PROPOSITION 5.4 (Random covering bound). *For every* $0 < \eta < \theta$, *we have*

$$N(S^{n-1}, g, \theta+\eta) \leqslant \left\lceil \frac{1}{V(\theta)} \log\left(\frac{V(\theta)}{V(\eta)}\right)\right\rceil + \frac{1}{V(\theta)}.$$

PROOF. Let $N = \lceil \frac{1}{V(\theta)} \log(V(\theta)/V(\eta))\rceil$. Choose $(x_i)_{1\leqslant i\leqslant N}$ randomly, independently according to σ, and denote $A = \bigcup\{C(x_i,\theta) \,:\, 1 \leqslant i \leqslant N\}$. The expected proportion of the sphere missed by A can be computed using the Fubini–Tonelli theorem

$$\mathbf{E}\sigma(S^{n-1}\backslash A) = (1 - V(\theta))^N \leqslant \exp(-NV(\theta)) \leqslant \frac{V(\eta)}{V(\theta)}. \tag{5.8}$$

In particular, there exist (x_i) such that $\sigma(S^{n-1}\backslash A) \leqslant V(\eta)/V(\theta)$. Let $\{C(y_j,\eta) \,:\, 1 \leqslant j \leqslant M\}$ be a maximal family of disjoint balls of radius η contained in $S^{n-1}\backslash A$. It follows from (5.8) that $M \leqslant 1/V(\theta)$. By construction, S^{n-1} is covered by the family

$$\{B(x_i,\theta+\eta) \,:\, 1 \leqslant i \leqslant N\} \cup \{B(y_j,2\eta) \,:\, 1 \leqslant j \leqslant M\}.$$

□

COROLLARY 5.5 (Neat random covering bound; see Exercise 5.8). *For every* $0 < \varepsilon < \pi/2$, *we have*

$$N(S^{n-1}, g, \varepsilon) \leqslant Cn\log n\, V(\varepsilon)^{-1} \tag{5.9}$$

for some absolute constant C.

It follows from (5.7), (5.9) and (5.4) that, for a fixed $\varepsilon \in (0,\pi/2)$, we have

$$\lim_{n\to\infty} \frac{1}{n}\log N(S^{n-1}, g, \varepsilon) = -\log(\sin\varepsilon). \tag{5.10}$$

We note for future reference the following fact.

PROPOSITION 5.6. *Let* $P \subset \mathbb{R}^n$ *be a polytope such that* $d_{BM}(P, B_2^n) \leqslant \lambda$. *Then* P *has at least* $2\exp((n-1)/2\lambda^2)$ *vertices and at least* $2\exp((n-1)/2\lambda^2)$ *facets.*

PROOF. Consider first the statement about vertices. Without loss of generality we may assume that $\lambda^{-1}B_2^n \subset P \subset B_2^n$, and that the vertices of P are unit vectors. Let V be the set of vertices of P. The hypothesis is equivalent to saying that V is a θ-net in (S^{n-1}, g) for $\cos\theta = 1/\lambda$ (see Exercise 5.7). Using (5.7), it follows that $\operatorname{card} V \geqslant 2(\sin\theta)^{-(n-1)} \geqslant 2\exp((n-1)/2\lambda^2)$, where we used the inequality $\sin\arccos t \leqslant \exp(-t^2/2)$ for $0 \leqslant t \leqslant 1$. Since $d_{BM}(P, B_2^n) = d_{BM}(P^\circ, B_2^n)$, and since vertices of P° are in bijection with facets of P, the statement about facets follows. □

We also point out that it is possible to approximate the sphere by polytopes with at most exponentially many vertices and, *simultaneously*, at most exponentially many facets (see Exercise 7.22).

EXERCISE 5.6. Check that the constant 2 cannot be replaced by a smaller number in the statement of Lemma 5.3.

EXERCISE 5.7 (Nets and convex hulls). Let $\mathcal{N} \subset S^{n-1}$ and $\theta \in (0,\pi/2)$. Prove that $\mathcal{N}$ is a θ-net in (S^{n-1}, g) if and only if $(\cos\theta)B_2^n \subset \operatorname{conv}\mathcal{N}$.

EXERCISE 5.8 (Proof of the neat random covering bound). Deduce Corollary 5.5 from Proposition 5.4.

EXERCISE 5.9 (On the optimality of Corollary 5.5). Let C_n be the smallest number such that the inequality $N(S^{n-1}, g, \varepsilon) \leqslant C_n V(\varepsilon)^{-1}$ holds for any $\varepsilon > 0$. By considering ε slightly smaller than $\pi/2$, show that $C_n \geqslant \frac{n+1}{2}$. A less trivial fact is that $C_n = \Omega(n)$ is also witnessed by taking ε very close to 0, see [**CFR59**] and Notes and Remarks.

EXERCISE 5.10 (Nets in the projective space). Prove the following result, which will be useful in Sections 8.1 and 9.4. Let $\varepsilon \in (0, \pi/2)$. If $\mathcal{N}$ is an ε-net in the projective space $\mathbf{P}(\mathbb{C}^d)$ (equipped with the Fubini-Study metric (B.5)), then $\operatorname{card} \mathcal{N} \geqslant (c/\varepsilon)^{2d-2}$ for some absolute positive constant c. In the opposite direction, there exists an ε-net of cardinality not exceeding $(C/\varepsilon)^{2d-2}$.

EXERCISE 5.11 (Volume of balls in $\mathbf{P}(\mathbb{C}^d)$). Consider the projective space $\mathbf{P}(\mathbb{C}^d)$ equipped with the Fubini-Study metric (B.5) and the invariant probability measure. If $\varepsilon \in (0, \pi/2]$, then the measure of any ball of radius ε in $\mathbf{P}(\mathbb{C}^d)$ is $\sin^{2d-2} \varepsilon$.

5.1.2.3. *Packing on the sphere.* Recall that $P(S^{n-1}, g, \varepsilon)$ is the maximal number of disjoint caps of geodesic radius $\varepsilon/2$. The exact value is known for $\pi/2 \leqslant \varepsilon < \pi$ (we have $P(S^{n-1}, g, \pi/2) = 2n$, see Exercise 5.12) and so we restrict our discussion to the range $0 < \varepsilon < \pi/2$.

Packing problems are usually harder than covering problems. For example, as opposed to (5.10), the exponential rate at which packing numbers increase, i.e., the value of

$$p(\varepsilon) = \limsup_{n \to \infty} \frac{1}{n} \log P(S^{n-1}, g, \varepsilon)$$

is not known for $\varepsilon \in (0, \pi/2)$. We know from (5.2) that $V(\varepsilon)^{-1} \leqslant P(S^{n-1}, g, \varepsilon) \leqslant V(\varepsilon/2)^{-1}$, and therefore

$$-\log \sin(\varepsilon) \leqslant p(\varepsilon) \leqslant -\log \sin(\varepsilon/2). \tag{5.11}$$

In this context the lower bound is known as the Chabauty–Shannon–Wyner bound and actually corresponds to using the trivial algorithm to produce packings: pick separated points, no matter how, as long as you can. It is an amazing fact that the lower bound $p(\varepsilon) \geqslant -\log \sin \varepsilon$ has never been improved: nobody knows how to substantially beat the worst possible choices!

On the other hand, the upper bound in (5.11) has received various improvements. It has been shown by Rankin that for $\varepsilon \in (0, \pi/2)$

$$p(\varepsilon) \leqslant -\log(\sqrt{2} \sin(\varepsilon/2)),$$

which matches the lower bound from (5.11) as ε increases to $\pi/2$. For small ε, further improvements due to Kabatjanskiĭ–Levenšteĭn are based on the so-called linear programming bound (see Notes and Remarks).

EXERCISE 5.12 (Packing large caps on the sphere). Suppose that (x_i) are N points in S^{n-1} such that $\langle x_i, x_j \rangle \leqslant t$ for $i \neq j$.
(i) Show that $N \leqslant 1 - 1/t$ if $t < 0$.
(ii) Show that $N \leqslant 2n$ if $t = 0$.
If $t > 0$ is fixed, we know from (5.11) that exponentially many points in the sphere may have pairwise inner products at most t. The situation when t tends to zero with n is investigated in the following exercise.

EXERCISE 5.13 (Coarse approximation of B_2^n by polytopes with few vertices). Suppose that (x_i) are N points in S^{n-1} such that $|\langle x_i, x_j\rangle| \leqslant t$ whenever $i \neq j$, for some $t > 0$.
(i) If $t < 1/\sqrt{n}$, show that $N \leqslant n/(1-nt^2)$.
(ii) By considering the family $(x_i^{\otimes k})_{1\leqslant i\leqslant N}$ for a suitable large k, show that if $t \leqslant 1/2$, then $N \leqslant (C/t)^{Ct^2 n}$ for some absolute constant C.
(iii) Deduce that, for $r \geqslant 2$, there is a polytope P with at most $(Cr)^{Cn/r^2}$ vertices such that $d_g(P, B_2^n) \leqslant r$.

5.1.3. Nets and packings in the discrete cube. Although the discussion from the previous sections dealt specifically with spheres, some ideas carry over directly to other settings. As an illustration we consider the case of the *discrete cube* $\{0,1\}^n$ (a.k.a. *Boolean cube*) equipped with the normalized *Hamming distance*

$$d_H(x,y) = \frac{1}{n}\,\mathrm{card}\{i \,:\, x_i \neq y_i\}. \tag{5.12}$$

We denote by $V(t)$ the volume (i.e., the cardinality) of a ball of radius $t \in (0,1)$. We have

$$V(t) = \mathrm{card}\left\{y \in \{0,1\}^n \,:\, d_H(x,y) \leqslant t\right\} = \sum_{k=0}^{\lfloor tn\rfloor} \binom{n}{k}.$$

The quantity $V(t)$ is governed by the binary entropy function H defined for $x \in (0,1)$ by $H(x) = -x\log_2 x - (1-x)\log_2(1-x)$. For $t \leqslant 1/2$ such that tn is an integer, we have (see Exercise 5.15)

$$\frac{1}{n+1}\,2^{nH(t)} \leqslant V(t) \leqslant 2^{nH(t)}. \tag{5.13}$$

Related estimates will be used when discussing concentration of measure, see (5.59).

As in the case of the sphere, the covering problem is simpler than the packing problem (at least in some asymptotic regimes). In particular (see Exercise 5.14), a random covering argument similar to Proposition 5.4—in combination with (5.13)—implies that, for $0 < \varepsilon < 1/2$,

$$\lim_{n\to\infty} \frac{1}{n}\log_2 N(\{0,1\}^n, d_H, \varepsilon) = 1 - H(\varepsilon). \tag{5.14}$$

On the other hand, the corresponding limit for packing is unknown; we only get from (5.2) the asymptotic bounds

$$1 - H(\varepsilon) \leqslant \limsup_{n\to\infty} \frac{1}{n}\log_2 P(\{0,1\}^n, d_H, \varepsilon) \leqslant 1 - H(\varepsilon/2) \tag{5.15}$$

for $0 < \varepsilon < 1/2$. As in the case of the sphere, the lower bound from (5.15) (known in this context as the Gilbert–Varshamov bound) has not been improved, while the upper bound has been subject to various enhancements.

For the q-ary version of the cube, i.e., the space $\{0,\dots,q-1\}^n$ (also equipped with normalized Hamming distance), the entropy function has to be replaced by

$$H_q(x) := -x\log_q x - (1-x)\log_q(1-x) + x\log_q(q-1).$$

Indeed, if $V_q(t)$ denotes the cardinality of a ball of radius t in $\{1,\dots,q-1\}^n$, for $t \in (0, 1-1/q)$ such that tn is an integer, then

$$\frac{1}{n+1}\,q^{nH_q(t)} \leqslant V_q(t) \leqslant q^{nH_q(t)}. \tag{5.16}$$

Estimates about the q-ary cube are useful when one wants to construct nets or separated sets in products of metric spaces. The following specific fact, which is an easy consequence of (5.16) and (5.1), will be used later.

PROPOSITION 5.7. *Let (K, d) be a metric space such that $P(K, d, \varepsilon) \geqslant q$. Given integer $n \in \mathbb{N}$, equip K^n with the distance*

$$d_n((x_1, \ldots, x_n), (y_1, \ldots, y_n)) = d(x_1, y_1) + \cdots + d(x_n, y_n).$$

Then, for $t \in (0, 1 - 1/q)$,

$$P(K^n, d^n, t\varepsilon n) \geqslant P(\{0, \ldots, q-1\}^n, d_H, t) \geqslant \frac{q^n}{V_q(t)} \geqslant q^{n(1-H_q(t))}. \tag{5.17}$$

EXERCISE 5.14 (Efficient random nets of the Boolean cube). Show (5.14) by adapting the random covering argument from Proposition 5.4.

EXERCISE 5.15 (Volume of balls in the q-ary discrete cube). Show (5.16) (which specified to $q = 2$ gives (5.13)).

5.1.4. Metric entropy for convex bodies. If the metric space (M, d) is actually a normed space with a unit ball B, we write $N(K, B, \varepsilon)$ or $N(K, \varepsilon B)$ instead of $N(K, d, \varepsilon)$. It is possible to come up with an alternative definition which does not refer to the norm, by saying that $N(K, B, \varepsilon)$ is the minimum number N such that there exist $x_1, \ldots, x_N$ in K with

$$K \subset \bigcup_{i=1}^{N} (x_i + \varepsilon B). \tag{5.18}$$

This alternative definition does not require the set B to be symmetric, or even convex, or to have nonempty interior, even though that is usually the case. In our context, the minimal reasonable hypothesis appears to be asking that B be *star-shaped* with respect to the origin, i.e., that $tB \subset B$ for $t \in [0, 1]$.

The technology for estimating covering/packing numbers of subsets (particularly convex subsets) of normed spaces is quite well-developed and frequently rather sophisticated. We quote here a simple well-known result that expresses $N(\cdot, \cdot)$ in terms of a "volume ratio".

LEMMA 5.8. *Let L be a symmetric convex body in $\mathbb{R}^n$ and let $K \subset \mathbb{R}^n$ be a Borel set. Then, for any $\varepsilon > 0$,*

$$\left(\frac{1}{\varepsilon}\right)^n \frac{\operatorname{vol}(K)}{\operatorname{vol}(L)} \leqslant N(K, L, \varepsilon) \leqslant \left(\frac{2}{\varepsilon}\right)^n \frac{\operatorname{vol}(K + \frac{\varepsilon}{2} L)}{\operatorname{vol}(L)}. \tag{5.19}$$

PROOF. If (x_i) is an ε-net in K with respect to $\|\cdot\|_L$, then the union of the sets $x_i + \varepsilon L$ contains K, and the left-hand side inequality in (5.19) follows from volume comparison. Consider now a family (x_i) of N elements of K which is ε-separated for $\|\cdot\|_L$. This means that the sets $x_i + \frac{\varepsilon}{2} L$ have disjoint interiors. Since they are all included in $K + \frac{\varepsilon}{2} L$, we have $N \operatorname{vol}(\frac{\varepsilon}{2} L) \leqslant \operatorname{vol}(K + \frac{\varepsilon}{2} L)$. Together with (5.1), this implies the right-hand side inequality in (5.19). □

When K is convex and the "regularizing" trick implicit in Exercise 5.17 below is applied, the lower and upper bounds are often as close as one can expect provided K and L are in the M-position (see Notes and Remarks). The case $K = L$ in Lemma 5.8 is related to the approximation of convex bodies by polytopes.

LEMMA 5.9. *Let $0 < \varepsilon < 1$, $K \subset \mathbb{R}^n$ be a symmetric convex body and $\mathcal{N}$ be an ε-net in K with respect to $\|\cdot\|_K$. Then* $\operatorname{conv} \mathcal{N} \supset (1-\varepsilon)K$.

PROOF. Let $P = \operatorname{conv} \mathcal{N}$ and denote $A = \sup\{\|y\|_P \ : \ y \in K\}$. One checks that P contains 0 in the interior, so that $A < \infty$. Given $x \in K$, there is $x' \in \mathcal{N}$ such that $\|x - x'\|_K \leqslant \varepsilon$, and therefore $\|x\|_P \leqslant \|x'\|_P + \|x - x'\|_P \leqslant 1 + \varepsilon A$. Taking supremum over x gives $A \leqslant 1 + \varepsilon A$, so that $A \leqslant (1-\varepsilon)^{-1}$, which is equivalent to the inclusion $P \supset (1-\varepsilon)K$. □

The following is an immediate consequence of Lemmas 5.8 and 5.9.

COROLLARY 5.10. *Let $\varepsilon \in (0,1)$. Any symmetric convex body in $\mathbb{R}^n$ is $(1-\varepsilon)^{-1}$-close, in the Banach–Mazur distance, to a polytope with at most $(1+2/\varepsilon)^n$ vertices.*

For an extension of Lemma 5.9 and 5.10 to not-necessarily-symmetric convex bodies, see Exercises 5.18–5.20. Note that the dependence on ε in Corollary 5.10 is not sharp (see Notes and Remarks). For the special case $K = B_2^n$, the conclusion of Lemma 5.9 can be easily improved to $\operatorname{conv} \mathcal{N} \supset (1 - \varepsilon^2/2)K$; see Exercise 5.7.

EXERCISE 5.16 (Covering with balls whose centers lie outside of the set). For convex bodies K, L in $\mathbb{R}^n$, let $N'(K, L)$ be the smallest number N such that there exist $x_1, \dots, x_N$ in $\mathbb{R}^n$ with $K \subset \bigcup_{1 \leqslant i \leqslant N} (x_i + L)$ (the difference with $N(K, L)$ is that x_i are not required to belong to K). Give an example with L symmetric for which $N'(K, L) < N(K, L)$. Can we have such an example with also K symmetric?

EXERCISE 5.17 (A regularizing trick). Let K, L be convex bodies in $\mathbb{R}^n$, with $0 \in L$. Show that $N(K, \varepsilon L) = N(K, (K-K) \cap \varepsilon L)$.

EXERCISE 5.18 (Approximating by polytopes with few vertices). Let $K \subset \mathbb{R}^n$ be a convex body with the centroid at the origin (K is not assumed to be symmetric). Using Lemma 5.8 and Proposition 4.18, show that for every $\varepsilon \in (0,1)$ we have $N(K, \varepsilon K) \leqslant (2 + 4/\varepsilon)^n$, where $N(K, \varepsilon K) = N(K, K, \varepsilon)$ is defined as in (5.18). By arguing as in the proof of Lemma 5.9, conclude that there exists a polytope P with at most $(2 + 4/\varepsilon)^n$ vertices such that $(1-\varepsilon)K \subset P \subset K$.

EXERCISE 5.19 (Approximating by polytopes with few facets). Let $\varepsilon \in (0,1)$ and $K \subset \mathbb{R}^n$ be a convex body with centroid at the origin. Show that there exists a polytope Q with at most $(2 + 4/\varepsilon)^n$ facets such that $(1-\varepsilon)Q \subset K \subset Q$.

EXERCISE 5.20 (Approximating by polytopes and the Santaló inequality). Let K be a convex body in $\mathbb{R}^n$ and let $\kappa = \operatorname{vrad}(K) \operatorname{vrad}(K^\circ) < \infty$ (i.e., K satisfies approximately the Santaló inequality, see Theorem 4.17 and the comments following it). If $\varepsilon \in (0,1)$, then K can be approximated up to ε (in the sense of Exercises 5.18 and 5.19) by a polytope P with at most $(C\kappa/\varepsilon)^n$ vertices (resp., facets).

EXERCISE 5.21 (Duality of metric entropy for ellipsoids). Let $\mathscr{E}$ and $\mathscr{F}$ be 0-symmetric ellipsoids in $\mathbb{R}^n$. Check that for every $\varepsilon > 0$, $N(\mathscr{E}, \mathscr{F}, \varepsilon) = N(\mathscr{F}^\circ, \mathscr{E}^\circ, \varepsilon)$.

EXERCISE 5.22 (Explicit nets in S^{n-1}). Here is an explicit construction of an ε-net in S^{n-1} with at most $(C/\varepsilon)^n$ elements, for some (suboptimal) constant C.
(i) Show that, if $\mathcal{N}$ is an ε-net in B_2^n (with $0 < \varepsilon < 1$), then the set $\{x/|x| \ : \ x \in \mathcal{N}\}$ is an η-net in $(S^{n-1}, |\cdot|)$ for $\eta = \sqrt{2 - 2\sqrt{1-\varepsilon^2}}$.
(ii) Let $\mathcal{N} = B_2^n \cap \frac{\varepsilon}{\sqrt{n}} \mathbb{Z}^n$. Show that $\mathcal{N}$ is an ε-net in B_2^n and that $\operatorname{card} \mathcal{N} \leqslant (C/\varepsilon)^n$.

5.1.5. Nets in Grassmann manifolds, orthogonal and unitary groups. We now extend the results given for the sphere to other classical manifolds, including unitary and orthogonal groups and Grassmann manifolds (which are introduced in Appendix B). Metric structures on such manifolds are induced by unitarily invariant norms on the corresponding matrix spaces, with Schatten p-norms being the most popular choices. While there are several natural ways (also discussed in detail in Appendix B) to define a metric on a manifold starting from a given Schatten norm, all such metrics—for a fixed p—differ at most by a multiplicative factor of $\pi/2$. Accordingly, the behavior of covering numbers in all such situations can be subsumed in the following single statement.

THEOREM 5.11 (Not proved here; see Exercise 5.23). *Let M be either $\mathsf{SO}(n)$, $\mathsf{U}(n)$, $\mathsf{SU}(n)$, $\mathsf{Gr}(k,\mathbb{R}^n)$ or $\mathsf{Gr}(k,\mathbb{C}^n)$, equipped with a metric generated by the Schatten norm $\|\cdot\|_p$ for some $1 \leqslant p \leqslant \infty$. Then for any $\varepsilon \in (0, \operatorname{diam} M]$,*

$$\left(\frac{c \operatorname{diam} M}{\varepsilon}\right)^{\dim M} \leqslant N(M,\varepsilon) \leqslant \left(\frac{C \operatorname{diam} M}{\varepsilon}\right)^{\dim M}, \tag{5.20}$$

where $C, c > 0$ are universal constants (independent of n, k, p and ε), $\dim M$ is the real dimension of M, and $\operatorname{diam} M$ the diameter of M with respect to the corresponding metric.

For easy reference, we list in Table 5.1 some of the values of the parameters (dimensions, diameters) that appear in (5.20).

TABLE 5.1. Real dimensions and diameters from the bounds (5.20) for covering numbers of a selection of classical manifolds. The distances used on $\mathsf{SO}(n)$ and $\mathsf{U}(n)$ are the extrinsic metrics obtained from the Schatten p-norm on M_n, and the distances on Grassmann manifolds are the corresponding quotient metrics. The restriction $k \leqslant n/2$ is imposed to reduce clutter (note that $\mathsf{Gr}(k,\mathbb{R}^n)$ and $\mathsf{Gr}(n-k,\mathbb{R}^n)$ are isometric).

M	$\dim M$	$\operatorname{diam} M$	comments
$\mathsf{SO}(n)$	$n(n-1)/2$	$2n^{1/p}$	
$\mathsf{U}(n)$	n^2	$2n^{1/p}$	
$\mathsf{Gr}(k,\mathbb{R}^n)$	$k(n-k)$	$2^{1/2}(2k)^{1/p}$	$k \leqslant n/2$
$\mathsf{Gr}(k,\mathbb{C}^n)$	$2k(n-k)$	$2^{1/2}(2k)^{1/p}$	$k \leqslant n/2$

EXERCISE 5.23 (Metric entropy of classical groups and manifolds). Prove Theorem 5.11 for $M = \mathsf{U}(n)$, $M = \mathsf{SU}(n)$ or $M = \mathsf{SO}(n)$ and for $p = \infty$, by appealing to Lipschitz properties of the exponential map with matrix argument (Exercise B.8).

EXERCISE 5.24. Derive the formula for the diameter of $\mathsf{Gr}(k,\mathbb{R}^n)$ in Table 5.1.

EXERCISE 5.25 (Volume of balls in classical groups and manifolds). Let M be either $\mathsf{SO}(n), \mathsf{U}(n)$, or $\mathsf{Gr}(k,\mathbb{R}^n)$, equipped with a metric as in Theorem 5.11. Denoting by σ the Haar probability measure on M, deduce from Theorem 5.11 a two-sided estimate for $\sigma(B(x,\varepsilon))$, where $B(x,\varepsilon)$ denotes the ball of radius ε centered at $x \in M$.

5.2. Concentration of measure

The classical isoperimetric inequality in $\mathbb{R}^n$ (Eq. (4.27), also known as Dido's problem) states that among all sets of a given volume, the Euclidean balls have the smallest surface area. As we already noticed in the setting of $\mathbb{R}^n$ in Section 4.3.1, an alternative methodology is to consider, instead of the surface area, the family of ε-enlargements of a given set. The latter approach makes sense in any metric space X equipped with a measure μ (a *metric measure space*, or a *metric probability space* if $\mu(X) = 1$, which will be assumed as a default): for a subset $A \subset X$ and $\varepsilon > 0$, we define

$$A_\varepsilon = \{x \in X \ : \ \operatorname{dist}(x, A) \leqslant \varepsilon\}.$$

The two viewpoints are roughly equivalent since the "surface area" relative to μ can be retrieved (when that makes sense) as the first-order variation of $\mu(A_\varepsilon)$ when ε goes to 0, cf. (4.23) and, conversely, the growth of the function $\varepsilon \mapsto \mu(A_\varepsilon)$ on the macroscopic scale can be recovered from the knowledge of its derivative. However, the enlargement-based approach seems simpler (a more flexible definition) and is often more fruitful since some otherwise useful bounds on $\mu(A_\varepsilon)$ may be meaningless for small ε, and/or may be available in the absence of any clue with regard to the nature of extremal sets.

Lower bounds for $\mu(A_\varepsilon)$ can be rephrased as deviation inequalities for Lipschitz functions. This leads, in some settings, to a remarkable phenomenon: every Lipschitz function concentrates strongly around some "central value". Statements to such and similar effect will be the focus of our presentation. Specifically, we will look for estimates of the form

$$\mu(f > M_f + t) \leqslant Ce^{-\lambda t^2} \tag{5.21}$$

and

$$\mu(f > \mathbf{E}f + t) \leqslant Ce^{-\lambda t^2}, \tag{5.22}$$

to be valid for any real-valued 1-Lipschitz function on X and all $t > 0$, where M_f and $\mathbf{E}f$ are the *median* and the *expected value* of f calculated with respect to μ. (A number M is said to be a median for a random variable X if $\mathbf{P}(X \geqslant M) \geqslant 1/2$ and $\mathbf{P}(X \leqslant M) \geqslant 1/2$.) Clearly, (5.21) and (5.22) formally imply then similar *two-sided* estimates for $\mu(|f - M_f| > t)$ and $\mu(|f - \mathbf{E}f| > t)$ with C replaced by $2C$. Concentration of this type is referred to as *subgaussian* (more on this terminology in Section 5.2.6). For the convenience of a casual reader—and for easy reference—we list in Table 5.2 the constants and the exponents that appear in subgaussian concentration inequalities for a selection of classical objects.

REMARK 5.12. We point out that if a function f is such that one of the inequalities (5.21) or (5.22) holds (for all $t > 0$) with constants C, λ, then the other inequality similarly holds (for the same function) with some other constants. For example, if (5.22) holds with $C \geqslant \frac{1}{2}$ and λ, then (5.21) holds with $2C^2$ and $\lambda/2$; if (5.21) holds with $C \geqslant e^{-1/3} \approx 0.717$ and λ, then (5.21) holds with eC^2 and $\lambda/2$ (see Proposition 5.29 and Remarks 5.30, 5.31.) Sharper results of this nature (i.e., with better dependence on C, λ) can sometimes be obtained if we assume that (5.21) (or (5.22)) holds for *all* real-valued 1-Lipschitz functions on X; some questions in that spirit are considered in [**Led01**] (see, e.g., Exercise 5.48).

TABLE 5.2. Constants and exponents in subgaussian concentration inequalities for a selection of classical objects. When applicable, the reference measure is the canonical invariant measure on the object in question. We made an effort to come up with reasonable values of constants/exponents, and some of them are optimal. Unless indicated otherwise, the metric used for manifolds is the Riemannian geodesic distance. d_H stands for the normalized Hamming distance (5.12). References: (a) Theorem 5.24. (b) The log-Sobolev inequality (LSI); see Table 5.4. (c) Corollary 5.17. (d) Proposition 5.20; what follows from the LSI is $\lambda = \frac{n-1}{2}$. (e) Ricci curvature; see Table 5.3. (f) Remark 5.12. (g) Corollary 5.52. (h) The LSI on the discrete cube; see Theorem 5.1 and Exercise 5.5 in [**BLM13**]. (i) Theorem 5.54; convex or concave functions only. (j) The constant in the exponent is $\frac{1}{8}$ and not $\frac{1}{2}$ due to rescaling ($\{-1,1\}$ vs. $\{0,1\}$). (k) Theorem 5.56; convex functions only. (l) Theorem 5.38. (m) Theorem 5.39. (n) $(C,\lambda) = (2,\frac{1}{4})$ if $n = 2$; see Remark 5.12. (o) Exercise 5.54. (p) Remark 5.19. (q) If we use instead the non-Riemannian metric (B.11), the parameter λ needs to be multiplied by 2 in view of (B.12). (r) Remark 5.53.

Object	C,λ in (5.21)–median	C,λ in (5.22)–mean	Comments
Gauss space $(\mathbb{R}^n, \lvert\cdot\rvert, \gamma_n)$	$\frac{1}{2}, \frac{1}{2}$ (a)	$1, \frac{1}{2}$ (b)	
Gauss space $(\mathbb{C}^n, \lvert\cdot\rvert, \gamma_n^{\mathbb{C}})$	$\frac{1}{2}, 1$ (a)	$1, 1$ (b)	
(S^{n-1}, g) or $(S^{n-1}, \lvert\cdot\rvert)$	$\frac{1}{2}, \frac{n}{2}$ (c)	$1, \frac{n}{2}$ (d)	$n > 2$ for (S^{n-1}, g) (p)
$\mathsf{SO}(n)$	$\frac{1}{2}, \frac{n-1}{8}$ (e)	$1, \frac{n-1}{8}$ (b)	metric (B.8)
$\mathsf{SU}(n)$	$\frac{1}{2}, \frac{n}{4}$ (e)	$1, \frac{n}{4}$ (b)	metric (B.8)
$\mathsf{U}(n)$	$2, \frac{n}{24}$ (f)	$1, \frac{n}{12}$ (b)	metric (B.8)
$\mathsf{Gr}(k, \mathbb{R}^n)$	$\frac{1}{2}, \frac{n-2}{4}$ (e)	$1, \frac{n-2}{4}$ (b)	metric (B.10) (q)
$\mathsf{Gr}(k, \mathbb{C}^n)$	$\frac{1}{2}, \frac{n}{2}$ (e)	$1, \frac{n}{2}$ (b)	metric (B.10) (q)
$(\{-1,1\}^n, d_H)$	$1, 2n$ (g)	$1, 2n$ (h)	$n \geqslant 3$ (r)
$(\{-1,1\}^n, \lvert\cdot\rvert)$	$2, \frac{1}{8}$ (i)(j)	$1, \frac{1}{8}$ (k)(j)	appropriate convexity hypotheses
Ricci curvature $\geqslant c$	$\frac{1}{2}, \frac{c}{2}$ (l)	$1, \frac{c}{2}$ (b)	
LSI with constant $\leqslant \alpha$	$2, \frac{1}{4\alpha}$ (f)	$1, \frac{1}{2\alpha}$ (m)	
$(S^{n-1})^k$	$\frac{1}{2}, \frac{n-2}{2}$ (e)(n)	$1, \frac{n-1}{2}$ (b)	ℓ_2 product metric
$[0,1]^k$	$\frac{1}{2}, \pi$ (o)	$1, \frac{\pi^2}{2}$ (b)	ℓ_2 product metric

In the next two subsections we will exemplify the concentration phenomenon and related techniques in the case of the Euclidean sphere and the Gaussian space. In subsequent subsections we will survey some general methods for proving isoperimetric/concentration results and present a selection of examples, in particular those listed in Table 5.2. We will concentrate on the objects that exhibit subgaussian concentration; more general settings will be addressed briefly in exercises and in Notes and Remarks (an exception is Section 5.2.6, which treats sums of independent subexponential random variables). A comprehensive presentation of diverse aspects and manifestations of the concentration phenomenon is beyond the scope of this work; we refer the interested reader to the monographs [**Led01**, **BLM13**] and/or to other sources listed in Notes and Remarks. Here we restrict our attention to highlighting several central techniques and, subsequently, to going over examples that appear to be of relevance to the quantum theory.

5.2.1. A prime example: concentration on the sphere. The settings of the Euclidean sphere and of the projective space are directly relevant to quantum information theory since the latter identifies canonically with the set of pure states. In the language of enlargements, the isoperimetric inequality on the sphere can be stated as follows.

THEOREM 5.13 (Spherical isoperimetric inequality, not proved here). *Equip the unit sphere $S^{n-1} \subset \mathbb{R}^n$ with the geodesic distance g and the uniform probability measure σ. If $A \subset S^{n-1}$ and if $C \subset S^{n-1}$ is a spherical cap such that $\sigma(A) = \sigma(C)$, then, for any $\varepsilon > 0$,*

$$\sigma(A_\varepsilon) \geqslant \sigma(C_\varepsilon). \tag{5.23}$$

Recall that the spherical cap with center $x \in S^{n-1}$ and radius ε is the set

$$C(x, \varepsilon) = \{y \in S^{n-1} \ : \ g(x, y) \leqslant \varepsilon\}.$$

Note that the class of spherical caps is stable under enlargements and that we have

$$C(x, \varepsilon)_\delta = C(x, \varepsilon + \delta) \quad \text{for any} \quad \delta, \varepsilon > 0. \tag{5.24}$$

In view of the simple relationship between g and the extrinsic (or chordal) distance inherited from the ambient Euclidean space (see Appendix B.1), Theorem 5.13 is valid also for the latter. However, it is traditionally stated for the geodesic distance. Also, the formula (5.24) for $C(x, \varepsilon)_\delta$ stated above would be more complicated if we used $|\cdot|$ to define caps.

The usefulness of Theorem 5.13 comes from the fact that there are explicit integral formulas and sharp bounds for the measure of spherical caps, which were explored in Section 5.1.2. However, while small caps seemed more interesting in the study of packing and covering, in the present context of concentration the radii close to $\pi/2$ are most relevant. This is because arguably the most useful instance of Theorem 5.13 is $\sigma(A) = \frac{1}{2}$, in which case the radius of the corresponding cap C is $\pi/2$ and the radius of its ε-enlargement, C_ε, is $\pi/2 + \varepsilon$. Taking into account the bound (5.5) leads then to

COROLLARY 5.14. *If $n > 2$ and if $A \subset S^{n-1}$ with $\sigma(A) \geqslant \frac{1}{2}$ and $\varepsilon > 0$, then*

$$\sigma(A_\varepsilon) \geqslant \sigma\left(C\left(x, \frac{\pi}{2} + \varepsilon\right)\right) \geqslant 1 - \frac{1}{2} e^{-n\varepsilon^2/2}. \tag{5.25}$$

There is no simple proof of the isoperimetric inequality on the sphere (Theorem 5.13) that we know of. However, a result just slightly weaker than Corollary 5.14 follows easily from the Brunn–Minkowski inequality (4.21). We have the following.

PROPOSITION 5.15. *If* $\varepsilon \in (0, \pi/2]$ *and* $K, L \subset S^{n-1}$ *are such that* $\operatorname{dist}(K, L) \geqslant \varepsilon$ *(in the geodesic distance), then* $\sigma(K)\sigma(L) \leqslant e^{-n\varepsilon^2/4}$. *In particular, if* $\sigma(K) \geqslant 1/2$, *then* $\sigma(K_\varepsilon) \geqslant 1 - 2e^{-n\varepsilon^2/4}$.

PROOF. The second statement follows by applying the first one with $L = K_\varepsilon^c$. It thus remains to prove the first statement.

Define $K' \subset B_2^n$ via $K' := \{tx : x \in K, t \in [0,1]\}$ and similarly for L'. Then $\operatorname{vol}(K') = \sigma(K)\operatorname{vol}(B_2^n)$ and $\operatorname{vol}(L') = \sigma(L)\operatorname{vol}(B_2^n)$. Consequently, by the Brunn–Minkowski inequality in the form (4.21),

$$\operatorname{vol}\left(\frac{K' + L'}{2}\right) \geqslant \sqrt{\operatorname{vol}(K')\operatorname{vol}(L')} = \sqrt{\sigma(K)\sigma(L)}\,\operatorname{vol}(B_2^n).$$

On the other hand, if $x, y \in S^{n-1}$ and the angle between x and y is at least ε, then $|(x+y)/2| \leqslant \cos(\varepsilon/2)$. If $\varepsilon \leqslant \pi/2$ (and so $\langle x, y\rangle \geqslant 0$), a simple calculation shows that the same is true if we replace x and y by $x' = sx$ and $y' = ty$, where $s, t \in [0,1]$ (in fact this is even true if $\varepsilon \leqslant 2\pi/3$). This means that we have then $\frac{K'+L'}{2} \subset \cos(\varepsilon/2)B_2^n$, and so $\sqrt{\sigma(K)\sigma(L)} \leqslant (\cos(\varepsilon/2))^n$. It remains to appeal to the (subtle but elementary) inequality $\cos u \leqslant e^{-u^2/2}$ (see Exercise 5.3). □

REMARK 5.16.
(1) Proposition 5.15 holds actually for the entire nontrivial range of ε, which is $[0, \pi]$; this follows *a posteriori* from the estimate in Lévy's lemma (see Exercise 5.26). The above proof fails for large ε; however, only the range $[0, \pi/2]$ is relevant to the second statement and to Corollary 5.14: if $\mu(K) \geqslant 1/2$, then no point x can verify $\operatorname{dist}(x, K) > \pi/2$.
(2) The estimate in the Proposition is fairly tight: if K, L are opposite (i.e., $K = -L$) caps with $\operatorname{dist}(K, L) = 2\varepsilon$, we conclude from the Proposition that $\mu(K) \leqslant e^{-n\varepsilon^2/2}$. This compares fairly well with the bound $\frac{1}{2}e^{-n\varepsilon^2/2}$ implicit in (5.25).

Corollary 5.14 readily implies a concentration result for Lipschitz functions, which is often referred to in quantum information circles as Lévy's lemma.

COROLLARY 5.17 (Lévy's lemma). *Let* $n > 2$. *If* $f : (S^{n-1}, g) \to \mathbb{R}$ *is a L-Lipschitz function and if* M_f *is a median for* f, *then, for any* $t > 0$,

$$\sigma(f > M_f + t) \leqslant \frac{1}{2}\exp(-nt^2/2L^2), \tag{5.26}$$

and therefore

$$\sigma(|f - M_f| > t) \leqslant \exp(-nt^2/2L^2). \tag{5.27}$$

PROOF. Let $A = \{x \in S^{n-1} : f(x) \leqslant M_f\}$ and set $\varepsilon = t/L$. Since $f \leqslant M_f$ on A and since f is L-Lipschitz (i.e., $|f(x) - f(y)| \leqslant Lg(x, y)$ for $x, y \in S^{n-1}$), it follows that for any $y \in S^{n-1}$ we have $f(y) \leqslant M_f + Lg(y, A)$. In particular, if $y \in A_\varepsilon$, then $g(y, A) \leqslant \varepsilon$ and so $f(y) \leqslant M_f + L\varepsilon = M_f + t$. In other words, we proved that $A_\varepsilon \subset \{f \leqslant M_f + t\} = \{f > M_f + t\}^c$. The first inequality in Corollary 5.17 follows now by observing that, by the definition of the median, $\sigma(A) \geqslant \frac{1}{2}$ and by appealing to Corollary 5.14.

The second inequality follows from the first one combined with an identical bound on $\sigma(f < M_f - t)$, which is shown either by the same argument applied to $A = \{x \in S^{n-1} : f(x) \geqslant M_f\}$, or by appealing to the first inequality with f replaced by $-f$. $\square$

REMARK 5.18. Both parts of the above proof are quite general. First, any lower bounds on measures of enlargements of sets of measure $\frac{1}{2}$ imply (in fact are equivalent to, see Exercise 5.27) bounds for deviation of Lipschitz function from their medians. Second, any one-sided bound for deviation from the median (or the expected value, or any other "symmetric" parameter) implies a two-sided bound, at the cost of a factor of 2.

REMARK 5.19. In Corollaries 5.14 and 5.17 we have to assume that $n > 2$ because the bound (5.5) is not valid in the entire nontrivial range $0 \leqslant t \leqslant \pi/2$. If $n = 2$, one needs to replace the function $\frac{1}{2}e^{-nt^2/2}$ by $\max\{\frac{1}{2} - \frac{t}{\pi}, 0\}$. However, no modifications are needed if the enlargements or the Lipschitz constants are calculated with respect to the ambient space metric, or if only small values of ε or t are of interest, say, $\varepsilon \leqslant 1$ or $t \leqslant L$.

Concentration around the median follows naturally from the isoperimetric inequality. As we mentioned in Remark 5.12, this formally implies concentration around the expectation with altered constants. In some situations, it is possible to obtain good constants with extra work.

PROPOSITION 5.20 (Lévy's lemma for the mean; not proved here). *Let $n > 2$. If $f : (S^{n-1}, g) \to \mathbb{R}$ is a 1-Lipschitz function, then for any $t > 0$,*

$$\sigma(f > \mathbf{E}f + t) \leqslant \exp(-nt^2/2). \tag{5.28}$$

As mentioned in Remark 5.18, the inequality $\sigma(|f - \mathbf{E}f| > t) \leqslant 2\exp(-nt^2/2)$ follows formally, but is probably not optimal. See Problem 5.26 for questions about possible better bounds in this and similar settings.

EXERCISE 5.26 (Proposition 5.15 holds for the full range of ε). Show that it follows *a posteriori* from Theorem 5.13 and the bound (5.5) that, for $n > 2$, in the notation and under the hypotheses of Proposition 5.15, we have $\sigma(K)\,\sigma(L) \leqslant \frac{1}{4}e^{-n\varepsilon^2/4}$. For $n = 2$, the optimal inequality is $\sigma(K)\,\sigma(L) \leqslant \frac{1}{4}\big(1 - \frac{\varepsilon}{\pi}\big)^2$ (cf. Remark 5.19).

EXERCISE 5.27 (Concentration implies isoperimetry). Show that, for a metric probability space (X, μ), concentration implies isoperimetry in the following sense: if $\mu(f > M_f + t) \leqslant \alpha$ for any 1-Lipschitz function f, then $\mu(A_t) \geqslant 1 - \alpha$ for any $A \subset X$ with $\mu(A) = \frac{1}{2}$.

EXERCISE 5.28 (A finer bound for the mean width of a union). Let K, L be two bounded sets in $\mathbb{R}^n$, and R the outradius of $K \cup L$. Show that $w(\mathrm{conv}(K \cup L)) \leqslant \max(w(K), w(L)) + \sqrt{\frac{2\pi}{n}}\, R$.

5.2.2. Gaussian concentration. Another classical setting where isoperimetry and concentration have been widely studied is the Gaussian space $(\mathbb{R}^n, |\cdot|, \gamma_n)$, where γ_n is the standard Gaussian measure on $\mathbb{R}^n$ (see Appendix A.2 for the notation, basic properties and relevant facts). It turns out that the extremal sets for the isoperimetric problem are then half-spaces, and since their enlargements are also half-spaces, the solution to the problem can be expressed simply in terms of the cumulative distribution function of an $N(0,1)$ variable, i.e., in terms of $\Phi(x) := \gamma_1((-\infty, x])$. We have

THEOREM 5.21 (Gaussian isoperimetric inequality; see Exercise 5.30). *Let $A \subset \mathbb{R}^n$, and let $a \in \mathbb{R}$ be defined by $\gamma_1\big((-\infty,a]\big) = \gamma_n(A)$. Then, for any $\varepsilon > 0$,*

$$\gamma_n(A_\varepsilon) \geqslant \gamma_1\big((-\infty, a+\varepsilon]\big) \tag{5.29}$$

or, equivalently,

$$\Phi^{-1}(\gamma_n(A_\varepsilon)) \geqslant \Phi^{-1}(\gamma_n(A)) + \varepsilon. \tag{5.30}$$

The solution to the Gaussian isoperimetric problem (Theorem 5.21) was originally derived from the spherical isoperimetric inequality (Theorem 5.13) via the following classical fact.

THEOREM 5.22 (Poincaré's lemma; see Exercise 5.29). *For $n, N \in \mathbb{N}$ with $N \geqslant n$, we consider $\mathbb{R}^n$ to be a subspace of $\mathbb{R}^N$. Next, fix n and let ν_N be the pushforward to $\mathbb{R}^n$, via the orthogonal projection, of the normalized uniform measure on $\sqrt{N}S^{N-1}$. Then, as $N \to \infty$, (ν_N) converges to γ_n, the standard Gaussian measure on $\mathbb{R}^n$.*

The convergence in Theorem 5.22 holds in a very strong sense, e.g., in total variation, or in uniform convergence of densities.

Another derivation of the Gaussian isoperimetric inequality is based on the following analogue of the Brunn–Minkowski inequality in the Gaussian setting.

THEOREM 5.23 (Ehrhard's inequality, not proved here). *Let A, B be Borel subsets of $\mathbb{R}^n$ and let $\lambda \in [0,1]$. Then*

$$\Phi^{-1}(\gamma_n((1-\lambda)A + \lambda B)) \geqslant (1-\lambda)\Phi^{-1}(\gamma_n(A)) + \lambda\Phi^{-1}(\gamma_n(B)). \tag{5.31}$$

Ehrhard's inequality is stronger than log-concavity of the Gaussian measure (Section 4.3.2), see Exercise 5.31. Assuming Ehrhard's inequality, the derivation of the Gaussian isoperimetric inequality goes as follows. Fix A, ε and let $\lambda \in (0,1)$. Since $A_\varepsilon = A + \varepsilon B_2^n = (1-\lambda)(1-\lambda)^{-1}A + \lambda\varepsilon\lambda^{-1}B_2^n$, we have, by (5.31),

$$\Phi^{-1}(\gamma_n(A_\varepsilon)) \geqslant (1-\lambda)\Phi^{-1}(\gamma_n((1-\lambda)^{-1}A)) + \lambda\Phi^{-1}(\gamma_n(\varepsilon\lambda^{-1}B_2^n)). \tag{5.32}$$

We now let $\lambda \to 0^+$. The first term on the right-hand side of (5.32) converges clearly to $\Phi^{-1}(\gamma_n(A))$, while the second term converges to ε (this is a little harder, but elementary, see Exercise 5.32), and so we proved the Gaussian isoperimetric inequality in the form (5.30).

The next theorem follows from Theorem 5.21 according to the general scheme indicated in Remark 5.18, with the explicit exponential bound being a consequence of Exercise A.1.

THEOREM 5.24. *If $f : \mathbb{R}^n \to \mathbb{R}$ is L-Lipschitz and M_f denotes its median (with respect to γ_n), then for any $t > 0$*

$$\gamma_n(f > M_f + t) \leqslant \gamma_1\big((t/L, \infty)\big) \leqslant \frac{1}{2}e^{-t^2/2L^2}, \tag{5.33}$$

$$\gamma_n(|f - M_f| > t) \leqslant e^{-t^2/2L^2}.$$

As we already noted in the setting of the sphere, concentration around the median formally implies similar concentration around the mean (see Remark 5.12). However, this approach leads to suboptimal constants. A more precise technique relies on the log-Sobolev inequality from Section 5.2.4.2, which specified to the Gaussian setting yields the following.

THEOREM 5.25 (See Theorem 5.39 and Proposition 5.42). *If $f : \mathbb{R}^n \to \mathbb{R}$ is L-Lipschitz and $\mathbf{E}f$ is the mean of f (with respect to γ_n), then for any $t > 0$*

$$\max\{\gamma_n(f > \mathbf{E}f + t), \gamma_n(f < \mathbf{E}f - t)\} \leqslant e^{-t^2/2L^2}. \tag{5.34}$$

There is some numerical evidence that the assertion of Theorem 5.25 can be further strengthened. We pose

PROBLEM 5.26. *If $f : \mathbb{R}^n \to \mathbb{R}$ is 1-Lipschitz and $\mathbf{E}f$ denotes its average with respect to γ_n, is it true that $\gamma_n(|f - \mathbf{E}f| > t) \leqslant e^{-t^2/2}$? The case $n = 1$ implies the general case and is probably not that hard to settle. Similarly, is it true that $\sigma(|f - \mathbf{E}f| > t) \leqslant \exp(-nt^2/2)$ if $f : (S^{n-1}, g) \to \mathbb{R}$ is a 1-Lipschitz function (and $n > 2$; see Remark 5.19 for comments on peculiarities of the case $n = 2$)?*

An example of a function for which Theorem 5.24 is meaningful is the Euclidean norm, which is trivially 1-Lipschitz. This gives the following (see also Exercise 5.37).

COROLLARY 5.27. *Let G be a standard Gaussian vector in $\mathbb{R}^n$. Then, for any $t > 0$,*

$$\mathbf{P}\big(|G| \geqslant \sqrt{n} + t\big) \leqslant \frac{1}{2}e^{-t^2/2} \quad \text{and} \quad \mathbf{P}\Big(|G| \leqslant \sqrt{n - \frac{2}{3}} - t\Big) \leqslant \frac{1}{2}e^{-t^2/2}.$$

The distribution of $|G|^2$ is commonly known as $\chi^2(n)$, the chi-squared distribution with n degrees of freedom. Denoting by m_n the median of $|G|$, what is required to deduce Corollary 5.27 from Theorem 5.24 are the inequalities $\sqrt{n - \frac{2}{3}} \leqslant m_n \leqslant \sqrt{n}$. The lower bound is proved in Exercise 5.34 and the upper bound follows from Proposition 5.34): we have $m_n \leqslant \kappa_n \leqslant \sqrt{n}$.

EXERCISE 5.29 (Weak convergence in Poincaré's lemma). In the context of Poincaré's lemma (Theorem 5.22), show without any computation that the sequence (ν_N) converges weakly towards γ_n.

EXERCISE 5.30 (Gaussian isoperimetric inequality via Poincaré lemma). Derive the Gaussian isoperimetric inequality (5.29) from the Poincaré lemma (Theorem 5.22) and the spherical isoperimetric inequality (Theorem 5.13).

EXERCISE 5.31 (Ehrhard's inequality implies log-concavity). Show that Theorem 5.23 (Ehrhard's inequality) formally implies that the Gaussian measure γ_n satisfies the log-concavity inequality (4.28).

EXERCISE 5.32 (Gaussian measure of large balls). Show that

$$\lim_{r \to +\infty} \frac{\Phi^{-1}\big(\gamma_n(rB_2^n)\big)}{r} = 1.$$

EXERCISE 5.33 (Ehrhard-like (a-)symmetrization). Show that the following statement is equivalent to the validity of Ehrhard's inequality for convex bodies. *Let $K \subset \mathbb{R}^n$ be a convex body and let $E \subset \mathbb{R}^n$ be a k-dimensional subspace with*

$0 < k < n$. *Identify* E *and* $E^\perp$ *with, respectively,* $\mathbb{R}^k$ *and* $\mathbb{R}^{n-k}$ *and define a set* $L \subset \mathbb{R}^{k+1}$ *by*

$$(x, s) \in L \iff s \leqslant \Phi^{-1}(\gamma_{n-k}(\{y \in E^\perp : (x, y) \in K\})),$$

where $x \in E, s \in \mathbb{R}$. *Then* L *is convex.*
In the case when $E = u^\perp$ is a hyperplane (i.e., $k = n - 1$) the transformation $K \mapsto L$ is called Ehrhard (a-)symmetrization in direction u.

EXERCISE 5.34 (Median of the chi-squared distribution, based on [**CR86**]). Let X be a random variable with distribution $\chi^2(n)$, and $V = \left(\frac{X}{n-2/3}\right)^{1/3}$. Show that the density h of V satisfies the inequality $h(1-t) \leqslant h(1+t)$ for $t \in [0,1]$, and conclude that the median of V is greater than 1, therefore the median of X is larger than $n - 2/3$. Higher order two-sided bounds for the median can be found in [**BS**].

5.2.3. Concentration tricks and treats. This section contains a selection of largely elementary facts related to the concentration phenomenon. It supplies a set of tools allowing for flexible applications of concentration results. As a rule, the facts are well known to experts in the area and are included here for future reference. Proofs are relegated to exercises.

5.2.3.1. *Laplace transform.* We mostly restrict ourselves to settings where concentration exhibits a subgaussian behaviour as in (5.21) or (5.22). Such behaviour can be proved via estimating the bilateral Laplace transform, using the *exponential Markov inequality* $\mathbf{P}(X > t) \leqslant e^{-st}\, \mathbf{E} \exp(sX)$ for $s > 0$.

LEMMA 5.28 (Laplace transform method). *Let* X *be a random variable such that* $\mathbf{E}\exp(sX) \leqslant A\exp(\beta s^2)$ *for every* $s \in \mathbb{R}$. *Then, for every* $t > 0$,

$$\max(\mathbf{P}(X > t), \mathbf{P}(-X > t)) \leqslant A\exp(-t^2/4\beta).$$

EXERCISE 5.35. Prove Lemma 5.28 about the Laplace transform method.

EXERCISE 5.36. Prove Hoeffding's lemma: if X is a mean zero random variable taking values in an interval $[a, b]$, then $\mathbf{E}\exp(sX) \leqslant \exp(\frac{1}{8}s^2(b-a)^2)$ for any $s \in \mathbb{R}$.

EXERCISE 5.37 (A large deviation bound for chi-squared variable, based on [**Vem04**]). Let X be a random variable with distribution $\chi^2(n)$, for example $X = |G|^2$ where G is a standard Gaussian vector in $\mathbb{R}^n$. Show that $\mathbf{E}\exp(sX) = (1 - 2s)^{-n/2}$ for any $s < 1/2$. Conclude that $\mathbf{P}(X \geqslant (1+\varepsilon)n) \leqslant \big((1+\varepsilon)\exp(-\varepsilon)\big)^{n/2}$ for any $\varepsilon > 0$ and that $\mathbf{P}(X \leqslant (1-\varepsilon)n) \leqslant \big((1-\varepsilon)\exp(\varepsilon)\big)^{n/2}$ for $\varepsilon \in (0, 1]$. (We known from Cramér's large deviations theorem that these bounds are sharp.) Conclude that

$$\mathbf{P}(|X - n| \geqslant \varepsilon n) \leqslant 2\exp\left(-\frac{n\varepsilon^2}{4 + 8\varepsilon/3}\right). \tag{5.35}$$

5.2.3.2. *Central values.* Once we know that a function is concentrated around some value, we can *a posteriori* infer that it also concentrates around the mean or the median, or any other particular quantile. This can be formalized by the concept of a central value. If Y is a real random variable, we will say that M is a *central value* of Y if M is either the mean of Y, or any number between the 1st and the 3rd quartile of Y (i.e., if $\min\{\mathbf{P}(Y \geqslant M), \mathbf{P}(Y \leqslant M)\} \geqslant \frac{1}{4}$; this happens in particular if M is the median of Y). The numbers $\frac{1}{4}$ and $\frac{3}{4}$ play no special role and can be changed to other numbers from $(0,1)$ at the cost of deteriorating (or improving) the constants in the statements that follow (See, e.g., Remark 5.31).

PROPOSITION 5.29 (See Exercises 5.38–5.40). *Let Y be a real random variable and let M be any central value for Y. Let $a \in \mathbb{R}$ and let constants $A \geqslant \frac{1}{2}, \lambda > 0$ be such that, for any $t > 0$,*

$$\max\{\mathbf{P}(Y > a+t), \mathbf{P}(Y < a-t)\} \leqslant A\exp(-\lambda t^2). \tag{5.36}$$

Then $|M-a| \leqslant \sqrt{\log(4A)}\,\lambda^{-1/2}$. Consequently, for any $t \geqslant \sqrt{\log(4A)}\,\lambda^{-1/2}$,

$$\max\{\mathbf{P}(Y > M+t), \mathbf{P}(Y < M-t)\} \leqslant 4A^2\exp(-\lambda t^2/2). \tag{5.37}$$

REMARK 5.30 (Improvements to Proposition 5.29). The expressions $\sqrt{\log(4A)}$ and $4A^2$ in the assertion of Proposition 5.29 can be replaced by $\sqrt{\log(\kappa A)}$ and κA^2, where $\kappa = 2$ when M is the median of Y and $\kappa = e$ when M is the expectation of Y; see Exercises 5.38, 5.39 and 5.40.

REMARK 5.31 (On the necessity of restrictions on t in Proposition 5.29). We point out that the bound on the first (resp., the second) probability appearing in (5.37) is valid under the formally weaker restriction $t > (M-a)^+$ (resp., $t > (M-a)^-$). The restriction $t \geqslant \sqrt{\log(4A)}\,\lambda^{-1/2}$, while annoying, cannot be completely avoided if we want to keep full generality because the hypothesis (5.36) does not necessarily supply any information about the probabilities appearing in the assertion if t is small. However, this is only a minor inconvenience since for such t the upper bound in (5.37) is never small and often holds for trivial reasons. In particular, (5.37) holds for all $t > 0$ if M is the mean or any quantile between the 27th and 73rd percentile, or if $A \geqslant 3^{2/3}/4 \approx 0.52$, and always if we replace the factor $4A^2$ by $3\sqrt{2}A^2$. If M is the median, we can go even further: no restrictions on t are needed even if we replace $4A^2$ by $2A^2$ on the right-hand side of (5.37); if M is the mean, similar improvement (i.e., eA^2 on the right-hand side) is possible when $A \geqslant e^{-1/3} \approx 0.717$ (these last observations were used in Remark 5.12).

COROLLARY 5.32 (Lévy's lemma for central values). *Let $f : (S^{n-1}, g) \to \mathbb{R}$ be an L-Lipschitz function and let M be any central value for f. Then $|M - M_f| \leqslant \sqrt{2\log 2}\, n^{-1/2}$ and, for any $\varepsilon > 0$,*

$$\mathbf{P}(f \geqslant M+\varepsilon) \leqslant \exp\Big(-\frac{n\varepsilon^2}{4L^2}\Big). \tag{5.38}$$

We sketch proofs and give more precise bounds and/or variations on the above results in Exercises 5.38–5.48. Note that while (5.38) follows from Proposition 5.29 and Corollary 5.17 for $n > 2$ and for ε not-too-small, a separate argument is needed to cover the remaining cases (cf. Remark 5.31). We also point out that while Proposition 5.29 is meant to give reasonably good estimates valid in the most general setting when concentration is present, better bounds are available in specific instances. For example, Corollary 5.32 can be improved when M is the mean (see Table 5.2 and Exercise 5.44), and similarly in the Gaussian case.

The heuristics behind Corollary 5.32 are as follows: *if we know that all sets of measure at least $\frac{1}{2}$ have large enlargements, then approximately the same is true for all sets of measure at least $\frac{1}{4}$.* Actually, almost the same is true for much smaller sets; here is a sample result.

PROPOSITION 5.33 (See Exercise 5.49). *Let (X, d, μ) be a metric probability space and let $\varepsilon > 0$. Suppose that any set $A \subset X$ with $\mu(A) \geqslant \frac{1}{2}$ verifies $\mu(A_\varepsilon) \geqslant 1 - Ce^{-\lambda\varepsilon^2}$. Then $\mu(B_{2\varepsilon}) \geqslant 1 - Ce^{-\lambda\varepsilon^2}$ for any set $B \subset X$ with $\mu(B) \geqslant Ce^{-\lambda\varepsilon^2}$.*

A common feature of concentration inequalities presented up to now is that in order to translate them to concrete bounds for concrete functions, we need to calculate—or at least reasonably estimate—the medians or expected values, or similar parameters of the functions under consideration. A selection of tools, some of them quite sharp, to handle expected values will be described in Section 6.1. The preceding three results tell us that it doesn't really matter which central value we employ, as long as we are willing to pay a small penalty in the form of an additional multiplicative constant in the exponent and in front of the exponential. The following observation shows that, in the Gaussian context, sometimes no penalty is needed at all.

PROPOSITION 5.34 (See Exercise 5.50). *Let $f : \mathbb{R}^n \to \mathbb{R}$ be a convex function. Denote by M_f (resp., $\mathbf{E}f$) the median (resp., the expectation) of f with respect to the standard Gaussian measure γ_n. Then $M_f \leqslant \mathbf{E}f$.*

EXERCISE 5.38. Show that a random variable Y_0 such that $P(Y_0 > t) \leqslant A\exp(-t^2)$ for $t > 0$ must verify $\mathbf{E}\,Y_0 \leqslant \mathbf{E}\,Y_0^+ \leqslant \min\{A\sqrt{\pi}/2, \sqrt{1 + \log^+ A}\}$. Deduce the first assertion of Proposition 5.29 and the corresponding improvement from Remark 5.30 if M is the mean of Y.

EXERCISE 5.39. Show that if Y_0 is a random variable such that $P(Y_0 > t) \leqslant A\exp(-t^2)$ for $t > 0$ and if $M_{3/4}$ is its 3rd quartile, then $M_{3/4} \leqslant \sqrt{\log^+(4A)}$. Deduce the first assertion of Proposition 5.29 if M is between the 1st or the 3rd quartile of Y, and the strengthening from Remark 5.30: $|M - a| \leqslant \sqrt{\log^+(2A)}\ \lambda^{-1/2}$ *if M is the median of Y.*

EXERCISE 5.40. Prove the inequality $e^{-s^2} \leqslant e^{\delta^2} e^{-(s+\delta)^2/2}$ for $s, \delta \in \mathbb{R}$. Use it and the last two exercises to show the second assertion of Proposition 5.29, and its strengthenings stated in Remark 5.30 when M is the median or the mean of Y.

EXERCISE 5.41. Verify the assertions in the last two sentences of Remark 5.31.

EXERCISE 5.42. Given $\alpha \in (0, 1)$, prove a version of (5.37) with the right-hand side of the form $B\exp(-\alpha\lambda t^2)$, where B depends only on A and α (and on κ from Remark 5.30, if applicable).

EXERCISE 5.43 (Lévy's lemma for central values). Let $n > 2$. Use Exercise 5.26 to derive Corollary 5.32 for any quantile between the 1st and the 3rd quartile.

EXERCISE 5.44 (The median and the mean on the sphere). Let f be a 1-Lipschitz function on (S^{n-1}, g) with $n > 2$. Show that the median and the mean of f differ at most by $\sqrt{\pi/8n}$ and describe the extremal function.

EXERCISE 5.45 (Variance of a Lipschitz function on the sphere). Let f be a 1-Lipschitz function on (S^{n-1}, g) with $n > 1$. Show that $\mathbf{Var}(f) \leqslant \frac{2}{n}$ and give an example with $\mathbf{Var}(f) \geqslant \frac{1}{n}$. What function gives the maximal variance?

EXERCISE 5.46 (Concentration around L_2 average). Let f be a 1-Lipschitz and positive function on (S^{n-1}, g) with $n > 1$. Set $q = (\mathbf{E}f^2)^{1/2}$. Show that for any $t > 0$, $\mathbf{P}(f \geqslant q + t) \leqslant \exp(-nt^2/2)$ and $\mathbf{P}(f \leqslant q - t) \leqslant e\exp(-nt^2/2)$.

EXERCISE 5.47 (The case of S^1). Using directly the solution to the isoperimetric problem on S^1, show that Corollary 5.32 holds also for $n = 2$.

EXERCISE 5.48. Let (X, d, μ) be a metric probability space and let $\alpha : [0, \infty) \to [0, \infty)$ be such that $\mu(f \geqslant \mathbf{E}f + t) \leqslant \alpha(t)$ for any bounded 1-Lipschitz function $f : X \to \mathbb{R}$ and for all $t > 0$. Then, for any such function f and for any $t > 0$, $\mu(f \geqslant M_f + t) \leqslant \alpha(t/2)$. Equivalently, $\mu(A_\varepsilon) \geqslant 1 - \alpha(\varepsilon/2)$ for any $A \subset X$ with $\mu(A) \geqslant 1/2$ and any $\varepsilon > 0$. The preceding argument can be iterated; see (1.18) in [**Led01**].

EXERCISE 5.49. Prove Proposition 5.33 about enlargements of fairly small sets.

EXERCISE 5.50 (Median vs. mean for convex functions of Gaussian variables). Prove Proposition 5.34 by showing first that the function $g : t \mapsto \Phi^{-1}(\gamma_n(\{f \leqslant t\}))$ is concave.

EXERCISE 5.51. Show that the following statement is a consequence of Proposition 5.34. If $(X_1, \ldots, X_N)$ are jointly Gaussian random variables and $f : \mathbb{R}^N \to \mathbb{R}$ is a convex function, then the median of the random variable $f(X_1, \ldots, X_N)$ does not exceed its expectation.

5.2.3.3. *Local versions.* It sometimes happens that a function defined on the sphere S^{n-1} has a poor global Lipschitz behaviour, while its restriction to a subset of large measure is much more regular. To take advantage of such situation, we formulate a "local" version of Lévy's lemma.

COROLLARY 5.35 (Lévy's lemma, local version). *Let $\Omega \subset S^{n-1}$ be a subset of measure larger than 3/4. Let $f : (S^{n-1}, g) \to \mathbb{R}$ be a function such that the restriction of f to Ω is L-Lipschitz. Then, for every $\varepsilon > 0$,*

$$\mathbf{P}(\{|f(x) - M_f| > \varepsilon\}) \leqslant \mathbf{P}(S^{n-1} \backslash \Omega) + 2\exp(-n\varepsilon^2/4L^2),$$

where M_f is the median of f.

One scenario under which the hypotheses of Corollary 5.35 may be satisfied is when we have an upper bound on some Sobolev norm of f (a "global" parameter, which suggests that "restricted version of Lévy's lemma" could have been better terminology). However, our applications of the Corollary will be rather straightforward and will not require any advanced notions.

EXERCISE 5.52. Prove Corollary 5.35, the local version of Lévy's lemma.

5.2.3.4. *Pushforward.* The following elementary result is very useful for establishing concentration phenomenon for many classical spaces. In a nutshell, it says that concentration results can be "pushed forward" by surjective contractions.

PROPOSITION 5.36 (Contraction principle). *Let (X, μ) and (Y, ν) be metric probability spaces. Assume that there exists a surjective contraction $\phi : X \to Y$ which pushes forward μ to ν (i.e., $\nu(B) = \mu(\phi^{-1}(B))$ and let $a \in (0, 1)$ and $\varepsilon > 0$. Then*

$$\inf_{B \subset Y,\, \nu(B) \geqslant a} \nu(B_\varepsilon) \geqslant \inf_{A \subset X,\, \mu(A) \geqslant a} \mu(A_\varepsilon). \tag{5.39}$$

Similarly, for any $t > 0$,

$$\sup_{g:Y \to \mathbb{R},\, g \text{ 1-Lipschitz}} \nu(g - \mathbf{E}g > t) \leqslant \sup_{f:X \to \mathbb{R},\, f \text{ 1-Lipschitz}} \mu(f - \mathbf{E}f > t). \tag{5.40}$$

Moreover, (5.40) *holds if expectation is replaced by median on both sides.*

EXERCISE 5.53. Prove Proposition 5.36, the contraction principle. State a more general version with $\phi : X \to Y$ assumed to be L-Lipschitz rather than a contraction.

EXERCISE 5.54 (Concentration on the solid cube via Gaussian pushforward). Let Y be the solid cube $[0,1]^n$ endowed with the Lebesgue measure and the Euclidean metric inherited from $\mathbb{R}^n$. Use Proposition 5.36 to show that Y verifies (5.21) with $(C,\lambda) = \left(\frac{1}{2},\pi\right)$ and (5.22) with $(C,\lambda) = (1,\pi)$.

5.2.3.5. *Direct products.* It is easy to see that the concentration phenomenon passes to direct products of metric probability spaces. Indeed, let X and Y be two such spaces that exhibit the concentration phenomenon and let $X \times Y$ be endowed with the product measure and some reasonable product metric, such as the ℓ_p *product metric* defined for (x_1, y_1) and (x_2, y_2) in $X \times Y$ as

$$d((x_1,y_1),(x_2,y_2)) = \left(d_X(x_1,x_2)^p + d_Y(x_1,x_2)^p\right)^{1/p}, \tag{5.41}$$

the limit case $p = \infty$ being interpreted as a maximum. If f is a 1-Lipschitz function on $X \times Y$, then $\phi(x) = M_{f(x,\cdot)}$ is 1-Lipschitz on X and hence concentrated around its median M_ϕ. Since, for each $x \in X$, $f(x,\cdot)$ is concentrated around $\phi(x)$, it follows that f is concentrated around M_ϕ. (See Exercise 5.55 for precise statements.) The above argument can be clearly iterated. Here is another elementary result involving product measures.

PROPOSITION 5.37 (Concentration on product spaces; see Exercise 5.55). *Let (X_i, d_i, μ_i), $1 \leqslant i \leqslant n$, be bounded metric probability spaces and denote $D_i = \operatorname{diam} X_i$. Let $X = X_1 \times \ldots \times X_n$ be endowed with the product measure μ and the ℓ_1 product metric d. Then, for every 1-Lipschitz function $f : X \to \mathbb{R}$ and for any $t \geqslant 0$,*

$$\mu(f \geqslant \mathbf{E}f + t) \leqslant e^{-2t^2/D^2}, \tag{5.42}$$

where $D = \left(\sum_{i=1}^n D_i^2\right)^{1/2}$.

Both approaches to products of metric probability spaces that are sketched above share an unsatisfactory feature: the constants deteriorate as the number of factors increases. In complete generality, this feature is unavoidable (see Section 5.2.5). However, in some natural settings (e.g., the Gaussian space) dimension-free results are possible.

EXERCISE 5.55 (Concentration on product spaces, a naive approach). For the purpose of this exercise the median of a random variable F is defined as $M_F = \frac{1}{2}(\sup\{t : \mathbf{P}(F \geqslant t) \geqslant 1/2\} + \inf\{t : \mathbf{P}(F \leqslant t) \geqslant 1/2\})$, but most other definitions would work if applied consistently and with sufficient care. Let (X, d_1, μ) and (Y, d_2, ν) be metric probability spaces. Consider the space $(X \times Y, d, \pi)$, where $\pi = \mu \otimes \nu$ and d is any metric verifying

$$d((x_1,y),(x_2,y)) = d_1(x_1,x_2) \quad \text{and} \quad d((x,y_1),(x,y_2)) = d_2(y_1,y_2)$$

for all $x, x_1, x_2 \in X$ and $y, y_1, y_2 \in Y$, and let $f : X \times Y \to \mathbb{R}$ be a 1-Lipschitz function with respect to d.

(i) Show that the function $\phi(x) = M_{f(x,\cdot)}$ is 1-Lipschitz on X.
(ii) If X and Y exhibit the concentration phenomenon in the sense of (5.21) for some C and λ, then $\pi(f > M_\phi + t) \leqslant 2Ce^{-\lambda t^2/4}$ for all $t > 0$, and similarly for $\pi(f < M_\phi - t)$.
(iii) Show that M_ϕ is a central value in the sense of Section 5.2.3.
(iv) Same as (ii) with (5.21) replaced by (5.22) and M_ϕ by $\mathbf{E}f$.

EXERCISE 5.56 (Concentration on product spaces, Laplace transform method). The Laplace functional of a probability metric space (X, d, μ) is defined for $\lambda \in \mathbb{R}$ as $E_{(X,d,\mu)}(\lambda) = \sup \int e^{\lambda f}\, d\mu$, where the supremum is taken over all 1-Lipschitz functions $f : X \to \mathbb{R}$ with mean 0.
(i) Show that if X has diameter D, then $E_{(X,d,\mu)}(\lambda) \leqslant \exp(\lambda^2 D^2/8)$ (use Exercise 5.36).
(ii) Show that if (X_1, d_1, μ_1) and (X_2, d_2, μ_2) are two metric probability spaces, if d denotes the ℓ_1 product metric on $X_1 \times X_2$ as defined in (5.41), then

$$E_{(X_1\times X_2,d,\mu_1\otimes\mu_2)}(\lambda) \leqslant E_{(X_1,d_1,\mu_1)}(\lambda)E_{(X_2,d_2,\mu_2)}(\lambda).$$

(iii) Show that in the context of Proposition 5.37, we have

$$E_{(X,d,\mu)}(\lambda) \leqslant \exp(\lambda^2 D^2/8).$$

(iv) Prove Proposition 5.37 using Lemma 5.28.

EXERCISE 5.57 (Hoeffding's inequality). Show that Proposition 5.37 implies Hoeffding's inequality: *if $X_1, \ldots, X_n$ are independent random variables such that X_i takes values in an interval of length l_i, then for any $t > 0$,*

$$\mathbf{P}(S \geqslant \mathbf{E}S + t) \leqslant e^{-2t^2/L^2}, \tag{5.43}$$

where $S = X_1 + \cdots + X_n$ and $L^2 = l_1^2 + \cdots + l_n^2$.

5.2.4. Geometric and analytic methods. Classical examples. In Sections 5.2.1 and 5.2.2 we sketched isoperimetric/concentration results on the Euclidean sphere and for the Gaussian measure. While these are admittedly very special situations, the fact of the matter is that, in high-dimensional settings, some form of concentration phenomenon is the rule rather than the exception.

5.2.4.1. *Gromov's comparison theorem.* The first result asserts that isoperimetric and concentration inequalities hold under geometric assumptions which significantly generalize the spherical case. The invariant that can be related to sphere-like behavior is the *Ricci curvature*, which describes the rate of growth of volume under geodesic flow on the manifold with the similar rate in the Euclidean space. For example (see Figure 5.3), the circumference of a circle of geodesic radius θ ($< \pi$) on the sphere S^2 is $2\pi \sin\theta$, and hence the length of the arc of the circle corresponding to an angle α (measured on the plane tangent at the center of the circle) is $\alpha \sin\theta \approx \alpha\big(\theta - \frac{\theta^3}{6}\big) = \alpha\theta\big(1 - \frac{\theta^2}{6}\big)$ compared to $\alpha\theta$ for the Euclidean plane. (Here and in the next paragraph $\approx$ means equality up to higher order terms.)

Repeating this calculation *mutatis mutandis* for an m-dimensional sphere (in $\mathbb{R}^{m+1}$) of radius R and a solid m-dimensional angle α, we get $\alpha\big(R\sin\frac{\theta}{R}\big)^{m-1} \approx \alpha\big(\theta - \frac{\theta^3}{6R^2}\big)^{m-1} \approx \alpha\theta^{m-1}\big(1 - \frac{m-1}{R^2}\frac{\theta^2}{6}\big)$ compared to $\alpha\theta^{m-1}$ in the Euclidean setting (i.e., in $\mathbb{R}^m$). This is subsumed by saying that the Ricci curvature of RS^m, the m-dimensional sphere of radius R, *at every point and in each direction* is $\frac{m-1}{R^2}$. The notion is generalized to an arbitrary point p on a Riemannian manifold X

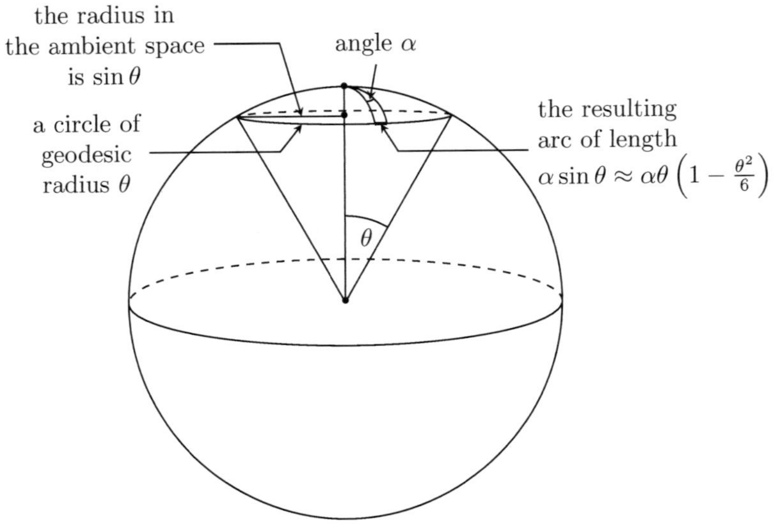

FIGURE 5.3. Volume growth on the sphere S^2 as a function of geodesic distance.

of dimension greater than or equal to 2 and to an arbitrary unit vector u in the tangent space at p by considering infinitesimal (solid) angles in the direction of u and finding the coefficient of $\frac{\theta^2}{6}$ in the corresponding expression for the volume on the geodesic sphere or radius θ centered at p; this coefficient is denoted by $\mathrm{Ric}_p(u)$. The minimum of $\mathrm{Ric}_p(u)$ over $p \in X$ and over direction u is denoted by $c(X)$.

Such straightforward calculation may be difficult to perform for more complicated manifolds. On a less elementary level, the Ricci curvature can be computed using the following formula expressed in the language of Riemannian geometry: whenever $(u_1, \ldots, u_m)$ is an orthonormal basis in the tangent space at p (thought of as a real inner product space), we have

$$\mathrm{Ric}_p(u_1) = \sum_{i=2}^{m} \sec(u_1, u_i), \tag{5.44}$$

where sec denotes the sectional curvature. This leads to an alternative explanation of the value of the Ricci curvature for the sphere, for other manifolds of constant sectional curvature such as the Euclidean space or the hyperbolic space, or for their quotients by discrete groups of symmetries (e.g., for tori or for the real projective space). In the case of Lie groups, sectional curvature can be expressed via Lie brackets. For examples of computations, see Exercises 5.58 and 5.59.

We are now ready to state the main result of this section. By RS^m we denote the sphere of radius R in $\mathbb{R}^{m+1}$.

THEOREM 5.38 (Gromov's comparison theorem, not proved here). *Let $m \geqslant 2$ and let X be an m-dimensional connected Riemannian manifold such that $c(X) \geqslant \frac{m-1}{R^2} = c(RS^m)$. Let $A \subset X$ and let $C \subset RS^m$ be a cap such that $\mu_X(A) = \mu_{RS^m}(C)$, where μ_X and μ_{RS^m} are normalized Riemannian volumes on, respectively, X and RS^m. Then, for every $\varepsilon > 0$, $\mu_X(A_\varepsilon) \geqslant \mu_{RS^m}(C_\varepsilon)$.*

It follows then (from the same proof as Corollary 5.17) that any 1-Lipschitz function $f : X \to \mathbb{R}$ with median M_f satisfies, for any $t > 0$,

$$\mu_X\left(\{f > M_f + t\}\right) \leqslant \frac{1}{2}\exp(-(m+1)t^2/2R^2).$$

As it turns out, the hypotheses of Theorem 5.38 are verified for many (but not all) manifolds that naturally appear in mathematics and that play a role in physics, notably for most classical Lie groups and their homogeneous spaces, see Table 5.3.

TABLE 5.3. Optimal bounds on Ricci curvature for a selection of classical manifolds. We restrict our attention to manifolds for which that curvature is nonnegative, which in particular excludes the hyperbolic space and its quotients. All the bounds concerning specific objects can be derived via formula (5.44) involving the (more standard) sectional curvatures. This is straightforward for spaces for which the sectional curvatures are constant (R^n, S^{n-1}, and $\mathsf{P}(\mathbb{R}^n)$); the remaining cases are covered by Exercises 5.58 and 5.59. Note that the values for the projective spaces $\mathsf{P}(V)$ and the corresponding $\mathsf{Gr}(1, V)$ do not coincide due to different normalization of the metric (an additional $\sqrt{2}$ factor in (B.10) when compared to (B.5)).

X	metric	$c(X)$	comments
$\mathbb{R}^n$	Euclidean	0	
S^{n-1}	geodesic	$n-2$	$n \geqslant 2$
$\mathsf{SO}(n)$	standard (B.8)	$\frac{n-2}{4}$	$n \geqslant 2$
$\mathsf{SU}(n)$	standard (B.8)	$\frac{n}{2}$	
$\mathsf{U}(n)$	standard (B.8)	0	
$\mathsf{Gr}(k, \mathbb{R}^n)$	quotient from $\mathsf{O}(n)$ (B.10)	$\frac{n-2}{2}$	$1 \leqslant k \leqslant n-1$
$\mathsf{Gr}(k, \mathbb{C}^n)$	quotient from $\mathsf{U}(n)$ (B.10)	n	$1 \leqslant k \leqslant n-1$
$\mathsf{P}(\mathbb{R}^n)$	Fubini–Study (B.5)	$n-2$	$n \geqslant 2$
$\mathsf{P}(\mathbb{C}^n)$	Fubini–Study (B.5)	$2n$	$n \geqslant 2$
$X_1 \times X_2$	ℓ_2 product metric (5.41)	$\min\{c(X_1), c(X_2)\}$	

EXERCISE 5.58 (Ricci curvature of Grassmannians). For $\mathsf{Gr}(k, \mathbb{R}^n)$ or $\mathsf{Gr}(k, \mathbb{C}^n)$, the tangent space at any point can be identified with $\mathsf{M}_{k,n-k}$. If $X, Y \in \mathsf{M}_{k,n-k}$ are orthogonal, one can show (see Section 8.2.1 in [**Pet06**]) that

$$\sec(X, Y) = \frac{1}{4}\left(\|XY^\dagger - YX^\dagger\|_{\mathrm{HS}}^2 + \|X^\dagger Y - Y^\dagger X\|_{\mathrm{HS}}^2\right). \tag{5.45}$$

Use this formula and (5.44) to compute the corresponding values from Table 5.3. In some references we find the coefficient $\frac{1}{2}$ instead of $\frac{1}{4}$ because of a different normalization of the metric.

EXERCISE 5.59 (Ricci curvature of classical groups). For $\mathsf{G} = \mathsf{SO}(n)$, $\mathsf{SU}(n)$, or $\mathsf{U}(n)$, the tangent space at I (or at any point) can be identified with the corresponding Lie algebra $\mathfrak{g}$ ($= \mathfrak{so}_n$, $\mathfrak{su}_n$ or $\mathfrak{u}_n$). If $X, Y \in \mathfrak{g}$ are orthonormal, one can show (see Exercise 2.19 in **[Pet06]**) that $\sec(X,Y) = \frac{1}{4}\|XY - YX\|_{\mathrm{HS}}^2$. Use this formula and (5.44) to compute the corresponding values from Table 5.3.

5.2.4.2. *Log-Sobolev inequalities* (LSI). The next technique that we present is of analytic nature. It is based on a class of inequalities which at the first sight seem irrelevant to the subject at hand. Let (X,μ) be a measure space and let f be a non-negative function on X. The (continuous Shannon) *entropy* is defined by

$$\mathrm{Ent}_\mu(f) := \int f \log f \,\mathrm{d}\mu \tag{5.46}$$

if $\int f\,\mathrm{d}\mu = 1$, where we used the convention $0\log 0 = 0$, and then extended to non-negative integrable functions by 1-homogeneity. An explicit formula that implements the extension is

$$\mathrm{Ent}_\mu(f) := \int f \log f \,\mathrm{d}\mu - \int f\,\mathrm{d}\mu\, \log\left(\int f\,\mathrm{d}\mu\right). \tag{5.47}$$

By Jensen's inequality, $\mathrm{Ent}_\mu(f) \geqslant 0$, with $+\infty$ being a possibility.

We now assume that X is a Riemannian manifold and that μ is a Borel measure on X. We say that (X,μ) verifies a logarithmic Sobolev inequality with parameter α if for every (sufficiently smooth) function $f : X \to \mathbb{R}$ we have

$$\mathrm{Ent}_\mu(f^2) \leqslant 2\alpha \int |\nabla f|^2 \,\mathrm{d}\mu. \tag{5.48}$$

The smallest constant α that works in (5.48) is called the *log-Sobolev constant* of (X,μ) and denoted by $\mathrm{LS}(X,\mu)$.

The relevance of this circle of ideas to the concentration phenomenon is explained by the following result.

THEOREM 5.39 (Herbst's argument). *Let X be a Riemannian manifold and let μ be a Borel probability measure on X such that $\mathrm{LS}(X,\mu) \leqslant \alpha$. Then every 1-Lipschitz function $F : X \to \mathbb{R}$ is integrable and satisfies, for every $t > 0$,*

$$\mu\Big(F > \int F\mathrm{d}\mu \,+\, t\Big) \leqslant e^{-t^2/2\alpha}. \tag{5.49}$$

REMARK 5.40. The above Theorem can be extended to the setting of general metric spaces, with essentially the same proof, once $|\nabla f|$ is properly defined. For example, we may use $|\nabla f|(x) = \limsup_{y\to x} \frac{|f(y)-f(x)|}{\mathrm{dist}(y,x)}$ if X has no isolated points; discrete spaces may also be handled with some care. However, for clarity of the exposition, we will assume for the rest of this subsection that the underlying spaces are (connected) Riemannian manifolds.

PROOF OF THEOREM 5.39. First, we may assume that F is smooth and that $\int F\,\mathrm{d}\mu = 0$; this may be achieved by replacing F with an appropriate approximation and subtracting a constant. The strategy is to show that the (bilateral) Laplace transform of F verifies

$$\int e^{\lambda F}\,\mathrm{d}\mu \leqslant e^{\alpha\lambda^2/2} \quad \text{for all } \lambda \in \mathbb{R}, \tag{5.50}$$

which by Lemma 5.28 implies that $\mu(F > t) \leqslant e^{-t^2/2\alpha}$, as needed. To establish (5.50), we introduce an auxiliary function $f = f_\lambda > 0$ defined via $f^2 = e^{\lambda F - \alpha\lambda^2/2}$. In other words, $f = e^{\lambda F/2 - \alpha\lambda^2/4}$, and it is readily checked that $\nabla f = \frac{\lambda}{2} f \nabla F$. Since

$|\nabla F| \leqslant 1$ (because F is 1-Lipschitz), it follows that $|\nabla f|^2 \leqslant \frac{\lambda^2}{4} f^2$. Consequently, by (5.48) (cf. (5.47)),

$$\text{(5.51)}\quad \operatorname{Ent}_\mu(f^2) = \int f^2\Big(\lambda F - \frac{\alpha\lambda^2}{2}\Big)\,\mathrm{d}\mu - \int f^2\,\mathrm{d}\mu\,\log\Big(\int f^2\,\mathrm{d}\mu\Big) \leqslant \frac{\alpha\lambda^2}{2}\int f^2\,\mathrm{d}\mu.$$

We now set $\phi(\lambda) = \int f^2\,\mathrm{d}\mu$ and note that differentiating under the integral sign gives

$$\phi'(\lambda) = \int f^2(F - \alpha\lambda)\,\mathrm{d}\mu.$$

This allows to rewrite (5.51) as

$$\lambda\phi'(\lambda) - \phi(\lambda)\,\log\big(\phi(\lambda)\big) \leqslant 0,$$

which, for $\lambda \neq 0$, is equivalent to

$$\text{(5.52)}\qquad \frac{\mathrm{d}}{\mathrm{d}\lambda}\Big(\frac{\log\big(\phi(\lambda)\big)}{\lambda}\Big) \leqslant 0.$$

On the other hand, given that $\phi(0) = 1$, l'Hôpital's rule yields

$$\text{(5.53)}\qquad \lim_{\lambda\to 0}\frac{\log\big(\phi(\lambda)\big)}{\lambda} = \lim_{\lambda\to 0}\frac{\phi'(\lambda)}{\phi(\lambda)} = \frac{\phi'(0)}{\phi(0)} = \frac{\int F\,\mathrm{d}\mu}{1} = 0.$$

Combining (5.52) and (5.53) we conclude that $\log\big(\phi(\lambda)\big)/\lambda \leqslant 0$ for $\lambda > 0$ and $\log\big(\phi(\lambda)\big)/\lambda \geqslant 0$ for $\lambda < 0$, which just means that $\phi(\lambda) \leqslant 1$ for all $\lambda \in \mathbb{R}$. In other words, $\int e^{\lambda F - \alpha\lambda^2/2}\,\mathrm{d}\mu \leqslant 1$ for $\lambda \in \mathbb{R}$, which is just a restatement of (5.50) and concludes the argument. □

Apart from the median being replaced by the expected value (which is largely a matter of convenience or elegance, see Proposition 5.29 in Section 5.2.3), the assertion of Theorem 5.39 closely resembles (5.26) and (5.33), which quantified the concentration phenomenon for Lipschitz functions in the spherical and Gaussian settings. However, its usefulness depends on availability of spaces (X, μ) verifying logarithmic Sobolev inequalities. The next few results ensure that the supply is indeed quite ample. For easy reference, the spaces and estimates on their log-Sobolev constants are cataloged in Table 5.4.

PROPOSITION 5.41 (Not proved here). *Let X be an m-dimensional Riemannian manifold such that $c(X) > 0$ and let μ be the normalized Riemannian volume. Then* $\operatorname{LS}(X, \mu) \leqslant \frac{m-1}{mc(X)}$.

PROPOSITION 5.42 (Not proved here). *Let μ be a measure on $\mathbb{R}^n$ whose density with respect to the Lebesgue measure is of the form e^{-U}, where U verifies* $\operatorname{Hess}(U) \geqslant \beta\mathrm{I}$ *for some $\beta > 0$. Then* $\operatorname{LS}(\mathbb{R}^n, \mu) \leqslant \beta^{-1}$. *In particular,* $\operatorname{LS}(\mathbb{R}^n, \gamma_n) \leqslant 1$ *and* $\operatorname{LS}(\mathbb{C}^n, \gamma_n^{\mathbb{C}}) \leqslant \frac{1}{2}$.

PROPOSITION 5.43 (Not proved here; see Exercise 5.61). *We have*

$$\operatorname{LS}(S^1, \sigma) = 1 \quad\text{and}\quad \operatorname{LS}([0,1], \mathrm{vol}_1) = \pi^{-2}.$$

PROPOSITION 5.44 (Tensorization property of LSI, not proved here). *Given (X_i, μ_i), $i = 1, \dots, k$, let $X = X_1 \times \cdots \times X_k$ be endowed with the ℓ_2 product metric as defined in (5.41) and the product measure $\mu = \mu_1 \otimes \cdots \otimes \mu_k$. Then* $\operatorname{LS}(X, \mu) = \max_{1\leqslant i\leqslant k}\operatorname{LS}(X_i, \mu_i)$.

REMARK 5.45 (Poincaré's inequality). Another related famous functional inequality is the Poincaré inequality, which reads as follows: for every smooth function $f : X \to \mathbb{R}$

$$\mathbf{Var}_\mu f \leqslant \alpha \int |\nabla f|^2 \, \mathrm{d}\mu, \tag{5.54}$$

where $\mathbf{Var}_\mu f$ denotes the quantity $\int f^2 \, \mathrm{d}\mu - \left(\int f \, \mathrm{d}\mu\right)^2$. The smallest α is called the Poincaré constant of (X, μ) and denoted $\mathrm{P}(X, \mu)$. Inequality (5.54) is implied by the LSI (5.48) (with the same constant α); it implies subexponential instead of subgaussian concentration. A list of Poincaré constants for common spaces can be found in Table 5.4. An example of a probability measure satisfying the Poincaré inequality but not the LSI is the (symmetric) exponential distribution on $\mathbb{R}$.

REMARK 5.46 (Contraction principle for LSI and Poincaré's inequality). If $\phi : (X, \mu) \to (Y, \nu)$ is a surjective contraction which pushes forward μ onto ν, then $\mathrm{LS}(Y, \nu) \leqslant \mathrm{LS}(X, \mu)$ and $\mathrm{P}(Y, \nu) \leqslant \mathrm{P}(X, \mu)$. This can be proved as in Exercise 5.53 and is especially transparent if we define $|\nabla f|$ as in Remark 5.40.

TABLE 5.4. Bounds on log-Sobolev and Poincaré constants for a selection of classical manifolds. We use the same metrics as in Table 5.3. Except as indicated, the estimates on log-Sobolev constants follow from estimates on the Ricci curvature (see Proposition 5.41). Most of the time we use the bound $\mathrm{LS}(X, \mu) < c(X)^{-1}$; the more precise expressions involving the dimension of X lead to slightly better but often cumbersome formulas. The upper bounds on the Poincaré constants of Grassmann manifolds follow from Remark 5.46. For more comments and references about Poincaré constants, see Notes and Remarks.

X or (X, μ)	$\mathrm{LS}(X, \mu)$	$\mathrm{P}(X, \mu)$	Comments
$\left([a, b], \frac{vol_1}{b-a}\right)$	$\frac{(b-a)^2}{\pi^2}$	$\frac{(b-a)^2}{\pi^2}$	Prop. 5.43
S^{n-1}	$\frac{1}{n-1}$	$\frac{1}{n-1}$	Prop. 5.43 for S^1
$\mathsf{P}(\mathbb{R}^n)$	$\leqslant \frac{1}{n-1}$	$\frac{1}{2n}$	
$\mathsf{P}(\mathbb{C}^n)$	$< \frac{1}{2n}$	$\frac{1}{4n}$	
$(\mathbb{R}^n, \gamma_n)$	1	1	Exercise 5.60
$\mathsf{SO}(n)$	$< \frac{4}{n-2}$	$\frac{2}{n-1}$	
$\mathsf{SU}(n)$	$< \frac{2}{n}$	$\frac{n}{n^2-1}$	
$\mathsf{U}(n)$	$\leqslant \frac{6}{n}$	$\frac{1}{n}$	**[MM13]**
$\mathsf{Gr}(k, \mathbb{R}^n)$	$< \frac{2}{n-2}$	$\leqslant \frac{2}{n-1}$	$1 \leqslant k \leqslant n-1$
$\mathsf{Gr}(k, \mathbb{C}^n)$	$< \frac{1}{n}$	$\leqslant \frac{1}{n}$	$1 \leqslant k \leqslant n-1$
$(X \times Y, \mu_X \otimes \mu_Y)$	$\max\{\mathrm{LS}(X), \mathrm{LS}(Y)\}$	$\max\{\mathrm{P}(X), \mathrm{P}(Y)\}$	ℓ_2 product metric

EXERCISE 5.60 (Log-Sobolev constant for the Gaussian space). Show that $\mathrm{LS}(\mathbb{R}^n, \gamma_n) \geqslant 1$ (we have actually equality, see Proposition 5.42).

EXERCISE 5.61 (Log-Sobolev constants for segments and circles). (i) Use the contraction principle from Remark 5.46 to show that $\mathrm{LS}([0,1],\mathrm{vol}_1) \leqslant \pi^{-2}\mathrm{LS}(S^1,\sigma)$ and $\mathrm{P}([0,1],\mathrm{vol}_1) \leqslant \pi^{-2}\mathrm{P}(S^1,\sigma)$. (ii) Verify that $\mathrm{P}(S^1,\sigma) = 1$. (iii) Verify that $\mathrm{P}([0,1],\mathrm{vol}_1) \geqslant \pi^{-2}$ (see Notes and Remarks for the reasons why there is actually an equality).

5.2.4.3. *Hypercontractivity, Gaussian polynomials.* We give a brief introduction to the concept of hypercontractivity and illustrate it to give an example of a concentration inequality for Gaussian polynomials.

We work on the probability space $(\mathbb{R}^n, \gamma_n)$. We define the Ornstein–Uhlenbeck semigroup of operators $(P_t)_{t\geqslant 0}$ as follows. For $f : \mathbb{R}^n \to \mathbb{R}$ is a bounded measurable function, and $x \in \mathbb{R}^n$, let

$$(P_t f)(x) = \mathbf{E}\, f\Big(e^{-t}x + \sqrt{1-e^{-2t}}G\Big), \tag{5.55}$$

where G is a standard Gaussian vector in $\mathbb{R}^n$. These operators satisfy the semigroup property $P_s P_t = P_{s+t}$. Moreover it is easily checked (Exercise 5.62) that for every $p \geqslant 1$ and $t \geqslant 0$,

$$\|P_t f\|_{L_p(\gamma_n)} \leqslant \|f\|_{L_p(\gamma_n)},$$

and therefore P_t extends to a bounded (contractive) operator on $L_p(\gamma_n)$. Remarkably, a stronger statement is true: provided $p > 1$ and $t > 0$, P_t is a contraction from $L_p(\gamma_n)$ to $L_q(\gamma_n)$ for some $q = q(t) > p$. This phenomenon is called hypercontractivity.

PROPOSITION 5.47 (Not proved here; see Exercise 5.63). *Let $1 \leqslant p \leqslant q < \infty$ and $t > 0$ such that $q \leqslant 1 + e^{2t}(p-1)$. Then*

$$\|P_t f\|_{L_q(\gamma_n)} \leqslant \|f\|_{L_p(\gamma_n)}.$$

The eigenvectors of P_t are the Hermite polynomials. In the one-dimensional case, denote by $(h_k)_{k\in\mathbb{N}}$ the sequence of polynomials obtained by orthonormalizing the sequence $(1, x, x^2, \dots)$ in the space $\mathscr{H}_1 := L_2(\mathbb{R}, \gamma_1)$. (In this context, we exceptionally mean $\mathbb{N} = \{0, 1, 2, 3, \dots\}$.) Given a multi-index $\alpha = (\alpha_1, \dots, \alpha_n) \in \mathbb{N}^n$, let h_α be the multivariate polynomial

$$h_\alpha(x_1, \dots, x_n) = h_{\alpha_1}(x_1)\cdots h_{\alpha_n}(x_n). \tag{5.56}$$

The family $(h_\alpha)_{\alpha\in\mathbb{N}^n}$ is an orthonormal basis in $\mathscr{H}_n := L_2(\mathbb{R}^n, \gamma_n)$, and we have

$$P_t h_\alpha = e^{-t|\alpha|}h_\alpha, \tag{5.57}$$

where $|\alpha| = \sum_{i=1}^n \alpha_i$ is the weight of the multi-index α, or the total degree of the polynomial h_α. Note that formula (5.57) allows us to define $P_t Q$ for any polynomial Q even when t is negative.

PROPOSITION 5.48. *Let Q be a polynomial in n variables of (total) degree at most k. Then, for every $q \geqslant 2$,*

$$\|Q\|_{L_q(\gamma_n)} \leqslant (q-1)^{k/2}\|Q\|_{L_2(\gamma_n)}.$$

PROOF. For any $t \geqslant 0$, we have $P_t P_{-t} Q = Q$ (see the remark following (5.57)). Choosing $t > 0$ such that $q - 1 = e^{2t}$, we may apply Proposition 5.47 to conclude that $\|Q\|_{L_q(\gamma_n)} \leqslant \|P_{-t}Q\|_{L_2(\gamma_n)}$. We may write the decomposition of Q in the basis of Hermite polynomials

$$Q = \sum_{|\alpha|\leqslant k} c_\alpha h_\alpha$$

for some coefficients (c_α). It follows that $\|Q\|_{L_2(\gamma_n)}^2 = \sum c_\alpha^2$, while

$$\|P_{-t}Q\|_{L_2(\gamma_n)}^2 = \sum_{|\alpha|\leqslant k} e^{2t|\alpha|}c_\alpha^2 \leqslant e^{2tk}\|Q\|_{L_2(\gamma_n)}^2,$$

and the result follows. □

COROLLARY 5.49 (Concentration inequality for Gaussian polynomials). *Let $Z_1, \dots, Z_n$ be independent $N(0,1)$ variables and let $X = Q(Z_1, \dots, Z_n)$, where Q is a polynomial of (total) degree at most k. Then, for any $t \geqslant (2e)^{k/2}$,*

$$\mathbf{P}\left(|X - \mathbf{E}X| \geqslant t\sqrt{\mathbf{Var}\, X}\right) \leqslant \exp\left(-\frac{k}{2e}t^{2/k}\right).$$

PROOF. There is no loss of generality in assuming that $Z_1, \dots, Z_n$ are defined as the coordinate functions on $(\mathbb{R}^n, \gamma_n)$, so that Proposition 5.48 applies. We may assume $\mathbf{E}X = 0$, $\mathbf{Var}\, X = 1$ and write by Markov's inequality, for any $q \geqslant 2$,

$$\mathbf{P}(|X| \geqslant t) \leqslant t^{-q}\,\mathbf{E}\,|X|^q \leqslant t^{-q}(q-1)^{kq/2} \leqslant (q^{k/2}/t)^q$$

where we used Proposition 5.48. The choice $q = t^{2/k}/e$ (which is larger than 2 provided $t \geqslant (2e)^{k/2}$) yields the result. □

REMARK 5.50. The phenomenon of hypercontractivity is not specific to the Gaussian case and is essentially equivalent to a log-Sobolev inequality (see Theorem 5.2.3 in [**BGL14**]). Similar concentration results are true for polynomials in binary random variables (see Theorem 9.21 in [**O'D14**]) and for polynomials on the sphere (cf. [**Mon12**]). Here is a precise statement of the latter. *If Q be a polynomial with total degree of at most k in $n_1 + \cdots + n_d$ variables and $X = (X_1, \dots, X_d)$ with X_i independent and uniformly distributed on S^{n_i-1}, then for every $q \geqslant 2$, $\|Q(X)\|_{L_q} \leqslant (q-1)^{k/2}\|Q(X)\|_{L_2}$.* (This is slightly more general than Corollary 12 in [**Mon12**], which assumes that $n_1 = \cdots = n_d$ and that the partial degrees in each variable are equal.) The argument is similar to the Gaussian case, using spherical harmonics instead of Hermite polynomials. Concentration estimates similar to Corollary 5.49 follow.

EXERCISE 5.62 (The Ornstein–Uhlenbeck semigroup is contractive). Show that P_t is a contraction on $L_p(\gamma_n)$ for any $t \geqslant 0$ and $p \geqslant 1$.

EXERCISE 5.63 (Sharpness of the hypercontractive inequality). When $n = 1$, compute $P_t f_\lambda$ when $f_\lambda(x) = e^{\lambda x}$. Conclude that Proposition 5.47 is sharp in the following sense: when $q > 1 + e^{2t}(p-1)$, there is no constant C such that the inequality $\|P_t f\|_{L_q(\gamma_1)} \leqslant C\|f\|_{L_p(\gamma_1)}$ holds.

5.2.5. Some discrete settings. All the *specific* instances of concentration we identified thus far involved manifolds. However, the phenomenon also occurs in the discrete case. We will exemplify it (and the issues that may arise) on the fundamental example of the *Boolean cube* $\{0,1\}^n$, or $\{-1,1\}^n$, endowed with the normalized counting measure μ and the normalized Hamming distance $d_H(x,y) := \frac{1}{n}\operatorname{card}\{i : x_i \neq y_i\}$, which up to normalization coincides with the ℓ_1 metric in the ambient space $\mathbb{R}^n$. (This setting was already studied in Section 5.1.3; other product measures, or metrics induced by ℓ_p-norms for other p, are also frequently considered, more about that later.)

A nearly optimal concentration result for the Boolean cube follows already from Proposition 5.37. However, we can do better: the exact solution to the isoperimetric problem on the cube is known. To describe it, we introduce a total order $<$ on $\{0,1\}^n$ (called the *simplicial order*) as follows: for $x = (x_i)$ and $y = (y_i)$ in $\{0,1\}^n$, declare that $x < y$ if either $x_1 + \cdots + x_n < y_1 + \cdots + y_n$ or $x_1 + \cdots + x_n = y_1 + \cdots + y_n$ and x precedes y in the lexicographic order. Then the initial segments for this order are isoperimetric sets. As opposed to the Gaussian and spherical case, the extremal sets are not unique in any reasonable sense (see Exercise 5.66)

THEOREM 5.51 (Harper's isoperimetric inequality, not proved here). *For any integer N with $1 \leqslant N < 2^n$, let $A \subset \{0,1\}^n$ be the set of the N smallest elements with respect to the simplicial order. Then A has the smallest ε-enlargements (for all $\varepsilon > 0$) among all sets of the same cardinality. The set A verifies*

$$B(x, k/2^n) \subset A \subset B(x, (k+1)/2^n) \tag{5.58}$$

for some $k \in \{0, \ldots, n-1\}$.

If we define the boundary of A as $\partial A := \{y \in \{0,1\}^n : \operatorname{dist}(y, A) = 1/n\}$, the sets from Theorem 5.51 also have the "smallest boundary" among subsets of $\{0,1\}^n$ of the same measure. In this language, the condition (5.58) says that A consists of a ball and a part of its boundary. If $N = \sum_{j=1}^{k} \binom{n}{j}$ for some k, the situation becomes simple: the optimal sets are balls, and so are their enlargements.

For example, if $n = 2m+1$ is odd, an example of an optimal set of measure $\frac{1}{2}$ is

$$A = \{y \in \{0,1\}^n : Y \leqslant m\},$$

where $Y = \sum_{j=1}^{n} y_j$. The enlargements of A are then clearly of the form $A_{s/n} = \{Y \leqslant m + s\}$ and, consequently,

$$\mu(A_{s/n}) = \frac{\sum_{j=1}^{m+s} \binom{n}{j}}{2^n} = 1 - \frac{\sum_{j>m+s} \binom{n}{j}}{2^n} \geqslant 1 - e^{-2s^2/n}, \tag{5.59}$$

where the inequality follows from Hoeffding's inequality (5.43). A similar analysis can be performed when n is even (see Exercise 5.64 for details). To summarize, we have

COROLLARY 5.52. *If $A \subset \{0,1\}^n$ with $\mu(A) \geqslant \frac{1}{2}$, $s \in \mathbb{N}$ and $\varepsilon = s/n$, then $\mu(A_\varepsilon) \geqslant 1 - e^{-2n\varepsilon^2}$. Consequently, if $f : \{0,1\}^n \to \mathbb{R}$ is a 1-Lipschitz function and M is its median, then $\mu(f > M + \varepsilon) \leqslant e^{-2n\varepsilon^2}$.*

REMARK 5.53. Some authors assert that the bound $\mu(A_\varepsilon) \geqslant 1 - e^{-2n\varepsilon^2}$ (for A satisfying $\mu(A) \geqslant \frac{1}{2}$) holds for all $\varepsilon > 0$. However, this may be false, but only if $n = 1$ or 2 and only for certain values of $\varepsilon \in (0, 1/n)$, see Exercise 5.65.

The setting of Corollary 5.52 is a special case of that of Proposition 5.37. (The differences include the mean being replaced by the median, and the numerical constants being better in the former, which is not surprising since it is a more specialized result.) The Corollary is an elegant and sharp result, but it exhibits the following unsatisfactory feature: if we use the standard Euclidean metric to define the 1-Lipschitz property of f or the expansions A_t, the exponential term in the estimates becomes $e^{-2t^2/n}$. This should be compared to the dimension-free (and differently scaled) term $\frac{1}{2}e^{-t^2/2}$ in Theorem 5.24, the Gaussian isoperimetric inequality. However, there is a fix to this difficulty due to Talagrand: if the function f is convex, its restriction to $\{0,1\}^n$ exhibits dimension-free subgaussian concentration. We have

THEOREM 5.54 (Talagrand's convex concentration inequality for the Boolean cube, not proved here). *Let A be a non-empty subset of $\{0,1\}^n \subset \mathbb{R}^n$ and set $\phi_A(x) := \operatorname{dist}(x, \operatorname{conv} A)$, where the distance is calculated with respect to the Euclidean metric. Then*

$$\mathbf{E}\, e^{\frac{1}{2}\phi_A^2} \leqslant 1/\mu(A) \tag{5.60}$$

and so $\mu(\phi_A > t) \leqslant e^{-t^2/2}/\mu(A)$ for $t > 0$. Consequently, if $f : [0,1]^n \to \mathbb{R}$ is a convex (or concave) 1-Lipschitz function and M is its median with respect to μ, then $\mu(f > M + t) \leqslant 2e^{-t^2/2}$ for $t > 0$.

In the statement of Theorem 5.54 we tacitly assume that μ is a measure on $\mathbb{R}^n$ supported on $\{0,1\}^n$. The second assertion of the Theorem follows from (5.60) by Markov's inequality. Some finer issues related to the derivation of the last assertion are addressed in Exercise 5.67. See also Exercise 5.68.

Theorem 5.54 turned out to be very useful (for example in the context of random matrices) and has been generalized in various ways. Here is one possible statement.

THEOREM 5.55 (Not proved here). *Let $V_1, V_2, \ldots, V_N$ be finite-dimensional normed spaces and let $V = \bigoplus_{j=1}^N V_j$ be their sum in the ℓ_q-sense (for some $q \geqslant 2$). For $j = 1, 2, \ldots, N$, let μ_j be a measure on V_j supported on a set of diameter at most 1 and let $\mu = \otimes_{j=1}^N \mu_j$. Further, assume that $F : V \to \mathbb{R}$ is 1-Lipschitz and quasiconvex (i.e., $F^{-1}\big((-\infty, a]\big)$ is convex for all $a \in \mathbb{R}$) or quasiconcave. Then*

$$\mu(F > M + t) \leqslant 2e^{-\frac{1}{4}t^q} \quad \textit{for all} \quad t > 0, \tag{5.61}$$

where M is the median of F with respect to μ.

We conclude this section with a result that is the counterpart of Theorem 5.54 with the median replaced by the mean, whose degree of generality is intermediate between those of Theorem 5.54 and Theorem 5.55.

THEOREM 5.56 (Convex concentration inequality for the mean, not proved here). *Let $\mu = \mu_1 \otimes \cdots \otimes \mu_k$ be a product measure on $[0,1]^n \subset \mathbb{R}^n$ and let $f : [0,1]^n \to \mathbb{R}$ be a function which is 1-Lipschitz with respect to the Euclidean distance and convex with respect to each variable. Then, for any $t \geqslant 0$,*

$$\mu(f > \mathbf{E}f + t) \leqslant e^{-t^2/2}. \tag{5.62}$$

While, by Remark 5.12 (which was based on the very general results from Section 5.2.3.2), statements about concentration around the median formally imply similar statements about the mean, we state Theorem 5.56 separately since it combines good constants with a different set of hypotheses.

EXERCISE 5.64 (Concentration on even-dimensional Boolean cube). If $n = 2m$ is even, an example of a set $A \subset \{0,1\}^n$ with $\mu(A) = \frac{1}{2}$ that is optimal in the sense of Theorem 5.51 is $A = \big\{\sum_{j=1}^n y_j < m\big\} \cup \big\{\sum_{j=1}^n y_j = m \text{ and } y_1 = 1\big\}$. Show that also in this case $\mu(A_{s/n}) \geqslant 1 - e^{-2s^2/n}$ for $s \in \mathbb{N}$.

EXERCISE 5.65. Show that the bound $\mu(A_\varepsilon) \geqslant 1 - e^{-2n\varepsilon^2}$ from Corollary 5.52 may fail for some $\varepsilon > 0$ if $n = 1$ or 2, but that it always holds if $n > 2$ or if $\varepsilon \geqslant 1/n$.

EXERCISE 5.66 (Non uniqueness in Harper's theorem). Give an example of a value N and two sets of N elements in $\{0,1\}^4$ with smallest ε-enlargements (for all values of ε) among sets with N elements, which are distinct up to symmetries of the hypercube. **Note**: it appears to be unknown whether uniqueness can be assured by insisting that both A and its complement are isoperimetric sets for all sizes of enlargement.

EXERCISE 5.67 (Talagrand's concentration inequality for concave functions). Derive the bound $\mu(f > M + t) \leqslant 2e^{-t^2/2}$ for concave f in Theorem 5.54 (or, equivalently, $\mu(f < M - t) \leqslant 2e^{-t^2/2}$ for convex f) from the inequalities preceding it.

EXERCISE 5.68 (Existence of convex Lipschitz extensions). Let $K \subset \mathbb{R}^n$ be a convex set and let $f : K \to \mathbb{R}$ be a convex 1-Lipschitz function. Then f admits a convex 1-Lipschitz extension to $\mathbb{R}^n$. Consequently, in Theorem 5.54 it doesn't matter whether we assume f to be convex and 1-Lipschitz on $\mathbb{R}^n$ or just on $[0,1]^n$.

EXERCISE 5.69 (No dimension-free subgaussian bound in absence of convexity). Here is an example showing that convexity is crucial in Theorem 5.54. Define $f : \{-1,1\}^n \to \mathbb{R}$ by $f(x_1,\dots,x_n) = \max(0, x_1 + \cdots + x_n)^{1/2}$. Show that f has median 0 and is $\frac{1}{\sqrt{2}}$-Lipschitz with respect to the Euclidean metric, while $\mu\left(f > cn^{1/4}\right) \geqslant c$ for some absolute constant $c > 0$.

5.2.6. Deviation inequalities for sums of independent random variables. In this section we gather some simple but useful facts about deviation inequalities for sum of independent mean zero random variables. We mostly focus on two families of random variables: subgaussian and subexponential variables.

In a probabilistic setting, the L_p-norm (for $p \geqslant 1$) of a random variable X is $\|X\|_p = (\mathbf{E}\,|X|^p)^{1/p}$. As a preliminary step, consider two prototypical examples: let Z be an $N(0,1)$ random variable and T be a symmetric exponential variable with parameter 1 (i.e., $\mathbf{P}(T > t) = \mathbf{P}(-T > t) = \frac{1}{2}e^{-t}$ for $t > 0$). A simple computation (cf. (A.1)) shows that

$$\|Z\|_p = \frac{\sqrt{2}}{\pi^{1/2p}}\Gamma\left(\frac{p+1}{2}\right)^{1/p} \sim \sqrt{\frac{p}{e}}, \tag{5.63}$$

$$\|T\|_p = \Gamma(p+1)^{1/p} \sim \frac{p}{e} \tag{5.64}$$

as p tends to infinity.

The growth of the L_p-norms motivates the following definitions: a random variable X is said to be *subgaussian* (or ψ_2) when

$$\|X\|_{\psi_2} := \sup_{p \geqslant 1} p^{-1/2}\|X\|_p < \infty. \tag{5.65}$$

This terminology is consistent with that introduced in the preamble to Section 5.2 and based on the tail behavior (cf. (5.21), (5.22); see Exercise 5.70 and Lemma 5.57 below). Similarly, X is said to be *subexponential* (or ψ_1) when

$$\|X\|_{\psi_1} := \sup_{p \geqslant 2} \frac{\|X\|_p}{\|T\|_p} < \infty. \tag{5.66}$$

The reader may be familiar with the arguably less *ad hoc* forms of ψ_r conditions, based on either the rate of growth of the (bilateral) Laplace transform or the appropriate Orlicz norms, or on the tail behavior of the type

$$\mathbf{P}(|X| > t) \leqslant Ce^{-\lambda t^r} \quad \text{for} \quad t \geqslant 0$$

(cf. (5.21) and (5.22)). There is no need to be alarmed, though: while not identical, all these approaches lead to quantities that are equivalent up to universal constants. The definitions (5.65)–(5.66) were chosen out of convenience in view of the sample applications we present. See Notes and Remarks for more details and references.

If follows from (5.63) and (5.64) that $\|T\|_{\psi_1} = 1$, $\|Z\|_{\psi_2} = \sqrt{2/\pi}$ and that $\|\cdot\|_{\psi_1} \leqslant \|\cdot\|_{\psi_2}$ (see Exercise 5.75). We have obviously $\|\cdot\|_{\psi_2} \leqslant \|\cdot\|_\infty$ and $\|\cdot\|_{\psi_1} \leqslant \|\cdot\|_\infty$, so the present discussion also applies to bounded variables. Another important example of subgaussian variables is obtained by taking the inner product with a fixed vector of a randomly chosen unit vector in $\mathbb{R}^d$ or $\mathbb{C}^d$. This has to be compared with Poincaré's lemma (Theorem 5.22), which says that the Gaussian measure appears at the limit $d \to \infty$.

LEMMA 5.57. *If X is uniformly distributed on S^{d-1} (resp., $S_{\mathbb{C}^d}$), then for every $u \in \mathbb{R}^d$ (resp., $u \in \mathbb{C}^d$), we have $\|\langle X, u\rangle\|_{\psi_2} \leqslant |u|/\sqrt{d}$.*

PROOF. We may assume by homogeneity that $|u| = 1$. Let G be a standard Gaussian vector in $\mathbb{R}^d$. The variable uniformly distributed on S^{d-1} can be then represented as $X = G/|G|$. Moreover, $|G|$ is independent of X and hence, for $p \geqslant 1$,

$$\|\langle G, u\rangle\|_p = \||G|\|_p \|\langle X, u\rangle\|_p.$$

We have $\||G|\|_p \geqslant \||G|\|_1 = \kappa_d$ (see Section 4.3.3). Since $\langle G, u\rangle$ has distribution $N(0,1)$, we know from (5.63) that $\|\langle X, u\rangle\|_{\psi_2} = \sqrt{2/\pi} = \kappa_1$. Therefore, using Proposition A.1(ii), we obtain $\|\langle X, u\rangle\|_{\psi_2} \leqslant \frac{\kappa_1}{\kappa_d} \leqslant \frac{1}{\sqrt{d}}$. The complex case is similar. □

We also note that the square of a subgaussian variable is subexponential, as follows easily from the definitions. We now consider the case of a sum of either subgaussian or subexponential mean zero random variables. If the random variables are bounded, we can apply Hoeffding's inequality (5.43). It turns out that essentially the same result holds for subgaussian variables.

PROPOSITION 5.58 (See Exercise 5.73). *Let $X_1, \dots, X_n$ be independent subgaussian real random variables with mean zero, and $S = X_1 + \cdots + X_n$. Define $K > 0$ by $K^2 = \|X_1\|_{\psi_2}^2 + \cdots + \|X_n\|_{\psi_2}^2$. Then for every $t > 0$,*

$$\mathbf{P}(|S| > t) \leqslant 2\exp\left(-\frac{t^2}{8eK^2}\right).$$

The proof actually yields a better bound $2\exp(-\frac{t^2}{2eK^2})$ when (X_i) are symmetric random variables (i.e., such that X_i and $-X_i$ have the same distribution for any fixed i).

In the case of ψ_1 variables, the situation is slightly more complicated since two tails enter the picture: subgaussian tails for moderate deviations (which are reminiscent of the central limit phenomenon) and subexponential tails for large deviations (which come from the tails of individual variables).

PROPOSITION 5.59 (Bernstein's inequalities; see Exercise 5.76). *Let $X_1, \ldots, X_n$ be independent real random variables with mean zero, and assume that $\|X_i\|_{\psi_1} \leqslant K$ for every index i. Then, for every vector $a = (a_1, \ldots, a_n) \in \mathbb{R}^n$ and every $t \geqslant 0$,*

$$\mathbf{P}\left(\left|\sum_{i=1}^{n} a_i X_i\right| > t\right) \leqslant 2\exp\left(-\min\left(\frac{t^2}{8K^2\|a\|_2^2}, \frac{t}{4K\|a\|_\infty}\right)\right).$$

REMARK 5.60. Propositions 5.58 and 5.59 readily generalize to the complex case (with possibly different numerical constants).

EXERCISE 5.70 (Lipschitz function on a Gaussian space is subgaussian). Let G be a standard Gaussian vector on $\mathbb{R}^n$ and $f : \mathbb{R}^n \to \mathbb{R}$ a 1-Lipschitz function such that $f(G)$ has mean zero. Deduce from the results of Section 5.2.2 that $\|f(G)\|_{\psi_2} \leqslant C$ for some absolute constant C. (Except for the value of the constant C, this is a generalization of Lemma 5.57.)

EXERCISE 5.71 (Khintchine inequalities). Let $X = \sum_{i=1}^n \varepsilon_i a_i$, where $a_1, \ldots, a_n$ are real numbers and (ε_i) is a sequence of independent random variables with $\mathbf{P}(\varepsilon_i = 1) = \mathbf{P}(\varepsilon_i = -1) = 1/2$. Show that, for any $p \geqslant 1$,

$$A_p\|X\|_{L_2} \leqslant \|X\|_{L_p} \leqslant B_p\|X\|_{L_2},$$

where $A_p > 0$ and B_p are constants depending only on p. Show that $B_p = O(\sqrt{p})$ as $p \to \infty$.

EXERCISE 5.72 (Khintchine–Kahane inequalities). Khintchine inequalities have a vector-valued generalization which is due to Kahane: *If $x_1, \ldots, x_n$ belong to some normed space Y and X' denotes the random variable $\|\sum_{i=1}^n \varepsilon_i x_i\|_Y$, then*

$$A'_p\|X'\|_{L_2} \leqslant \|X'\|_{L_p} \leqslant B'_p\|X'\|_{L_2}$$

where $A'_p > 0$ and B'_p are constants depending only on p. Prove this. Moreover, we have $A_1 = A'_1 = 1/\sqrt{2}$ and $B'_p = \Theta(\sqrt{p})$ as $p \to \infty$.

EXERCISE 5.73. Prove Proposition 5.58 by following the outline given below.
(i) If X is symmetric, show that $\mathbf{E}\exp(\lambda X) \leqslant \exp(\frac{e}{2}\|X\|_{\psi_2}^2\lambda^2)$ for any $\lambda > 0$.
(ii) Let Y be an independent copy of a mean zero random variable X. Show that $\mathbf{E}\exp(\lambda X) \leqslant \mathbf{E}\exp(\lambda(X - Y))$. Using this symmetrization trick, deduce from (i) that the inequality $\mathbf{E}\exp(\lambda X) \leqslant \exp(2e\|X\|_{\psi_2}^2\lambda^2)$ holds for any mean zero random variable X.
(iii) Deduce Proposition 5.58 using Lemma 5.28.

EXERCISE 5.74 (Linear combinations of subgaussian random variables are subgaussian). Show the following variant of Proposition 5.58: if $X_1, \ldots, X_n$ are independent and mean zero, then $\|X_1 + \cdots + X_n\|_{\psi_2} \leqslant C(\|X_1\|_{\psi_2}^2 + \cdots + \|X_n\|_{\psi_2}^2)$ for some absolute constant C.

EXERCISE 5.75. Verify that $\|Z\|_{\psi_2} = \sqrt{2/\pi}$ and that, for any variable X, $\|X\|_{\psi_1} \leqslant \|X\|_{\psi_2}$.

EXERCISE 5.76 (Bernstein's inequalities). (i) Show that if $\mathbf{E}X = 0$ and $\|X\|_{\psi_1} \leqslant 1$, then $\mathbf{E}\exp(\lambda X) \leqslant 1 + 2\lambda^2 \leqslant \exp(2\lambda^2)$ for $|\lambda| < 1/2$ (cf. Lemma 5.28).
(ii) Under the hypotheses of Proposition 5.59, assuming $K = 1$ and denoting $S = a_1X_1 + \cdots + a_nX_n$, prove that $\mathbf{E}\exp(\lambda S) \leqslant \exp(2\lambda^2 \sum a_i^2)$ for $|\lambda| \leqslant 1/(2\|a\|_\infty)$.
(iii) Prove Proposition 5.59.

Notes and Remarks

Section 5.1. An encyclopedic reference for sphere packings is the book [**CS99**]. Other valuable and historically significant references are [**Rog64, Bör04, FT97**].

Packing and covering on the Euclidean sphere and the discrete cube. To complement Proposition 5.1, it has been proved in [**BGK$^+$01**] that for $0 \leqslant t \leqslant \arccos\sqrt{2/n}$, we have $V(t) \geqslant (6\sqrt{n}\cos t)^{-1}(\sin t)^{n-1}$ (similar estimates appear in [**Bör04**], Lemma 6.8.6). For some values of n, t (roughly for $t > 1.14$ and for large n), this is better than the lower bound from (5.4), and similarly superior to the improved bound from Exercise 5.4 if $t > 1.221$.

The random covering argument from Proposition 5.4 is due to Rogers [**Rog57, Rog63**]. The factor $Cn\log n$ from Corollary 5.5 is usually referred to as the *density of the covering*, even though calling it "the overlap" or "the redundancy" would seem more logical. Both the original Rogers's argument, and the one presented here, allow achieving $C = 1$ at the expense of additional lower order terms (see Exercise 5.8 and its hint). Recent advances by Dumer [**Dum07**] improve the bound on the density to $(\frac{1}{2} + o(1))n\log n$. The paper [**Dum07**] establishes also a density bound $\frac{1}{2}n\log n + 2n\log\log n + 5n$, valid for all $\varepsilon \in (0,1)$ and all $n \geqslant 4$. It should be noted, however, that the latter result deals with a slightly easier problem, covering the sphere $S^{n-1} \subset \mathbb{R}^n$ by balls whose centers are not required to belong to S^{n-1} (i.e., with the parameter N' from Exercise 5.1). Finally, at the price of increasing the constant C, the result from Corollary 5.5 can be strengthened as follows: for any dimension n and angle ε, there is a covering of S^{n-1} by caps of radius ε such that any point belongs to at most $400n\log n$ caps [**BW03**].

Since the sphere looks locally like a Euclidean space, as the radii of the caps tend to 0, the packing/covering problems for S^{n-1} converge to the corresponding problems for $\mathbb{R}^{n-1}$. (The original random covering argument of Rogers [**Rog57**] considered an even more general question, economical coverings of $\mathbb{R}^n$ by translates of an arbitrary convex body—the spherical variant being an afterthought—and led to an upper bound of $n\log n + n\log\log n + 5n$ for the appropriately defined asymptotic density.) In that setting, a lower bound on density of optimal coverings by Euclidean balls is $\Omega(n)$ [**CFR59**] and this estimate can be transferred back to S^{n-1} if the radius is small enough; see Example 6.3 in [**BW03**] for an argument that works if $\varepsilon \leqslant \arcsin(1/\sqrt{n})$.

References for the results mentioned about packing are [**Ran55**] (Rankin) and [**KL78**] (Kabatjanskiĭ–Levenšteĭn); we refer to [**CS99**] for more information (see also [**BN06a**]). Again, when the radius of the cap tends to 0, the problem becomes the classical sphere packing problem in $\mathbb{R}^n$. In this context, a classical result due to Minkowski–Hlawka shows the existence of lattice packings of Euclidean balls (or actually, of any symmetric convex body) in $\mathbb{R}^n$ which cover a proportion $1/2^{n-1}$ of the space (a.k.a. *packing density*). Remarkably, this result has been only marginally improved in the past century [**Rog47, DR47, Bal92b**] and is exponentially far from Kabatjanskiĭ–Levenšteĭn upper bound—which is approximately of order 0.66^n—for the proportion covered by a (non-necessarily) lattice packing (see [**Gru07**] for more on this topic).

Covering and particularly packing in the Hamming cube is of fundamental importance in coding theory; see, e.g., [**Rot06, CHLL97**]. The case of (very small) balls of radius $1/n$ in $\{0, \dots, q-1\}^n$ is treated in [**KP88**].

The Gilbert–Varshamov bound has been improved in the q-ary cube for certain large values of q in [**TVZ82**], using a link with modular curves.

Packing and covering for convex bodies. For early references on metric entropy of convex bodies see [**CS90**], [**Pis89b**].

The arguments from [**Bar14**] imply the following improvement on the volumetric bound from Corollary 5.10: for $\varepsilon \in (0, 1)$, any symmetric convex body in $\mathbb{R}^n$ is $(1+\varepsilon)$-close in Banach–Mazur distance to a polytope with $(C/\sqrt{\varepsilon})^n$ vertices. (This is sharp: consider the case of the sphere.) To the best of our knowledge, it is not known whether an analogous statement holds for not-necessarily symmetric bodies and the affine version (4.2) of the Banach–Mazur distance. Similar questions can be considered for large ε, or even ε growing with the dimension. In the case of the sphere, this is essentially the problem considered in Exercise 5.13. Again, [**Bar14**] contains good estimates in the general case. However, the bounds from [**Bar14**] deteriorate as the *asymmetry* of the body (defined, for example, as the minimal distance d_{BM} to a symmetric body) increases. Estimates that are superior for some ranges of parameters can be found in [**Sza**].

Let us also mention an important open problem, known as the duality conjecture: *do there exist absolute constants $c, C > 0$ such that for every two symmetric convex bodies $K, L \subset \mathbb{R}^n$ we have*

$$\log N(L^\circ, K^\circ) \leqslant C \log N(K, cL)? \tag{5.67}$$

This was proved when K or L is the Euclidean ball [**AMS04**] and extended to the case when a bound on the K-convexity constant (as defined in Section 7.1.2) is present in [**AMSTJ04**]. Another possible generalization to the setting of non-symmetric convex bodies is more tricky: in that case, even the proper formulation of (5.67) is not entirely clear.

A deep fact about covering numbers is the following ([**Mil86**]; see also the discussion in [**Pis89b**]): there is an absolute constant C such that, for every symmetric convex body $K \subset \mathbb{R}^n$, there is an 0-symmetric ellipsoid $\mathscr{E}$ such that

$$\max\left(N(K, \mathscr{E}), N(\mathscr{E}, K)\right) \leqslant C^n. \tag{5.68}$$

Note that since metric entropy duality (5.67) is known to hold when one of the bodies is an ellipsoid, it follows then that similar bounds automatically hold also for $N(K^\circ, \mathscr{E}^\circ)$ and $N(\mathscr{E}^\circ, K^\circ)$. (In the original definitions, all four quantities were included explicitly or implicitly.) Such an ellipsoid $\mathscr{E}$ is called an *M-ellipsoid for K*, and K is said to be in the *M-position* when B_2^n is an M-ellipsoid for K. The M-ellipsoids are discussed in detail in [**AAGM15**].

Metric entropy of classical manifolds. Theorem 5.11 is from [**Sza82**], which covers the case of all metrics induced by unitarily invariant norms (see also [**Sza83, Sza98**] and [**Paj99**]). Examples of packings in some Grassmannians (mostly low-dimensional), some of them optimal, can be found in [**CHS96, SS98**]. More recent references, motivated by information transmission issues and concentrated on different asymptotics (k fixed and n tending to infinity), are [**BN02, BN05, BN06b**]. It appears that the theoretical computer science community is not aware that questions of that nature were considered in AGA already in 1980s.

Section 5.2. Classical general references about concentration of measure are [**Led01**] and [**Sch03**]. We particularly recommend the recent monograph [**BLM13**]. For a presentation directed towards applications to data science, see [**Ver**].

Isoperimetry and concentration. A geometry-oriented reference about isoperimetric inequalities is [**BZ88**]. The paternity of the isoperimetric inequality on the sphere (Theorem 5.13) is usually attributed to Lévy [**Lév22, Lév51**] although the arguments he presented were not fully rigorous; [**Sch48**] is usually cited as the first rigorous proof. Remarkably, the functional version (Lévy's lemma, in the language of our Corollary 5.17) appears explicitly in [**Lév22**] (see p. 279) and is therefore almost one century old!

A self-contained proof of the isoperimetric inequality on S^{n-1}, based on the concept of spherical symmetrization, appears in [**FLM77**]. Another symmetrization procedure (the two-point symmetrization) is applied in [**Ben84**]. The simple proof of the non-sharp inequality from Proposition 5.15 is based on [**AdRBV98**]. Proposition 5.20 is from [**JS**].

The Gaussian isoperimetric inequality was proved independently by Borell [**Bor75b**] and Sudakov–Tsireslon [**SC74**]. For a proof of Poincaré's lemma (Theorem 5.22) going beyond the weak convergence version from Exercise 5.29, we refer to [**DF87**] (which also advocates that the statement was first formulated by Borel and not by Poincaré). See also [**Led96**] and references therein. For a direct proof of concentration of measure on Gauss space, see [**Pis86**].

Ehrhard's inequality (5.31) was proved in [**Ehr83**] for convex sets, then extended in [**Lat96**] to the case where only one of the sets is convex, with the general case being treated in [**Bor03**]. *A priori*, deriving an isoperimetric inequality such as (5.29) requires validity of (5.31) for an arbitrary Borel set and a ball; the paper [**Ehr83**], however, contains a direct application of the technique to prove (5.29). A general reference for this circle of ideas is [**Lat02**].

The concept of central values was formalized and applied in the context of QIT in [**ASW11**], which also contains versions of Corollaries 5.32 and 5.35. However, instances of the arguments can be found in [**Has09**] and in AGA literature dating to (at least) the 1980s.

Proposition 5.34 appears in [**Dmi90, Kwa94, Fer97**]. Exercise 5.48 appears as Proposition 1.7 in [**Led01**]. Proposition 5.37 is Corollary 1.17 from [**Led01**].

There are various generalizations of Hoeffding's inequality appearing in Exercise 5.57, notably due to Azuma [**Azu67**] and McDiarmid [**McD89**] in the context of martingales.

Geometric and analytical methods. General references for Section 5.2.4 are [**MS86, Sch03, DS01, GM00, BLM13, BGL14, GZ03**].

Gromov's comparison theorem (Theorem 5.38) appeared first in the preprint [**Gro80**]. A proof can be found in an appendix in [**MS86**]. A new proof and an extension to non-Riemannian spaces was proposed recently in [**CM15**]. While the theorem is sharp as stated, there is a reason to suspect that a more precise result should be available: the proof proceeds via a local/variational argument and the *globally* normalized volume appears only *a posteriori*. A more satisfactory variant appears in [**Mil15**]. In addition to the curvature, it takes into account the actual diameter of the manifold in question, which may be strictly smaller than the bound following indirectly from the curvature. However, since the results in [**Mil15**] necessarily involve model manifolds more complicated than spheres, their statements are somewhat technical.

The case of manifolds of dimension 1 is a little special. First, while the definition of Ricci curvature in dimension 1 needs to be properly construed, the only sensible value is 0 since every such manifold looks locally like a segment. Accordingly, Proposition 5.41 is then vacuously true. Next, the solution to the isoperimetric problem in S^1 (resp., in $\mathbb{R}$) is very simple: among sets of any (positive, but not full) measure, the boundary is the smallest if it consists of exactly two points. Consequently, the solutions, both for the "smallest boundary" and the "smallest enlargement" problems, are arcs (resp., segments). However, finer analytic statements (including but not limited to LSI) are interesting and highly nontrivial already in dimension 1. For example, in view of Proposition 5.44, the validity of (5.48) for the 1-dimensional Gaussian measure implies the same inequality in any dimension (with the same constant α, which, in view of Proposition 5.42, can be taken to be 1, which is optimal). Indeed, even statements about spaces consisting of only two points can be deep; as for example in the elementary proof of the Gaussian isoperimetric inequality presented in [**Bob97**]. We will return to the same theme further when reporting on developments directly related to LSI and hypercontractivity.

Log-Sobolev inequalities (LSI) were introduced in a seminal paper by Gross [**Gro75**]. Again, the case of manifolds of dimension 1 (segments, circles) is a little special; see [**GMW14**] for an elementary overview of this aspect of the subject and for references. The link with concentration of measure (the Herbst argument) originates in an unpublished letter from Herbst to Gross. The connection between LSI, Ricci curvature, and the Hessian of the density was put forward in [**BÉ85, Bak94**]. For a comprehensive treatment of functional inequalities (including complete references), see [**BGL14**]. Another fruitful approach is the connection between LSI and the quadratic transportation cost inequalities; see Chapter 6 in [**Led01**].

As exemplified in Table 5.4, the values of the Poincaré constants can often be computed exactly. Indeed, the Poincaré inequality (5.54) can be rewritten as $\mathbf{Var}_\mu f \leqslant \alpha \int (-\Delta f) f \, \mathrm{d}\mu$, where Δ is the Laplace–Beltrami operator on $L_2(X, \mu)$. It follows that the optimal α is equal to the reciprocal of the "spectral gap," i.e., the smallest nonzero eigenvalue of $-\Delta$. In some examples the eigenfunctions of the Laplace–Beltrami operator can be explicitly described: for the Gauss space they are the Hermite polynomials, for the sphere they are the spherical harmonics (see the elementary [**See66**], or [**BGM71**] which covers also the case of the projective spaces). On S^{n-1}, equality in (5.54) is achieved for functions of the form $x \mapsto \langle x, y \rangle$ with $y \in \mathbb{R}^n$. For Lie groups there is a connection with the spectrum of the Casimir operator and representations of the associated Lie algebra (see Proposition 10.6 in [**Hal15**]), which allows to derive the entire spectrum of $-\Delta$. The case of $\mathsf{SO}(n)$ and $\mathsf{SU}(n)$ appears in [**SC94**] (for $\mathsf{U}(n)$, see [**Voi91**]). Note that in these examples there is equality in (5.54) when f is a function of the form $M \mapsto \mathrm{Tr}(AM)$ for $A \in \mathsf{M}_n$. For a complete list of semisimple Lie algebras, see [**Rot86**]. The spectrum of Grassmann manifolds is considered in [**Tsu81, EC04, TK04, Hal07**], which allows in principle to retrieve the value of the Poincaré constant for specific dimensions if needed.

Hypercontractivity for the Ornstein–Uhlenbeck semigroup (Proposition 5.47) has been first established by Nelson [**Nel73**]. The connection with log-Sobolev inequalities was put forward by Gross [**Gro75**].

In many situations, the Gaussian case can be treated as a limit case from the case of the hypercube via the central limit theorem. By the tensorization property (Proposition 5.44), this amounts ultimately to verifying statements about the two-point space $\{-1, 1\}$ (see [**Gro75**] for a proof of the Gaussian LSI along these lines). The hypercontractivity inequality on the discrete cube is known as the Bonami–Beckner inequality [**Bon70, Bec75**]. Some variants of Proposition 5.48 appear in [**Jan97**]. For a more sophisticated technology giving sharp estimations on the moments of Gaussian polynomials (or Gaussian chaoses) see [**Lat06**]. The statement about concentration on polynomials on products of spheres appearing in Remark 5.50 follows from the proof of Corollary 12 in [**Mon12**].

Discrete settings. A reference focusing on the case of the hypercube is [**O'D14**] (it contains in particular the versions of Proposition 5.48 and Corollary 5.49 for the hypercube alluded to in Remark 5.50). In addition to [**O'D14**], general references for Section 5.2.5 are [**Mat02, McD98**]. The main statement of Theorem 5.51 was proved in [**Har66**] and rediscovered in [**Kat75**]. A short proof may be found in [**FF81**]; we also recommend the reference [**Lea91**]. Theorem 5.51 deals with vertex-isoperimetry. If we consider instead edge-isoperimetry (minimizing the number of edges joining A to A^c), the optimal sets are no longer Hamming balls but subcubes.

Theorem 5.54 is taken from [**Tal88**] (Note that [**Tal88**] states the result for the cube $\{-1,1\}^n$ and so the coefficient in the exponent in the estimate corresponding to (5.60) is there $\frac{1}{8}$.) Theorem 5.55 appears in [**JS91**] and [**Mec04**]. The latter paper addresses general unconditional direct sums and not only ℓ_q-sums; see also [**Mec03**]. Similar results with quite different proofs were presented in [**Mau91**] and [**Dem97**]. The most abstract (and most flexible) statements are arguably in [**Tal95, Tal96b, Tal96a**]. The arguments addressing settings more general than that of Theorem 5.54 usually led to a coefficient $\frac{1}{4}$ in the exponent as in (5.61), except for [**Tal95**], which includes a statement (Theorem 4.2.4) featuring coefficient $\frac{1}{2}$, but at the cost of introducing additional factors of lower order and restricting the range of t. A clean proof of Theorem 5.56 (which also has coefficient $\frac{1}{2}$ in the exponent) can be found in [**BLM13**]; the argument is attributed to [**Led97**] and the result itself to [**Tal96b**].

Deviation inequalities. Some references for Section 5.2.6 are [**Ver12**] and [**CGLP12**] (the latter treats also the case of intermediate growth between subgaussian and subexponential). As pointed out in the main text, there are several possible forms of ψ_r conditions and of definitions of the ψ_r-norms. The original ones were (presumably) in terms of Orlicz/Young functions: given an increasing convex function $\psi : \mathbb{R}_+ \to \mathbb{R}_+$ with $\psi(0) = 0$ and $\psi(x) \to \infty$ as $x \to \infty$, we may define a the ψ-norm of a random variable X as (for example)

$$\|X\|_\psi = \inf\{c > 0 \ : \ \mathbf{E}\,\psi(|X|/c) \leqslant \psi(1)\}.$$

If one considers $\psi_r(x) = \exp(x^r) - 1$ ($r \geqslant 1$), then, for $r = 1, 2$, one gets norms which are equivalent (although not equal) to the ones defined in (5.66) and (5.65). For precise statements and proofs, see Theorem 1.1.5 in [**CGLP12**], which also covers the link to (the rate of growth of) the Laplace transforms mentioned in the main text; cf. Lemma 5.28 and Exercise 5.76. Overall, Section 1.1 of [**CGLP12**] is an excellent reference for ψ_r conditions/norms, which are otherwise difficult to extract from books/surveys on the more general Orlicz spaces.

For a historical account of Bernstein's contributions, we refer to pp. 126–128 in [**AAGM15**]. For more precise results about moments of sums of independent variables, see [**Lat97**]. For non-commutative analogues of these inequalities (i.e., for sums of random matrices), see [**Tro12**].

Finally, among other techniques to prove concentration of measure, we mention the so-called martingale method which implies for example concentration on permutation groups (see [**Sch82, Mau79, MS86**]): *If we equip the symmetric group* $\mathfrak{S}_n$ *with the uniform probability measure and the distance* $d(\sigma,\tau) = \frac{1}{n}\,\mathrm{card}\{i : \sigma(i) \neq \tau(i)\}$, *then any* 1*-Lipschitz function* f *on* $(\mathfrak{S}_n, d)$ *satisfies* $\mathbf{P}(f \geqslant \mathbf{E}f + t) \leqslant \exp(-nt^2/8)$ *for any* $t \geqslant 0$.

The best constants in Khintchine inequalities (see Exercise 5.72) have been found in [**Sza76**] (who proved $A_1 = 1/\sqrt{2}$) and in [**Haa81**] (for $p > 1$). The Khintchine–Kahane inequalities from Exercise 5.72 were first proved in [**Kah85**]. The correct asymptotic order of the constants as $p \to \infty$ was found in [**Kwa76**], while the value $A_1' = 1/\sqrt{2}$ is from [**LO94**]. A complete proof of the Khintchine–Kahane inequalities can be found by consulting Theorem 3.5.2 of [**AAGM15**].

CHAPTER 6

Gaussian processes and random matrices

This chapter is devoted to the development of probabilistic techniques which, along the concentration of measure from Chapter 5, constitute our most powerful tools. Specifically, we will consider stochastic processes (mostly, but not exclusively, Gaussian) and present deep results permitting their quantitative study. The key insights are the link between suprema of Gaussian processes and the mean width of convex bodies, and the use of comparison theorems for Gaussian processes to the analysis of spectral behavior of random matrices.

6.1. Gaussian processes

This section deals with Gaussian processes (widely used in mathematical modeling and in statistics) and presents several tools for estimating various parameters related to such processes. A *Gaussian process* $\mathbf{X} = (X_t)_{t\in T}$ is simply a family of *jointly* Gaussian variables, normally with mean zero, defined on some probability space Ω, which may or may not be specified. See Appendix A for more on the terminology and for basic and not-so-basic facts about Gaussian variables.

We especially focus on studying the supremum of Gaussian processes, e.g., computing (or estimating) $\mathbf{E}\sup\{X_t : t \in T\}$. In our context, suprema of Gaussian processes appear when considering the Gaussian mean width of a convex body (and this is essentially the general case; see Section 6.1.1) and therefore can be used to estimate other geometric parameters such as volume. There are essentially three levels of sophistication when investigating the supremum of a Gaussian process.

(i) Discretize the problem by using an ε-net and appealing to the union bound.

(ii) Use a recursive version of (i) by considering a whole hierarchy of ε-nets (for example $\varepsilon = 2^{-k}$ for every integer k). This is called a "chaining argument".

(iii) Use a further sophistication of (ii), where instead of using nets whose resolution parameter is uniform across the index set, we allow more general partition schemes. This is called the "generic chaining" or the "majorizing measure" approach.

A deep result due to Talagrand asserts that (iii) provides an estimate on the supremum of any Gaussian process which is always sharp up to a multiplicative constant. However, we mostly consider the situations (i) and (ii) since they are much simpler and sufficient for our purposes.

We note for the record that without any assumptions on regularity of $\mathbf{X}$, which will be implicitly made in what follows, measurability issues and other complications may in principle arise, particularly when T is uncountable. For the benefit of a non-specialist reader we sketch examples of possible pathologies in Exercise 6.1. However, such potential difficulties are not relevant in our context and we will henceforth largely ignore them. For example, in all the settings we are interested in we will have enough regularity so that

$$\mathbf{E}\sup\{X_t \,:\, t \in T\} = \sup_{F\subset T,\, F \text{ finite}} \mathbf{E}\max\{X_t \,:\, t \in F\}, \tag{6.1}$$

and other questions can similarly be reduced to considering instances of the problem with finite index sets. As usual, the crucial point will be that the constants that may appear in the statements do not depend on $\mathbf{X}$ and, in particular, on the size of T.

EXERCISE 6.1. Give examples of processes $(X_t)_{t\in T}$ such that, for every $t \in T$, $X_t = 0$ a.s., but (a) $\mathbf{E}\sup\{X_t \,:\, t \in T\} = \infty$ (b) $\sup\{X_t \,:\, t \in T\}$ is not measurable.

6.1.1. Key example and basic estimates. We start with a simple—but crucial—observation that if $G : \Omega \to \mathbb{R}^n$ is a standard Gaussian vector, then $(\langle G, x\rangle)_{x\in\mathbb{R}^n}$ is a Gaussian process. Recalling the definition of the Gaussian mean width of a (bounded nonempty) set $K \subset \mathbb{R}^n$, as introduced in Section 4.3.3,

$$w_G(K) = \mathbf{E}\,\sup\{\langle G, x\rangle \,:\, x \in K\}, \tag{6.2}$$

we see that calculating $w_G(K)$ is equivalent to finding the expectation of the supremum of a certain Gaussian process, a subprocess of $(\langle G, x\rangle)_{x\in\mathbb{R}^n}$.

This instance is actually, more or less, the general case. This follows by combining two facts:
(i) the map $x \mapsto \langle G, x\rangle$ is an isometry from $(\mathbb{R}^n, |\cdot|)$ to $L_2(\Omega)$,
(ii) the joint distribution of $\mathbf{X} = (X_t)_{t\in T}$ is uniquely determined by the covariances $\big(\mathbf{E}X_sX_t\big)_{s,t\in T}$ and so all the stochastically relevant information about the process is encoded in the geometry of $\mathbf{X}$, considered as a subset of $L_2(\Omega)$.
Consequently, if E is a Euclidean space and vectors $x_t \in E$ (for $t \in T$) are such that $\langle x_s, x_t\rangle = \mathbf{E}X_sX_t$ for all $s, t \in T$, and if G_E is a standard Gaussian vector on E, then the Gaussian process $(\langle G_E, x_t\rangle)_{t\in T}$ is a faithful copy of $\mathbf{X}$.

For a finite process $\mathbf{X} = (X_k)_{1\leqslant k\leqslant N}$ this is easily realized: we can choose $E := \operatorname{span}\{X_k\} \subset L_2(\Omega)$ and $x_k = X_k$. We then have in particular

$$\mathbf{E}\max_{1\leqslant k\leqslant N} X_k = w_G(\mathbf{X}) = w_G(K_{\mathbf{X}}), \tag{6.3}$$

where $K_{\mathbf{X}} := \operatorname{conv}\{X_k \,:\, 1 \leqslant k \leqslant N\}$ is a convex set in E. (This effectively covers any situation where (6.1) applies.)

The above construction shows that the two (classes of) problems, namely calculating (1) the mean width of a convex set and (2) the expectation of the supremum of a Gaussian process, are essentially equivalent. This equivalence will turn out to be very fruitful. Recall that if $0 \in K$, then $\sup\{\langle y, x\rangle \,:\, x \in K\} = \|y\|_{K^\circ}$ and so $w_G(K) = \mathbf{E}\,\|G\|_{K^\circ}$. It may happen that the set $K_{\mathbf{X}}$ does not contain 0, but this can be remedied by considering instead $\mathbf{X}' = (X_k - X_0)_{1\leqslant k\leqslant N}$ for some $X_0 \in \operatorname{conv}\{X_k\}$. We have then

$$\mathbf{E}\,\max\{X_k \,:\, 1 \leqslant k \leqslant N\} = \mathbf{E}\,\max\{X_k - X_0 \,:\, 1 \leqslant k \leqslant N\},$$

which is reminiscent of the fact that the mean width does not depend on the choice of the origin. Note that if we select X_0 belonging to the relative interior of $\operatorname{conv}\{X_k\}$, we will even be able to stay in the category of convex bodies with the origin in the interior.

We next state a simple upper bound on the expectation of the supremum of a finite Gaussian process.

LEMMA 6.1. *Let $(X_k)_{1 \leqslant k \leqslant N}$ be Gaussian random variables with mean zero and variance bounded by* 1. *Then*

$$\mathbf{E} \max_{1 \leqslant k \leqslant N} X_k \leqslant \sqrt{2 \log N}. \tag{6.4}$$

Moreover, if $(X_k)_{1 \leqslant k \leqslant N}$ are independent $N(0,1)$ random variables, then

$$\mathbf{E} \max_{1 \leqslant k \leqslant N} X_k \geqslant (1 - o(1))\sqrt{2 \log N}. \tag{6.5}$$

PROOF. We use the following elementary computation: if X has distribution $N(0, \sigma^2)$ with $\sigma^2 \leqslant 1$ then $\mathbf{E}\, e^{tX} = \exp(t^2\sigma^2/2) \leqslant \exp(t^2/2)$ for any real t. For $\beta > 0$ to be determined, we have (the second inequality being Jensen's inequality)

$$\mathbf{E} \max_{1 \leqslant k \leqslant N} X_k \leqslant \mathbf{E}\, \frac{1}{\beta} \log \sum_{k=1}^{N} e^{\beta X_k} \leqslant \frac{1}{\beta} \log \mathbf{E} \sum_{k=1}^{N} e^{\beta X_k} \leqslant \frac{1}{\beta} \log(N \exp(\beta^2/2)),$$

and the optimal choice $\beta = \sqrt{2 \log N}$ yields (6.4).

This completes the proof of the first inequality. A slightly weaker, but more general estimate, based on the simple (and not-so-optimal, see Appendix A.1) upper bound

$$\mathbf{P}(Z \geqslant t) \leqslant \frac{1}{2} e^{-t^2/2} \quad \text{if } t \geqslant 0 \tag{6.6}$$

for the tail of a standard normal variable Z (see Exercise A.1) is given in Lemma 6.16. We relegate the proof of the second inequality (based on a *lower* bound for the tail of Z) to Exercise 6.2, which also gives an explicit expression for the $o(1)$ quantity. □

We note that the estimate from (6.4) also holds for the expected maximum of the *absolute values* of Gaussian variables.

LEMMA 6.2 (See Exercise 6.3). *Let $N \geqslant 2$ and let $(X_k)_{1 \leqslant k \leqslant N}$ be jointly Gaussian random variables with variance bounded by* 1. *Then*

$$\mathbf{E} \max_{1 \leqslant k \leqslant N} |X_k| \leqslant \sqrt{2 \log N}.$$

*When $N \geqslant 4$, the inequality holds for any Gaussian random variables (that is, not necessarily **jointly** Gaussian).*

As an application, we have a bound on the volume of a polytope, given its number of vertices.

PROPOSITION 6.3. *Let $K \subset \mathbb{R}^n$ be a polytope with (no more than) N vertices and whose outradius is at most* 1. *Then*

$$\operatorname{vrad} K \leqslant \kappa_n^{-1} \sqrt{2 \log N} \sim \sqrt{\frac{2 \log N}{n}},$$

where κ_n is defined by (A.8).

PROOF. Let $x_1, \dots x_N$ be the vertices of K. Without loss of generality we may assume that $K \subset B_2^n$. We can now apply the first part of Lemma 6.1 with $X_k = \langle G, x_k \rangle$ to obtain

$$\mathbf{E} \max_{1 \leqslant k \leqslant N} \langle G, x_k \rangle \leqslant \sqrt{2 \log N}.$$

Since, for any $y \in \mathbb{R}^n$, $\sup\{\langle y, x\rangle : x \in K\} = \max_{1 \leqslant k \leqslant N}\langle y, x_k\rangle$, the above bound is (cf. (6.2)) equivalent to $w_G(K) \leqslant \sqrt{2\log N}$. It remains to appeal to the relation (4.32) between the Gaussian mean width and the usual mean width, and to Urysohn's inequality (4.34). $\square$

REMARK 6.4 (Sharp bound on volume of polytopes with few vertices). The bound in Proposition 6.3 can be improved to $O\left(\sqrt{\frac{\log(N/n)}{n}}\right)$. This improvement is meaningful only when N is not much larger than n. For example, if $K = B_1^n$ (the unit ball of ℓ_1^n), then $K = \operatorname{conv}\{\pm e_1, \ldots, \pm e_n\}$, where $(e_k)_{k=1}^n$ is the standard unit vector basis in $\mathbb{R}^n$. Consequently, Proposition 6.3 used with $N = 2n$ leads to the bound $\operatorname{vrad} B_1^n = O\big(\sqrt{\log(n)/n}\big)$, while the correct value (cf. Table 4.1) is $O(1/\sqrt{n})$. Some of these issues are explored in Exercise 6.4.

REMARK 6.5 (Conjectured extremal property of the regular simplex). It is conjectured that the polytope with N vertices and outradius 1 that has the largest Gaussian mean width is the regular simplex inscribed in the unit ball. This is known (and easy) for $N \leqslant 3$. By the argument used in the proof of Proposition 6.3, this is equivalent to characterizing the instances giving the extremal value of $\mathbf{E}\max_{1 \leqslant k \leqslant N} X_k$ in the context of Lemma 6.1 (with $(X_k)_{1 \leqslant k \leqslant N}$ jointly Gaussian).

EXERCISE 6.2. Show that, in the context of the second part of Lemma 6.1, we have

$$\mathbf{E}\max_{1 \leqslant k \leqslant N} X_k \geqslant \sqrt{2\log N} - O\left(\frac{\log\log N}{\sqrt{\log N}}\right)$$

by using the lower bound from (A.4).

EXERCISE 6.3. Prove Lemma 6.2 for $N \geqslant 4$ as follows: if Z is an $N(0,1)$ random variable, then

$$\mathbf{E}\max_{1 \leqslant k \leqslant N} |X_k| \leqslant T + 2N\int_T^\infty \mathbf{P}(Z > t)\,dt = T + \frac{2N}{\sqrt{2\pi}}e^{-T^2/2} - 2NT\,\mathbf{P}(Z > T),$$

and check numerically that the choice $T = \sqrt{2\log N - 3/2}$ gives the needed inequality. Note that this proof does not use the hypothesis that the variables are *jointly* Gaussian. For 2 or 3 jointly Gaussian variables, use Proposition 6.9 to identify extremal configurations.

EXERCISE 6.4 (Volume of polytopes with very few vertices). Show that if, in the notation of Proposition 6.3, $N = O(n)$, then $\operatorname{vrad} K = O(1/\sqrt{n})$, which yields the better bound stated in Remark 6.4 for that range of N.

EXERCISE 6.5 (Volume of symmetric polytopes with few vertices). Show that if $K \subset B_2^n$ is a **symmetric** polytope with N vertices, the conclusion of Proposition 6.3 can be slightly improved to the inequality $\operatorname{vrad}(K) \leqslant \sqrt{2\log N}/\sqrt{n}$.

EXERCISE 6.6 (Mean widths of standard sets). Prove the estimates involving mean width from Table 4.1.

6.1.2. Comparison inequalities for Gaussian processes. The following fundamental inequality is known as Slepian's lemma. It expresses the fact that strengthening correlations of a Gaussian process decreases the supremum.

PROPOSITION 6.6 (Slepian's lemma, not proved here). *Let $(X_k)_{1\leqslant k\leqslant N}$ and $(Y_k)_{1\leqslant k\leqslant N}$ be Gaussian processes, and assume that*

$$\mathbf{E}\left[(X_k - X_j)^2\right] \leqslant \mathbf{E}\left[(Y_k - Y_j)^2\right]$$

for every $1 \leqslant j, k \leqslant N$. Then,

$$\mathbf{E} \sup_{1\leqslant k\leqslant N} X_k \leqslant \mathbf{E} \sup_{1\leqslant k\leqslant N} Y_k. \tag{6.7}$$

Moreover, if also $\mathbf{E}\, X_k^2 = \mathbf{E}\, Y_k^2$ for all k, then for any $\lambda_1, \dots, \lambda_N \in \mathbb{R}$,

$$\mathbf{P}\left(X_k \geqslant \lambda_k \text{ for some } k\right) \leqslant \mathbf{P}\left(Y_k \geqslant \lambda_k \text{ for some } k\right). \tag{6.8}$$

Slepian's lemma can be reformulated in geometric language: contractions decrease the mean width. More precisely, if $T \subset \mathbb{R}^n$ and if $\phi : T \to \mathbb{R}^m$ is a contraction (with respect to the Euclidean distance, not necessarily linear), then

$$w_G\big(\operatorname{conv}(\phi(T))\big) = w_G(\phi(T)) \leqslant w_G(T) = w_G(\operatorname{conv}(T)). \tag{6.9}$$

If $m = n$, we can immediately deduce from (4.32) that also $w(\phi(T)) \leqslant w(T)$. This property seems intuitively obvious, but we know a simple proof only if ϕ is linear (or affine, see Exercise 4.46).

Slepian's lemma admits a number of variants and generalizations, and the following one has been quite useful. In particular, it leads to elegant proofs of various statements about random matrices (see Section 6.2) and versions of Dvoretzky's theorem (Section 7.2).

PROPOSITION 6.7 (Gordon's lemma, not proved here). *Let $(X_t)_{t\in T}$ and $(Y_t)_{t\in T}$ be Gaussian processes. Assume further that $T = \bigcup_{s\in S} T_s$ and that*

(i) $\|X_t - X_{t'}\|_2 \leqslant \|Y_t - Y_{t'}\|_2$ if $t \in T_s, t' \in T_{s'}$ with $s \neq s'$,
(ii) $\|X_t - X_{t'}\|_2 \geqslant \|Y_t - Y_{t'}\|_2$ if $t, t' \in T_s$ for some s.

Then

$$\mathbf{E} \max_{s\in S} \min_{t\in T_s} X_t \leqslant \mathbf{E} \max_{s\in S} \min_{t\in T_s} Y_t\,.$$

Moreover, if also $\mathbf{E}\, X_t^2 = \mathbf{E}\, Y_t^2$ for all $t \in T$, then for any choice of real numbers $(\lambda_t)_{t\in T}$,

$$\mathbf{P}\left(\bigcup_{s\in S} \bigcap_{t\in T_s} \{X_t \geqslant \lambda_t\}\right) \leqslant \mathbf{P}\left(\bigcup_{s\in S} \bigcap_{t\in T_s} \{Y_t \geqslant \lambda_t\}\right).$$

REMARK 6.8. (1) When all T_s are singletons, Gordon's lemma reduces to the Slepian version. Accordingly, Proposition 6.7 is sometimes referred to as the *Slepian–Gordon lemma.* (2) Replacing X_t, Y_t with $-X_t, -Y_t$ we get analogous statements for min max in place of max min, and similarly for the Slepian's lemma and for the statements about probabilities. (3) Further generalizations to min and max applied alternatively more than twice are possible.

Another fundamental comparison inequality is the Khatri–Šidák lemma.

PROPOSITION 6.9 (Khatri–Šidák; see Exercise 6.9). *Consider two Gaussian processes $(X_k)_{1\leqslant k\leqslant N}$ and $(Y_k)_{1\leqslant k\leqslant N}$, and assume that*

(1) for every $1 \leqslant k \leqslant N$, $\mathbf{E}\, X_k^2 = \mathbf{E}\, Y_k^2$,
(2) the random variables $(Y_k)_{1\leqslant k\leqslant N}$ are independent.

Then

$$\mathbf{E} \sup_{1\leqslant k\leqslant N} |X_k| \leqslant \mathbf{E} \sup_{1\leqslant k\leqslant N} |Y_k|. \tag{6.10}$$

Moreover, for any $t_1, \ldots, t_N \geqslant 0$,

$$\mathbf{P}(|X_k| \geqslant t_k \text{ for some } k) \leqslant \mathbf{P}(|Y_k| \geqslant t_k \text{ for some } k) \tag{6.11}$$

or equivalently

$$\mathbf{P}(|X_k| \leqslant t_k \text{ for all } k) \geqslant \mathbf{P}(|Y_k| \leqslant t_k \text{ for all } k) = \prod_{k=1}^{N} \mathbf{P}(|Y_k| \leqslant t_k). \tag{6.12}$$

Similarly to Slepian's lemma, both (6.10) and (6.12) have nice geometric interpretations. Consider n bands in $\mathbb{R}^n$ of the form $B_i = \{x \in \mathbb{R}^n : |\langle x, u_i\rangle| \leqslant a_i\}$, where $u_1, \ldots, u_n \in S^{n-1}$ are unit vectors and $a_1, \ldots, a_n$ are positive numbers. Then the mean width of $B_1^\circ \cap \cdots \cap B_n^\circ$ is minimal when the directions of the bands (i.e., the normal vectors u_i) are pairwise orthogonal. Similarly, the (Gaussian) measure of the intersection of the bands is minimal if the bands are orthogonal.

A remarkable statement that generalizes (6.12) and that has been a long-standing open problem is the *Gaussian correlation conjecture*. It was answered affirmatively very recently by Royen, who proved the following inequality: *given* 0*-symmetric convex sets* $K, L \subset \mathbb{R}^n$ *and a centered Gaussian measure* $\mathbf{P}$ *on* $\mathbb{R}^n$, *we have*

$$\mathbf{P}(K \cap L) \geqslant \mathbf{P}(K)\mathbf{P}(L). \tag{6.13}$$

EXERCISE 6.7 (Comparison of tails implies comparison of expectations). Deduce the first part (6.7) of Slepian's lemma from the second part (6.8). To get rid of the "equal variance" assumption, approximate the space by a sphere of large radius.

EXERCISE 6.8. Show that it is enough to verify (6.13) when $\mathbf{P}$ is the standard Gaussian measure.

EXERCISE 6.9 (Proof of the Khatri–Šidák inequality). Prove the correlation conjecture (6.13) in the special case where L is a band by using the fact that the Gaussian measure is log-concave and therefore satisfies (4.28). Then deduce the Khatri–Šidák inequality (Proposition 6.9).

6.1.3. Sudakov and dual Sudakov inequalities. Given a Gaussian process $\mathbf{X} = (X_t)_{t\in T}$, we may identify $\mathbf{X}$ with a subset of the Hilbert space $L_2(\Omega)$ (cf. (6.3) and the comments in the paragraph containing it). Since the joint distribution of $(X_t)_{t\in T}$ is uniquely determined by the covariances $(\mathbf{E}X_sX_t)_{s,t\in T}$, it follows that all the stochastically relevant information about the process is encoded in the geometry of $\mathbf{X}$. As it turns out, the value of the expected supremum of $\mathbf{X}$ is intimately related to the behavior of covering numbers $N(\mathbf{X}, \varepsilon)$. The first result in this direction is the Sudakov inequality.

PROPOSITION 6.10 (Sudakov minoration). *Let* $\mathbf{X} = (X_t)_{t\in T}$ *be a Gaussian process. Then*

$$c \sup_{\varepsilon>0} \varepsilon\sqrt{\log N(\mathbf{X}, \varepsilon)} \leqslant \mathbf{E} \sup_{t\in T} X_t \tag{6.14}$$

for some absolute constant $c > 0$.

PROOF. By (5.1), we may equivalently work with the packing number $P(\mathbf{X}, \varepsilon)$. Let $\varepsilon > 0$ and let $S \subset T$ be a subset which is ε-separated in the L_2-norm, that is, verifying $\|X_s - X_t\|_2 \geqslant \varepsilon$ whenever $s, t \in S$ and $s \neq t$. Let $(Y_s)_{s \in S}$ be a Gaussian process such that Y_s are independent $N(0, \varepsilon^2/2)$ random variables. By construction, we have

$$\|Y_s - Y_t\|_2 = \varepsilon \leqslant \|X_s - X_t\|_2$$

for any $s, t \in S$ with $s \neq t$. Accordingly, by Slepian's lemma and Lemma 6.1, we can conclude that

$$\varepsilon\sqrt{\log(\operatorname{card} S)} \sim \mathbf{E} \sup_{s \in S} Y_s \leqslant \mathbf{E} \sup_{s \in S} X_s \leqslant \mathbf{E} \sup_{t \in T} X_t,$$

as needed. □

In view of the comments in Section 6.1.1 (cf. (6.2), (6.3)), Sudakov's inequality (6.14) is really a statement about Gaussian mean widths of subsets of a Hilbert space. Since $w_G(K) \sim \sqrt{n}\, w(K)$ for $K \subset \mathbb{R}^n$ (see Section 4.3.3), the inequality (6.14) may be restated as follows: *for every bounded set (or, equivalently, for every convex body) $K \subset \mathbb{R}^n$ we have*

$$\log N(K, \varepsilon B_2^n) \lesssim w_G(K)^2/\varepsilon^2 \sim n w(K)^2/\varepsilon^2. \tag{6.15}$$

In general, Sudakov's inequality is not tight (see Exercise 6.11). However, in combination with the equally simple-minded bound (6.5) (applied at the appropriate "level of resolution"), it often leads to surprisingly precise estimates for $\mathbf{E} \sup_{t \in T} X_t$. We will elaborate on this point in the next section, in which we prove the companion bound, Dudley's inequality (Proposition 6.13).

When information about the mean width of K is available, (6.15) can be used to upper-bound covering/packing numbers of K. As a rule of thumb, this yields a reasonable estimate when $\log N(K, \varepsilon) = O(n)$. For smaller ε, i.e., when $\log N(K, \varepsilon) \gg n$, the volumetric approach from Lemma 5.8 is generally more precise. We exemplify these phenomena in Exercise 6.12.

A dual version of the Sudakov inequality also holds.

PROPOSITION 6.11 (Dual Sudakov minoration). *For any bounded set $K \subset \mathbb{R}^n$, we have*

$$\log N(B_2^n, K^\circ, \varepsilon) = \log N(B_2^n, \varepsilon K^\circ) \lesssim w_G(K)^2/\varepsilon^2 \sim n w(K)^2/\varepsilon^2. \tag{6.16}$$

Modulo minor issues related primarily to possible lack of symmetry, Proposition 6.11 follows from Proposition 6.10, and vice versa, by the (known) Euclidean case of the duality conjecture of covering numbers (5.67). However, there is a simple self-contained argument.

PROOF OF PROPOSITION 6.11. First, we may assume that K is a convex body since replacing K with its closed convex hull and passing to a subspace (if K was not of full dimension) does not change any of the quantities involved. Next, we may assume that 0 is an interior point of K since otherwise K° contains a half-space and the left-hand side is 0. Further, we may assume that K is symmetric since while replacing K by $K - K$ increases both sides, the right-hand side changes precisely by a factor of 4. The last "trivial" reduction is a rescaling. Since $\log N(B_2^n, \varepsilon K^\circ) = \log N(rB_2^n, r\varepsilon K^\circ)$, using $r = \frac{4w_G(K)}{\varepsilon}$ we reduce the problem to the following: *If $L \subset \mathbb{R}^n$ is a symmetric convex body with $w_G(L) = 1$, then $\log(N(rB_2^n, 4L^\circ)) \lesssim r^2$ for $r > 0$.*

As in the previous argument, it is more handy to argue via packings. Let $x_1, x_2, \ldots, x_N \in rB_2^n$ be such that $x_i + 2L^\circ$ are disjoint and let γ_n be the standard Gaussian measure on $\mathbb{R}^n$. The remainder of the proof depends on two simple observations.

(a) Since $1 = w_G(L) = \int \|x\|_{L^\circ}\,\mathrm{d}\gamma_n$, it follows by Markov's inequality that $\gamma_n(2L^\circ) \geqslant \frac{1}{2}$.

(b) Since $|x_i| \leqslant r$ for all $i \leqslant N$, the measure of each translation $x_i + 2L^\circ$ cannot be "too small" and since the translations are disjoint, there cannot be too many of them.

Here are details of the calculation behind the second observation. First, by symmetry of L°,

$$\gamma_n(x_i + 2L^\circ) = \frac{\gamma_n(x_i + 2L^\circ) + \gamma_n(-x_i + 2L^\circ)}{2} = \int_{2L^\circ} \frac{\phi(x + x_i) + \phi(x - x_i)}{2}\,\mathrm{d}x,$$

where $\phi(x) = (2\pi)^{-n/2}e^{-|x|^2/2}$ is the density of γ_n. Next, by convexity of the exponential function and by the parallelogram identity,

$$\begin{aligned}\frac{\phi(x + x_i) + \phi(x - x_i)}{2} &= (2\pi)^{-n/2}\,\frac{e^{-|x+x_i|^2/2} + e^{-|x-x_i|^2/2}}{2}\\ &\geqslant (2\pi)^{-n/2}e^{-(|x+x_i|^2+|x-x_i|^2)/4}\\ &= (2\pi)^{-n/2}e^{-(|x|^2+|x_i|^2)/2}\\ &= e^{-|x_i|^2/2}\phi(x)\\ &\geqslant e^{-r^2/2}\phi(x).\end{aligned}$$

Inserting this estimate into the preceding formula, we get

$$\gamma_n(x_i + 2L^\circ) \geqslant e^{-r^2/2}\gamma_n(2L^\circ) \geqslant \frac{1}{2}e^{-r^2/2}$$

and so $N \leqslant 2e^{r^2/2}$. This is exactly what we needed, except in the case when r is small, which can be handled separately by an elementary argument showing that the left-hand side of (6.16) is then 0; see Exercise 6.13. ☐

REMARK 6.12. In the setting of observation (a) in the proof above, a stronger statement is actually true: if $w_G(L) = 1$, then $\gamma_n(L^\circ) \geqslant \frac{1}{2}$; see Exercise 6.14.

EXERCISE 6.10 (Optimal constant in Sudakov's inequality). Show that the optimal constant in (6.14) is $c = (2\pi \log 2)^{-1/2} > 0.479$.

EXERCISE 6.11 (The gap in Sudakov's inequality). Show that the gap in Sudakov's inequality, i.e., the ratio between $w_G(K)$ and $\sup_{\varepsilon>0} \varepsilon\sqrt{\log N(K,\varepsilon)}$, can be arbitrarily large. For example, let $(d_j)_{j=1}^n$ be a "sufficiently fast" increasing sequence of positive integers and consider $K = K_1 \times K_2 \times \ldots \times K_n$, where K_j is a Euclidean sphere of dimension d_j and radius $1/\sqrt{d_j}$.

EXERCISE 6.12 (Metric entropy of B_1^n). Let $K = n^{1/2}B_1^n$. It is known (see Theorem 1 in [**Sch84**]) that then

$$\log N(K,\varepsilon) \simeq \begin{cases} n\,\frac{\log(2\varepsilon)}{\varepsilon^2} & \text{if} \quad 1 \leqslant \varepsilon \leqslant \frac{1}{2}n^{1/2},\\ n\log(2/\varepsilon) & \text{if} \quad 0 < \varepsilon \leqslant 1.\end{cases} \tag{6.17}$$

Compare the performance/facility of application of (6.15) to that of Lemma 5.8 when estimating $\log N(K,\varepsilon)$.

EXERCISE 6.13 (Gaussian measure and the inradius). Let γ_n be the standard Gaussian measure on $\mathbb{R}^n$. Show that if a symmetric convex body $K \subset \mathbb{R}^n$ satisfies $\gamma_n(K) \geqslant \gamma_1([-r,r])$, then $K \supset rB_2^n$. In particular, if $\gamma_n(K) \geqslant .683$, then $N(B_2^n, K) = 1$. Conclude that the left-hand side of (6.16) is 0 whenever $w(K)/\varepsilon \leqslant .317$.

EXERCISE 6.14 (Gaussian measure and the mean width). Show that if a symmetric convex body $L \subset \mathbb{R}^n$ satisfies $w_G(L) \leqslant 1$, then $\gamma_n(L^\circ) \geqslant \frac{1}{2}$.

EXERCISE 6.15 (Metric entropy of B_∞^n). Use one of the Sudakov inequalities to show that, for every $0 < \varepsilon < 1$, $N(B_2^n, B_\infty^n, \varepsilon)$ grows (at most) polynomially with the dimension n.
It is actually known (see Theorem 1 in [**Sch84**]) that

$$\log N(B_2^n, B_\infty^n, \varepsilon) \simeq \begin{cases} \frac{\log(2n\varepsilon^2)}{\varepsilon^2} & \text{if} \quad n^{-1/2} \leqslant \varepsilon \leqslant 1/2, \\ n \log \frac{2}{n\varepsilon^2} & \text{if} \quad 0 < \varepsilon \leqslant n^{-1/2}. \end{cases} \tag{6.18}$$

The similarity of the estimates (6.17) and (6.18) is not a coincidence; see (5.67). (Note that (6.17) could have been equivalently stated with $\log(2\varepsilon^2)$ and $\log(2/\varepsilon^2)$ instead of $\log(2\varepsilon)$ and $\log(2/\varepsilon)$, making the similarity even more apparent.)

6.1.4. Dudley's inequality and the generic chaining. The preceding section presented lower bounds for expected suprema of a Gaussian process in terms of the related covering/packing numbers. In this section we will present similar upper bounds in a slightly more general setting.

Let (S, ρ) be a compact metric space and let $(X_s)_{s\in S}$ be a family of random variables (a stochastic process indexed by S). We say that (X_s) is *centered* if $\mathbf{E}\, X_s = 0$ for all $s \in S$, and that it is *subgaussian* if, for all $s, t \in S$ with $s \neq t$ and for all $\lambda > 0$,

$$\mathbf{P}(X_s - X_t > \lambda) \leqslant A \exp\left(-\alpha \frac{\lambda^2}{\rho(s,t)^2}\right), \tag{6.19}$$

where A, α are positive parameters (independent of λ, s, t). The motivation for the terminology is that if the process is Gaussian, then (6.19) holds with $A = \alpha = \frac{1}{2}$ and with respect to the metric $\rho(s,t) = \|X_s - X_t\|_2$, and the bound is then essentially tight (see Exercise A.1).

PROPOSITION 6.13 (Dudley's inequality). *If $(X_s)_{s\in S}$ is centered and satisfies (6.19) with $A \geqslant \frac{1}{2}$, then*

$$\mathbf{E} \sup_{s\in S} X_s \leqslant 6\alpha^{-1/2} \int_0^{R/2} \sqrt{1 + 2\log\left(A^{1/2} N(S,\eta)\right)}\, d\eta, \tag{6.20}$$

where R is the radius of S.

COROLLARY 6.14. *If $(X_s)_{s\in S}$ satisfies (6.19) with $A \geqslant \frac{1}{2}$, but is not necessarily centered, then*

$$\mathbf{E} \sup_{s\in S} X_s \leqslant \sup_{s\in S} \mathbf{E}\, X_s + B \quad \textit{and} \quad \mathbf{E} \sup_{s\in S} |X_s| \leqslant \sup_{s\in S} \mathbf{E}\, |X_s| + B \tag{6.21}$$

where B is the quantity on the right-hand side of (6.20).

The first bound in the Corollary follows immediately by considering $X_s' = X_s - \mathbf{E}\, X_s$, and the second by noticing that if (X_s) verifies (6.19), then so does $(|X_s|)$.

REMARK 6.15. (1) Most formulations of Dudley's inequality involve the expression $\int \sqrt{\log N(S,\eta)}\,\mathrm{d}\eta$. In that case, the integrand is 0 if η is larger than the radius of S, and so one may as well integrate over $[0,\infty)$. In our formulation, the integrand is never 0; this is the price we are paying for having good dependence of the bound on A and, to a lesser extent, for Lemma 6.16 being stated for not-necessarily-centered variables.
(2) Some applications require majorizing the expected value of $\sup_{s,t}|X_s - X_t| = \sup_{s,t}(X_s - X_t)$; the proof below yields then (in the notation of Corollary 6.14) the bound $2B$, without having to assume that (X_s) is centered.
(3) When comparing Dudley's inequality to Sudakov's inequality (6.14), we notice that the former involves the L_1-norm of the function $\phi(\eta) = \sqrt{\log N(S,\eta)}$, while the latter the weak L_1-quasinorm (see [**Gra14**] for the definition). This explains why the two bounds are often of the same order and even if they are not, their ratio depends rather weakly on the dimension and other parameters.

PROOF OF DUDLEY'S INEQUALITY. Observe first that both sides of the inequality change in the same way if we rescale the process and/or the metric (i.e., replace (X_t) by (aX_t) and/or ρ by $b\rho$ for some $a, b > 0$) and appropriately adjust the parameter α. Accordingly, we may assume that both α and the radius of S are equal to 1. For every integer $k \geqslant 0$, let $\mathcal{N}_k$ be a 2^{-k}-net of minimal cardinality for (S,ρ). By hypothesis, the net $\mathcal{N}_0$ consists of a single element s_0. For every k and for every $s \in S$, denote by $\pi_k(s)$ an element of $\mathcal{N}_k$ satisfying $\rho(s,\pi_k(s)) \leqslant 2^{-k}$. The *chaining equation* reads for every $s \in S$

$$X_s = X_{s_0} + \sum_{k\geqslant 0} \big(X_{\pi_{k+1}(s)} - X_{\pi_k(s)}\big). \tag{6.22}$$

It follows that

$$\sup_{s\in S} X_s \leqslant X_{s_0} + \sum_{k\geqslant 0} \sup_{s\in S} \big(X_{\pi_{k+1}(s)} - X_{\pi_k(s)}\big) \leqslant X_{s_0} + \sum_{k\geqslant 0} \sup_{u,u'} (X_u - X_{u'}), \tag{6.23}$$

where the last supremum is taken over couples $(u,u') \in \mathcal{N}_{k+1} \times \mathcal{N}_k$ satisfying $\rho(u,u') \leqslant 2^{-k} + 2^{-(k+1)} = 3\cdot 2^{-(k+1)}$. Since $\mathbf{E}\, X_{s_0} = 0$, it remains to bound the expectation of each term in the sum, using the following fact

LEMMA 6.16. *If $A \geqslant \frac{1}{2}, \beta > 0$ and if $Y_1,\dots,Y_N$ are random variables satisfying $\mathbf{P}(Y_i > t) \leqslant A\exp(-t^2/\beta^2)$ for all $t \geqslant 0$, then*

$$\mathbf{E} \max_{1\leqslant i\leqslant N} Y_i \leqslant \beta\sqrt{1+\log(AN)}. \tag{6.24}$$

To bound $\mathbf{E}\sup(X_u - X_{u'})$, we apply the above Lemma with $\beta = 3\cdot 2^{-(k+1)}$ and $N = \operatorname{card}(\mathcal{N}_k)\cdot\operatorname{card}(\mathcal{N}_{k+1}) \leqslant N(S,2^{-(k+1)})^2$. This gives

$$\mathbf{E}\sup_{s\in S} X_s \leqslant 3\sum_{k\geqslant 1} 2^{-k}\sqrt{1+2\log\big(A^{1/2}N(S,2^{-k})\big)}\,. \tag{6.25}$$

The result follows now by majorizing the last series with an integral. □

PROOF OF LEMMA 6.16. We may assume that $\beta = 1$ by working with Y_i/β and that the variables Y_i are non-negative by working with the positive parts Y_i^+.

If $N \geqslant 2$, then $AN \geqslant 1$ and so

$$\begin{aligned}\mathbf{E}\max_i Y_i &= \int_0^\infty \mathbf{P}(\max_i Y_i \geqslant t)\,\mathrm{d}t \\ &\leqslant \sqrt{\log(AN)} + AN\int_{\sqrt{\log(AN)}}^\infty \exp(-t^2)\,\mathrm{d}t \\ &\leqslant \sqrt{\log(AN)+1}.\end{aligned}$$

The first inequality is the union bound; the second one is the upper bound in Komatu's inequality (A.4) which can be rewritten as $\int_u^\infty e^{-t^2}\,\mathrm{d}t \leqslant (\sqrt{u^2+1}-u)e^{-u^2}$ (valid for $u \geqslant -0.3893$ and applied with $u = \sqrt{\log(AN)}$).

If $N = 1$, the inequality is trivial if the variable has mean 0 and can be checked directly otherwise; see Exercise 6.17, which also treats in detail the case of small A. □

Although Dudley's inequality is not sharp in general (see Exercises 6.19 and 6.20, which exhibit two different reasons for a possible gap), it does become sharp when sufficiently many symmetries are present; such situation is referred to as the *stationary case* in probability literature. Here is a statement demonstrating this principle expressed in the language of convex sets and their Gaussian mean widths.

PROPOSITION 6.17 (Not proved here). *Let $K \subset \mathbb{R}^n$ be a nonempty compact convex set and let F be the set of extreme points of K. If the isometry group of K acts transitively on F, then*

$$w_G(K) = w_G(F) \simeq \int_0^{\operatorname{outrad}(F)} \sqrt{1+\log\left(N(F,\eta)\right)}\,\mathrm{d}\eta.$$

In the most general situation, the chaining argument used in the proof of Proposition 6.13 (which is based on a decomposition along consecutive "levels of resolutions") can be improved by using a generic version of the chaining.

THEOREM 6.18 (Generic chaining, not proved here). *Let $(X_t)_{t\in T}$ be a centered subgaussian process and let ρ be the distance on T defined by $\rho(s,t) = \|X_s - X_t\|_{L_2}$. Let $(T_k)_{k\in\mathbb{N}}$ be an increasing family of subsets of T such that $\operatorname{card}(T_0) = 1$ and $\operatorname{card}(T_k) \leqslant 2^{2^k}$ for $k \geqslant 1$. Then*

$$\mathbf{E}\sup_{t\in T} X_t \leqslant C\sup_{s\in T}\sum_{k=0}^\infty 2^{k/2}\rho(s,T_k) \tag{6.26}$$

for some absolute constant C. Conversely, if the process $(X_t)_{t\in T}$ is Gaussian, this bound is always sharp in the following sense: if $\gamma_2(T)$ denotes the infimum of $\sup_{s\in T}\sum_{k=0}^\infty 2^{k/2}\rho(s,T_k)$ over all such families (T_k), then we have

$$\mathbf{E}\sup_{t\in T} X_t \geqslant c\,\gamma_2(T)$$

for some absolute constant c.

To grasp the difference between Dudley's integral and the generic chaining bound, it is useful to rephrase the former in the language of Theorem 6.18. One checks (see Exercise 6.22) that, for any compact metric space (T,ρ),

$$\int_0^{\operatorname{diam} T} \sqrt{\log N(T,\eta)}\,\mathrm{d}\eta \simeq \inf_{(T_k)}\sum_{k=0}^\infty 2^{k/2}\sup_{s\in T}\rho(s,T_k), \tag{6.27}$$

where the infimum is taken over families (T_k) as in Theorem 6.18. Note that the right-hand sides of (6.26) and (6.27) differ in the relative position of the summation and the supremum.

EXERCISE 6.16 (The constant in Dudley's inequality). Show that the constant 6 in Dudley's inequality (6.20) can be improved to $3 + 2\sqrt{2} \approx 5.83$ if we repeat the proof with $\mathcal{N}_k$ being a θ^k-net, and optimize over $\theta \in (0, 1)$.

EXERCISE 6.17. The argument in the proof of Lemma 6.16 works if $AN \geqslant 1$. Show that when $AN < 1$, then the optimal majorant is $\frac{\sqrt{\pi}}{2}\beta AN$ and check that, consequently, the bound from Lemma 6.16 holds whenever $AN \geqslant 0.4236$.

EXERCISE 6.18 (Median of the maximum of a subgaussian process). Show that under the hypotheses of Lemma 6.16 the median of $\max_i Y_i$ is at most $\beta\sqrt{\log(2AN)}$.

EXERCISE 6.19 (The gap in Dudley's inequality). Let $(Z_k)_{k=1}^n$ be an i.i.d. sequence of $N(0, 1)$ variables and let $X_k = Z_k/\sqrt{1 + \log k}$. Check that $\mathbf{E}\max_k X_k < 3$ for any $n \in \mathbb{N}$, but that the integral on the right-hand side of (6.20) is $\Theta(\log\log n)$.

EXERCISE 6.20 (The gap in Dudley's inequality via B_1^n). Let $K = B_1^n$. Show that $\int_0^1 \sqrt{\log N(K, \eta)}\, \mathrm{d}\eta \simeq (\log n)^{3/2}$ while $w_G(K) \sim \sqrt{2\log n}$. Interpret this discrepancy as a gap in Dudley's inequality.

EXERCISE 6.21 (Law of the iterated logarithm via Dudley's inequality). Here is a rough version of the law of the iterated logarithm. Let $(Z_i)_{1\leqslant i\leqslant n}$ be independent $N(0, 1)$ random variables and consider the Gaussian process $\mathbf{X} = (X_k)_{1\leqslant k\leqslant n}$ defined by $X_k = \frac{1}{\sqrt{k}}(Z_1 + \cdots + Z_k)$. Estimate the covering numbers of $\mathbf{X}$ and conclude that $\mathbf{E}\max\{X_k : 1 \leqslant k \leqslant n\} = \Theta(\log\log n)$.

EXERCISE 6.22 (Dudley integral as a chaining bound). Prove (6.27).

EXERCISE 6.23 (Generic chaining improves on Dudley's inequality). Show that the processes from Exercise 6.19 can be shown to be uniformly bounded via generic chaining.

6.2. Random matrices

Random matrix theory (RMT) studies spectral properties of large-dimensional matrices generated by some random procedure. We present in this chapter a very small selection of results from RMT, which will be useful to analyze random constructions of interest in QIT. In particular, while we focus mostly on the Gaussian setting, most of the limit theorems are valid for a much wider class of random matrices; this principle is known as universality. We study primarily (but not exclusively) matrices with complex entries since these are the most relevant to QIT. In contrast, much of the original motivation for RMT research came from statistics, the setting in which the real case is more usual.

For $A \in \mathsf{M}_n^{\mathrm{sa}}$, we denote by $(\lambda_i(A))_{1\leqslant i\leqslant n}$ or simply $(\lambda_i)_{1\leqslant i\leqslant n}$ the eigenvalues of A, listed with multiplicities and arranged so that

$$\lambda_1(A) \geqslant \lambda_2(A) \geqslant \cdots \geqslant \lambda_n(A). \tag{6.28}$$

The *empirical spectral distribution* of A, denoted by $\mu_{\mathrm{sp}}(A)$, is the probability measure obtained as the uniform measure over the spectrum of A. More formally

$$\mu_{\mathrm{sp}}(A) = \frac{1}{n}\sum_{i=1}^{n} \delta_{\lambda_i(A)}, \tag{6.29}$$

which is clearly independent of the order of eigenvalues. Obviously, if the matrix A is random, the corresponding empirical spectral distribution is also random. We are interested in giving a description of the typical shape of this random measure.

6.2.1. ∞-Wasserstein distance. At least two kinds of RMT limit theorems are relevant for quantum information theory: fine information about the extreme eigenvalues (or about the operator norm) and large-scale information about the entire spectrum. These two possible perspectives are known in RMT as "local" vs. "global" regimes. In order to encompass both aspects, we find it convenient to introduce the ∞-Wasserstein distance between probability measures on $\mathbb{R}$.

DEFINITION 6.19. Let μ_1, μ_2 be probability measures on $\mathbb{R}$. The ∞-Wasserstein distance is defined as

$$d_\infty(\mu_1, \mu_2) := \inf \|X_1 - X_2\|_{L_\infty}, \tag{6.30}$$

with infimum over all couples (X_1, X_2) of random variables with (marginal) laws μ_1 and μ_2, defined on a common probability space. Similarly, if Y_1, Y_2 are real random variables, we will mean by $d_\infty(Y_1, Y_2)$ the ∞-Wasserstein distance between the laws of Y_1 and Y_2.

The definition of ∞-Wasserstein distance immediately extends to probability measures on a metric space (E, d) if we interpret in (6.30) the quantity $\|X_1 - X_2\|_{L_\infty}$ as the smallest Δ such that $\mathbf{P}(d(X_1, X_2) \leqslant \Delta) = 1$. Similarly, replacing the L_∞-norm by the L_p-norm leads to the p-Wasserstein distance d_p, with the "finite p" case (and particularly $p = 1, 2$) being much more intensively studied than $p = \infty$. The metric d_1 is also known, particularly in the computer science community, as the *Earth Mover's distance*.

We note the following inequality (cf. Exercise 6.24): *whenever $f : \mathbb{R} \to \mathbb{R}$ is an L-Lipschitz function and X, Y are random variables, then*

$$|\mathbf{E}\, f(X) - \mathbf{E}\, f(Y)| \leqslant L\, d_\infty(X, Y). \tag{6.31}$$

The ∞-Wasserstein distance can be computed from cumulative distribution functions: if $F_X(t) = \mathbf{P}(X \leqslant t)$, then

$$d_\infty(X, Y) = \inf\{\varepsilon > 0 \ :\ F_X(t - \varepsilon) \leqslant F_Y(t) \leqslant F_X(t + \varepsilon) \text{ for all } t \in \mathbb{R}\}. \tag{6.32}$$

Note the similarity with the definition of Lévy distance d_L, which metrizes the weak convergence

$$d_L(X, Y) = \inf\{\varepsilon > 0 \ :\ F_X(t - \varepsilon) - \varepsilon \leqslant F_Y(t) \leqslant F_X(t + \varepsilon) + \varepsilon \text{ for all } t \in \mathbb{R}\}.$$

The following lemma is elementary, but it will be crucial for our purposes.

LEMMA 6.20. *Let Z be a random variable distributed according to a measure ν_Z, with support* **equal** *to some bounded interval $[a, b]$. If (Y_n) is a sequence of random variables, the following are equivalent:*
(1) $d_\infty(Y_n, Z) \to 0$,
(2) $Y_n \to Z$ *weakly and* $\sup Y_n \to b$, $\inf Y_n \to a$.

By inf and sup we really mean here *essential* inf and sup. Note that the hypothesis on the support is vital: the equivalence fails if the support is not connected (see Exercise 6.29).

PROOF. Since $d_L \leqslant d_\infty$, convergence in ∞-Wasserstein distance implies weak convergence. Moreover we have $|\sup Y_n - \sup Z| \leqslant d_\infty(Y_n, Z)$ and similarly for the infima, and therefore (1) implies (2).

Conversely, assume (2). Given $\varepsilon > 0$, choose $a = x_0 < x_1 < \cdots < x_r = b$ such that $x_{j+1} - x_j < \varepsilon$ and such that, for $0 < j < r$, x_j is a continuity point of F_Z (such points are dense in $\mathbb{R}$). The hypothesis on the support of ν_Z implies that F_Z is strictly increasing on $[a, b]$, so that there exists $\alpha > 0$ with the property that $F_Z(x_j) \geqslant F_Z(x_{j-1}) + \alpha$ for $0 < j \leqslant r$. For n large enough, we have $\inf Y_n > a - \varepsilon$, $\sup Y_n < b + \varepsilon$ and $|F_{Y_n}(x_j) - F_Z(x_j)| < \alpha$ for any $0 < j < r$ (using the fact that F_Z is continuous at x_j). These conditions imply that, for any real number t,

$$F_Z(t - 2\varepsilon) \leqslant F_{Y_n}(t) \leqslant F_Z(t + 2\varepsilon),$$

and therefore $d_\infty(Y_n, Z) \leqslant 2\varepsilon$. □

REMARK 6.21. The proof of Lemma 6.20 gives actually the following: a neighbourhood basis around ν_Z for the topology induced by d_∞ is given by $(V_\varepsilon)_{\varepsilon>0}$, where V_ε is the set of probability measures μ satisfying the condition

$$\max\Big(d_L(\mu, \nu_Z), |\sup \mu - \sup \nu_Z|, |\inf \mu - \inf \nu_Z|\Big) < \varepsilon,$$

where by $\inf \nu$ and $\sup \nu$ we denote the infimum and supremum of the support of a measure ν.

EXERCISE 6.24 (∞-Wasserstein distance and Lipschitz functions). Show the stronger version of (6.31) : *If $f : \mathbb{R} \to \mathbb{R}$ is an L-Lipschitz function, then* $|\mathbf{E} f(X) - \mathbf{E} f(Y)| \leqslant L\, d_1(X, Y)$.

EXERCISE 6.25. Show that if $f : \mathbb{R} \to \mathbb{R}_+$ is an L-Lipschitz function and $d_\infty(X, Y) \leqslant \varepsilon$, then $\mathbf{E} f(Y) \geqslant \mathbf{E} g(X)$, where $g = (f - L\varepsilon)^+$.

EXERCISE 6.26 (∞-Wasserstein distance via cumulative distribution functions). Prove the alternate formula (6.32) for the ∞-Wasserstein distance.

EXERCISE 6.27 (∞-Wasserstein distance and weak convergence). Show *directly* that $d_\infty(Y_n, Z) \to 0$ implies the weak convergence $Y_n \to Z$, i.e., the convergence $\mathbf{E} f(Y_n) \to \mathbf{E} f(Z)$ for any bounded continuous function $f : \mathbb{R} \to \mathbb{R}$.

EXERCISE 6.28. Show that under the hypotheses of Lemma 6.20, $d_\infty(Y_n, Z) \to 0$ implies the convergence $\mathbf{E} f(Y_n) \to \mathbf{E} f(Z)$ for any continuous function $f : \mathbb{R} \to \mathbb{R}$ (bounded or not). Show, by example, that this may be false when Z is unbounded.

EXERCISE 6.29. Give an example showing that connectedness is important in Lemma 6.20.

EXERCISE 6.30. Show that if $A, B \in \mathsf{M}_n^{\mathrm{sa}}$, then $d_\infty(\mu_{\mathrm{sp}}(A), \mu_{\mathrm{sp}}(B)) \leqslant \|A - B\|_{\mathrm{op}}$.

6.2.2. The Gaussian Unitary Ensemble (GUE).

6.2.2.1. *Definition of* GUE. Recall that the space $\mathsf{M}_n^{\mathrm{sa}}$ of complex Hermitian $n \times n$ matrices can be considered as a real Euclidean space when equipped with the Hilbert–Schmidt inner product. We denote by GUE(n) the distribution of the standard Gaussian vector in $\mathsf{M}_n^{\mathrm{sa}}$ (see Appendix A). When a random matrix A has distribution GUE(n), we say simply that A is a GUE(n) matrix. Here are some other equivalent descriptions of GUE(n) matrices (see Exercise 6.31).

(1) The density of GUE(n) is $c_n e^{-\operatorname{Tr} X^2/2}$ for $X \in \mathsf{M}_n^{\mathrm{sa}}$, where c_n is the appropriate normalization constant.
(2) Let $C \in \mathsf{M}_n$ be a random matrix with independent $N_{\mathbb{C}}(0,1)$ entries (see Appendix A). Then the matrix $A = (C + C^\dagger)/\sqrt{2}$ is a GUE(n) matrix.
(3) $A = (a_{ij}) \in \mathsf{M}_n^{\mathrm{sa}}$ is a GUE(n) matrix if and only if the random variables $(a_{ij})_{1 \leqslant i \leqslant j \leqslant n}$ are independent, the random variable a_{ij} having distribution $N_{\mathbb{C}}(0,1)$ when $i \neq j$ and $N_{\mathbb{R}}(0,1)$ when $i = j$.

The GUE has the property of unitary invariance: if $A \in \mathsf{M}_n$ is a GUE(n) matrix, then, for any fixed $U \in \mathsf{U}(n)$, the random matrix $UAU^\dagger$ is also a GUE(n) matrix.

Although it plays almost no role in our approach, an important feature of natural unitarily invariant models is that there are explicit formulas for the density of eigenvalues (see also Exercise 6.32).

PROPOSITION 6.22 (Ginibre formula, not proved here). *Let A be a GUE(n) matrix, and $\lambda(A) = (\lambda_i)_{1 \leqslant i \leqslant n}$ be the spectrum of A, arranged in the non-increasing order. Then the density of the random vector $\lambda(A)$ is given by*

$$c_n \mathbf{1}_{\{\lambda_1 \geqslant \cdots \geqslant \lambda_n\}}\, e^{-\frac{1}{2}\sum_{i=1}^n \lambda_i^2} \prod_{1 \leqslant i < j \leqslant n} (\lambda_i - \lambda_j)^2,$$

where c_n is the appropriate normalization constant.

The real-valued companion to the GUE is the Gaussian Orthogonal Ensemble or GOE, which corresponds to the standard Gaussian vector in the space of self-adjoint real matrices (up to normalization, see Section 6.2.4). The Gaussian Symplectic Ensemble (GSE) similarly corresponds to the standard Gaussian vector in the space of quaternionic Hermitian matrices.

For some arguments, it is important to introduce what we call the GUE$_0$(n) ensemble, which is the GUE ensemble conditioned to have trace zero. In other words, G_0 is a GUE$_0$(n) matrix if it has the distribution of a standard Gaussian vector in the hyperplane $H \subset \mathsf{M}_n^{\mathrm{sa}}$ of trace zero Hermitian matrices. If A is a GUE(n) matrix, then $A_0 := A - \frac{\operatorname{Tr} A}{n} \mathrm{I}$ is a GUE$_0$(n) matrix. Note that the coefficient $\frac{\operatorname{Tr} A}{n}$ has distribution $N(0, 1/n)$ and is independent of A_0.

EXERCISE 6.31. Show that (1), (2) and (3) provide equivalent definitions of GUE(n).

EXERCISE 6.32 (Characterization of GUE(n)). Show that GUE(n) is the only unitarily invariant distribution on $\mathsf{M}_n^{\mathrm{sa}}$ for which the formula from Proposition 6.22 holds.

6.2.2.2. *Limit theorems.* The probability distribution that is the non-commutative analogue of the Gaussian distribution is the *semicircular distribution* (or *semicircle law*). The *standard* semicircular distribution μ_{SC} is the probability distribution on $\mathbb{R}$ with support $[-2, 2]$ and with density

$$\frac{1}{2\pi}\sqrt{4 - x^2} \tag{6.33}$$

with respect to the Lebesgue measure. The even moments of the semicircular distribution are the Catalan numbers: for a nonnegative integer p, we have

$$\int_{-2}^{2} x^{2p}\, \mathrm{d}\mu_{\mathrm{SC}}(x) = \frac{1}{p+1}\binom{2p}{p}. \tag{6.34}$$

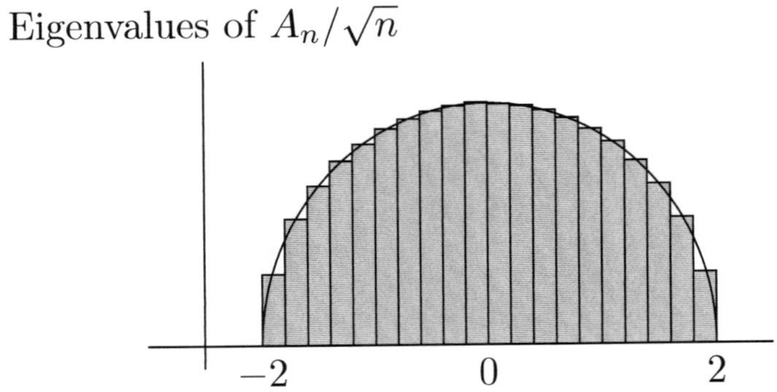

FIGURE 6.1. The empirical eigenvalue distribution of a GUE(n) matrix A_n for $n = 10000$ approaches the semicircular distribution.

In particular the variance equals 1. If X is a random variable with distribution μ_{SC}, then for any $m \in \mathbb{R}$ and $\sigma \geqslant 0$, we denote by $\mu_{\text{SC}(m,\sigma^2)}$ the distribution of $m + \sigma X$, called the semicircular distribution with mean m and variance σ.

The semicircular distribution appears as the limit spectral distribution of GUE random matrices (see Figure 6.1).

THEOREM 6.23 (Convergence of GUE spectrum towards the semicircular distribution, not proved here). *For each n, let A_n be a* GUE(n) *or* GUE$_0$(n) *matrix. After normalization, the sequence of empirical spectral distributions $(\mu_{\text{sp}}(A_n))$ converges towards the semicircular distribution (with respect to the ∞-Wasserstein distance) in the following sense: for any $\varepsilon > 0$,*

$$\lim_{n\to\infty} \mathbf{P}(d_\infty(\mu_{\text{sp}}(n^{-1/2}A_n), \mu_{\text{SC}}) > \varepsilon) = 0.$$

Using Lemma 6.20 (see also Remark 6.21), one checks that Theorem 6.23 brings together two facts, usually presented (and proved) independently in the RMT literature:

(1) The fact that the sequence $(\mu_{\text{sp}}(n^{-1/2}A_n))$ of random empirical measures converges (weakly, in probability) towards the semicircle law, a result going back to Wigner.
(2) The convergence (in probability) of the largest and smallest eigenvalues of $n^{-1/2}A_n$ towards ± 2. This requires a different and finer analysis, which we sketch in what follows.

Since GUE(n) is the standard Gaussian vector in M_n^{sa}, and by the duality between Schatten norms (see Proposition 1.17) the quantity $\mathbf{E}\,\|A_n\|$ is exactly the Gaussian mean width of $S_1^{n,\text{sa}}$, the self-adjoint part of the unit ball for the trace norm. Although the order of magnitude of $\mathbf{E}\,\|A_n\|$ can be readily deduced from general principles (see Exercise 6.33), the derivation of the precise constant 2 requires more specialized arguments. However, once an appropriate bound such as (6.37) below is established, concentration of $\|A_n\|$ around its expectation is provided by Theorem 5.24 and gives the following estimates.

PROPOSITION 6.24. *Let A_n be a* GUE(n) *or* GUE$_0$(n) *matrix. Then, for any $\varepsilon > 0$,*

$$\mathbf{P}\left(\lambda_1(n^{-1/2}A_n) \geqslant 2+\varepsilon\right) \leqslant \mathbf{P}\left(\left\|n^{-1/2}A_n\right\|_\infty \geqslant 2+\varepsilon\right) \leqslant \frac{1}{2}\exp\left(-\frac{n\varepsilon^2}{2}\right). \tag{6.35}$$

PROOF. Since $\|\cdot\|_\infty \leqslant \|\cdot\|_{\mathrm{HS}}$, the function $\|\cdot\|_\infty$ is a 1-Lipschitz function. By Theorem 5.24 (recall that $\mathrm{GUE}(n)$ is the standard Gaussian vector in the space $\mathsf{M}_n^{\mathrm{sa}}$, and similarly for $\mathrm{GUE}_0(n)$ and the hyperplane of trace zero matrices), it follows that

$$\mathbf{P}\left(\|A_n\|_\infty \geqslant M+t\right) \leqslant \frac{1}{2}\exp(-t^2/2), \tag{6.36}$$

where M is the median of the random variable $\|A_n\|_\infty$. We claim that $M < 2\sqrt{n}$. This follows from two facts. First, we have the inequality

$$\mathbf{E}\,\|A_n\|_\infty < 2\sqrt{n} - 0.6n^{-1/6} < 2\sqrt{n}, \tag{6.37}$$

which was derived in Appendix F in [**Sza05**] (note that this inequality extends to the case of $\mathrm{GUE}_0(n)$ via Jensen's inequality). Second, it follows from Proposition 5.34 that the median of the random variable $\|A_n\|_\infty$ is smaller than its mean. Once we know that $M \leqslant 2\sqrt{n}$, (6.35) follows by setting $t = \varepsilon\sqrt{n}$ and appealing to (6.36).

An alternative proof is to directly use (6.37) in combination with Theorem 5.25, but we opted for the argument above since, in our approach, concentration around the median is more elementary than that around the mean. □

Similar estimates also hold for the GOE. For example, if A_n is a $\mathrm{GOE}(n)$ matrix, we have $\mathbf{E}\,\lambda_1(A_n) \leqslant 2\sqrt{n}$ (see Exercise 6.48) and therefore

$$\mathbf{P}\Big(\lambda_1(n^{-1/2}A_n) \geqslant 2+\varepsilon\Big) \leqslant \frac{1}{2}\exp\Big(-\frac{n\varepsilon^2}{2}\Big).$$

We next note that if $A \in \mathsf{M}_n^{\mathrm{sa}}$, then $\|A\|_\infty = \max\{\lambda_1(A), -\lambda_n(A)\}$, and that, by symmetry of $\mathrm{GOE}(n)$, the distribution of $-\lambda_n(A_n)$ is the same as that of $\lambda_1(A_n)$. Combining these observations with the bound above yields

$$\mathbf{P}\Big(\|n^{-1/2}A_n\|_\infty \geqslant 2+\varepsilon\Big) \leqslant \exp\Big(-\frac{n\varepsilon^2}{2}\Big). \tag{6.38}$$

The bound from Proposition 6.24 can be improved for small values of ε (the *Tracy–Widom effect*).

PROPOSITION 6.25 (Not proved here). *Let A_n be a $\mathrm{GUE}(n)$ or a $\mathrm{GOE}(n)$ matrix. Then for any $\varepsilon \in (0,1)$,*

$$\mathbf{P}\Big(\lambda_1(n^{-1/2}A_n) \geqslant 2+\varepsilon\Big) \leqslant C\exp(-cn\varepsilon^{3/2})$$

and

$$\mathbf{P}\Big(\lambda_1(n^{-1/2}A_n) \leqslant 2-\varepsilon\Big) \leqslant C\exp(-cn^2\varepsilon^3),$$

for some absolute constants $C, c > 0$.

The main result of this section, Theorem 6.23, is formulated as an asymptotic statement. One can ask for a more quantitative version, or for a fixed–dimension bound.

PROBLEM 6.26. *If A_n is a $\mathrm{GUE}(n)$, a $\mathrm{GUE}_0(n)$, or a $\mathrm{GOE}(n)$ matrix, what is the rate of convergence in $d_\infty(\mu_{\mathrm{sp}}(n^{-1/2}A_n), \mu_{\mathrm{SC}}) \to 0$? Proposition 6.25 suggests that the answer may be $\Theta(n^{-2/3})$. The convergence cannot be faster than $n^{-2/3}$ due to the Tracy–Widom effect; see Notes and Remarks. The same question can be asked about the Wishart matrices considered in the next section.*

EXERCISE 6.33 (An elementary proof of boundedness of GUE(n)). Using a net argument, show that if A_n is a GUE(n) matrix, then $\|A_n\|_\infty \leqslant C\sqrt{n}$ with large probability, where $C > 2$ is some universal constant.

EXERCISE 6.34. Show that the GUE(n) version of Theorem 6.23 implies the $\mathrm{GUE}_0(n)$ version.

6.2.3. Wishart matrices.

6.2.3.1. *Definition of the Wishart ensemble.* Let n, s be nonzero integers. Let $B \in \mathsf{M}_{n,s}$ be a random matrix with independent $N_{\mathbb{C}}(0,1)$ entries. The random matrix $W = BB^\dagger \in \mathsf{M}_n^{\mathrm{sa}}$ is called a (complex) Wishart matrix and its distribution is denoted by Wishart(n, s). We often say simply that B is a Wishart(n, s) matrix. The eigenvalues of W are the squares of the singular values of B, so that statements about the spectrum of Wishart matrices are equivalent to statements about singular values of a random (rectangular) Gaussian matrix.

Here is an equivalent description: let $(G_1, \ldots, G_s)$ be s independent copies of a standard complex Gaussian vector in the space $\mathbb{C}^n$. Then the matrix

$$W = \sum_{i=1}^{s} |G_i\rangle\langle G_i| \tag{6.39}$$

has distribution Wishart(n, s).

The rank of a Wishart(n, s) matrix is almost surely equal to $\min(n, s)$. In the following we often assume that $s \geqslant n$, i.e., that the Wishart matrices are almost surely positive definite. This is not really a restriction since the case $s < n$ can be covered by the following observation: if $B \in \mathsf{M}_{n,s}$ is a random matrix with independent $N_{\mathbb{C}}(0,1)$ entries, then $W_1 = BB^\dagger$ is a Wishart(n, s) matrix while $W_2 = B^\dagger B$ is a Wishart(s, n) matrix (because the $N_{\mathbb{C}}(0,1)$ distribution is invariant under complex conjugation), and the matrices W_1 and W_2 share the same non-zero eigenvalues.

One can also consider the real version of Wishart matrices by starting with $G_1, \ldots, G_s$ that are standard Gaussian vectors on $\mathbb{R}^n$ rather than on $\mathbb{C}^n$ (see Section 6.2.4). This setup has a long history due to the fact that it is frequently encountered in statistics.

6.2.3.2. *Limit theorems.* What does the spectrum of large Wishart matrices look like? Before answering this question, it might be useful to have in mind the following elementary result from probability theory, which can be considered as the commutative analogue of a Wishart(n, s) matrix (think of p as $1/n$).

Let X be a random variable following a binomial distribution of parameters $s \in \mathbb{N}$ and $p \in (0, 1)$ (this means that X has the same distribution as the sum of s independent Bernoulli random variables taking values 1 with probability p and 0 with probability $1 - p$). We then have

FACT 6.27 (Easy). *When s tends to infinity and p tends to 0, then*
(i) if $\alpha = \lim sp$ exists in $(0, \infty)$, then X converges (weakly) towards a Poisson distribution of parameter α.
(ii) If $\lim sp = \infty$, then $(X - sp)/\sqrt{sp}$ converges (weakly) towards a standard Gaussian distribution.

In the non-commutative context, we replace independent Bernoulli variables by *free* Bernoulli variables. The resulting limit laws are the so-called *free Poisson* distribution and, again, the semi-circular distribution given by (6.33). Free probability theory is beyond the scope of this book (see Section 6.2.5 for a brief introduction) and so, rather than defining *freeness*, we will explain the heuristics relating it to RMT.

In noncommutative probability theory, a Bernoulli variable with parameter $p = \frac{1}{n}$ can be represented as a random rank 1 projection on $\mathbb{C}^n$ (i.e., uniformly distributed on $\mathsf{Gr}(n,1)$; more generally, we may consider a random rank $p \dim \mathcal{H}$ projection on $\mathcal{H}$). According to a fundamental paradigm of free probability, freeness is realized as a large dimension limit of independent matrix ensembles. Accordingly, the RMT model to consider is

$$X = \sum_{i=1}^{s} |\psi_i\rangle\langle\psi_i|, \tag{6.40}$$

where the vectors ψ_i are i.i.d. and uniformly distributed on the sphere in $\mathbb{C}^n$ and $n, s \to \infty$. Since, for large n, the standard Gaussian vector on $\mathbb{C}^n$ is close to being uniformly distributed on the sphere of radius $n^{1/2}$ (see Corollary 5.27), it follows that X is close to the appropriately rescaled Wishart random matrix given by (6.39) (see Exercise 6.37). Consequently, the limiting behavior that is the noncommutative analogue of Fact 6.27 can be retrieved from the results on spectral properties of Wishart(n, s) as $n, s \to \infty$. Such results have been known for quite a while, even if the full extent of the analogy and the identification of the limit laws as the free analogues of the Poisson and normal distributions had to await the development of the language of free probability.

To make the limit results for Wishart matrices more tangible, we need to describe explicitly what the free Poisson distributions are. They originally appeared in RMT as *Marčenko–Pastur distributions*. First, for $\lambda > 0$, we let $x_\pm = (1 \pm \sqrt{\lambda})^2$ and define a function supported on $[x_-, x_+]$ by

$$f_\lambda(x) = \frac{\sqrt{(x - x_-)(x_+ - x)}}{2\pi x} \mathbf{1}_{[x_-, x_+]}(x).$$

The Marčenko–Pastur (a.k.a. free Poisson) distribution with parameter λ, denoted $\mu_{\mathrm{MP}(\lambda)}$, is then defined by

$$\mu_{\mathrm{MP}(\lambda)} = (1 - \lambda)^+ \delta_0 + f_\lambda \, dx, \tag{6.41}$$

where δ_0 denotes a Dirac mass at 0 and $f \, dx$ is the measure whose density (with respect to the Lebesgue measure) is f (see Figure 6.2).

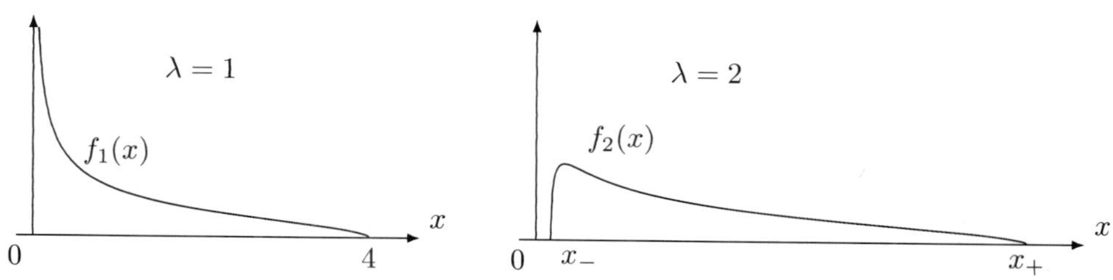

FIGURE 6.2. Marčenko–Pastur densities for $\lambda = 1$ and $\lambda = 2$.

THEOREM 6.28 (Not proved here). *Consider a sequence of indices (n, s) which tend to infinity in such a way that $\lambda = \lim s/n \in [1, \infty)$ exists. For each (n, s),* *let* $W_{n,s}$ *be a* Wishart(n,s) *matrix. After renormalization, the sequence of random empirical spectral distributions* $(\mu_{\mathrm{sp}}(W_{n,s}))$ *converges in probability towards the Marčenko–Pastur distribution* $\mu_{\mathrm{MP}(\lambda)}$ *with respect to the* ∞*-Wasserstein distance: for any* $\varepsilon > 0$,

$$\lim_{(n,s)\to\infty} \mathbf{P}(d_\infty(\mu_{\mathrm{sp}}(n^{-1}W_{n,s}), \mu_{\mathrm{MP}(\lambda)}) > \varepsilon) = 0.$$

The alternative normalization $s^{-1}W_{n,s} = (s^{-1/2}B)(s^{-1/2}B)^\dagger$ converges towards a rescaled Marčenko–Pastur distribution with support $[(1-1/\sqrt{\lambda})^2, (1+1/\sqrt{\lambda})^2]$. For large λ, this shows that the matrix $s^{-1/2}B$ is an almost isometric embedding from $\mathbb{C}^n$ into $\mathbb{C}^s$, all singular values being close to 1.

As explained earlier, a similar result follows formally in the case $\lambda \in (0,1)$. However, some care is needed in the formulation, since the atomic part in the Marčenko–Pastur distribution is supported outside of the continuous part, and this lack of connectedness may prevent convergence with respect to the ∞-Wasserstein distance (cf. Lemma 6.20 and Exercises 6.29 and 6.36).

In the case where the ratio s/n tends to infinity, the limiting Marčenko–Pastur distribution degenerates into a semicircular distribution, in the same way that a Poisson distribution with a large parameter is almost Gaussian.

THEOREM 6.29 (Not proved here). *Consider a sequence of indices* (n,s) *which both tend to infinity in such a way that* $\lim s/n = \infty$*. For each* (n,s)*, let* $W_{n,s}$ *be a* Wishart(n,s) *matrix. After renormalization and recentering, the sequence of empirical spectral distributions* $(\mu_{\mathrm{sp}}(W_{n,s}))$ *converges in probability towards the semicircular distribution* μ_{SC} *with respect to the* ∞*-Wasserstein distance, in the following sense: for any* $\varepsilon > 0$,

$$\lim_{n\to\infty} \mathbf{P}(d_\infty(\mu_{\mathrm{sp}}(A_{n,s}), \mu_{\mathrm{SC}}) > \varepsilon) = 0,$$

where $A_{n,s}$ *stands for* $\frac{1}{\sqrt{ns}}(W_{n,s} - s\,\mathrm{I})$.

As in Theorem 6.23, the assertion of convergence in ∞-Wasserstein distance in Theorems 6.28 and 6.29 subsumes both global convergence of the spectrum towards the limit distribution, and convergence of the extreme eigenvalues towards the edges of the limit distribution (see Proposition 6.33).

Our last limit theorem deals with partial transposition of Wishart matrices. As we shall see, the partial transposition dramatically changes the limit behavior. Note that the distributions MP(λ) and SC(λ,λ) which appear in Theorems 6.28 and 6.30 have the same mean and the same variance (see Exercise 6.35). This was to be expected since the partial transposition preserves both the trace and the Hilbert–Schmidt norm.

THEOREM 6.30 (Not proved here). *Consider a sequence of indices* (d,s) *which tend to infinity in such a way that* $\lambda = \lim s/d^2 \in (0,\infty)$ *exists. For each* d, s*, let* $W_{d^2,s}$ *be a* Wishart(d^2,s) *random matrix (considered as an operator on* $\mathbb{C}^d \otimes \mathbb{C}^d$*) and* $W^\Gamma_{d^2,s}$ *its partial transpose. Then, for any* $\varepsilon > 0$,

$$\lim_{(d,s)\to\infty} \mathbf{P}(d_\infty(\mu_{\mathrm{sp}}(d^{-2}W^\Gamma_{d^2,s}), \mu_{\mathrm{SC}(\lambda,\lambda)}) > \varepsilon) = 0,$$

where $\mu_{\mathrm{SC}(\lambda,\lambda)}$ *denotes the semicircular distribution with mean* λ *and variance* λ.

EXERCISE 6.35. Verify that (6.41) does indeed define a probability distribution both for $\lambda \geqslant 1$ and for $0 < \lambda < 1$, and that the expected value and the variance of the corresponding random variable are both equal to λ.

EXERCISE 6.36. Check that $f_\lambda(\lambda x) = f_{1/\lambda}(x)$. Use this to deduce from Theorem 6.28 that the weak convergence of $\mu_{\mathrm{sp}}(\frac{1}{n}W_{n,s})$ towards $\mu_{\mathrm{MP}(\lambda)}$ holds for any $\lambda > 0$.

EXERCISE 6.37 (Spherical variant of Wishart ensemble). Deduce from Theorem 6.28 the following variant: if $X_{n,s}$ is defined as in (6.40) and n, s tend to infinity with $\lim s/n = \lambda$, then $\mu_{\mathrm{sp}}(X_{n,s})$ converge towards $\mathrm{MP}(\lambda)$ (in probability, in ∞-Wasserstein distance).

EXERCISE 6.38 (The quartercircular distribution). Check that if X has a standard semicircular distribution, then X^2 has a $\mathrm{MP}(1)$ distribution. In what sense can we say that the singular value distribution of a large random (non-Hermitian) square matrix B with independent $N_{\mathbb{C}}(0,1)$ entries is given by a quartercircular distribution?

EXERCISE 6.39 (Free Poisson variables in the large λ limit).
(a) Show that if X_λ has a $\mathrm{MP}(\lambda)$ distribution, then $(X_\lambda - \lambda)/\sqrt{\lambda}$ converges to the standard semicircular distribution with respect to the ∞-Wasserstein distance as $\lambda \to \infty$.
(b) Find a gap in the following argument, which purports to show that part (a) in combination with Theorem 6.28 implies Theorem 6.29.
By Theorem 6.28, the empirical spectral distribution of $W_{n,s}/n$ is approximately X_λ (in the sense of the ∞-Wasserstein distance) if $s/n \approx \lambda$ and n, s are large. Consequently $(X_\lambda - \lambda)/\sqrt{\lambda} \approx (X_\lambda - s/n)/\sqrt{s/n}$ is approximately the empirical spectral distribution of $(W_{n,s}/n - s/n)/\sqrt{s/n} = (W_{n,s} - s)/\sqrt{sn}$, which is exactly the assertion of Theorem 6.29.

6.2.3.3. *Concentration of spectrum.* In view of Theorem 6.28, it is natural to expect that the spectrum of a typical Wishart(n,s) matrix lies close to the interval $[(\sqrt{s}-\sqrt{n})^2, (\sqrt{s}+\sqrt{n})^2]$ (for $s \geqslant n$), or equivalently that all singular values of an $n \times s$ matrix with i.i.d. $N_{\mathbb{C}}(0,1)$ entries lie close to the interval $[\sqrt{s}-\sqrt{n}, \sqrt{s}+\sqrt{n}]$. A first result in this direction is a precise bound (without any multiplicative constants or error terms) for the expected largest singular value, i.e., the operator norm.

PROPOSITION 6.31. *Let B be an $n \times s$ random matrix with independent $N_{\mathbb{C}}(0,1)$ entries. Then*

$$\mathbf{E}\,\|B\|_{\mathrm{op}} \leqslant \sqrt{n} + \sqrt{s}.$$

Proposition 6.31 will be deduced from its analogue for real Wishart matrices, which requires methods specific to that setting. Accordingly, we postpone its proof until Section 6.2.4.2.

In view of Proposition 6.31, it is natural to ask the following question, the answer to which is known to be affirmative in the real case (see Corollary 6.38). Recall that $s_n(B)$ denotes the smallest singular value of B.

PROBLEM 6.32. *Let $s \geqslant n$, and let B be an $n \times s$ random matrix with independent $N_{\mathbb{C}}(0,1)$ entries. Do we have the inequality $\mathbf{E}\, s_n(B) \geqslant \sqrt{s} - \sqrt{n}$?*

We now state a concentration result for the spectrum of Wishart matrices.

PROPOSITION 6.33. *Let B be a random $n \times s$ matrix with independent $N_{\mathbb{C}}(0,1)$ entries. For every $t > 0$,*

$$\mathbf{P}\left(\|B\|_{\mathrm{op}} \geqslant \sqrt{n} + \sqrt{s} + t\right) \leqslant \frac{1}{2}\exp(-t^2). \tag{6.42}$$

If $s > n$, then for every $t > 4\sqrt{2\log n}/(\sqrt{s/n} - 1)$,

$$\mathbf{P}\left(s_n(B) \leqslant \sqrt{s} - \sqrt{n} - t\right) \leqslant \exp(-t^2/4), \tag{6.43}$$

where C and c denote absolute constants.

The above result is closely related to Proposition 6.24 and shares many of the ramifications of the latter. For example, while we know from the general theory of Gaussian concentration that the quantities in question are concentrated around some value, identifying that value requires a separate argument and may be hard. In particular, a positive answer to Problem 6.32 would imply the validity of (6.43) for all $t \geqslant 0$ and with the bound $\exp(-t^2)$.

PROOF. The functions $\|\cdot\|_{\mathrm{op}}$ and s_n are 1-Lipschitz with respect to the Hilbert–Schmidt norm on $\mathsf{M}_{n,s}$. Let M be the median of $\|B\|_{\mathrm{op}}$. By combining Propositions 6.31 and 5.34, it follows that $M \leqslant \sqrt{n} + \sqrt{s}$, and we deduce (6.42) by using the values from Table 5.2.

Next, let M' be the median of $s_n(B)$. We claim that

$$M' \geqslant \sqrt{s} - \sqrt{n} - \frac{2\sqrt{s+n}\sqrt{\log 2n}}{\sqrt{s} - \sqrt{n}}. \tag{6.44}$$

Once this is established, we argue as before, using the values from Table 5.2, and get for $t > 4\sqrt{2\log n}/(\sqrt{s/n} - 1)$

$$\mathbf{P}(s_n(B) \leqslant \sqrt{s} - \sqrt{n} - t) \leqslant \mathbf{P}(s_n(B) \leqslant M' - t/2) \leqslant \frac{1}{2}\exp(-t^2/4).$$

In turn, we can obtain (6.44) as a consequence of the following inequality valid for any $t > 0$:

$$\frac{1}{2}\exp(-tM'^2) \leqslant \mathbf{E}\,\mathrm{Tr}\exp(-tBB^\dagger) \leqslant n\exp\left(-(\sqrt{s} - \sqrt{n})^2 t + (s+n)t^2\right). \tag{6.45}$$

(The second inequality in (6.45) is not at all immediate to prove; it appears as Lemma 7.2 in [**HT03**].) We then use the optimal choice $t = \sqrt{(s+n)\log(2n)}$ and the inequality $\sqrt{a-b} \geqslant \sqrt{a} - b/\sqrt{a}$ (valid for $a \geqslant b$). $\square$

6.2.3.4. *Random induced states.* Wishart matrices are of interest in quantum theory since they lead to a very natural model of random quantum states. One possible way to generate a random state on $\mathbb{C}^n$ is to take independent unit vectors $(\psi_i)_{1 \leqslant i \leqslant s}$ distributed uniformly on the sphere and to consider the average of corresponding pure states, i.e.,

$$\rho = \frac{1}{s}\sum_{i=1}^{s} |\psi_i\rangle\langle\psi_i|.$$

This is exactly (6.40) up to normalization. However a closely related and often better model is to consider the partial trace of a Haar-distributed pure state on $\mathbb{C}^n \otimes \mathbb{C}^s$. We call states obtained that way *random induced states.*

Let us denote by $\mu_{n,s}$ the distribution of the induced state $\mathrm{Tr}_{\mathbb{C}^s}|\psi\rangle\langle\psi|$ when ψ is uniformly distributed on the unit sphere in $\mathbb{C}^n \otimes \mathbb{C}^s$. The measure $\mu_{n,s}$ is a probability measure on the set $\mathrm{D}(\mathbb{C}^n)$ of states on $\mathbb{C}^n$. As the following simple fact shows, this measure is just a renormalization of $\mathrm{Wishart}(n, s)$.

PROPOSITION 6.34 (Wishart matrices as induced states). *Let W be a random matrix with distribution* $\mathrm{Wishart}(n, s)$. *Then $\frac{W}{\operatorname{Tr} W}$ has distribution $\mu_{n,s}$ and is independent of* $\operatorname{Tr} W$.

PROOF. The Proposition follows from the combination of two facts. First, if G is a standard Gaussian vector in any given Euclidean or Hilbert space V (in our case $V = \mathbb{C}^n \otimes \mathbb{C}^s$), then the vector $\frac{G}{|G|}$ is uniformly distributed on the unit sphere of V and is independent of $|G|$. Second, when we identify a tensor $\psi \in \mathbb{C}^n \otimes \mathbb{C}^s$ with a matrix $A \in \mathsf{M}_{n,s}$, we have (see Section 0.8)

$$\operatorname{Tr}_{\mathbb{C}^s} |\psi\rangle\langle\psi| = AA^\dagger. \qquad \square$$

The normalization factor $\operatorname{Tr} W$ is very strongly concentrated around the value ns (see Exercise 6.40). Therefore, it can be virtually treated as a constant when translating the results for Wishart matrices in the language of induced states.

We have the following (recall that $\mu_{\mathrm{sp}}(A)$ is the empirical spectral distribution of a self-adjoint matrix A, see (6.29)).

THEOREM 6.35. *Given integers n, s, let $\rho_{n,s}$ be a random induced state with distribution $\mu_{n,s}$.*
(i) If n is fixed and s tends to infinity, then $\sqrt{n(n-1)s}\,(\rho_{n,s} - \frac{\mathrm{I}}{n})$ converges in distribution towards a $\mathrm{GUE}_0(n)$ matrix.
(ii) If n tends to infinity and $\lim s/n = \lambda \in (0, \infty)$, then $\mu_{\mathrm{sp}}(s\rho_{n,s})$ converges weakly in probability towards $\mu_{\mathrm{MP}(\lambda)}$. Moreover, if $\lambda \geqslant 1$, then the convergence also holds in ∞-Wasserstein distance.
(iii) If both n and s/n tend to infinity, then $\mu_{\mathrm{sp}}(\sqrt{ns}(\rho_{n,s} - \mathrm{I}/n))$ converges in probability in ∞-Wasserstein distance towards μ_{SC}.

Recall that the empirical spectral distributions of a *rescaled* GUE_0 matrix is almost semicircular (see Theorem 6.23), so that (i) and (iii) are indeed consistent. To deduce (ii) from Theorem 6.28 and (iii) from Theorem 6.29, use Proposition 6.34 and the bounds from Exercise 6.40. The statement (i) is more elementary (see Exercise 6.41).

Similarly, Proposition 6.33 can be restated as a result about spectra of random induced states or as a result about Schmidt coefficients of random pure states. We single it out as a separate statement since it will be used several times. Alternatively, a weaker statement follows from an elementary net argument (see Exercise 6.43).

PROPOSITION 6.36. *For $n \leqslant s$, let ψ be a random vector uniformly distributed on the unit sphere of $\mathbb{C}^n \otimes \mathbb{C}^s$ and let $\lambda_1(\psi) \geqslant \cdots \geqslant \lambda_n(\psi)$ be its Schmidt coefficients. Then, for any $\varepsilon > 0$,*

$$\mathbf{P}\left(\lambda_1(\psi) \geqslant \frac{1}{\sqrt{n}} + \frac{1+\varepsilon}{\sqrt{s}}\right) \leqslant \exp(-n\varepsilon^2) \tag{6.46}$$

and, for any $\varepsilon \geqslant C\sqrt{s \log n}/(\sqrt{ns} - n)$,

$$\mathbf{P}\left(\lambda_n(\psi) \leqslant \frac{1}{\sqrt{n}} - \frac{1+\varepsilon}{\sqrt{s}}\right) \leqslant \exp(-cn\varepsilon^2),$$

where c and C are absolute constants.

Proposition 6.36 can be deduced from Proposition 6.33 or proved in the same way using concentration of measure on the sphere (cf. Exercise 6.42). We also note that Proposition 6.36 can be equivalently restated using matrices instead of tensors: *if $M \in \mathsf{M}_{n,s}$ is uniformly distributed on the Hilbert–Schmidt sphere, then with large probability all its singular values belong to the interval* $\left[\frac{1}{\sqrt{n}} - \frac{1+\varepsilon}{\sqrt{s}}, \frac{1}{\sqrt{n}} + \frac{1+\varepsilon}{\sqrt{s}}\right]$.

When $s \geqslant n$, the probability measure $\mu_{n,s}$ has a density with respect to the Lebesgue measure on $\mathrm{D}(\mathbb{C}^n)$, which has a simple form

$$\frac{\mathrm{d}\mu_{n,s}}{\mathrm{d}\,\mathrm{vol}}(\rho) = \frac{1}{Z_{n,s}}(\det \rho)^{s-n}, \tag{6.47}$$

where $Z_{n,s}$ is a normalization factor. Note that formula (6.47) allows us to define the measure $\mu_{n,s}$ (in particular) for every real $s \geqslant n$, while the partial trace construction makes sense only for integer values of s. The explicit formula (6.47) will not be used in this book.

In the important special case where $s = n$, the density of the measure $\mu_{n,n}$ is constant: a random state distributed according to $\mu_{n,n}$ is distributed with respect to the uniform (Lebesgue) measure on $\mathrm{D}(\mathbb{C}^n)$. This can be seen as a non-commutative version of the following classical fact: if $\psi = (\psi_1, \dots, \psi_n)$ is uniformly distributed on the unit sphere in $\mathbb{C}^n$, the vector $(|\psi_1|^2, \dots, |\psi_n|^2)$ is uniformly distributed on the $(n-1)$-dimensional simplex.

Exercise 6.40 (Trace of a Wishart matrix). Let W be a Wishart(n, s) matrix. Check that $2 \operatorname{Tr} W$ has distribution $\chi^2(2ns)$ and deduce from Exercise 5.37 that for any $t > 0$,

$$\mathbf{P}(|\operatorname{Tr} W - ns| > tns) \leqslant 2 \exp\left(-\frac{nst^2}{2 + 4t/3}\right).$$

Exercise 6.41. Use the multivariate central limit theorem to prove part (i) of Theorem 6.35.

Exercise 6.42 (Mean of the largest Schmidt coefficient). Let ψ be a random vector uniformly distributed on $S_{\mathbb{C}^n \otimes \mathbb{C}^s}$. Deduce from (6.49) that $\mathbf{E}\,\lambda_1(\psi) \leqslant \frac{\kappa_{2n}+\kappa_{2s}}{\kappa_{2ns}} \leqslant \frac{1}{\sqrt{n}} + \frac{1}{\sqrt{s}}$. Then prove (6.46).

Exercise 6.43 (Elementary bounds on the spectra of random induced states). Let ρ be a random induced state with distribution $\mu_{n,s}$, i.e., $\rho = \operatorname{Tr}_{\mathbb{C}^s} |\psi\rangle\langle\psi|$ with ψ uniformly distributed on $S_{\mathbb{C}^n \otimes \mathbb{C}^s}$.
(i) For any $y \in S_{\mathbb{C}^n}$, show that the function f defined on $S_{\mathbb{C}^n \otimes \mathbb{C}^s}$ by $f(\psi) = \sqrt{\langle y|\rho|y\rangle}$ is 1-Lipschitz and that $\mathbf{E}\, f^2 = 1/n$. Conclude from Exercise 5.46 that, for any $t > 0$, $\mathbf{P}(|f - 1/\sqrt{n}| > t) \leqslant (1 + e)\exp(-nst^2)$.
(ii) Let $\mathcal{N}$ be a δ-net in $S_{\mathbb{C}^n}$ for $\delta < 1/2$. Denote $\Delta = \rho - \mathrm{I}/n$ and show that

$$\|\Delta\|_\infty \leqslant \frac{1}{1 - 2\delta} \sup_{y \in \mathcal{N}} |\langle y|\Delta|y\rangle| .$$

(iii) Let $s \geqslant n$. Conclude that $\|\Delta\|_\infty \leqslant C/\sqrt{ns}$ with high probability for some constant C.

Exercise 6.44 (The limit distribution of the partial transpose). Let ν be the law of XY, where X and Y are independent random variables following the standard semicircular distribution. Let $\psi \in S_{\mathbb{C}^d \otimes \mathbb{C}^d}$ be a uniformly distributed random vector, and $A = d|\psi\rangle\langle\psi|^\Gamma$. (The partial transposition Γ was defined in Section 2.2.6.) Show that, when d tends to infinity, $\mu_{\mathrm{sp}}(A)$ converges in probability, in ∞-Wasserstein distance, towards ν.

EXERCISE 6.45 (Low moments of Wishart matrices and expected purity of random induced states). For a quantum state σ, the quantity $\operatorname{Tr}\sigma^2$ is called the *purity* of σ.
(i) Let G be an $n \times s$ random matrix with independent $N_{\mathbb{C}}(0,1)$ entries. Show that $\mathbf{E}\operatorname{Tr}(GG^\dagger GG^\dagger) = n^2 s + s^2 n$ and that $\mathbf{E}(\operatorname{Tr} GG^\dagger)^2 = ns(ns+1)$.
(ii) Let ρ be a random induced state with distribution $\mu_{n,s}$. Show that $\mathbf{E}\operatorname{Tr}\rho^2 = \frac{n+s}{ns+1}$.

6.2.4. Real RMT models and Chevet–Gordon inequalities. We consider now variants of the random matrix models introduced before, where the entries are real instead of complex. All the theorems stated for the GUE and for complex Wishart matrices carry over *mutatis mutandis* to the real case. One important modification that is worth pointing out is that in the density formula from Proposition 6.22 the factors $\lambda_i - \lambda_j$ are not squared, which makes certain arguments harder. However, the formulas in question play almost no role in our approach. On the other hand, some other tools—most notably the analysis via Gaussian processes—are more adapted to the real setting.

The GOE is the real version of the GUE. A random matrix A has the $\mathrm{GOE}(n)$ distribution if the random variables $(a_{ij})_{1 \leqslant i \leqslant j \leqslant n}$ are independent, with a_{ii} having the $N(0,2)$ distribution and a_{ij} (for $i \neq j$) having the $N(0,1)$ distribution. This normalization is chosen so that the distribution is invariant under conjugacy by an orthogonal matrix. Note also that $A/\sqrt{2}$ is a standard Gaussian vector in the space $\mathsf{M}_n^{\mathrm{sa}}$.

Real Wishart matrices are then defined exactly as their complex analogues: if B is an $n \times s$ random matrix with independent $N(0,1)$ entries, the distribution of $W = BB^\dagger$ is denoted by $\mathrm{Wishart}_{\mathbb{R}}(n,s)$.

In both settings, an argument based on Gordon's lemma (Proposition 6.7) allows for concise proofs of precise inequalities. This scheme actually allows obtaining sharp bounds on the norm of a random matrix as an operator between any two real normed spaces. The basic ingredient is a contraction property of the tensor product map which holds only in the real case (Exercise 6.47).

6.2.4.1. *Chevet–Gordon inequalities.*

PROPOSITION 6.37 (Chevet–Gordon inequalities). *Let $B \in \mathsf{M}_{n,s}$ be a random matrix with independent $N(0,1)$ entries. Let $K \subset \mathbb{R}^s$ and $L \subset S^{n-1}$ be compact sets, and $r_K \geqslant 0$ such that $K \subset r_K B_2^s$. Then*

$$w_G(K) - r_K w_G(L) \leqslant \mathbf{E} \min_{u \in L} \max_{t \in K} \langle Bt, u\rangle \leqslant \mathbf{E} \max_{u \in L} \max_{t \in K} \langle Bt, u\rangle \leqslant w_G(K) + r_K w_G(L).$$

Note that the upper bound in Proposition 6.37 is always sharp up to a factor of 2 (see Exercise 6.46).

PROOF. Let G be a standard Gaussian vector in $\mathbb{R}^s \oplus \mathbb{R}^n$. We are going to compare the following Gaussian processes indexed by $(t,u) \in K \times L$,

$$X_{t,u} = \langle Bt, u\rangle,$$

$$Y_{t,u} = \langle G, t \oplus r_K u\rangle.$$

One checks (see Exercise 6.47(ii)) that for $(t,u),(t',u')$ in $K \times L$

$$\mathbf{E}(X_{t,u} - X_{t',u'})^2 \leqslant \mathbf{E}(Y_{t,u} - Y_{t',u'})^2. \tag{6.48}$$

We may now apply Slepian's lemma (Proposition 6.6; as usual, the fact that the supremum is presently taken over an infinite set can be circumvented by considering all finite subfamilies, see (6.1)) to conclude that

$$\mathbf{E} \max_{(t,u)\in K\times L} X_{t,u} \leqslant \mathbf{E} \max_{(t,u)\in K\times L} Y_{t,u} = w_G(K) + r_K w_G(L).$$

To prove the other inequality we use the Slepian–Gordon lemma (Proposition 6.7, in the min max version; see Remark 6.8) with the partition $K \times L = \bigcup_{u\in L} T_u$, where $T_u = K \times \{u\}$. The hypotheses are satisfied since there is equality in (6.48) when $u = u'$. Consequently,

$$\mathbf{E} \min_{u\in L} \max_{t\in K} X_{t,u} \geqslant \mathbf{E} \min_{u\in L} \max_{t\in K} Y_{t,u} = w_G(K) - r_K w_G(L). \qquad \square$$

As a corollary we obtain sharp bounds on the extreme singular values of a rectangular Gaussian matrix, or equivalently on the extreme eigenvalues of a Wishart matrix. These bounds match the support of the Marčenko–Pastur distribution from Theorem 6.28. It is then routine to derive concentration estimates.

COROLLARY 6.38. *Let $n \leqslant s$, let $B \in \mathsf{M}_{n,s}$ be a random matrix with independent $N(0,1)$ entries, and denote by $s_n(B)$ its smallest singular value. Then*

$$\sqrt{s} - \sqrt{n} \leqslant \kappa_s - \kappa_n \leqslant \mathbf{E}\, s_n(B) \leqslant \mathbf{E}\, \|B\|_{\mathrm{op}} \leqslant \kappa_s + \kappa_n \leqslant \sqrt{s} + \sqrt{n}.$$

Consequently, for any $t \geqslant 0$,

$$\mathbf{P}(\|B\|_{\mathrm{op}} \geqslant \sqrt{s} + \sqrt{n} + t) \leqslant \frac{1}{2}\exp(-t^2/2),$$

$$\mathbf{P}(s_n(B) \leqslant \sqrt{s} - \sqrt{n} - t) \leqslant \exp(-t^2/2).$$

PROOF. We apply Proposition 6.37 with $K = S^{s-1}$ and $L = S^{n-1}$. Note that $w_G(K) = \kappa_s$ and $w_G(L) = \kappa_n$. The leftmost and rightmost inequalities follow from Proposition A.1 (iv) and (i). The concentration estimates are proved as in Proposition 6.33. $\square$

EXERCISE 6.46 (Sharpness of Chevet's inequality). In the notation of Proposition 6.37, show that

$$\mathbf{E} \max_{u\in L} \max_{t\in K} \langle Bt, u\rangle \geqslant \max(w_G(K), \mathrm{outrad}(K) w_G(L)).$$

EXERCISE 6.47 (The contractions underlying the Gordon–Chevet inequality). Let m, n be integers.
(i) If $\delta : \mathbb{R}^m \times \mathbb{R}^n \to \mathbb{R}^m \times \mathbb{R}^n$ is defined by $\delta(x,y) = (|y|x, |x|y)$, show that for any (x,y) and (x',y') in $\mathbb{R}^m \times \mathbb{R}^n$,

$$|x \otimes y - x' \otimes y'| \leqslant |\delta(x,y) - \delta(x',y')|.$$

(ii) Fix $r > 0$ and consider the map $\delta_r : \mathbb{R}^m \times \mathbb{R}^n \to \mathbb{R}^m \times \mathbb{R}^n$ defined by $\delta_r(x,y) = (rx, |x|y)$. Show that for any (x,y) and (x',y') in $\mathbb{R}^m \times rB_2^n$,

$$|x \otimes y - x' \otimes y'| \leqslant |\delta_r(x,y) - \delta_r(x',y')|.$$

(iii) Show that the analogues of (i) and (ii) fail in the complex setting.

EXERCISE 6.48 (Sharp bounds on the largest eigenvalue of GOE(n) and GUE(n) matrices). Let A be a GOE(n) or GUE(n) random matrix. By arguing along the lines of the proofs of Proposition 6.37 and Corollary 6.38, show that $\mathbf{E}\,\lambda_1(A) \leqslant 2\sqrt{n}$.

EXERCISE 6.49 (Mean width of the projective tensor product). Let $K \subset \mathbb{R}^m$ and $L \subset \mathbb{R}^n$ be convex bodies. Assume that $K \subset r_K B_2^m$ and $L \subset r_L B_2^n$. Prove that $w_G(K \hat{\otimes} L) \leqslant w_G(K) r_L + w_G(L) r_K$ and $w(K \hat{\otimes} L) \leqslant w(K)\frac{r_L}{\sqrt{n}} + w(L)\frac{r_K}{\sqrt{m}}$.

6.2.4.2. *A coupling argument.* We prove here Proposition 6.31. Let B be an $n \times s$ random matrix with independent $N_{\mathbb{C}}(0,1)$ entries and let A be a $2n \times 2s$ random matrix with independent $N(0,1)$ entries. We show that

$$\mathbf{E}\,\|B\|_{\mathrm{op}} \leqslant \frac{1}{\sqrt{2}}\,\mathbf{E}\,\|A\|_{\mathrm{op}} \leqslant \frac{\kappa_{2n} + \kappa_{2s}}{\sqrt{2}} \leqslant \sqrt{n} + \sqrt{s}. \tag{6.49}$$

We use representations of A and B via χ-distributed random variables. If G is a standard Gaussian vector in $\mathbb{R}^n$, the distribution of $|G|$ is denoted by $\chi(n)$ (the square of a $\chi(n)$-distributed variable has distribution $\chi^2(n)$). If G is a standard Gaussian vector in $\mathbb{C}^n$, then $\sqrt{2}|G|$ has distribution $\chi(2n)$.

LEMMA 6.39 (See Exercise 6.50). *Let $n \leqslant s$ and let A be an $n \times s$ random matrix with independent $N(0,1)$ entries. There exist random matrices $U \in \mathsf{O}(n)$ and $V \in \mathsf{O}(s)$, such that, denoting $R = UAV$,*
(i) the random variables $\{r_{i,j} \,:\, 1 \leqslant i \leqslant n, 1 \leqslant j \leqslant s\}$ are independent,
(ii) for $1 \leqslant i \leqslant n$, $r_{i,i}$ has distribution $\chi(s+1-i)$,
(iii) for $2 \leqslant i \leqslant n$, $r_{i,i-1}$ has distribution $\chi(n+1-i)$,
(iv) other entries of R are almost surely zero.

LEMMA 6.40 (See Exercise 6.50). *Let $n \leqslant s$ and let B be an $n \times s$ random matrix with independent $N_{\mathbb{C}}(0,1)$ entries. There exist random matrices $U' \in \mathsf{U}(n)$ and $V' \in \mathsf{U}(s)$, such that, denoting $S = \sqrt{2}U'BV'$,*
(i) the random variables $\{s_{i,j} \,:\, 1 \leqslant i \leqslant n, 1 \leqslant j \leqslant s\}$ are independent,
(ii) for $1 \leqslant i \leqslant n$, $s_{i,i}$ has distribution $\chi(2s+2-2i)$,
(iii) for $2 \leqslant i \leqslant n$, $s_{i,i-1}$ has distribution $\chi(2n+2-2i)$,
(iv) other entries of S are almost surely zero.

We apply Lemmas 6.39 (with dimensions $2n \times 2s$ instead of $n \times s$) and 6.40 to the matrices A and B appearing in (6.49). Since $2s + 2 - 2i \leqslant 2s + 1 - i$ for $1 \leqslant i \leqslant n$ and $2n + 2 - 2i \leqslant 2n + 1 - i$ for $2 \leqslant i \leqslant n$, the initial matrices A and B can be coupled (i.e., both defined on a single probability space) in such a way that, almost surely, $s_{ij} \leqslant r_{ij}$ for any $1 \leqslant i \leqslant n$ and $1 \leqslant j \leqslant s$. Since R and S have positive entries, this implies that (almost surely) $\|S\|_{\mathrm{op}} \leqslant \|R\|_{\mathrm{op}}$. Since $\|A\|_{\mathrm{op}} = \|R\|_{\mathrm{op}}$ and $\|B\|_{\mathrm{op}} = \frac{1}{\sqrt{2}}\|S\|_{\mathrm{op}}$, it follows that $\mathbf{E}\,\|B\|_{\mathrm{op}} \leqslant \frac{1}{\sqrt{2}}\,\mathbf{E}\,\|A\|_{\mathrm{op}}$. The remaining inequalities in (6.49) are proved in Corollary 6.38.

PROBLEM 6.41. *Does there exist an argument along similar lines (i.e., using Slepian's lemma and coupling) that yields inequalities in the spirit of (6.49), but involving* GUE *and* GOE *matrices (say, $\mathbf{E}\,\|B\|_{\mathrm{op}} \leqslant \frac{1}{\sqrt{2}}\,\mathbf{E}\,\|A\|_{\mathrm{op}} \leqslant 2\sqrt{n}$, with B being a* GUE(n) *matrix and A being a* GOE($2n$) *matrix)?*

EXERCISE 6.50 (Representation of Wishart matrices via χ-distributed variables). Prove Lemmas 6.39 and 6.40. Show also that the matrices U, V and R can be chosen to be independent, with U, V Haar-distributed (same for U', V' and S).

EXERCISE 6.51 (Neat bounds on the norms of Wishart matrices). Let A be a random $n \times s$ matrix with independent $N(0,1)$ entries with $n \leqslant s$.
(i) Show that $\mathbf{E}\,\|A\| \geqslant \|M\|$ where $M = (m_{i,j})$ is the $n \times s$ matrix such that $m_{i,i} = \kappa_{s+1-i}$ for $1 \leqslant i \leqslant n$ and $m_{i,i-1} = \kappa_{n+1-i}$ for $2 \leqslant i \leqslant n$ (other entries being zero).
(ii) Conclude that $\mathbf{E}\,\|A\| \geqslant (\sqrt{n-k} + \sqrt{s-k})\sqrt{1-1/k}$ for any $1 \leqslant k \leqslant n$. Show that this inequality also holds when A is defined using $N_{\mathbb{C}}(0,1)$ variables.

6.2.4.3. *The escape phenomenon.* Another consequence of the Chevet–Gordon inequalities is the fact that a subset of the sphere which is small (when measured using mean width) typically does not intersect a subspace of large dimension: a generic subspace "escapes" from any small set. This is made very precise in the following proposition.

PROPOSITION 6.42. *Let $L \subset S^{n-1}$ a closed subset, $k \in \{1, \dots, n-1\}$ such that $w_G(L) < \kappa_{n-k}$, and let $E \subset \mathbb{R}^n$ be a random k-dimensional subspace. Then*

$$\mathbf{P}\,(E \cap L \neq \varnothing) \leqslant \exp(-(\kappa_{n-k} - w_G(L))^2/2).$$

Proposition 6.42 will give a direct proof of the low-M^* estimate (Theorem 7.45).

PROOF OF PROPOSITION 6.42. Let $s = n-k$, let B be an $n \times s$ random matrix with i.i.d. $N(0,1)$ entries, and let $E = \ker B$. One checks that E is distributed according to the Haar measure on $\mathsf{Gr}(k, \mathbb{R}^n)$. (This follows from the characterization of the Haar measure as the only measure invariant under the action of $\mathsf{O}(n)$.) Moreover, since L is closed, the condition $E \cap L = \varnothing$ is equivalent to $\min_{x \in L} |Bx| > 0$. We apply the Chevet–Gordon inequalities (Proposition 6.37) with $K = S^{s-1}$ to conclude that

$$\mathbf{E} \min_{x \in L} |Bx| \geqslant \kappa_s - w_G(L).$$

Since the function $g : B \mapsto \min_{x \in L} |Bx|$ is 1-Lipschitz with respect to the Hilbert–Schmidt distance, we may apply Gaussian concentration of measure (see Table 5.2) to conclude that

$$\begin{aligned} \mathbf{P}(E \cap L \neq \varnothing) &= \mathbf{P}(g(B) = 0) \\ &= \mathbf{P}\Big(g(B) \leqslant \mathbf{E}\, g(B) - (\kappa_s - w_G(L))\Big) \\ &\leqslant \exp(-(\kappa_s - w_G(L))^2/2). \end{aligned}$$

□

6.2.5. A quick initiation to free probability. We now mention briefly deeper results about high-dimensional random matrices that touch upon the connection with free probability. A rigorous introduction to free probability is beyond the scope of this book, so we instead illustrate, in an example, the kind of conclusions that can be derived from the general theory.

Free probability describes limit objects towards which large-dimensional random matrices converge. Here is a typical statement about polynomials in independent GUE matrices.

THEOREM 6.43 (Not proved here). *Let P be a non-commutative self-adjoint polynomial in N variables. For every n, let $A_n^{(1)}, \dots, A_n^{(N)}$ be N independent random matrices with* GUE(n) *distribution, and let $X_n = P(A_n^{(1)}/\sqrt{n}, \dots, A_n^{(N)}/\sqrt{n})$. Then, as $n \to \infty$, the empirical spectral distributions $(\mu_{\mathrm{sp}}(X_n))$ converge weakly,* *in probability, towards the distribution of* $P(a_1, \ldots, a_N)$*, where* $a_1, \ldots, a_N$ *are free semicircular variables. Moreover,* $\|X_n\|_\infty$ *converges in probability towards the value* $\|P(a_1, \ldots, a_N)\|$.

Let us explain the meaning of the concepts and notions that appear in Theorem 6.43. First, a polynomial P is self-adjoint if $P(M_1, \ldots, M_N) \in \mathsf{M}_n^{\mathrm{sa}}$ whenever $M_1, \ldots, M_N \in \mathsf{M}_n^{\mathrm{sa}}$; an example is $P(x_1, x_2) = x_1 x_2 x_1$. Second, a family of N "free semicircular random variables" can be concretely realized as follows: let $\mathcal{F} = \bigoplus_{k \in \mathbb{N}} (\mathbb{C}^N)^{\otimes k}$ be the Fock space over $\mathbb{C}^N$, with the usual convention that $(\mathbb{C}^N)^{\otimes 0}$ is a one-dimensional space spanned by a unit vector Ω. Let $|1\rangle, \ldots, |N\rangle$ be the canonical basis of $\mathbb{C}^N$, and let $h_1, \ldots, h_N \in B(\mathcal{F})$ be the corresponding creation operators, defined by $h_i(x) = |i\rangle \otimes x \in (\mathbb{C}^N)^{\otimes(k+1)}$ for every $x \in (\mathbb{C}^N)^{\otimes k}$. Set $a_i := h_i + h_i^\dagger$; then the operators $a_1, \ldots, a_N$ are an example of "free semicircular variables." The quantity $\|P(a_1, \ldots, a_N)\|$ appearing in Theorem 6.43 is simply the operator norm, and the *distribution* of a self-adjoint operator $Y \in B(\mathcal{H})$ is defined as the unique probability measure μ on $\mathbb{R}$ such that, for every bounded continuous function $f : \mathbb{R} \to \mathbb{R}$,

$$\langle \Omega | f(Y) | \Omega \rangle = \int_{\mathbb{R}} f \, d\mu$$

(it is enough to consider the case where f is a polynomial). The unfamiliar reader is invited to check that this formalism is consistent with Theorem 6.23 (see Exercise 6.52).

The phenomenon behind Theorem 6.43 is called "asymptotic freeness of random matrices" and is not limited to the case of GUE matrices (see Notes and Remarks for more references). Here is another example involving unitary matrices. The "free additive convolution" is a binary operation (denoted by $\boxplus$ and not defined here) on probability measures (say, with compact support) on $\mathbb{R}$.

THEOREM 6.44 (Not proved here). *Let* μ *and* ν *be two compactly supported probability measures on* $\mathbb{R}$*. For every* n*, let* $A_n, B_n \in \mathsf{M}_n^{\mathrm{sa}}$ *be real (resp., complex) self-adjoint matrices such that the sequences of empirical measures* $(\mu_{\mathrm{sp}}(A_n))$ *and* $(\mu_{\mathrm{sp}}(B_n))$ *converge weakly towards* μ *and* ν *as* $n \to \infty$*. Let* U_n *be a Haar-distributed random orthogonal (resp., unitary) matrix. Then (weakly, in probability)*

$$\lim_{n \to \infty} \mu_{\mathrm{sp}}(A_n + U_n B_n U_n^\dagger) = \mu \boxplus \nu.$$

The usefulness of Theorem 6.44 comes from the fact that in many situations the free additive convolution of probability measures can be computed using the so-called R-transform, a non-commutative analogue of the Fourier transform (see for example Lecture 12 in [**NS06**]). Here is an example of conclusions that can be derived from Theorem 6.44.

COROLLARY 6.45 (Not proved here). *Fix* $0 \leqslant t \leqslant 1/2$*. For any* n*, let* E_n *and* F_n *be subspaces which are independent and Haar-distributed on the Grassmann manifold* $\mathsf{Gr}(\lfloor tn \rfloor, \mathbb{C}^n)$ *and denote* $A_n = P_{E_n} + P_{F_n}$ *the sum of the corresponding projectors. Then the sequence of empirical measures* $(\mu_{\mathrm{sp}}(A_n))$ *converges weakly in probability towards the deterministic measure*

$$(1-2t)\delta_0 + \frac{\sqrt{4t(1-t) - (x-1)^2}}{\pi x(2-x)} \mathbf{1}_{\left[1-2\sqrt{t(1-t)},\, 1+2\sqrt{t(1-t)}\right]}(x)\, dx. \tag{6.50}$$

Moreover, the sequence $(\|A_n\|_\infty)$ *converges in probability towards* $1 + 2\sqrt{t(1-t)}$.

An analogous statement for $t \geqslant 1/2$ follows by applying 6.45 to $E_n^\perp$ and $F_n^\perp$. The measure defined in (6.50) is the free additive convolution of the measure $(1-t)\delta_0 + t\delta_1$ with itself (for $t = 1/2$ we recover the arcsine distribution).

EXERCISE 6.52 (Semicircular variables via creation operators). Show that the distribution of the operators a_i defined on the Fock space in the paragraph following Theorem 6.43 is indeed the semicircular distribution.

Notes and Remarks

Section 6.1. The elegant proof of Lemma 6.1 is due to Talagrand (see [**Tal11**]). Denote by u_n the expectation of the maximum of n independent $N(0,1)$ variables. For small n, explicit formulas are known: $u_1 = 0$, $u_2 = 1/\sqrt{\pi}$, $u_3 = 3/(2\sqrt{\pi})$, $u_4 = \frac{3}{\sqrt{\pi}}(\frac{1}{2} + \frac{1}{\pi}\arcsin(\frac{1}{3})))$ and $u_5 = \frac{5}{2\sqrt{\pi}}(\frac{1}{2} + \frac{3}{\pi}\arcsin(\frac{1}{3})))$ (numbers are from the website [**@2**]). Moreover, an asymptotic expansion of u_n can be obtained from the convergence of the maximum of independent Gaussian samples to the Gumbel distribution (see [**LLR83**], Theorem 1.5.3 and also [**Pic68**] to justify convergence in expectation)

$$u_n = \sqrt{2\log n} - \frac{\log\log n}{2\sqrt{2\log n}} + O\left(\frac{1}{\sqrt{\log n}}\right).$$

Inequalities in a similar spirit, but also for fixed n, appear in [**DLS14**].

References for the result from Remark 6.4 are [**Glu88**], [**CP88**] and [**BF88**]. The conjecture from Remark 6.5 appears in [**HS05**].

The second part of Proposition 6.6 was originally proved by Slepian [**Sle62**] and is usually referred to as Slepian's lemma. The first assertion, which follows from the second one, is sometimes called the Sudakov–Fernique inequality and appears in [**Fer75**]. Several proofs of Proposition 6.6 are available in addition to the original one; see, e.g., Kahane [**Kah86**] and Gromov [**Gro87**].

We also mention a well-known open problem related to Slepian's lemma which is known as the Kneser–Poulsen conjecture. Suppose that $x_1, \dots, x_N$ and $y_1, \dots, y_N$ are points in $\mathbb{R}^d$ with the property that $|x_i - x_j| \leqslant |y_i - y_j|$ for any $1 \leqslant i, j \leqslant N$. The conjecture asks whether for every radii $r_1, \dots, r_N > 0$, we have $\mathrm{vol}(\bigcup B(x_i, r_i)) \leqslant \mathrm{vol}(\bigcup B(y_i, r_i))$. Under the same hypotheses, a sister conjecture is whether the inequality $\mathrm{vol}(\bigcap B(x_i, r_i)) \geqslant \mathrm{vol}(\bigcap B(y_i, r_i))$ holds. Similar questions can be asked for the spherical, hyperbolic and projective spaces. Note that, in the spherical case, the two conjectures are equivalent since the complement of a cap is also a cap. Also, since all Riemannian manifolds are asymptotically flat as distances go to 0, the Euclidean case (in any particular dimension) would be a formal consequence of a positive answer in any other setting. The answers were shown to be affirmative when $k \leqslant d+1$ in the spherical setting (see [**Gro87**]) and when $k \leqslant d+3$ or when $d = 2$ (for arbitrary k) in the Euclidean setting (see [**BC02**]). Both conjectures are known to be true for spherical caps of angle $\pi/2$, see [**Bez08**], which also surveys partial results and specific open problems in the hyperbolic setting. In the setting of projective spaces the question about unions appears to have a negative answer, as indicated by counterexamples in section 4 of [**Šid68**], which show that a full two-sided analogue of Slepian's lemma (in the spirit of Proposition 6.9) does not hold.

Proposition 6.9 was proved independently by Khatri [**Kha67**] and Šidák [**Šid67**] (see also [**Šid68, Glu88**]). The Gaussian correlation conjecture was proved by

Royen in [**Roy14**]. A more accessible and more detailed exposition can be found in [**LM17**], to which we also refer for more background and references.

The Sudakov minoration (Proposition 6.10) appears in [**Sud71**]. The dual Sudakov inequality (Proposition 6.11) is due to Pajor–Tomczak-Jaegermann [**PTJ85**]. The proof presented here is due to Talagrand (see [**LT91**]). Some refinements of both inequalities appear in [**MTJ87**].

Dudley's inequality (Proposition 6.13) goes back to [**Dud67**] and was generalized to the subgaussian setting in [**JM78**]. A version of Proposition 6.17 in the language of stationary Gaussian processes can be found in [**Fer97**]. The first part of Theorem 6.18 is due to Fernique [**Fer75**] and the second part (which is much harder) is due to Talagrand [**Tal87**] (a later paper [**Tal01**] contains a more transparent exposition). For more information about the "generic chaining" principle (which is a reincarnation of the "majorizing measures"), we refer to the books [**Tal05**] and [**Tal14**] by Talagrand, the latter one being more accessible.

Section 6.2. Two recent and excellent references about RMT are [**AGZ10**] and [**Tao12**], and we direct the reader to them for the background, further information, and bibliography. In particular, a huge branch of RMT which is not considered here revolves around the universality principle and aims at extending convergence results to models with less symmetries and/or with weaker integrability properties. Random matrices drawn from classical compact groups are the topic of the forthcoming monograph [**Mec**].

In the context of empirical measures, the ∞-Wasserstein distance was introduced in [**ASY14**]. The ∞-Wasserstein distance is much less popular than its "finite p" cousins; for example, in [**Vil09**] it appears only in the bibliographical notes to the entire chapter devoted to the topic. However, it has a few interesting applications, see for example [**McC06**]. We refer to [**Vil09**] for a thorough discussion of why the terminology "Wasserstein distance" is as highly questionable as it is predominant. For a proof that the Lévy distance metrizes weak convergence, see Section 4.3 in [**Gal95**]. Knowing that the weak convergence is metrizable gives unambiguous meaning to statements asserting that a sequence of random measures "converges weakly in probability", which are ubiquitous in RMT. A long list of conditions equivalent to weak convergence is known as the *Portmanteau lemma* and can be found, along many other facts about convergence of probability measures, in [**Bil99**].

Wigner's theorem about convergence to the semicircle distribution originates from [**Wig55, Wig58**] and has been extended and strengthened in various directions, notably to matrices with independent (but not necessarily Gaussian) entries (see, e.g., references in [**Tao12**]).

The "small deviation" inequalities from Proposition 6.25 are from [**Aub05, Led03, LR10**]. The perhaps surprising normalization is sharp and reflects the fact that fluctuations of large random matrices are asymptotically smaller than the upper bound given by the Gaussian concentration. For example, the quantity $\lambda_1(\mathrm{GUE}(n)) - 2\sqrt{n}$ is of order $n^{-1/6}$ (as opposed to $O(1)$ following from the Gaussian isoperimetric inequality), and it converges, after normalization, to the Tracy–Widom distribution [**TW94**] (resp., $\mathrm{GOE}(n)$, [**TW96**]).

The Marčenko–Pastur distribution appearing in Theorem 6.28 was introduced in [**MP67**], where the weak convergence was proved. Convergence of extreme eigenvalues was obtained in [**Gem80, Sil85**]. A reference for Theorem 6.29 is

[**BY88**]. Theorem 6.30 about partial transposition appears in [**Aub12**] (see also [**BN13, FŚ13**] for a slightly different setting). We also refer to [**CN16**] for a survey of RMT techniques in quantum information theory.

Proposition 6.31 seems new, but it is likely that—similarly to its special case, (6.37)—it can be derived from the subtle inequalities contained in [**HT03, HT05**]. The proof is, to the best of our knowledge, new; however, Lemma 6.39 appears in [**Sil85**].

The formula (6.47) has been derived in [**ŻS01**] (and probably independently in many other sources). Proposition 6.37 is from [**Gor85**] and improves on [**Che78**]. The argument leading to Corollary 6.38 is taken from [**DS01**]. Proposition 6.42 is from [**Gor88**].

Free probability. The very interesting and fruitful link between free probability and large random matrices mentioned in Section 6.2.5 goes back to [**Voi91**]. The monograph [**NS06**] gives an accessible and comprehensive approach to the subject with an emphasis on its combinatorial aspects. A highly readable exposition of many aspects of the subject relevant to quantum information theory can be found in [**HP00**].

The weak convergence in Theorem 6.43 was proved by Voiculescu. The extension to the convergence of the operator norm is a difficult result which was derived later by Haagerup–Thorbjørnsen [**HT05**].

Free additive convolution was introduced by Voiculescu in [**Voi85**] and the statement of Theorem 6.44 is from [**Voi90**]. The needed convergence of the operator norms required for the last part of Corollary 6.45 was supplied recently in [**CM14**]. A formula for the sum of more than two projectors can also be derived, see [**FN15**].

Finally, we mention that some concentration estimates for polynomials in random matrices can be found in [**MS12**].

CHAPTER 7

Some tools from asymptotic geometric analysis

This chapter contains a selection of results from asymptotic geometric analysis that we believe to be of interest to quantum information theory. The most famous of them is arguably Dvoretzky's theorem which asserts that, roughly speaking, every convex body of sufficiently large dimension admits sections which are arbitrarily close to Euclidean balls. There are actually several variations on this statement and they are studied in detail in Section 7.2. We also introduce the ℓ-position of convex bodies and use it to deduce the MM^*-estimate, an important result that allows appealing to duality when studying mean widths.

7.1. ℓ-position, K-convexity and the MM^*-estimate

7.1.1. ℓ-norm and ℓ-position. Let $K \subset \mathbb{R}^n$ be a convex body containing 0 in the interior. For $T \in \mathsf{M}_n$, we define the quantity $\ell_K(T)$ as

$$\ell_K(T) = \mathbf{E}\,\|T(G)\|_K,$$

where G denotes a standard Gaussian vector in $\mathbb{R}^n$. If there is no ambiguity about the underlying convex body, we write ℓ instead of ℓ_K. The following proposition collects elementary properties of this concept.

PROPOSITION 7.1 (See Exercise 7.1). *If $K \subset \mathbb{R}^n$ is a convex body containing 0 in the interior, then*
(i) $\ell_K(\cdot)$ is a norm on M_n,
(ii) ℓ_K obeys the ideal property: for $S, T \in \mathsf{M}_n$, we have $\ell_K(ST) \leqslant \ell_K(S)\|T\|_{\mathrm{op}}$,
(iii) $\ell_K(\mathrm{I}) = w_G(K^\circ) = \kappa_n w(K^\circ) \sim \sqrt{n}\, w(K^\circ)$,
(iv) for $T \in \mathsf{GL}(n, \mathbb{R})$, $\ell_K(T) = \ell_{T^{-1}K}(\mathrm{I})$,
(v) if P_E denotes the orthogonal projection on a subspace $E \subset \mathbb{R}^n$, then

$$\ell_K(P_E) = w_G((K \cap E)^\circ) = w_G(P_E K^\circ),$$

where by $(K \cap E)^\circ$ we mean the polar of $K \cap E$ inside E.

We now introduce the concept of ℓ-position via the following lemma.

LEMMA 7.2. *For any convex body $K \subset \mathbb{R}^n$ containing 0 in the interior, there is a unique $T_0 \in \mathcal{PSD}$ that is a solution to the maximization problem*

$$\max\{\det T \ : \ T \in \mathcal{PSD}(\mathbb{R}^n), \ell_K(T) \leqslant 1\}. \tag{7.1}$$

If T_0 is a multiple of the identity, we say that K is in the ℓ-position.

PROOF OF LEMMA 7.2. The maximum is attained by compactness and is obviously strictly positive. Assume that $T_0, T_1 \in \mathcal{PSD}(\mathbb{R}^n)$ are both solutions of the maximization problem. If $T_0 \neq T_1$, it would follow that $T = (T_0 + T_1)/2$ verifies, on the one hand, $\ell(T) \leqslant 1$ and, on the other hand, $\det T > (\det T_0)^{1/2}(\det T_1)^{1/2} =$

$\det T_0$ (by strict log-concavity of det over $\mathcal{PSD}$, see Exercise 1.42), a contradiction. □

Note that the ℓ-position of a convex body is unique up to homotheties and rotations. It follows from Proposition 4.8 that convex bodies with enough symmetries are automatically in the ℓ-position.

LEMMA 7.3. *Let K be a convex body in the ℓ-position. Then $w_G(K^\circ)\operatorname{Tr}(A) \leqslant n\ell_K(A)$ for any $A \in \mathsf{M}_n$.*

PROOF. We may assume $A \in \mathcal{PSD}$. Indeed, any $B \in \mathsf{M}_n$ can be written as AO for $A \in \mathcal{PSD}$ and $O \in \mathsf{O}(n)$, and we have $\ell_K(A) = \ell_K(B)$ by rotational invariance of the Gaussian measure, while $\operatorname{Tr} B \leqslant \|B\|_1 = \operatorname{Tr} A$.

Since K is in ℓ-position, the solution of the variational problem (7.1) is $\lambda\,\mathrm{I}$ with $\lambda = \ell_K(\mathrm{I})^{-1} = w_G(K^\circ)^{-1}$. Consider $A \in \mathcal{PSD}$ and $\varepsilon > 0$ small enough such that $\mathrm{I} + \varepsilon A \in \mathcal{PSD}$. Let $B = (\ell_K(\mathrm{I} + \varepsilon A))^{-1}(\mathrm{I} + \varepsilon A)$. Since $\ell_K(B) = 1$ it follows that $\det(B) \leqslant \det(\lambda\,\mathrm{I}) = \lambda^n$. Consequently, using the triangle inequality,

$$(\det(\mathrm{I} + \varepsilon A))^{1/n} \leqslant \lambda \ell_K(\mathrm{I} + \varepsilon A) \leqslant 1 + \varepsilon\lambda\ell_K(A).$$

Since $\det(\mathrm{I} + \varepsilon A)^{1/n} = 1 + \frac{1}{n}\varepsilon \operatorname{Tr}(A) + o(\varepsilon)$ as ε goes to 0, the result follows. □

REMARK 7.4. Before proceeding, let us point out that the more common definition of the ℓ-norm (and of the ℓ-position) is via the second moment, namely $(\mathbf{E}\,\|T(G)\|_K^2)^{1/2}$. Using the second moment leads to nicer duality relations, but we prefer to use the first moment to make the connection to the mean width more transparent. The next proposition shows that the two quantities are equivalent; however, they are not equal or proportional, and so the corresponding two maximization problems lead to two slightly different notions of ℓ-position.

PROPOSITION 7.5 (Not proved here). *For any symmetric convex body $K \subset \mathbb{R}^n$ and for any linear operator $T : \mathbb{R}^n \to \mathbb{R}^n$, we have*

$$\mathbf{E}\,\|T(G)\|_K \leqslant (\mathbf{E}\,\|T(G)\|_K^2)^{1/2} \leqslant \sqrt{\frac{\pi}{2}}\,\mathbf{E}\,\|T(G)\|_K.$$

EXERCISE 7.1. Prove the properties of the ℓ-norm listed in Proposition 7.1.

EXERCISE 7.2 (The left ideal property). In the setting of Proposition 7.1, is it true that $\ell_K(ST) \leqslant \|S\|\ell_K(T)$?

7.1.2. K-convexity and the MM^*-estimate. Consider the Hilbert space $\mathscr{H}_k := L_2(\mathbb{R}^k, \gamma_k)$. It is useful to write the norm of an element $f \in \mathscr{H}_k$ as

$$\left(\mathbf{E}\,|f(G)|^2\right)^{1/2},$$

where G is a standard Gaussian vector in $\mathbb{R}^k$. Recall that the Hermite polynomials $(h_\alpha)_{\alpha\in\mathbb{N}^k}$ defined in (5.56) form an orthonormal basis in $\mathscr{H}_k$. For an integer $d \geqslant 0$, denote by $R_d : \mathscr{H}_k \to \mathscr{H}_k$ the orthogonal projection onto the subspace of homogeneous polynomials of total degree d : $R_d(h_\alpha) = h_\alpha$ if $|\alpha| = d$ and $R_d(h_\alpha) = 0$ if $|\alpha| \neq d$. For $f \in \mathscr{H}_k$, $R_0(f)$ is a constant function and $R_1(f)$ has the form $x \mapsto \langle x, a\rangle$ for some $a \in \mathbb{R}^k$.

Given $n \in \mathbb{N}$, let $\mathscr{H}_{k,n}$ be the space of Borel functions $\Theta = (f_1, \dots, f_n) : \mathbb{R}^k \to \mathbb{R}^n$ such that $f_i \in \mathscr{H}_k$ for each i. The space $\mathscr{H}_{k,n}$ is a Hilbert space for the inner product

$$\langle\langle \Theta, \Theta' \rangle\rangle := \mathbf{E}\langle \Theta(G), \Theta'(G)\rangle \tag{7.2}$$

and can be identified with the Hilbert space tensor product $\mathscr{H}_k \otimes \mathbb{R}^n$. (This is the canonical identification of the space of $\mathcal{H}$-valued L_2 functions on Ω with $L_2(\Omega)\otimes\mathcal{H}$; if $\dim \mathcal{H} < \infty$, no completion of the latter is needed.) The projections R_d induce extensions $\tilde{R}_d := R_d \otimes \mathrm{I}_{\mathbb{R}^n} : \mathscr{H}_{k,n} \to \mathscr{H}_{k,n}$. More concretely, for $\Theta \in \mathscr{H}_{k,n}$, we have $\tilde{R}_d(\Theta) := (R_d f_1, \ldots, R_d f_n)$. Similarly as for $n = 1$, the function $\tilde{R}_1(\Theta) : \mathbb{R}^k \to \mathbb{R}^n$ is linear, i.e., it has the form $x \mapsto Ax$ for some $A \in \mathsf{M}_{k,n}$ (depending on Θ), and the operator $\tilde{R}_1$ is the orthogonal projection onto the subspace of $\mathscr{H}_{k,n}$ formed by such linear functions.

Let K be a convex body in $\mathbb{R}^n$ containing 0 in the interior. For $\Theta \in \mathscr{H}_{k,n}$, define

$$\|\!|\Theta|\!\|_K = \left(\mathbf{E}\,\|\Theta(G)\|_K^2\right)^{1/2} \tag{7.3}$$

(this quantity is a norm when K is symmetric; again, we have here X-valued L_2 functions on $(\mathbb{R}^k, \gamma_k)$, where $X = (\mathbb{R}^n, \|\cdot\|_K))$. It is easily checked that, for $\Theta \in \mathscr{H}_{k,n}$,

$$\|\!|\Theta|\!\|_K = \sup\{\langle\!\langle \Theta, \Xi\rangle\!\rangle \;:\; \Xi \in \mathscr{H}_{k,n}, \|\!|\Xi|\!\|_{K^\circ} \leqslant 1\}. \tag{7.4}$$

The K-*convexity constant* of the convex body K, denoted by $\mathbf{K}(K)$, is the smallest constant C such that the inequality

$$\|\!|\tilde{R}_1(\Theta)|\!\|_K \leqslant C\|\!|\Theta|\!\|_K \tag{7.5}$$

holds for every k and for all $\Theta \in \mathscr{H}_{k,n}$. It is not hard to show that $\mathbf{K}(K) < \infty$ (see Exercise 7.3). Moreover, rather surprisingly, $\mathbf{K}(\cdot)$ is often *uniformly* bounded for large classes of bodies (for example, for balls in all commutative or non-commutative ℓ_p spaces for a fixed $p \in (1, \infty)$). For general *symmetric* convex bodies, the sharp estimate $\mathbf{K}(K) = O(\log n)$ appears in Corollary 7.9.

We now connect the K-convexity constant with mean width estimates.

PROPOSITION 7.6. *Let $K \subset \mathbb{R}^n$ be a convex body containing 0 in the interior which is in the ℓ-position. Then $w_G(K)w_G(K^\circ) \leqslant n\mathbf{K}(K)$.*

PROOF. To each $x \in \mathbb{R}^n$ associate $\Theta(x) \in K$ with the property that $\langle x, \Theta(x)\rangle = \|x\|_{K^\circ}$; we can also ensure that the map Θ is Borel (see Exercise 1.12), so that $\Theta \in \mathscr{H}_{n,n}$. Since Θ takes values in K, it follows that $\|\!|\Theta|\!\|_K \leqslant 1$. (Actually, since $x \neq 0$ implies $\Theta(x) \in \partial K$, we necessarily have $\|\!|\Theta|\!\|_K = 1$.) We have

$$w_G(K) = \mathbf{E}\,\|G\|_{K^\circ} = \mathbf{E}\langle G, \Theta(G)\rangle = \langle\!\langle \mathrm{I}_{\mathbb{R}^n}, \Theta\rangle\!\rangle.$$

Given that $\tilde{R}_1$ is an orthogonal projection onto a subspace containing $\mathrm{I}_{\mathbb{R}^n}$, we have $\langle\!\langle \mathrm{I}_{\mathbb{R}^n}, \Theta\rangle\!\rangle = \langle\!\langle \mathrm{I}_{\mathbb{R}^n}, \tilde{R}_1(\Theta)\rangle\!\rangle$. Recalling that $\tilde{R}_1(\Theta)$ has the form $x \mapsto Ax$ for some $A \in \mathsf{M}_n$, we can write

$$w_G(K) = \mathbf{E}\langle G, AG\rangle.$$

Since an elementary computation shows that $\mathbf{E}\langle G, AG\rangle = \operatorname{Tr} A$, a straightforward application of Lemma 7.3 yields

$$w_G(K)w_G(K^\circ) \leqslant n\ell_K(A). \tag{7.6}$$

It remains to unscramble the meaning of the quantity $\ell_K(A)$. We have

$$\begin{aligned} \ell_K(A) &= \mathbf{E}\,\|A(G)\|_K \leqslant \left(\mathbf{E}\,\|A(G)\|_K^2\right)^{1/2} = \|\!|A|\!\|_K \\ &= \|\!|\tilde{R}_1(\Theta)|\!\|_K \leqslant \mathbf{K}(K)\|\!|\Theta|\!\|_K \leqslant \mathbf{K}(K) \end{aligned}$$

as needed, the only significant step in the above chain of equalities/inequalities being the application of (7.5), the definition of the K-convexity constant. $\square$

An upper bound on the K-convexity constant, whose importance cannot be overstated, is the following result due to Pisier.

THEOREM 7.7. *There is a universal constant C such that*

$$\mathbf{K}(K) \leqslant C(1 + \log d(K, B_2^n))$$

for any $n \in \mathbb{N}$ and any ***symmetric*** *convex body $K \subset \mathbb{R}^n$.*

REMARK 7.8. If $K \subset \mathbb{R}^n$ is unconditional, the bound in Theorem 7.7 can be improved to $C\big(1 + \log d(K, B_2^n)\big)^{1/2}$.

Before proving Theorem 7.7, we derive some of its consequences. First, since $d(K, B_2^n) \leqslant \sqrt{n}$ for every symmetric convex body $K \subset \mathbb{R}^n$ (see Exercise 4.20; actually, the weaker result from Exercise 4.2 would suffice), we first have

COROLLARY 7.9. *There is a universal constant C such that $\mathbf{K}(K) \leqslant C \log n$ for any* ***symmetric*** *convex body $K \subset \mathbb{R}^n, n \geqslant 2$.*

Combined with Proposition 7.6, this implies the following result known in asymptotic geometric analysis as the "MM^*-estimate".

THEOREM 7.10 (The MM^*-estimate). *Let $n \geqslant 2$ and let $K \subset \mathbb{R}^n$ be a* ***symmetric*** *convex body which is in the ℓ-position. Then*

$$1 \leqslant w(K)\, w(K^\circ) \leqslant C \log n. \tag{7.7}$$

We point out that the lower bound $w(K)w(K^\circ) \geqslant 1$ is elementary (see Exercise 4.37). As a corollary, we obtain the fact that, in the ℓ-position, the Urysohn inequality (4.34) is sharp up to a logarithmic factor.

COROLLARY 7.11 (Reverse Urysohn's inequality). *Let $n \geqslant 2$ and let $K \subset \mathbb{R}^n$ be a symmetric convex body. Then there exists a linear transformation $T \in \mathsf{GL}(n, \mathbb{R})$ such that*

$$w(T(K)) \leqslant C \log n \operatorname{vrad}(T(K)).$$

Moreover, T can be chosen to commute with the group of isometries of K. In particular, if K has enough symmetries, one may take $T = \mathrm{I}$.

Note that since both $w(T(K))$ and $\operatorname{vrad}(T(K))$ are 1-homogeneous in T, one may require in Corollary 7.11 that $T \in SL(\mathbb{R}^n)$, in which case $\operatorname{vrad}(T(K)) = \operatorname{vrad}(K)$.

For the proof of Theorem 7.7 we need two auxiliary lemmas, the first of which requires recalling some notation. Fix $k \geqslant 1$ and let $(P_t)_{t \geqslant 0}$ be the Ornstein–Uhlenbeck semigroup introduced in (5.55). Then each P_t is a contraction on $\mathscr{H}_k$ (Exercise 5.62). Moreover, the operator P_t extends to an operator $\tilde{P}_t$ on $\mathscr{H}_{k,n}$ by the formula

$$\tilde{P}_t(f_1, \ldots, f_k) = (P_t f_1, \ldots, P_t f_k)$$

(or, more abstractly, $\tilde{P}_t = P_t \otimes \mathrm{I}_{\mathbb{R}^n}$), and this extension is also a contraction with respect to any "reasonable functional norm".

LEMMA 7.12. *For any $\Theta \in \mathscr{H}_{k,n}$ and for any convex body $K \subset \mathbb{R}^n$ containing 0 in the interior, we have $|||\tilde{P}_t \Theta|||_K \leqslant |||\Theta|||_K$.*

PROOF. For $x \in \mathbb{R}^k$, denote $g(x) = \|\Theta(x)\|_K$, so that $|||\Theta|||_K = \|g\|_{\mathscr{H}_k}$. Then, for any $z \in K^\circ$ and any $x \in \mathbb{R}^k$, we have $\langle \Theta(x), z\rangle \leqslant g(x)$. Since P_t preserves positivity (this is clear from (5.55)), it follows that

$$\langle (\tilde{P}_t\Theta)(x), z\rangle = P_t(\langle \Theta(x), z\rangle) \leqslant (P_t g)(x).$$

Taking supremum over $z \in K^\circ$ yields $\|(\tilde{P}_t\Theta)(x)\|_K \leqslant (P_t g)(x)$. Squaring and integrating against γ_k we obtain (cf. (7.3))

$$|||\tilde{P}_t\Theta|||_K \leqslant ||P_t g||_{\mathscr{H}_k} \leqslant ||g||_{\mathscr{H}_k} = |||\Theta|||_K,$$

the second inequality following from P_t being a contraction on $\mathscr{H}_k$ (see Exercise 5.62). □

The second lemma that we need for the proof of Theorem 7.7 is the following.

LEMMA 7.13 (See Exercise 7.6). *Let p be a polynomial such that (1) $|p(x)| \leqslant 1$ for any $x \in [-1,1]$ and (2) for some $\lambda \geqslant e$, $|p(z)| \leqslant \lambda$ for any complex number z with $|z| \leqslant 1$. Then $|p'(0)| \leqslant \frac{4e}{\pi}\log\lambda$.*

PROOF OF THEOREM 7.7. Fix $k \geqslant 1$ and let $\lambda = d(K, B_2^n)$. Since the K-convexity constant is linearly invariant (see Exercise 7.3), we may assume that $B_2^n \subset K \subset \lambda B_2^n$ and therefore

$$|||\cdot|||_K \leqslant |||\cdot|||_{B_2^n} \leqslant \lambda|||\cdot|||_K. \tag{7.8}$$

Further, since $\mathbf{K}(K) \leqslant d(K, B_2^n)$ (again, by Exercise 7.3, or directly from (7.8)), we may assume that $\lambda \geqslant e$. Note that the Hilbert space norm on $\mathscr{H}_{k,n}$ corresponding to the inner product (7.2) is exactly $|||\cdot|||_{B_2^n}$.

Our objective is to show that if $\Theta \in \mathscr{H}_{k,n}$ satisfies $|||\Theta|||_K \leqslant 1$, then $|||\tilde{R}_1(\Theta)|||_K \leqslant \frac{4e}{\pi}\log\lambda$. (This will imply the Theorem with $C = \frac{4e}{\pi}$.) By density, we may assume that Θ is a polynomial; denote by m its degree. Consider the $\mathscr{H}_{k,n}$-valued polynomial defined for $z \in \mathbb{C}$ by

$$\pi(z) = \sum_{j=1}^{m} z^j \tilde{R}_j(\Theta).$$

For $|z| \leqslant 1$, we have

$$|||\pi(z)|||_K \leqslant |||\pi(z)|||_{B_2^n} \leqslant |||\Theta|||_{B_2^n} \leqslant \lambda$$

where the middle inequality uses the Pythagorean theorem. If $x = \exp(-t) > 0$, then $\pi(x) = \tilde{P}_t\Theta$ (by (5.57)) and therefore $|||\pi(x)|||_K \leqslant 1$. Similarly, if $y = -\exp(-t) < 0$, then $\pi(y) = \tilde{P}_t\Psi$ where Ψ is defined by $\Psi(x) = -\Theta(-x)$. Because K is symmetric, we have $|||\Psi|||_K = |||\Theta|||_K$ and therefore $|||\pi(y)|||_K \leqslant 1$. For any $\Xi \in \mathscr{H}_{k,n}$ with $|||\Xi|||_{K^\circ} \leqslant 1$, the polynomial $p(z) = \langle\langle \pi(z), \Xi\rangle\rangle$ satisfies the hypotheses of Lemma 7.13. It follows that $|p'(0)| = |\langle\langle \tilde{R}_1(\Theta), \Xi\rangle\rangle| \leqslant \frac{4e}{\pi}\log\lambda$ and the conclusion follows from the duality formula (7.4). □

REMARK 7.14. An alternative definition of K-convexity is obtained if we replace the Gauss space by the discrete cube. Given a function $f : \{-1,1\}^k \to \mathbb{R}^n$, consider its decomposition $f = \sum_{A\subset\{1,\ldots,k\}} w_A x_A$, where w_A is the Walsh function $(\varepsilon_1,\ldots,\varepsilon_k) \mapsto \prod_{i\in A}\varepsilon_i$. Define then $Rf := \sum_{i=1}^k w_{\{i\}}x_{\{i\}}$, the orthogonal projection onto the space of linear functions (the Rademacher projection). Given a convex body K in $\mathbb{R}^n$, let $\mathbf{K}'(K)$ be the smallest constant C such that the inequality

$$\left(\mathbf{E}\,\|Rf(\varepsilon)\|_K^2\right)^{1/2} \leqslant C\left(\mathbf{E}\,\|f(\varepsilon)\|_K^2\right)^{1/2}$$

holds for every k and every $f : \{-1,1\}^k \to \mathbb{R}^n$, where ε is uniformly distributed on $\{-1,1\}^k$. It can be shown (see Section 6.6 in [**AAGM15**] for a detailed argument) that for any symmetric convex body K,

$$\frac{2}{\pi}\mathbf{K}'(K) \leqslant \mathbf{K}(K) \leqslant \mathbf{K}'(K).$$

This definition allows for a derivation of the estimate from Theorem 7.7 that parallels the one presented above, with the Hermite polynomials being replaced by the Walsh functions, and Lemma 7.13 replaced by a careful application of Bernstein's inequality: *If p is a polynomial of degree at most m such that $|p(x)| \leqslant 1$ for $x \in [-1,1]$, then $|p'(0)| \leqslant m$.*

EXERCISE 7.3 (A rough bound for the K-convexity constant). (i) Show that $\mathbf{K}(B_2^n) = 1$. (ii) Show that if K, L are symmetric convex bodies in $\mathbb{R}^n$, then $\mathbf{K}(K) \leqslant d_{BM}(K,L)\mathbf{K}(L)$. (iii) Conclude that $\mathbf{K}(K) \leqslant \sqrt{n}$ for symmetric convex bodies $K \subset \mathbb{R}^n$.

EXERCISE 7.4 (K-convexity and duality). Show that $\mathbf{K}(K) = \mathbf{K}(K^\circ)$ for every convex body K containing 0 in the interior.

EXERCISE 7.5 (The K-convexity constant for B_1^n and for the cube). Let $N = 2^k$ and write the canonical basis of $\mathbb{R}^N$ as $(e_\varepsilon)_{\varepsilon \in \{-1,1\}^k}$. Define a map $\Theta \in \mathscr{H}_{k,N}$ by $\Theta(x) = e_\varepsilon$ if the signs of the coordinates of $x \in \mathbb{R}^k$ match the sequence $\varepsilon \in \{-1,1\}^k$. Show that $|||\tilde{R}_1(\Theta)|||_{B_1^N} \geqslant c\sqrt{k}$ for some $c > 0$ and conclude that $\mathbf{K}(B_1^n) = \Omega(\sqrt{\log n}) = \mathbf{K}(B_\infty^n)$.

EXERCISE 7.6 (A Bernstein-like inequality). Prove Lemma 7.13 by using the conformal transformation $z \mapsto \tanh(\pi z/4)$ mapping the strip $S = \{z : |\operatorname{Im} z| < 1\}$ onto the open unit disk; reformulate the question as an inequality about holomorphic functions on S and use the Hadamard three-lines lemma.

7.2. Sections of convex bodies

7.2.1. Dvoretzky's theorem for Lipschitz functions. We start with the simple but crucial observation that concentration of measure for Lipschitz functions defined on the unit sphere (Corollary 5.17) implies that such functions are actually almost constant on a typical (randomly chosen) subspace of large dimension.

Throughout this section, whenever we consider a "random" k-dimensional subspace $E \subset \mathbb{R}^n$ (resp., $E \subset \mathbb{C}^n$), it is tacitly assumed that E is distributed uniformly with respect to the Haar measure (as defined in Appendix B.4) on the Grassmann manifold $\mathsf{Gr}(k,\mathbb{R}^n)$ (resp., $\mathsf{Gr}(k,\mathbb{C}^n)$), for example by setting $E = U(\mathbb{R}^k)$ (resp., $E = U(\mathbb{C}^k)$) where U is Haar-distributed on $\mathsf{O}(n)$ or $\mathsf{SO}(n)$ (resp., on $\mathsf{U}(n)$ or $\mathsf{SU}(n)$) and $\mathbb{R}^k \subset \mathbb{R}^n$ (resp., $\mathbb{C}^k \subset \mathbb{C}^n$) is the canonical inclusion.

It will be convenient to use the following concept: given a function $f : X \to \mathbb{R}$, the *oscillation* of f around the value μ on a subset $A \subset X$ is defined as

$$\operatorname{osc}(f,A,\mu) = \sup_{x\in A} |f(x) - \mu|.$$

In the following we consider the space $S^{n-1} \subset \mathbb{R}^n$ equipped with the geodesic metric g. The objective is to show that, for a Lipschitz function $f : S^{n-1} \to \mathbb{R}$ and a random k-dimensional subspace $E \subset \mathbb{R}^n$, the oscillation of f around a central value on the subsphere $S_E := S^{n-1} \cap E$ is small (and similarly for $S_{\mathbb{C}^n} \subset \mathbb{C}^n$). We first present a straightforward ε-net argument, which gives easily a result that is only slightly worse than Theorem 7.15 below. We focus on the real case, but the same argument applies in the complex setting. Note, however, that the latter does not follow formally from the former: while $\mathbb{C}^n, S_{\mathbb{C}^n}$ can be identified with $\mathbb{R}^{2n}, S^{2n-1}$ as metric spaces, not every $2k$-dimensional $\mathbb{R}$-linear subspace of $\mathbb{R}^{2n}$ corresponds to k-dimensional $\mathbb{C}$-linear subspace of $\mathbb{C}^n$.

Let $f : (S^{n-1}, g) \to \mathbb{R}$ be a 1-Lipschitz function, let μ_f be a central value for f, and let $E = U(\mathbb{R}^k)$ be a random k-dimensional subspace of $\mathbb{R}^n$, with U Haar-distributed on $\mathsf{O}(n)$. Let $\varepsilon \in (0,1)$ and let $\mathcal{N}$ be an ε-net in (S^{k-1}, g). First, since the function $f \circ U$ is 1-Lipschitz, we have

$$\operatorname{osc}(f \circ U, S^{k-1}, \mu_f) \leqslant \varepsilon + \operatorname{osc}(f \circ U, \mathcal{N}, \mu_f).$$

We know from Corollary 5.32 that for any $x \in \mathcal{N}$,

$$\mathbf{P}\left(|f(U(x)) - \mu_f| > \varepsilon\right) \leqslant 2\exp(-n\varepsilon^2/4).$$

By the union bound, it follows that

$$\mathbf{P}\left(\operatorname{osc}(f \circ U, \mathcal{N}, \mu_f) > \varepsilon\right) \leqslant \operatorname{card}(\mathcal{N}) \cdot 2\exp(-n\varepsilon^2/4). \tag{7.9}$$

By Lemma 5.3, we may choose $\mathcal{N}$ with $\operatorname{card}\mathcal{N} \leqslant (\pi/\varepsilon)^k$, so that the bound from (7.9) is substantially smaller than 1 provided $k \leqslant c'n\varepsilon^2/\log(1/\varepsilon)$. In that case we have $\operatorname{osc}(f, S_E, \mu_f) \leqslant 2\varepsilon$ with high probability. We will slightly improve the dependence on ε in Theorem 7.15 below; this improvement turns out to be crucial for some applications.

A function $f : S_{\mathbb{C}^n} \to \mathbb{R}$ is said to be *circled* if it satisfies $f(e^{i\theta}x) = f(x)$ for every $x \in S_{\mathbb{C}^n}$ and $\theta \in \mathbb{R}$. Circled functions are the complex counterpart of even functions.

THEOREM 7.15 (Dvoretzky–Milman theorem for Lipschitz functions). *There are constants $c, c' > 0$ such that the following holds. Let $f : S_{\mathbb{C}^n} \to \mathbb{R}$ be a 1-Lipschitz circled function, let μ_f be a central value for f (with respect to the uniform measure), and let $0 < \varepsilon < 1$. Assume that $k \leqslant cn\varepsilon^2$, and let $E \subset \mathbb{C}^n$ be a random k-dimensional subspace. Then, with probability larger than $1 - \exp(-c'n\varepsilon^2)$,*

$$\operatorname{osc}(f, S_E, \mu_f) \leqslant \varepsilon.$$

The same conclusion holds for any 1-Lipschitz function $f : S^{n-1} \to \mathbb{R}$ and a random subspace $E \subset \mathbb{R}^n$. In both cases the dimension changes to $cn\varepsilon^2/L^2$ if the function f is L-Lipschitz.

REMARK 7.16. The proof given below gives for example the value $c = 1/400$, which is certainly far from optimal. (The argument actually works provided $k+1 \leqslant n\varepsilon^2/200$.) While the bound can be undoubtedly improved, the use of Dudley's inequality inevitably results in poor constants. In the real case, the use of Slepian–Gordon inequalities gives a constant of order $1/6$ (see Exercise 7.7), and even better when the function f is the restriction of a norm (see Remark 7.23). It would be desirable to come up with a complex version of that argument, the difficulty being that the inequalities from Exercise 6.47 do not carry over to the complex case.

PROOF OF THEOREM 7.15. We consider the complex case and note that the same argument applies *mutatis mutandis* in the real setting. We may also assume that $\mu_f = 0$ (otherwise consider $f - \mu_f$).

Let $E = U(\mathbb{C}^k)$, with $U \in \mathsf{SU}(n)$ a random Haar-distributed unitary matrix (we could use equivalently the Haar measure on $\mathsf{U}(n)$, but this would lead to worse constants in (7.10) below, see Table 5.2). Consider the function $F : \mathsf{SU}(n) \to \mathbb{R}$ defined by

$$F(U) = \sup_{S_E} |f| = \sup_{x \in S_{\mathbb{C}^k}} |f(U(x))|.$$

For $U, V \in \mathsf{SU}(n)$ and $x \in S_{\mathbb{C}^k}$, we have (see Exercise B.5 for the last inequality)

$$|f(Ux) - f(Vx)| \leqslant |Ux - Vx| \leqslant \|U - V\|_{\mathrm{op}} \leqslant \|U - V\|_{\mathrm{HS}} \leqslant g_2(U, V),$$

where g_2 denotes the geodesic distance on $\mathsf{SU}(n)$, defined in (B.8). It follows that F is 1-Lipschitz on $(\mathsf{SU}(n), g_2)$. Using concentration of measure (see Table 5.2) gives then, for any $t > 0$,

$$\mathbf{P}(F \geqslant \mathbf{E}\, F + t) \leqslant \exp(-nt^2/4). \tag{7.10}$$

The remaining part of the proof consists in bounding $\mathbf{E}\, F$. We will rely on the following lemma.

LEMMA 7.17. *Let $f : S_{\mathbb{C}^n} \to \mathbb{R}$ be a 1-Lipschitz circled function and $U \in \mathsf{SU}(n)$ be a Haar-distributed random unitary matrix. Then for any $x, y \in S_{\mathbb{C}^n}$ with $x \neq y$ and for any $\lambda > 0$,*

$$\mathbf{P}(f(Ux) - f(Uy) > \lambda) \leqslant \exp\left(-\frac{(n-1)\lambda^2}{2|x-y|^2}\right),$$

where A and c are absolute constants.

PROOF. Fix $x, y \in S_{\mathbb{C}^n}$. Since f is circled (and U is $\mathbb{C}$-linear), we may replace y by $e^{i\theta} y$ and choose θ so that $\langle x|y\rangle$ is real nonnegative; note that this choice of θ minimizes $|x - y|$ and ensures that $x + y$ and $x - y$ are orthogonal. Set $z = \frac{x+y}{2}$ and $w = \frac{x-y}{2}$, then $x = z + w$ and $y = z - w$. Further, set $\beta = |w| = \frac{1}{2}|x - y|$ (we may assume that $\beta \neq 0$) and $w' = \beta^{-1} w$. Then, conditionally on $u = U(z)$, $U(w')$ is distributed uniformly on the sphere $S_{u^\perp} := S_{\mathbb{C}^n} \cap u^\perp$. Since $U(x) = u + \beta U(w')$ and $U(y) = u - \beta U(w')$, it follows that the conditional (on $u = U(z)$) distribution of $f(Ux) - f(Uy)$ is the same as that of $f_u : S_{u^\perp} \to \mathbb{R}$ defined by

$$f_u(v) = f(u + \beta v) - f(u - \beta v).$$

As is readily seen, f_u is 2β-Lipschitz and its mean is 0. From Lévy's lemma (Corollary 5.32) applied to f_u and to the $(2n-3)$-dimensional sphere $S_{u^\perp}$, we deduce that, conditionally on $u = U(z)$,

$$\mathbf{P}(f(Ux) - f(Uy) > \lambda) \leqslant \exp(-(2n-2)\lambda^2/4|x-y|^2),$$

and hence the same inequality holds also without the conditioning. □

We now return to the proof of Theorem 7.15. Lemma 7.17 asserts that the process $(X_s)_{s \in S_{\mathbb{C}^k}}$ defined by $X_s = f(Us)$ is subgaussian (a notion defined in (6.19)) with constants $A = 1$ and $\alpha = (n-1)/2$. We apply Dudley's inequality in the form given in Corollary 6.14 to obtain

$$\mathbf{E} \sup_{s \in S_{\mathbb{C}^k}} X_s \leqslant \sup_{s \in S_{\mathbb{C}^k}} \mathbf{E}\, X_s + \frac{6\sqrt{2}}{\sqrt{n-1}} \int_0^{1/2} \sqrt{1 + 2\log(N(S_{\mathbb{C}^k}, |\cdot|, \eta))}\, d\eta. \tag{7.11}$$

For any $s \in S$, $\mathbf{E}\, X_s$ is equal to the mean of f. Since 0 is a central value for f, it follows from Corollary 5.32 that $\mathbf{E}\, X_s \leqslant \sqrt{2\log 2}/\sqrt{2n}$. We know from Lemma 5.3 that $N(S_{\mathbb{C}^k}, |\cdot|, \eta) \leqslant (2/\eta)^{2k}$. Using the bound $\sqrt{1+t} \leqslant 1 + \sqrt{t}$ gives

$$\mathbf{E}\, F = \mathbf{E} \sup_{s \in S_{\mathbb{C}^k}} X_s \leqslant \frac{\sqrt{\log 2}}{\sqrt{n}} + \frac{3\sqrt{2}}{\sqrt{n-1}} + \frac{12\sqrt{2k}}{\sqrt{n-1}} \int_0^{1/2} \sqrt{\log(2/\eta)}\, \mathrm{d}\eta.$$

The numerical value $\int_0^{1/2} \sqrt{\log(2/\eta)}\, \mathrm{d}\eta \leqslant 0.759$ leads to

$$\mathbf{E}\, F = \mathbf{E} \sup_{s \in S} X_s \leqslant \frac{5.08 + 12.89\sqrt{k}}{\sqrt{n-1}}.$$

This quantity is smaller than $\varepsilon/2$ provided $k \leqslant cn\varepsilon^2$ for some constant c, and the conclusion follows by applying (7.10) for $t = \varepsilon/2$.

To obtain the constant $c = 1/400$, one checks the inequality $5.08 + 12.89\sqrt{k} \leqslant \sqrt{200(k+1) - 1}$. It follows that $\mathbf{E}\, F \leqslant \varepsilon$ provided $\sqrt{200(k+1)-1} \leqslant \varepsilon\sqrt{n-1}$, or (since $\varepsilon < 1$) when $k + 1 \leqslant n\varepsilon^2/200$. Since we may assume that $n\varepsilon^2 \geqslant 400$ (otherwise there is nothing to prove), this inequality is implied by the condition $k \leqslant n\varepsilon^2/400$. □

EXERCISE 7.7 (An alternative argument for Theorem 7.15 in the real case). Let $f : S^{n-1} \to \mathbb{R}$ be a 1-Lipschitz function. Denote by M_f the median of f, and consider $T = \{f = M_f\}$. Let $\varepsilon > 0$ be such that $n\varepsilon^2 \geqslant 12$, and let k be an integer such that $k + 1 \leqslant \frac{1}{6}\varepsilon^2 n$.
(i) For $\alpha \in (0, \pi/2)$, let $T_\alpha = \{x \in S^{n-1} : \operatorname{dist}(x, T) \leqslant \alpha\}$, where distance refers to the geodesic metric. Show that $\sigma(S^{n-1} \backslash T_\alpha) \leqslant \exp(-n\alpha^2/2)$. We now set $\alpha = \sqrt{2\log 2}/\sqrt{n}$, so that $\sigma(T_\alpha) \geqslant 1/2$.
(ii) Show that if $B \subset S^{n-1}$ satisfies $\sigma(B) \geqslant 1/2$, then $w(S^{n-1} \backslash B_\beta) \leqslant \frac{1+\cos\beta}{2}$.
(iii) Let $A = S^{n-1} \backslash T_\varepsilon$. Check that the assumptions on n, k, ε imply the inequality $\frac{1+\cos(\varepsilon-\alpha)}{2} \leqslant \sqrt{1 - (k+1)/n}$, and conclude from (ii) that $w_G(A) < \kappa_{n-k}$.
(iv) Using Proposition 6.42, conclude that with positive probability, a random k-dimensional subspace $E \subset \mathbb{R}^n$ satisfies $E \cap S^{n-1} \subset T_\varepsilon$, and thus $\operatorname{osc}(f, S_E, M_f) \leqslant \varepsilon$.

EXERCISE 7.8 (Removing the circledness assumption in Theorem 7.15). Show that the following holds for some constants $C, c > 0$. Let $f : S_{\mathbb{C}^n} \to \mathbb{R}$ be a 1-Lipschitz function with mean μ_f and $\varepsilon \geqslant C\sqrt{\log n}/\sqrt{n}$. Then, for $k \leqslant c\varepsilon^2 n$, a random k-dimensional subspace $E \subset \mathbb{C}^n$ satisfies $\operatorname{osc}(f, S_E, \mu_f) \leqslant \varepsilon$ with high probability. (Start by introducing the auxiliary circled functions $g(x) = \frac{1}{2\pi}\int_0^{2\pi} f(e^{i\theta}x)\, \mathrm{d}\theta$ and $h(x) = \max\{|f(e^{i\theta}x) - f(x)| : \theta \in [0, 2\pi]\}$.)
We do not know whether the assumption $\varepsilon \geqslant C\sqrt{\log n}/\sqrt{n}$ can be dropped.

7.2.2. The Dvoretzky dimension. A convex body K is said to be C-Euclidean if $d_{BM}(K, B_2^{\dim K}) \leqslant C$ (the Banach–Mazur distance d_{BM} and the geometric distance d_g were defined in (4.2) and (4.1)). It is customary to separate the situation where C is controlled, but possibly large (the "isomorphic" theory), from the situation where $C = 1 + \varepsilon$ with $\varepsilon \ll 1$ or at least "sufficiently small" (the "almost isometric" theory). Still another aspect is when $\varepsilon = 0$ (the "isometric" theory), which is quite different in nature and hardly mentioned in this book (with the exception of Section 11.1).

The goal of this section, and of the following ones, is to give upper and lower bounds on the maximal possible dimension of a subspace $E \subset \mathbb{R}^n$ such that $K \cap E$

is C-Euclidean (when K is symmetric, we restrict ourselves to subspaces through the origin, so that the results can be translated in terms of subspaces of normed spaces); see Figure 7.1. It is remarkable that, up to an absolute multiplicative constant, this maximal dimension can be computed via a simple formula, which we now introduce under the name of the *Dvoretzky dimension*.

DEFINITION 7.18 (Dvoretzky dimension). Let K be either a convex body in $\mathbb{R}^n$ containing 0 in the interior, or a circled convex body in $\mathbb{C}^n$. The *Dvoretzky dimension* of K is defined as

$$k_*(K) = (w(K^\circ)\,\mathrm{inrad}(K))^2 n.$$

If $\|\cdot\|$ is a norm on $\mathbb{R}^n$ (or $\mathbb{C}^n$), the Dvoretzky dimension of $X = (\mathbb{R}^n, \|\cdot\|)$ is defined as the Dvoretzky dimension of its unit ball, or equivalently as

$$k_*(X) = (M/b)^2 n,$$

where b is the smallest number such that $\|\cdot\| \leqslant b|\cdot|$ and $M = \mathbf{E}\,\|X\|$, where X is a random variable uniformly distributed on S^{n-1}.

We note that b corresponds to the maximum value of $\|\cdot\|$ over the Euclidean sphere, while M is the average value. Hence we always have $M \leqslant b$, thus $k_* \leqslant n$. Note also the inequality $k_*(K) \leqslant d_g(K,L)k_*(L)$ for a pair of convex bodies K, L. We should think of k_* as a quantity meaningful only up to an (absolute) multiplicative constant. Likewise, in order to not to obscure the arguments, we will sometimes pretend in what follows that k_* and similar expressions are integers.

The Dvoretzky dimension of a convex body $K \subset \mathbb{R}^n$ depends on the choice of the underlying Euclidean structure. The remarkable fact is that the following two quantities are equivalent up to multiplicative universal constants (see Exercise 7.10 and Theorem 7.19).

(i) The supremum of $k_*(K)$ over all Euclidean structures on $\mathbb{R}^n$.

(ii) The largest k such that $K \cap E$ is 2-Euclidean for some $E \in \mathsf{Gr}(k, \mathbb{R}^n)$.

The usefulness of this concept comes from the fact that, for standard norms, the Dvoretzky dimension is usually easily computed. We illustrate this in the case of ℓ_p spaces and Schatten norms in Section 7.2.4. However, the following Theorem 7.19 is also of interest when applied to abstract norms. For example, it implies the celebrated fact that any high-dimensional convex body has sections which are arbitrarily close to a Euclidean ball (see Corollary 7.40).

THEOREM 7.19 (Tangible Dvoretzky–Milman theorem). *There are absolute constants $c, c' > 0$ such that the following holds. Let K be either a convex body in $\mathbb{R}^n$ containing 0 in the interior, or a circled convex body in $\mathbb{C}^n$. Let M and $k_* = k_*(K)$ (the Dvoretzky dimension of K) be as in Definition 7.18. Fix $0 < \varepsilon \leqslant 1$, and let $k = c\varepsilon^2 k_*$. Then a random k-dimensional subspace E satisfies the following: with probability larger than $1 - \exp(-c'\varepsilon^2 k_*)$, we have*

$$\forall x \in E, \quad (1-\varepsilon)M|x| \leqslant \|x\|_K \leqslant (1+\varepsilon)M|x| \tag{7.12}$$

and consequently

$$d_g(K \cap E, B_2^E) \leqslant \frac{1+\varepsilon}{1-\varepsilon},$$

where d_g denotes the geometric distance as defined in (4.1) and $B_2^E = B_2^n \cap E$.

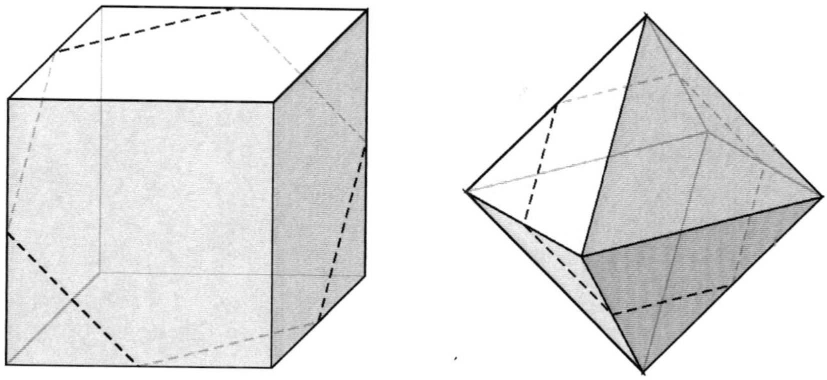

FIGURE 7.1. Low-dimensional illustration of Dvoretzky's theorem: the regular hexagon appears as a section of B_1^3 and B_∞^3.

PROOF. This is a straightforward consequence of Theorem 7.15, applied to the function $f(x) = \|x\|_K$, which is a b-Lipschitz function (and, moreover, is circled in the complex case). Indeed, provided $k \leqslant cn(\varepsilon M/b)^2$, we obtain with probability larger than $1 - \exp(-c'(\varepsilon M)^2 n)$ that $\operatorname{osc}(\|\cdot\|_K, S_E, M) \leqslant \varepsilon M$, which is equivalent to (7.12). □

REMARK 7.20. A simple ε-net argument combined with a little trick (see Exercise 7.11) gives a version of the complex case of Theorem 7.19 with a slightly worse dependence on ε, but without the assumption that K is circled.

REMARK 7.21 (about the dependence on ε). We now comment about the sharpness of Theorem 7.19. First, the isomorphic version (for macroscopic ε) is always sharp: the dimension of generic 2-Euclidean sections can never exceed $k_*(K)$ (see Exercise 7.12). Second, one can construct norms for which the dependence on ε is sharp (see Exercise 7.20). However, for some natural and interesting instances the dependence on ε can be improved (we will see a very important example in Chapter 8, connected to the additivity conjecture; see Remark 8.21).

REMARK 7.22. If K is A-Euclidean, then $k_*(K) \geqslant n/A^2$. Consequently, by Theorem 7.19, for any fixed $\varepsilon > 0$, K admits sections of proportional dimension which are $(1+\varepsilon)$-Euclidean. Therefore any result about isomorphically Euclidean sections implies a counterpart about almost isometric sections, the dimension of the section being affected only by a multiplicative $\Omega(\varepsilon^2)$ constant.

REMARK 7.23. In the real case, the conclusion of Theorem 7.19 also holds for a gauge (i.e., without the symmetry assumption of K). Moreover, a derivation from the Chevet–Gordon inequalities allows for a more direct proof and gives a better constant. For any $k \leqslant k_*$, we show the existence of a k-dimensional subspace $E \subset \mathbb{R}^n$ such that

$$d_{BM}(K \cap E, B_2^k) \leqslant \frac{1 + \sqrt{k/k_*}}{1 - \sqrt{k/k_*}}. \tag{7.13}$$

To achieve this, we consider a random matrix $B \in \mathsf{M}_{k,n}$, which we interpret as an operator from $\mathbb{R}^k$ to $\mathbb{R}^n$. By the Chevet–Gordon inequalities (Proposition 6.37),

$$w_G(K^\circ) - b\kappa_k \leqslant \mathbf{E} \min_{x \in S^{k-1}} \max_{y \in K^\circ} \langle Bx, y\rangle \leqslant \mathbf{E} \max_{x \in S^{k-1}} \max_{y \in K^\circ} \langle Bx, y\rangle \leqslant w_G(K^\circ) + b\kappa_k.$$

Using the inequality $\kappa_k/\sqrt{k} \leqslant \kappa_n/\sqrt{n}$ (Proposition A.1), we are led to

$$\begin{aligned}\kappa_n\left(M - b\sqrt{\frac{k}{n}}\right) &\leqslant \mathbf{E} \min_{x\in S^{k-1}} \max_{y\in K^\circ}\langle Bx, y\rangle \\ &\leqslant \mathbf{E} \max_{x\in S^{k-1}} \max_{y\in K^\circ}\langle Bx, y\rangle \leqslant \kappa_n\left(M + b\sqrt{\frac{k}{n}}\right),\end{aligned}$$

and the existence of a subspace $E = B(\mathbb{R}^k)$ satisfying (7.13) follows.

Due to the duality between sections and projections of convex bodies (see (1.12) and (1.13)), Theorem 7.19 admits a dual formulation via projections onto subspaces.

COROLLARY 7.24. *Let K be a convex body in $\mathbb{R}^n$, and $\varepsilon > 0$. Provided $k \leqslant c\varepsilon^2 k_*(K^\circ)$, a random k-dimensional subspace E satisfies with large probability*

$$(1-\varepsilon)w(K)B_2^E \subset P_E K \subset (1+\varepsilon)w(K)B_2^E.$$

REMARK 7.25 (Geometric interpretation of the MM^*-estimate). Let $K \subset \mathbb{R}^n$ be a symmetric convex body and let $k \leqslant c\varepsilon^2 \min(k_*(K), k_*(K^\circ))$. We know then from Theorem 7.19 and Corollary 7.24 that, for a random subspace $E \in \mathsf{Gr}(k, \mathbb{R}^n)$, the section $K \cap E$ is $(1+\varepsilon)$-close to a Euclidean ball of radius $w(K^\circ)^{-1}$ while the projection $P_E K$ is $(1+\varepsilon)$-close to a Euclidean ball of radius $w(K)$; the ratio of these radii is the quantity $w(K)w(K^\circ)$ which appears in Theorem 7.10. In particular, if K is in the ℓ-position, the radius of a typical k-dimensional projection only exceeds the radius of a typical k-dimensional section by a $\log(n)$ factor. However it is not clear whether the ℓ-position is always compatible with the conditions $k_*(K) \gg 1$ and $k_*(K^\circ) \gg 1$ (see Problem 7.26).

PROBLEM 7.26. *Does there exist, for every symmetric convex body $K \subset \mathbb{R}^n$, a subspace E of dimension $c \log n$ such that*

$$r_1 B_2^E \subset K \cap E \subset 2r_1 B_2^E \quad \text{and} \quad r_2 B_2^E \subset P_E K \subset 2r_2 B_2^E,$$

with $r_2/r_1 = O(\log n)$? Without the constraint on the radii this follows from classical facts, see Exercise 7.27.

EXERCISE 7.9 (Dvoretzky dimension and duality). Let $K \subset \mathbb{R}^n$ be a convex body such that $d_g(K, B_2^n) \leqslant A$. Show that $k_*(K)k_*(K^\circ) \geqslant n^2/A^2$. In particular, if K is symmetric and is in John or Löwner position, then $k_*(K)k_*(K^\circ) \geqslant n$.

EXERCISE 7.10. Let $K \subset \mathbb{R}^n$ be a symmetric convex body, and suppose that $d_{BM}(K \cap E, B_E) < A$ for some k-dimensional subspace E, where $B_E = B_2^n \cap E$. Show that there is a linear transformation $T \in \mathsf{GL}(n, \mathbb{R})$ such that $k_*(T(L)) \geqslant (k-1)/A^2$.

EXERCISE 7.11 (Almost spherical sections discretized).
(i) Let $\mathcal{N}$ be a δ-net in $(S_{\mathbb{C}^n}, |\cdot|)$, and $\|\cdot\|$ a norm on $\mathbb{C}^n$ such that

$$\forall x \in \mathcal{N}, \quad 1 - \alpha \leqslant \|x\| \leqslant 1 + \beta.$$

Show that

$$\forall x \in S_{\mathbb{C}^n}, \quad 1 - \alpha - \frac{\delta(1+\beta)}{1-\delta} \leqslant \|x\| \leqslant \frac{1+\beta}{1-\delta}. \tag{7.14}$$

(ii) Use (i) to show that, when $k \leqslant c\varepsilon^2 \log^{-1}(1/\varepsilon) k_*(K)$, the conclusion from Theorem 7.15 can be derived via the elementary net argument that led to (7.9).

EXERCISE 7.12 (Sharpness of the Dvoretzky–Milman theorem for random subspaces). Consider a norm $\|\cdot\|$ on $\mathbb{R}^n$, and let M, b, k_* be as in Definition 7.18. The goal of this exercise is to show that, in Theorem 7.19, the value k_* is always sharp for macroscopic values of ε (say, $\varepsilon = 1$ so that the lower bound in (7.12) is vacuous). Assume that k is an integer with the following property: with probability larger than $1 - 1/n$, a random k-dimensional subspace E satisfies

$$\forall x \in E, \quad \|x\| \leqslant 2M|x|. \tag{7.15}$$

(i) Show that there is an orthogonal decomposition of $\mathbb{R}^n$ as the direct sum of $\lceil n/k \rceil$ subspaces, each of them satisfying (7.15).
(ii) Show that for every $x \in \mathbb{R}^n$, $\|x\| \leqslant 2M\sqrt{\lceil n/k \rceil}|x|$.
(iii) Conclude that $k \leqslant C(M/b)^2 n$ for some absolute constant C.

7.2.3. The Figiel–Lindenstrauss–Milman inequality. In this section we will derive, as a consequence of Theorem 7.19, a useful inequality due to Figiel–Lindenstrauss–Milman which can be interpreted as follows: *complexity (of any convex body) must lie somewhere.*

Fix a convex body $K \subset \mathbb{R}^n$ containing the origin in the interior. Define the *verticial dimension* of K as

$$\dim_V(K) = \log \inf\{N \ : \ \text{there is a polytope } P \text{ with } N \text{ vertices s.t. } K \subset P \subset 4K\}$$

and the *facial dimension* of K as

$$\dim_F(K) = \log \inf\{N \ : \ \text{there is a polytope } Q \text{ with } N \text{ facets s.t. } K \subset Q \subset 4K\}.$$

The number 4 plays no special role in these definitions; all the results below are only affected in the values of the constants if 4 is replaced by another number larger than 1 (see Exercise 7.15). The basic properties of these concepts are gathered in Proposition 7.27.

PROPOSITION 7.27. *Let $K \subset \mathbb{R}^n$ a convex body containing the origin in the interior. Then*
(i) for any $T \in \mathsf{GL}(n, \mathbb{R})$, we have $\dim_V(TK) = \dim_V(K)$ and $\dim_F(TK) = \dim_F(K)$,
(ii) we have $\dim_V(K^\circ) = \dim_F(K)$ and $\dim_F(K^\circ) = \dim_V(K)$,
(iii) for any subspace $E \subset \mathbb{R}^n$, we have $\dim_F(K \cap E) \leqslant \dim_F K$ and $\dim_V(P_E K) \leqslant \dim_V K$,
(iv) if K has centroid at the origin, then $\dim_V(K) \leqslant Cn$ and $\dim_F(K) \leqslant Cn$ for some absolute constant C.

We note that the verticial and facial dimensions are linearly invariant but not affinely invariant (see Exercise 7.13).

PROOF. (i) is obvious and (ii) follows from the fact that polarity exchanges vertices and facets for polytopes (see (1.16)). The two dual inequalities in (iii) hold since projections do not increase the number of vertices of polytopes, while sections do not increase the number of facets. For (iv), see Exercises 5.18 and 5.19. □

Define also the *asphericity* of a convex body $K \subset \mathbb{R}^n$ as

$$a(K) = \inf\left\{\frac{R}{r} \ : \ \text{there is a 0-symmetric ellipsoid } \mathscr{E} \text{ with } r\mathscr{E} \subset K \subset R\mathscr{E}\right\}. \tag{7.16}$$

We have $a(K) = d_{BM}(K, B_2^n)$ if K is centrally symmetric. The following lemma gives a simple connection between asphericity and verticial (resp., facial) dimension. It is an immediate consequence of Proposition 5.6.

LEMMA 7.28. *Let $K \subset \mathbb{R}^n$ be a convex body containing the origin in the interior. Then*

$$\dim_V(K)\, a(K)^2 \geqslant \frac{n-1}{32}, \quad \dim_F(K)\, a(K)^2 \geqslant \frac{n-1}{32}.$$

When combined with Dvoretzky's theorem, the inequalities from Lemma 7.28 give a much sharper result.

THEOREM 7.29 (Figiel–Lindenstrauss–Milman inequality). *For any convex body $K \subset \mathbb{R}^n$ containing the origin in the interior we have*

$$\dim_F(K)\, \dim_V(K)\, a(K)^2 \geqslant cn^2, \tag{7.17}$$

where $c > 0$ is an absolute constant.

PROOF. We may assume that $rB_2^n \subset K \subset RB_2^n$ with $R/r = a(K)$. Let $M = \mathbf{E}\,\|X\|_K$ and $M^* = \mathbf{E}\,\|X\|_{K^\circ}$, where X is a random vector uniformly distributed on the unit sphere.

We apply Theorem 7.19 to K for $\varepsilon = 1/2$ (say). This yields a subspace $E \subset \mathbb{R}^n$ of dimension $c(rM)^2 n$ such that

$$\frac{M}{2} B_2^E \subset K \cap E \subset \frac{3M}{2} B_2^E.$$

It follows (using Proposition 7.27(iii) and Lemma 7.28) that $\dim_F(K) \geqslant \dim_F(K \cap E) \geqslant c(rM)^2 n$ for an absolute constant $c > 0$. We apply the same argument to K° (note that $R^{-1}B_2^n \subset K^\circ$) and obtain that $\dim_F(K^\circ) = c(M^*/R)^2 n$. Since $\dim_V(K) = \dim_F(K^\circ)$, it follows that

$$\dim_F(K) \dim_V(K) \geqslant c^2 n^2 (MM^*)^2 (r/R)^2 = c^2 n^2 / a(K)^2$$

as needed, where we used the fact (see Exercise (4.37)) that $MM^* \geqslant 1$. □

A consequence is a remarkable combinatorial result about symmetric polytopes. Indeed, we know from Exercise 4.20 that $a(K) \leqslant \sqrt{n}$ for any symmetric convex body $K \subset \mathbb{R}^n$.

COROLLARY 7.30. *Let $P \subset \mathbb{R}^n$ be a symmetric polytope with n_1 vertices and n_2 faces. Then*

$$(\log n_1)(\log n_2) \geqslant cn.$$

The conclusion of the Corollary fails dramatically for non-symmetric polytopes (consider the simplex).

EXERCISE 7.13 (The position of the origin matters). Give examples of planar convex bodies containing the origin in the interior, whose verticial or facial dimension is arbitrarily high.

EXERCISE 7.14. Give examples of symmetric polytopes in $\mathbb{R}^n$ with $\exp(o(n))$ vertices and (simultaneously) $\exp(o(n))$ facets.

EXERCISE 7.15 (Isomorphic facial and verticial dimension). Let $K \subset \mathbb{R}^n$ be a convex body containing the origin in the interior. For $A \geqslant 1$, define $\dim_F(K, A)$

(resp., $\dim_V(K, A)$) as $\log N$, where N is the minimal number of facets (resp., vertices) of a polytope P such that $K \subset P \subset AK$. Show that, for any $A, B \geqslant 1$,

$$A^2 \dim_F(K, A) \cdot B^2 \dim_V(K, B) \cdot a(K)^2 \geqslant cn^2,$$

where $c > 0$ is an absolute constant.

7.2.4. The Dvoretzky dimension of standard spaces. In this section we compute the Dvoretzky dimension for the unit balls with respect to the most standard norms: the commutative and non-commutative p-norms. Unless specified otherwise, the statements refer to both the real and the complex case.

7.2.4.1. *ℓ_p norms.* Let B_p^n denote the unit ball (in either $\mathbb{R}^n$ or $\mathbb{C}^n$) for the norm $\|\cdot\|_p$, where $p \in [1, \infty]$. We also define the conjugate exponent $q \in [1, \infty]$ by the relation $p^{-1} + q^{-1} = 1$. Recall that $(B_p^n)^\circ = B_q^n$.

THEOREM 7.31. *The Dvoretzky dimension of B_p^n is of the following order:*

$$k_*(B_p^n) \simeq \begin{cases} n & \text{if } 1 \leqslant p \leqslant 2, \\ pn^{2/p} & \text{if } 2 \leqslant p \leqslant \log n, \\ \log n & \text{if } \log n \leqslant p \leqslant \infty. \end{cases}$$

REMARK 7.32. We emphasize that the constants implicit in the relations "$\simeq$" *do not* depend on p (in addition to not depending on n). The proof actually shows that, for fixed p and as n tends to ∞, $w(B_q^n) \sim n^{1/p-1/2}\|g\|_{L_p}$, where g is a standard $N(0,1)$ (real or complex, accordingly) Gaussian random variable ("$\sim$" is uniform in p on bounded sets, but not globally). A closed expression for $\|g\|_{L_p}$ is given in (5.63).

PROOF. We treat the real case, the complex case being similar. Let $q \in [1, \infty]$ be such that $1/p + 1/q = 1$. By Definition 7.18, we have

$$k_*(B_p^n) = n \operatorname{inrad}(B_p^n)^2 w(B_q^n)^2.$$

Accordingly, the theorem will follow from the estimates

$$\operatorname{inrad}(B_p^n) = \begin{cases} n^{1/2-1/p} & \text{if } 1 \leqslant p \leqslant 2, \\ 1 & \text{if } 2 \leqslant p \leqslant \infty, \end{cases}$$

and

$$\mathbf{E}\,\|x\|_p = w(B_q^n) \simeq \begin{cases} \sqrt{p}\, n^{1/p-1/2} & \text{if } 1 \leqslant p \leqslant \log n, \\ \sqrt{\log n}/\sqrt{n} & \text{if } \log n \leqslant p \leqslant \infty, \end{cases} \tag{7.18}$$

where x is a random vector uniformly distributed on S^{n-1}.

The value of $\operatorname{inrad}(B_p^n)$ is a reformulation of the corresponding inequality (1.4) between ℓ_p-norms. We estimate the mean width by introducing a standard Gaussian vector $G = (g_1, \dots, g_n)$ in $\mathbb{R}^n$, so that $w(B_q^n) = \kappa_n^{-1}\,\mathbf{E}\,\|G\|_p$, where $\kappa_n \sim n^{1/2}$ was defined in (A.8). Consider also the random variable $X = \|G\|_p$. What is easy to compute is the pth moment (or the L_p-norm) of X:

$$\|X\|_{L_p} = (\mathbf{E}\,X^p)^{1/p} = (n\,\mathbf{E}\,|g_1|^p)^{1/p} \sim \sqrt{p/e}\; n^{1/p} \tag{7.19}$$

(see (5.63) for the last relation). Since we are interested in the value of $\mathbf{E}\,X$, we will use Gaussian concentration to relate the expectation of X to its pth moment.

Consider first the case $p \geqslant 2$, then $\|\cdot\|_p$ is 1-Lipschitz (with respect to the Euclidean metric) and so by Proposition 5.34 and Theorem 5.24

$$\mathbf{P}(X - \mathbf{E}X > t) \leqslant \mathbf{P}(X - M > t) \leqslant \mathbf{P}(g_1 > t) \qquad \text{for all } t > 0,$$

where M is the median of X. In particular, we have

$$\mathbf{E}\left((X - \mathbf{E}X)^+\right)^p = \int_0^\infty pt^{p-1}\mathbf{P}(|X - \mathbf{E}X| > t)\,\mathrm{d}t \leqslant \mathbf{E}(g_1^+)^p \leqslant \left(\frac{p}{e}\right)^{p/2}$$

(see (A.1) or (5.63)) and so

$$\|(X - \mathbf{E}X)^+\|_{L_p} \leqslant \sqrt{p/e}. \tag{7.20}$$

Since $\|X\|_{L_p} - \|(X - \mathbf{E}X)^+\|_{L_p} \leqslant \mathbf{E}X \leqslant \|X\|_{L_p}$, it follows from (7.19) and (7.20) that $w((B_p^n)^\circ) = \Theta(\sqrt{p}n^{1/p-1/2})$ whenever $2 \leqslant p \leqslant \log n$. For $\log n \leqslant p \leqslant \infty$, we have $\|\cdot\|_\infty \leqslant \|\cdot\|_p \leqslant e\|\cdot\|_\infty$, so that it suffices to prove the second part of (7.18) for $p = \infty$. This is exactly (modulo the relation between the spherical and Gaussian means) the content of Lemma 6.1, which asserts that, in the present notation, $\mathbf{E}\|G\|_\infty \sim \sqrt{2\log n}$.

If $1 \leqslant p < 2$, $\|\cdot\|_p$ is $n^{1/p-1/2}$-Lipschitz and an argument along the same lines yields

$$\|(X - \mathbf{E}X)^+\|_{L_p} \leqslant n^{1/p-1/2}\sqrt{p/e}. \tag{7.21}$$

Combining this with (7.19) shows that $\mathbf{E}X = \Theta(n^{1/p})$ for $1 \leqslant p < 2$, whence (7.18) for that range of p readily follows. □

While the above argument relies heavily on tools specific to the Gaussian case, most of its elements can be carried over to a much more general setting. An example of a more robust calculation is given in Exercise 7.17.

REMARK 7.33 (Sharpness of Theorem 7.31). It can be shown that the estimates for the dimension of nearly Euclidean subspaces implied by Theorem 7.31 are sharp in the following sense: for $2 < p < \infty$, if some k-dimensional subspace $E \subset \mathbb{R}^n$ is such that $d_{BM}(B_p^n \cap E, B_2^k) \leqslant 2$, then $k \leqslant Cpn^{2/p}$, where C is an absolute constant (see Exercise 7.19).

REMARK 7.34 (Euclidean sections of ℓ_∞^n). The case of ℓ_∞^n deserves a special mention since almost Euclidean subspaces of ℓ_∞^n are closely related to ε-nets in the unit Euclidean sphere. It is easily checked (see Exercise 7.18) that the following two statements are equivalent:
(i) There is a k-dimensional subspace $E \subset \mathbb{R}^n$ such that $d_{BM}(B_2^k, B_\infty^n \cap E) \leqslant 1+\varepsilon$.
(ii) There exist n points $x_1, \dots, x_n$ in S^{k-1} such that

$$(1+\varepsilon)^{-1}B_2^k \subset \operatorname{conv}\{\pm x_i \;:\; 1 \leqslant i \leqslant n\}. \tag{7.22}$$

Moreover, (7.22) is also equivalent to $(\pm x_i)$ being a θ-net in (S^{k-1}, g) with $\cos\theta = (1+\varepsilon)^{-1}$ (Exercise 5.7). Since the smallest cardinality of such a net is essentially of order $V(\theta)^{-1}$ (Corollary 5.5), it follows from Proposition 5.1 that the largest dimension of a $(1+\varepsilon)$-Euclidean subspace in ℓ_∞^n is $\Theta(\log(n)/\log(1/\varepsilon))$.

The Dvoretzky dimension of ℓ_1^n is of order n, and consequently (by Theorem 7.19), for $0 < \varepsilon < 1$, a typical subspace of dimension $c\varepsilon^2 n$ of ℓ_1^n is $(1+\varepsilon)$-close to the Euclidean space.Remarkably, this phenomenon persists even when the dimension of the subspace approaches n. We have the following theorem.

THEOREM 7.35 (See Section 7.2.6.2). *Let $0 < \alpha < 1$. With large probability, a typical $\lfloor (1-\alpha)n \rfloor$-dimensional subspace $E \subset \mathbb{R}^n$ has the property that for every $x \in E$,*

$$A(\alpha)^{-1}\sqrt{n}|x| \leqslant \|x\|_1 \leqslant \sqrt{n}|x|, \tag{7.23}$$

where $A(\alpha)$ is a constant depending only on α.

For a more general result in this direction, see Theorem 7.42.

REMARK 7.36. (i) The optimal dependence as α tends to 0 in Theorem 7.35 is $A(\alpha) = \Theta((\log(2/\alpha)/\alpha)^{1/2})$. The upper bound will be shown in Section 7.2.6.2, where the Theorem is proved; see Exercise 7.16 for a slightly weaker lower bound. An alternative approach to Theorem 7.35 (with a simpler proof, but worse dependence on α) is via Theorem 7.42. (ii) In the context of Theorem 7.35, the parameter A is often called the distortion (of the ℓ_1-norm over E). However, an alternative (and arguably better, see Section 7.2.7) definition of the distortion of a subspace $E \subset \mathbb{R}^n$ is the ratio between the maximum and the minimum of the function $\|x\|_1$ over $S^{n-1} \cap E$. This is because for $A < \sqrt{\pi/2}$ the inequality $\|x\|_1 \geqslant A^{-1}\sqrt{n}|x|$ may hold for all $x \in E$ only if $\dim E$ is small (depending on A) [**SW**].

EXERCISE 7.16 (Simple lower bound on ℓ_1 distortion). Let $a \in (0,1]$ and let $E \subset \mathbb{R}^n$ be a subspace such that the inequality $\|x\|_1 \geqslant a\sqrt{n}|x|$ is satisfied for all $x \in E$. Show that the codimension of E is at least $a^2 n - 1$. (This is elementary.) Conclude that the optimal $A(\alpha)$ in Theorem 7.35 satisfies $A(\alpha) = \Omega(1/\sqrt{\alpha})$.

EXERCISE 7.17 (p-norms of subgaussian vectors). Let $(Y_1, \dots, Y_n)$ be independent random variables satisfying $\|Y_i\|_{\psi_2} \leqslant A$ for some $A \geqslant 0$. Denote $Y = (Y_1, \dots, Y_n)$. Show that $\mathbf{E}\,\|Y\|_p \leqslant A\sqrt{p}\,n^{1/p}$ for $1 \leqslant p < +\infty$ and that $\mathbf{E}\,\|Y\|_\infty \leqslant CA\sqrt{\log n}$.

EXERCISE 7.18 (Optimal almost spherical sections of the cube). Show the equivalence (i) $\iff$ (ii) in Remark 7.34.

EXERCISE 7.19 (Sharpness of Dvoretzky–Milman theorem for B_p^n). We show here that for $2 < p < \infty$, if some k-dimensional subspace $E \subset \mathbb{R}^n$ is such that $d_{BM}(B_p^n \cap E, B_2^k) \leqslant 2$ then $k \leqslant Cpn^{2/p}$, where C is an absolute constant. (i) Prove that $k_*(T(B_p^n)) \leqslant Cpn^{2/p}$ for any $T \in \mathsf{GL}(n,\mathbb{R})$. (ii) Conclude using Exercise 7.10.

EXERCISE 7.20 (Isomorphic vs. almost isometric Euclidean subspaces). Given $0 < \varepsilon < 1$ and n large enough, here is an example of a norm $\|\cdot\|$ on $\mathbb{R}^n$ such that $|\cdot| \leqslant \|\cdot\| \leqslant 2|\cdot|$, while if $E \subset \mathbb{R}^n$ is a subspace such that (for some A)

$$\forall x \in E \quad A|x| \leqslant \|x\| \leqslant (1+\varepsilon)A|x|, \tag{7.24}$$

then necessarily $\dim E = O(\varepsilon^2 n)$. We define the norm as $\|\cdot\| = |\cdot| + \|\cdot\|_p$, where $p \in [2,4]$ is given by the relation $n^{1/p-1/2} = 2\varepsilon$ (this is possible for n large enough). Suppose that a subspace E satisfies (7.24).
(i) Show that $A \geqslant 1 + \varepsilon/2$ and that $\varepsilon A \leqslant 4(A-1)$.
(ii) Show that for every $x \in E$, we have $B|x| \leqslant \|x\|_p \leqslant 5B|x|$ with $B = A - 1$.
(iii) Using the result of Exercise 7.19, conclude that $\dim E \leqslant C\varepsilon^2 n$ for some absolute constant C.

EXERCISE 7.21. Fix integers $m, n \geqslant 1$ and consider the convex body obtained as the ℓ_1-sum of m copies of B_2^n

$$K = \{(x_1, \dots, x_m) \in (\mathbb{R}^n)^m \ : \ |x_1| + \cdots + |x_m| \leqslant 1\}.$$

Show that $k_*(K) \geqslant cnm$ for some absolute constant $c > 0$.

EXERCISE 7.22. Show that, for every $\varepsilon > 0$, there is a polytope P with at most $\exp(Cn/\varepsilon^2)$ vertices and at most $\exp(Cn/\varepsilon^2)$ facets, such that $(1-\varepsilon)B_2^n \subset P \subset B_2^n$.

7.2.4.2. *Schatten norms.* We now consider the Schatten p-norms, for $p \in [1, \infty]$. Recall that $S_p^{m,n}$ is the corresponding unit ball in the space of (real or complex) $m \times n$ matrices, and $S_p^{n,\mathrm{sa}}$ is its analogue for the space of (real or complex) self-adjoint $n \times n$ matrices. Also recall (see Corollary 1.18) that $(S_p^{m,n})^\circ = S_q^{m,n}$ and $(S_p^{n,\mathrm{sa}})^\circ = S_q^{n,\mathrm{sa}}$, where $q \in [1, \infty]$ is defined by $p^{-1} + q^{-1} = 1$.

THEOREM 7.37 (Dvoretzky dimension for Schatten norms). *Consider two integers $m \leqslant n$, and $p \in [1, \infty]$. The Dvoretzky dimension of $S_p^{m,n}$ satisfies*

$$k_*(S_p^{m,n}) \simeq \begin{cases} mn & \text{if } 1 \leqslant p \leqslant 2, \\ m^{2/p}n & \text{if } 2 \leqslant p \leqslant \infty. \end{cases}$$

Moreover, in the case $m = n$, the same estimates are true for $k_*(S_p^{n,\mathrm{sa}})$.

REMARK 7.38. We emphasize again that the constants implicit in the $\simeq$ notation are absolute and do not depend on p, m, n. Moreover, the proof allows us to describe the precise asymptotic behavior of $k_*(S_p^{m,n})$ and $k_*(S_p^{n,\mathrm{sa}})$ (i.e., relations "$\sim$" in place of "$\simeq$," with reasonably explicit constants), see Exercise 7.23.

PROOF. We focus primarily on the real case, the complex case being similar. Let $q \in [1, \infty]$ be such that $1/p + 1/q = 1$. We have (see Definition 7.18)

$$k_*(S_p^{m,n}) = nm \operatorname{inrad}(S_p^{m,n})^2 w(S_q^{m,n})^2.$$

Accordingly, the Theorem will follow from the estimates

$$\operatorname{inrad}(S_p^{m,n}) = \begin{cases} m^{1/2-1/p} & \text{if } 1 \leqslant p \leqslant 2, \\ 1 & \text{if } 2 \leqslant p \leqslant \infty \end{cases}$$

and

$$\mathbf{E}\,\|A\|_p = w(S_q^{m,n}) = \Theta(m^{1/p-1/2}), \tag{7.25}$$

where A is a random matrix uniformly distributed on the Hilbert–Schmidt unit sphere in $\mathsf{M}_{m,n}$. The inradius is the same as in the commutative case: we are just comparing the ℓ_p-norm and the ℓ_2-norm of the sequence of singular values of a matrix (see (1.29); the comparison is formalized in (1.31)). In turn, (7.25) will be obtained by combining well-known properties of random matrices with the relation (A.7) between the spherical and the Gaussian mean. To that end, we note first that once we show the following one-sided bounds for the extreme values of p,

$$\text{(i) } \mathbf{E}\,\|A\|_\infty \lesssim m^{-1/2} \quad \text{and} \quad \text{(ii) } \mathbf{E}\,\|A\|_1 \gtrsim m^{1/2},$$

the remaining cases will follow by appealing again to the inequalities (1.31) relating different Schatten p-norms,

$$m^{1/p-1}\|\cdot\|_1 \leqslant \|\cdot\|_p \leqslant m^{1/p}\|\cdot\|_\infty.$$

Next, we know from the duality $(S_1^{m,n})^\circ = S_\infty^{m,n}$ and from Exercise 4.37 that

$$w(S_1^{m,n})w(S_\infty^{m,n}) \geqslant 1,$$

so that (ii) follows from (i) with $c = 1/C$. Finally, to justify (i), introduce a standard Gaussian vector B in $\mathsf{M}_{m,n}$, so that $\mathbf{E}\,\|A\|_\infty = \kappa_{mn}^{-1}\,\mathbf{E}\,\|B\|_\infty$ (in the complex case replace κ_{mn} by $\kappa_{mn}^{\mathbb{C}}$). Note that the random matrix $W = BB^\dagger$ is a Wishart matrix, allowing us to use the results from Section 6.2.3. We know (see Proposition 6.31 for the complex case and Corollary 6.38 for the real case) that $\mathbf{E}\,\|B\|_\infty \leqslant \sqrt{m} + \sqrt{n}$. Since $\kappa_{mn} \sim \sqrt{mn}$, this shows (i), completes the proof of (7.25) and, consequently, of the part of the Theorem concerning $k_*(S_p^{m,n})$.

The self-adjoint version can be treated exactly the same way, using estimates on the norm of GOE/GUE matrices (Proposition 6.24); recall that the GOE (resp., GUE) is essentially the standard Gaussian vector in the space of real symmetric (resp., complex self-adjoint) matrices. □

EXERCISE 7.23 (Sharp bounds for mean widths of Schatten balls). We consider either the real or the complex case. (i) Fix $p \in [1, \infty]$ and let n, s tend to infinity in such a way that $\lim \frac{s}{n} = \lambda \in [1, \infty)$. Show that the quantity $\mathbf{E}\,\|A\|_p = w(S_q^{n,s})$ appearing in (7.25) is equivalent to $\alpha_p \lambda^{-1/2} n^{1/p-1/2}$, where α_p is defined by $\alpha_p^p = \int |x|^{p/2}\, \mathrm{d}\mu_{\mathrm{MP}(\lambda)}(x)$ for $1 \leqslant p < \infty$, and $\alpha_\infty = 1 + \sqrt{\lambda}$. (One can check that the product $\alpha_p \lambda^{-1/2}$ is bounded away from 0 and $+\infty$.) (ii) Fix $p \in [1, \infty]$. Show that, as n tends to infinity, the quantity $w(S_q^{n,\mathrm{sa}})$ is equivalent to $\beta_p n^{1/p-1/2}$, where β_p is defined by $\beta_p^p = \int_{-2}^2 |x|^p\, \mathrm{d}\mu_{\mathrm{SC}}(x)$ for $1 \leqslant p < \infty$, and $\beta_\infty = 2$.

EXERCISE 7.24 (Uniformly bounded volume ratio for Schatten balls, $1 \leqslant p \leqslant 2$). Using the (reverse) Santaló inequality, show that for $m \leqslant n$ and $1 \leqslant p \leqslant \infty$,

$$cm^{1/2-1/p} \leqslant \mathrm{vrad}(S_p^{m,n}) \leqslant w(S_p^{m,n}) \leqslant Cm^{1/2-1/p}.$$

Deduce that the convex bodies $S_p^{m,n}$ have a (uniformly) bounded volume ratio if $1 \leqslant p \leqslant 2$. (See Section 7.2.6.1 for the definition.)

EXERCISE 7.25 (Sharpness of Dvoretzky–Milman theorem for Schatten spaces). Let $m \leqslant n$ be integers, let $p \in [1, \infty]$, and suppose that $E \subset \mathsf{M}_{m,n}$ is a k-dimensional subspace such that $d_{BM}(E \cap S_p^{m,n}, B_2^k) \leqslant 2$. The goal of this exercise is to show that

$$k \leqslant Ck_*(S_p^{m,n}), \tag{7.26}$$

where C is an absolute constant. This shows that, for isomorphically Euclidean sections, the Dvoretzky dimension gives a sharp bound. (Note, however, the hypothesis $d_{BM}(E \cap S_p^{m,n}, B_2^k) \leqslant 1 + \varepsilon$ *does not* imply that $k \leqslant C\varepsilon^2 k_*(S_p^{m,n})$; exploiting this "room for improvement" will be crucial in Chapter 8, see Remark 8.21.) Note that (7.26) holds trivially when $1 \leqslant p \leqslant 2$.
(i) Show that there is a constant C_0 and a polytope P with at most C_0^{m+n} vertices such that $P \subset S_1^{m,n} \subset 2P$.
(ii) Using (i) and Remark 7.34, show that (7.26) holds when $p = \infty$.
(iii) Assume now that $2 \leqslant p < \infty$, and suppose that $d_{BM}(E \cap S_p^{m,n}, B_2^k) \leqslant 2$. Show that $k_*(E \cap S_\infty^{m,n}) \geqslant ck/n^{2/p}$, and (using the previous question) that $k \leqslant Ck_*(S_p^{m,n})$.

7.2.5. Dvoretzky's theorem for general convex bodies. A famous consequence of the Dvoretzky–Milman theorem is the fact that any convex body of sufficiently large dimension admits almost Euclidean sections (see Corollary 7.40 below). It is based on the fact that n-dimensional convex bodies, which are in John position, have Dvoretzky dimensions that are $\Omega(\log n)$.

PROPOSITION 7.39. *Let $K \subset \mathbb{R}^n$ be a convex body in John position. Then the Dvoretzky dimension of K satisfies $k_*(K) \geqslant c \log n$ for some absolute constant $c > 0$.*

COROLLARY 7.40 (Dvoretzky's theorem). *There is a constant $c > 0$ such that the following holds. Let K be a symmetric convex body in $\mathbb{R}^n$ (for some $n \in \mathbb{N}$) and let $\varepsilon > 0$. Then there exists a subspace $E \subset \mathbb{R}^n$ of dimension at least $c\varepsilon^2 \log n$ such that*

$$d_g(K \cap E, B_2^E) \leqslant 1 + \varepsilon,$$

where B_2^E is the Euclidean unit ball in E.

If K is a non-symmetric convex body, the same conclusion holds for some k-dimensional affine subspace E and the corresponding notion of the distance.

PROOF OF COROLLARY 7.40. If K is in John position, the conclusion follows immediately from Proposition 7.39 and Theorem 7.19. For a general convex body $K \subset \mathbb{R}^n$, we know from Proposition 4.7 that there is a linear map T such that TK is in John position. Therefore there exists a subspace E with dimension $c\varepsilon^2 \log n$ such that $d_g(T(K) \cap E, B_2^E) \leqslant 1 + \varepsilon$. It follows that there is an ellipsoid $\mathscr{E} \subset T^{-1}(E)$ such that

$$\mathscr{E} \subset K \cap E \subset (1 + \varepsilon)\mathscr{E}.$$

We now use the result from Exercise 1.25 to conclude that $\mathscr{E}$ can be replaced by a multiple of the Euclidean ball if we replace E by a subspace $F \subset E$ with $\dim F = \lceil \frac{1}{2} \dim E \rceil$.

Finally, the same argument works for non-symmetric convex bodies, except that the subspace E is affine. □

The key estimate needed for the proof of Proposition 7.39 is the following lemma, known as the Dvoretzky–Rogers lemma.

LEMMA 7.41 (Dvoretzky–Rogers lemma). *Let $K \subset \mathbb{R}^n$ be a convex body which is in John position. Then there exists an orthonormal basis $(x_k)_{1 \leqslant k \leqslant n}$ such that, for any $1 \leqslant k \leqslant n$,*

$$\|x_k\|_K \geqslant \sqrt{k/n}.$$

PROOF. The Lemma is a consequence of the following claim: *under the hypotheses of the Lemma, any m-dimensional subspace $F \subset \mathbb{R}^n$ contains a vector x with $|x| = 1$ and $\|x\|_K \geqslant \sqrt{m/n}$.* Indeed, we construct successively $x_n, \ldots, x_1$ and obtain x_k by applying the claim to the subspace orthogonal to $\{x_i \,:\, i > k\}$.

To prove the claim, consider a resolution of identity (c_i, x_i) given by Proposition 4.7. Recall that $x_i \in \partial K \cap B_2^n$ are contact points; in particular it follows that K is contained in each half-space $\{\langle \cdot, x_i \rangle \leqslant 1\}$, or that $\| \cdot \|_K \geqslant \langle \cdot, x_i \rangle$. Given an m-dimensional subspace $F \subset \mathbb{R}^n$, we have

$$P_F = \sum c_i P_F |x_i\rangle\langle x_i|.$$

Taking the trace gives $m = \sum c_i |P_F x_i|^2$. Since $\sum c_i = n$, there exists an index j with $|P_F x_j| \geqslant \sqrt{m/n}$. Let $x = P_F x_j / |P_F x_j|$. We have

$$\|x\|_K \geqslant \langle x, x_j \rangle = |P_F x_j| \geqslant \sqrt{m/n}. \qquad \square$$

We can now complete the proof of Proposition 7.39.

PROOF OF PROPOSITION 7.39. Let K be a convex body in John position, and let X be a random vector uniformly distributed on S^{n-1}. Since $\operatorname{inrad}(K) = 1$, it suffices to prove in view of Definition 7.18 that

$$\mathbf{E}\,\|X\|_K \geqslant c\sqrt{\log n / n}.$$

for some constant c.

We know from Lemma 7.41 that there exists an orthonormal family of $n/4$ vectors (x_i) with $\|x_i\|_K \geqslant 1/2$. In particular, we have $\|\cdot\|_K \geqslant \frac{1}{2}\max\{\langle \cdot, x_i\rangle \ : \ 1 \leqslant i \leqslant n/4\}$. Consequently, if G denotes a standard Gaussian vector in $\mathbb{R}^n$, then

$$\mathbf{E}\,\|X\|_K = \frac{1}{\kappa_n}\mathbf{E}\,\|G\|_K \geqslant \frac{1}{2\kappa_n}\mathbf{E}\max\{\langle G, x_i\rangle \ : \ 1 \leqslant i \leqslant n/4\}.$$

The random variables $\langle G, x_i \rangle$ are i.i.d. standard normal variables, and therefore the expectation of their maximum is of order $\sqrt{\log n}$ by Lemma 6.1, as needed. $\square$

EXERCISE 7.26 (Complex version of Dvoretzky's theorem). Check that Corollary 7.40 remains valid for a circled convex body $K \subset \mathbb{C}^n$.

EXERCISE 7.27 (Simultaneous spherical sections for a set and its polar). (i) Show that the following holds for some constant $c > 0$: for every symmetric convex body $K \subset \mathbb{R}^n$ there is a k-dimensional subspace $E \subset \mathbb{R}^n$ with $k = c\log n$ such that both $K \cap E$ and $P_E K$ (or, equivalently, $K^\circ \cap E$) are 2-Euclidean. (ii) Can we choose a position of K such that the conclusion is valid for most subspaces E?

7.2.6. Related results.

7.2.6.1. *Volume ratio.* Define the *volume ratio* of a convex body $K \subset \mathbb{R}^n$ as

$$\operatorname{vr}(K) = \left(\frac{\operatorname{vol}(K)}{\operatorname{vol}(\operatorname{John}(K))}\right)^{1/n}.$$

The quantity $\operatorname{vr}(K)$ is an affine invariant. Consequently, if $K = B_X$, it makes sense to denote $\operatorname{vr}(X) = \operatorname{vr}(K)$. Examples of convex bodies with "bounded volume ratio" (i.e., bounded by a dimension-independent constant) include B_p^n, $S_p^{m,n}$ and $S_p^{n,\mathrm{sa}}$ for $1 \leqslant p \leqslant 2$. For B_p^n, this is a consequence of the computations from Section 4.3.3 (Table 4.1, Exercises 4.39 and 4.40). For the Schatten spaces, the boundedness follows from the proof of Theorem 7.37 (see also Exercise 7.24). The following theorem asserts that bodies (resp., spaces) with bounded volume ratio always have nearly Euclidean sections (resp., subspaces) of proportional dimension, for arbitrary proportion $\alpha \in (0,1)$.

THEOREM 7.42 (Not proved here). *Let $K \subset \mathbb{R}^n$ a convex body in John position and denote $A = \operatorname{vr}(K)$. Let $E \subset \mathbb{R}^n$ be a random k-dimensional subspace. Then, with probability larger than $1 - e^{-n}$,*

$$B_2^E \subset K \cap E \subset (CA)^{\frac{n}{n-k}} B_2^E,$$

where C is an absolute constant.

In general, the bounded volume ratio property is inherited by subspaces only if the dimension of the space and that of the subspace are comparable (Exercise 7.28). However, *all* subspaces of the classical and noncommutative L_p-spaces alluded to above do have uniformly bounded volume ratio. This is due to the fact that they possess the so-called cotype 2 property, which is clearly inherited by subspaces and which is known to imply the bounded volume ratio (see Notes and Remarks).

An important instance of Theorem 7.42 is the following striking fact.

COROLLARY 7.43 (See Exercise 7.29). *Let $n = 2k$; there exist $E_1, E_2 \subset \mathbb{R}^n$ with* $\dim E_1 = \dim E_2 = k$ *and $E_1 \perp E_2$ such that*

$$c|x| \leqslant n^{-1/2}\|x\|_1 \leqslant |x| \quad \text{for } x \in E_i,\ i = 1, 2, \tag{7.27}$$

where $c > 0$ is a universal constant. Similarly, if $n = 3k$, there exist mutually orthogonal k-dimensional subspaces E_1, E_2, E_3 such that the bounds from (7.27) hold for $x \in E_i + E_j$, for any $\{i, j\} \subset \{1, 2, 3\}$.

The property expressed by (7.27) is usually referred to as *the Kashin decomposition of ℓ_1^n*. Another statement closely related to Theorem 7.42 is the following.

THEOREM 7.44 (Not proved here). *Let $K \subset \mathbb{R}^n$ a convex body in John position and denote $A =$* $\mathrm{vr}(K)$. *There is an orthogonal transformation $U \in \mathsf{O}(n)$ such that $K \cap UK \subset 8A^2 B_2^n$.*

EXERCISE 7.28 (Volume ratio of subspaces). (a) Let $K \subset \mathbb{R}^n$ be a symmetric convex body and let $E \subset \mathbb{R}^n$ be a k-dimensional subspace. Show that $\mathrm{vr}(K \cap E) \leqslant (C\,\mathrm{vr}(K))^{n/k}$. (b) Give examples of symmetric convex bodies $K \subset \mathbb{R}^n$ and subspaces $E \subset \mathbb{R}^n$ such that the ratio $\mathrm{vr}(K \cap E)/\mathrm{vr}(K)$ is arbitrarily large.

EXERCISE 7.29 (Kashin decomposition via volume ratio). (i) Derive Corollary 7.43 from Theorem 7.35. (ii) Show that the assertion of Corollary 7.43 holds for spaces X with uniform bound on their volume ratios (i.e., with constant c depending only on $\mathrm{vr}(X)$).

EXERCISE 7.30 (A dual Kashin decomposition). Show that, for any $n \in \mathbb{N}$, there is an orthogonal transformation $U \in \mathsf{O}(n)$ such that $c\sqrt{n}B_2^n \subset \mathrm{conv}(B_\infty^n, UB_\infty^n)$, where $c > 0$ is an absolute constant. Note that we always have $\mathrm{conv}(B_\infty^n, UB_\infty^n) \subset \sqrt{n}B_2^n$.

7.2.6.2. *The low-M^* estimate and the proof of Theorem* 7.35. Let $K \subset \mathbb{R}^n$ be a symmetric convex body. The argument from Exercise 7.12 shows that sections of K of dimension larger than $k_*(K)$ cannot be isomorphically Euclidean. Remarkably, "one half" of the estimates (7.12) persists: an avatar of the lower bound remains valid for subspaces of proportional dimension.

THEOREM 7.45 (Low-M^* estimate). *Let K be either a convex body in $\mathbb{R}^n$ containing 0 in the interior or a circled convex body in $\mathbb{C}^n$, and $M^* = w(K)$. Let $0 < \alpha < 1$ and $k = n(1 - \alpha)$. Then, with probability larger than $1 - \exp(-c\alpha n)$, a random k-dimensional subspace E satisfies*

$$\forall x \in E, \quad \frac{c\sqrt{\alpha}}{M^*}|x| \leqslant \|x\|_K, \tag{7.28}$$

where $c > 0$ is an absolute constant.

If we denote (as in Theorem 7.19) $w(K^\circ)$ by M, we recall that $MM^* \geqslant 1$ (see Exercise 4.37), so that the lower bound in (7.28) is always worse than the lower bound in (7.12). However, when a good upper bound on the product MM^* is present (which is always the case for some choice of the Euclidean structure, see Theorem 7.10), both estimates become comparable.

PROOF. We give a proof (valid only in the real case) based on Proposition 6.42. Consider $L = S^{n-1} \cap tK$ for $t > 0$ to be chosen later. We have $w_G(L) \leqslant w_G(tK) = tw_G(K) = t\kappa_n M^*$. We now chose t such that $t\kappa_n M^* = \frac{1}{2}\kappa_{n-k}$; this implies $t \geqslant c\sqrt{\alpha}/M_*$ for some $c > 0$ because $\kappa_m \sim \sqrt{m}$. Proposition 6.42 implies then that, with high probability, a random subspace $E \in \mathsf{Gr}(k, \mathbb{R}^n)$ does not intersect L. This is equivalent to the fact that the inequality $\|\cdot\|_K > t|\cdot|$ holds on E. □

PROOF OF THEOREM 7.35. We argue as in the proof of Theorem 7.45 specified to $K = B_1^n$; the only modification comes in upper-bounding $w_G(L)$. Denote $\tilde{L} = B_2^n \cap tB_1^n$ ($t \in [1, \sqrt{n}]$ to be chosen later), then clearly

$$w_G(L) \leqslant w_G(\tilde{L}).$$

(We actually have equality since $\tilde{L} = \operatorname{conv} L$; this is a fairly easy consequence of the fact that no extreme point of tB_1^n lies inside B_2^n.) Next, $\tilde{L}^\circ = \operatorname{conv}\left(B_2^n \cup t^{-1}B_\infty^n\right)$ by (1.15) and so we have

$$w_G(\tilde{L}^\circ) \geqslant w_G(t^{-1}B_\infty^n) = t^{-1} \times \sqrt{\frac{2}{\pi}}\, n,$$

see Table 4.1 and Exercise 6.6 for the equality. Given that $\tilde{L}$ is permutationally symmetric, it has enough symmetries and hence it is in the ℓ-position (see Section 4.2.2 and particularly Proposition 4.8). Accordingly, Proposition 7.6 applies and shows that

$$w_G(\tilde{L})w_G(\tilde{L}^\circ) \leqslant n\mathbf{K}(\tilde{L}).$$

Further, since $\tilde{L}$ is unconditional, it follows from Remark 7.8 that

$$\mathbf{K}(\tilde{L}) \leqslant C\big(1 + \log d(\tilde{L}, B_2^n)\big)^{1/2} = C\big(1 + \log(t/\sqrt{n})\big)^{1/2}.$$

Combining the above inequalities yields

$$w_G(L) \leqslant w_G(\tilde{L}) \leqslant C\sqrt{\frac{\pi}{2}}\, t\big(1 + \log(\sqrt{n}/t)\big)^{1/2}.$$

As in the proof of Theorem 7.45, we now choose t so that

$$2C\sqrt{\frac{\pi}{2}}\, t\big(1 + \log(\sqrt{n}/t)\big)^{1/2} = \kappa_{n-k} \sim \sqrt{\alpha n},$$

which can be rewritten as $g(\lambda) \sim c\alpha^{-1/2}$, where $g(x) = x(1 + \log x)^{-1/2}$, $\lambda = \sqrt{n}/t$ and $c = \sqrt{2\pi}C$. Solving for λ we obtain $\lambda \simeq \alpha^{-1/2}\big(\log(2/\alpha)\big)^{1/2}$, and therefore $t = \sqrt{n}/\lambda \simeq (\alpha/\log(2/\alpha))^{1/2}\sqrt{n}$, as needed. (We are using here the fact that if $\beta \in \mathbb{R}$ is fixed and if $y = g(x) := x(1 + \log x)^\beta$, then the inverse function—which is defined for sufficiently large y—satisfies $g^{-1}(y) \sim y(1 + \log y)^{-\beta}$ as $y \to \infty$.) □

7.2.6.3. *The quotient of a subspace theorem.* It follows from Corollary 7.40 that any convex body $K \subset \mathbb{R}^n$ admits isomorphically Euclidean sections of dimension $\Omega(\log n)$. Dually, any convex body admits orthogonal projections of the same dimension which are isomorphically Euclidean. The bound $\Omega(\log n)$ cannot be improved, as shown by the case of the cube (for sections) or of the ℓ_1^n ball (for projections). However, it turns out that combining both operations leads to a surprising phenomenon: every convex body admits a projection of a section of proportional dimension which is isomorphically Euclidean.

THEOREM 7.46 (Quotient of a subspace theorem, not proved here). *Given a symmetric convex body $K \subset \mathbb{R}^n$ and $\alpha \in (0,1)$, there exist subspaces $E \subset F \subset \mathbb{R}^n$ with* $\dim E \geqslant (1-\alpha)n$ *such that*

$$d_{BM}(P_E(K \cap F), B_2^{\dim E}) \leqslant C\alpha^{-1}\log(C\alpha^{-1}).$$

We note that an "almost isometric" version of the quotient of a subspace theorem follows then by appealing to Remark 7.22.

EXERCISE 7.31 (Quotient of a subspace $=$ subspace of a quotient). Show that given a decomposition $\mathbb{R}^n = E \oplus F \oplus G$ into orthogonal subspaces, we have, for $K \subset \mathbb{R}^n$

$$(P_{E\oplus F}K) \cap E = P_E(K \cap (E \oplus G)).$$

Conclude that the class of sections of projections of K coincides with the class of projections of sections of K.

EXERCISE 7.32 (Combining quotient and subspace operations is necessary). Give an example of a family of convex bodies of growing dimension which has neither sections nor projections of proportional dimension which are isomorphically Euclidean. Therefore, in general, combining both operations is needed for Theorem 7.46 to be valid.

EXERCISE 7.33 (Quotient of a subspace implies reverse Santaló). We show here how to derive the reverse Santaló inequality from the quotient of a subspace theorem (Theorem 7.46). Let $K \subset \mathbb{R}^n$ be a symmetric convex body, and $\mathbb{R}^n = E_1 \oplus E_2 \oplus E_3$ be an orthogonal decomposition. Let $n_i = \dim E_i$. Denote $K_1 = P_{E_1}(K \cap (E_1 \oplus E_2))$, $K_2 = K \cap E_2$, and $K_3 = P_{E_3}K$; these are convex bodies in, respectively E_1, E_2, and E_3.
(i) Check by applying Lemma 4.20 twice that $\mathrm{vol}(K) \geqslant \frac{1}{4^n}\mathrm{vol}(K_1)\mathrm{vol}(K_2)\mathrm{vol}(K_3)$ and $\mathrm{vol}(K^\circ) \geqslant \frac{1}{4^n}\mathrm{vol}(K_1^\circ)\mathrm{vol}(K_2^\circ)\mathrm{vol}(K_3^\circ)$.
(ii) Given convex body $L \subset \mathbb{R}^k$, define $\alpha(L) = \mathrm{vrad}(L)\,\mathrm{vrad}(L^\circ)$. Show that, for some constant c, $\alpha(K)^n \geqslant c^n \alpha(K_1)^{n_1}\alpha(K_2)^{n_2}\alpha(K_3)^{n_3}$.
(iii) By Theorem 7.46, we may assume that $n_1 = n/2$, and that K_1 is A-Euclidean for some absolute constant A. Show that $\alpha(K_1) \geqslant A^{-1}$. If β_N denotes the infimum of $\alpha(K)$ over all symmetric convex bodies of dimension at most N, conclude that $\beta_N \geqslant c^2/A$.

7.2.6.4. *Approximation of zonoids by zonotopes.* We first state a reformulation of Dvoretzky's theorem for ℓ_1^n.

THEOREM 7.47 (See Exercise 7.34). *For any $n \in \mathbb{N}$, $\varepsilon > 0$, there exists an integer $N \leqslant Cn/\varepsilon^2$ and vectors $x_1, \dots, x_N \in \mathbb{R}^n$ such that $Z \subset B_2^n \subset (1+\varepsilon)Z$, where Z denotes the zonotope*

$$Z = [-x_1, x_1] + \cdots + [-x_N, x_N]. \tag{7.29}$$

It is natural to ask whether a version of Theorem 7.47 holds when the Euclidean ball is replaced by an arbitrary zonoid. The best result in this direction is the following.

THEOREM 7.48 (Not proved here). *For any* 0*-symmetric zonoid* $Y \subset \mathbb{R}^n$ *and* $\varepsilon > 0$, *there exists an integer* $N \leqslant Cn\log(n)/\varepsilon^2$ *and vectors* $x_1, \ldots, x_N \in \mathbb{R}^n$ *such that* $Z \subset Y \subset (1+\varepsilon)Z$, *where* Z *denotes the zonotope* (7.29). *Moreover, we can ensure that* $\operatorname{supp}\mu_Z \subset \operatorname{supp}\mu_Y$, *where the measures* μ_Y, μ_Z *are defined in* (4.8).

EXERCISE 7.34 (Approximating balls by zonotopes via Dvoretzky's theorem). Prove Theorem 7.47 using the fact that the Dvoretzky dimension of B_1^n is of order n (Theorem 7.31).

7.2.6.5. *The Johnson–Lindenstrauss lemma.*

THEOREM 7.49 (The Johnson–Lindenstrauss lemma). *Let* A *be a finite subset of* $\mathbb{R}^n$, $m = \operatorname{card} A$, *and* $\varepsilon \in (0,1)$. *If* $k \geqslant 4\varepsilon^{-2}\log m$, *there exists a linear map* $f : \mathbb{R}^n \to \mathbb{R}^k$ *such that, for every* $x, y \in A$,

$$(1-\varepsilon)|x-y| \leqslant |f(x)-f(y)| \leqslant (1+\varepsilon)|x-y|. \tag{7.30}$$

PROOF. We show that a random choice for f satisfies (7.30) with high probability. Let $B : \mathbb{R}^n \to \mathbb{R}^k$ be a random matrix with i.i.d. $N(0,1)$ entries. For every unit vector $u \in \mathbb{R}^n$, Bu is a standard Gaussian vector in $\mathbb{R}^k$, and the random variable $|Bu|$ follows the $\chi^2(k)$ distribution. Denoting by M_k^2 the median of the $\chi^2(k)$ distribution, it follows from Theorem 5.24 that for any $t > 0$,

$$\mathbf{P}\left(\big||Bu| - M_k\big| > t\right) \leqslant \exp(-t^2/2).$$

Define f as $\frac{1}{M_k}B$. Given $x, y \in A$, we apply the above inequality for $u = (x-y)/|x-y|$ and $t = \varepsilon M_k$ to obtain

$$\mathbf{P}\left(\big||f(x)-f(y)| - |x-y|\big| > \varepsilon|x-y|\right) \leqslant \exp(-\varepsilon^2 M_k^2/2).$$

We now take the union bound over the $\binom{m}{2} \leqslant m^2/2$ pairs of points from A. It follows that (7.30) is satisfied whenever $m^2/2 \cdot \exp(-\varepsilon^2 M_k^2/2) < 1$, i.e., $2\log m < \varepsilon^2 M_k^2/2 + \log 2$. Since $M_k^2 \geqslant k - 2/3$ (see Exercise 5.34), this condition is satisfied provided $k \geqslant 4\varepsilon^{-2}\log m$. □

7.2.7. Constructivity. A general feature of the proofs of most of the theorems in this chapter is a heavy use of the probabilistic method. For example, the existence of a subspace satisfying the conclusion of Dvoretzky's theorem or its variants is proved by selecting it at random according to the unitarily invariant measure on the corresponding Grassmannian (after a suitable Euclidean structure has been chosen) or by using random matrices. Random constructions benefit from the *blessing of dimensionality*, as opposed to the *curse of dimensionality*, which renders an exhaustive search (and many deterministic algorithms) nonfeasible.

However, for theoretical and practical reasons, existence results are often unsatisfactory. For example, to write a computer code implementing an error-correcting algorithm, one needs a specific encoding matrix. This leads to the class of problems asking for *explicit* versions of, or pseudo-random models for, objects whose constructions involve probabilistic arguments. By "explicit" we mean here an algorithm, whose complexity is manageable (say, with running time being polynomial in the dimension). Individual constructions are often "more explicit" than that; they may involve, e.g., closed formulas. An alternative to an explicit solution may be a guarantee that we can efficiently check whether a randomly generated object actually does the job.

When the initial setting is completely abstract, it seems unrealistic to expect any meaningful statement. We therefore mostly concentrate on standard convex bodies. Here is an example of a satisfactory result.

THEOREM 7.50 (Explicit quotient of a subspace theorem for the simplex, not proved here). *Given $n \in \mathbb{N}$, there exists a set $S \subset \mathbb{R}^n$ which is an explicit affine image of an explicit section of the $5n$-dimensional simplex and which verifies*

$$B_2^n \subset S \subset CB_2^n.$$

Moreover, C can be replaced by $1+\varepsilon$ for $\varepsilon \in (0,1)$, if we use a simplex of dimension $\geqslant C_1 n \log(2/\varepsilon)$.

Another result, for which substantial efforts has been devoted to derandomization, is Dvoretzky's theorem for B_1^n (or ℓ_1^n). Recall that the (ℓ_1-)*distortion* of a subspace $E \subset \mathbb{R}^n$ is the ratio between the maximum and the minimum of the function $\|x\|_1$ over $S^{n-1} \cap E$. We already showed, via the probabilistic method, the existence of subspaces of proportional dimension with arbitrarily small distortion (Theorem 7.31) and the existence of subspaces of arbitrarily large proportional dimension with bounded distortion (Theorem 7.42). The randomness relied on the Haar measure on Grassmann manifold, which requires an infinite amount of random bits to be exactly simulated. However, a careful look at the arguments shows that the same conclusion can be derived using only $O(n^2)$ random bits.

A natural step towards explicit examples is randomness reduction: can we match, or approach, the optimal dimension and distortion bounds using fewer random bits? We point out that constructions using $O(\log n)$ random bits are very close to being explicit, since we can then perform an exhaustive search among the polynomially many possible bit strings. However, it is not clear whether the distortion of a given subspace can be efficiently estimated; the following seems to be unknown.

PROBLEM 7.51. *Is the problem of calculating (or approximating well enough) the ℓ_1-distortion of a general subspace $E \subset \mathbb{R}^n$ NP-hard?*

The best results known to the authors and directed towards constructing explicit subspaces of ℓ_1^n (going in several different directions) are gathered in Table 7.1. One result that "doesn't fit" in the table is the following.

THEOREM 7.52 (Not proved here). *Given $n \in \mathbb{N}$, $p \in (1,2)$ and $\eta \in (0,1)$, there is an explicitly defined subspace $E \subset \mathbb{R}^n$ of dimension $(1-\eta)n$ such that*

$$d_g(B_1^n \cap E, B_p^n \cap E) \leqslant (1/\eta)^{O\left((2-p)^{-1}\right)}. \tag{7.31}$$

In the language of this section, (7.31) gives a bound on the distortion of the ℓ_1-norm on the sphere of ℓ_p^n intersected with E.

In a different direction, we state a result which derandomizes Dvoretzky's theorem (Corollary 7.40) *simultaneously* for a wide class of convex bodies.

THEOREM 7.53 (Not proved here). *Given $n \in \mathbb{N}$ and $\varepsilon \in (0,1)$, there is an explicitly defined subspace $E \subset \mathbb{R}^n$ of dimension $k = c\log n/\log(1/\varepsilon)$ such that the*

TABLE 7.1. The best known results concerning derandomization/randomness reduction when constructing almost-Euclidean sections of B_1^n. The parameters $\epsilon, \eta, \gamma \in (0,1)$ are assumed to be constants, although we explicitly point out when the dependence on them is subsumed by the big-Oh or the little-oh notation. The last column indicates the number of random bits used in the construction.

Reference	Distortion	Subspace dimension	Randomness
[**Ind07**]	$1+\epsilon$	$n^{1-o_\epsilon(1)}$	explicit
[**GLR10**]	$(\log n)^{O_\eta(\log\log\log n)}$	$(1-\eta)n$	explicit
[**Ind00**]	$1+\epsilon$	$\Omega(\epsilon^2/\log(1/\epsilon))n$	$O(n\log^2 n)$
[**AAM06, LS08**]	$O_\eta(1)$	$(1-\eta)n$	$O(n)$
[**GLW08**]	$2^{O_\eta(1/\gamma)}$	$(1-\eta)n$	$O(n^\gamma)$
[**IS10**]	$1+\epsilon$	$(\gamma\epsilon)^{O(1/\gamma)}n$	$O(n^\gamma)$

following holds. If $K \subset \mathbb{R}^n$ is a convex body invariant under the isometry group of the cube (i.e., permutation of coordinates and sign flips) then

$$d_g(K \cap E, B_2^E) \leqslant 1+\varepsilon.$$

Notes and Remarks

A recent and comprehensive reference for the material presented in this chapter (and much more) is [**AAGM15**]. Older standard and valuable references include [**MS86, Pis89b, TJ89, Ver**].

Section 7.1. Proposition 7.5 is a special case of Corollary 3 in [**LO99**]. If we do not insist on obtaining the optimal constant $\sqrt{\pi/2}$, the result is more elementary: it is an instance of the Gaussian version of the Khintchine–Kahane inequality, Exercise 5.72, whose proof carries over to the present context (modulo replacing an application of Theorem 5.23 with that of Theorem 5.51) and extends to non-symmetric convex bodies (see, e.g., [**BLPS99**], Lemma 3.3).

The K-convexity constant is more frequently defined in the literature for a normed space Y and corresponds, on our notation, to $\mathbf{K}(B_Y)$.

Proposition 7.6 is due to Figiel and Tomczak-Jaegermann [**FTJ79**] (where the ℓ-norm is also introduced), whereas Theorem 7.7 and the bound stated in Remark 7.8 are due to Pisier (see [**Pis80, Pis81**]). The proof of Theorem 7.7 that is presented here is based on Lemma 7.13, which is from [**Mau03**]. The bound on the K-convexity constant from Theorem 7.7 is sharp: there is an example due to Bourgain [**Bou84**] of a symmetric convex body $K \subset \mathbb{R}^n$ (for an arbitrarily large n) with $\mathbf{K}(K) = \Omega(\log n)$; this example is presented in detail in [**AAGM15**, Section 6.7]. Besides unconditional bodies, the improved bound $\mathbf{K}(K) = O(\sqrt{\log n})$ holds if $K \subset \mathbb{R}^n$ is, for example, a zonoid (see [**Pis80**]; or Theorem IV.5 in [**LQ04**] for a detailed proof).

In is unknown if the MM^*-estimate is sharp, i.e., whether $\log n$ can be replaced by a smaller function in Theorem 7.10. The pair (B_1^n, B_∞^n) gives an example of a (sequence of) symmetric convex bodies, for which $w(K)w(K^\circ) = \Theta(\sqrt{\log n})$, and one may conjecture that the MM^*-estimate holds (for symmetric bodies) with a bound $O(\sqrt{\log n})$. In the non-symmetric case, the n-dimensional simplex Δ is an example with $w(\Delta)w(\Delta^\circ) \simeq \log n$. While it is conceivable that the MM^*-estimate holds with a bound that is polynomial in $\log n$ also for non-symmetric bodies, the known general upper bounds in that setting are much weaker [**BLPS99, Rud00**]. This question is related to the problem of determining the diameter of the Banach–Mazur compactum of not-necessarily-symmetric convex bodies.

Section 7.2. The history around Dvoretzky's theorem starts with a conjecture by Grothendieck [**Gro53b**]: does every n-dimensional normed space contain a $k(\varepsilon, n)$-dimensional subspace which is $(1+\varepsilon)$-Euclidean, for some function $k(\varepsilon, n)$ tending to infinity with n? This was shown affirmatively by Dvoretzky [**Dvo61**], and later refined by [**Mil71**] using crucially concentration of measure. Other early proofs include [**Sza74**] and [**Fig76**].

Theorem 7.15 with the dependence on ε as stated appears in [**Gor88**] in the real case (see Exercise 7.7). The proof via Lemma 7.17 is from [**Sch89**], and it was noticed in [**ASW11**] that it carries over to the complex case.

When asking about the dependence on ε in Dvoretzky's theorem, it is important to keep in mind that there are two different questions, depending whether we ask if $(1+\varepsilon)$-Euclidean subspaces either (i) exist or (ii) have measure $1 - o(1)$ in the Grassmann manifold equipped with the standard Haar measure.

For example, one may ask: given $\varepsilon > 0$ and k, for which values of n can we guarantee that every n-dimensional symmetric convex body has a k-dimensional section which is $(1+\varepsilon)$-Euclidean? If we believe that the worst case is the cube, it is natural to conjecture that this holds for $n \geqslant C(k)\varepsilon^{-(k-1)/2}$. This conjecture is confirmed for $k = 2$ (see [**Mil88**]). For $k > 2$ the problem is wide open and a good dependence would follow from a positive answer to a weak version of the Knaster problem, see [**KS03**]. In a related direction, the random version of the Dvoretzky theorem for the cube has been studied in [**Sch07, Tik14**] and the dependence on ε in Theorem 7.19 for $K = B_\infty^n$ is $c(\varepsilon) = \Theta(\varepsilon/\ln(1/\varepsilon))$.

Most of the material from Sections 7.2.2 through 7.2.4 is based on the very influential paper [**FLM77**]. The concepts of the verticial and facial dimensions of a convex body were formally defined in [**AS17**].

Exercise 7.12 about the sharpness of the Dvoretzky dimension is an observation due to Milman–Schechtman [**MS97**] (see [**HW17**] for a sharper statement). The paper [**MS97**] also introduces global versions of Dvoretzky's theorem, of which here is a sample: *for any symmetric convex body $K \subset \mathbb{R}^n$, there is an integer $t \leqslant C\varepsilon^{-2}n/k_*(K)$ and $U_1, \dots, U_t \in \mathsf{O}(n)$ such that the Minkowski sum $U_1(K) + \cdots + U_t(K)$ is $(1+\varepsilon)$-Euclidean.* For other similar results, see [**AAGM15**]. The result from Exercise 7.19 appears in [**BDG$^+$77**] (for another proof, see [**AAGM15**, Theorem 5.4.3]). The construction from Exercise 7.20 is due to Figiel.

The estimate from Exercise 7.21 is relevant to [**FHS13**]. Theorem 7.35 is from [**Kaš77**]; the correct order of magnitude of the distortion constant $A(\alpha)$ was determined in [**GG84**]; the proof of the upper bound presented in Section 7.2.6.2 follows [**PTJ90**]. We also refer to [**FR13**, Chapter 10] for a detailed presentation focusing on applications to compressed sensing.

The Dvoretzky–Rogers lemma was first proved in [**DR50**]. The proof presented comes from [**Pel80**]. It has been realized since [**BS88**] that actually a stronger property holds: *There is a function $f : (0,1] \to [1,\infty)$ such that, for any n-dimensional normed space X there exist $m \geqslant (1-\delta)n$ and operators $\alpha : \mathbb{R}^m \to X$, $\beta : X \to \mathbb{R}^m$ verifying $\beta \circ \alpha = \mathrm{I}$ and $\|\alpha : \ell_1^m \to X\| \cdot \|\beta : X \to \ell_2^m\| \leqslant f(\delta)$.* The above is often referred to as a proportional *Dvoretzky–Rogers factorization.* It is known that $f(\delta) = O(\delta^{-1})$ and $f(\delta) = \Omega(\delta^{-1/2})$ [**Gia96, Rud97**]. Variants for nonsymmetric bodies were also shown, see [**You14**]. For more information and references see the website [**@3**].

Regarding Proposition 7.39, it has been proved in [**Bal89, Bal91**] that the cube (resp., the simplex) has the smallest mean width among all symmetric (resp., non-necessarily symmetric) convex bodies in John position.

The relevance of the concept of volume ratio to Dvoretzky-like theorems was realized in [**Sza78, ST80**], which were inspired by the important work [**Kaš77**] that in particular established the existence of the Kashin decomposition of ℓ_1^n (see Corollary 7.43). This concept is related to the notion of cotype 2. Let (ε_n) be a sequence of independent variables such that $\mathbf{P}(\varepsilon_i = 1) = \mathbf{P}(\varepsilon_i = -1) = 1/2$. The cotype 2 constant of a normed space X is the smallest number $C_2(X)$ such that, for every vectors $x_1, \ldots, x_n \in X$, we have

$$\sum_{i=1}^{n} \|x_i\|^2 \leqslant C_2(X)\, \mathbf{E} \left\| \sum_{i=1}^{n} \varepsilon_i x_i \right\|^2 .$$

The estimate $\mathrm{vr}(X) = O(C_2(X) \log C_2(X))$ connecting volume ratio and cotype 2 was proved in [**BM87, MP86**] (see [**Mil87**] for a simpler proof and [**DS85**] for an earlier argument yielding Kashin's decompositions under cotype 2 assumptions). Any bound on the cotype 2 constant is obviously inherited by subspaces. For more information about the type and cotype theory, see [**Mau03**]. The formulation of Theorem 7.44 appears in [**Bal97**].

The low-M^* estimate (Theorem 7.45) was proved originally by Milman with a worse dependence on α; the proof we present is due to Gordon [**Gor88**]. Another proof giving the correct dependence, and valid also in the complex setting, is due to Pajor and Tomczak–Jaegermann [**PT86**]. See [**AAGM15**] for a presentation of several different proofs. We also point that in some cases the upper bound in the Dvoretzky–Milman theorem (Theorem 7.19) holds for dimensions larger than the Dvoretzky dimension: see [**KV07**].

The quotient of a subspace theorem is due to Milman [**Mil85**]. The simple argument to deduce the reverse Santaló inequality sketched in Exercise 7.33 is due to Pisier. Another related result due to Milman [**Mil86**] is the *reverse Brunn–Minkowski inequality*, which asserts the following: *for any symmetric convex body $B \subset \mathbb{R}^n$ there is a volume-preserving linear map $T_B \in \mathsf{SL}(n, \mathbb{R})$ such that, if K, L are symmetric convex bodies, then*

$$\mathrm{vol}\left(T_K(K) + T_L(L)\right)^{1/n} \leqslant C \left(\mathrm{vol}(K)^{1/n} + \mathrm{vol}(L)^{1/n}\right). \tag{7.32}$$

There is a close link with the M-ellipsoid and M-position introduced in (5.68), since (7.32) is easily seen to hold when $T_K(K)$ and $T_L(L)$ admit multiples of Euclidean balls as M-ellipsoids.

The results from AGA are classically presented in the real setting, but typically remain valid for complex spaces (or circled convex bodies) as well. This is the case for Theorems 7.42, 7.45 and 7.46. Often the proofs can be translated verbatim, with the notable exception of the Chevet–Gordon inequalities, for which no complex analogue is known. We also note that Pisier [**Pis89a**] obtained a proof of (7.32) via interpolation which works primarily in the complex setting (see Chapter 7 in [**Pis89b**]).

The theme of the approximation of zonoids by zonotopes with few summands attracted attention in the late nineteen eighties. The best result (Theorem 7.48) is due to Talagrand [**Tal90**] and improves on [**Sch87, BLM89**]. It is an open question whether Theorem 7.48 holds without the factor $\log n$, i.e., with $N \leqslant C(\varepsilon)n$.

The Johnson–Lindenstrauss lemma appeared in [**JL84**]. It was announced in [**LN16**] that the dependence on ε in the version presented here is optimal.

Theorem 7.50 is due to Ben-Tal and Nemirovski [**BTN01b**]; see also [**KTJ09**]. Problem 7.51 appears to be folklore. An analogous question for a vaguely similar restricted invertibility property (RIP), important in the theory of compressed sensing, was answered in the affirmative, see [**BDMS13**]. Table 7.1 comes from [**IS10**].

Theorem 7.52 is a special case of a result from [**Kar11**], which deals with the distortion of the ℓ_r-norm on the sphere of ℓ_p^n for any $0 < r < p < 2$. Theorem 7.53 is from [**Fre14**], which contains also a version of the Theorem for convex bodies that are only assumed to be invariant under permutation of coordinates.

Part 3

The Meeting: AGA and QIT

CHAPTER 8

Entanglement of pure states in high dimensions

Throughout this chapter, we consider a multipartite Hilbert space

$$\mathcal{H} = \mathbb{C}^{d_1} \otimes \cdots \otimes \mathbb{C}^{d_k}$$

and study the entanglement of pure states on $\mathcal{H}$. We will *always* assume that $k \geqslant 2$ and that $d_1, \ldots, d_k \geqslant 2$.

We identify pure states on $\mathcal{H}$ with elements of $\mathsf{P}(\mathcal{H})$, the projective space on $\mathcal{H}$. The set of product vectors forms the Segré variety Seg $\subset \mathsf{P}(\mathcal{H})$ (see (B.6) in Appendix B.2). A simple remark, on which we will elaborate, is that most pure states are entangled. Indeed, since the variety Seg $\subset \mathsf{P}(\mathcal{H})$ has lower dimension and measure zero, it follows that a randomly chosen—in any reasonable sense—pure state in $\mathcal{H}$ is almost surely entangled.

A problem which turns out to be fundamental to several constructions in QIT is to show the existence of large-dimensional subspaces of $\mathcal{H}$, in which every unit vector corresponds to an entangled pure state. There are several variations on this question. We may consider the qualitative version of the problem, where we require the subspace simply to contain no nonzero product vector (see Theorem 8.1). Alternatively, we may insist that the subspace contains only very entangled vectors, once it is specified how to quantify entanglement; for pure states this may be done via the von Neumann or Rényi entropy of the partial trace.

The versions of Dvoretzky's theorem that were discussed in Section 7.2 are obviously relevant to such questions, since they show the existence of large subspaces on which a given function is almost constant. This approach allows us to give a complete presentation of Hastings's counterexample to the additivity problem (Section 8.4.4).

Much of our exposition will be focused on detailed study of the bipartite case $\mathcal{H} = \mathbb{C}^k \otimes \mathbb{C}^d$ (we will always assume that $k \leqslant d$). One reason for such emphasis is the fact that subspaces of a bipartite Hilbert space can provide a convenient description of quantum channels through the Stinespring representation, as we explain in Section 8.2.2. Fine aspects of pure state entanglement in multipartite systems are dealt with in the last part of the chapter (Section 8.5).

8.1. Entangled subspaces: Qualitative approach

Let $\mathcal{H} = \mathbb{C}^{d_1} \otimes \mathbb{C}^{d_2}$. A fundamental *qualitative* question we may ask about entangled subspaces is: "What is the maximal dimension of a subspace of $\mathcal{H}$ in which every unit vector corresponds to an entangled pure state?" The answer to this question is $(d_1 - 1)(d_2 - 1)$, as shown by the following theorem, which also settles the multipartite case.

THEOREM 8.1. *Let* $\mathcal{H} = \mathbb{C}^{d_1} \otimes \cdots \otimes \mathbb{C}^{d_k}$, *and let* $n_0 = d_1 \cdots d_k - (d_1 + \cdots + d_k) + k - 1$. *Then*

(1) *If* $m > n_0$, *then any* m*-dimensional subspace of* $\mathcal{H}$ *contains a (nonzero) product vector.*
(2) *If* $m \leqslant n_0$, *a generic* m*-dimensional subspace of* $\mathcal{H}$ *contains no (nonzero) product vector.*

PROOF. We only give an argument for the second part of the Theorem (the first assertion can be proved via the projective dimension theorem from algebraic geometry). The proof is based on dimension counting, and we find it instructive to give a "probabilistic" version of dimension counting, which naturally fits in the general framework of this book. For simplicity, we only consider the case $\mathcal{H} = \mathbb{C}^d \otimes \mathbb{C}^d$ (so that $n_0 = (d-1)^2$), the general case being similar.

We work in the projective space $\mathsf{P}(\mathcal{H})$, which we equip with the Fubini–Study distance given by (B.5). The ball of center ψ and radius r is denoted by $B(\psi, r)$. We use bounds on the size of ε-nets and on the measure of ε-balls in $\mathsf{P}(\mathcal{H})$, which are extracted from Theorem 5.11 (and from Exercise 5.25; however, the more elementary results from Section 5.1.2 would actually suffice, cf. Exercise 5.10 and (5.2)). In this proof, as opposed to most material in this book, the dependence of constants on dimension *is* allowed, and we will denote by C, C' etc. positive constants which may depend on d and m, but are independent of the parameter ε.

Let F be a random m-dimensional subspace of $\mathcal{H}$, chosen with respect to the Haar measure on the Grassmann manifold. More concretely, we may realize F as $F = U(F_0)$, where F_0 is any fixed m-dimensional subspace, and U is a Haar-distributed unitary matrix. Recall that $\mathrm{Seg} \subset \mathsf{P}(\mathcal{H})$ is the set of product vectors (the Segré variety).

We are going to show that the event $\mathrm{Seg} \cap F = \varnothing$ has probability 1. Given $\varepsilon > 0$, let $\mathcal{M}_\varepsilon$ be an ε-net inside the projective space $\mathsf{P}(F_0)$ with $\mathrm{card}(\mathcal{M}_\varepsilon) \leqslant (C'/\varepsilon)^{2m-2}$. Next, let $\mathcal{N}_\varepsilon$ be an ε-net inside $\mathsf{P}(\mathbb{C}^d)$ with $\mathrm{card}(\mathcal{N}_\varepsilon) \leqslant (C'/\varepsilon)^{2d-2}$. One checks that $\mathcal{N}_\varepsilon^{\otimes 2} := \{x \otimes y \,:\, x, y \in \mathcal{N}_\varepsilon\}$ is a 2ε-net inside Seg. We use the union bound in the following way:

$$\begin{aligned}
\mathbf{P}(\mathrm{Seg} \cap F \neq \varnothing) &\leqslant \mathbf{P}\left(\bigcup_{\varphi \in \mathcal{N}_\varepsilon^{\otimes 2}} B(\varphi, 2\varepsilon) \cap U\left(\bigcup_{\psi \in \mathcal{M}_\varepsilon} B(\psi, \varepsilon)\right) \neq \varnothing \right) \\
&\leqslant \sum_{\varphi \in \mathcal{N}_\varepsilon^{\otimes 2}, \psi \in \mathcal{M}_\varepsilon} \mathbf{P}\left(B(\varphi, 2\varepsilon) \cap U(B(\psi, \varepsilon)) \neq \varnothing\right) \\
&\leqslant \sum_{\varphi \in \mathcal{N}_\varepsilon^{\otimes 2}, \psi \in \mathcal{M}_\varepsilon} \mathbf{P}\left(d(\varphi, U\psi) < 3\varepsilon\right).
\end{aligned}$$

The quantity $\mathbf{P}(d(\varphi, U\psi) < 3\varepsilon)$ does not depend on the particular points $\varphi, \psi \in \mathsf{P}(\mathcal{H})$, and is equal to the normalized measure of a ball of radius 3ε in $\mathsf{P}(\mathcal{H})$, which is bounded from above by $(C''\varepsilon)^{2d^2-2}$ (or see Exercise 5.11 for the exact value). Consequently,

$$\begin{aligned}
\mathbf{P}(\mathrm{Seg} \cap U(F_0) \neq \varnothing) &\leqslant \mathrm{card}(\mathcal{N}_\varepsilon^{\otimes 2})\, \mathrm{card}(\mathcal{M}_\varepsilon) (C''\varepsilon)^{2d^2-2} \\
&\leqslant C\varepsilon^{2d^2-2-(2m-2)-2(2d-2)}.
\end{aligned}$$

Provided $m \leqslant (d-1)^2$, the last quantity tends to 0 as ε tends to 0. This shows that the event $\{F \text{ intersects Seg}\}$ has probability 0, so that F contains no nonzero product vector. □

EXERCISE 8.1 (Universal entanglers). Show that whenever $d \geqslant 4$, a generic unitary matrix $U \in \mathsf{U}(d^2)$ has the property that for every product unit vector $\psi \in \mathbb{C}^d \otimes \mathbb{C}^d$, $U|\psi\rangle\langle\psi|U^\dagger$ is entangled.

8.2. Entropies of entanglement and additivity questions

8.2.1. Quantifying entanglement for pure states. The most common way to quantify the entanglement of a bipartite pure state is to use the entropy of entanglement (for operational meanings of the entropy of entanglement, we refer to Notes and Remarks).

Let $\psi \in \mathbb{C}^k \otimes \mathbb{C}^d$ be a unit vector. The *entropy of entanglement* of ψ, denoted by $E(\psi)$, is defined as the von Neumann entropy of the reduced matrix $\rho = \operatorname{Tr}_{\mathbb{C}^d} |\psi\rangle\langle\psi|$.

$$E(\psi) = S(\rho) = -\operatorname{Tr} \rho \log \rho. \tag{8.1}$$

Both parties play a symmetric role since the two reduced matrices $\operatorname{Tr}_{\mathbb{C}^d} |\psi\rangle\langle\psi|$ and $\operatorname{Tr}_{\mathbb{C}^k} |\psi\rangle\langle\psi|$ have the same von Neumann entropy (in the matrix formalism, a consequence of the fact that $MM^\dagger$ and $M^\dagger M$ have the same nonzero eigenvalues for $M \in \mathsf{M}_{k,d}$). If $\psi = \sum \lambda_i \varphi_i \otimes \chi_i$ is a Schmidt decomposition of ψ, then

$$E(\psi) = -\sum \lambda_i^2 \log \lambda_i^2 = -2\sum \lambda_i^2 \log \lambda_i. \tag{8.2}$$

For any $p \in [0, \infty]$, we introduce the p-entropy of entanglement, defined as

$$E_p(\psi) = S_p(\rho), \tag{8.3}$$

where $\rho = \operatorname{Tr}_{\mathbb{C}^d} |\psi\rangle\langle\psi|$ and S_p is the p-Rényi entropy introduced in Section 1.3.3. Recall that the case $p = 1$ corresponds to the von Neumann entropy, i.e., $E_1(\psi) = E(\psi)$ (as given by (8.1)). The limit cases $p = 0$ and $p = \infty$ should be interpreted as $E_0(\psi) = \log \operatorname{rank}(\psi)$ and $E_\infty(\psi) = -2 \log \max \lambda_1$, where $\operatorname{rank} \psi$ is the Schmidt rank of ψ and λ_1 its largest Schmidt coefficient.

Rényi entropies for $p > 1$ are easier to manipulate since they are closely related to Schatten norms. If we identify a vector $\psi \in \mathbb{C}^k \otimes \mathbb{C}^d$ with a matrix $M \in \mathsf{M}_{k,d}$ as explained in Section 0.8, we obtain (see (2.12))

$$\|\rho\|_p = \|M\|_{2p}^2, \tag{8.4}$$

and therefore

$$E_p(\psi) = \frac{p}{1-p} \log \|\rho\|_p = \frac{2p}{1-p} \log \|M\|_{2p}. \tag{8.5}$$

In all this chapter we assume that $k \leqslant d$, and therefore (for any $p \in [0, \infty]$) the p-entropy of entanglement varies between 0 and $\log k$. Moreover, a pure state ψ satisfies $E_p(\psi) = 0$ if and only if it is a product vector, and satisfies $E_p(\psi) = \log k$ if and only if it is a maximally entangled vector.

These definitions make sense only in the bipartite case, as they rely on the Schmidt decomposition of a bipartite pure state, which has no canonical analogue for the multipartite case. The limit case $p = \infty$ is different: E_∞ depends only on the largest Schmidt coeficient, which can be defined in a multipartite system as the maximal modulus of inner product (or the maximal overlap) with a product vector (cf. (2.13)). We elaborate on this in Section 8.5.

One of the goals of this chapter is to find subspaces $\mathcal{W} \subset \mathbb{C}^k \otimes \mathbb{C}^d$ which are very entangled, in the sense that the quantity $E(\psi)$ (or $E_p(\psi)$) has a uniform lower bound over all unit vectors $\psi \in \mathcal{W}$.

8.2.2. Channels as subspaces. A crucial insight allowing to relate analysis of quantum channels to high-dimensional convex geometry is the observation that there is an essentially one-to-one correspondence between channels and linear subspaces of composite Hilbert spaces. Specifically, let $\mathcal{W}$ be a subspace of $\mathbb{C}^k \otimes \mathbb{C}^d$ of dimension m. Then $\Phi : B(\mathcal{W}) \to \mathsf{M}_k$ defined by $\Phi(\rho) = \mathrm{Tr}_{\mathbb{C}^d}(\rho)$ is a quantum channel. Alternatively, and perhaps more properly, we could identify $\mathcal{W}$ with $\mathbb{C}^m$ via an isometry $V : \mathbb{C}^m \to \mathbb{C}^k \otimes \mathbb{C}^d$ whose range is $\mathcal{W}$ and define, for $\rho \in \mathsf{M}_m$, the corresponding channel $\Phi : \mathsf{M}_m \to \mathsf{M}_k$ by

$$\Phi(\rho) = \mathrm{Tr}_{\mathbb{C}^d}(V\rho V^\dagger). \tag{8.6}$$

There is no limitation in considering quantum channels of the form (8.6): by the Stinespring representation theorem (Theorem 2.24), any quantum channel $\Phi : \mathsf{M}_m \to \mathsf{M}_k$ can be represented via (8.6) for some subspace $\mathcal{W} \subset \mathbb{C}^k \otimes \mathbb{C}^d$, with $d = km$.

It is now easy to define a natural family of random quantum channels. They will be associated, via the above scheme, to random m-dimensional subspaces $\mathcal{W}$ of $\mathbb{C}^k \otimes \mathbb{C}^d$, distributed according to the Haar measure on the corresponding Grassmann manifold (for some fixed positive integers m, d, k that will be specified later). Note that most interesting parameters of a channel defined by (8.6) depend only on the subspace $\mathcal{W} = V(\mathbb{C}^m)$ and not on a particular choice of the isometry V (see, e.g., Lemma 8.2). In this sense, the language of "random m-dimensional subspaces of $\mathbb{C}^k \otimes \mathbb{C}^d$" is equivalent to that of "random isometries from $\mathbb{C}^m$ to $\mathbb{C}^k \otimes \mathbb{C}^d$", with the corresponding mathematical objects being, respectively, the closely related Grassmann manifolds and Stiefel manifolds (see Appendix B.4).

8.2.3. Minimal output entropy and additivity problems. Given a quantum channel $\Phi : \mathsf{M}_m \to \mathsf{M}_k$, we define its *minimum output entropy* as

$$S^{\min}(\Phi) = S_1^{\min}(\Phi) = \min_{\rho \in \mathrm{D}(\mathbb{C}^m)} S(\Phi(\rho)), \tag{8.7}$$

as well as the p-entropy variant for $p \geqslant 0$,

$$S_p^{\min}(\Phi) = \min_{\rho \in \mathrm{D}(\mathbb{C}^m)} S_p(\Phi(\rho)).$$

The following lemma shows that, for channels defined via (8.6), the minimum output entropy depends only on the range of the isometry V.

LEMMA 8.2. *Let $\Phi : \mathsf{M}_m \to \mathsf{M}_k$ a random channel, obtained by (8.6) from a Haar-distributed isometry $V : \mathbb{C}^m \to \mathbb{C}^k \otimes \mathbb{C}^d$. Then, for any $0 \leqslant p \leqslant \infty$,*

$$S_p^{\min}(\Phi) = \min_{\psi \in \mathcal{W}, |\psi|=1} E_p(\psi),$$

where $\mathcal{W} \subset \mathbb{C}^k \otimes \mathbb{C}^d$ is the range of V.

PROOF. Since the function S_p is concave (see Section 1.3.3), the minimum is achieved on a pure state (pure states are extreme points of $\mathrm{D}(\mathbb{C}^m)$). Consequently,

$$S_p^{\min}(\Phi) = \min_{\varphi \in S_{\mathbb{C}^m}} S_p(\Phi(|\varphi\rangle\langle\varphi|)) = \min_{\psi \in \mathcal{W} \,:\, |\psi|=1} S_p(\mathrm{Tr}_{\mathbb{C}^d} |\psi\rangle\langle\psi|),$$

and the result follows. □

For some time, an important open problem in quantum information theory was to decide whether the quantity $S^{\min}$ is additive, i.e., whether every pair (Φ, Ψ) of quantum channels satisfies

$$S^{\min}(\Phi \otimes \Psi) \stackrel{?}{=} S^{\min}(\Phi) + S^{\min}(\Psi). \tag{8.8}$$

The problem admits several equivalent formulations with operational meaning, notably whether entangled inputs can increase the capacity of a quantum channel to transmit classical information. (Note that the inequality "$\leqslant$" in (8.8) always holds and is easy, see Exercise 8.2.)

A similar question can be asked for the quantities $S_p^{\min}$, the motivation being that a positive answer to the $p > 1$ question would have implied a positive answer to the (arguably more important) $p = 1$ problem. However, it turns out that *all* these equalities do not hold, at least for sufficiently large dimensions.

THEOREM 8.3. *For any $p \geqslant 1$, there exist quantum channels Φ, Ψ such that*

$$S_p^{\min}(\Phi \otimes \Psi) < S_p^{\min}(\Phi) + S_p^{\min}(\Psi). \tag{8.9}$$

Theorem 8.3 will be a consequence of Proposition 8.6 (for $p > 1$) and Proposition 8.24 (for $p = 1$).

EXERCISE 8.2 ($S_p^{\min}$ is always subadditive). Show that the inequality $S_p^{\min}(\Phi \otimes \Psi) \leqslant S_p^{\min}(\Phi) + S_p^{\min}(\Psi)$ is satisfied for any channels Φ, Ψ and any $p \geqslant 0$.

EXERCISE 8.3 (Reduction of the additivity problem to the case $\Phi = \Psi$). A trick based on direct sums (as defined in (2.42)) allows a reduction to the case $\Phi = \Psi$ in questions such as (8.8).
(i) Given quantum channels Φ, Ψ, show that $S_p^{\min}(\Phi \oplus \Psi) = \min(S_p^{\min}(\Phi), S_p^{\min}(\Psi))$.
(ii) Assume that there is a pair of channels Φ, Ψ such that (8.9) holds for some p. Deduce formally the existence of a channel Ξ such that $S_p^{\min}(\Xi \otimes \Xi) < 2S_p^{\min}(\Xi)$.

8.2.4. On the $1 \to p$ norm of quantum channels. The $p > 1$ version of the additivity problem has a nice functional-analytic interpretation. If $p > 1$ and ρ is a state, then $S_p(\rho) = \frac{p}{1-p} \log \|\rho\|_p$, and so the study of $S_p^{\min}(\Phi)$ is replaced by that of $\max\{\|\Phi(\rho)\|_p \, : \, \rho \in \mathsf{D}(\mathbb{C}^m)\}$, or the *maximum output p-norm*. The latter quantity equals $\|\Phi\|_{1 \to p}$, i.e., the norm of Φ as an operator from $(\mathsf{M}_m^{\mathrm{sa}}, \|\cdot\|_1)$ to $(\mathsf{M}_k^{\mathrm{sa}}, \|\cdot\|_p)$. Therefore (8.9) is equivalent to

$$\|\Phi \otimes \Psi\|_{1 \to p} > \|\Phi\|_{1 \to p} \|\Psi\|_{1 \to p}. \tag{8.10}$$

A remarkable fact is that for completely positive maps (and even for 2-positive maps), the norm $\|\cdot\|_{1 \to p}$ is unchanged if we drop the self-adjointness constraint.

PROPOSITION 8.4. *Let $\Phi : \mathsf{M}_m \to \mathsf{M}_k$ be a 2-positive map, and $p \geqslant 1$. Then*

$$\sup_{X \in \mathsf{M}_m, \|X\|_1 = 1} \|\Phi(X)\|_p = \sup_{X \in \mathsf{M}_m^{\mathrm{sa}}, \|X\|_1 = 1} \|\Phi(X)\|_p. \tag{8.11}$$

We first show the following fact.

LEMMA 8.5. *If $A, B, C \in \mathsf{M}_k$ are such that the block matrix $M = \begin{bmatrix} A & B \\ B^\dagger & C \end{bmatrix}$ is positive semi-definite, then for every $p \geqslant 1$ we have $\|B\|_p^2 \leqslant \|A\|_p \|C\|_p$.*

PROOF. From the singular value decomposition, there exist unitary matrices $U, V \in \mathsf{U}(k)$ such that $UBV^\dagger$ is a diagonal matrix with nonnegative diagonal entries. Denote $W = U \oplus V \in \mathsf{U}(2k)$. We have

$$WMW^\dagger = \begin{bmatrix} UAU^\dagger & UBV^\dagger \\ VB^\dagger U^\dagger & VCV^\dagger \end{bmatrix}.$$

Since the Schatten norms are invariant under multiplication by unitaries, this shows that to prove the Lemma it is enough to treat the case when the matrix B is diagonal with nonnegative entries, which we consider now.

We first note that $b_{ii}^2 \leqslant a_{ii}c_{ii}$, which follows from the matrix $\begin{bmatrix} a_{ii} & b_{ii} \\ b_{ii} & c_{ii} \end{bmatrix}$ being positive as a submatrix of M. Consequently, we have

$$\|B\|_p^p = \sum_{i=1}^k b_{ii}^p \leqslant \sum_{i=1}^k a_{ii}^{p/2} c_{ii}^{p/2} \leqslant \left(\sum_{i=1}^k a_{ii}^p\right)^{1/2} \left(\sum_{i=1}^k c_{ii}^p\right)^{1/2} \leqslant \|A\|_p^{p/2} \|C\|_p^{p/2},$$

where the last inequality uses the fact that the diagonal is majorized by the spectrum (Lemma 1.14). □

PROOF OF PROPOSITION 8.4. For $\varphi, \psi \in S_{\mathbb{C}^m}$, consider $u = \varphi \otimes |1\rangle + \psi \otimes |2\rangle \in \mathbb{C}^m \otimes \mathbb{C}^2$. By direct calculation

$$\Phi \otimes \mathrm{Id}_{\mathsf{M}_2}(|u\rangle\langle u|) = \begin{bmatrix} \Phi(|\varphi\rangle\langle\varphi|) & \Phi(|\psi\rangle\langle\varphi|) \\ \Phi(|\varphi\rangle\langle\psi|) & \Phi(|\psi\rangle\langle\psi|) \end{bmatrix}.$$

Since Φ is 2-positive, the resulting matrix is block-positive and thus, by Lemma 8.5,

$$\|\Phi(|\psi\rangle\langle\varphi|)\|_p^2 \leqslant \|\Phi(|\psi\rangle\langle\psi|)\|_p \|\Phi(|\varphi\rangle\langle\varphi|)\|_p.$$

Taking supremum over unit vectors gives the required result (recall that extreme points of S_1^d and $S_1^{d,\mathrm{sa}}$ are rank 1 operators). □

EXERCISE 8.4 (The equality (8.11) does not always hold). Define $\Phi : \mathsf{M}_2 \to \mathsf{M}_2$ by $\Phi(X) = X - \mathrm{Tr}(X)\frac{\mathrm{I}}{2}$. Show that for $p > 1$, Φ fails to satisfy the equality (8.11). Known examples where (8.11) fails for $p = 1$ are more complicated; see [**Wat05**].

8.3. Concentration of E_p for $p > 1$ and applications

8.3.1. Counterexamples to the multiplicativity problem.

We first consider the case of the p-entropy of entanglement with $p > 1$, and show that the Dvoretzky theorem can be used to produce counterexamples to the multiplicativity problem as announced in Theorem 8.3.

PROPOSITION 8.6. *There is a constant c such that the following holds. Let $p > 1$, and $\Phi : \mathsf{M}_m \to \mathsf{M}_k$ be a random channel, obtained by (8.6) from a Haar-distributed isometry $V : \mathbb{C}^m \to \mathbb{C}^k \otimes \mathbb{C}^d$. Denote $\Psi = \overline{\Phi}$, the channel obtained from $\overline{V}$, the complex conjugate of V. Assume that $k = d$ and that $m = cd^{1+1/p}$. Then, for d large enough, with high probability,*

$$\|\Phi \otimes \Psi\|_{1\to p} > \|\Phi\|_{1\to p} \|\Psi\|_{1\to p}. \tag{8.12}$$

As in other similar situations, in order to not to obscure the arguments, above and in what follows we pretend that $cd^{1+1/p}$ and similar expressions are integers.

PROOF. Denote by $\mathcal{W} \subset \mathsf{M}_d$ the range of V (we may consider $\mathcal{W}$ as a subspace of M_d after we identify tensors and matrices). From (8.4) and Lemma 8.2, we have

$$\|\Phi\|_{1\to p} = \max_{A\in\mathcal{W}\,:\,\|A\|_{\mathrm{HS}}=1} \|A\|_{2p}^2. \tag{8.13}$$

We remark that $\|\Phi\|_{1\to p} = \|\Psi\|_{1\to p}$ since the Schatten norms are invariant under complex conjugation. We now appeal to Dvoretzky's theorem for the Schatten norm $\|\cdot\|_q$ with $q = 2p$. Provided that $m \leqslant cd^{1+2/q}$ for an appropriate universal constant $c > 0$, it follows from Theorem 7.37 that, with large probability,

$$d^{1/q-1/2}\|A\|_{\mathrm{HS}} \leqslant \|A\|_q \leqslant Cd^{1/q-1/2}\|A\|_{\mathrm{HS}}$$

for all $A \in \mathcal{W}$. We have therefore, by (8.13),

$$d^{1/p-1} \leqslant \|\Phi\|_{1\to p} = \|\Psi\|_{1\to p} \leqslant \left(Cd^{1/q-1/2}\right)^2 = C^2 d^{1/p-1}. \tag{8.14}$$

The reason for choosing $\overline{\Phi}$ as a second channel is that the channel $\Phi\otimes\overline{\Phi}$ necessarily has at least one output with at least one large eigenvalue, as shown by the following lemma.

LEMMA 8.7. *Let $\Phi : \mathsf{M}_m \to \mathsf{M}_k$ be a quantum channel obtained from an isometry $V : \mathbb{C}^m \to \mathbb{C}^k \otimes \mathbb{C}^d$, as in (8.6). Denote by $\psi \in \mathbb{C}^m \otimes \mathbb{C}^m$ the maximally entangled state*

$$\psi = \frac{1}{\sqrt{m}}\left(|1\rangle\otimes|1\rangle + \cdots + |m\rangle\otimes|m\rangle\right).$$

Then

$$\left\|(\Phi\otimes\overline{\Phi})(|\psi\rangle\langle\psi|)\right\|_\infty \geqslant \frac{m}{dk}$$

and consequently, for any $p > 1$,

$$\|\Phi\otimes\overline{\Phi}\|_{1\to p} \geqslant \|\Phi\otimes\overline{\Phi}\|_{1\to\infty} \geqslant \frac{m}{dk}.$$

In our setting, $d = k$ and $m = cd^{1+1/p}$, so we obtain from Lemma 8.7 the lower bound $\|\Phi\otimes\overline{\Phi}\|_{1\to p} = \Omega(d^{1/p-1})$. Since we have, by (8.14),

$$\|\Phi\|_{1\to p}\|\overline{\Phi}\|_{1\to p} = \|\Phi\|_{1\to p}^2 = \Theta(d^{2(1/p-1)}),$$

we conclude that the inequality (8.12) holds for d large enough (*a priori* depending on $p > 1$). □

REMARK 8.8. The proof shows that, for any fixed $p > 1$, both the multiplicative violation in (8.10) and the additive violation in (8.9) tend to infinity as the dimension of the problem increases (at the rates $\Omega(d^{1-1/p})$ and $\Omega(\log d)$ respectively).

PROOF OF LEMMA 8.7. We work in the matrix formalism. Identify the range of V with an m-dimensional subspace $\mathcal{W} \subset \mathsf{M}_{k,d}$. Let $(A_1, \ldots, A_m)$ be the orthonormal basis in $\mathcal{W}$ (with respect to the Hilbert–Schmidt inner product) obtained as the image under V of the canonical basis in $\mathbb{C}^m$, and

$$M = \frac{1}{\sqrt{m}}\sum_{i=1}^m A_i \otimes \overline{A_i} \in \mathcal{W}\otimes\overline{\mathcal{W}}.$$

The conclusion of the Lemma is equivalent to the inequality $\|M\|_\infty \geqslant \sqrt{m/kd}$.

Let $(\varphi_j)_{1\leqslant j\leqslant k}$ and $(\psi_{j'})_{1\leqslant j'\leqslant d}$ be orthonormal bases in $\mathbb{C}^k$ and $\mathbb{C}^d$, respectively. We consider the maximally entangled states

$$\varphi = \frac{1}{\sqrt{k}}\sum_{j=1}^{k}\varphi_j\otimes\overline{\varphi_j}, \quad \psi = \frac{1}{\sqrt{d}}\sum_{j'=1}^{d}\psi_{j'}\otimes\overline{\psi_{j'}}$$

and compute

$$\begin{aligned}\|M\|_\infty &\geqslant \left|\langle\psi|M|\varphi\rangle\right| \\ &= \frac{1}{\sqrt{mkd}}\sum_{i=1}^{m}\sum_{j=1}^{k}\sum_{j'=1}^{d}\left|\langle\psi_{j'}\otimes\overline{\psi_{j'}}|A_i\otimes\overline{A_i}|\varphi_j\otimes\overline{\varphi_j}\rangle\right| \\ &= \frac{1}{\sqrt{mkd}}\sum_{i=1}^{m}\sum_{j=1}^{k}\sum_{j'=1}^{d}\left|\langle\psi_{j'}|A_i|\varphi_j\rangle\right|^2 \\ &= \frac{\sqrt{m}}{\sqrt{kd}},\end{aligned}$$

where we used the fact that $\|X\|_{\mathrm{HS}}^2 = \sum_{j,j'}\left|\langle\psi_{j'}|X|\varphi_j\rangle\right|^2$. □

EXERCISE 8.5 (Non-random counterexamples for $p > 2$). Let $\mathcal{W} \subset \mathsf{M}_d$ be the subspace of anti-symmetric matrices, i.e., such that $A^T = -A$.
(i) Show that for any $A \in \mathcal{W}$ we have $\|A\|_\infty \leqslant \frac{1}{\sqrt{2}}\|A\|_{\mathrm{HS}}$.
(ii) Let Φ be the quantum channel constructed from $\mathcal{W}$ as in (8.6) and fix $p > 2$. Using Lemma 8.7, show that the pair $(\Phi, \overline{\Phi})$ is an example for which (8.10) holds for d large enough.

8.3.2. Almost randomizing channels.

A variant of the construction used in the proof of Proposition 8.6 for $p = +\infty$ gives the following: a channel $\Phi : \mathsf{M}_d \to \mathsf{M}_d$ constructed from a generic random embedding $V : \mathbb{C}^d \to \mathbb{C}^d \otimes \mathbb{C}^N$ with $N = O(d)$ has the property that $\|\Phi(\rho)\|_{\mathrm{op}} \leqslant C/d$ for any state $\rho \in \mathrm{D}(\mathbb{C}^d)$. In other words, all output states have small eigenvalues. It is natural to ask whether similar lower bounds of the eigenvalues of output states can also be achieved; showing that this is indeed the case is the content of this section. Recall also (see Section 2.3.3) that the dimension N of the environment in the Stinespring representation is an upper bound on the Kraus rank of Φ.

Let $0 < \varepsilon < 1$. A quantum channel $\Phi : \mathsf{M}_d \to \mathsf{M}_d$ is said to be ε-randomizing if for all states $\rho \in \mathrm{D}(\mathbb{C}^d)$

$$\|\Phi(\rho) - \rho_*\|_{\mathrm{op}} \leqslant \varepsilon/d.$$

Recall that $\rho_* = \mathrm{I}/d$ denotes the maximally mixed state. These channels can be thought as approximations of the completely randomizing channel R, which is defined by the property $R(\rho) = \rho_*$ for any $\rho \in \mathrm{D}(\mathbb{C}^d)$. The completely randomizing channel rank has Kraus rank equal to d^2 (see Exercise 8.6). On the other hand, it turns out that there exist ε-randomizing channels with a substantially smaller Kraus rank, as shown by the following theorem. The dependence on d is optimal since any ε-randomizing channel has Kraus rank at least d, which is due to the fact that rank one states must be mapped to full rank states.

THEOREM 8.9. *Let $(U_i)_{1 \leqslant i \leqslant N}$ be independent random matrices Haar-distributed on the unitary group* $\mathsf{U}(d)$. *Let $\Phi : \mathsf{M}_d \to \mathsf{M}_d$ be the quantum channel defined by*

$$\Phi(\rho) = \frac{1}{N} \sum_{i=1}^{N} U_i \rho U_i^\dagger .$$

Assume that $0 < \varepsilon < 1$ and $N \geqslant Cd/\varepsilon^2$. Then the channel Φ is ε-randomizing with high probability.

The proof of Theorem 8.9 is based on the following two lemmas.

LEMMA 8.10. *Let ρ and σ be pure states on $\mathbb{C}^d$ and let $(U_i)_{1 \leqslant i \leqslant N}$ be independent Haar-distributed random unitary matrices. Then, for every $0 < \delta < 1$,*

$$\mathbf{P}\left(\left| \frac{1}{N} \sum_{i=1}^{N} \operatorname{Tr}(U_i \rho U_i^\dagger \sigma) - \frac{1}{d} \right| \geqslant \frac{\delta}{d} \right) \leqslant 2\exp(-c\delta^2 N).$$

PROOF. Write $\rho = |\varphi\rangle\langle\varphi|$ and $\sigma = |\psi\rangle\langle\psi|$. Denote $X_i = d \operatorname{Tr}(U_i \rho U_i^\dagger \sigma) = \left|\sqrt{d}\langle\psi|U_i|\varphi\rangle\right|^2$. We know from Lemma 5.57 that this variable is subexponential (as the square of a subgaussian variable) and satisfies $\|X_i\|_{\psi_1} \leqslant C$. The conclusion follows now directly from Bernstein's inequalities (Proposition 5.59). □

LEMMA 8.11. *Let $\Delta : \mathsf{M}_d^{\mathrm{sa}} \to \mathsf{M}_d^{\mathrm{sa}}$ be a linear map. Let A be the quantity*

$$A = \sup_{\rho \in \mathrm{D}(\mathbb{C}^d)} \|\Delta(\rho)\|_{\mathrm{op}} = \sup_{\rho,\sigma \in \mathrm{D}(\mathbb{C}^d)} |\operatorname{Tr} \sigma \Delta(\rho)| .$$

Let $0 < \delta < 1/4$ and let $\mathcal{N}$ be a δ-net in $(S_{\mathbb{C}^d}, |\cdot|)$. Then $A \leqslant (1-4\delta)^{-1} B$, where

$$B = \sup_{\varphi,\psi \in \mathcal{N}} |\operatorname{Tr} |\psi\rangle\langle\psi| \Delta(|\varphi\rangle\langle\varphi|)| .$$

PROOF OF LEMMA 8.11. First note that for any $X, Y \in \mathsf{M}_d^{\mathrm{sa}}$ we have

$$|\operatorname{Tr} Y \Delta(X)| \leqslant A \|X\|_1 \|Y\|_1. \tag{8.15}$$

By a convexity argument, the supremum in A can be restricted to pure states. Given unit vectors $\varphi, \psi \in S_{\mathbb{C}^d}$, let $\varphi_0, \psi_0 \in \mathcal{N}$ so that $|\varphi - \varphi_0| \leqslant \delta$ and $|\psi - \psi_0| \leqslant \delta$. Given $\chi \in S_{\mathbb{C}^d}$, we write P_χ for $|\chi\rangle\langle\chi|$. We have

$$\|P_\varphi - P_{\varphi_0}\|_1 \leqslant \|P_\varphi - |\varphi\rangle\langle\varphi_0|\|_1 + \||\varphi\rangle\langle\varphi_0| - P_{\varphi_0}\|_1 \leqslant 2\delta,$$

and similarly $\|P_\psi - P_{\psi_0}\|_1 \leqslant 2\delta$ (this simple bound is not optimal). We now write

$$|\operatorname{Tr} P_\psi \Delta(P_\varphi)| \leqslant |\operatorname{Tr}(P_\psi - P_{\psi_0})\Delta(P_\varphi)| + |\operatorname{Tr} P_{\psi_0} \Delta(P_\varphi - P_{\varphi_0})| + |\operatorname{Tr} P_{\psi_0} \Delta(P_{\varphi_0})| .$$

Using twice (8.15) and taking supremum over φ, ψ gives $A \leqslant 2\delta A + 2\delta A + B$, hence the result. □

PROOF OF THEOREM 8.9. Fix a $\frac{1}{8}$-net $\mathcal{N} \subset (S_{\mathbb{C}^d}, |\cdot|)$ with $\operatorname{card} \mathcal{N} \leqslant 16^{2d}$, as provided by Lemma 5.3. Let $\Delta = R - \Phi$ and A, B as in Lemma 8.11. Here A and B are random quantities and it follows from Lemma 8.11 that

$$\mathbf{P}\left(A \geqslant \frac{\varepsilon}{d}\right) \leqslant \mathbf{P}\left(B \geqslant \frac{\varepsilon}{2d}\right).$$

Using the union bound and Lemma 8.10, we get

$$\mathbf{P}\left(B \geqslant \frac{\varepsilon}{2d}\right) \leqslant 16^{4d} \cdot 2\exp(-c\varepsilon^2 N/4).$$

This is less than 1 if $N \geqslant Cd/\varepsilon^2$, for some constant C. □

EXERCISE 8.6 (Kraus decomposition of the completely randomizing channel). (i) Show that the Kraus rank of the completely randomizing channel R is d^2.
(ii) Let $\omega = \exp(2i\pi/d)$ and A, B be the unitary operators defined by their action on the canonical basis by

$$A|j\rangle = |j+1 \bmod d\rangle, \quad B|j\rangle = \omega^j |j\rangle. \tag{8.16}$$

Show that the operators $(B^j A^k)_{1 \leqslant j,k \leqslant d}$ give a Kraus decomposition of R. These operators are sometimes called the Heisenberg–Weyl operators.

8.4. Concentration of von Neumann entropy and applications

8.4.1. The basic concentration argument.

We now consider the von Neumann entropy (instead of the p-Rényi entropy) as the invariant quantifying entanglement. Since the von Neumann entropy is not naturally associated with a norm, we are going to use the version of the Dvoretzky theorem for Lipschitz functions (Theorem 7.15). The relevant function is the entropy of entanglement $\psi \mapsto E(\psi)$, defined (via (8.1)) on the unit sphere in $\mathbb{C}^k \otimes \mathbb{C}^d$. As usual in such situations, we need two pieces of information: the Lipschitz constant of $E(\cdot)$ and a central value. They are provided by the next two lemmas.

LEMMA 8.12. *The Lipschitz constant of the function $\psi \mapsto E(\psi)$, defined on $(S_{\mathbb{C}^k\otimes\mathbb{C}^d}, |\cdot|)$, is bounded from above by $C \log k$ for some absolute constant C.*

This is clearly optimal up to the value of the constant C, since the function E maps $S_{\mathbb{C}^k\otimes\mathbb{C}^d}$ (which has diameter π, or $\pi/2$ if we consider E as a function on $\mathsf{P}(\mathbb{C}^k\otimes\mathbb{C}^d)$) onto the segment $[0, \log k]$. (Remember that in this chapter we always assume $k \leqslant d$.) Note that, in view of (B.1), it doesn't matter—apart from the value of the constant—whether we use the geodesic distance or the extrinsic distance. For a discussion of the optimal values of the constants see Exercise 8.7.

PROOF. We first check the commutative case by considering the function $f : S^{k-1} \to [0, \log k]$ defined by

$$f(x) = -\sum x_i^2 \log(x_i^2), \tag{8.17}$$

i.e., the Shannon entropy of the probability distribution $(x_i^2) \in \Delta_k$. In the terminology of (8.2), this is equivalent to restricting attention to vectors ψ whose Schmidt decompositions use fixed sequences (φ_i), (χ_i). One computes

$$|\nabla f(x)|^2 = 4\sum_{i=1}^k x_i^2(1+\log(x_i^2))^2 \leqslant C \log^2 k, \tag{8.18}$$

where the last inequality can be obtained by observing that the function $t \mapsto t(1+\log t)^2$ is concave on $[0, e^{-2}]$, and so the quantity $|\nabla f(x)|$ increases when we replace the coordinates of x smaller than e^{-1} by their ℓ_2 average. It follows that if L is the Lipschitz constant of f with respect to the geodesic distance on S^{k-1}, then $L \leqslant C^{1/2} \log k$. Our objective is to show is that the same constant works for the function $\psi \mapsto E(\psi)$.

To that end, we will consider an auxiliary function which is defined as follows. Let $(u_i)_{1\leqslant i\leqslant k}$ be an orthonormal basis of $\mathbb{C}^k$. If $\psi \in S_{\mathbb{C}^k\otimes\mathbb{C}^d}$, set $\rho = \operatorname{Tr}_{\mathbb{C}^d} |\psi\rangle\langle\psi|$ and let

$$\tilde{f}(\psi) = -\sum_{i=1}^k \langle u_i|\rho|u_i\rangle \log(\langle u_i|\rho|u_i\rangle). \tag{8.19}$$

In other words, $\tilde{f}(\psi)$ is the entropy of the diagonal part of ρ, calculated in the basis (u_i). An important property of $\tilde{f}$ is that $\tilde{f}(\psi) = S(\rho)$ if (u_i) is a basis which diagonalizes ρ (which is obvious from the definitions) and $\tilde{f}(\psi) \geqslant S(\rho)$ in general (which is a consequence of concavity of S and is the content of Exercise 1.50). Next, one verifies that $\langle u_i|\rho|u_i\rangle = |P_i\psi|$, where P_i is the orthogonal projection onto the subspace $u_i \otimes \mathbb{C}^d \subset \mathbb{C}^k \otimes \mathbb{C}^d$. Since the map $\psi \mapsto (|P_1\psi|, \dots, |P_k\psi|)$ is a contraction, it follows that the Lipschitz constant of $\tilde{f}$ (with respect to g, the geodesic distance on $S_{\mathbb{C}^k\otimes\mathbb{C}^d}$) is at most L.

We now return to the original question. Let $\psi_1, \psi_2 \in S_{\mathbb{C}^k\otimes\mathbb{C}^d}$; set $\rho_k = \operatorname{Tr}_{\mathbb{C}^d} |\psi_k\rangle\langle\psi_k|$ and let $\tilde{f}$ be defined by (8.19) using a basis (u_i) which diagonalizes ρ_2. Then

$$E(\psi_1) - E(\psi_2) = S(\rho_1) - S(\rho_2) = S(\rho_1) - \tilde{f}(\psi_2) \leqslant \tilde{f}(\psi_1) - \tilde{f}(\psi_2) \leqslant L\, g(\psi_1, \psi_2).$$

Since the roles of ψ_1 and ψ_2 can be reversed, it follows that the Lipschitz constant of E with respect to g is at most L (and hence exactly L), as claimed. □

LEMMA 8.13 (Not proved here; see Remark 8.14). *For $k \leqslant d$, the expectation of the function $\psi \mapsto E(\psi)$ (with respect to the uniform measure on the unit sphere in $\mathbb{C}^k \otimes \mathbb{C}^d$) satisfies*

$$\mathbf{E}\, E(\psi) = \left(\sum_{j=d+1}^{kd} \frac{1}{j} \right) - \frac{k-1}{2d} \geqslant \log k - \frac{1}{2}\frac{k}{d}. \tag{8.20}$$

REMARK 8.14 (An easy bound on the entropy of entanglement). An inequality slightly weaker than (8.20) follows readily from Proposition 6.36 (or Exercise 6.43, which is even more elementary). First, with large probability, all Schmidt coefficients of ψ belong to the interval

$$\left[\frac{1}{\sqrt{k}} - \frac{C}{\sqrt{d}}, \frac{1}{\sqrt{k}} + \frac{C}{\sqrt{d}} \right]$$

for some constant C. It follows that all the eigenvalues of the $\operatorname{Tr}_{\mathbb{C}^d} |\psi\rangle\langle\psi|$ lie then in an interval $\left[\frac{1-\varepsilon}{k}, \frac{1+\varepsilon}{k}\right]$ for some $\varepsilon = O\big(\sqrt{k/d}\big)$, and Lemma 1.20 yields the bound $E(\psi) = S(\rho) \geqslant \log k - C'k/d$. (The use of Lemma 1.20 requires $\varepsilon \leqslant 1$, for larger ε we may use the simpler bound $S(\rho) \geqslant S_\infty(\rho) = -\log \|\rho\|_\infty$.)

An immediate consequence of Dvoretzky's theorem (in the form from Theorem 7.15) follows.

THEOREM 8.15. *Let $\varepsilon > 0$ and $m \leqslant c\varepsilon^2 kd/\log^2 k$. Then most m-dimensional subspaces $\mathcal{W} \subset \mathbb{C}^k \otimes \mathbb{C}^d$ have the property that any unit vector $x \in \mathcal{W}$ satisfies*

$$E(x) \geqslant \log k - \frac{1}{2}\frac{k}{d} - \varepsilon.$$

In some cases the result given by Theorem 8.15 can be improved. In particular, in order to obtain violations for the additivity of $S_{\min}$ we will need to produce "extremely entangled subspaces", in which every state has entropy $\log(k) - o(1)$ (see Section 8.4.3).

In the opposite direction, Exercise 8.9 shows an upper bound on the minimal entropy inside *any* subspace of given dimension.

EXERCISE 8.7 (Sharp bounds for the Lipschitz constant of E). In the notation of Lemma 8.12, assume $k \leqslant d$ and let $L = L_k$ be the Lipschitz constant of the function $\psi \mapsto E(\psi)$, calculated with respect to the geodesic distance on $S_{\mathbb{C}^k \otimes \mathbb{C}^d}$ (or on $\mathsf{P}(\mathbb{C}^k \otimes \mathbb{C}^d)$). Show that $L_k \sim \log k$.

EXERCISE 8.8. Show that any s-dimensional subspace $F \subset \mathbb{C}^n$ contains a unit vector x satisfying $\|x\|_\infty \geqslant \sqrt{s/n}$.

EXERCISE 8.9 (An upper bound on the minimal entropy for general subspaces). Let $\mathcal{W} \subset \mathbb{C}^k \otimes \mathbb{C}^d$ be a subspace of dimension αkd, with $\alpha \geqslant 1/k$. (i) Using the previous exercise, show that $\mathcal{W}$ contains a unit vector ψ satisfying $E(\psi) \leqslant h(\alpha) + (1-\alpha)\log(k-1)$, where $h(t) = -t\log t - (1-t)\log(1-t) \leqslant \log 2$ is the binary entropy function. (ii) Conclude that if $\lambda \geqslant 1$ and $E(\psi) \geqslant \log k - \lambda/k$ for all $\psi \in \mathcal{W}$, then $\dim \mathcal{W} = O\big(\lambda d/(1+\log\lambda)\big)$.

8.4.2. Entangled subspaces of small codimension. The argument from the previous section gives nothing for subspaces of dimension cdk or larger: if $\varepsilon = \log d$, the conclusion of Theorem 8.15 does not even imply nonnegativity of $E(x)$. However, in view of Theorem 8.1, it seems plausible to quantify entanglement on subspaces of larger dimension. This can be achieved provided we use a suitable measure of entanglement.

One possibility is to use the p-Rényi entropy for $p = 1/2$. Recall from (8.5) that if we identify a unit vector $x \in \mathbb{C}^k \otimes \mathbb{C}^d$ with $A \in \mathsf{M}_{k,d}$, then

$$E_{1/2}(x) = 2\log\|A\|_1,$$

and our problem becomes a question about the behavior of $\|\cdot\|_1$ vs. $\|\cdot\|_2$ on subspaces of $\mathsf{M}_{k,d}$.

THEOREM 8.16. *Let $k \leqslant d$, and $\mathcal{W} \subset \mathbb{C}^k \otimes \mathbb{C}^d$ be a random subspace of dimension m. The following holds with large probability: for every unit vector $x \in \mathcal{W}$ we have*

$$E_{1/2}(x) \geqslant \log(k - m/d) - C.$$

The conclusion of Theorem 8.16 yields nontrivial quantitative information for subspaces of codimension larger than $C_1 d$, for some constant C_1. This compares well with Theorem 8.1, which asserts that subspaces of codimension smaller than $d+k-1$ are *never* fully entangled.

PROOF. We identify $\mathbb{C}^k \otimes \mathbb{C}^d$ with $\mathsf{M}_{k,d}$, and apply the low M^*-estimate (Theorem 7.45) to the norm $\|\cdot\|_1$. One needs the value of $M^* := \mathbf{E}\,\|X\|_{\mathrm{op}}$, where X is uniformly distributed on the Hilbert–Schmidt sphere in $\mathsf{M}_{k,d}$. The inequality $M^* \leqslant C/\sqrt{k}$ follows Proposition 6.36. Denoting $\alpha = 1 - m/kd$, we conclude that, for every $A \in \mathcal{W}$,

$$\|A\|_1 \geqslant c\sqrt{k}\sqrt{\alpha}\|A\|_{\mathrm{HS}}$$

and therefore, for every unit vector $x \in \mathcal{W}$ (now seen as a subspace of $\mathbb{C}^k \otimes \mathbb{C}^d$),

$$E_{1/2}(x) = 2\log\|A\|_1 \geqslant \log(k - m/d) - C. \qquad \square$$

8.4.3. Extremely entangled subspaces. In a different direction, we might seek subspaces of not-so-large dimension, but with near-maximal entropy of entanglement, say $\log k - o(1)$ for example. In view of Lemma 8.13, this requires $k = o(d)$. For simplicity, we will focus on the case $d = k^2$. This choice of dimensions allows us to produce an example of a pair of channels violating the additivity relation (8.8), although the method is applicable to a wider range of parameters.

PROPOSITION 8.17. *There are absolute constants c, C such that the following holds. Let k be an integer and set $d = k^2$, $m = ck^2$. With large probability, a random m-dimensional subspace $\mathcal{W} \subset \mathbb{C}^k \otimes \mathbb{C}^d$ has the property that any unit vector $\psi \in \mathcal{W}$ satisfies*

$$E(\psi) \geqslant \log k - \frac{C}{k}.$$

REMARK 8.18. Proposition 8.17 is optimal in the following sense. First, we cannot hope for larger values of $E(\psi)$ on a random subspace since (by Lemma 8.13) the *global* average value is precisely of order $\log k - \frac{C}{k}$. Second, subspaces of dimension larger than Ck^2 cannot have this property, as shown by Exercise 8.9 (ii).

We start by relating the entropy of very mixed states to their Hilbert–Schmidt distance to the maximally mixed state ρ_* (cf. Lemma 1.20, which leads to a slightly stronger conclusion under stronger hypothesis).

LEMMA 8.19. *If ρ is any state on $\mathbb{C}^k$, then*

$$S(\rho) \geqslant \log k - k \left\|\rho - \rho_*\right\|_{\mathrm{HS}}^2 .$$

PROOF. The following inequality compares the entropy with a second order approximation: for every $x, t \in [0, 1]$,

$$-x\log x \geqslant -t\log t - (1+\log t)(x-t) - \frac{1}{t}(x-t)^2. \tag{8.21}$$

To check inequality (8.21), notice that it can be rewritten (after some work) as $\log(y) \leqslant y - 1$ with $y = x/t$. Given a state $\rho \in \mathrm{D}(\mathbb{C}^k)$ with eigenvalues $(p_i)_{1\leqslant i\leqslant k}$, we apply (8.21) with $x = p_i$ and $t = 1/k$. Summing over i, we obtain the announced inequality. □

It will be more convenient to work with a random matrix $\mathsf{M} \in \mathsf{M}_{k,d}$ of Hilbert–Schmidt norm 1, rather than with a random unit vector $\psi \in \mathbb{C}^k \otimes \mathbb{C}^d$ (both approaches are equivalent, see Section 0.8). Also recall that when a vector ψ is identified with a matrix M, we have $\mathrm{Tr}_{\mathbb{C}^d} |\psi\rangle\langle\psi| = MM^\dagger$, see (2.12).

Here is a proposition which (via Lemma 8.19) immediately implies Proposition 8.17.

PROPOSITION 8.20. *There are absolute constants c, C such that the following holds. Let k be an integer, $d = k^2$, $m = ck^2$ and let S_{HS} be the Hilbert–Schmidt sphere in $\mathsf{M}_{k,d}$. Consider the function $g : S_{\mathrm{HS}} \to \mathbb{R}$ defined by*

$$g(M) = \left\| MM^\dagger - \frac{\mathrm{I}}{k} \right\|_{\mathrm{HS}} .$$

With large probability, a random m-dimensional subspace $\mathcal{W} \subset \mathsf{M}_{k,d}$ has the property that

$$\sup_{M \in S_{\mathrm{HS}} \cap \mathcal{W}} g(M) \leqslant C/k. \tag{8.22}$$

REMARK 8.21. We wish to point out that while Proposition 8.20 will be *derived from* Dvoretzky's theorem for Lipschitz functions, it can be *rephrased* in the language of the standard Dvoretzky theorem. Indeed, its assertion says that for every $M \in \mathcal{W}$ with $\|M\|_{\mathrm{HS}} = 1$ we have

$$\frac{C^2}{k^2} \geqslant \left\| MM^\dagger - \frac{\mathrm{I}}{k} \right\|_{\mathrm{HS}}^2 = \operatorname{Tr}|M|^4 - \frac{2\operatorname{Tr} MM^\dagger}{k} + \frac{\operatorname{Tr}\mathrm{I}}{k^2} = \operatorname{Tr}|M|^4 - \frac{1}{k} \geqslant 0. \tag{8.23}$$

Consequently,

$$k^{-1/4}\|M\|_{\mathrm{HS}} \leqslant \|M\|_4 \leqslant k^{-1/4}\Big(1 + \frac{C^2}{k}\Big)^{1/4}\|M\|_{\mathrm{HS}} \leqslant k^{-1/4}\Big(1 + \frac{C^2}{4k}\Big)\|M\|_{\mathrm{HS}} \tag{8.24}$$

for all $M \in \mathcal{W}$. In other words, $\mathcal{W}$ is $(1+\delta)$-Euclidean, with $\delta = \frac{C^2}{4k}$, when considered as a subspace of the Schatten normed space $(\mathsf{M}_{k,d}, \|\cdot\|_4)$. On the other hand, the Dvoretzky dimension of $(\mathsf{M}_{k,d}, \|\cdot\|_4)$ equals $k^{1/2}d$ (see Theorem 7.37) and therefore the general theory (such as Theorem 7.19) gives only $\delta = O(k^{-1/4})$ for m-dimensional subspaces. Although the Dvoretzky dimension is sharp for the size of isomorphically Euclidean subspaces (in the sense exemplified in Exercises 7.12 and 7.25), (8.24) supplies an instance where it can be beaten for almost isometrically Euclidean subspaces.

Before embarking on the proof of Proposition 8.20 we offer some preliminary remarks. We know from Proposition 6.36 (the elementary argument from Exercise 6.43 would actually be sufficient) that all singular values of a typical $M \in S_{\mathrm{HS}}$ belong to the interval

$$\left[\frac{1}{\sqrt{k}} - \frac{C}{\sqrt{d}}, \frac{1}{\sqrt{k}} + \frac{C}{\sqrt{d}}\right]. \tag{8.25}$$

It follows that $\|MM^\dagger - \mathrm{I}/k\|_\infty = O(k^{-3/2})$ and thus the median M_g of g satisfies $M_g \leqslant C/k$. We next estimate the Lipschitz constant of g. The inequality

$$\|MM^\dagger - NN^\dagger\|_{\mathrm{HS}} \leqslant \|M(M^\dagger - N^\dagger) + (M-N)N^\dagger\|_{\mathrm{HS}} \leqslant (\|M\|_{\mathrm{op}} + \|N\|_{\mathrm{op}})\|M-N\|_{\mathrm{HS}}$$

has the following immediate consequence.

LEMMA 8.22. *Let $\Omega_t = \{M \in S_{\mathrm{HS}} : \|M\|_{\mathrm{op}} \leqslant t\}$ for some $t \geqslant 0$. The function defined on Ω_t by $M \mapsto MM^\dagger$ is $2t$-Lipschitz with respect to the Hilbert–Schmidt norm.*

In particular, the function g is 2-Lipschitz on $\Omega_1 = S_{\mathrm{HS}}$. However, a direct application of Theorem 7.15 yields only a bound of order $1/\sqrt{k}$ in (8.22). (This calculation parallels the one from Remark 8.21 that was expressed in the alternative language of the Dvoretzky dimension.) The trick is to apply concentration of measure twice: to the function g itself, and to the function $f : M \mapsto \|M\|_{\mathrm{op}}$, which is used to control the Lipschitz constant of g.

The function f is 1-Lipschitz on S_{HS}. By (8.25), its median equals $1/\sqrt{k} + O(1/k)$; in particular it is bounded by $2/\sqrt{k}$ for k large enough. Consequently, Lévy's lemma (Corollary 5.17) implies that

$$\mathbf{P}\left(f(M) \geqslant 3/\sqrt{k}\right) \leqslant \frac{1}{2}\exp(-k^2). \tag{8.26}$$

Similarly, an application of the standard Dvoretzky's theorem (Theorem 7.19) to the norm $\|\cdot\|_\infty$ with $\varepsilon = 1/\sqrt{k}$ (note that the dimension of the ambient space is $n = kd$ and that the Dvoretzky dimension is of order d, see Theorem 7.37) shows that a random ck^2-dimensional subspace $\mathcal{W}$ satisfies $S_{\mathrm{HS}} \cap \mathcal{W} \subset \Omega_{3/\sqrt{k}}$ with high probability.

Starting from this point, we will present two possible paths to complete the proof of Proposition 8.20. The first argument uses twice the general Dvoretzky theorem for Lipschitz functions (Theorem 7.15) with the optimal dependence on ε. The second argument is based on a trick due to Fukuda making the overall argument more elementary. In terms of the hierarchy discussed at the beginning of Section 6.1, the first proof we give uses principles from level (ii), namely the Dudley inequality, whereas the second argument uses a single ε-net, staying at level (i).

PROOF #1 OF PROPOSITION 8.20. We know from Lemma 8.22 that the function g is $2t$-Lipschitz on Ω_t. Let $\tilde{g}$ be a $2t$-Lipschitz extension of $g_{|\Omega}$ to S_{HS}. Note that, in any metric space X, it is possible to extend any L-Lipschitz function h defined on a subset Y without increasing the Lipschitz constant; use, e.g., the formula

$$\tilde{h}(x) = \inf_{y \in Y} \left[h(y) + L \operatorname{dist}(x, y) \right].$$

This formula also guarantees that the extended function $\tilde{g}$ is circled. Since $\tilde{g} = g$ on most of S_{HS}, the median of g (resp., $\tilde{g}$) is a central value of $\tilde{g}$ (resp., g). We apply Theorem 7.19 to $\tilde{g}$ with $\varepsilon = 1/k$, $\mu = M_g$ and $L = 2t = 6k^{-1/2}$ to get

$$\sup_{S_{\mathrm{HS}} \cap \mathcal{W}} |\tilde{g} - \mu| \leqslant 1/k$$

on a random subspace $\mathcal{W} \subset \mathsf{M}_{k,d}$ of dimension $m = c_0 \cdot kd \cdot (k^{-1}/(6k^{-1/2}))^2 = cd$. We then have

$$\sup_{S_{\mathrm{HS}} \cap \mathcal{W}} \tilde{g} \leqslant \mu + \frac{1}{k} \leqslant \frac{C'}{k}.$$

If $S_{\mathrm{HS}} \cap \mathcal{W} \subset \Omega$ (which, as noticed before, holds with large probability), g and $\tilde{g}$ coincide on $S_{\mathrm{HS}} \cap \mathcal{W}$ and therefore $g \leqslant C'/k$ on $S_{\mathrm{HS}} \cap \mathcal{W}$, proving (8.22). □

PROOF #2 OF PROPOSITION 8.20. We use the following lemma which allows to discretize the supremum in (8.22).

LEMMA 8.23. *Let $\mathcal{N}$ be an ε-net in $(S_{\mathrm{HS}} \cap \mathcal{W}, |\cdot|)$ with $\varepsilon < \sqrt{2} - 1$. Then*

$$\sup_{M \in S_{\mathrm{HS}} \cap \mathcal{W}} g(M) \leqslant \frac{1}{1 - \varepsilon^2 - 2\varepsilon} \sup_{M \in \mathcal{N}} g(M).$$

PROOF OF LEMMA 8.23. Let $M \in S_{\mathrm{HS}} \cap \mathcal{W}$. There exists $M_0 \in \mathcal{N}$ such that $\delta := \|M - M_0\|_{\mathrm{HS}} \leqslant \varepsilon$. We write $M = M_0 + \delta N$ with $N \in S_{\mathrm{HS}}$, and consider also $A = M_0 + N$ and $B = M_0 - N$ (note that the operators N, A and B all belong to $\mathcal{W}$). One checks that $\|A\|_{\mathrm{HS}}^2 = 2 - \delta$ and $\|B\|_{\mathrm{HS}}^2 = 2 + \delta$. We then set

$$\Delta := MM^\dagger - MM_0^\dagger = \frac{\delta}{2} \left(AA^\dagger - BB^\dagger + 2\delta NN^\dagger \right),$$

and the triangle inequality implies

$$\|\Delta\|_{\mathrm{HS}} \leqslant \frac{\delta}{2} \left(\|AA^\dagger - (2-\delta)\rho_*\|_{\mathrm{HS}} - \|BB^\dagger - (2+\delta)\rho_*\|_{\mathrm{HS}} + \|2\delta NN^\dagger - 2\delta\rho_*\|_{\mathrm{HS}} \right).$$

We can thus estimate

$$\begin{aligned} g(M) &\leqslant g(M_0) + \|MM^\dagger - M_0M_0^\dagger\|_{\mathrm{HS}} \\ &\leqslant g(M_0) + \frac{\delta}{2}\left((2-\delta)g(A/\|A\|_{\mathrm{HS}}) + (2+\delta)g(B/\|B\|_{\mathrm{HS}}) + 2\delta g(N)\right) \\ &\leqslant g(M_0) + (2\delta + \delta^2) \sup_{X \in S_{\mathrm{HS}} \cap \mathcal{W}} g(X) \\ &\leqslant g(M_0) + (2\varepsilon + \varepsilon^2) \sup_{X \in S_{\mathrm{HS}} \cap \mathcal{W}} g(X), \end{aligned}$$

and taking supremum over $M \in S_{\mathrm{HS}}$ gives the result. □

We now return to the proof of the Proposition. The random subspace is realized as $\mathcal{W} = V(\mathbb{C}^m)$ where $V : \mathbb{C}^m \to \mathsf{M}_{k,d}$ is a Haar-distributed isometry. If $\mathcal{M}$ is an ε-net in $(S_{\mathbb{C}^m}, |\cdot|)$, then $\mathcal{N} = V(\mathcal{M})$ is an ε-net in $(S_{\mathrm{HS}} \cap \mathcal{W}, |\cdot|)$. Let us choose (for example) $\varepsilon = 1/3$; by Lemma 5.3, we can ensure that $\operatorname{card} \mathcal{N} \leqslant 36^m$.

We apply the "local Lévy lemma" (Corollary 5.35) to the function g with the subset $\Omega = \Omega_{3/\sqrt{k}} \subset S_{\mathrm{HS}}$ and $\varepsilon = 1/k$. The function $g_{|\Omega}$ is $6/\sqrt{k}$-Lipschitz, and therefore, using (8.26),

$$\mathbf{P}(\{g > M_g + 1/k\}) \leqslant \mathbf{P}(S_{\mathrm{HS}} \subset \Omega) + 2\exp(-d/36) \leqslant C\exp(-cd).$$

Using the union bound and Lemma 8.23, this gives

$$\mathbf{P}\left(\sup_{M \in S_{\mathrm{HS}} \cap \mathcal{W}} g(M) \geqslant \frac{9}{2}(M_g + 1/k) \right) \leqslant 36^m C \exp(-cd),$$

and this quantity is (much) smaller than 1 provided $m \leqslant c'd$, for sufficiently small $c' > 0$. Since $M_g = O(1/k)$, this concludes the proof. □

8.4.4. Counterexamples to the additivity problem. Using Proposition 8.17 and the approach used in Proposition 8.6 for the p-Rényi entropy, we can show the following.

PROPOSITION 8.24. *There is a constant c such that the following holds. Let $d = k^2$, $m = ck^2$ and $\Phi : \mathsf{M}_m \to \mathsf{M}_d$ be a random channel, obtained by (8.6) from a Haar-distributed isometry $V : \mathbb{C}^m \to \mathbb{C}^d \otimes \mathbb{C}^d$. Set $\Psi = \overline{\Phi}$, the channel obtained from $\overline{V}$, the complex conjugate of V. If k is large enough, then with large probability,*

$$S^{\min}(\Phi \otimes \Psi) < S^{\min}(\Phi) + S^{\min}(\Psi).$$

PROOF. Denote by $\mathcal{W} \subset \mathbb{C}^k \otimes \mathbb{C}^d$ the range of V. From Lemma 8.2, we have

$$S_{\min}(\Phi) = \min_{\psi \in \mathcal{W}, |\psi|=1} E(\psi).$$

Note that $S_{\min}(\Phi) = S_{\min}(\Psi)$. From Proposition 8.17, we have with large probability

$$S_{\min}(\Phi) \geqslant \log k - \frac{C}{k}.$$

On the other hand, we know from Lemma 8.7 that applying $\Phi \otimes \overline{\Phi}$ to the maximally entangled state yields an output state with an eigenvalue greater than or equal to $\frac{\dim \mathcal{W}}{\dim \mathsf{M}_{k,d}} = \frac{m}{kd} = \frac{c}{k}$. Then, a simple argument using just concavity of S (see

Proposition 1.19) reduces the problem to calculating the entropy of the state with one eigenvalue *equal* to $\frac{c}{k}$ and all the remaining ones identical, which yields

$$S_{\min}(\Phi \otimes \Psi) \leqslant 2\log k - \frac{c\log k}{k} + \frac{1}{k}.$$

We have therefore $S^{\min}(\Phi\otimes\Psi) < S^{\min}(\Phi)+S^{\min}(\Psi)$ provided k is large enough. □

8.5. Entangled pure states in multipartite systems

8.5.1. Geometric measure of entanglement. The definition of the p-entropy of entanglement relies on the Schmidt decomposition, which is specific to the bipartite case. However, the case $p = \infty$ is different since its definition only involves the largest Schmidt coefficient, and this quantity can be defined in a multipartite setting as the square of the maximal overlap with a product vector. In the multipartite setting, the corresponding "∞-entropy of entanglement" has been introduced in the QIT literature via the *geometric measure of entanglement*.

Let $\mathcal{H} = \mathcal{H}_1 \otimes \cdots \otimes \mathcal{H}_k$ be a multipartite real or complex Hilbert space. Given a unit vector $\psi \in \mathcal{H}$, the geometric measure of entanglement of ψ is defined as

$$g(\psi) = \max\left\{\left|\langle\psi, \psi_1 \otimes \cdots \otimes \psi_k\rangle\right| \;:\; \psi_i \text{ unit vector in } \mathcal{H}_i, 1 \leqslant i \leqslant k\right\} \tag{8.27}$$

(cf. (2.13)) and the ∞-entropy of entanglement is

$$E_\infty(\psi) = -2\log g(\psi). \tag{8.28}$$

We always have $E_\infty(\psi) \geqslant 0$, and $E_\infty(\psi)$ is equal to 0 if and only if ψ is a product vector. Therefore, it makes sense to call unit vectors ψ which maximize $E_\infty(\psi)$ "maximally entangled" vectors. In the bipartite case $\mathbb{C}^d \otimes \mathbb{C}^d$, one recovers the usual notion of a maximally entangled state (see Section 2.2.4). However, in the multipartite case it seems hard to describe the maximally entangled vectors. The problem has an immediate geometric reformulation.

PROPOSITION 8.25 (Easy). *Let $\mathcal{H} = \mathcal{H}_1\otimes\cdots\otimes\mathcal{H}_k$. The following numbers are equal.*
(i) The minimal value of $g(\psi)$ over all unit vectors $\psi \in \mathcal{H}$.
(ii) The inradius of $B_{\mathcal{H}_1} \widehat{\otimes} \cdots \widehat{\otimes} B_{\mathcal{H}_k}$, where $B_{\mathcal{H}_i}$ denotes the unit ball in $\mathcal{H}_i$.
(iii) The largest constant c such that any k-linear map $\phi : \mathcal{H}_1 \times \cdots \times \mathcal{H}_k \to \mathbb{C}$ satisfies

$$c|||\phi||| \leqslant \max\{|\phi(x_1,\ldots,x_k)| \;:\; |x_1| \leqslant 1,\ldots,|x_k| \leqslant 1\},$$

where $|||\cdot|||$ denotes the norm

$$|||\phi\,||^2 = \sum_{x_1\in\mathscr{B}_1} \cdots \sum_{x_k\in\mathscr{B}_k} |\phi(x_1,\ldots,x_k)|^2$$

with $\mathscr{B}_i$ an orthonormal basis in $\mathcal{H}_i$ (the value of $|||\cdot|||$ does not depend on the choice of the bases).

Denote by $g_{\min}(\mathcal{H})$ the common value of the numbers appearing in Proposition 8.25. There is a simple lower bound on $g_{\min}(\mathcal{H})$.

LEMMA 8.26. *If $\mathcal{H} = \mathbb{C}^{d_1}\otimes\cdots\otimes\mathbb{C}^{d_k}$ or $\mathcal{H} = \mathbb{R}^{d_1}\otimes\cdots\otimes\mathbb{R}^{d_k}$ with $d_1 \leqslant \cdots \leqslant d_k$, then $g_{\min}(\mathcal{H}) \geqslant 1/\sqrt{d_1\cdots d_{k-1}}$. Equivalently, for every unit vector $\psi \in \mathcal{H}$,*

$$E_\infty(\psi) \leqslant \log(d_1) + \cdots \log(d_{k-1}).$$

PROOF OF LEMMA 8.26. The same argument works for the real case and the complex case; we prove the Lemma by induction on k. For $k = 2$, we have

$$g_{\min}(\mathbb{C}^{d_1} \otimes \mathbb{C}^{d_2}) = \frac{1}{\min(\sqrt{d_1}, \sqrt{d_2})},$$

which is a restatement on the inequalities between the trace norm and the Hilbert–Schmidt norm on the space of $d_1 \times d_2$ matrices. For the induction step, we use the bound (which is again the $k = 2$ case)

$$g_{\min}(\mathbb{C}^{d_1} \otimes \mathcal{H}) \geqslant \frac{1}{\sqrt{d_1}} g_{\min}(\mathcal{H}).$$ □

8.5.2. The case of many qubits. We will now focus, for simplicity, on the particular case of k qubits, i.e., $d_1 = d_2 = \cdots = d_k = 2$ in the complex case.

In this section it is convenient to define entropy via logarithm to the base 2 and so we will exceptionally use $E_\infty^{(2)}(\psi) := -2\log_2 g(\psi)$ (cf. (8.28)). In this notation, the conclusion of Lemma 8.26 can be rewritten as follows: for any pure state $\psi \in (\mathbb{C}^2)^{\otimes k}$, we have $E_\infty^{(2)}(\psi) \leqslant k - 1$. The following seems to be unknown.

PROBLEM 8.27. *Does there exist a constant C, and for each k a unit vector $\psi \in (\mathbb{C}^2)^{\otimes k}$, such that*

$$E_\infty^{(2)}(\psi) \geqslant k - C\ ?$$

The next proposition shows that random states are typically very entangled, but not entangled enough to give a positive answer to Problem 8.27.

PROPOSITION 8.28. *There exist absolute constants c, C such that a uniformly distributed random unit vector $\psi \in (\mathbb{C}^2)^{\otimes k}$ satisfies with high probability*

$$c\frac{\sqrt{k \log k}}{2^{k/2}} \leqslant g(\psi) \leqslant C\frac{\sqrt{k \log k}}{2^{k/2}}.$$

The conclusion of Proposition 8.28 can be equivalently rewritten as

$$k - \log(k) - \log\log(k) - C' \leqslant E_\infty^{(2)}(\psi) \leqslant k - \log(k) - \log\log(k) + C'.$$

PROOF OF PROPOSITION 8.28. The average of g over the unit sphere is exactly the mean width of $K = (B_{\mathbb{C}^2})^{\hat{\otimes} k}$ (we think of $(\mathbb{C}^2)^{\otimes k}$ as a 2^{k+1}-dimensional real space). The concentration of the functional g around its mean follows from Lévy's lemma (see Table 5.2). Indeed, since K is contained in the unit ball, the functional $g = w(K, \cdot)$ is 1-Lipschitz and therefore

$$\mathbf{P}(|g(\psi) - w(K)| > t) \leqslant 2\exp(-2^k t^2).$$

It remains to show that $w(K) = \Theta(\sqrt{k \log k}\, 2^{-k/2})$, or equivalently that $w_G(K) = \Theta(\sqrt{k \log k})$. The upper bound follows from a standard ε-net argument: let $\mathcal{N}$ be an ε-net in $(S_{\mathbb{C}^2}, |\cdot|)$ with $\operatorname{card}\mathcal{N} \leqslant (2/\varepsilon)^4$ (see Lemma 5.3). From Exercise 5.7 (the weaker result from Lemma 5.9 would be enough here), it follows that $\operatorname{conv}\mathcal{N} \supset (1 - \varepsilon^2/2)B_{\mathbb{C}^2}$. Consequently, denoting by $\mathcal{N}^{\otimes k}$ the set

$$\mathcal{N}^{\otimes k} = \{\psi_1 \otimes \cdots \otimes \psi_k \ :\ \psi_i \in \mathcal{N} \text{ for } 1 \leqslant i \leqslant k\},$$

we have

$$\operatorname{conv}(\mathcal{N}^{\otimes k}) \supset (1 - \varepsilon^2/2)^k K.$$

Using Lemma 6.1, we conclude that

$$w_G(\mathrm{conv}(\mathcal{N}^{\otimes k})) \leqslant \sqrt{2\,\mathrm{card}(\mathcal{N}^{\otimes k})} \leqslant \sqrt{8k\log(2/\varepsilon)}.$$

Choosing $\varepsilon = 1/\sqrt{k}$ gives the upper bound $w_G(K) = O(\sqrt{k\log k})$.

To show that this argument is sharp, we are going to construct large separated sets in K. Start with a set $\mathcal{M} = \{x_1, \ldots, x_N\}$ which is $1/\sqrt{k}$-separated in the projective space over $\mathbb{C}^2$, with $N = \mathrm{card}(\mathcal{M}) \geqslant ck$. (The estimate on the size of separated sets in $\mathsf{P}(\mathbb{C}^2)$ is an elementary special case of Theorem 5.11 or Exercise 5.10; note that $\mathsf{P}(\mathbb{C}^2)$ identifies with the Bloch sphere, a 2-dimensional Euclidean sphere of radius 1/2, if we use the metric (B.5).) This means that, for $i \neq j$, we have $|\langle x_i, x_j\rangle| \leqslant 1 - 1/2k$.

We claim that a large subset of $\mathcal{M}^{\otimes k}$ is separated. To construct it, introduce $Q = \{1, \ldots, N\}^k$, equipped with the normalized Hamming metric, defined for $\alpha, \beta \in Q$ by

$$d(\alpha, \beta) = \frac{1}{k}\,\mathrm{card}\{i \;:\; \alpha_i \neq \beta_i\}.$$

To each element $\alpha = (\alpha_1, \ldots, \alpha_k) \in Q$ we associate the vector

$$x_\alpha = x_{\alpha_1} \otimes \cdots \otimes x_{\alpha_k} \in K.$$

When $\alpha, \beta \in Q$ are such that $d(\alpha, \beta) \geqslant k/10$, we have

$$|\langle x_\alpha, x_\beta\rangle| = \prod_{j=1}^{k} |\langle x_{\alpha_j}, x_{\beta_j}\rangle| \leqslant (1 - 1/2k)^{k/10} \leqslant c$$

for some constant $c < 1$. We then have $|x_\alpha - x_\beta| \geqslant c' := \sqrt{2 - 2c} > 0$. If we start from a subset $\mathcal{Q} \subset Q$ which is $k/10$-separated, the set $\{x_\alpha \;:\; \alpha \in \mathcal{Q}\}$ is c'-separated in $(\mathbb{C}^2)^{\otimes k}$. By the Sudakov inequality (Proposition 6.10), we have then

$$w_G(K) \geqslant c\sqrt{\log \mathrm{card}\,\mathcal{Q}}.$$

It remains to give a lower bound on the size of $\mathcal{Q}$. Using the inequality (5.17) from Chapter 5 (which was obtained by the greedy packing algorithm), we obtain $\mathrm{card}\,\mathcal{Q} \geqslant N^{k(1-H_N(1/5))} \geqslant N^{c''k}$ for some constant $c'' > 0$. It follows that $w_G(K) \geqslant c\sqrt{k\log k}$. □

8.5.3. Multipartite entanglement in real Hilbert spaces. It turns out that, in the real case, Lemma 8.26 is surprisingly sharp, so that the real version of Problem 8.27 has a positive answer with $C = 1$. The construction from Proposition 8.29 seems to be specific to the real case. For variants related to Clifford algebras, see Exercise 8.10.

PROPOSITION 8.29. *For any integers $k \geqslant 1$, we have*

$$g_{\min}((\mathbb{R}^2)^{\otimes k}) = 2^{-(k-1)/2}.$$

PROOF OF PROPOSITION 8.29. The inequality $g_{\min}((\mathbb{R}^2)^{\otimes k}) \geqslant 2^{-(k-1)/2}$ is a consequence of Lemma 8.26. Using Proposition 8.25(iii), the converse inequality will follow provided we show the existence of a k-linear form $\phi : (\mathbb{R}^2)^k \to \mathbb{R}$ such that $|\phi(x_1, \ldots, x_k)| \leqslant 1$ for unit vectors $x_1, \ldots, x_k$, and $|||\phi||| = 2^{(k-1)/2}$. Let $\theta : \mathbb{R}^2 \to \mathbb{C}$ be the canonical isomorphism. It is easily verified that

$$\phi : (x_1, \ldots, x_k) \mapsto \mathrm{Re}\prod_{i=1}^{k} \theta(x_i)$$

(where $\prod$ means complex multiplication) satisfies the desired conclusion. $\square$

EXERCISE 8.10 (Clifford matrices and multipartite maximally entangled states). Given $d \geqslant 2$, let N such that $\mathsf{M}_N(\mathbb{R})$ contains a d-dimensional subspace E in which every matrix is a multiple of an isometry (the smallest possible N is described in Theorem 11.4). Show that

$$g_{\min}((\mathbb{R}^d)^{\otimes k}) \leqslant \frac{\sqrt{N}}{d^{k/2}}. \tag{8.29}$$

When $d \in \{2, 4, 8\}$, one can achieve $N = d$ and the upper bound (8.29) matches the lower bound from Lemma 8.26.

Notes and Remarks

Section 8.1. Theorem 8.1 was proved in [**Wal02, Par04, WS08**]. The statement from Exercise 8.1 is taken from [**CDJ$^+$08**].

Section 8.2. There are multiple operational motivations to use the von Neumann entropy when defining the entropy of entanglement in (8.1). This is because, given a bipartite state ρ, there are several ways to quantify how much entanglement it contains. Two approaches that are in some sense extremal and dual to each other are the entanglement of distillation (the rate at which one can LOCC-transform copies of ρ into Bell states, see also Chapter 12) and the entanglement cost (the rate at which one can LOCC-transform Bell states into copies of ρ). For a general survey on entanglement measures we refer to [**PV07**]. If we restrict ourselves to **pure** states as we do in this chapter, all these entanglement measures coincide with the entropy of entanglement (see Chapter 12.5.2 in [**NC00**]).

The "additivity conjecture" (8.8) has been a major open problem in QIT, particularly since work by Shor [**Sho04**], who showed that the additivity of the minimum output von Neumann entropy was equivalent to the additivity of several other quantities, including the capacity of quantum channels to carry classical information and the entanglement of formation (defined later in Section 10.3.1). For example, the entire ICM 2006 talk by A. Holevo [**Hol06**] was devoted to this circle of ideas. A positive answer would have greatly simplified the theory, leading to a "single letter" formula for the aforementioned capacity, see, e.g., [**Hol06**]. However, the answer to the conjecture was shown to be negative by Hastings [**Has09**].

Exercise 8.3 is based on [**FW07**].

Proposition 8.4 was proved in [**Wat05, Aud09, Sza10**]. We follow here the argument from [**Sza10**].

Our presentation in this chapter barely scratches the surface of the topic of quantum channel capacities. In the quantum context, there are many notions of capacity (see, e.g., [**Wil17**]) and each of them leads to its own class of mathematical questions. For a recent overview of applications of operator space theory to the problem of estimating quantum capacity (i.e., the capacity to carry quantum information), see [**LJL15**].

Section 8.3. The question of the multiplicativity of $\|\cdot\|_{1\to p}$ (8.10) has been considered in [**WH02**] and solved in [**HW08**]. The presentation in the text is based on [**ASW10**], where the connection to Dvoretzky's theorem was noticed. It is also known that $\|\cdot\|_{1\to p}$ is not multiplicative for p close to 0 [**CHL$^+$08**], but part of the range $0 \leqslant p < 1$ is not covered by any approach. The explicit example from Exercise 8.5 comes from [**GHP10**].

Modulo the optimal dependence on the dimension, Theorem 8.9 concerning ε-randomizing channels has been proved in [**HLSW04**]; the parasitic logarithmic factor has been removed in [**Aub09**]. A step towards derandomization has also been made in [**Aub09**], where it was shown that the unitaries in question can be sampled from any Kraus decomposition of the completely randomizing channel.

Section 8.4. Lemma 8.12 appears in [**HLW06**] with the value $C = \sqrt{8}/\log 2$. The argument leading to a better constant ($C_k \sim 1$) in Lemma 8.12 that is sketched in Exercise 8.7 was an unpublished byproduct of the work on [**ASW11**]. For various aspects of continuity of the von Neumann entropy, see [**Win16**].

The exact formula (8.20) from Lemma 8.13 has been conjectured in [**Pag93**] and proved in [**FK94, SR95, Sen96**]. Having the precise form (as opposed to the weaker version stated in Remark 8.14) results in better constants in Theorem 10.16 in Section 10.3.1.

Theorem 8.16 appears to be new.

After Hastings's counterexample to the additivity conjecture [**Has09**] appeared, several papers tried to simplify and extend the original approach, including [**BH10, FKM10, FK10, ASW11, Fuk14**]. We follow mostly [**ASW11**]; Lemma 8.23 and the second proof of Proposition 8.20 are from [**Fuk14**].

A completely different strategy was used in a series of papers initiated by Collins–Nechita [**CN10, CN11**] via free probability and allows us to derive results which are more precise in some regimes. Here is a sample theorem from [**BCN12, CFN15**]. *Fix an integer k and $t \in (0,1)$. There is a deterministic convex set $K_{k,t} \subset \mathrm{D}(\mathbb{C}^k)$ with the following property: if $\Phi : \mathsf{M}_m \to \mathsf{M}_k$ is a quantum channel obtained from a random embedding $V : \mathbb{C}^m \to \mathbb{C}^k \otimes \mathbb{C}^d$ with $m = tkd$, then, almost surely as $d \to \infty$, the set $\Phi(\mathrm{D}(\mathbb{C}^m))$ converges to $K_{k,t}$.* This allows us, at least in principle, to answer any question about minimal output entropies of generic channels in this range of parameters. It was subsequently shown in [**BCN16**] that generic channels violating additivity can be obtained by following this strategy if and only if $k \geqslant 183$. Moreover, the defect of non-additivity, i.e., the difference between the two sides of (8.9) is generically almost $\log 2$ for large k (or 1 bit if we use $\log_2$ to define entropy). This improves on the preceding arguments—including the one presented in the text—which showed a violation that was minuscule. Still, in contrast with the Hayden–Winter example [**HW08**] (cf. Remark 8.8), the demonstrated violation does not go to infinity as the dimensions increase. A drawback of the free probability-based method is that the results are valid only when the environment dimension d goes to infinity, and obtaining explicit values of d, for which these asymptotic phenomena hold, requires extra analysis, which is not supplied in [**BCN16**]. For more information on this approach we refer to the survey [**CN16**]. Still another approach, due to Collins [**Col16**] and perhaps more conceptual, relies on the Haagerup inequality about the norms of convolutions on the free group.

In the opposite direction, it is proved in [**Mon13**] that random quantum channels satisfy a weak form of multiplicativity.

Section 8.5. The geometric measure of entanglement was considered under a different terminology in [**Shi95, BL01**]; see also [**WG03**]. Lemma 8.26 is well-known and appears for example in [**AS06, JHK$^+$08, Arv09**].

We could not locate Problem 8.27 in the literature although it seems a very natural question. It is known that $E_\infty(\psi) < k-1$ for any unit vector $\psi \in (\mathbb{C}^2)^{\otimes k}$ whenever $k \geqslant 3$ (see [**JHK$^+$08**]). The fact that random states are very entangled (the upper bound from Proposition 8.28) has been noticed and used in [**GFE09**, **BMW09**].

The argument behind Proposition 8.29 and Exercise 8.10 was communicated to us by Mikael de la Salle (see also Theorem 3.3 in [**Hil07a**]). The papers [**Hil06**, **Hil07a**] compute also the exact values $g_{\min}((\mathbb{R}^3)^{\otimes 4}) = 1/\sqrt{7}$ and $g_{\min}((\mathbb{R}^3)^{\otimes 4}) = 1/\sqrt{21}$.

CHAPTER 9

Geometry of the set of mixed states

Let $\mathcal{H} = \mathcal{H}_1 \otimes \cdots \otimes \mathcal{H}_k$ be a multipartite Hilbert space. We are interested in the geometry of the set of separable states on $\mathcal{H}$, and related questions. To simplify the exposition we are going to focus on two specific cases: the bipartite case $\mathcal{H} = \mathbb{C}^{d_1} \otimes \mathbb{C}^{d_2}$ (we may restrict ourselves to the *balanced* case $d_1 = d_2 = d$ in order to keep notation simple) and the case of k qubits $\mathcal{H} = (\mathbb{C}^2)^{\otimes k}$. However, essentially all the methods carry over to the general case, except that the formulas may sometimes become not very elegant (see, for example, Theorem 9.12). The sets $\mathrm{D} = \mathrm{D}(\mathcal{H})$, $\mathrm{Sep} = \mathrm{Sep}(\mathcal{H})$ and $\mathrm{PPT} = \mathrm{PPT}(\mathcal{H})$ were defined in Chapter 2. Recall that $\mathrm{Sep} \subset \mathrm{PPT} \subset \mathrm{D}$. One of the main goals of this chapter is to produce a table (Table 9.1) which contains radii estimates for these sets of states, similar to Table 4.1 for the classical examples of convex bodies. Table 9.2 matches estimates from Table 9.1 to the corresponding theorems in the text.

TABLE 9.1. Radii estimates for sets of quantum states. In each row n denotes the dimension of the corresponding Hilbert space. The first columns reads as $\mathrm{D} = \mathrm{D}(\mathbb{C}^n)$, $\mathrm{Sep} = \mathrm{Sep}(\mathbb{C}^d \otimes \mathbb{C}^d)$, $\mathrm{PPT} = \mathrm{PPT}(\mathbb{C}^d \otimes \mathbb{C}^d)$ and $\mathrm{Sep}' = \mathrm{Sep}((\mathbb{C}^2)^{\otimes k})$. The notation Θ^* indicates a two-sided estimate up to multiplicative factors polynomial in $\log n$. References to precise statements can be found in Table 9.2. Quantities in each row are non-decreasing from left to right, see Exercise 4.51, Proposition 2.5, and Proposition 2.18. (This gives in particular non-matching two-sided bounds for the missing entry in the last row.)

K	n	$\mathrm{inrad}(K)$	$w(K^\circ)^{-1}$	$\mathrm{vrad}(K)$	$w(K)$	$\mathrm{outrad}(K)$
D	n	$\frac{1}{\sqrt{n(n-1)}}$	$\sim \frac{1}{2\sqrt{n}}$	$\sim \frac{\exp(-1/4)}{\sqrt{n}}$	$\sim \frac{2}{\sqrt{n}}$	$\sqrt{\frac{n-1}{n}}$
Sep	d^2	$\frac{1}{\sqrt{n(n-1)}}$	$\Theta^*\left(n^{-3/4}\right)$	$\Theta\left(n^{-3/4}\right)$	$\Theta\left(n^{-3/4}\right)$	$\sqrt{\frac{n-1}{n}}$
PPT	d^2	$\frac{1}{\sqrt{n(n-1)}}$	$\Theta(n^{-1/2})$	$\Theta\left(n^{-1/2}\right)$	$\Theta\left(n^{-1/2}\right)$	$\sqrt{\frac{n-1}{n}}$
Sep′	2^k	$\Theta\left(n^{-1.292\ldots}\right)$	??	$\Theta^*\left(n^{-1.094\ldots}\right)$	$\Theta^*\left(n^{-1}\right)$	$\sqrt{\frac{n-1}{n}}$

We next clarify the statements about the radii appearing in Table 9.1. They are all computed with respect to the Hilbert–Schmidt Euclidean structure. Both inradii and outradii are computed for Hilbert–Schmidt balls centered at the maximally mixed state ρ_*. This choice of a center is optimal: one may argue that the optimal center can be chosen to be invariant under isometries of the convex set, and this property characterizes ρ_* (see Propositions 2.5 and 2.18, cf. Exercise 4.51 and its hint). Statements referred to as trivial in Table 9.2 follow from (2.7).

TABLE 9.2. References for proofs of the results from Table 9.1.

K	$\operatorname{inrad}(K)$	$w(K^\circ)^{-1}$	$\operatorname{vrad}(K)$, $w(K)$	$\operatorname{outrad}(K)$
$\mathrm{D}(\mathbb{C}^n)$	trivial	use (1.26)	Theorem 9.1	trivial
$\mathrm{Sep}(\mathbb{C}^d \otimes \mathbb{C}^d)$	Theorem 9.15	Theorem 9.6	Theorem 9.3	trivial
$\mathrm{PPT}(\mathbb{C}^d \otimes \mathbb{C}^d)$	trivial	Theorem 9.13	Theorem 9.13	trivial
$\mathrm{Sep}((\mathbb{C}^2)^{\otimes k})$	Theorem 9.21	unknown	Theorem 9.11	trivial

Some arguments require us to consider the affine space H_1 of trace one Hermitian matrices as a vector space with ρ_* as the origin. In order to emphasize this point of view we use a specialized notation: if $\rho \in H_1$ and $t \in \mathbb{R}$, then we write

$$t \bullet \rho := t\rho + (1-t)\rho_*. \tag{9.1}$$

If $K \subset H_1$, we denote $t \bullet K = \{t \bullet x \,:\, x \in K\}$. A similar caveat applies to polarity calculated inside the space H_1.

It is a remarkable fact that, despite sharing the same inradii and outradii, the sets Sep and D behave so differently with respect to volume radius. In particular, the proportion of states on $\mathbb{C}^d \otimes \mathbb{C}^d$ which are separable, when measured in terms of volume, is extremely small, of order $\exp(-cd^4 \log d)$. We will return to such considerations in Chapter 10.

9.1. Volume and mean width estimates

In this section, we prove the volume radius and mean width estimates from Table 9.1. In particular, we compute (up to a logarithmic factor) the mean width of Sep° (Theorem 9.6), which will play a crucial role in Chapter 10.

9.1.1. Symmetrization.

We heavily use the symmetrization operations defined in Section 4.1.2. Recall that, on a multipartite Hilbert space $\mathcal{H} = \mathcal{H}_1 \otimes \cdots \otimes \mathcal{H}_k$, we have

$$\mathrm{Sep}(\mathcal{H})_{\mathcal{O}} = \mathrm{D}(\mathcal{H}_1)_{\mathcal{O}} \,\hat{\otimes} \cdots \hat{\otimes}\, \mathrm{D}(\mathcal{H}_k)_{\mathcal{O}},$$

and that $\mathrm{D}(\mathcal{H}_i)_{\mathcal{O}}$ is the unit ball for the space $(B^{\mathrm{sa}}(\mathcal{H}_i), \|\cdot\|_1)$.

The Rogers–Shephard inequality (Theorem 4.22) controls how much the volume changes after symmetrization. In our context (i.e., $\mathcal{H} = \mathbb{C}^d \otimes \mathbb{C}^d \leftrightarrow \mathbb{C}^n$, $\dim \mathrm{D} = \dim \mathrm{Sep} = n^2 - 1$), it implies the inequalities

$$\frac{2}{\sqrt{n}} \operatorname{vol}_{n^2-1}(\mathrm{D}) \leqslant \operatorname{vol}_{n^2}(\mathrm{D}_{\mathcal{O}}) \leqslant \frac{2^{n^2}}{n^{5/2}} \operatorname{vol}_{n^2-1}(\mathrm{D}), \tag{9.2}$$

$$\frac{2}{\sqrt{n}} \operatorname{vol}_{n^2-1}(\mathrm{Sep}) \leqslant \operatorname{vol}_{n^2}(\mathrm{Sep}_{\mathcal{O}}) \leqslant \frac{2^{n^2}}{n^{5/2}} \operatorname{vol}_{n^2-1}(\mathrm{Sep}). \tag{9.3}$$

9.1.2. The set of all quantum states.

THEOREM 9.1. *Let* $\mathrm{D} = \mathrm{D}(\mathbb{C}^n)$ *be the set of states on* $\mathbb{C}^n$. *The volume of* D *equals*

$$\operatorname{vol}(\mathrm{D}) = \sqrt{n}\,(2\pi)^{n(n-1)/2} \frac{\prod_{j=1}^n \Gamma(j)}{\Gamma(n^2)}, \tag{9.4}$$

and satisfies the two-sided estimates

$$\frac{1}{2\sqrt{n}} \leqslant \operatorname{vrad}(\mathrm{D}) \leqslant \frac{1}{\sqrt{n}}. \tag{9.5}$$

The mean width of D *satisfies the asymptotic estimate* $w(\mathrm{D}) \sim 2/\sqrt{n}$ *when* $n \to \infty$. *Moreover, the upper bound* $w(\mathrm{D}) \leqslant 2/\sqrt{n}$ *holds for every dimension* n.

PROOF. We do not derive the exact value (9.4). From there, a tedious but routine calculation based on the Stirling formula gives then the asymptotic behavior of $\operatorname{vrad}(\mathrm{D})$ in Table 9.1.

Alternatively, we present a "soft" way to prove (9.5). First, we know from the Santaló inequality (Theorem 4.17) that $\operatorname{vrad}(\mathrm{D})\operatorname{vrad}(\mathrm{D}^\circ) \leqslant 1$. On the other hand, $\mathrm{D}^\circ = (-n) \bullet \mathrm{D}$ (see (1.26), recall that polarity is with respect to ρ_*). This gives the upper bound in (9.5).

For the lower bound, consider the symmetrization $\mathrm{D}_\oslash = S_1^{n,\mathrm{sa}}$, the unit ball with respect to the trace norm. Since $\|\cdot\|_1 \leqslant \sqrt{n}\|\cdot\|_{\mathrm{HS}}$, the inradius of $\mathrm{D}_\oslash$ equals $1/\sqrt{n}$ and therefore $\operatorname{vrad}(\mathrm{D}_\oslash) \geqslant 1/\sqrt{n}$. We may now appeal to the Rogers–Shephard inequality (9.2) to obtain the lower bound $\operatorname{vrad}(\mathrm{D}) \geqslant \frac{1}{2\sqrt{n}}$ (this requires some numerical verification since the convex bodies D and $\mathrm{D}_\oslash$ live in different dimensions, leading to different powers in the definition of the volume radii).

We now compute the Gaussian mean width of D. If A_n is a $\mathrm{GUE}_0(n)$ random matrix, then

$$w_G(\mathrm{D}) = \mathbf{E} \sup_{\rho \in \mathrm{D}} \operatorname{Tr}(A_n\rho) = \mathbf{E} \sup_{\psi \in \mathcal{H}, |\psi|=1} \operatorname{Tr}(A_n|\psi\rangle\langle\psi|) = \mathbf{E}\,\lambda_1(A_n), \tag{9.6}$$

since $\operatorname{Tr}(B|\psi\rangle\langle\psi|) = \langle\psi|B|\psi\rangle$. Given that $w(\mathrm{D}) = \kappa_{n^2-1}^{-1} w_G(\mathrm{D})$, the asymptotic estimate follows from the facts that $\kappa_{n^2-1} \sim n$ and $\mathbf{E}\,\lambda_1(A_n) \sim 2\sqrt{n}$ (Theorem 6.23). To show that the inequality $w(\mathrm{D}) \leqslant 2/\sqrt{n}$ holds in every dimension, we use the refined bounds from Proposition A.1(i) and from (6.37). □

It is possible to give a more direct proof of the upper bound $w(\mathrm{D}) = O(1/\sqrt{n})$ using a discretization lemma, which we state for future reference (see Exercise 9.1).

LEMMA 9.2. *Let* $\mathcal{H} = \mathbb{C}^d$, *and let* $\mathcal{N}$ *be an* α*-net in* $(S_{\mathbb{C}^d}, g)$, *with* $\alpha < \pi/4$. *Then*

$$\cos(2\alpha)\mathrm{D}_\oslash \subset \operatorname{conv}\{\pm|\psi\rangle\langle\psi| \ : \ \psi \in \mathcal{N}\} \subset \mathrm{D}_\oslash. \tag{9.7}$$

Equivalently, $\mathcal{N}$ *is an* ε*-net in* $(S_{\mathbb{C}^d}, |\cdot|)$ *for* $\varepsilon = 2\sin(\alpha/2)$, *and* $\cos(2\alpha) = 1 - 2\varepsilon^2 + \varepsilon^4/2$.

PROOF OF LEMMA 9.2. Set $P = \operatorname{conv}\{\pm|\psi\rangle\langle\psi| \ : \ \psi \in \mathcal{N}\}$. The inclusion $P \subset \mathrm{D}_\oslash$ is trivial. Let us check the other inclusion through the corresponding dual (polar) norms

$$\|A\|_{(\mathrm{D}_\oslash)^\circ} = \max_{\varphi \in S_{\mathbb{C}^d}} |\langle\varphi|A|\varphi\rangle| = \|A\|_{\mathrm{op}},$$

$$\|A\|_{P^\circ} = \max_{\psi \in \mathcal{N}} |\langle\psi|A|\psi\rangle| .$$

We need to show that $\|A\|_{P^\circ} \geqslant \cos(2\alpha)\|A\|_{\mathrm{op}}$ for every $A \in \mathsf{M}_d^{\mathrm{sa}}$. We may assume by homogeneity and symmetry that $\|A\|_{\mathrm{op}}$ and the largest eigenvalue of A are both equal to 1. Let $\varphi \in \mathbb{C}^d$ be a unit vector such that $A\varphi = \varphi$. Choose $\psi \in \mathcal{N}$ verifying $g(\varphi, \psi) \leqslant \alpha$. By adjusting the phase of φ (i.e., replacing φ with an appropriate element of $[\varphi]$), we may write $\psi = \cos(\beta)\varphi + \sin(\beta)\chi$ for a unit vector $\chi \perp \varphi$, and $0 \leqslant \beta \leqslant \alpha$. We have then (since $\langle\varphi|A|\chi\rangle = 0$ and $\langle\chi|A|\chi\rangle \geqslant -1$)

$$\langle\psi|A|\psi\rangle = \cos^2(\beta)\langle\varphi|A|\varphi\rangle + \sin^2(\beta)\langle\chi|A|\chi\rangle \geqslant \cos^2\beta - \sin^2\beta = \cos(2\beta) \geqslant \cos(2\alpha),$$

as needed. □

EXERCISE 9.1 (An easy upper bound on the mean width of D). Using Lemma 9.2, give an alternate proof of the relation

$$w(\mathrm{D}(\mathbb{C}^n)) = O(1/\sqrt{n}).$$

EXERCISE 9.2. Show that Lemma 9.2 is sharp on $\mathbb{C}^2$, i.e., that $\cos(2\alpha)$ cannot be replaced by a larger number in (9.7).

9.1.3. The set of separable states (the bipartite case).

THEOREM 9.3. *If* $\mathrm{Sep} = \mathrm{Sep}(\mathbb{C}^d \otimes \mathbb{C}^d)$, *we have the two-sided estimates*

$$\frac{1}{6d^{3/2}} \leqslant \mathrm{vrad}(\mathrm{Sep}) \leqslant w(\mathrm{Sep}) \leqslant \frac{4}{d^{3/2}}. \tag{9.8}$$

The inequality $\mathrm{vrad}(\mathrm{Sep}) \leqslant w(\mathrm{Sep})$ is the Urysohn inequality (Proposition 4.15). We first give an elementary argument showing that $w(\mathrm{Sep}) = O(d^{-3/2})$, and then prove separately the more precise bounds from (9.8).

PROOF THAT $w(\mathrm{Sep}) = O(d^{-3/2})$. We proceed through a net argument. It is easier to work with the Gaussian mean width, and therefore we prove the equivalent statement $w_G(\mathrm{Sep}) = O(\sqrt{d})$. Since $w_G(\mathrm{Sep}) \leqslant w_G(\mathrm{Sep}_{\oslash})$, it is enough to give an upper bound on $w_G(\mathrm{Sep}_{\oslash})$. Let P be the polytope given by Lemma 9.4 below. Then

$$w_G(\mathrm{Sep}_{\oslash}) \leqslant 2w_G(P) \leqslant 2\sqrt{2\log(C^d)} = O(\sqrt{d}),$$

where we used Proposition 6.3 (note that vertices of P have Hilbert–Schmidt norm 1). □

LEMMA 9.4. *There is a constant* $C > 0$ *such that, for every dimension* d, *there is a family* $\mathcal{N}$ *of product pure states on* $\mathcal{H} = \mathbb{C}^d \otimes \mathbb{C}^d$ *with* $\mathrm{card}\,\mathcal{N} \leqslant C^d$ *and such that, if we denote by* P *the polytope* $\mathrm{conv}\{\pm|\varphi\otimes\psi\rangle\langle\varphi\otimes\psi| \;:\; \varphi\otimes\psi \in \mathcal{N}\}$, *then we have*

$$\frac{1}{2}\,\mathrm{Sep}_{\oslash} \subset P \subset \mathrm{Sep}_{\oslash}.$$

The constant $1/2$ appearing in Lemma 9.4 could be replaced by $1-\epsilon$ for any $\varepsilon > 0$, affecting only the value of C. Interestingly, the analogous statement for Sep (i.e., without symmetrization) is false, see Proposition 9.31.

PROOF. Let $\mathcal{M}$ be an α-net in $(S_{\mathbb{C}^d}, g)$ and $P_0 = \mathrm{conv}\{\pm|\psi\rangle\langle\psi| \;:\; \psi \in \mathcal{M}\}$. We write D for $\mathrm{D}(\mathbb{C}^d)$ and Sep for $\mathrm{Sep}(\mathbb{C}^d \otimes \mathbb{C}^d)$. We know from Lemma 9.2 that

$$\cos(2\alpha)\mathrm{D}_{\oslash} \subset P_0 \subset \mathrm{D}_{\oslash}.$$

Since $\mathrm{Sep}_{\oslash} = \mathrm{D}_{\oslash} \mathbin{\widehat{\otimes}} \mathrm{D}_{\oslash}$, it follows that

$$\cos^2(2\alpha)\mathrm{Sep}_{\oslash} \subset P_0 \mathbin{\widehat{\otimes}} P_0 \subset \mathrm{Sep}_{\oslash}. \tag{9.9}$$

It remains to choose $\alpha = \pi/8$, so that $\cos^2(2\alpha) = 1/2$. We choose $\mathcal{N}$ to be the set $\{\varphi \otimes \psi \;:\; \varphi, \psi \in \mathcal{M}\}$, so that $P = P_0 \mathbin{\widehat{\otimes}} P_0$. We bound the cardinality of $\mathcal{M}$ using Lemma 5.3, yielding $\mathrm{card}\,\mathcal{N} \leqslant C^d$ for some absolute constant C. □

PROOF THAT $w(\mathrm{Sep}) \leqslant 4d^{-3/2}$. We have $\mathrm{Sep} = \mathrm{D} \hat{\otimes} \mathrm{D}$, where D means $\mathrm{D}(\mathbb{C}^d)$. We use the Chevet–Gordon inequality in the form of Exercise 6.49 to obtain that $w_G(\mathrm{Sep}) \leqslant 2w_G(\mathrm{D})$. The bound $w(\mathrm{D}) \leqslant 2/\sqrt{d}$ from Theorem 9.1 implies only $w(\mathrm{Sep}) \leqslant (4 + o(1))d^{-3/2}$. However, using the refined bound (6.37) (cf. the proof of Theorem 9.1), we can obtain

$$w(\mathrm{Sep}) = \frac{1}{\kappa_{d^4-1}} w_G(\mathrm{Sep}) \leqslant \frac{4\sqrt{d} - 1.2d^{-1/6}}{\sqrt{d^4 - 1}} \leqslant 4d^{-3/2}. \qquad \square$$

PROOF THAT $\mathrm{vrad}(\mathrm{Sep}) \geqslant \frac{1}{6}d^{-3/2}$. We first give a lower bound on $\mathrm{vrad}(\mathrm{Sep}_{\varnothing})$ by estimating from below the inradius of $\mathrm{Sep}_{\varnothing}$. We are going to compare $\mathrm{Sep}_{\varnothing}$ with a simpler convex body which we now define. Let $K \subset B(\mathcal{H})$ be the convex hull of rank one product operators (not necessarily self-adjoint).

$$K := \mathrm{conv}\left\{|x_1 \otimes x_2\rangle\langle y_1 \otimes y_2| \ : \ x_1, y_1, x_2, y_2 \in B_{\mathbb{C}^d}\right\}.$$

The convex body K is most naturally seen as $S_1^d \mathbin{\hat{\otimes}} S_1^d$, it can also be identified with $(B_{\mathbb{C}^d})^{\hat{\otimes} 4}$ up to identification with dual space. The next lemma (the proof we postpone for a moment) relates K to $\mathrm{Sep}_{\varnothing}$.

LEMMA 9.5. *Let $\mathcal{H} = \mathbb{C}^d \otimes \mathbb{C}^d$. Let $\pi : B(\mathcal{H}) \to B^{\mathrm{sa}}(\mathcal{H})$ be the projection onto self-adjoint part, $\pi(A) := \frac{1}{2}(A + A^\dagger)$. Then*

$$\mathrm{Sep}_{\varnothing} \subset \pi(K) \subset 3\,\mathrm{Sep}_{\varnothing}.$$

Lemma 9.5 implies that $\mathrm{inrad}(\mathrm{Sep}_{\varnothing}) \geqslant \frac{1}{3}\,\mathrm{inrad}(K)$. We also know from Lemma 8.26 that

$$\mathrm{inrad}(K) = \mathrm{inrad}\left((B_{\mathbb{C}^d})^{\hat{\otimes} 4}\right) \geqslant \frac{1}{d^{3/2}}.$$

Therefore,

$$\mathrm{vrad}(\mathrm{Sep}_{\varnothing}) \geqslant \mathrm{inrad}(\mathrm{Sep}_{\varnothing}) \geqslant \frac{1}{3d^{3/2}}.$$

We conclude using (9.3) that $\mathrm{vrad}(\mathrm{Sep}) \geqslant \frac{1}{6d^{3/2}}$. (As in the proof of Theorem 9.1, this requires a somewhat tedious verification due to the fact that $\mathrm{Sep}_{\varnothing}$ and Sep live in different dimensions.) $\square$

PROOF OF LEMMA 9.5. The factor 3 appears as an upper bound on the geometric distance between the sets $\mathrm{D}_{\varnothing}$ and $\mathrm{Sep}_{\varnothing}$ corresponding to 2 qubits, i.e., the smallest positive number λ such that $\mathrm{D}(\mathbb{C}^2 \otimes \mathbb{C}^2)_{\varnothing} \subset \lambda \mathrm{Sep}(\mathbb{C}^2 \otimes \mathbb{C}^2)_{\varnothing}$. The upper bound $\lambda \leqslant 3$ follows from Proposition 9.17, or by noting that any state ρ can be decomposed as

$$\rho = 2\underbrace{\frac{\rho + \mathrm{I}/2}{3}}_{\text{separable}} - \underbrace{\frac{\mathrm{I} - \rho}{3}}_{\text{separable}},$$

where separability can be checked, e.g., using the Peres criterion (see Theorem 2.15).

It is enough to show that extreme points of $\pi(K)$ are contained in $3\mathrm{Sep}_{\varnothing}$. Any extreme point A of $\pi(K)$ can be written as

$$A = \frac{1}{2}\left(|x_1 \otimes x_2\rangle\langle y_1 \otimes y_2| + |y_1 \otimes y_2\rangle\langle x_1 \otimes x_2|\right).$$

It may appear at the first sight that the above representation shows that A is separable. However, while the two terms in the parentheses are indeed product operators, they are not self-adjoint and we can only conclude that $A \in \mathrm{D}(\mathcal{H})_{\otimes}$ (as a self-adjoint operator whose trace norm is $\leqslant 1$).

Let $\mathcal{H}_i$ be the 2-dimensional subspace of $\mathbb{C}^d$ spanned by x_i and y_i (if the vectors are proportional, add any vector to get a 2-dimensional space) and let $\mathcal{H}' := \mathcal{H}_1 \otimes \mathcal{H}_2$. Then A can be considered as an operator on $\mathcal{H}'$; more precisely, as an element of $\mathrm{D}(\mathcal{H}')_{\otimes}$ (and, conversely, any operator acting on $\mathcal{H}'$ can be canonically lifted to one acting on $\mathcal{H}$). Since A belongs to $\mathrm{D}(\mathcal{H}')_{\otimes}$, it also belongs to $3\mathrm{Sep}(\mathcal{H}')_{\otimes}$, and thus to $3\mathrm{Sep}(\mathcal{H})_{\otimes}$. □

9.1.4. The set of block-positive matrices. Let $\mathrm{Sep} = \mathrm{Sep}(\mathbb{C}^d \otimes \mathbb{C}^d)$. In Theorem 9.3 we computed the order of magnitude of the mean width of Sep. We now focus on the dual quantity: the mean gauge of Sep, or the mean width of Sep° (recall that polarity is taken with maximally mixed state $\rho_* = \mathrm{I}/d^2$ being the origin).

THEOREM 9.6. *Let* Sep *be the set of separable states on* $\mathbb{C}^d \otimes \mathbb{C}^d$. *Then for some absolute constants* c, C,

$$cd^{3/2} \leqslant \mathrm{vrad}(\mathrm{Sep}^\circ) \leqslant Cd^{3/2},$$

$$cd^{3/2} \leqslant w(\mathrm{Sep}^\circ) \leqslant Cd^{3/2}\log(d).$$

Since the cone $\mathcal{BP}$ of block-positive operators is dual to the cone $\mathcal{SEP}$ of separable operators (see Section 2.4), we obtain the following corollary.

COROLLARY 9.7. *Let* BP *be the set of trace one block-positive operators on* $\mathbb{C}^d \otimes \mathbb{C}^d$. *Then, for some absolute constants* c, C

$$cd^{-1/2} \leqslant \mathrm{vrad}(\mathrm{BP}) \leqslant Cd^{-1/2},$$

$$cd^{-1/2} \leqslant w(\mathrm{BP}) \leqslant Cd^{-1/2}\log(d).$$

PROOF. Since $\mathrm{BP} = -d^{-2}\mathrm{Sep}^\circ$ (see (2.47)), the derivation of Corollary 9.7 from Theorem 9.6 is immediate. □

The Santaló and reverse Santaló inequalities (Theorem 4.17) allow us to directly estimate $\mathrm{vrad}(\mathrm{Sep}^\circ)$ from $\mathrm{vrad}(\mathrm{Sep})$, so the first part of Theorem 9.6 follows from Theorem 9.3. However the analogous result for the mean width, the MM^*-estimate (Theorem 7.10), is more demanding. Since we already know that $w(\mathrm{Sep}) = \Theta(d^{-3/2})$ (again from Theorem 9.3), the conclusion of Theorem 9.6 follows after we prove the MM^*-estimate (7.7) for the pair $(\mathrm{Sep}, \mathrm{Sep}^\circ)$, i.e.,

$$w(\mathrm{Sep})w(\mathrm{Sep}^\circ) = O(\log d). \tag{9.10}$$

Recall that the lower bound $w(\mathrm{Sep})w(\mathrm{Sep}^\circ) \geqslant 1$ is elementary and holds for any pair of polar bodies (see Exercise 4.37). However, (9.10) does not follow immediately from the general theory: Theorem 7.10 is known to hold only for symmetric convex bodies which are in a specific position (the ℓ-position). In our situation Sep is not symmetric and there is no reason to think that it is in the ℓ-position.

The first step towards proving Theorem 9.6 is to introduce the following symmetrization of Sep:

$$\mathrm{Sep}_\cap = -\mathrm{Sep} \cap \mathrm{Sep},$$

where $-\mathrm{Sep} = (-1) \bullet \mathrm{Sep}$, see (9.1). We check that the relevant geometric parameters are essentially unchanged by this symmetrization procedure.

PROPOSITION 9.8. *The convex bodies* Sep *and* $\mathrm{Sep}_\cap$ *have comparable volume radius, mean width and dual mean with, as show by the following formulas, where* $\mathrm{Sep}_\cap^\circ$ *means* $(\mathrm{Sep}_\cap)^\circ$*:*

$$w(\mathrm{Sep}^\circ) \leqslant w(\mathrm{Sep}_\cap^\circ) \leqslant 2w(\mathrm{Sep}^\circ), \tag{9.11}$$

$$\frac{1}{2}\,\mathrm{vrad}(\mathrm{Sep}) \leqslant \mathrm{vrad}(\mathrm{Sep}_\cap) \leqslant \mathrm{vrad}(\mathrm{Sep}), \tag{9.12}$$

$$w(\mathrm{Sep}) \simeq w(\mathrm{Sep}_\cap) \simeq d^{-3/2}. \tag{9.13}$$

Moreover, Sep *and* $\mathrm{Sep}_\cap$ *have the same inradius, equal to* $(d^2(d^2-1))^{-1/2}$*. However, the outradius of* $\mathrm{Sep}_\cap$ *is bounded by* $1/d$*, while the outradius of* Sep *is of order* 1.

PROOF. We have, for any self-adjoint A with zero trace,

$$\|\rho_* + A\|_{\mathrm{Sep}_\cap} = \max(\|\rho_* + A\|_{\mathrm{Sep}}, \|\rho_* - A\|_{\mathrm{Sep}}) \leqslant \|\rho_* + A\|_{\mathrm{Sep}} + \|\rho_* - A\|_{\mathrm{Sep}}.$$

When averaging A over the Hilbert–Schmidt sphere, using the fact that A and $-A$ have the same distribution, we obtain (9.11). Inequalities (9.12) follow from Proposition 4.18. For (9.13), we already know (cf. Theorem 9.3) that

$$\mathrm{vrad}(\mathrm{Sep}) \simeq w(\mathrm{Sep}) \simeq d^{-3/2}.$$

We therefore have the following chain of inequalities: the first is trivial, the third is (9.12) and the last is Urysohn's inequality (Proposition 4.15):

$$w(\mathrm{Sep}_\cap) \leqslant w(\mathrm{Sep}) \simeq \mathrm{vrad}(\mathrm{Sep}) \simeq \mathrm{vrad}(\mathrm{Sep}_\cap) \leqslant w(\mathrm{Sep}_\cap).$$

Therefore all these quantities are comparable, and (9.13) follows.

The statement about the inradius is trivial. On the other hand, any matrix A such that $\rho_* + A \in \mathrm{Sep}$ satisfies $A \geqslant -\mathrm{I}/d^2$. Consequently, any A such that $\rho_* + A \in \mathrm{Sep}_\cap$ satisfies $-\mathrm{I}/d^2 \leqslant A \leqslant \mathrm{I}/d^2$, or $\|A\|_\infty \leqslant 1/d^2$. It follows that the outradius of $\mathrm{Sep}_\cap$, which is measured with respect to the Hilbert–Schmidt norm, is bounded by $1/d$. □

We are now going to prove that the $MM^\star$-estimate holds for $\mathrm{Sep}_\cap$.

PROPOSITION 9.9. *There is an absolute constant* C *such that*

$$d^4 \sim \kappa_{d^4-1}^2 \leqslant w_G(\mathrm{Sep}_\cap) w_G(\mathrm{Sep}_\cap^\circ) \leqslant C d^4 \log d. \tag{9.14}$$

It is now easy to deduce Theorem 9.6. Indeed, using the relation (4.32) between spherical and Gaussian widths, Proposition 9.9 implies that $w(\mathrm{Sep}_\cap)w(\mathrm{Sep}_\cap^\circ) = O(\log d)$, and (9.10) follows from (i) and (iii) of Proposition 9.8.

PROOF OF PROPOSITION 9.9. Denote $K = \mathrm{Sep}_\cap - \rho_*$, so that K is a symmetric convex body in the space H_0 of self-adjoint trace zero operators on $\mathbb{C}^d \otimes \mathbb{C}^d$. The lower bound in (9.14) is a reformulation of the inequality $w(K)w(K^\circ) \geqslant 1$, which is elementary (see Exercise 4.37). Using the ℓ-norms introduced in Section 7.1.1 (especially Proposition 7.1(iii)), we may reformulate (9.14) as

$$d^4 \lesssim \ell_K(\mathrm{I}_{H_0}) \ell_{K^\circ}(\mathrm{I}_{H_0}) \leqslant C d^4 \log d. \tag{9.15}$$

To prove the upper bound in (9.15), let $T : H_0 \to H_0$ be a linear map such that TK is in the ℓ-position. We will take advantage of the symmetries of K. The set Sep (hence also K) is invariant under local unitaries, and the decomposition of H_0 into irreducible subspaces is (see Lemma 2.19) $H_0 = E \oplus F_1 \oplus F_2$, where

$$E = \operatorname{span}\{\sigma_1 \otimes \sigma_2 \ : \ \operatorname{Tr} \sigma_1 = \operatorname{Tr} \sigma_2 = 0\},$$
$$F_1 = \operatorname{span}\{\sigma_1 \otimes \mathrm{I} \ : \ \operatorname{Tr} \sigma_1 = 0\},$$
$$F_2 = \operatorname{span}\{\mathrm{I} \otimes \sigma_2 \ : \ \operatorname{Tr} \sigma_2 = 0\}.$$

By Proposition 4.8, we may assume that $T = \alpha P_E + \lambda_1 P_{F_1} + \lambda_2 P_{F_2}$ for some positive numbers $\alpha, \lambda_1, \lambda_2$. We may also assume $\alpha = 1$ without loss of generality. The ideal property of the ℓ-norm (Proposition 7.1(ii)) implies that

$$\ell_K(P_E) = \ell_K(TP_E) \leqslant \ell_K(T),$$

and similarly for $\ell_{K^\circ}(P_E)$. By the MM^*-estimate (Theorem 7.10), we know that

$$\ell_K(T)\ell_{K^\circ}(T^{-1}) = O(d^4 \log d).$$

Noting that $T^{-1} = P_E + \lambda_1^{-1} P_{F_1} + \lambda_2^{-1} P_{F_2}$, it follows that

$$\ell_K(P_E)\ell_{K^\circ}(P_E) = O(d^4 \log d). \tag{9.16}$$

The ℓ-norms of the projections P_{F_1}, P_{F_2} can be upper-bounded in a rather straightforward fashion, mostly due to the fact that their ranks are relatively small. We have the following.

LEMMA 9.10. *Let $F = F_1 \oplus F_2$. Then $\ell_K(P_F) = O(d^3)$ and $\ell_{K^\circ}(P_F) = O(1)$.*

We now postpone the proof of Lemma 9.10 and show how it allows to complete the proof of Proposition 9.9. To that end, we compare the estimates from Lemma 9.10 with the bounds $\ell_{K^\circ}(\mathrm{I}_{H_0}) \simeq \sqrt{d}$ (a reformulation of Theorem 9.3) and $\ell_K(\mathrm{I}_{H_0}) \gtrsim d^{7/2}$ (which follows from the already proved lower bound in (9.15)). For $L = K$ or $L = K^\circ$, we have therefore $\ell_L(P_F) \leqslant \frac{1}{2}\ell_L(\mathrm{I}_{H_0})$ for d large enough. Using the triangle inequality $\ell_L(\mathrm{I}_{H_0}) \leqslant \ell_L(P_E) + \ell_L(P_F)$, it follows that $\ell_L(\mathrm{I}_{H_0}) \leqslant 2\ell_L(P_E)$ for d large enough. Combined with (9.16), this gives the upper bound in (9.15), as needed. □

PROOF OF LEMMA 9.10. We have $\dim F = 2(d^2 - 1)$. We use Proposition 7.1(v) and the estimates on the inradius and the outradius of $\mathrm{Sep}_\cap$ from Proposition 9.8 to deduce the following inequalities (recall that κ_n is of order $\sqrt{n}$, see Proposition A.1):

$$\ell_K(P_F) = w_G((K \cap F)^\circ) \leqslant d^2 \kappa_{\dim F} \lesssim d^3,$$
$$\ell_{K^\circ}(P_F) = w_G(P_F K) \leqslant d^{-1} \kappa_{\dim F} \lesssim 1.$$ □

9.1.5. The set of separable states (multipartite case). We first note that an iteration of the arguments from the bipartite case (Theorem 9.3) can be used to show the following estimates (where the constants c_k, C_k depend *a priori* on k), for $\mathcal{H} = \mathbb{C}^{d_1} \otimes \cdots \otimes \mathbb{C}^{d_k}$.

$$c_k \frac{\max(\sqrt{d_1}, \ldots, \sqrt{d_k})}{d_1 \cdots d_k} = \operatorname{vrad}(\mathrm{Sep}) \leqslant w(\mathrm{Sep}) \leqslant C_k \frac{\sqrt{d_1} + \cdots + \sqrt{d_k}}{d_1 \cdots d_k}.$$

These estimates are reasonably sharp as long as k remains bounded (few subsystems, each of them being possibly large), but deteriorate very quickly once k grows. However, it is also possible to obtain fairly sharp bounds valid for large values of k (many small subsystems). For simplicity, we first consider the case of k qubits.

THEOREM 9.11. *Let $k \geqslant 1$, $n = 2^k$, and $\mathcal{H} = (\mathbb{C}^2)^{\otimes k}$. Then*

$$\frac{c\sqrt{\log n \log\log n}}{n} \leqslant w(\mathrm{Sep}) \leqslant \frac{C\sqrt{\log n \log\log n}}{n} \tag{9.17}$$

and

$$\frac{c}{n^{1+\alpha}} \leqslant \mathrm{vrad}(\mathrm{Sep}) \leqslant \frac{C\sqrt{\log n \log\log n}}{n^{1+\alpha}}, \tag{9.18}$$

where c, C are absolute constants, and $\alpha = \frac{1}{8}\log_2(27/16) \approx 0.094$.

PROOF OF THEOREM 9.11. We write $\mathrm{D} = \mathrm{D}(\mathbb{C}^2)$ and $\mathrm{Sep} = \mathrm{Sep}(\mathcal{H})$. Since $\mathrm{Sep}_{\mathcal{O}} = \mathrm{D}_{\mathcal{O}}^{\hat{\otimes} k}$, it follows from Lemma 9.2 that, if $\mathcal{N}$ is an ε-net in $(S_{\mathbb{C}^2}, g)$, then

$$\cos(2\varepsilon)^k \,\mathrm{Sep}_{\mathcal{O}} \subset P \subset \mathrm{Sep}_{\mathcal{O}},$$

where

$$P := \mathrm{conv}\{\pm|\psi_1 \otimes \cdots \otimes \psi_k\rangle\langle\psi_1 \otimes \cdots \otimes \psi_k| \;:\; \psi_1, \ldots, \psi_k \in \mathcal{N}\}.$$

We choose ε such that $\cos(2\varepsilon)^k = 1/2$, i.e., $\varepsilon \simeq 1/\sqrt{k}$. The polytope P is contained in the Hilbert–Schmidt unit ball, and (using Lemma 5.3) can be chosen with at most $2(\operatorname{card}\mathcal{N})^k \leqslant \exp(Ck\log k)$ vertices. The first idea would be to apply directly Proposition 6.3. This approach yields the bound

$$\mathrm{vrad}(\mathrm{Sep}) \leqslant w(\mathrm{Sep}) \leqslant w(\mathrm{Sep}_{\mathcal{O}}) \leqslant \frac{C\sqrt{\log n \log\log n}}{n},$$

which is the upper bound in (9.17). For the lower bound in (9.17), see Exercise 9.3.

The reason for the extra factor n^α in (9.18) comes from the fact that the Hilbert–Schmidt Euclidean structure is not the most adapted to the present problem. When we apply Proposition 6.3 in the Euclidean structure induced by some ellipsoid $\mathcal{E}$, we actually obtain the following result: if P is a polytope with v vertices contained in an ellipsoid $\mathcal{E} \subset \mathbb{R}^N$, then we have

$$\left(\frac{\mathrm{vol}\,P}{\mathrm{vol}\,\mathcal{E}}\right)^{1/N} \leqslant \sqrt{\frac{2\log v}{N}}. \tag{9.19}$$

In this inequality, for a fixed polytope P, the best choice of ellipsoid is given by the Löwner ellipsoid of P. Accordingly, we are going to consider the Löwner ellipsoid associated to the set $\mathrm{Sep}_{\mathcal{O}}$. By Lemma 4.9, we have

$$\mathrm{Löw}(\mathrm{Sep}_{\mathcal{O}}) = \mathrm{Löw}(\mathrm{D}_{\mathcal{O}})^{\otimes_2 k}. \tag{9.20}$$

The set $\mathrm{D}_{\mathcal{O}}$ is a cylinder. To compute its Löwner ellipsoid, we use Lemma 4.3 with $n = 3$, $h = 1/\sqrt{2}$, $a = 0$ and $S = \mathrm{I}/\sqrt{2} \in \mathsf{M}_2$. It follows that $\mathrm{Löw}(\mathrm{D}_{\mathcal{O}}) = T(B_{\mathrm{HS}})$, where B_{HS} denotes the Hilbert–Schmidt unit ball in $\mathsf{M}_2^{\mathrm{sa}}$ and T is the matrix $\mathrm{Diag}(\sqrt{2}, \sqrt{2/3}, \sqrt{2/3}, \sqrt{2/3})$ in the basis of Pauli matrices (2.3). Consequently,

$$\frac{\mathrm{vol}\,\mathrm{Löw}(\mathrm{D}_{\mathcal{O}})}{\mathrm{vol}\,B_{\mathrm{HS}}} = \det T = \sqrt{\frac{16}{27}}$$

or, equivalently, $\mathrm{vrad}(\mathrm{Löw}(\mathrm{D}_{\mathcal{O}})) = (16/27)^{1/8}$. From the formula

$$\mathrm{vrad}(\mathrm{Löw}(\mathrm{Sep}_{\mathcal{O}})) = \mathrm{vrad}(\mathrm{Löw}(\mathrm{D}_{\mathcal{O}}))^k$$

(which follows from (9.20), see Exercise 4.32), we conclude that

$$\mathrm{vrad}\,\mathrm{Löw}(\mathrm{Sep}_{\mathcal{O}}) = (16/27)^{k/8} = n^{-\alpha}$$

with $\alpha = \frac{1}{8}\log_2(27/16)$. If we use the (inner product induced by the) Löwner ellipsoid of $\mathrm{Sep}_{\wp}$ as the reference Euclidean structure to apply (9.19), we obtain the upper bound

$$\mathrm{vrad}(\mathrm{Sep}_{\wp}) \leqslant C\frac{\sqrt{k\log k}}{n}\,\mathrm{vrad}(\mathrm{L\ddot{o}w}(\mathrm{Sep}_{\wp})) = C\frac{\sqrt{\log n \log\log n}}{n^{1+\alpha}}.$$

To show the lower bound in (9.18), we use the fact (see Exercise 4.20) that for every symmetric convex body $K \subset \mathbb{R}^N$, the inclusion $K \supset \frac{1}{\sqrt{N}}\,\mathrm{L\ddot{o}w}(K)$ holds. We apply this for $K = \mathrm{Sep}_{\wp}$ (so that $N = n^2$) to conclude that

$$\mathrm{vrad}(\mathrm{Sep}_{\wp}) \geqslant \frac{1}{n}\,\mathrm{vrad}(\mathrm{L\ddot{o}w}(\mathrm{Sep}_{\wp})) = \frac{1}{n^{1+\alpha}}.$$

Finally, an application of the Rogers–Shephard inequality (9.3) shows that $\mathrm{vrad}(\mathrm{Sep})$ and $\mathrm{vrad}(\mathrm{Sep}_{\wp})$ are of the same order. □

A similar argument allows us to estimate the size of the set of separable states on k "qudits", i.e., on $(\mathbb{C}^d)^{\otimes k}$.

THEOREM 9.12 (See Exercise 9.5). *Let $d \geqslant 2$, $k \geqslant 1$, $n = d^k$, and $\mathcal{H} = (\mathbb{C}^d)^{\otimes k}$. Then*

$$\frac{c_d\sqrt{\log n \log\log n}}{n} \leqslant w(\mathrm{Sep}) \leqslant \frac{C_d\sqrt{\log n \log\log n}}{n} \tag{9.21}$$

and

$$\frac{c_d}{n^{1+\alpha_d}} \leqslant \mathrm{vrad}(\mathrm{Sep}) \leqslant \frac{C_d\sqrt{\log n \log\log n}}{n^{1+\alpha_d}}, \tag{9.22}$$

where $\alpha_d = \frac{1}{2}\log_d(1+\frac{1}{d}) - \frac{1}{2d^2}\log_d(d+1)$.

EXERCISE 9.3 (Lower bound on the mean width of Sep). Show that, for some constant $c > 0$, $\mathrm{Sep}\left((\mathbb{C}^2)^{\otimes k}\right)$ contains k^{ck} elements which are c-separated with respect to the Hilbert–Schmidt distance. Then, use the Sudakov minoration (Proposition 6.10) to establish the lower bound in (9.17).

EXERCISE 9.4 (Löwner ellipsoid and the Killing form). Check that the Löwner ellipsoid of $\mathrm{D}(\mathbb{C}^2)_{\wp}$ induces on $\mathsf{M}_2^{\mathrm{sa}}$ the inner product

$$\langle u, v\rangle_L = \frac{3}{2}\,\mathrm{Tr}(uv) - \frac{1}{2}\,\mathrm{Tr}(u)\,\mathrm{Tr}(v).$$

EXERCISE 9.5 (The size of of Sep for k qudits). Complete the proof of Theorem 9.12.

9.1.6. The set of PPT states. We present estimates for the volume and mean width of PPT. For asymptotic versions improving some of the constants, see Exercise 9.6.

THEOREM 9.13 (Volume and mean width of PPT). *For $\mathcal{H} = \mathbb{C}^d \otimes \mathbb{C}^d$, we have*

$$\frac{1}{4d} \leqslant w(\mathrm{PPT}^{\circ})^{-1} \leqslant \mathrm{vrad}(\mathrm{PPT}) \leqslant w(\mathrm{PPT}) \leqslant \frac{2}{d}.$$

PROOF. The upper bound on the mean width follows from the obvious inequality $w(\mathrm{PPT}) \leqslant w(\mathrm{D})$ and from the bound $w(\mathrm{D}) \leqslant 2/d$ (Theorem 9.1). To prove the lower bound, we use the dual Urysohn inequality (Proposition 4.16), where polarity is taken with respect to ρ_*

$$\mathrm{vrad}(\mathrm{PPT}) \geqslant \frac{1}{w(\mathrm{PPT}^\circ)}.$$

If Γ denotes the partial transposition on $\mathcal{H}$, then $\mathrm{PPT} = \mathrm{D} \cap \Gamma(\mathrm{D})$ and therefore

$$\mathrm{PPT}^\circ = \mathrm{conv}(\mathrm{D}^\circ \cup \Gamma(\mathrm{D})^\circ) \subset \mathrm{D}^\circ + \Gamma(\mathrm{D})^\circ. \tag{9.23}$$

Geometrically, the transformation Γ is an isometry with respect to the Hilbert–Schmidt norm (cf. Exercise 2.22; the argument we present actually works for any Hilbert–Schmidt isometry). Using the fact that $\mathrm{D}^\circ = -d^2\mathrm{D}$ and the upper bound from Theorem 9.1, we obtain

$$w(\mathrm{PPT}^\circ) \leqslant w(\mathrm{D}^\circ) + w(\Gamma(\mathrm{D})^\circ) \leqslant 2w(\mathrm{D}^\circ) = 2d^2 w(\mathrm{D}) \leqslant 4d. \qquad \square$$

It follows from Theorem 9.13 that D and PPT have comparable volume radii, up to an absolute constant. An interesting question is whether this constant approaches 1 as the dimension increases.

PROBLEM 9.14. *Is there an absolute constant $c < 1$ such that, for every $d \geqslant 3$,*

$$\mathrm{vrad}(\mathrm{PPT}(\mathbb{C}^d \otimes \mathbb{C}^d)) \leqslant c\, \mathrm{vrad}(\mathrm{D}(\mathbb{C}^d \otimes \mathbb{C}^d))?$$

EXERCISE 9.6 (Sharper asymptotic bounds on the size of PPT). Prove that $w(\mathrm{PPT}^\circ) \leqslant (2 + o(1))d$ and conclude that

$$w(\mathrm{PPT}) \geqslant \mathrm{vrad}(\mathrm{PPT}) \geqslant \frac{1 - o(1)}{2d}.$$

EXERCISE 9.7 (Volume radius of PPT as a large deviation problem). Show that Problem 9.14 can be reformulated as follows: *does there exist a constant $c > 0$ such that, if B is a $d^2 \times d^2$ matrix with independent $N_{\mathbb{C}}(0,1)$ entries, then*

$$\mathbf{P}\left((BB^\dagger)^\Gamma \text{ is positive}\right) \leqslant \exp(-cd^4)? \tag{9.24}$$

This recasts the problem as a large deviation estimate for some random matrix ensemble. Note that the same ensemble appears in Theorem 6.30, which asserts that it is asymptotically semicircular with appropriate parameters. It is worth pointing out that bounds in the spirit of (9.24) hold for the GUE ensembles and Wishart ensembles, see [**BAG97**, **HP98**] and [**AGZ10**].

9.2. Distance estimates

In this section we gather known estimates for the geometric distance (defined in (4.1)) between D, Sep, and the Hilbert–Schmidt ball B_{HS}. Computing the distance to the Hilbert–Schmidt ball is equivalent to computing the inradius and outradius. In particular, it follows from the results of Table 9.1 that, for $\mathcal{H} = \mathbb{C}^d \otimes \mathbb{C}^d$,

$$d_g(\mathrm{D}, B_{\mathrm{HS}}) = d_g(\mathrm{Sep}, B_{\mathrm{HS}}) = d^2 - 1. \tag{9.25}$$

9.2.1. The Gurvits–Barnum theorem. A remarkable fact, which is implicit in (9.25) above, is that—in the bipartite case—not only the outradii, but also the inradii of Sep and D are the same.

THEOREM 9.15. *Let* $\mathcal{H} = \mathbb{C}^{d_1} \otimes \mathbb{C}^{d_2}$, $n = d_1 d_2$ *and* ρ *be a state on* $\mathcal{H}$ *such that*

$$\left\| \rho - \frac{\mathrm{I}}{n} \right\|_{\mathrm{HS}} \leqslant \frac{1}{\sqrt{n(n-1)}}.$$

Then ρ *is separable.*

An elementary geometric argument shows that Theorem 9.15 is equivalent to the following statement: if $A \in B^{\mathrm{sa}}(\mathbb{C}^{d_1} \otimes \mathbb{C}^{d_2})$ satisfies $\|A\|_{\mathrm{HS}} \leqslant 1$, then $\mathrm{I} + A \in \mathcal{SEP}$.

PROOF. Let $K \subset \mathrm{D}(\mathcal{H})$ be the set of states ρ such that $\|\rho - \mathrm{I}/n\|_{\mathrm{HS}} \leqslant 1/\sqrt{n(n-1)}$ and $\mathcal{C} = \mathbb{R}_+ K$ be the Lorentz-like cone generated by K. The assertion of Theorem 9.15 is equivalent to the cone inclusion $\mathcal{C} \subset \mathcal{SEP}$. By cone duality (see Section 1.2.1), this is also equivalent to $\mathcal{SEP}^* \subset \mathcal{C}^*$. Recall that $\mathcal{SEP}^*$ is the cone of block-positive operators, see (2.46).

Let $M \in B^{\mathrm{sa}}(\mathcal{H})$. One checks that

$$M \in \mathcal{C} \iff \|M\|_{\mathrm{HS}} \leqslant \frac{1}{\sqrt{n-1}} \operatorname{Tr} M.$$

It follows (see Exercise 1.31) that

$$M \in \mathcal{C}^* \iff \|M\|_{\mathrm{HS}} \leqslant \operatorname{Tr} M.$$

We thus reduced the proof of Theorem 9.15 to the following problem: *for a block-positive matrix* $M \in B^{\mathrm{sa}}(\mathbb{C}^{d_1} \otimes \mathbb{C}^{d_2})$, *prove that* $\operatorname{Tr} M^2 \leqslant (\operatorname{Tr} M)^2$. We will need the following lemma.

LEMMA 9.16. *Let* $M = \begin{pmatrix} A & B \\ B^\dagger & C \end{pmatrix}$ *be a block-positive operator in* $B^{\mathrm{sa}}(\mathbb{C}^{d_1} \otimes \mathbb{C}^2)$. *Then*

$$\|B\|_2^2 \leqslant \|A\|_1 \|C\|_1.$$

By assuming the lemma, we can complete the proof of Theorem 9.15. Denote by $M_{kl} \in B(\mathbb{C}^{d_1})$ the blocks of M. For $k, l \in \{1, \ldots, d_2\}$, we then have

$$\|M_{kl}\|_2^2 \leqslant \|M_{kk}\|_1 \|M_{ll}\|_1$$

(if $k = l$ this is obvious; if $k \neq l$ this is the content of the Lemma). Noting that the diagonal blocks M_{kk} are positive semi-definite and summing over k, l gives the needed inequality $\|M\|_{\mathrm{HS}}^2 \leqslant (\operatorname{Tr} M)^2$. □

PROOF OF LEMMA 9.16. Let $B = HU$ be the polar decomposition of B, with H positive and U unitary. We may choose an orthonormal basis in $\mathbb{C}^{d_1}$ which makes U diagonal. From the inequalities $|B_{ii}|^2 \leqslant A_{ii} C_{ii}$ and $|H_{ij}|^2 \leqslant H_{ii} H_{jj}$, we get

$$|B_{ij}|^2 = |H_{ij}|^2 \leqslant H_{ii} H_{jj} = |B_{ii} B_{jj}| \leqslant \sqrt{A_{ii} C_{ii} A_{jj} C_{jj}} \leqslant \frac{1}{2}(A_{ii} C_{jj} + A_{jj} C_{ii}).$$

Summing over i, j proves the Lemma. □

EXERCISE 9.8 (Another proof of the Gurvits–Barnum theorem). Here is an alternative argument for Theorem 9.15.
(i) Show that for any operators $A_{ij} \in \mathsf{M}_{d_1}$,

$$\left\| \sum_{i,j=1}^{d_2} A_{ij} \otimes |i\rangle\langle j| \right\|_{\mathrm{op}}^2 \leqslant \sum_{i,j=1}^{d_2} \|A_{ij}\|_{\mathrm{op}}^2.$$

(ii) Use (i), Theorem 2.34 and Exercise 2.30 to give an alternate proof of Theorem 9.15.

9.2.2. Robustness in the bipartite case. We now compute the geometric distance between D and Sep in the bipartite case.

PROPOSITION 9.17. *Let $\mathcal{H} = \mathbb{C}^{d_1} \otimes \mathbb{C}^{d_2}$ for $d_1, d_2 \geqslant 2$, and denote $n = d_1 d_2$. We have*

$$d_g(\mathrm{D}, \mathrm{Sep}) = d_g(\mathrm{D}, \mathrm{PPT}) = \frac{n}{2} + 1.$$

An equivalent way to describe the geometric distance is to define the *robustness* of a state ρ as follows (the notation $\bullet$ was defined in (9.1)):

$$R(\rho) = \inf \left\{ s \geqslant 0 \ : \ \frac{1}{1+s} \bullet \rho \in \mathrm{Sep} \right\}. \tag{9.26}$$

Proposition 9.17 asserts that the maximal robustness of a state on $\mathbb{C}^{d_1} \otimes \mathbb{C}^{d_2}$ equals $n/2$. Since $\mathrm{Sep} \subset \mathrm{PPT} \subset \mathrm{D}$, it suffices to prove that $d_g(\mathrm{D}, \mathrm{PPT}) \geqslant \frac{n}{2} + 1$ and $d_g(\mathrm{D}, \mathrm{Sep}) \leqslant \frac{n}{2} + 1$.

PROOF THAT $d_g(\mathrm{D}, \mathrm{PPT}) \geqslant \frac{n}{2} + 1$. Let $\chi = \frac{1}{\sqrt{2}}(|1\rangle \otimes |1\rangle + |2\rangle \otimes |2\rangle) \in \mathbb{C}^{d_1} \otimes \mathbb{C}^{d_2}$ and $\rho = |\chi\rangle\langle\chi|$. Now, for $0 < t < 1$, consider the state $\rho_t = t \bullet \rho$. The non-zero eigenvalues of ρ^Γ are $(1/2, 1/2, 1/2, -1/2)$. It follows that ρ_t is not PPT whenever $-t/2 + (1-t)/n < 0$, or equivalently $t > 2/(2+n)$. Therefore $d_g(\mathrm{D}, \mathrm{PPT}) \geqslant \frac{n}{2} + 1$. □

PROOF THAT $d_g(\mathrm{D}, \mathrm{Sep}) \leqslant \frac{n}{2} + 1$. We have to show that for any state ρ, the state $t_0 \bullet \rho$ is separable when $t_0 := 2/(2+n)$. By convexity, we may assume that ρ is a pure state $|\chi\rangle\langle\chi|$. Consider the Schmidt decomposition of χ,

$$\chi = \sum_{j=1}^{d} \lambda_j \varphi_j \otimes \psi_j,$$

for some $d \leqslant \min(d_1, d_2)$ and orthonormal bases (φ_j) in $\mathbb{C}^{d_1}$ and (ψ_j) in $\mathbb{C}^{d_2}$.

Let $\theta = (\theta_1, \dots, \theta_d)$ a d-tuple of complex numbers with modulus one. Consider the vectors

$$\varphi(\theta) = \sum_{j=1}^{d} \sqrt{\lambda_j} \theta_j \varphi_j \in \mathbb{C}^{d_1},$$

$$\psi(\theta) = \sum_{j=1}^{d} \sqrt{\lambda_j} \theta_j \psi_j \in \mathbb{C}^{d_2}.$$

We compute $\mathbf{E}\, |\varphi(\theta) \otimes \psi(\bar{\theta})\rangle\langle\varphi(\theta) \otimes \psi(\bar{\theta})|$, where $\theta_1, \dots, \theta_d$ are independent and uniformly distributed on the unit circle, and $\bar{\theta}$ denotes the coordinatewise complex conjugate of θ. The resulting operator B, which belongs to the separable cone $\mathcal{SEP}$ by construction, equals

$$B = \sum_{j,k,l,m=1}^{d} \sqrt{\lambda_j \lambda_k \lambda_l \lambda_m}\, \mathbf{E}\left[\theta_j \bar{\theta}_k \theta_l \bar{\theta}_m\right] |\varphi_j \otimes \psi_k\rangle\langle \varphi_l \otimes \psi_m|.$$

The quantity $\mathbf{E}[\theta_j \bar{\theta}_k \theta_l \bar{\theta}_m]$ vanishes unless either (1) $j = k$ and $l = m$, or (2) $j = m$ and $k = l$. The non-vanishing terms can be gathered as $B = |\chi\rangle\langle\chi| + A$, where

$$A = \sum_{j \neq k} \lambda_j \lambda_k |\varphi_j \otimes \psi_k\rangle\langle \varphi_j \otimes \psi_k|.$$

Denote $\alpha = \max\{\lambda_j \lambda_k \ : \ j \neq k\}$. It is easily checked that $\alpha \mathrm{I} - A \in \mathcal{SEP}$ since it can be written as a positive combination of the operators

$$\{|\varphi_j \otimes \psi_k\rangle\langle \varphi_j \otimes \psi_k| \ : \ 1 \leqslant j \leqslant d_1, 1 \leqslant k \leqslant d_2\}.$$

Note that $\alpha \leqslant \frac{1}{2}$ since $\lambda_j \lambda_k \leqslant \frac{1}{2}(\lambda_j^2 + \lambda_k^2) \leqslant \frac{1}{2}$. It follows that $\frac{1}{2}\mathrm{I} - A \in \mathcal{SEP}$, and therefore that

$$t_0 \bullet \rho = t_0 \left(|\chi\rangle\langle\chi| + \frac{n}{2}\rho_* \right) = t_0 \left(B - A + \frac{1}{2}\mathrm{I} \right)$$

is a separable state, as needed. □

9.2.3. Distances involving the set of PPT states. We consider the case of a balanced bipartite Hilbert space $\mathcal{H} = \mathbb{C}^d \otimes \mathbb{C}^d$. Another relevant quantity—not covered by Proposition 9.17—is the geometric distance between PPT and Sep. This quantity is of interest since it quantifies the degree to which PPT is a poor substitute for separability in large dimensions. However, even the order of magnitude of the distance seems unknown. Actually, we are not aware of any upper bound improving substantially on the obvious estimate $d_g(\mathrm{Sep}, \mathrm{PPT}) \leqslant d_g(\mathrm{Sep}, \mathrm{D})$.

PROPOSITION 9.18. *Let $\mathcal{H} = \mathbb{C}^d \otimes \mathbb{C}^d$. We have*

$$\frac{\sqrt{d}}{16} \leqslant d_g(\mathrm{Sep}, \mathrm{PPT}).$$

PROOF. We use the lower bound on the distance that comes from volume comparison

$$d_g(\mathrm{Sep}, \mathrm{PPT}) \geqslant \frac{\operatorname{vrad} \mathrm{PPT}}{\operatorname{vrad} \mathrm{Sep}},$$

together with the lower bound $\operatorname{vrad} \mathrm{PPT} \geqslant \frac{1}{4d}$ (Theorem 9.13) and the upper bound $\operatorname{vrad}(\mathrm{Sep}) \leqslant 4d^{-3/2}$ (Theorem 9.3). □

Proposition 9.18 asserts that there are PPT states that are far from the set of separable states. Another way of quantifying this phenomenon is as follows. Given a state ρ on $\mathbb{C}^d \otimes \mathbb{C}^d$, we introduce

$$d_{\mathrm{Sep}}(\rho) = \min_{\sigma \in \mathrm{Sep}(\mathbb{C}^d \otimes \mathbb{C}^d)} \|\rho - \sigma\|_1.$$

THEOREM 9.19 (Not proved here). *For every $\varepsilon > 0$, for d large enough, there is a PPT state ρ on $\mathbb{C}^d \otimes \mathbb{C}^d$ such that $d_{\mathrm{Sep}}(\rho) \geqslant 2 - \varepsilon$.*

The proof of Theorem 9.19 involves tricks that are beyond the scope of this book. However, we present an argument showing that a weaker lower bound on the distance to separable states (1/4 instead of 2) is achieved in a generic direction.

PROPOSITION 9.20. *Let S denote the unit sphere in the space of trace zero Hermitian operators on $\mathbb{C}^d \otimes \mathbb{C}^d$. For most directions $u \in S$, there exists a PPT state ρ such that*

$$d_{\mathrm{Sep}}(\rho) \geqslant \|u\|_\infty^{-1} \min_{\sigma \in \mathrm{Sep}(\mathbb{C}^d \otimes \mathbb{C}^d)} \mathrm{Tr}((\rho - \sigma)u) \geqslant \frac{1}{4} - o(1).$$

PROOF. We consider the support functions $w(\mathrm{PPT}, \cdot)$ and $w(\mathrm{Sep}, \cdot)$, as defined in (4.29). Since the outradii of PPT and Sep are less than 1, these functions are 1-Lipschitz on S. Note also that the average of these functions on S is exactly the mean width of the corresponding set. Using the values from Table 5.2, we conclude that, for $K = \mathrm{PPT}$ or $K = \mathrm{Sep}$ and for $\varepsilon > 0$,

$$\mathbf{P}(|w(K, \cdot) - w(K)| > \varepsilon) \leqslant 2\exp(-\varepsilon^2(d^4 - 1)/2).$$

We next use the bounds $w(\mathrm{PPT}) \geqslant (\frac{1}{2} - o(1))d^{-1}$ (Exercise 9.6) and $w(\mathrm{Sep}) \leqslant 4d^{-3/2}$ (Theorem 9.3) to conclude that, for most directions $u \in S$, we have

$$w(\mathrm{PPT}, u) \geqslant \left(\frac{1}{2} - o(1)\right) d^{-1}, \quad w(\mathrm{Sep}, u) \leqslant 5d^{-3/2}. \tag{9.27}$$

Moreover (see Proposition 6.24), most directions u also satisfy

$$\|u\|_\infty \leqslant (2 + o(1))d^{-1}. \tag{9.28}$$

Choose $u \in S$ satisfying both (9.27) and (9.28), and let $\rho \in \mathrm{PPT}$ be such that $\mathrm{Tr}(\rho u) = w(\mathrm{PPT}, u)$. We then have

$$\sup_{\sigma \in \mathrm{Sep}} \mathrm{Tr}((\rho - \sigma)u) = w(\mathrm{PPT}, u) - w(\mathrm{Sep}, u) \geqslant \left(\frac{1}{2} - o(1)\right) d^{-1}.$$

Using the inequality $\mathrm{Tr}((\rho - \sigma)u) \leqslant \|u\|_\infty \|\rho - \sigma\|_1 \leqslant (2 + o(1))d^{-1}\|\rho - \sigma\|_1$, we obtain

$$d_{\mathrm{Sep}}(\rho) \geqslant \frac{1}{4} - o(1). \qquad \square$$

Any improvement on the lower bound (9.27) for the mean width of PPT would improve the lower bound in Proposition 9.20.

9.2.4. Distance estimates in the multipartite case.

We now focus on the case of k qubits, i.e., the Hilbert space $\mathcal{H} = (\mathbb{C}^2)^{\otimes k}$. Recall that the inradius of Sep is witnessed by balls centered at ρ_* (see Proposition 2.18 and the discussion in the preamble to the present chapter). The inradius of Sep is known up to a universal (not too large) multiplicative constant.

THEOREM 9.21 (Not proved here; see Exercise 9.9). *For $\mathcal{H} = (\mathbb{C}^2)^{\otimes k}$, we have*

$$\sqrt{54/17} \times 6^{-k/2} \leqslant \mathrm{inrad}(\mathrm{Sep}) \leqslant 2 \times 6^{-k/2}.$$

We next turn to the problem of estimating the geometric distance between D and Sep in the case of many qubits, for which even the asymptotic order is not known.

PROPOSITION 9.22 (Robustness for many qubits). *For $\mathcal{H} = (\mathbb{C}^2)^{\otimes k}$, we have*

$$2^{k-1} + 1 \leqslant d_g(\mathrm{Sep}, \mathrm{D}) \leqslant (\sqrt{6})^k.$$

PROOF. The upper bound is fairly straightforward: it follows by comparing the two sets with the Hilbert–Schmidt ball. Specifically, we use the elementary inequality $d_g(\mathrm{Sep}, \mathrm{D}) \leqslant \mathrm{outrad}(\mathrm{D})/\mathrm{inrad}(\mathrm{Sep})$, combined with Theorem 9.21 and with the obvious fact that $\mathrm{outrad}(\mathrm{D}(\mathbb{C}^n)) = \sqrt{(n-1)/n}$.

To prove the lower bound, consider any decomposition $(\mathbb{C}^2)^{\otimes k} = \mathcal{A} \otimes \mathcal{B}$, where $\mathcal{A} = (\mathbb{C}^2)^{\otimes j}$ and $\mathcal{B} = (\mathbb{C}^2)^{\otimes(k-j)}$ for some $0 < j < k$. A separable state on $(\mathbb{C}^2)^{\otimes k}$ is also separable along the $\mathcal{A} : \mathcal{B}$ cut, and therefore

$$d_g(\mathrm{D}((\mathbb{C}^2)^{\otimes k}), \mathrm{Sep}((\mathbb{C}^2)^{\otimes k})) \geqslant d_g(\mathrm{D}(\mathcal{A} \otimes \mathcal{B}), \mathrm{Sep}(\mathcal{A} \otimes \mathcal{B})) = 2^{k-1} + 1,$$

where the last equality comes from Proposition 9.17. □

EXERCISE 9.9 (A bound on the inradius of Sep on k qubits via mean width). Let $P : \mathsf{M}_2^{\mathrm{sa}} \to \mathsf{M}_2^{\mathrm{sa}}$ be the orthogonal projection onto the hyperplane of trace zero matrices, and let $\Pi = P^{\otimes k}$.
(i) Check that $\Pi(\mathrm{Sep}((\mathbb{C}^2)^{\otimes k})_{\wp}) = (P(\mathrm{D}(\mathbb{C}^2)_{\wp}))^{\hat{\otimes} k}$.
(ii) Show that

$$\mathrm{inrad}\left(\mathrm{Sep}((\mathbb{C}^2)^{\otimes k})\right) \leqslant \mathrm{inrad}\left(\left(2^{-1/2} B_2^3\right)^{\hat{\otimes} k}\right) = O(\sqrt{k \log k} \cdot 6^{-k/2}).$$

9.3. The super-picture: Classes of maps

Up to now, we focused on determining volumes and other geometric parameters for various classes of states. Due to the Choi–Jamiołkowski isomorphism (see Section 2.3.1), these results can be translated into statements about the corresponding classes of quantum maps, or *superoperators*. However, there are some fine points that need to be addressed for such translation to be rigorous.

To exemplify the fine points, consider the cone $\boldsymbol{CP} = \boldsymbol{CP}(\mathsf{M}_m, \mathsf{M}_n)$ of completely positive maps $\Phi : \mathsf{M}_m \to \mathsf{M}_n$, which can be identified via the Choi isomorphism $\Phi \mapsto C(\Phi)$ (see Section 2.4, especially Table 2.2) with the positive semi-definite cone $\mathcal{PSD}(\mathbb{C}^n \otimes \mathbb{C}^m)$. So far, so good. However, if we restrict our attention to the subset of trace-preserving maps Φ (denoted by $\boldsymbol{CP}_{\mathrm{TP}}$), the set of the corresponding Choi matrices $C(\Phi)$ forms a *proper* subset of the rescaled set of states $m\mathrm{D}(\mathbb{C}^n \otimes \mathbb{C}^m)$. This is due to the fact that the trace-preserving condition $\mathrm{Tr}\,\Phi(\rho) = \mathrm{Tr}\,\rho$ (for $\rho \in \mathsf{M}_m$) translates into $\mathrm{Tr}_{\mathbb{C}^n} C(\Phi) = \mathrm{I}_{\mathbb{C}^m}$ (which implies $\mathrm{Tr}\, C(\Phi) = m$, hence the rescaling factor m), which represents m^2 independent (real linear) scalar constraints. On the other hand, membership in $m\mathrm{D}(\mathbb{C}^n \otimes \mathbb{C}^m)$ is represented by just one scalar constraint $\mathrm{Tr}(\cdot) = m$ (in addition to the positive semi-definiteness constraint common to both settings).

In other words, if we denote by $H \subset B^{\mathrm{sa}}(\mathbb{C}^n \otimes \mathbb{C}^m)$ the affine subspace $\{\mathrm{Tr}_{\mathbb{C}^n}(\cdot) = \mathrm{I}_{\mathbb{C}^m}\}$, then the rescaled set of states $K = m\mathrm{D}(\mathbb{C}^n \otimes \mathbb{C}^m)$ is a base of the positive semi-definite cone, which is an $m^2n^2 - 1$-dimensional convex set, while the set of Choi matrices corresponding to completely positive trace-preserving maps is $K \cap H$, a section of that base of relative codimension $m^2 - 1$, i.e., a convex set of dimension $m^2n^2 - m^2$.

The problem of relating the size of a convex set to that of its (central) sections is in general nontrivial, and two-sided bounds are only possible if the set is isotropic (in the technical sense defined in Section 4.4; see especially Proposition 4.26). The set D of all states actually *is* isotropic (see Proposition 4.25). While not all natural sets of states have this property, they are all sufficiently balanced so that the more robust Proposition 4.28 leads to reasonable estimates. For notational simplicity, we restrict ourself to superoperators $\Phi : \mathsf{M}_d \to \mathsf{M}_d$ in the following theorem.

THEOREM 9.23. *Let $\boldsymbol{C} = \boldsymbol{C}(\mathsf{M}_d, \mathsf{M}_d)$ be one of the cones of superoperators appearing in Table 9.3, and $\boldsymbol{C}_{\mathrm{TP}} := \{\Phi \in \boldsymbol{C} : \Phi \text{ is trace-preserving}\}$. Denote also by $\mathcal{C} = \{C(\Phi) : \Phi \in \boldsymbol{C}\}$ the corresponding cone of Choi matrices and by $\mathcal{C}^{\mathrm{b}} = \{A \in \mathcal{C} : \operatorname{Tr} A = 1\}$ its base. Then, as $d \to \infty$,*

$$\operatorname{vrad}\big(\boldsymbol{C}_{\mathrm{TP}}\big) \sim d \operatorname{vrad}\big(\mathcal{C}^{\mathrm{b}}\big). \tag{9.29}$$

TABLE 9.3. Each cone $\boldsymbol{C}$ of superoperators is a nondegenerate cone in $B(\mathsf{M}_d^{\mathrm{sa}}, \mathsf{M}_d^{\mathrm{sa}})$ and the subset $\boldsymbol{C}_{\mathrm{TP}}$ of trace-preserving elements is a convex set of dimension $d^4 - d^2$. The cone $\mathcal{C} \subset B^{\mathrm{sa}}(\mathbb{C}^d \otimes \mathbb{C}^d)$ is the image of $\boldsymbol{C}$ under the map $\Phi \mapsto C(\Phi)$, see Section 2.4.

Cone of superoperators $\boldsymbol{C}$		Cone $\mathcal{C}$	Base $\mathcal{C}^{\mathrm{b}}$	$\operatorname{vrad}(\boldsymbol{C}_{\mathrm{TP}})$
Positivity-preserving	$\boldsymbol{P}$	$\mathcal{BP}$	BP	$\Theta(\sqrt{d})$
Decomposable	$\boldsymbol{DEC}$	co-$\mathcal{PSD}$ +$\mathcal{PSD}$	$\operatorname{conv}(\mathrm{D} \cup \Gamma(\mathrm{D}))$	$\Theta(1)$
Completely positive	$\boldsymbol{CP}$	$\mathcal{PSD}$	D	$\sim e^{-1/4}$
PPT-inducing	$\boldsymbol{PPT}$	$\mathcal{PPT}$	PPT	$\Theta(1)$
Entanglement breaking	$\boldsymbol{EB}$	$\mathcal{SEP}$	Sep	$\Theta(1/\sqrt{d})$

PROOF OF THEOREM 9.23. Denote $K = d\mathcal{C}^{\mathrm{b}}$ and $n = \dim K = d^4 - 1$. Since K is invariant under local unitaries, it follows (see Proposition 2.18) that the centroid of K equals I/d.

As explained earlier, $\boldsymbol{C}_{\mathrm{TP}}$ identifies with a section of K (through the centroid) of codimension $k = d^2 - 1$. It follows from Proposition 4.28 that

$$R^{-\theta}\, b(n,k) \leqslant \frac{\operatorname{vrad}(\boldsymbol{C}_{\mathrm{TP}})^{1-\theta}}{\operatorname{vrad}(K)} \leqslant r^{-\theta}\, b(n,k) \binom{n}{k}^{\frac{1}{n}}, \tag{9.30}$$

where $\theta = \frac{k}{n} = \frac{d^2-1}{d^4-1} < \frac{1}{d^2}$ and r, R denote respectively the inradius and outradius of K. The constants $b(n,k)$ were defined in (4.51); in our setting the bounds (4.55) can be sharpened (see Exercise 9.10) to

$$b(n,k) = 1 - O\left(\frac{\log d}{d^2}\right), \quad b(n,k)\binom{n}{k}^{\frac{1}{n}} = 1 + O\left(\frac{\log d}{d^2}\right). \tag{9.31}$$

Since all the cones we consider have the property that $\mathrm{Sep} \subset \mathcal{C}^{\mathrm{b}} \subset \mathrm{BP} = -d^2\mathrm{Sep}^\circ$, we know from Table 9.1 that $r = 1/\sqrt{d^2-1}$ and $R = \sqrt{d^2-1}$, so $r^{-\theta} = R^{\theta} = 1 + O\big(\frac{\log d}{d^2}\big)$. Combining (9.30) and (9.31) yields $\operatorname{vrad}(\boldsymbol{C}_{\mathrm{TP}})^{1-\theta} \sim d \operatorname{vrad}\big(\mathcal{C}^{\mathrm{b}}\big)$, and it remains to again notice that since θ is small, the exponent $1-\theta$ does not make much of a difference (this uses very weakly the estimates on the volume radii from Table 9.1, or just rough bounds given by r and R). □

The same argument leads to non-asymptotic bounds (i.e., stated for a fixed dimension) and to bounds for maps from M_m to M_n. We also state explicitly a version of Theorem 9.23 for the mean width. As we shall see in Chapter 10, the latter may also be of independent importance.

PROPOSITION 9.24. *In the same notation as in Theorem 9.23, we have*

$$w(C_{\mathrm{TP}}) \leqslant (1 + d^{-2})\, d\, w(\mathcal{C}^{\mathrm{b}}). \tag{9.32}$$

PROOF. This is a consequence of the following inequality: for an n-dimensional convex body K and a k-codimensional affine subspace H, we have

$$w(K \cap H) \leqslant \sqrt{\frac{n}{n-k-1}}\, w(K). \tag{9.33}$$

Inequality (9.33) follows from the link (4.32) between the Gaussian mean width and the standard mean width, from the fact that the Gaussian mean width of a subset does not exceed that of the entire set, and from the inequality $\sqrt{n-1} \leqslant \kappa_n \leqslant \sqrt{n}$ (Proposition A.1(i)). □

Deriving meaningful lower bounds for $w(K \cap H)$ in terms of $w(K)$ in a general setting (such as Proposition 4.28 for the volume radius) is not that easy. However, when K is one of the sets $\mathcal{C}^{\mathrm{b}}$ from Table 9.3, nontrivial lower bounds for the mean width follow from the estimates on the volume radii contained in the Table and from Urysohn's inequality.

EXERCISE 9.10. Prove the bounds (9.31).

EXERCISE 9.11 (Cones of channels are not self-dual). Let $\mathcal{H} = \mathbb{C}^m \otimes \mathbb{C}^n$.
(i) Consider the affine subspace $H = \{A \in B(\mathcal{H}) \ : \ \mathrm{Tr}_{\mathbb{C}^n} A = \frac{\mathrm{I}}{m}\}$. Show that $\mathrm{D} \cap H \subsetneq P_H\mathrm{D}$ and $\mathrm{Sep} \cap H \subsetneq P_H\mathrm{Sep}$.
(ii) Conclude in particular that $(\mathrm{D} \cap H)^\circ \neq -mn(\mathrm{D} \cap H)$: the self-duality of D is destroyed by the partial trace condition.
(iii) Consider the affine subspace $F = \{\frac{\mathrm{I}}{m} \otimes \sigma \ : \ \sigma \in \mathsf{M}_n^{\mathrm{sa}} \ : \ \mathrm{Tr}\, \sigma = 1\} \subsetneq H$. Show that $\mathrm{D} \cap F = P_F\mathrm{D} = \mathrm{Sep} \cap F = P_F\mathrm{Sep}$.

9.4. Approximation by polytopes

The proofs of volume and mean width estimates given in Section 9.1 proceed through a symmetrization argument, by showing that the symmetrized sets ($\mathrm{D}_{\varnothing}$ or $\mathrm{Sep}_{\varnothing}$) are close, with respect to the geometric distance, to a polytope with not-too-many vertices. It is natural to wonder whether similar effect can be achieved without symmetrization. This problem is also of independent interest since approximating convex sets by polytopes with not-too-many vertices, or with not-too-many faces, has important algorithmic implications, and is a much studied question in computational geometry. It is convenient to formulate the results using the notion of verticial and facial dimensions introduced in Section 7.2.3. For an easy overview and reference, we list the results in Table 9.4; the proofs can be found in the next two sections.

9.4.1. Approximating the set of all quantum states. We first show that it is possible to approximate D by a polytope whose number of vertices is exponential in the dimension of the underlying Hilbert space. Recall that the notation $t \bullet K$ was defined in (9.1).

PROPOSITION 9.25. *For every $\varepsilon \in (0,1)$, there is a constant $C(\varepsilon)$ such that the following holds: for every dimension $d \geqslant 2$, there exists a family $\mathcal{N} = (\varphi_i)_{1 \leqslant i \leqslant N}$ of unit vectors in $\mathbb{C}^d$, with $N \leqslant \exp(C(\varepsilon)d)$, such that*

$$(1-\varepsilon) \bullet \mathrm{D}(\mathbb{C}^d) \subset \mathrm{conv}\{|\varphi_i\rangle\langle\varphi_i| \ : \ \varphi_i \in \mathcal{N}\}. \tag{9.34}$$

TABLE 9.4. Verticial and facial dimensions of the set of states $\mathrm{D}(\mathbb{C}^m)$ and of the set of separable states $\mathrm{Sep}(\mathbb{C}^d \otimes \mathbb{C}^d)$. They are proved respectively in Sections 9.4.1 (Corollary 9.26) and 9.4.2 (Proposition 9.31 and Corollary 9.32). We also include the values of asphericities $a(\mathrm{D})$ and $a(\mathrm{Sep})$ (see Exercises 9.12 and 9.14), which are important in some applications and derivations. For all these notions, the maximally mixed state ρ_* plays the role of the origin.

K	dimension	$a(K)$	$\dim_V(K)$	$\dim_F(K)$
$\mathrm{D}(\mathbb{C}^m)$	m^2-1	$m-1$	$\Theta(m)$	$\Theta(m)$
$\mathrm{Sep}(\mathbb{C}^d \otimes \mathbb{C}^d)$	d^4-1	d^2-1	$\Theta(d\log d)$	$\Omega(d^3/\log d)$

The result from Proposition 9.25 can be rephrased as estimates on the verticial (or facial) dimension of $\mathrm{D}(\mathbb{C}^d)$.

COROLLARY 9.26. *There are absolute constants c, C such that, for any $d \geqslant 2$,*

$$cd \leqslant \dim_V(\mathrm{D}(\mathbb{C}^d)) = \dim_F(\mathrm{D}(\mathbb{C}^d)) \leqslant Cd.$$

PROOF. Since $\mathrm{D}^\circ = (-d) \bullet \mathrm{D}$, the facial and verticial dimensions are equal. The upper bound follows from Proposition 9.25. Using the value $a(\mathrm{D}) = d-1$ (see Table 9.1 and Exercise 9.12), one can deduce the lower bound from Theorem 7.29. Alternatively, an elementary argument is sketched in Exercise 9.13. □

It may seem reasonable to expect that choosing as $\mathcal{N}$ a δ-net in $S_{\mathbb{C}^d}$ (for some δ depending only on ε) would be enough for the conclusion of Proposition 9.25 to hold. This is the case for $\mathrm{D}_{\wp}$ (see Lemma 9.2). However, this approach fails for D. Indeed, given δ, for d large enough, a δ-net $\mathcal{N}$ may have the property that, for some fixed unit vector ψ, we have $|\langle \varphi_i, \psi\rangle| > 1/\sqrt{d}$ for every $\varphi_i \in \mathcal{N}$. It follows that $\langle\psi|\rho|\psi\rangle > 1/d$ for every $\rho \in \mathrm{conv}\{|\varphi_i\rangle\langle\varphi_i|\}$. However, this inequality fails for $\rho = \rho_*$, which shows that even the maximally mixed state does not belong to the convex hull of the net! Elements of the net may somehow conspire towards the direction ψ.

Yet, this approach can be salvaged if we use a balanced δ-net to avoid such conspiracies. The idea is to use, instead of an arbitrary net, a family of random points independently and uniformly distributed on the unit sphere, and to show that these points satisfy the conclusion of Proposition 9.25 with high probability. This is reminiscent of the random covering argument used in Proposition 5.4.

We start with a lemma which gives a rough bound on the number of unit vectors that are needed.

LEMMA 9.27. *Let $\mathcal{M}$ be a δ-net in $(S_{\mathbb{C}^d}, |\cdot|)$. Then*

$$(1-2d\delta) \bullet \mathrm{D}(\mathbb{C}^d) \subset \mathrm{conv}\{|\psi_i\rangle\langle\psi_i| \;:\; \psi_i \in \mathcal{M}\} \subset \mathrm{D}(\mathbb{C}^d). \tag{9.35}$$

The reader will notice that the proof given below can be fine-tuned to yield a slightly better (but more complicated) factor $\big(1-2(d-1)\delta\big)$ in (9.35).

PROOF. We have to show that, for any trace zero Hermitian matrix A,

$$\lambda_1(A) = \sup_{\psi \in S_{\mathbb{C}^d}} \langle\psi|A|\psi\rangle \leqslant (1-2\delta d)^{-1} \sup_{\psi_i \in \mathcal{N}} \langle\psi_i|A|\psi_i\rangle.$$

Since A has zero trace, we have $\|A\|_\infty \leqslant d\lambda_1(A)$. Given $\psi \in S_{\mathbb{C}^d}$, there is $\psi_i \in \mathcal{M}$ with $|\psi - \psi_i| \leqslant \delta$. By the triangle inequality, we have

$$\langle\psi|A|\psi\rangle \leqslant \delta\|A\|_\infty + \langle\psi|A|\psi_i\rangle \tag{9.36}$$

$$\leqslant 2\delta\|A\|_\infty + \langle\psi_i|A|\psi_i\rangle \tag{9.37}$$

$$\leqslant 2\delta d\lambda_1(A) + \langle\psi_i|A|\psi_i\rangle. \tag{9.38}$$

Taking supremum over ψ, we get $\lambda_1(A) \leqslant 2\delta d\,\lambda_1(A) + \sup\{\langle\psi_i|A|\psi_i\rangle \ : \ \psi_i \in \mathcal{M}\}$ and the result follows. □

Lemma 9.27 is not sufficient to directly imply Proposition 9.25, but it can be "bootstrapped" to yield the needed estimate.

Proof of Proposition 9.25. The conclusion (9.34) can be equivalently reformulated as follows: *For any self-adjoint trace zero matrix A we have*

$$\lambda_1(A) = \sup_{\psi\in S_{\mathbb{C}^d}} \langle\psi|A|\psi\rangle \leqslant \frac{1}{1-\varepsilon} \sup_{\varphi_i\in\mathcal{N}} \langle\varphi_i|A|\varphi_i\rangle. \tag{9.39}$$

Let $\mathcal{M}$ be a $\frac{\varepsilon}{4d}$-net in $(S_{\mathbb{C}^d}, |\cdot|)$. By Lemma 5.3, we may enforce $\operatorname{card}\mathcal{M} \leqslant (8d/\varepsilon)^{2d}$. By Lemma 9.27, we have

$$\sup_{\psi\in S_{\mathbb{C}^d}} \langle\psi|A|\psi\rangle \leqslant \frac{1}{1-\varepsilon/2} \sup_{\psi\in\mathcal{M}} \langle\psi|A|\psi\rangle. \tag{9.40}$$

Set $\eta = \sqrt{\varepsilon/8}$. For $\psi \in S_{\mathbb{C}^d}$, denote by $C(\psi,\eta) \subset S_{\mathbb{C}^d}$ the cap with center ψ and radius η with respect to the geodesic distance. By symmetry, there is a number α (depending on d and ε) such that

$$\frac{1}{\sigma(C(\psi,\eta))}\int_{C(\psi,\eta)} |\varphi\rangle\langle\varphi|\,\mathrm{d}\sigma(\varphi) = (1-\alpha)\bullet|\psi\rangle\langle\psi|. \tag{9.41}$$

Taking (Hilbert–Schmidt) inner product with $|\psi\rangle\langle\psi|$, we obtain

$$1-\alpha+\frac{\alpha}{d} = \frac{1}{\sigma(C(\psi,\eta))}\int_{C(\psi,\eta)} |\langle\psi,\varphi\rangle|^2\,\mathrm{d}\sigma(\varphi) \geqslant \cos^2\eta \geqslant 1-\eta^2$$

so that

$$\alpha \leqslant \eta^2\frac{d}{d-1} \leqslant \varepsilon/4. \tag{9.42}$$

Denote $L := \sigma(C(\psi,\eta))^{-1}$ and let $\mathcal{N} = \{\varphi_i \ : \ 1 \leqslant i \leqslant 2L^3\}$ be a family of $N = \lceil 2L^3\rceil$ independent random vectors uniformly distributed on $S_{\mathbb{C}^d}$. (To not to obscure the argument, we will pretend in what follows that $2L^3$ is an integer and so $N = 2L^3$.) We will rely on the following lemma.

Lemma 9.28. *Let $S_\infty^d = \{\Delta \in \mathsf{M}_d \ : \ \|\Delta\|_{\mathrm{op}} \leqslant 1\}$ be the unit ball for the operator norm. For $\psi \in S_{\mathbb{C}^d}$ and $t \geqslant 0$, the event*

$$E_{\psi,t} = \Big\{(\varphi_i) \ : \ (1-\alpha)\bullet|\psi\rangle\langle\psi| \in tS_\infty^d + \operatorname{conv}\{|\varphi_i\rangle\langle\varphi_i| \ : 1 \leqslant i \leqslant 2L^3\}\Big\}$$

satisfies

$$1-\mathbf{P}(E_{\psi,t}) \leqslant \exp(-L) + 2d\exp\left(-t^2L^2/8\right).$$

We apply Lemma 9.28 with $t = \varepsilon/8d$. When the event $E_{\psi,t}$ holds, we have

$$(1-\alpha)\langle\psi|A|\psi\rangle \leqslant t\|A\|_1 + \sup_{\varphi_i\in\mathcal{N}}\langle\varphi_i|A|\varphi_i\rangle. \tag{9.43}$$

If the events $E_{\psi,t}$ hold simultaneously for every $\psi \in \mathcal{M}$, we can conclude from (9.40) and (9.43) that

$$(1-\varepsilon/2)(1-\alpha)\lambda_1(A) \leqslant t\|A\|_1 + \sup_{\varphi_i\in\mathcal{N}}\langle\varphi_i|A|\varphi_i\rangle. \tag{9.44}$$

Since A has zero trace, we have $\|A\|_1 \leqslant 2d\lambda_1(A)$, and (9.44) combined with (9.42) implies that

$$(1-\varepsilon)\lambda_1(A) \leqslant \big((1-\varepsilon/2)(1-\alpha)-2td\big)\lambda_1(A) \leqslant \sup_{\varphi_i\in\mathcal{N}}\langle\varphi_i|A|\varphi_i\rangle,$$

yielding (9.39). The Proposition will follow once we show that the events $E_{\psi,t}$ hold simultaneously for every $\psi \in \mathcal{M}$ with positive probability. To that end, we use Lemma 9.28 and the union bound

$$\mathbf{P}\left(\bigcap_{\psi\in\mathcal{M}} E_{\psi,t}\right) \geqslant 1-\sum_{\psi\in\mathcal{M}}(1-\mathbf{P}(E_{\psi,t})) \tag{9.45}$$

$$\geqslant 1-\left(\frac{8d}{\varepsilon}\right)^{2d}\Big(\exp(-L)+2d\exp\big(-\varepsilon^2d^{-2}L^2/512\big)\Big). \tag{9.46}$$

We know from Proposition 5.1 that $\exp(c_1(\varepsilon)d) \leqslant L \leqslant \exp(C_1(\varepsilon)d)$ for some constants $c_1(\varepsilon)$, $C_1(\varepsilon)$ depending only on ε. It follows that the quantity in (9.45)–(9.46) is positive for d large enough (depending on ε), yielding a family of $2L^3 \leqslant 2\exp(3C_1(\varepsilon)d)$ vectors satisfying the conclusion of Proposition 9.25. Small values of d are taken care of by adjusting the constant $C(\varepsilon)$ if necessary. □

PROOF OF LEMMA 9.28. Let $M_\psi = \operatorname{card}(\mathcal{N} \cap C(\psi,\eta))$. The random variable M_ψ follows the binomial distribution $B(N,p)$ for $N = 2L^3$ and $p = 1/L$. It follows from Hoeffding's inequality (5.43) that

$$\mathbf{P}\left(B(N,p) \leqslant \frac{Np}{2}\right) \leqslant \exp\left(-\frac{p^2N}{2}\right).$$

Specialized to our situation, this yields

$$\mathbf{P}\left(M_\psi \leqslant L^2\right) \leqslant \exp\left(-L\right). \tag{9.47}$$

Moreover, conditionally on the value of M_ψ, the points from $\mathcal{N} \cap C(\psi,\eta)$ have the same distribution as $(\varphi_k)_{1\leqslant k\leqslant M_\psi}$, where (φ_k) are independent and uniformly distributed inside $C(\psi,\eta)$. The random matrices

$$X_k = |\varphi_k\rangle\langle\varphi_k| - \mathbf{E}\,|\varphi_1\rangle\langle\varphi_1| = |\varphi_k\rangle\langle\varphi_k| - (1-\alpha)\bullet|\psi\rangle\langle\psi|$$

are independent mean zero matrices. We now use the *matrix Hoeffding inequality* (see, e.g., Theorem 1.3 in [**Tro12**]) to conclude that, for any $t \geqslant 0$,

$$\mathbf{P}\left(\left\|\frac{1}{M_\psi}\sum_{k=1}^{M_\psi}X_k\right\|_\infty \geqslant t\right) \leqslant 2d\exp(-M_\psi t^2/8) \tag{9.48}$$

(the factor 2 appears because we want to control the operator norm rather than the largest eigenvalue). Define a random matrix Δ by the relation

$$\frac{1}{M_\psi} \sum_{k=1}^{M_\psi} |\varphi_k\rangle\langle\varphi_k| + \Delta = (1-\alpha) \bullet |\psi\rangle\langle\psi|.$$

The bound (9.48) translates into $\mathbf{P}(\|\Delta\|_\infty \geqslant t) \leqslant 2d\exp(-M_\psi t^2/8)$. If we remove the conditioning on M_ψ and take (9.47) into account, we are led to

$$\mathbf{P}(\|\Delta\|_\infty \geqslant t) \leqslant \exp(-L) + 2d\exp\left(-L^2t^2/8\right),$$

which is exactly the content of Lemma 9.28. □

EXERCISE 9.12 (Asphericity of D). By comparing the values of the inradius and the outradius of $\mathrm{D}(\mathbb{C}^m)$ from Table 9.1, we see that the asphericity of $\mathrm{D}(\mathbb{C}^m)$ is at most $m-1$. Prove that it actually equals $m-1$.

EXERCISE 9.13 (An elementary bound for verticial dimension of D). Let P be a polytope such that $\frac{1}{4} \bullet \mathrm{D}(\mathbb{C}^d) \subset P \subset \mathrm{D}(\mathbb{C}^d)$. Use Proposition 6.3 to prove that P has at least $\exp(cd)$ vertices for some $c > 0$.

9.4.2. Approximating the set of separable states. For simplicity, we only consider the case $\mathcal{H} = \mathbb{C}^d \otimes \mathbb{C}^d$ and denote $\mathrm{Sep} = \mathrm{Sep}(\mathbb{C}^d \otimes \mathbb{C}^d)$. As in the case of D, a simple net argument (Lemma 9.29) shows that the verticial dimension of Sep is $O(d\log d)$. However there is no analogue of the random construction used in Proposition 9.25: this upper bound is sharp (see Proposition 9.31 below). Here are the precise statements and the proofs.

LEMMA 9.29. *Let* $\mathrm{Sep} = \mathrm{Sep}(\mathbb{C}^d \otimes \mathbb{C}^d)$. *If* $\mathcal{N}$ *is a* $\frac{\varepsilon}{4d^2}$*-net in* $(S_{\mathbb{C}^d}, |\cdot|)$, *then*

$$(1-\varepsilon) \bullet \mathrm{Sep} \subset \mathrm{conv}\left\{|\psi_\alpha \otimes \psi_\beta\rangle\langle\psi_\alpha \otimes \psi_\beta| \ : \ \psi_\alpha, \psi_\beta \in \mathcal{N}\right\}.$$

In particular, $\dim_V(\mathrm{Sep}) \leqslant Cd\log d$ *for some constant* C.

PROOF. We have to show that for any trace zero Hermitian matrix A we have

$$W := \sup_{\psi,\varphi \in S_{\mathbb{C}^d}} \langle\psi\otimes\varphi|A|\psi\otimes\varphi\rangle \leqslant (1-\varepsilon)^{-1} \sup_{\psi_\alpha,\psi_\beta\in\mathcal{N}} \langle\psi_\alpha\otimes\psi_\beta|A|\psi_\alpha\otimes\psi_\beta\rangle.$$

First, note using Theorem 9.15 that

$$W \geqslant \frac{1}{d^2}\|A\|_2 \geqslant \frac{1}{d^2}\|A\|_\infty.$$

Let $\delta = \varepsilon/4d^2$. Given $\varphi, \psi \in S_{\mathbb{C}^d}$, there are $\psi_\alpha, \psi_\beta \in \mathcal{N}$ with $|\varphi - \psi_\alpha| \leqslant \delta$ and $|\psi - \psi_\beta| \leqslant \delta$. Using the triangle inequality as in (9.36)–(9.37), we obtain

$$\langle\varphi\otimes\psi|A|\varphi\otimes\psi\rangle \leqslant 4\delta\|A\|_\infty + \langle\psi_\alpha\otimes\psi_\beta|A|\psi_\alpha\otimes\psi_\beta\rangle \leqslant \varepsilon W + \langle\psi_\alpha\otimes\psi_\beta|A|\psi_\alpha\otimes\psi_\beta\rangle.$$

Taking supremum over ψ, φ gives the result. The estimate on the verticial dimension follows from Lemma 5.3. □

REMARK 9.30. A closer examination of the above proof shows that the bound $O(d\log d)$ on the verticial dimension allows for an approximation more precise than the default "up to factor 4" implicit in the definitions found in Section 7.2.3. For example, the argument gives that (in the notation from Exercise 7.15) $\dim_V(\mathrm{Sep}, 1+d^{-\kappa}) \leqslant Cd\log d$, where the constant C depends on $\kappa > 0$.

We will show next that the upper bound obtained in Lemma 9.29 is sharp. This is in contrast with the case of the symmetrized set $\mathrm{Sep}_{\oslash}$, whose verticial dimension is of order d (see Lemma 9.4).

PROPOSITION 9.31. *Let* $\mathrm{Sep} = \mathrm{Sep}(\mathbb{C}^d \otimes \mathbb{C}^d)$. *Then* $\dim_V(\mathrm{Sep}) \geqslant cd\log d$ *for some constant* $c > 0$.

PROOF. Let P be a polytope with N vertices such that $\frac{1}{4} \bullet \mathrm{Sep} \subset P \subset \mathrm{Sep}$. By Carathéodory's theorem, we may write each vertex of P as a combination of d^4 extreme points of Sep (which are pure product states, i.e., of the form $|\psi\otimes\varphi\rangle\langle\psi\otimes\varphi|$ for unit vectors $\psi, \varphi \in \mathbb{C}^d$). We obtain therefore a polytope Q which is the convex hull of $N' \leqslant Nd^4$ pure product states, and such that $\frac{1}{4} \bullet \mathrm{Sep} \subset Q \subset \mathrm{Sep}$. Let $(|\psi_i \otimes \varphi_i\rangle\langle\psi_i \otimes \varphi_i|)_{1\leqslant i\leqslant N'}$ be the vertices of Q. Fix $\chi \in S_{\mathbb{C}^d}$ arbitrarily. For any $\varphi \in S_{\mathbb{C}^d}$, let $\alpha = \max\{|\langle\varphi,\varphi_i\rangle|^2 \ : \ 1 \leqslant i \leqslant N'\}$. Consider the linear form

$$g(\rho) = \mathrm{Tr}\left[\rho\left(|\chi\rangle\langle\chi| \otimes (\alpha\,\mathrm{I}_{\mathbb{C}^d} - |\varphi\rangle\langle\varphi|)\right)\right].$$

For any $1 \leqslant i \leqslant N'$ we have

$$g(|\psi_i \otimes \varphi_i\rangle\langle\psi_i \otimes \varphi_i|) = |\langle\chi,\psi_i\rangle|^2(\alpha - |\langle\varphi,\varphi_i\rangle|^2) \geqslant 0,$$

and therefore g is nonnegative on Q. Since $Q \supset \frac{1}{4} \bullet \mathrm{Sep}$, we have

$$\begin{aligned} 0 \leqslant g\left(\frac{1}{4} \bullet |\chi\otimes\varphi\rangle\langle\chi\otimes\varphi|\right) &= \frac{1}{4} g(|\chi\otimes\varphi\rangle\langle\chi\otimes\varphi|) + \frac{3}{4} g(\rho_*) \\ &= \frac{1}{4}(\alpha - 1) + \frac{3}{4} \times \frac{1}{d}\left(\alpha - \frac{1}{d}\right) \\ &= \alpha\left(\frac{1}{4} + \frac{3}{4d}\right) - \left(\frac{1}{4} + \frac{3}{4d^2}\right). \end{aligned}$$

It follows that

$$\alpha \geqslant \frac{1 + \frac{3}{d^2}}{1 + \frac{3}{d}} \geqslant 1 - \frac{3}{d}.$$

In other words, we proved that for every $\varphi \in S_{\mathbb{C}^d}$ there is an index $i \in \{1, \ldots, N'\}$ such that $|\langle\varphi,\varphi_i\rangle|^2 \geqslant 1 - 3/d$. This means that $(\varphi_i)_{1\leqslant i\leqslant N'}$ is a $(C/\sqrt{d})$-net in the projective space $\mathsf{P}(\mathbb{C}^d)$ equipped with the quotient metric from $(S_{\mathbb{C}^d}, |\cdot|)$. By Theorem 5.11 (or Exercise 5.10), this implies that $N' \geqslant (c'\sqrt{d})^{2(d-1)}$, and therefore $\log N \geqslant cd\log d$ for some constant $c > 0$. □

We conclude this section by stating an estimate on the facial dimension of Sep.

COROLLARY 9.32. *Let* $\mathrm{Sep} = \mathrm{Sep}(\mathbb{C}^d \otimes \mathbb{C}^d)$. *Then*

$$cd^3/\log d \leqslant \dim_F(\mathrm{Sep}) \leqslant Cd^4 \tag{9.49}$$

for some absolute constants $C, c > 0$.

PROOF. We use the Figiel–Lindenstrauss–Milman inequality (Theorem 7.29). Recall that the dimension of Sep equals $d^4 - 1$. The asphericity of Sep is bounded from above by the ratio outrad(Sep)/inrad(Sep) (see Table 9.1 for the values of the radii; as indicated in Table 9.4, there is actually equality, see Exercise 9.14). Since the value of this ratio is $d^2 - 1$, it follows that

$$\dim_F(\mathrm{Sep})\dim_V(\mathrm{Sep}) \geqslant cd^4. \tag{9.50}$$

The lower bound in (9.49) is now immediate from (9.50) and Lemma 9.29, while the upper bound follows from Proposition 7.27(iv). □

PROBLEM 9.33. *Is it true that* $\dim_F(\mathrm{Sep}(\mathbb{C}^d \otimes \mathbb{C}^d)) = \Theta(d^4)$*?*

EXERCISE 9.14 (Asphericity of Sep). Prove that the asphericity of $\mathrm{Sep}(\mathbb{C}^d \otimes \mathbb{C}^d)$ equals $d^2 - 1$.

9.4.3. Exponentially many entanglement witnesses are necessary. We conclude this chapter by showing that Dvoretzky's theorem (applied via the Figiel–Lindenstrauss–Milman inequality (7.17)) implies that the set of separable states is complex in the following sense: super-exponentially many entanglement witnesses are necessary to approximate it within a constant factor.

In this section we write D, $\mathcal{PSD}$ and Sep for $\mathrm{D}(\mathbb{C}^d \otimes \mathbb{C}^d)$, $\mathcal{PSD}(\mathbb{C}^d \otimes \mathbb{C}^d)$ and $\mathrm{Sep}(\mathbb{C}^d \otimes \mathbb{C}^d)$. We denote by $\boldsymbol{P}(\mathbb{C}^d)$ the cone of positivity-preserving operators from M_d to M_d. Recall the statement of Theorem 2.34: *a state* $\rho \in \mathrm{D}$ *is entangled if and only if there exists an entanglement witness, i.e.,* $\Phi \in \boldsymbol{P}(\mathbb{C}^d)$ *such that* $(\Phi \otimes \mathrm{Id})(\rho)$ *is not positive.* In other words,

$$\mathrm{Sep} = \bigcap_{\Phi \in \boldsymbol{P}(\mathbb{C}^d)} \{\rho \in \mathrm{D} \ : \ (\Phi \otimes \mathrm{Id})(\rho) \in \mathcal{PSD}\}. \tag{9.51}$$

It is natural to wonder whether the intersection in (9.51) can be taken over a smaller subfamily. For $d = 2$, two superoperators suffice, namely Id and T; this is the content of Størmer's theorem. It is known that for $d \geqslant 3$ an infinite family is needed. If we consider instead the isomorphic version of the problem, the following theorem shows that super-exponentially (in the dimension of the underlying Hilbert space) many witnesses are necessary.

THEOREM 9.34. *There is a constant* $c > 0$ *such that the following holds: if* $\Phi_1, \ldots, \Phi_N \in \boldsymbol{P}(\mathbb{C}^d)$ *are such that*

$$\bigcap_{i=1}^{N} \{\rho \in \mathrm{D} \ : \ (\Phi_i \otimes \mathrm{Id})(\rho) \in \mathcal{PSD}\} \subset 2 \bullet \mathrm{Sep}, \tag{9.52}$$

then $N + 1 \geqslant \exp(cd^3/\log(d))$.

The following variant of the above theorem also holds.

THEOREM 9.35 (See Exercise 9.16). *There are universal constants* $c_0, c > 0$ *such that the following holds: if* $\Phi_1, \ldots, \Phi_N \in \boldsymbol{P}(\mathbb{C}^d)$ *are such that*

$$\bigcap_{i=1}^{N} \{\rho \in \mathrm{D} \ : \ (\Phi_i \otimes \mathrm{Id})(\rho) \in \mathcal{PSD}\} \subset \frac{c_0\sqrt{d}}{\log d} \bullet \mathrm{Sep}, \tag{9.53}$$

Then $N + 1 \geqslant \exp(cd^2 \log d)$.

In other words, even being able to detect very robust entanglement requires super-exponentially many witnesses. It would be of some interest to determine the maximal robustness level (defined in (9.26)) at which this phenomenon still persists. Note that, by Proposition 9.17, $\mathrm{D} \subset \left(1 + \frac{d^2}{2}\right) \bullet \mathrm{Sep}$ for states on $\mathbb{C}^d \otimes \mathbb{C}^d$, so the question is nontrivial only if a threshold for the robustness level is smaller than $\frac{d^2}{2}$.

PROOF OF THEOREM 9.34. Without loss of generality, we may assume that each superoperator Φ_i is unital (see Exercise 9.15). We use the following lemma.

LEMMA 9.36. *Let* $\Phi \in \boldsymbol{P}(\mathbb{C}^d)$ *be unital. Then for any* $\rho \in \mathrm{D}$,

$$0 \leqslant \mathrm{Tr}\left[(\Phi \otimes \mathrm{Id})\rho\right] \leqslant d.$$

PROOF. Since linear forms achieve their extrema on extreme points of convex compact sets, we may assume that $\rho = |\psi\rangle\langle\psi|$ is pure. Let $\psi = \sum \lambda_i e_i \otimes f_i$ be the Schmidt decomposition of ψ. We compute

$$\mathrm{Tr}\,[(\Phi \otimes \mathrm{Id})\rho] = \sum_{i=1}^{d} \lambda_i^2 \,\mathrm{Tr}\,\Phi(|e_i\rangle\langle e_i|) \leqslant d,$$

where the last inequality follows from the facts that $\sum \lambda_i^2 = 1$ and $\Phi(|e_i\rangle\langle e_i|) \leqslant \Phi(\mathrm{I}) = \mathrm{I}$. □

Let $\varepsilon = 1/(1+d)$. Let P be a polytope with at most $\exp(C_0 d^2 \log(d))$ facets such that

$$(1-\varepsilon) \bullet \mathrm{D} \subset P \subset \mathrm{D}. \tag{9.54}$$

The existence of P is guaranteed by Lemma 9.27, by the relation $\mathrm{D}^\circ = (-d^2) \bullet \mathrm{D}$, and by the fact that facets of P are in bijection with vertices of P° (see Section 1.1.5). Introduce the convex body

$$K_i = \{\rho \in \mathrm{D} \,:\, (\Phi_i \otimes \mathrm{Id})(\rho) \in \mathcal{PSD}\} = \mathrm{D} \cap (\Phi_i \otimes \mathrm{Id})^{-1}(\mathcal{PSD})$$

(note that $\mathrm{Sep} \subset K_i$) and the polyhedral cone

$$\mathcal{C}_i := \left\{A \in B^{\mathrm{sa}}(\mathbb{C}^d \otimes \mathbb{C}^d) \,:\, (\Phi_i \otimes \mathrm{Id})(A) \in \mathbb{R}_+ P\right\}. \tag{9.55}$$

We claim that

$$\frac{1}{2} \bullet K_i \subset P \cap \mathcal{C}_i \subset K_i. \tag{9.56}$$

Before proving the claim, let us first show how it implies the Theorem. Combining (9.56) and (9.52) we obtain

$$\frac{1}{2} \bullet \mathrm{Sep} \subset \bigcap_{i=1}^{N} \left(\frac{1}{2} \bullet K_i\right) \subset \bigcap_{i=1}^{N} (P \cap \mathcal{C}_i) = P \cap \bigcap_{i=1}^{N} \mathcal{C}_i \subset \bigcap_{i=1}^{N} K_i \subset 2 \bullet \mathrm{Sep}.$$

The polytope $R = P \cap \bigcap_{1 \leqslant i \leqslant N} \mathcal{C}_i$ has at most $f := (N+1)\exp(C_0 d^2 \log d)$ facets. Consequently, by the definition of the facial dimension (see Section 7.2.3), we must have $\log f \geqslant \dim_F(\mathrm{Sep})$ and so

$$\log(N+1) + Cd^2 \log d \geqslant \dim_F(\mathrm{Sep}).$$

Since we know from Corollary 9.32 that $\dim_F(\mathrm{Sep}) = \Omega(d^3/\log d)$, it follows that $\log(N+1) \geqslant cd^3/\log d$ for d large enough. Since small values of d can be taken into account by adjusting the constant c if necessary, this implies the Theorem.

It remains to prove the claimed inclusions (9.56). The second inclusion is immediate from the definitions and from (9.54). For the first inclusion, it is clearly enough to show that $\frac{1}{2} \bullet K_i \subset \mathcal{C}_i$. To that end, let $\rho \in K_i$ and denote $t = \mathrm{Tr}\,[(\Phi_i \otimes \mathrm{Id})\rho] \geqslant 0$. We now consider two cases. First, if $t = 0$, then (since $(\Phi_i \otimes \mathrm{Id})(\rho)$ is a positive operator) we must have $(\Phi_i \otimes \mathrm{Id})(\rho) = 0$. Hence trivially $\rho \in \mathcal{C}_i$ and, *a fortiori*, $\frac{1}{2} \bullet \rho \in \mathcal{C}_i$. If $t > 0$, we note that $t^{-1}(\Phi_i \otimes \mathrm{Id})(\rho) \in \mathrm{D}$ and that, by Lemma 9.36, we have $t \leqslant d$, and therefore $\frac{t}{1+t} = 1 - \frac{1}{1+t} \leqslant 1 - \frac{1}{1+d} = 1 - \varepsilon$. It thus follows from (9.54) that

$$\frac{t}{1+t} \bullet t^{-1}(\Phi_i \otimes \mathrm{Id})(\rho) \in \frac{t}{1+t} \bullet D \subset (1-\varepsilon) \bullet D \subset P.$$

It remains to notice that

$$\frac{t}{1+t} \bullet t^{-1}(\Phi_i \otimes \mathrm{Id})(\rho) = \frac{(\Phi_i \otimes \mathrm{Id})(\rho) + \rho_*}{1+t} = \frac{2}{1+t}(\Phi_i \otimes \mathrm{Id}) \left(\frac{\rho + \rho_*}{2} \right),$$

which means that we showed that $(\Phi_i \otimes \mathrm{Id}) \left(\frac{1}{2} \bullet \rho \right) \in \frac{1+t}{2} P$. In particular (cf. (9.55)), $\frac{1}{2} \bullet \rho \in \mathcal{C}_i$, as needed. □

EXERCISE 9.15 (Unital witnesses suffice). Let $\Phi \in \boldsymbol{P}(\mathbb{C}^d)$. Show that there is a unital map $\Psi \in \boldsymbol{P}(\mathbb{C}^d)$ with the property that, for any $\rho \in \mathrm{D}(\mathbb{C}^d \otimes \mathbb{C}^d)$,

$$(\Phi \otimes \mathrm{Id})(\rho) \in \mathcal{PSD} \iff (\Psi \otimes \mathrm{Id})(\rho) \in \mathcal{PSD}.$$

EXERCISE 9.16 (Detecting very robust entanglement is also hard). (i) Show that, in the notation of Exercise 7.15, we have $\dim_F(\mathrm{Sep}(\mathbb{C}^d \otimes \mathbb{C}^d), A) \geqslant d^3 A^{-2}/\log d$ for every $A > 1$, where $c > 0$ is an absolute constant. (ii) Prove Theorem 9.35.

Notes and Remarks

Section 9.1. The exact formula (9.4) for the volume of D appears in [**ŻS03**]. The question of computing exactly the volume of Sep was asked in [**ŻHSL98**] and seems challenging already in the bipartite case. A conjecture by Slater [**Sla12**], strongly supported by numerical evidence, is that for $\mathcal{H} = \mathbb{C}^2 \otimes \mathbb{C}^2$, one has $\mathrm{vol}(\mathrm{Sep})/\mathrm{vol}(\mathrm{D}) = 8/33$.

Theorems 9.3, 9.12 and 9.13 are from [**AS06**]; Theorem 9.11 appeared earlier in [**Sza05**]. Theorem 9.6 and its corollary about block-positive matrices is from [**ASY14**], and will be crucial in Chapter 10. The same question for multipartite Hilbert spaces or unbalanced bipartite Hilbert spaces was also studied in [**ASY14**]; an extra ingredient needed is the fact that $P_F \mathrm{Sep} = \mathrm{Sep} \cap F$ for certain subspaces F, see Exercise 9.11(iii).

Volume and mean width estimates for the hierarchies of states introduced in Section 2.2.5 are also known. For $1 \leqslant k \leqslant d$, denote by Ent_k the set of k-entangled states in $\mathbb{C}^d \otimes \mathbb{C}^d$. It is proved in [**SWŻ11**] that

$$\frac{ck^{1/2}}{d^{3/2}} \leqslant \mathrm{vrad}(\mathrm{Ent}_k) \leqslant w(\mathrm{Ent}_k) \leqslant \frac{Ck^{1/2}}{d^{3/2}}, \tag{9.57}$$

which is of course compatible with the extreme cases $\mathrm{Ent}_1 = \mathrm{Sep}$ and $\mathrm{Ent}_d = \mathrm{D}$.

Similarly, if Ext_k denotes the set of k-extendible states on $\mathbb{C}^d \otimes \mathbb{C}^d$, it is proved in [**Lan16**] that for each fixed k, as $d \to \infty$

$$w(\mathrm{Ext}_k) \sim \frac{2}{d\sqrt{k}}. \tag{9.58}$$

Note that $\mathrm{D} = \mathrm{Ext}_1$ and $\mathrm{Sep} = \bigcap \{\mathrm{Ext}_k \,:\, k \geqslant 1\}$. However the implicit dependence on k in (9.58) does not allow to recover Theorem 9.3 as $k \to \infty$.

Section 9.2. Theorem 9.15 was proved in [**GB02**]. The proof we present is due to Hans-Jürgen Sommers and appears is [**Som09**]; the equivalence between Theorem 9.15 and the inequality $\mathrm{Tr}(M^2) \leqslant (\mathrm{Tr}\, M)^2$ for a block-positive matrix M has been noted in [**SWŻ08**]. The alternative argument from Exercise 9.8 is from [**Wat**].

Proposition 9.17 (in the language of robustness) has been proved by Vidal and Tarrach [**VT99**]. Proposition 9.18 is from [**Jen13**]. The result from Theorem

9.19 is due to Beigi and Shor [**BS10**] and relies on the quantum de Finetti theorem. Another argument, yielding better quantitative estimates, was presented in [**BHH$^+$14**] and was based on the concept of private states. Proposition 9.20 is also from [**BHH$^+$14**].

Both inequalities from Theorem 9.21 are due to Hildebrand ([**Hil06**] for the lower bound and [**Hil07a**] for the upper bound), improving on previous results by Gurvits and Barnum [**GB03, GB05**] (the lower bound) and [**AS06**] (the upper bound, cf. the proof of Proposition 9.22).

The question of determining the exact order of $d_g(\mathrm{Sep}, \mathrm{D})$ for many qubits (cf. Proposition 9.22) deserves attention since it can be connected to feasibility of nuclear magnetic resonance (NMR) quantum information protocols (see, e.g., [**GB05**]).

Section 9.3. Theorem 9.23 was derived in [**SWŻ08**], to which we refer for precise estimates for the constants implicit in the $\Theta(\cdot)$ notation from Table 9.3.

Another class of superoperators for which volume estimates are known is the class of k-positive maps. Indeed, this class is essentially dual to the class of k-entangled operators (see Exercise 2.48). It was proved in [**SWŻ11**]—as a consequence of (9.57)—that if $\boldsymbol{P}_{k,\mathrm{TP}}$ denotes the set of k-positive trace-preserving maps from M_d to itself, then

$$c\sqrt{k/d} \leqslant \mathrm{vrad}(\boldsymbol{P}_{k,\mathrm{TP}}) \leqslant C\sqrt{k/d}.$$

Section 9.4. The results from this section are from [**AS17**]. The fact that for $d \geqslant 3$ the intersection in (9.51) cannot be restricted to a finite subfamily has been proved in [**Sko16**] and is based on [**HK11**].

CHAPTER 10

Random quantum states

The main goal of this chapter is to prove the following result. Consider a system of N identical particles (e.g., N qubits) in a random pure state. For some $k \leqslant N/2$, let A and B be two subsystems, each consisting of k particles. There exists a threshold function $k_0(N)$ which satisfies $k_0(N) \sim N/5$ as $N \to \infty$ and such that the following holds. *If $k < k_0(N)$, then with high probability the two subsystems A and B share entanglement. Conversely, if $k > k_0(N)$, then with high probability the two subsystems A and B do not share entanglement.*

If the Hilbert space associated with a single particle is $\mathbb{C}^q$ (e.g., $q = 2$ for qubits), the dimension of the system $A \otimes B$ equals q^{2k} and the state ρ describing the $A \otimes B$ subsystem is obtained as a partial trace over an environment of dimension q^{N-2k} (the remaining $N - 2k$ particles). If the global system is in a random and uniformly distributed pure state, the state ρ is a random induced state as introduced in Section 6.2.3.4, where its distribution was denoted by $\mu_{q^{2k},q^{N-2k}}$. The central result of the chapter (Theorem 10.12) answers the question of whether a random induced state on $\mathbb{C}^d \otimes \mathbb{C}^d$ with distribution $\mu_{d^2,s}$ is separable or entangled. It relies on the volume and mean width estimates from Chapter 9.

Section 10.3 contains results about other thresholds for random induced states: for the PPT vs. non-PPT dichotomy (Theorem 10.17) and for the value of the entanglement of formation being close to maximal or close to minimal (Theorem 10.16).

10.1. Miscellaneous tools

The first sections of this chapter lead to an auxiliary result (a kind of a quantitative central limit theorem) about approximation of random induced states by Gaussian matrices (Proposition 10.6). As tools, we will use some majorization inequalities, which we present in Section 10.1.1.

10.1.1. Majorization inequalities. Majorization was introduced in Section 1.3.1. We first state a technical result that ascertains that "flat" vectors (i.e., vectors with a large ℓ_1-norm and small ℓ_∞-norm) majorize many other vectors. Since we need to consider homotheties, it is natural to work in $\mathbb{R}^{n,0}$, the hyperplane of $\mathbb{R}^n$ consisting of vectors whose coordinates add up to 0.

LEMMA 10.1. *Let $x, y \in \mathbb{R}^{n,0}$. Assume that $\|y\|_\infty \leqslant 1$ and $\|y\|_1 \geqslant \alpha n$ for some $\alpha \in (0, 1]$. Then*

$$x \prec (2/\alpha - 1)\|x\|_\infty \, y. \tag{10.1}$$

PROOF OF LEMMA 10.1. By homogeneity, it is enough to verify that the condition $\|x\|_\infty \leqslant 1$ implies $x \prec (2/\alpha - 1)y$. Moreover, it is enough to check this for

x being an extreme point of the set $A := \{x \in \mathbb{R}^{n,0} : \|x\|_\infty \leqslant 1\}$, since the set $\{x \in \mathbb{R}^{n,0} : x \prec z\}$ is convex for any $z \in \mathbb{R}^{n,0}$.

Extreme points of A are of the following form: $\lfloor n/2 \rfloor$ coordinates are equal to 1 and $\lfloor n/2 \rfloor$ coordinates equal to -1. In the case of odd n there is one remaining coordinate, which is necessarily equal to 0. It is thus enough to verify that if x is of that form, and if y satisfies $\|y\|_\infty \leqslant 1$ and $\|y\|_1 = \alpha n$, then $x \prec (2/\alpha - 1)y$. This is shown by establishing that an average of permutations of y is a multiple of x.

First, average separately the positive and the negative coordinates of y to obtain a vector y' whose coordinates take only two values, one positive and one negative. Since the ℓ_1-norm of the positive and the negative part of y' is equal and amounts to $\alpha n/2$, the support of each part must be at least $\alpha n/2$ and at most $(1 - \alpha/2)n$, and the absolute value of each coordinate at least $\alpha/(2 - \alpha)$.

Assume now that n is even. Next, select a set of $n/2$ equal coordinates (positive or negative, depending on which part has larger support) and average the remaining ones. The obtained vector is a multiple of an extreme point, as needed. If n is odd, select $\lfloor n/2 \rfloor$ equal coordinates (from the dominant sign) and average the remaining ones to produce one zero and $\lfloor n/2 \rfloor$ equal coordinates. The resulting vector is also a multiple of an extreme point. □

A simpler but less precise version of Lemma 10.1 can be obtained without any hypothesis on $\|y\|_\infty$.

LEMMA 10.2. *Let $x, y \in \mathbb{R}^{n,0}$ with $y \neq 0$. Then*

$$x \prec \frac{2n\|x\|_\infty}{\|y\|_1} y. \tag{10.2}$$

PROOF. By homogeneity, we may assume that $\|y\|_\infty = 1$ and the result follows from Lemma 10.1. □

As a consequence, we obtain the fact that if two vectors from $\mathbb{R}^{n,0}$ are flat and close to each other, one is majorized by a small perturbation of the other one.

PROPOSITION 10.3. *Let $x, y \in \mathbb{R}^{n,0}$. Assume that $\|x - y\|_\infty \leqslant \varepsilon$ and $\|y\|_1 \geqslant \alpha n$ for some $\alpha > 0$. Then*

$$x \prec \left(1 + \frac{2\varepsilon}{\alpha}\right) y.$$

PROOF. We use the following elementary property of majorization: if $x_1 \prec \lambda_1 y$ and $x_2 \prec \lambda_2 y$ for some positive λ_1, λ_2, then $x_1 + x_2 \prec (\lambda_1 + \lambda_2)y$. We apply this fact with $x_1 = y$, $\lambda_1 = 1$ and $x_2 = x - y$. Lemma 10.2 shows that we can choose $\lambda_2 = 2\varepsilon/\alpha$, and the Proposition follows. □

EXERCISE 10.1. Provide an alternative proof of Lemma 10.2 by using directly the definition of majorization.

10.1.2. Spectra and norms of unitarily invariant random matrices. A lot of information about a self-adjoint matrix can be retrieved from its spectrum; for example, all unitarily invariant norms can be computed if one knows the eigenvalues (see Section 1.3.2). In contrast, computing the values of other norms or gauges (e.g., the gauge associated to the set of separable states) usually requires some knowledge about the eigenvectors.

However, if the matrix is random and if its distribution is unitarily invariant, it is possible to circumvent this difficulty. Heuristically, the principle we are going to establish and use is as follows: *if A and B are two unitarily invariant random matrices with similar spectra, then, for any norm or gauge $\|\cdot\|$, the typical values of $\|A\|$ and of $\|B\|$ are comparable.*

It is convenient to work in the hyperplane $\mathsf{M}_n^{\mathrm{sa},0}$ of self-adjoint complex $n \times n$ matrices with trace zero. One says that a $\mathsf{M}_n^{\mathrm{sa},0}$-valued random variable A is *unitarily invariant* if, for any $U \in \mathsf{U}(n)$, the random matrices A and $UAU^\dagger$ have the same distribution. Recall also that μ_{SC} is the standard semicircular distribution, that $\mu_{\mathrm{sp}}(A)$ is the empirical spectral distribution of a self-adjoint matrix A, and that d_∞ denotes the ∞-Wasserstein distance. All these concepts were introduced in Section 6.2.

PROPOSITION 10.4. *Let A and B be two $\mathsf{M}_n^{\mathrm{sa},0}$-valued random variables which are unitarily invariant and satisfy the conditions*

$$\mathbf{P}(d_\infty(\mu_{\mathrm{sp}}(A), \mu_{\mathrm{SC}}) \leqslant \varepsilon) \geqslant 1-p \quad \text{and} \quad \mathbf{E}\, d_\infty(\mu_{\mathrm{sp}}(A), \mu_{\mathrm{SC}}) \leqslant \varepsilon \tag{10.3}$$

for some $\varepsilon, p \in (0,1)$, and similarly for B. Then, for any convex body $K \subset \mathsf{M}_n^{\mathrm{sa},0}$ containing the origin in its interior,

$$\frac{1-p}{1+C\varepsilon}\,\mathbf{E}\,\|A\|_K \leqslant \mathbf{E}\,\|B\|_K \leqslant \frac{1+C\varepsilon}{1-p}\,\mathbf{E}\,\|A\|_K$$

for some absolute constant C.

PROOF OF PROPOSITION 10.4. Note that possible relations between A and B (such as independence) are irrelevant in the present situation. Consider the following function on $\mathbb{R}^{n,0}$ (recall that $\mathbb{R}^{n,0}$ denotes the hyperplane of vectors of sum zero in $\mathbb{R}^n$):

$$\phi(x) = \mathbf{E}\,\|U\,\mathrm{Diag}(x)U^\dagger\|_K,$$

where $U \in \mathsf{U}(n)$ denotes a Haar-distributed random unitary matrix (independent of everything else) and $\mathrm{Diag}(x)$ is the diagonal matrix whose ii-th entry is x_i. Unitary invariance implies that

$$\mathbf{E}\,\|A\|_K = \mathbf{E}\,\phi(\mathrm{spec}(A)) \tag{10.4}$$

and similarly for B (see Exercise 10.2). Let E be the event $\{d_\infty(\mu_{\mathrm{sp}}(B), \mu_{\mathrm{SC}}) \leqslant \varepsilon\}$. Assume for the moment that E holds, we have then (see Exercise 6.25)

$$\begin{aligned} \|B\|_1 &= n\int |x|\,\mathrm{d}\mu_{\mathrm{sp}}(B)(x) \geqslant n\int_{-2}^{2} (|x|-\varepsilon)^+\,\mathrm{d}\mu_{\mathrm{SC}}(x) \\ &\geqslant n\int_{-2}^{2} (|x|-1)^+\,\mathrm{d}\mu_{\mathrm{SC}}(x) = \alpha n, \end{aligned}$$

$\alpha \approx 0.16$ being a numerical constant. Applying Proposition 10.3 to the vectors $\mathrm{spec}(A)$ and $\mathrm{spec}(B)$, we conclude that (with $C = 2/\alpha$)

$$\mathrm{spec}(A) \prec (1 + C d_\infty(\mu_{\mathrm{sp}}(A), \mu_{\mathrm{sp}}(B)))\,\mathrm{spec}(B).$$

Since ϕ is convex and permutationally invariant, it follows that

$$\phi(\mathrm{spec}(A)) \leqslant (1 + C d_\infty(\mu_{\mathrm{sp}}(A), \mu_{\mathrm{sp}}(B)))\phi(\mathrm{spec}(B)).$$

Using the fact that $d_\infty(\mu_{\mathrm{sp}}(A), \mu_{\mathrm{sp}}(B)) \leqslant \varepsilon + d_\infty(\mu_{\mathrm{sp}}(A), \mu_{\mathrm{SC}})$ and taking expectation over A yields

$$\mathbf{E}\,\phi(\mathrm{spec}(A)) \leqslant (1 + 2C\varepsilon)\phi(\mathrm{spec}(B)).$$

Recall that the above inequality is true conditionally on E. Consequently,

$$\mathbf{E}\,\phi(\mathrm{spec}(B)) \geqslant \mathbf{E}\,\phi(\mathrm{spec}(B))\mathbf{1}_E \geqslant (1+2C\varepsilon)^{-1}\mathbf{P}(E)\,\mathbf{E}\,\phi(\mathrm{spec}(A)).$$

In view of (10.4) and since $\mathbf{P}(E) \geqslant 1-p$ by hypothesis, this shows that

$$\mathbf{E}\,\|A\|_K \leqslant \frac{1+2C\varepsilon}{1-p}\,\mathbf{E}\,\|B\|_K.$$

The other inequality follows by symmetry. □

If ε is large (2 or larger), the hypothesis $d_\infty(\mu_{\mathrm{sp}}(A), \mu_{\mathrm{SC}}) \leqslant \varepsilon$ does not prevent A from being identically zero. However, an isomorphic version of Proposition 10.4 can be similarly obtained under the hypothesis that the spectra of A and B are reasonably flat.

PROPOSITION 10.5 (See Exercise 10.3). *Let A and B be two $\mathsf{M}_n^{\mathrm{sa},0}$-valued random variables which are unitarily invariant. Assume that*

$$\mathbf{P}(\|A\|_1 \geqslant c_1 n) \geqslant 1-p \quad \text{and} \quad \mathbf{E}\,\|A\|_\infty \leqslant C_2, \tag{10.5}$$

and similarly for B. Then, for any convex body $K \subset \mathsf{M}_n^{\mathrm{sa},0}$ containing the origin in the interior,

$$C^{-1}\,\mathbf{E}\,\|A\|_K \leqslant \mathbf{E}\,\|B\|_K \leqslant C\,\mathbf{E}\,\|A\|_K$$

with $C = (1-p)^{-1}(2C_2/c_1)$.

EXERCISE 10.2 (Retrieving unitarily invariant distributions from the spectrum). Let A be a $\mathsf{M}_n^{\mathrm{sa},0}$-valued random variable which is unitarily invariant. Recall that $\mathrm{Diag}(\mathrm{spec}(A))$ is the diagonal matrix whose diagonal entries are the eigenvalues of A arranged in the non-increasing order. Let $U \in \mathsf{U}(n)$ be a Haar-distributed random unitary matrix independent of A. Show that the random matrix $U\,\mathrm{Diag}(\mathrm{spec}(A))U^\dagger$ has the same distribution as A.

EXERCISE 10.3 (All flat unitarily invariant distributions look alike). Prove Proposition 10.5.

10.1.3. Gaussian approximation to induced states. We are going to investigate typical properties of random induced states, in the large dimension regime. Their spectral properties were discussed in Section 6.2.3, and are described either by the Marčenko–Pastur distribution (when s is proportional to n) or by the semicircular distribution (when $s \gg n$).

However, we are also interested in properties that cannot be inferred from the spectrum (the main example being separability vs. entanglement on a bipartite system). In this context, it is useful to compare induced states with their Gaussian approximation. Indeed, the Gaussian model allows us to connect with tools from convex geometry, such as the mean width.

It is convenient to work in the hyperplane $\mathsf{M}_n^{\mathrm{sa},0}$ and to consider the shifted operators $\rho - \mathrm{I}/n$, which we compare with a GUE_0 random matrix (see Section 6.2.2). The following proposition compares the expected value of any norm (or gauge) computed for both models.

PROPOSITION 10.6. *Given integers n, s, denote by $\rho_{n,s}$ a random induced state on $\mathbb{C}^n$ with distribution $\mu_{n,s}$, and by G_n an $n \times n$ GUE_0 random matrix. Let $C_{n,s}$ be* *the smallest constant such that the following holds: for any convex body* $K \subset \mathsf{M}_n^{\mathrm{sa},0}$ *containing* 0 *in the interior,*

$$C_{n,s}^{-1}\,\mathbf{E}\left\|\frac{G_n}{n\sqrt{s}}\right\|_K \leqslant \mathbf{E}\left\|\rho_{n,s}-\frac{\mathrm{I}}{n}\right\|_K \leqslant C_{n,s}\,\mathbf{E}\left\|\frac{G_n}{n\sqrt{s}}\right\|_K. \tag{10.6}$$

Then
(i) For any sequences (n_k) *and* (s_k) *such that* $\lim_{k\to\infty} n_k = \lim_{k\to\infty} s_k/n_k = \infty$, *we have* $\lim_{k\to\infty} C_{n_k,s_k} = 1$.
(ii) For any $a > 0$, *we have* $\sup\{C_{n,s} \ : \ s \geqslant an\} < \infty$.

REMARK 10.7. We emphasize that the quantity $\mathbf{E}\,\|G_n\|_K$ appearing in (10.6) is exactly the Gaussian mean width of the polar set K°. Indeed, the standard Gaussian vector in the space $\mathsf{M}_n^{\mathrm{sa},0}$ (equipped with the Hilbert–Schmidt scalar product, as always) is exactly a GUE_0 random matrix. In view of (4.32), we could have equivalently formulated Proposition 10.6 using the usual mean width: if $\tilde{C}_{n,s}$ denotes the smallest constant such that the inequalities

$$\tilde{C}_{n,s}^{-1}\frac{w(K^\circ)}{\sqrt{s}} \leqslant \mathbf{E}\left\|\rho_{n,s}-\frac{\mathrm{I}}{n}\right\|_K \leqslant \tilde{C}_{n,s}\frac{w(K^\circ)}{\sqrt{s}} \tag{10.7}$$

are true for every convex body containing 0 in the interior, then the conclusions of Proposition 10.6 hold for $\tilde{C}_{n,s}$ instead of $C_{n,s}$.

PROOF. It is easy to check that (10.6) holds for some $C_{n,s} < +\infty$ if n and s are fixed (see Exercise 10.4). Moreover, we know from Theorem 6.35(i) that, for every fixed n,

$$\sup\{C_{n,s} \ : \ s \in \mathbb{N}\} < +\infty. \tag{10.8}$$

(i) Assume that $n = n_k$ and $s = s_k$, with n_k and s_k/n_k both tending to infinity, and denote $A_k = \sqrt{ns}(\rho_{n,s} - \mathrm{I}/n)$ and $B_k = G_n/\sqrt{n}$. Consider the random variables $X_k = d_\infty(\mu_{\mathrm{sp}}(A_k), \mu_{\mathrm{SC}})$ and $Y_k = d_\infty(\mu_{\mathrm{sp}}(B_k), \mu_{\mathrm{SC}})$. We know from Theorem 6.23 and Theorem 6.35(iii) that X_k and Y_k converge to zero in probability. We also claim that $\lim \mathbf{E}\,X_k = \lim \mathbf{E}\,Y_k = 0$; this follows from the fact that $X_k \leqslant 2 + \|A_k\|$, $Y_k \leqslant 2 + \|B_k\|$ and from Proposition 6.24 and Proposition 6.33. Part (i) follows now from Proposition 10.4.
(ii) Let A_k and B_k be as before, but now we only assume that $s_k \geqslant a n_k$ for some $a > 0$. We argue by contradiction: suppose that C_{n_k,s_k} tends to infinity. We know from (10.8) that the sequence (n_k) cannot be bounded, so we may assume $\lim_k n_k = +\infty$. Similarly, using part (i), we may assume that s_k/n_k is bounded, and therefore (by passing to a subsequence) that $\lim s_k/n_k = \lambda \in [a,\infty)$. We know from Theorem 6.35(ii) and Theorem 6.23 that $\mu_{\mathrm{sp}}(A_k)$ and $\mu_{\mathrm{sp}}(B_k)$ converge in probability towards a nontrivial deterministic limit, and therefore satisfy the hypotheses of Proposition 10.5 for some constants p, c_1, C_2. □

EXERCISE 10.4. Let X and Y be two $\mathbb{R}^n$-valued random vectors with the property that, for any $\theta \in S^{n-1}$, we have $0 < \mathbf{E}\,|\langle X,\theta\rangle| < +\infty$ and $0 < \mathbf{E}\,|\langle Y,\theta\rangle| < +\infty$. Show that there exists a constant C (depending on n, X, Y) such that, for any convex body K containing the origin in the interior, we have $\mathbf{E}\,\|X\|_K \leqslant C\,\mathbf{E}\,\|Y\|_K$.

10.1.4. Concentration for gauges of induced states. We present a concentration result valid for any gauge evaluated on random induced states.

PROPOSITION 10.8. *Let $s \geqslant n$, let $K \subset \mathrm{D}(\mathbb{C}^n)$ be a convex body with inradius r, and let ρ be a random state with distribution $\mu_{n,s}$. Let M be the median of $\|\rho - \mathrm{I}/n\|_{K_0}$, with $K_0 = K - \mathrm{I}/n$. Then, for every $\eta > 0$,*

$$\mathbf{P}\left(\left|\left\|\rho - \frac{\mathrm{I}}{n}\right\|_{K_0} - M\right| \geqslant \eta\right) \leqslant \exp(-s) + 2\exp(-n^2 s r^2 \eta^2/72).$$

PROOF OF PROPOSITION 10.8. We know that ρ has the same distribution as $AA^\dagger$, where A is an $n \times s$ matrix uniformly distributed on the Hilbert–Schmidt sphere S_{HS}. Consider the function $f : S_{\mathrm{HS}} \to \mathbb{R}$ defined by

$$f(A) = \left\|AA^\dagger - \frac{\mathrm{I}}{n}\right\|_{K_0}. \tag{10.9}$$

For every $t > 0$, denote by Ω_t the subset $\Omega_t = \{A \in S_{\mathrm{HS}} \, : \, \|A\|_\infty \leqslant t\}$. The function f is the composition of several operations:
(a) The map $A \mapsto \|A\|_{K_0}$, which is $1/r$-Lipschitz with respect to the Hilbert–Schmidt norm.
(b) The map $A \mapsto A - \mathrm{I}/n$, which is an isometry for the Hilbert–Schmidt norm.
(c) The map $A \mapsto AA^\dagger$, which is $2t$-Lipschitz on Ω_t (see Lemma 8.22).
It follows that the Lipschitz constant of the restriction of f to Ω_t is bounded by $2t/r$. We now apply the local version of Lévy's lemma (Corollary 5.35) and obtain that, for every $\eta > 0$,

$$\mathbf{P}(|f - M| \geqslant \eta) \leqslant \mathbf{P}(S_{\mathrm{HS}} \backslash \Omega_t) + 2\exp(-nsr^2\eta^2/8t^2).$$

If we choose $t = 3/\sqrt{n}$, then $\mathbf{P}(S_{\mathrm{HS}} \backslash \Omega_t) \leqslant \exp(-s)$ (apply Proposition 6.36 with $\varepsilon = \sqrt{s/n}$) and the result follows. □

REMARK 10.9. Taking $t = 1$ in the argument above, one obtains that the global Lipschitz constant of f is bounded by $2/r$. This implies (see Proposition 5.29) that any two central values for f differ by at most $C/(r\sqrt{ns})$.

10.2. Separability of random states

Assume now that we work in a bipartite Hilbert space, and for simplicity consider the case of $\mathbb{C}^d \otimes \mathbb{C}^d$ where both parties play a symmetric role. Throughout this section we write Sep for $\mathrm{Sep}(\mathbb{C}^d \otimes \mathbb{C}^d)$ and consider random induced states on $\mathbb{C}^d \otimes \mathbb{C}^d$ with distribution $\mu_{d^2,s}$.

10.2.1. Almost sure entanglement for low-dimensional environments. Since the maximally mixed state lies in the interior of the set of separable states, and since the measures $\mu_{d^2,s}$ converge weakly towards the Dirac mass at the maximally mixed state (see Section 6.2.3.4), it follows that $\mu_{d^2,s}(\mathrm{Sep})$ tends to 1 when s tends to infinity (d being fixed). Conversely, the following result shows that random induced states are entangled with probability one when $s \leqslant (d-1)^2$.

PROPOSITION 10.10. *Let d, s be integers with $s \leqslant (d-1)^2$. Then $\mu_{d^2,s}(\mathrm{Sep}) = 0$.*

PROOF. Let $S \subset \mathbb{C}^d \otimes \mathbb{C}^d$ be the range of ρ. The random subspace S is Haar-distributed on the Grassmann manifold $\mathsf{Gr}(s, \mathbb{C}^d \otimes \mathbb{C}^d)$. We use the following simple fact which is an immediate consequence of the definition of separability: if ρ is separable, then S is spanned by product vectors. The Proposition now follows from Theorem 8.1: when $s \leqslant (d-1)^2$, S almost surely contains no nonzero product vector. □

PROBLEM 10.11. *For which values of d, s do we have $\mu_{d^2,s}(\mathrm{Sep}) = 0$?*

EXERCISE 10.5. Let d, s be integers with $s \geqslant d^2$. Show that $0 < \mu_{d^2,s}(\mathrm{Sep}) < 1$.

EXERCISE 10.6. Let d, s be integers such that $\mu_{d^2,s}(\mathrm{Sep}) > 0$. Show that $\mu_{d^2,t}(\mathrm{Sep}) > 0$ for every $t \geqslant s$. (Cf. Problem 10.14.)

10.2.2. The threshold theorem. From the two extreme cases, $s \leqslant (d-1)^2$ and $s = \infty$, we may infer that induced states are more likely to be separable when the environment has larger dimension. As it turns out, a phase transition takes place (at least when d is sufficiently large): the generic behavior of ρ "flips" to the opposite one when s changes from being a little smaller than a certain threshold dimension s_0 to being larger than s_0. More precisely, we have the following theorem.

THEOREM 10.12. *Define a function $s_0(d)$ as $s_0(d) = w(\mathrm{Sep}(\mathbb{C}^d \otimes \mathbb{C}^d)^\circ)^2$. This function satisfies*

$$cd^3 \leqslant s_0(d) \leqslant Cd^3 \log^2 d \tag{10.10}$$

for some constants c, C and is the threshold between separability and entanglement in the following sense. If ρ is a random state on $\mathbb{C}^d \otimes \mathbb{C}^d$ induced by the environment $\mathbb{C}^s$, then, for any $\varepsilon > 0$,

(i) if $s \leqslant (1-\varepsilon)s_0(d)$, we have

$$\mathbf{P}(\rho \text{ is entangled}) \geqslant 1 - 2\exp(-c(\varepsilon)d^3), \tag{10.11}$$

(ii) if $s \geqslant (1+\varepsilon)s_0(d)$, we have

$$\mathbf{P}(\rho \text{ is separable}) \geqslant 1 - 2\exp(-c(\varepsilon)s), \tag{10.12}$$

where $c(\varepsilon)$ is a constant depending only on ε.

As a corollary, we recover the result mentioned in the preamble of the chapter: given N identical particles in a generic pure state, if we assign k of them to Alice and k of them to Bob, their shared state suddenly jumps from typically entangled to typically separable when k crosses a certain threshold value $k_N \sim N/5$. We state the result for qubits only, but both the statement and the proof easily generalize to D-level particles for $D > 2$.

COROLLARY 10.13 (See Exercise 10.8). *Given an integer N, there is $k_N \sim N/5$ with the following property. For some integer $k \leqslant N/2$, decompose $\mathcal{H} = (\mathbb{C}^2)^{\otimes N}$ as $\mathcal{A} \otimes \mathcal{B} \otimes \mathcal{E}$ with $\mathcal{A} = \mathcal{B} = (\mathbb{C}^2)^{\otimes k}$ and $\mathcal{E} = (\mathbb{C}^2)^{\otimes (N-2k)}$, and consider a unit vector $\psi \in \mathcal{H}$ chosen uniformly at random. Let $\rho = \mathrm{Tr}_{\mathcal{E}} |\psi\rangle\langle\psi|$ be the induced state on $\mathcal{A} \otimes \mathcal{B}$. Then*

(1) for $k < k_N$, $\mathbf{P}(\rho$ is entangled$) \geqslant 1 - 2\exp(-\alpha^N)$,
(2) for $k > k_N$, $\mathbf{P}(\rho$ is separable$) \geqslant 1 - 2\exp(-\alpha^N)$,

where $\alpha > 1$ is a constant independent of N.

PROOF OF THEOREM 10.12. The inequalities (10.10) are a direct consequence of Theorem 9.6.

We next present a detailed proof of part (ii). Let $\rho_{d^2,s}$ be a random state with distribution $\mu_{d^2,s}$. Denote $\mathrm{Sep}_0 = \mathrm{Sep} - \mathrm{I}/d^2$. Consider also the function $f(\rho) = \|\rho - \frac{\mathrm{I}}{d^2}\|_{\mathrm{Sep}_0}$ and the quantity $E_{d,s} := \mathbf{E}\, f(\rho_{d^2,s})$.

Fix $\varepsilon > 0$, and let s, d be such that $s \geqslant (1+\varepsilon)s_0(d)$. Appealing to Proposition 10.6 (in the version given in Remark 10.7), we obtain

$$E_{d,s} \leqslant \tilde{C}_{n,s} \frac{w(K^\circ)}{\sqrt{s}} \leqslant \frac{\tilde{C}_{n,s}}{\sqrt{1+\varepsilon}}, \tag{10.13}$$

where $\tilde{C}_{n,s}$ is the constant appearing in (10.7). The constants $\tilde{C}_{n,s}$ tend to 1 as d and s tend to infinity under the constraint $s \geqslant (1+\varepsilon)s_0(d)$.

Let $M_{d,s}$ be the median of $f(\rho_{d^2,s})$. We know from Proposition 10.8 (the inradius of Sep being $\Theta(1/d^2)$, see Table 9.1) that

$$\mathbf{P}\left(f(\rho_{d^2,s}) > M_{d,s} + \eta\right) \leqslant \exp(-s) + 2\exp(-cs\eta^2). \tag{10.14}$$

Remark 10.9 implies that $|M_{d,s} - E_{d,s}| \leqslant Cd/\sqrt{s}$. It follows then from (10.13) that there is an $\eta > 0$ (depending only on ε) with the property that $M_{d,s} + \eta \leqslant 1$ for all d large enough and $s \geqslant (1+\varepsilon)s_0(d)$. The inequality (10.12) follows now from (10.14) and from the obvious remark that a state ρ is entangled if and only if $f(\rho) > 1$. Small values of d can be taken into account by adjusting the constants if necessary. Note that the argument yields *a priori* a bound $C' \exp(-c'(\varepsilon)s)$, possibly with $C' > 2$, but the bound (10.12) follows then with $c(\varepsilon) = c'(\varepsilon)/\log_2 C'$.

The proof of part (i) goes along similar lines, particularly if we do not care about the exact power of d appearing in the exponent of the probability bound in (10.11); this is because Proposition 10.8 yields an estimate parallel to (10.14) for $\mathbf{P}\left(f(\rho_{d^2,s}) < M_{d,s} - \eta\right)$. There are some fine points which emerge when s is relatively small, but they can be handled using inequalities from Exercise 10.7; see [**ASY14**] for details. See also Remark 10.15. □

The fine points in the proof of part (i) of Theorem 10.12 would disappear if the answer to the following natural problem was positive (cf. Exercise 10.6).

PROBLEM 10.14 (As environment increases, entanglement decreases). *Fix an integer $d \geqslant 2$. Is it true that the function $s \mapsto \mu_{d^2,s}(\mathrm{Sep})$ is non-decreasing?*

REMARK 10.15. An alternative and simpler argument to prove part (i) of Theorem 10.12 is sketched in Exercise 10.9. That argument also has the advantage that it produces explicitly an entanglement witness certifying that the induced state is entangled. However, the argument works only in the range $s \leqslant cd^3$ for some constant $c > 0$; while this does not cover the entire range, it handles the case of relatively small s that does not readily follow from Proposition 10.8.

EXERCISE 10.7 (Partial results on monotonicity of entanglement). Set $\pi_{d,s} := \mu_{d^2,s}(\mathrm{Sep}(\mathbb{C}^d \otimes \mathbb{C}^d))$.
(i) Show that the function $d \mapsto \pi_{d,s}$ is non-increasing for any integer $s \geqslant 1$.
(ii) Show the inequality $\pi_{2d,s} \leqslant \pi_{d,4s}$.

EXERCISE 10.8 (Proof of the $N/5$ threshold result). Prove Corollary 10.13 by combining Theorem 10.12 (applied with $\varepsilon = 1/2$) and Exercise 10.7.

EXERCISE 10.9 (The induced state is its own witness). Let ρ be a random state on $\mathbb{C}^d \otimes \mathbb{C}^d$ with distribution $\mu_{d^2,s}$, and $W = \rho - \mathrm{I}/d^2$.
(i) Show that $\mathrm{Tr}(W\rho)$ is of order $1/s$ with high probability.
(ii) Show that for any unit vector $x \in \mathbb{C}^d \otimes \mathbb{C}^d$ and $0 < \eta < 1$, we have

$$\mathbf{P}\left(|\langle x|W|x\rangle| > \frac{\eta}{d^2}\right) \leqslant C\exp(-cs\eta^2).$$

(iii) Conclude that, with high probability, $\sup\{\operatorname{Tr}(\sigma W) : \sigma \in \mathrm{Sep}\} \leqslant Cd^{-3/2}s^{-1/2}$.
(iv) Conclude that in the regime $s \leqslant cd^3$, with high probability, W witnesses the fact that ρ is entangled.

10.3. Other thresholds

10.3.1. Entanglement of formation. Theorem 10.12 settles the "entanglement vs. separability" dichotomy for random induced states. In the generic entanglement regime, we could be more precise and ask about quantitative estimates: *how strongly* is a random state entangled?

To address the above question we need a method to quantify the amount of entanglement present in a quantum state. The approach from the preceding section allows us to use the value of the gauge $\|\rho - \mathrm{I}/d^2\|_{\mathrm{Sep}_0}$ as a measure of the strength of entanglement. In this section we will work with invariants that are more "native" to quantum information theory.

For a pure state ψ, the entropy of entanglement $E(\psi)$ was introduced in (8.1). A possible way to extend this definition to mixed states is to use a "convex roof" construction. For a state ρ on $\mathbb{C}^d \otimes \mathbb{C}^d$, define its *entanglement of formation* $E_F(\rho)$ as

$$E_F(\rho) = \inf\left\{\sum p_i E(\psi_i) \ : \ \rho = \sum p_i |\psi_i\rangle\langle\psi_i|\right\}, \tag{10.15}$$

the infimum being taken over all decompositions of ρ as convex combinations of pure states. Equivalently, the entanglement of formation is the largest convex function which coincides with the entropy of entanglement on pure states.

Entanglement of pure states was studied in Chapter 8. In particular, for a random pure state ψ (which corresponds to the case $s = 1$), we typically have $E_F(|\psi\rangle\langle\psi|) = E(\psi) = \log d - \frac{1}{2} + o(1)$; see Lemma 8.13. Here is a statement describing a "behavior shift" which takes place as s increases.

THEOREM 10.16 (Entanglement of formation for random induced states). *Let ρ be a random state on $\mathbb{C}^d \otimes \mathbb{C}^d$ with distribution $\mu_{d^2,s}$.*

(1) If $s \leqslant cd^2/\log^2 d$, then with high probability $E_F(\rho) \geqslant \log(d) - 1$.
(2) If $0 < \varepsilon < 1$ and $s \geqslant C\varepsilon^{-2}d^2\log^2 d$, then with high probability $E_F(\rho) \leqslant \varepsilon$.

PROOF. Assume $s \leqslant d^2$. If S denotes the range of ρ, then S is a random Haar-distributed s-dimensional subspace of $\mathbb{C}^d \otimes \mathbb{C}^d$. We use the following relaxation:

$$E_F(\rho) \geqslant \inf\{E(\psi) \ : \ \psi \in S\}.$$

We then conclude using Theorem 8.15 that, with high probability, $E_F(\rho) \geqslant \log(d) - 1$ provided $s \leqslant cd^2/\log^2 d$.

For the second part, denote by a the smallest eigenvalue of ρ and consider the convex combination

$$\rho = (\rho - a\,\mathrm{I}) + a\,\mathrm{I} = (1 - d^2a)\sigma + d^2a\frac{\mathrm{I}}{d^2}$$

for some state σ. Using the convexity of E_F and the obvious facts that $E_F(\sigma) \leqslant \log d$ and $E_F(\mathrm{I}/d^2) = 0$, we obtain $E_F(\rho) \leqslant (1 - d^2a)\log d$. However, we know from Proposition 6.36 (or Exercise 6.43) that $a \geqslant \frac{1}{d^2} - \frac{C}{d\sqrt{s}}$ with large probability. It follows that as long as $s \geqslant C^2\varepsilon^{-2}d^2\log^2 d$, then

$$E_F(\rho) \leqslant \frac{Cd\log(d)}{\sqrt{s}} \leqslant \varepsilon. \qquad \square$$

EXERCISE 10.10. Check that $E_F(\rho) = 0$ if and only if ρ is separable.

10.3.2. Threshold for PPT. The machinery developed in this chapter can be applied to any property instead of separability and allows us to reduce the estimation of threshold dimensions to the estimation of a geometric quantity (the mean width for the polar set).

One natural example is the PPT property. Since $\mathrm{PPT} = \mathrm{D} \cap \Gamma(\mathrm{D})$, where Γ is the partial transpose, it follows easily (arguing as in the first part of the proof of Proposition 9.8) that $w(\mathrm{PPT}_0^\circ) \leqslant 2w(\mathrm{D}_0^\circ) \simeq d$. The threshold s_1 appearing in this approach then satisfies

$$s_1(d) = w(\mathrm{PPT}_0^\circ)^2 = \Theta(d^2).$$

However, we know that the spectrum of large-dimensional partially transposed random states is described by a non-centered semicircular distribution (see Theorem 6.30). A more precise estimation of the threshold follows (note that the distribution $SC(\lambda, \lambda)$ appearing in Theorem 6.30 has support $[\lambda - 2\sqrt{\lambda}, \lambda + 2\sqrt{\lambda}]$, which is contained in $[0, +\infty)$ if and only if $\lambda \geqslant 4$).

THEOREM 10.17 (Threshold for the PPT property). *Define $s_1(d) = 4d^2$. Let ρ be a random state on $\mathbb{C}^d \otimes \mathbb{C}^d$ with distribution $\mu_{d^2,s}$. Then*

(i) if $s \leqslant (1-\varepsilon)s_1(d)$, we have

$$\mathbf{P}(\rho \text{ is PPT}) \leqslant 2\exp(-c(\varepsilon)d^2),$$

(ii) if $s \geqslant (1+\varepsilon)s_1(d)$, we have

$$\mathbf{P}(\rho \text{ is PPT}) \geqslant 1 - 2\exp(-c(\varepsilon)s).$$

Here $c(\varepsilon)$ is a constant depending only on ε.

The comparison between Theorems 10.12, 10.16 and 10.17 is instructive: if s is sufficiently larger than d^2, but sufficiently smaller than d^3, random states are typically PPT and entangled (in particular they cannot be distilled, see Chapter 12), but have an amount of entanglement extremely small when measured via the entanglement of formation.

EXERCISE 10.11. Explain the presence of expressions of the form $\Omega_\varepsilon(d^2)$ and $\Omega_\varepsilon(s)$ in the exponents in Theorem 10.17.

Notes and Remarks

Theorem 10.12, as well as the preliminary results from Section 10.1, are from [**ASY14**]. A high-level non-technical overview can be found in [**ASY12**]. In particular, the existence of a separability threshold around the value $s = d^3$ was proved in [**ASY14**]; previously only the cases $s \leqslant d^2$ or $s \geqslant d^4$ were covered (see e.g [**HLW06**]).

The answer to Problem 10.11 is known for qubits: we have $\mu_{4,2}(\mathrm{Sep}(\mathbb{C}^2 \otimes \mathbb{C}^2)) = 0$ and $\mu_{4,3}(\mathrm{Sep}(\mathbb{C}^2 \otimes \mathbb{C}^2)) > 0$. As explained in section 7.1 of [**ASY14**], this follows from results of [**RW09**] and [**SBŻ06**], respectively.

The entanglement of formation is only one of the many possible ways to quantify entanglement of mixed states. However, other measures are harder to manipulate. For a survey of the subject of entanglement measures see [**PV07**].

The threshold for the entanglement of formation (Theorem 10.16) is essentially from [**HLW06**], and the threshold for the PPT property (Theorem 10.17) is from [**Aub12**] (see also [**ASY12**]).

Other threshold functions have been computed or estimated: for the realignment criterion see [**AN12**], for the k-extendibility property see [**Lan16**], and for still other properties see [**CNY12, JLN14, JLN15**] (including the absolute PPT property and the reduction criterion).

CHAPTER 11

Bell inequalities and the Grothendieck–Tsirelson inequality

In this chapter we briefly sketch the connection (originally made by Tsirelson) between the celebrated Bell inequalities from the quantum theory, and the equally celebrated Grothendieck inequality from functional analysis. The presentation is anything but comprehensive: it has been unequivocally established in the last dozen or so years that the proper "mathematical home" of Bell inequalities is in the theories of *operator spaces* and *operator systems*, which are beyond the scope of this book. An excellent survey that addresses these aspects of the topic in much greater detail is [**PV16**].

11.1. Isometrically Euclidean subspaces via Clifford algebras

In Section 7.2.4 we studied in detail the almost Euclidean subspaces of M_n, i.e., on which a given Schatten p-norm is $(1+\varepsilon)$-equivalent to the Hilbert–Schmidt norm. For the purposes of the present chapter it is useful to focus on the case of *exactly* or *isometrically* Euclidean subspaces, i.e., $\varepsilon = 0$.

We first note that for a rank one matrix, all Schatten p-norms are equal. It follows that there are subspaces of dimension n in M_n (e.g., the space of all matrices with zero coefficients outside the first row) in which the ratio $\|\cdot\|_{\mathrm{op}}/\|\cdot\|_{\mathrm{HS}}$ is constant and equal to 1. However, such a subspace is not at the "correct level": for subspaces produced by Dvoretzky's theorem—which are also of dimension $\Theta(n)$—the same ratio is $\Theta(1/\sqrt{n})$ (or, more precisely, $\sim 2/\sqrt{n}$, see Exercise 7.23).

A less trivial construction, based on Clifford algebras (or, in more elementary terms, on Pauli matrices), gives isometrically Euclidean subspaces of M_n (and even of $\mathsf{M}_n^{\mathrm{sa}}$), at the correct level, of dimension $\Theta(\log n)$, at least when n is a power of 2. It is a natural question whether it is possible to interpolate between that construction and the subspaces given by Dvoretzky's theorem (see Problem 11.27 in Notes and Remarks).

LEMMA 11.1. *For every $k \geqslant 2$, there is a $(2k-2)$-dimensional subspace of the space of $2^k \times 2^k$ real self-adjoint matrices in which every matrix is a multiple of an orthogonal matrix.*

This result is specific to subspaces over the real field: any 2-dimensional complex subspace of complex matrices (or, more to the point, every 1-dimensional complex affine subspace not containing 0) contains a singular matrix since the polynomial $\lambda \mapsto \det(A + \lambda B)$ must vanish (see also Exercise 11.1). However, a similar phenomenon holds for complex *self-adjoint* matrices.

LEMMA 11.2. *For every $k \geqslant 1$, there is a $2k$-dimensional (real vector) subspace of the space of $2^k \times 2^k$ complex Hermitian matrices in which every matrix is a multiple of a unitary matrix.*

Lemma 11.2 immediately implies Lemma 11.1 since an $n \times n$ unitary matrix can be considered as a $2n \times 2n$ orthogonal matrix when one disregards the complex structure.

PROOF OF LEMMA 11.2. Consider the following elements $U_1, \dots, U_{2k}$ of $\mathsf{M}_2^{\otimes k}$:

$$U_i = \mathrm{I}^{\otimes(i-1)} \otimes \sigma_x \otimes \sigma_y^{\otimes(k-i)},$$

$$U_{k+i} = \mathrm{I}^{\otimes(i-1)} \otimes \sigma_z \otimes \sigma_y^{\otimes(k-i)},$$

where $\sigma_x, \sigma_y, \sigma_z$ are the Pauli matrices introduced in (2.2). It is easily checked (cf. Exercise 2.4) that the operators $(U_i)_{1 \leqslant i \leqslant 2k}$ are self-adjoint and are anticommuting reflections: $U_i^2 = \mathrm{I}$ and $U_i U_j = -U_j U_i$ for $i \neq j$. It follows that for any $\xi \in \mathbb{R}^{2k}$, the matrix $X = \xi_1 U_1 + \cdots + \xi_{2k} U_{2k}$ satisfies $XX^\dagger = |\xi|^2 \mathrm{I}$ and therefore is a multiple of a unitary matrix. □

REMARK 11.3. The subspaces in Lemmas 11.1 and 11.2 consist of trace zero matrices.

The dimensions appearing in Lemma 11.1 are not optimal. Finding the minimal possible dimension is related to the Radon–Hurwitz problem and involves more advanced analysis of Clifford algebras.

THEOREM 11.4 (Not proved here). *Given an integer $k \geqslant 1$, consider*
(i) $\alpha(k)$, the minimal integer n such that $\mathsf{M}_n(\mathbb{R})$ contains a k-dimensional subspace in which every matrix is a multiple of an orthogonal matrix.
(ii) $\beta(k)$, the minimal integer n such that $\mathsf{M}_n(\mathbb{R})$ contains a k-dimensional subspace in which every nonzero matrix is invertible.
Then

$$\alpha(k) = \beta(k) = \begin{cases} 2^{(k-2)/2} & \text{if } k = 0 \mod 8, \\ 2^{(k-1)/2} & \text{if } k = 1 \text{ or } k = 7 \mod 8, \\ 2^{k/2} & \text{if } k = 2 \text{ or } k = 4 \text{ or } k = 6 \mod 8, \\ 2^{(k+1)/2} & \text{if } k = 3 \text{ or } k = 5 \mod 8. \end{cases}$$

EXERCISE 11.1 (Isometrically Euclidean subspaces and parity of the dimension). Show that $\mathsf{M}_n(\mathbb{R})$ contains a 2-dimensional subspace in which every matrix is a multiple of an orthogonal matrix if and only if n is even.

11.2. Local vs. quantum correlations

Ever since the seminal 1935 paper [**EPR35**] by Einstein, Podolsky and Rosen it has been apparent that quantum theory leads to predictions which are incompatible with the classical understanding of physical reality. Specifically, the outcomes of some experiment may be correlated in a way contradicting common sense ("spooky action at a distance"). In this section we formalize the concept of correlations, which will lead to the famous Bell inequalities discovered in [**Bel64**].

11.2.1. Correlation matrices. Let us start by defining what we mean by correlation matrices in the classical and the quantum worlds. As we shall see, comparing the two naturally involves the Grothendieck constant.

DEFINITION 11.5. A $m \times n$ real matrix $A = (a_{ij})$ is called a *classical* (or *local*) *correlation matrix* if there exist random variables $(X_i)_{1 \leqslant i \leqslant m}$ and $(Y_j)_{1 \leqslant j \leqslant n}$ defined on a common probability space, satisfying $|X_i| \leqslant 1$, $|Y_j| \leqslant 1$ (almost surely), and such that, for any $1 \leqslant i \leqslant m$, $1 \leqslant j \leqslant n$,

$$a_{ij} = \mathbf{E}\, X_i Y_j. \tag{11.1}$$

We write $\mathsf{LC}_{m,n}$ (or simply LC) for the set of $m \times n$ local correlation matrices.

We emphasize that this notion does not coincide with the correlation or covariance matrices from statistics. In that context, covariance matrices are square and positive semi-definite, corresponding to the scenario when $(X_i) = (Y_i)$ and $\mathbf{E}\, X_i = 0$ (see, e.g., Appendix A.2), while the correlation matrix of (X_i) is the covariance matrix of the standardized variables $(\tilde{X}_i) = (X_i/\|X\|_2)$. When $\mathbf{E}\, X_i = \mathbf{E}\, Y_j = 0$, (11.1) coincides with the somewhat less frequently used notion of *cross-covariance*.

The set $\mathsf{LC}_{m,n}$ is a polytope with 2^{n+m-1} vertices (see Proposition 11.7) and appears in the literature under various names such as correlation polytope, Bell polytope, local hidden variable polytope, and local polytope. (The reader should be forewarned, though, that sometimes the same names are used for sets of the more general objects, the so-called *boxes*, defined in Section 11.3.2.) The reasons for the adjective "local" will become clear later on. The facial structure of $\mathsf{LC}_{m,n}$ is rather complicated (except in very low dimensions, see Exercises 11.4, 11.12, and 11.15).

DEFINITION 11.6. A $m \times n$ real matrix $A = (a_{ij})$ is called a *quantum correlation matrix* if there is a state $\rho \in \mathsf{D}(\mathbb{C}^{d_1} \otimes \mathbb{C}^{d_2})$ (for some d_1, d_2), self-adjoint operators $(X_i)_{1 \leqslant i \leqslant m}$ on $\mathbb{C}^{d_1}$ and $(Y_j)_{1 \leqslant j \leqslant n}$ on $\mathbb{C}^{d_2}$ satisfying $\|X_i\|_\infty \leqslant 1$, $\|Y_j\|_\infty \leqslant 1$, and such that, for any $1 \leqslant i \leqslant m$ and $1 \leqslant j \leqslant n$,

$$a_{ij} = \operatorname{Tr} \rho(X_i \otimes Y_j). \tag{11.2}$$

We write $\mathsf{QC}_{m,n}$ (or simply QC) for the set of $m \times n$ quantum correlation matrices.

It turns out that both sets LC and QC have simple descriptions.

PROPOSITION 11.7. *The set $\mathsf{LC}_{m,n}$ can be alternatively described as*

$$\mathsf{LC}_{m,n} = \operatorname{conv}\{(\xi_i \eta_j)_{1 \leqslant i \leqslant m, 1 \leqslant j \leqslant n} \ : \ \xi \in \{-1,1\}^m, \eta \in \{-1,1\}^n\} = B_\infty^m \,\hat{\otimes}\, B_\infty^n.$$

PROPOSITION 11.8. *The set $\mathsf{QC}_{m,n}$ is convex and can be alternatively described as*

$$\mathsf{QC}_{m,n} = \left\{ (\langle x_i, y_j \rangle)_{1 \leqslant i \leqslant m, 1 \leqslant j \leqslant n} \ : \ x_i, y_j \in \mathbb{R}^{\min(m,n)}, |x_i| \leqslant 1, |y_j| \leqslant 1 \right\}.$$

It is obvious from the Propositions that $\mathsf{LC} \subset \mathsf{QC}$. (This can also be established directly from the definitions, without appealing to the results of Section 11.1.) The crucial point—which is simple, but not entirely trivial, and will be studied in detail in the next section—is that this inclusion is *strict*. This is one mathematical manifestation of the fact that the quantum description of reality is different from the classical one. Correlation matrices that do not belong to LC will be called *nonclassical* or *nonlocal*.

PROOF OF PROPOSITION 11.7. We first prove the inclusion $\supset$. It is clear that given $\xi \in \{-1,1\}^m$ and $\eta \in \{-1,1\}^n$, we have $(\xi_i\eta_j) \in \mathsf{LC}_{m,n}$ (consider constant random variables taking values ± 1), so it suffices to show that $\mathsf{LC}_{m,n}$ is convex. If $a_{ij}^{(1)} = \mathbf{E}\, X_i^{(1)} Y_j^{(1)}$ and $a_{ij}^{(2)} = \mathbf{E}\, X_i^{(2)} Y_j^{(2)}$ are two classical correlation matrices (without loss of generality we may assume that all random variables are defined on the same probability space), define random variables $X_i = X_i^{(\alpha)}$ and $Y_j = Y_j^{(\alpha)}$, where α is an independent random index, equal to 1 with probability p and equal to 2 with probability $1-p$. Then $\mathbf{E}\, X_i Y_i = p a_{ij}^{(1)} + (1-p) a_{ij}^{(2)}$ and this shows that $\mathsf{LC}_{m,n}$ is convex.

Conversely, note that any vector $X \in [-1,1]^d$ can be written as a convex combination of elements of $I_d := \{-1,1\}^d$

$$X = \sum_{\xi \in I_d} \lambda_\xi^d(X)\xi$$

with the functions $\lambda_\xi^d : [-1,1]^d \to [0,1]$ being measurable (or even continuous) and adding to 1. If $A \in \mathsf{LC}_{m,n}$ is a classical correlation matrix with $a_{ij} = \mathbf{E}\, X_i Y_j$, we may write (denoting $X = (X_1, \dots, X_m)$ and $Y = (Y_1, \dots, Y_n)$)

$$a_{ij} = \mathbf{E}\Big(\sum_{\xi \in I_m} \lambda_\xi^m(X)\xi_i\Big)\Big(\sum_{\eta \in I_n} \lambda_\eta^n(Y)\eta_j\Big) = \sum_{\xi \in I_m, \eta \in I_n} \mathbf{E}\left[\lambda_\xi^m(X)\lambda_\eta^n(Y)\right]\xi_i\eta_j,$$

which shows that $A \in \operatorname{conv}\{(\xi_i\eta_j)_{i,j} \ : \ \xi \in I_m, \eta \in I_n\}$. □

PROOF OF PROPOSITION 11.8. Let us first prove the direct inclusion. Let $(a_{ij}) \in \mathsf{QC}_{m,n}$. There is a Hilbert space $\mathcal{H} = \mathbb{C}^{d_1} \otimes \mathbb{C}^{d_2}$, a state $\rho \in \mathrm{D}(\mathcal{H})$ and self-adjoint contractions (X_i) and (Y_j) such that $a_{ij} = \operatorname{Tr} \rho(X_i \otimes Y_j)$. We define a bilinear form on the space $B^{\mathrm{sa}}(\mathcal{H})$,

$$\beta(S,T) = \operatorname{Re} \operatorname{Tr}(\rho S T).$$

This bilinear form is positive semi-definite (to check symmetry, use the fact that $\operatorname{Re}\operatorname{Tr} X = \operatorname{Re}\operatorname{Tr} X^\dagger$) and therefore, after possibly passing to a quotient, it makes $B^{\mathrm{sa}}(\mathcal{H})$ into a real Euclidean space. The conclusion follows since $a_{ij} = \beta(X_i \otimes \mathrm{I}, \mathrm{I} \otimes Y_j)$ while $\beta(X_i \otimes \mathrm{I}, X_i \otimes \mathrm{I}) \leqslant 1$ and $\beta(\mathrm{I} \otimes Y_j, \mathrm{I} \otimes Y_j) \leqslant 1$. To obtain the dimension $\min(m,n)$ as claimed, note that we may *a posteriori* project the vectors $(x_i)_{1 \leqslant i \leqslant m}$ onto $\operatorname{span}\{y_j \ : \ 1 \leqslant j \leqslant n\}$, or vice versa.

Conversely, let $(x_i)_{1 \leqslant i \leqslant m}$ and $(y_j)_{1 \leqslant j \leqslant n}$ be vectors of Euclidean norm at most 1 in $\mathbb{R}^{\min(m,n)}$. By Lemma 11.2, there exist $d \times d$ complex Hermitian matrices A_i, B_j (for some d), with Hilbert–Schmidt norm at most 1 and such that $\operatorname{Tr} A_i B_j = \langle x_i, y_j \rangle$. Moreover, A_i, B_j are multiples of unitaries. Set $X_i = d^{1/2} A_i$ and $Y_j = d^{1/2} B_j^T$; then X_i, Y_j are unitaries and in particular $\|X_i\|_\infty \leqslant 1$ and $\|Y_j\|_\infty \leqslant 1$. Finally, if $\rho = |\psi\rangle\langle\psi|$, where $\psi = \frac{1}{\sqrt{d}} \sum_{i=1}^d |ii\rangle \in \mathbb{C}^d \otimes \mathbb{C}^d$ is a maximally entangled vector, then we have

$$\operatorname{Tr} \rho\big(X_i \otimes Y_j\big) = \frac{1}{d} \operatorname{Tr} X_i Y_j^T = \operatorname{Tr} A_i B_j = \langle x_i, y_j \rangle,$$

where the first equality follows by direct calculation (see Exercise 2.12). □

REMARK 11.9. As a by-product of the proof, we obtain the following extra information: Definition 11.6 is unchanged if we require the operators X_i, Y_j to satisfy $X_i^2 = \mathrm{I}$, $Y_j^2 = \mathrm{I}$ and $\operatorname{Tr} X_i = \operatorname{Tr} Y_j = 0$ (cf. Remark 11.3). Moreover, the latter reduction can be performed in a "functorial" way which preserves many properties of ρ, see Exercise 11.7.

REMARK 11.10. Definitions 11.5 and 11.6 can be readily extended to the multipartite setting. One defines $\mathsf{LC}_{n_1,\dots,n_k} \subset \mathbb{R}^{n_1\cdots n_k}$ as the set of arrays $(a_{i_1,\dots,i_k})$ of the form $a_{i_1,\dots,i_k} = \mathbf{E}\left[X^{(1)}_{i_1}\cdots X^{(k)}_{i_k}\right]$ where all the $X^{(j)}_{i_j}$ are random variables with $|X^{(j)}_{i_j}| \leqslant 1$ a.s., and $\mathsf{QC}_{n_1,\dots,n_k} \subset \mathbb{R}^{n_1\cdots n_k}$ as the set of arrays $(a_{i_1,\dots,i_k})$ of the form $a_{i_1,\dots,i_k} = \operatorname{Tr}\left[\rho(X^{(1)}_{i_1}\otimes\cdots\otimes X^{(k)}_{i_k})\right]$ where all the $X^{(j)}_{i_j} \in B(\mathcal{H}_j)$ are self-adjoint operators with $\|X^{(j)}_{i_j}\|_\infty \leqslant 1$, and $\rho \in \mathrm{D}(\mathcal{H}_1\otimes\cdots\otimes\mathcal{H}_k)$.

EXERCISE 11.2 (Convexity of the set of quantum correlations). Show (directly from the definition) that the set QC is convex.

EXERCISE 11.3 (Unit vectors suffice). Show that

$$\mathsf{QC}_{m,n} = \left\{(\langle x_i, y_j\rangle)_{1\leqslant i\leqslant m, 1\leqslant j\leqslant n} \ : \ x_i, y_j \in \mathbb{R}^d, d\in\mathbb{N}, |x_i| = 1, |y_j| = 1\right\}.$$

EXERCISE 11.4 (The 2×2 local correlation polytope is an ℓ_1-ball). Show that $\mathsf{LC}_{2,2}$, considered as a subset of $\mathbb{R}^4$, is congruent to $2B_1^4$ (a ball of radius 2 in the ℓ_1-norm).

EXERCISE 11.5 (Local correlation polytope and the cut-norm). The cut-norm of a matrix $B \in \mathsf{M}_{m,n}$ is defined as

$$\|B\|_{cut} = \sup\left\{\left|\sum_{i\in I}\sum_{j\in J} b_{ij}\right| \ : \ I\subset\{1,\dots,m\}, J\subset\{1,\dots,n\}\right\}.$$

Show that $\|B\|_{cut} \leqslant \|B\|_{\mathsf{LC}^\circ_{m,n}} = \sup\{\operatorname{Tr} AB \ : \ A\in\mathsf{LC}_{m,n}\} \leqslant 4\|B\|_{cut}$.

EXERCISE 11.6 (Correlation polytopes and operator norms). Let $M \in \mathsf{M}_{m,n}$ be a real matrix. Verify that $\|M\|_{\mathsf{LC}^\circ_{m,n}}$ equals $\|M : \ell^n_\infty \to \ell^m_1\|$. Similarly, $\|M\|_{\mathsf{QC}^\circ_{m,n}}$ equals $\|M : \ell^n_\infty(\mathcal{H}) \to \ell^m_1(\mathcal{H})\|$, where $\mathcal{H}$ is any real Hilbert space of dimension at least $\min\{m,n\}$.

EXERCISE 11.7 (Trace zero measurements suffice). Let (a_{ij}) be a quantum correlation matrix defined by (11.2). Show that (a_{ij}) can be realized with a state $\tilde\rho = \rho\otimes\sigma\otimes\tau \in \mathrm{D}(\mathbb{C}^{d_1}\otimes\mathbb{C}^{d_2}\otimes\mathbb{C}^2\otimes\mathbb{C}^2)$ (so that, in particular, $\tilde\rho$ is separable or PPT if ρ is) and with operators $\tilde X_i \in B^{\mathrm{sa}}(\mathbb{C}^{d_1}\otimes\mathbb{C}^2)$, $\tilde Y_j \in B^{\mathrm{sa}}(\mathbb{C}^{d_2}\otimes\mathbb{C}^2)$ such that, in addition to $\|\tilde X_i\|_\infty \leqslant 1$ and $\|\tilde Y_j\|_\infty \leqslant 1$, we have also $\operatorname{Tr}\tilde X_i = \operatorname{Tr}\tilde Y_j = 0$ for all i,j. Moreover, it can be arranged that all $\tilde X_i$ and $\tilde Y_j$ are multiples of isometries if X_i, Y_j are.

EXERCISE 11.8 (Local correlation polytope on k qubits is also an ℓ_1-ball). Show that the set $\mathsf{LC}_{2,2,\dots,2} \subset \mathbb{R}^{2^k}$ (as defined in Remark 11.10) is a convex polytope with 2^{k+1} vertices and 2^{2^k} facets.

EXERCISE 11.9. Find the inradius and the outradius of the sets LC and QC.

EXERCISE 11.10. Show that the sets LC and QC have enough symmetries (in the sense of Section 4.2.2).

11.2.2. Bell correlation inequalities and the Grothendieck constant. In the context of correlation matrices, a *Bell correlation inequality* is a linear functional $\varphi : \mathsf{M}_{m,n} \to \mathbb{R}$ with the property that $\varphi(A) \leqslant 1$ for any classical correlation matrix $A \in \mathsf{LC}_{m,n}$. (We will discuss a more general setup in Section 11.3.) If we identify $\mathsf{M}_{m,n}$ with its dual space, the set of Bell correlation inequalities becomes the polytope $\mathsf{LC}_{m,n}^{\circ}$ (the polar of $\mathsf{LC}_{m,n}$) and can be identified with $B_1^m \check{\otimes} B_1^n$. Of particular interest are the extreme (or optimal) inequalities or, equivalently, the facets of $\mathsf{LC}_{m,n}$ (cf. Section 1.1.5).

A famous example of a Bell correlation inequality in the 2×2 case is the Clauser–Horne–Shimony–Holt or *CHSH inequality* φ_{CHSH}, which is the linear functional $A \mapsto \frac{1}{2}\operatorname{Tr}(AM_{CHSH})$, where

$$M_{CHSH} = \big(m_{ij}\big)_{i,j=1}^{2} := \begin{bmatrix} 1 & 1 \\ 1 & -1 \end{bmatrix}. \tag{11.3}$$

It is easily checked that $\frac{1}{2}M_{CHSH} \in \mathsf{LC}_{2,2}^{\circ}$ since, for any choice of $\xi, \eta \in \{-1,1\}^2$,

$$\xi_1\eta_1 + \xi_1\eta_2 + \xi_2\eta_1 - \xi_2\eta_2 \leqslant 2. \tag{11.4}$$

Moreover, 8 of the 16 possible choices of (ξ,η) saturate this bound.

Since, as we mentioned, the inclusion $\mathsf{LC}_{m,n} \subset \mathsf{QC}_{m,n}$ is strict (provided $m,n \geqslant 2$) it may happen that for a Bell correlation inequality φ and a quantum correlation matrix $A \in \mathsf{QC}_{m,n}$, we have $\varphi(A) > 1$. In that case, we say that the Bell correlation inequality φ is violated by A and the quantity $\varphi(A)$ is called the *violation* or, more precisely, the *quantum violation*. This is, in particular, the case for the CHSH inequality. We have

PROPOSITION 11.11 (CHSH violations, see Exercises 11.11–11.13). *The maximal quantum violation of the CHSH inequality is $\sqrt{2}$, and no Bell correlation inequality for 2×2 correlation matrices yields a larger violation.*

A remarkable fact is that violations of Bell correlation inequalities *of arbitrary size* cannot exceed a universal constant called the Grothendieck constant.

THEOREM 11.12 (Grothendieck–Tsirelson, not proved here). *There exists an absolute constant $K \geqslant 1$ such that, for any positive integers m, n, the following three equivalent conditions hold:*
1° *We have the inclusion*

$$\mathsf{QC}_{m,n} \subset K\mathsf{LC}_{m,n}. \tag{11.5}$$

2° *For any $m \times n$ real matrix $\big(m_{ij}\big)$ and for any ρ, X_i, Y_j verifying the conditions of Definition 11.6 we have*

$$\sum_{i,j} m_{ij} \operatorname{Tr} \rho(X_i \otimes Y_j) \leqslant K \max_{\xi\in\{-1,1\}^m, \eta\in\{-1,1\}^n} \sum_{i,j} m_{ij}\xi_i\eta_j. \tag{11.6}$$

3° *For any $m \times n$ real matrix $\big(m_{ij}\big)$ and for any (real) Hilbert space vectors x_i, y_j with $|x_i| \leqslant 1, |y_j| \leqslant 1$ we have*

$$\sum_{i,j} m_{ij}\langle x_i, y_j\rangle \leqslant K \max_{\xi\in\{-1,1\}^m, \eta\in\{-1,1\}^n} \sum_{i,j} m_{ij}\xi_i\eta_j. \tag{11.7}$$

The traditional version of Grothendieck's inequality is (11.7), the point being the existence of K independent of m and n (not proved here). The equivalence of 3° with 2° (the *Tsirelson's bound*) is the content of Proposition 11.8. Finally, the equivalence 1° $\iff$ 2° is just duality combined with the "classical" Proposition 11.7.

The best constant K such that (11.5)–(11.7) hold for any m, n is called the (real) Grothendieck constant and denoted by K_G. The precise value of K_G is not known; as of this writing, the best estimates are $1.6769 < K_G < \frac{\pi}{2\ln(1+\sqrt{2})} \approx 1.7822$. We also denote by $K_G^{(m,n)}$ the best constant in (11.5)–(11.7) for fixed m, n, and $K_G^{(n)} = K_G^{(n\,n)}$. This should not be confused with the optimal constant in (11.7) under the restriction that x_i, y_j live in an n-dimensional Hilbert space, which is denoted similarly by some authors. The values of all these and related "Grothendieck constants" are discussed in Exercises 11.13–11.17 and in Notes and Remarks.

One sees immediately that the maximum on the right-hand side of (11.7) is the norm of the bilinear form

$$M = \big(m_{ij}\big) : \ell_\infty^m \times \ell_\infty^n \to \mathbb{R}. \tag{11.8}$$

Thus Proposition 11.7 is really an instance of the duality between the projective and injective tensor products (see Section 4.1.4 and particularly Exercise 4.18). Similarly, the maximum on the left-hand side of (11.7) is the norm of M as a bilinear form on $\ell_\infty^m(\mathcal{H}) \times \ell_\infty^n(\mathcal{H})$. In the setting of operator spaces, the latter quantity may be interpreted as the so-called *completely bounded* norm of the bilinear form (11.8) or, equivalently, the minimal tensor norm of M in that category. In other words, the values of the Grothendieck constants and of the maximal violations of Bell correlation inequalities may be obtained by comparing two norms which naturally appear in the context of operator spaces. We will not go into the details of that theory (or even define precisely the concepts we mentioned above) since to do that at a reasonable level of diligence would require (at least) another chapter. Instead, we refer the interested reader to the excellent survey [**PV16**].

An important question, which has attracted lots of attention over the last twenty or so years, is the characterization of states ρ that may lead to nonlocal correlations. It is easy to see that if a state ρ is separable, then any correlation matrix (11.2) belongs to the local polytope LC. (A more general fact of this nature is discussed in Exercise 11.25.) In other words, entanglement is necessary—at least in the present context—for nonlocality. However, it is known to be insufficient [**Wer89**] and, with the goal of clarifying these issues, Peres asked in 1998 whether there is a link between locality and the PPT property. Various variants of the question have been answered, but the following most basic version is apparently still open (see also Remark 11.21 and Notes and Remarks on Section 11.3).

PROBLEM 11.13 (Peres conjecture for correlation matrices). *Can nonlocal correlations be obtained, in the sense of Definition* 11.6, *from a PPT state?*

As we mentioned earlier, the facial structure of the polytope $\mathsf{LC}_{m,n}$ is, for large m, n, rather complicated. For example, we could not find in the literature an answer to the following simple question.

PROBLEM 11.14 (How many Bell correlation inequalities are there?). *How does the number of facets of* $\mathsf{LC}_{n,n}$ *grow with* n*? By general arguments (see Exercises* 11.18 *and* 11.19*) it follows that* $\mathsf{LC}_{n,n}$ *has at least* $\exp(\Omega(n))$ *facets. An upper bound of* $n^{O(n^2)}$ *facets can be derived from the theory of* 0/1 *polytopes, i.e., of polytopes which are the convex hull of a subset of* $\{0,1\}^n$. *(See* [**Zie00, BP01**].*)*

Of course, an even more important problem is to characterize all facets/optimal Bell correlation inequalities modulo symmetries of $\mathsf{LC}_{n,n}$. However, most experts appear to think that, for large n, a satisfactory answer to such question is unlikely.

Let us conclude this section with a result giving volume and mean width estimates for the sets of correlation matrices. We state them for classical correlations only, since similar estimates for quantum correlations follow formally via Theorem 11.12 (see, however, Problem 11.16).

PROPOSITION 11.15. *For $m, n \in \mathbb{N}$ we have*

$$\left(\frac{1}{\sqrt{2}} - o(1)\right) \max(\sqrt{m}, \sqrt{n}) \leqslant \operatorname{vrad}(\mathsf{LC}_{m,n}) \leqslant w(\mathsf{LC}_{m,n}) \leqslant \sqrt{\frac{2}{\pi}}(\sqrt{m} + \sqrt{n})\frac{\sqrt{mn}}{\kappa_{mn}}, \tag{11.9}$$

where $o(1)$ indicates the behavior as $m, n \to \infty$. (Recall that the ratio $\sqrt{k}/\kappa_k$ decreases from $\sqrt{\pi/2}$ to 1 as k increases from 1 to ∞, see Proposition A.1.*)*

PROOF. The middle inequality is the Urysohn inequality (Proposition 4.15). To get the upper bound on the mean width, we use the Chevet–Gordon inequality (see Section 6.2.4.1) in the form from Exercise 6.49:

$$w_G(\mathsf{LC}_{m,n}) \leqslant \sqrt{n}\, w_G(B_\infty^m) + \sqrt{m}\, w_G(B_\infty^n) = \sqrt{2/\pi}(m\sqrt{n} + n\sqrt{m}).$$

For the lower bound on the volume radius, we may assume $m \geqslant n$. We claim that (with the identification $\mathsf{M}_{m,n} \leftrightarrow \mathbb{R}^{mn} \leftrightarrow (\mathbb{R}^n)^m$) we have

$$\frac{1}{\sqrt{2}}(B_2^n)^m \subset \mathsf{LC}_{m,n}. \tag{11.10}$$

Since the volume radius of $(B_2^n)^m$ is easy to calculate, namely

$$\operatorname{vrad}\left((B_2^n)^m\right) = \frac{\operatorname{vol}(B_2^n)^{1/n}}{\operatorname{vol}(B_2^{mn})^{1/mn}} = \frac{\Gamma\left(\frac{mn}{2}+1\right)^{1/mn}}{\Gamma\left(\frac{n}{2}+1\right)^{1/n}}$$

by (B.3), the lower bound in (11.9) follows then readily from Stirling's formula (as does an explicit nonasymptotic bound, should it be needed).

To establish (11.10), we note that, for $B \in \mathsf{M}_{m,n}$,

$$\begin{aligned} \sup_{A \in \mathsf{LC}_{m,n}} \operatorname{Tr}(AB) &= \sup_{\xi \in \{-1,1\}^m} \sum_{i=1}^m \left| \sum_{j=1}^n b_{ij}\xi_i \right| \\ &\geqslant \operatorname*{Ave}_{\xi \in \{-1,1\}^m} \sum_{i=1}^m \left| \sum_{j=1}^n b_{ij}\xi_i \right| \\ &\overset{(*)}{\geqslant} \frac{1}{\sqrt{2}} \sum_{i=1}^m \left(\sum_{j=1}^n b_{ij}^2 \right)^{1/2}, \end{aligned} \tag{11.11}$$

where $(*)$ denotes an application of the optimal Khintchine inequality (Exercise 5.71, with the value $A_1 = 1/\sqrt{2}$ from [**Sza76**]). It remains to observe that the inequality between the first and the last term in (11.11) is an equivalent dual version of (11.10). $\square$

While Theorem 11.12 implies that $\operatorname{vrad}(\mathsf{LC}_{n,n}) \simeq \operatorname{vrad}(\mathsf{QC}_{n,n})$ uniformly in $m, n \in \mathbb{N}$, it is not clear how different the two volume radii can be. Here is a question whose flavor is similar to that of Problem 9.14.

PROBLEM 11.16. *Is there an absolute constant $c < 1$ such that, for every $n \geqslant 2$,*

$$\mathrm{vrad}(\mathsf{LC}_{n,n}) \leqslant c\, \mathrm{vrad}(\mathsf{QC}_{n,n})?$$

Even showing that the ratio $\mathrm{vol}(\mathsf{LC})/\mathrm{vol}(\mathsf{QC})$ *tends to* 0 *does not seem straightforward.*

EXERCISE 11.11 (The CHSH bound). Show that $\sup\{\varphi_{CHSH}(A) \; : \; A \in \mathsf{QC}_{2,2}\} = \sqrt{2}$.

EXERCISE 11.12 (CHSH is the only 2×2 Bell correlation inequality). By Exercise 11.4, the polytope $\mathsf{LC}_{2,2}$ has 16 facets. Show that the unit normals to these facets are (up to the sign) exactly the matrices that can be obtained by permuting the entries of either $\frac{1}{2}M_{CHSH}$ or of $\begin{bmatrix} 1 & 0 \\ 0 & 0 \end{bmatrix}$. Conclude that, up to the obvious symmetries, φ_{CHSH} is the only nontrivial 2×2 Bell correlation inequality.

EXERCISE 11.13 (The Grothendieck–Tsirelson bound). Show that the sequence $\big(K_G^{(n)}\big)_n$ increases to K_G and that $K_G^{(2)} = \sqrt{2}$.

EXERCISE 11.14 (CHSH is the only $2 \times n$ Bell correlation inequality). Show that $K_G^{(2,n)} = \sqrt{2}$ for any $n \geqslant 2$.

EXERCISE 11.15 (CHSH is the only 3×3 Bell correlation inequality). Using the MATLAB multi-parametric toolbox (or other software, or lots of time), it is routine to establish that $\mathsf{LC}_{3,3}$ has 90 facets. Using this information, show that, up to the obvious symmetries, φ_{CHSH} is the only nontrivial 3×3 Bell correlation inequality and deduce that $K_G^{(3)} = \sqrt{2}$.

EXERCISE 11.16. Show that $K_G^{(2)}$ coincides with the maximal ratio of $\|M : \ell_\infty^2(\mathbb{C}) \to \ell_1^2(\mathbb{C})\|$ and $\|M : \ell_\infty^2(\mathbb{R}) \to \ell_1^2(\mathbb{R})\|$, where M varies over the set of real 2×2 matrices.

EXERCISE 11.17. Show that the complex Grothendieck constant (see (11.37) in Notes and Remarks for the definition) for 2×2 matrices equals 1.

EXERCISE 11.18 (Facial dimension of the local correlation polytope). Using Corollary 7.30, show that $\mathsf{LC}_{n,n}$ has $\exp(\Omega(n))$ facets. Moreover, for any fixed $\lambda > 1$, show that any polytope P such that $P \subset \mathsf{LC}_{n,n} \subset \lambda P$ or $P \subset \mathsf{QC}_{n,n} \subset \lambda P$ has $\exp(\Omega(n))$ facets.

EXERCISE 11.19 (Facial dimension of the local correlation polytope, take #2). Combine Proposition 6.3, Theorem 4.17, Proposition 11.15, and Exercise 11.9 to show that $\mathsf{LC}_{n,n}$ has $\exp(\Omega(n))$ facets.

11.3. Boxes and games

This section outlines more general Bell inequalities described in the language of boxes and games. It includes an explanation of how the original Grothendieck–Bell setup fits into the broader framework, the CHSH inequality as a game, and a presentation of several examples and special features such as no-signaling, PR-boxes, and bounded or unbounded violations.

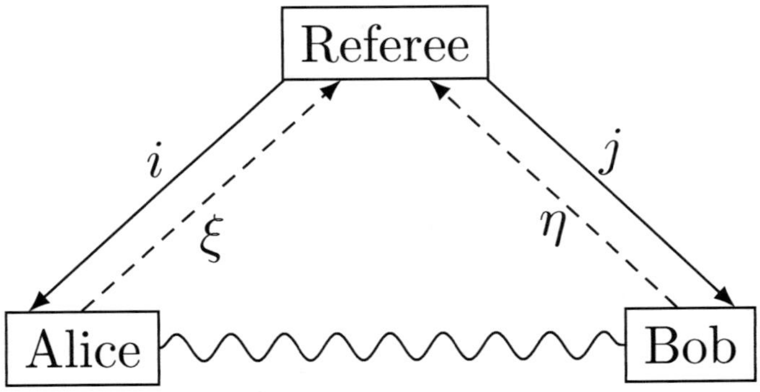

FIGURE 11.1. Diagrammatic representation of a quantum game. Prior to the game, Alice and Bob can agree on some strategy which, in the quantum variant, may involve sharing a bipartite quantum state (as depicted by the wavy line). Once the game starts, they are no longer allowed to communicate. The referee sends private input i to Alice and input j to Bob; Alice and Bob answer him privately with their outputs, respectively ξ and η.

11.3.1. Bell inequalities as games. We start by rephrasing the CHSH inequality (11.4) as a game. The game involves two cooperating players, Alice and Bob, and a fair but tough referee. The players may use a strategy agreed upon in advance and may share some resources, but are not allowed to communicate during the game. At each round of the game, the referee provides Alice and Bob with inputs (or settings) i and j, which can be 1 or 2, and each of them must respond with an output (respectively ξ and η) which can be 1 or -1. Alice and Bob win if the product $\xi\eta$ equals m_{ij}, the (i, j)th entry of the CHSH matrix (11.3), and lose otherwise. The difficulty is that while Alice knows her setting $i \in \{1, 2\}$, she doesn't know Bob's setting j, and similarly with the roles reversed.

A *deterministic* strategy consists of two vectors $(\xi_i)_{i=1}^2, (\eta_j)_{j=1}^2 \in \{-1, 1\}^2$ indicating players' responses for all values of the inputs. If the amount won or lost in each round is 1, the winnings per round, averaged over all possible inputs i, j (the *value* of the game), are

$$\frac{1}{4}\sum_{i,j} m_{ij}\,\xi_i\eta_j \leqslant \frac{1}{2} \tag{11.12}$$

(this is the same as the bound of 2 from (11.4) after renormalization), and half of deterministic strategies saturate this bound. Consequently, the same bound holds, and is optimal, for *random* strategies involving choosing at each round a random pair of vectors $(\xi(\omega), \eta(\omega))$ according to some distribution $p(\omega)$ (this requires *shared randomness* if the choices of Alice and Bob are not to be independent; such strategies, deterministic or random, are usually called *local* or *classical*). If we are interested instead in the probability of winning, the quantity to consider is the average of $\frac{1}{2}(1 + m_{ij}\,\xi_i\eta_j)$, which yields a bound of $\frac{3}{4}$.

The reader may wonder whether the uniform distribution on the set of inputs that is implicit in (11.12) is rather arbitrary. However, it is not hard to verify that such distribution faces the players with the toughest challenge. Similarly, there is a random strategy that yields game value $\frac{1}{2}$ for any probability distribution on the set of inputs that the referee may be using. (See Exercise 11.20.)

The quantum version of the CHSH game (see Figure 11.1) is very similar, except that rather than being deterministic or using shared randomness, the responses of Alice and Bob are based on measurements performed (locally on their respective sites $\mathcal{H}_A$ and $\mathcal{H}_B$) on a shared quantum state $\rho \in \mathrm{D}(\mathcal{H}_A \otimes \mathcal{H}_B)$. More precisely, for every setting i of Alice (resp., j of Bob) there is a pair of complementary projections $E_i^\xi, \xi = \xi_i \in \{-1, 1\}$ on $\mathcal{H}_A$ (resp., $F_j^\eta, \eta = \eta_j \in \{-1, 1\}$ on $\mathcal{H}_B$). If Alice receives from the referee the input i, she performs the projective measurement corresponding to $(E_i^\xi)_{\xi=\pm 1}$ and responds with the value of ξ supplied by the outcome of the measurement, and similarly for Bob. According to the Born rule (3.8), if the referee provides Alice and Bob with inputs (i, j), the probability of a pair of responses (ξ, η) will be $\operatorname{Tr} \rho\big(E_i^\xi \otimes F_j^\eta\big)$. Consequently, for these inputs, the expected value of the CHSH game will be m_{ij} (the corresponding entry of the payoff matrix M_{CHSH} from (11.3)) times

$$\sum_{\xi,\eta=\pm 1} \xi\eta \operatorname{Tr} \rho(E_i^\xi \otimes F_j^\eta) = \operatorname{Tr} \rho(X_i \otimes Y_j), \tag{11.13}$$

where $X_i = \sum_{\xi=\pm 1} \xi E_i^\xi = E_i^{+1} - E_i^{-1}$ and, similarly, $Y_j = F_j^{+1} - F_j^{-1}$. Averaging over all inputs i and j, we obtain the value

$$\frac{1}{4} \sum_{i,j} m_{ij} \operatorname{Tr} \rho(X_i \otimes Y_j). \tag{11.14}$$

Comparing (11.14) with (11.12) and appealing to Proposition 11.11 we conclude that there exists a *quantum game strategy* (i.e., ρ, $\big(E_i^\xi\big)$, $\big(F_j^\eta\big)$), which yields the value of $\frac{\sqrt{2}}{2}$ (which is also optimal). This is substantially better than the value of $\frac{1}{2}$ that can be achieved with classical strategies (deterministic or random). Similarly, if we want to focus on the probability of winning the game, the quantum strategy yields $\frac{2+\sqrt{2}}{4} \approx 0.8536$, which needs to be compared to the upper bound of $\frac{3}{4}$ for classical strategies that was calculated earlier. For a discussion of fine points of the optimality of this strategy see Exercise 11.21.

EXERCISE 11.20 (Optimality of the classical CHSH game strategies). (a) Show that if, in the CHSH game, the referee uses a non-uniform distribution on the set of inputs, then Alice and Bob have a deterministic strategy which gives a value strictly larger than $\frac{1}{2}$. (b) Describe all classical strategies of Alice and Bob that yield $\frac{1}{2}$ as the value of the CHSH game, irrespectively of the probability distribution on the set of inputs used by the referee.

EXERCISE 11.21 (Optimality of the quantum CHSH game strategies). State and prove a quantum version of the preceding exercise.

11.3.2. Boxes and the nonsignaling principle. The scheme that we described above via the example of the CHSH game can be conceptualized and generalized using the language of *boxes*. A box is a family of joint probability distributions

$$P = \{p(\cdot, \cdot | i, j) \ : \ 1 \leqslant i \leqslant m, \ 1 \leqslant j \leqslant n\}. \tag{11.15}$$

In the context of the two-player games described earlier, $p(\xi, \eta | i, j)$ is the probability that Alice and Bob respond with outputs ξ, η when presented with inputs i, j. If the payoff corresponding to this scenario is $v(\xi,\eta,i,j)$, the (average) value of the game is

$$V = \frac{1}{mn}\sum_{\xi,\eta,i,j} p(\xi,\eta|i,j)v(\xi,\eta,i,j). \tag{11.16}$$

For the CHSH game (classical or quantum), we had

$$v(\xi,\eta,i,j) = m_{ij}\xi\eta \tag{11.17}$$

with m_{ij}'s given by (11.3) and ξ,η taking values in $\{-1,1\}$. In the general case, ξ and η are no longer binary and we will not require that they take the same number of values. We will assume throughout this section that $\xi \in \{1,\ldots,k\}$ and $\eta \in \{1,\ldots,l\}$. (In fact, in some scenarios it may even be natural to consider boxes with the number of possible outputs dependent on the particular input.)

While the payoff function v can be *a priori* arbitrary, the probabilities implicit in the box P reflect the players' strategy and the resources available to them.

• Deterministic strategies (i.e., $\xi = f(i)$ and $\eta = g(j)$ for some functions f and g) result in a *deterministic* box:

$$p(\xi,\eta|i,j) = \mathbf{1}_{\{\xi=f(i)\}}\mathbf{1}_{\{\eta=g(j)\}}. \tag{11.18}$$

• Random strategies result in *product* boxes:

$$p(\xi,\eta|i,j) = p(\xi|i)p(\eta|j), \tag{11.19}$$

where $p(\cdot|i) = p_A(\cdot|i)$ and $p(\cdot|j) = p_B(\cdot|j)$ are the (independent) marginals of the distribution $p(\cdot,\cdot|i,j)$.

• Random strategies with shared randomness result in *local* (or *classical*) boxes:

$$p(\xi,\eta|i,j) = \int_\Lambda p(\xi|i,\lambda)\, p(\eta|j,\lambda)\, \mathrm{d}\mu(\lambda), \tag{11.20}$$

where $\lambda \in \Lambda$ is the (shared, knowingly or not) hidden variable, and μ a probability distribution on Λ.

• Quantum strategies result in *quantum* boxes:

$$p(\xi,\eta|i,j) = \operatorname{Tr}\rho\big(E_i^\xi \otimes F_j^\eta\big), \tag{11.21}$$

where ρ is a quantum state shared by Alice and Bob and, for each i (resp., j), $\big(E_i^\xi\big)_\xi$ (resp., $\big(F_j^\eta\big)_\eta$) is a POVM on Alice's space $\mathcal{H}_A$ (resp., Bob's space $\mathcal{H}_B$).

Let us denote the corresponding sets of boxes by DB, RB, LB and QB. If there is a need to specify the dimensions involved, we use expressions such as $\mathsf{QB}_{k,l|m,n}$. Since the number of values taken by ξ and η is, respectively, k and l, every box can be thought of as an element of $\mathbb{R}_+^{klmn}$ and we have

$$\mathsf{DB} \subset \mathsf{RB} \subset \mathsf{LB} \subset \mathsf{QB}. \tag{11.22}$$

The first inclusion is trivial and it is clear from the definition that $\mathsf{LB} = \operatorname{conv}\mathsf{RB}$; in particular LB is convex. (A moment of reflection—see Exercise 11.22—shows also that every product box is a mixture of deterministic boxes and so in fact $\mathsf{LB} = \operatorname{conv}\mathsf{DB}$.) The convexity of QB and the last inclusion in (11.22), which follows from it, are slightly less obvious (see Exercises 11.23, 11.25, 11.26 and Notes and Remarks for a discussion of these points and related issues). Except in trivial cases, the inclusion $\mathsf{LB} \subset \mathsf{QB}$ is strict; this follows, for example, from the fact that correlations can be retrieved from boxes (as in (11.16)–(11.17)) and from the inclusion $\mathsf{LC}_{m,n} \subset \mathsf{QC}_{m,n}$ being strict. Boxes that do not belong to LB are called *nonclassical* or *nonlocal*.

We next present a description of LB in the language of projective tensor products. First, consider the set of conditional marginal probability distributions

$$\mathrm{K}_{k,m} := \{p(\xi|i) \ : \ 1 \leqslant \xi \leqslant k, 1 \leqslant i \leqslant m\}, \tag{11.23}$$

i.e., of matrices $M = (m_{\xi,i}) \in \mathsf{M}_{k,m}$ with nonnegative coefficients and columns summing to 1. Then $\mathrm{K}_{k,m}$ is a convex compact set that canonically identifies with $(\Delta_{k-1})^m \subset (\mathbb{R}^k)^m = \mathbb{R}^{km}$ and one sees (directly from the definitions) that

$$\mathsf{LB}_{k,l|m,n} = \mathrm{K}_{k,m} \mathbin{\widehat{\otimes}} \mathrm{K}_{l,n}. \tag{11.24}$$

Due to the requirement that $p(\cdot,\cdot|i,j)$ be probability distributions, it is evident that the sets DB, RB, LB and QB are not full-dimensional in $\mathbb{R}^{klmn}$. The description (11.24) allows us to deduce that

$$\begin{aligned} \dim \mathsf{LB}_{k,l|m\,n} &= (\dim \mathrm{K}_{k,m} + 1)(\dim \mathrm{K}_{l,n} + 1) - 1 \\ &= mn(k-1)(l-1) + m(k-1) + n(l-1) \end{aligned}$$

(see Exercise 11.27). The geometry of QB is not as transparent as that of LB. To shed some light on it, let us consider a quantum box $P = \{p(\xi,\eta|i,j)\} = \{\operatorname{Tr}\rho(E_i^\xi \otimes F_j^\eta)\} \in \mathsf{QB}$ and, for given i,j, let us calculate the marginal density $p(\xi|i,j)$ of $p(\xi,\eta|i,j)$. We then obtain

$$p(\xi|i,j) = \sum_\eta p(\xi,\eta|i,j) = \sum_\eta \operatorname{Tr}\rho(E_i^\xi \otimes F_j^\eta) = \operatorname{Tr}\rho(E_i^\xi \otimes \mathrm{I}_{\mathcal{H}_B}) = \operatorname{Tr}\left(\rho_A E_i^\xi\right),$$

which doesn't depend on j (here $\rho_A = \operatorname{Tr}_{\mathcal{H}_B}\rho$ is the partial trace, cf. (3.10)). Similarly, the marginal densities $p(\eta|i,j)$ do not depend on i. In other words, there exist distributions $p(\cdot|i) = p_A(\cdot|i)$, $i = 1,\dots,m$, and $p(\cdot|j) = p_B(\cdot|j)$, $j = 1,\dots,n$ such that, for every i,j, $p_A(\xi|i)$ and $p_B(\eta|j)$ are the marginals of $p(\xi,\eta|i,j)$, i.e.,

$$\sum_\eta p(\xi,\eta|i,j) = p_A(\xi|i) \quad \text{and} \quad \sum_\xi p(\xi,\eta|i,j) = p_B(\eta|j). \tag{11.25}$$

Let us reflect now on the operational significance of (11.25). If, for some i, the distributions $p(\xi|i,j)$ depended on j, then (by implementing the procedure determining her response ξ to the input i obtained from the referee) Alice would gain information about the input j sent by the referee to Bob (*complete* information if the distributions $p(\cdot|i,j)$ were disjointly supported for distinct j, and *some* information if they were just different). This hypothetical event is usually interpreted as instant—or at least faster than light—signaling or communication and, consequently, the constraint (11.25) is usually referred to as the *nonsignaling principle*. (Actually, the arguably more appropriate interpretation may be that of *precognition* as nothing seems to forbid Alice from determining her response before—in the sense of being inside the past light cone—Bob determines his or, indeed, before Bob or even the referee knows the value of j. Note that while, in that case, Alice could in principle communicate her response to Bob, this has no effect on the statistics of *her* outputs.)

The set of boxes verifying (11.25) is called the *nonsignaling polytope* and we will denote it by NSB. (It is indeed a polytope, being the intersection of an affine subspace of $\mathbb{R}^{klmn}$ with the cube $[0,1]^{klmn}$.) An analysis of the constraints shows that LB and NSB (and hence the intermediate set QB) have the same dimension;

see Exercises 11.28 and 11.29. For 2-output nonsignaling boxes (i.e., if $k = l = 2$, in which case one may assume that ξ, η take values ± 1) one can still define the corresponding correlation matrices by the formula that is (modulo different normalization) implicit in (11.16)–(11.17), namely

$$a_{ij} = \sum_{\xi,\eta=\pm 1} \xi \eta\, p(\xi, \eta | i, j). \tag{11.26}$$

We will denote the set of such matrices by $\mathsf{NSC}_{m,n}$.

Here is an important example of elements of NSB, the so called Popescu–Rohrlich boxes, or PR-boxes. Let $m = n = k = l = 2$ (a bipartite 2×2 system with binary outputs: $\xi, \eta = \pm 1; i, j = 1, 2$) and consider the box P given by

$$p(\xi, \eta | i, j) = \begin{cases} \frac{1}{2}\mathbf{1}_{\{\xi \neq \eta\}} & \text{if } i = j = 2, \\ \frac{1}{2}\mathbf{1}_{\{\xi = \eta\}} & \text{otherwise.} \end{cases} \tag{11.27}$$

In other words, the joint distributions $p(\cdot, \cdot | i, j)$ are, respectively, either $\begin{bmatrix} 0 & \frac{1}{2} \\ \frac{1}{2} & 0 \end{bmatrix}$ or $\begin{bmatrix} \frac{1}{2} & 0 \\ 0 & \frac{1}{2} \end{bmatrix}$. Since all marginals $p_A(\cdot|i)$ and $p_B(\cdot|j)$ are identical, with probabilities of both outputs equal to $\frac{1}{2}$, it is immediately clear that P is nonsignaling. It is also apparent that for each combination (i, j) of inputs we have $\sum_{\xi,\eta=\pm 1} \xi\eta p(\xi, \eta | i, j) = m_{ij}$, where (m_{ij}) is given by (11.3). Accordingly, the value of the CHSH game (as given by (11.16)–(11.17)) is 1, as is the probability of winning. Since the analysis from Section 11.3.1 (based on Proposition 11.11) shows that the best value that can be achieved by a quantum strategy is $\frac{\sqrt{2}}{2}$, it follows that the PR-box cannot be realized as a quantum box via (11.21). This implies that the inclusion $\mathsf{QB} \subset \mathsf{NSB}$ is always proper.

We will conclude this section by giving volume estimates for the sets of nonsignaling boxes NSB and sets of nonsignaling correlation matrices $\mathsf{NSC}_{m,n}$.

PROPOSITION 11.17. *For $k, l, m, n \in \mathbb{N}$ we have*

$$\operatorname{vrad}(\mathsf{NSB}_{k,l|m,n}) = \Theta(\sqrt{mn}) \quad \text{and} \quad \operatorname{vrad}(\mathsf{NSC}_{m,n}) = \Omega(\sqrt{mn}). \tag{11.28}$$

PROOF. Since $\mathsf{NSB} = [0,1]^{klmn} \cap H$, where $H \subset \mathbb{R}^{klmn}$ is the nonsignaling affine subspace (in the notation of Exercises 11.28–11.27, $H = V_{k,m} \widehat{\otimes} V_{l,n}$), the first relation follows almost immediately from Proposition 4.27. The only two additional points that need to be made are as follows. First, while H doesn't contain the center of the cube $[0,1]^{klmn}$, it does contain the point whose coordinates are $\frac{1}{4}$, the center of the cube $[0, \frac{1}{2}]^{klmn}$. Accordingly, $\operatorname{vol}_N(\mathsf{NSB}) \geqslant \left(\frac{1}{2}\right)^N$, where $N = \dim H = mn + m + n$ (by Exercise 11.27). Since, by the Brunn–Minkowski inequality, central sections are at least as large as (parallel) non-central sections, the upper bound from Proposition 4.27 works without change and yields $\operatorname{vol}_N(\mathsf{NSB}) \leqslant 2^{(klmn-N)/2}$. The second point is that the dimension and the codimension of H are of the same order, and so $\left(2^{(klmn-N)/2}\right)^{1/N} = \Theta(1)$. It remains to combine the above estimates with the well-known asymptotic expression $\operatorname{vol}(B_2^N)^{1/N} \sim \sqrt{2\pi e/N}$ (as $N \to \infty$, see Appendix B.1 and particularly Exercise B.1).

The second relation can be analyzed in a similar way. By definition, $\mathsf{NSC}_{m,n}$ is a linear image of NSB, essentially a projection of a section of $[0,1]^{klmn}$. Since a projection of a section is larger than a section of a section, we get a lower bound.

(The reason for "essentially" is that the vector $\xi\eta = (1,-1)\otimes(1,-1) \in \mathbb{R}^2\otimes\mathbb{R}^2 \leftrightarrow \mathbb{R}^4$ is of norm 2 rather than 1.) $\square$

PROBLEM 11.18 (Volume radius and mean width of sets of boxes). *In the assertion of Proposition* 11.17, *can* Ω *be replaced by* Θ? *The argument given above (combined with, say, Proposition* 4.28*) runs into complications if* m *and* n *are of very different orders. More generally, what are the asymptotic orders of the volume radii and mean widths of sets of boxes of the sets* LB, QB, NSB *for arbitrary values of* k, l? *Some of the cases (e.g.,* LB, *because of* (11.24)*) appear fairly straightforward consequences of the methods presented in this book, but some of other ones seem to require further analysis.*

EXERCISE 11.22. Show that every product box is a convex combination of deterministic boxes.

EXERCISE 11.23 (Convexity of the set of quantum boxes). Show that the set QB of quantum boxes is a convex subset of $\mathbb{R}_+^{klmn}$.

EXERCISE 11.24 (Pure states suffice). Show that in the definition of quantum boxes (11.21) we can require the state ρ to be pure.

EXERCISE 11.25. Show that (i) $\mathsf{LB} \subset \mathsf{QB}$ and (ii) moreover, that every $P \in \mathsf{LB}$ can be realized as a quantum box (11.21) with ρ separable.

EXERCISE 11.26. Show that if a quantum box P can be written as $p(\xi,\eta|i,j) = \operatorname{Tr}(\rho(E_i^\xi \otimes F_j^\eta))$ with $\rho \in \mathrm{Sep}$, then $P \in \mathsf{LB}$.

EXERCISE 11.27 (The dimension of the set of local boxes). Show that $\dim \mathsf{LB} = mn(k-1)(l-1) + m(k-1) + n(l-1)$.

EXERCISE 11.28 (All sets of boxes have the same dimension). Show that

$$\dim \mathsf{QB} = \dim \mathsf{NSB} = \dim \mathsf{LB}.$$

EXERCISE 11.29. Deduce the equality $\dim \mathsf{QB} = \dim \mathsf{LB}$ from the fact that $\dim \mathrm{D} = \dim \mathrm{Sep}$ (shown in Section 2.2.3).

11.3.3. Bell violations. Consider a linear functional V on $\mathbb{R}^{mnkl}$ (sometimes called a "Bell functional" or a "Bell expression"). It can be written as

$$V(P) = \sum_{\xi,\eta,i,j} p(\xi,\eta|i,j)v(\xi,\eta,i,j). \tag{11.29}$$

Except for the normalizing factor, which was removed to reduce the clutter, this is the same as the average value of a game defined in (11.16). The local (or classical) optimal value of P is defined as

$$\omega_{\mathrm{L}}(V) = \max\{|V(P)| \ : \ P \in \mathsf{LB}\}. \tag{11.30}$$

(We will always tacitly assume that $\omega_{\mathrm{L}}(V) > 0$, i.e., that $V \notin \mathsf{LB}^\perp = \mathsf{NSB}^\perp$.) In this context, a Bell inequality is an inequality of the kind $|V(\cdot)| \leqslant \omega_{\mathrm{L}}(V)$. If a (necessarily nonlocal) box P satisfies $|V(P)| > \omega_{\mathrm{L}}(V)$, one says that the Bell inequality is violated and the ratio $|V(P)|/\omega_{\mathrm{L}}(V)$ is called the violation. Similarly, the quantum and nonsignaling optimal values of V are defined as

$$\omega_{\mathrm{Q}}(V) = \sup\{|V(P)| \ : \ P \in \mathsf{QB}\}, \quad \omega_{\mathrm{NS}}(V) = \max\{|V(P)| \ : \ P \in \mathsf{NSB}\}. \tag{11.31}$$

Finally, $\max_V \omega_{\mathrm{Q}}(V)/\omega_{\mathrm{L}}(V)$ is called the maximal quantum violation (for the particular values of m, n, k, l; more precisely, quantum-to-classical or quantum-to-local violation), and similarly for violations involving nonsignaling boxes. For example, the discussion following the definition (11.27) of PR-boxes shows that, for the CHSH game, nonsignaling-to-classical violations can as large as 2 (see Exercise 11.33 and cf. Proposition 11.24). All these parameters have nice functional-analytic interpretations, see Exercise 11.31.

As in the case of the CHSH game, the reader may wonder whether the uniform distribution on the set of inputs implicit in the definition of $V(P)$, and hence indirectly in (11.30)–(11.31), is justified. While for some "balanced" Bell functionals it will be true that—as for the CHSH game, see Exercise 11.20—the von Neumann–Nash-type equilibrium indeed involves the uniform distribution, this will not be universally the case. However, there is a simple trick that allows us to sidestep this issue: a game with the distribution $\pi(i, j)$ on input settings and the payoff function $v(\xi, \eta, i, j)$ is equivalent to the game with the payoff function $mn\pi(i, j)v(\xi, \eta, i, j)$ and the uniform distribution. In other words, considering uniform distributions on sets of inputs covers all possible scenarios: it is just one of many *essentially* equivalent ways of parameterizing the set of all possible Bell functionals. However, in some situations a moment of reflection will be needed; since, for example, the optimal $\pi(i, j)$'s for the local, quantum and nonsignaling strategies may be different, one has to be sure that one does not compare "apples to oranges".

As we will see later, measurement schemes involving boxes may lead to arbitrarily large violations. However, this is not the case for boxes with 2-outcomes (i.e., when $k = l = 2$). The reason is that sets of 2-outcome boxes are closely related to sets of correlations introduced in Section 11.2. This is particularly clear when one compares the set $\mathsf{LC}_{m,n}$ of classical/local correlations, which, by Proposition 11.7, identifies canonically with $B_\infty^m \,\widehat{\otimes}\, B_\infty^n \subset \mathbb{R}^m \otimes \mathbb{R}^n \leftrightarrow \mathbb{R}^{mn}$, and the corresponding set $\mathsf{LB}_{2,2|m,n}$ of local boxes, which, by (11.24), identifies with $\mathrm{K}_{2,m} \,\widehat{\otimes}\, \mathrm{K}_{2,n} = (\Delta_1)^m \,\widehat{\otimes}\, (\Delta_1)^n \subset \mathbb{R}^{2m} \otimes \mathbb{R}^{2n} \leftrightarrow \mathbb{R}^{4mn}$. In other words, $\mathsf{LC}_{m,n}$ is the projective tensor product of two 0-symmetric cubes, while $\mathsf{LB}_{2,2|m,n}$ is the projective tensor product of two similar cubes, but contained in spaces twice their dimension and centered at the point whose coordinates are $\frac{1}{2}$.

PROPOSITION 11.19. *If $k = l = 2$ then for any Bell expression V, we have $\omega_{\mathrm{Q}}(V) \leqslant K_G\, \omega_{\mathrm{L}}(V)$, where K_G is the Grothendieck constant. If, additionally, $m = n = 2$, then K_G can be replaced by $\sqrt{2}$.*

PROOF. Assume that the labels ξ and η belong to $\{-1, 1\}$ rather than $\{1, 2\}$. The maximum in $\omega_{\mathrm{L}}(V)$ is achieved on an extreme point of LB, i.e., on a deterministic box that is of the form (cf. (11.18))

$$p(\xi, \eta | i, j) = \mathbf{1}_{\{\xi = x_i, \eta = y_j\}} = \frac{1}{4}(1 + \xi x_i)(1 + \eta y_j)$$

for some vectors $x \in \{-1, 1\}^m$ and $y \in \{-1, 1\}^n$. We can then write

$$V(P) = \sum_{i=1}^{m} \sum_{j=1}^{n} \alpha_{i,j} x_i y_j + \sum_{i=1}^{m} \beta_i x_i + \sum_{j=1}^{n} \gamma_j y_j + \delta$$

with $\alpha_{i,j} = \mathrm{Ave}[\xi\eta\, v(\xi, \eta, i, j)]$, $\beta_i = \mathrm{Ave}[\xi\, v(\xi, \eta, i, j)]$, $\gamma_j = \mathrm{Ave}[\eta\, v(\xi, \eta, i, j)]$ and $\delta = \mathrm{Ave}[v(\xi, \eta, i, j)]$. (In each formula, Ave is a shortcut for $\frac{1}{4} \sum$ over all indices among i, j, ξ, η not appearing on the left of the equation.) We can gather all these quantities in a single $(m+1)\times(n+1)$ matrix by defining $\alpha_{i,n+1}=\beta_i$, $\alpha_{n+1,j}=\gamma_j$ and $\alpha_{n+1,n+1}=\delta$ and obtain

$$(11.32)\qquad \omega_{\mathrm{L}}(V)=\left|V(P)\right|=\max\left\{\left|\sum_{i=1}^{m+1}\sum_{j=1}^{n+1}\alpha_{ij}a_{ij}\right| \;:\; (a_{ij})\in \mathsf{LC}_{m+1,n+1}\right\}.$$

Consider now a quantum box $P'\in \mathsf{QB}_{2,2|m,n}$, of the form $p'(\xi,\eta|i,j)=\operatorname{Tr}\rho(E_i^\xi\otimes F_j^\eta)$. Using the same notation as before and setting $X_i=E_i^1-E_i^{-1}$ and $Y_j=F_i^1-F_j^{-1}$ as in (11.13)–(11.14), we can write

$$\begin{aligned} V(P') &= \sum_{i=1}^{m}\sum_{j=1}^{n}\alpha_{i,j}\operatorname{Tr}\rho(X_i\otimes Y_j)+\sum_{i=1}^{m}\beta_i\operatorname{Tr}\rho(X_i\otimes \mathrm{I})+\sum_{j=1}^{n}\gamma_j\operatorname{Tr}\rho(\mathrm{I}\otimes Y_j)+\delta\\ &= \sum_{i=1}^{m+1}\sum_{j=1}^{n+1}\alpha_{i,j}\operatorname{Tr}\rho(X_i\otimes Y_j), \end{aligned}$$

where in the last sum we defined $X_{m+1}=\mathrm{I}$ and $Y_{n+1}=\mathrm{I}$. It now follows that

$$(11.33)\qquad \left|V(P')\right| \leqslant \max\left\{\left|\sum_{i=1}^{m+1}\sum_{j=1}^{n+1}\alpha_{ij}a_{ij}\right| \;:\; (a_{ij})\in \mathsf{QC}_{m+1,n+1}\right\}.$$

Since $P'\in\mathsf{QB}$ was arbitrary, the first statement of the Proposition follows by comparing (11.33) with (11.32) and appealing to Theorem 11.12. For the second statement, we note that if $m=n=2$, then Theorem 11.12 will be used for 3×3 matrices and so K_G may be replaced by $K_G^{(3)}=\sqrt{2}$, see Exercises 11.13–11.15. □

REMARK 11.20. The argument shows that the violations of bipartite n-input, 2-output boxes do not exceed $K_G^{(n+1)}$ (and similarly for "rectangular" boxes, i.e., $m\neq n$). Still, the matrices (a_{ij}) that appear in (11.33) have a special structure and so it is conceivable that the bound $K_G^{(n)}$ works, too. However, this is unlikely to be a matter of a formal algebraic reduction since, for example, the setting of 3-input, 2-output boxes leads to optimal Bell inequalities, which are not present in the context of 3×3 correlation matrices (the so-called I_{3322} inequalities).

REMARK 11.21. Since the proof of Proposition 11.19 translates violations for 2-output quantum boxes to quantum violations for correlation matrices associated with *the same* state ρ, it follows that Problem 11.13, i.e., the Peres conjecture for correlation matrices, is formally equivalent to the analogous problem for boxes. However, if we allow three outputs in one of the boxes, the answer is known: there is an example of a PPT state producing violations, even with $\dim\mathcal{H}_A=\dim\mathcal{H}_B=3$.

As we mentioned earlier, measurement schemes involving more general boxes may lead to arbitrarily large violations. This may happen for two reasons: either the system is not bipartite (i.e., it involves three or more parties) or the outputs are not binary. These two situations are exemplified by the following pair of results. Recall that $\mathsf{LC}_{n_1,\ldots,n_k}$ and $\mathsf{QC}_{n_1,\ldots,n_k}$ are the k-partite generalizations of the sets of classical and quantum correlation matrices, see Remark 11.10 for details.

PROPOSITION 11.22 (Not proved here). *Denote by $K_G^{(n_1,\ldots,n_k)}$ the best constant K such that the inclusion $\mathsf{QC}_{n_1,\ldots,n_k}\subset K\,\mathsf{LC}_{n_1,\ldots,n_k}$ holds. Then $K_G^{(n,n,n)}=\Omega(n^{1/4}(\log n)^{-3/2})$ and $K_G^{(n_1,n_2,n_3)}\leqslant K_G\min\{n_1,n_2,n_3\}^{1/2}$.*

PROPOSITION 11.23 (Not proved here). *For any Bell functional V we have $\omega_{\mathrm{Q}}(V)/\omega_{\mathrm{L}}(V) \leqslant K_G^{\mathbb{C}}\sqrt{kl}$ (independently of the values of m,n). On the other hand, if $l=2$ and $m,n=2^k$, then there exists V such that $\omega_{\mathrm{Q}}(V)/\omega_{\mathrm{L}}(V) = \Omega\big(\sqrt{k}/\log^2 k\big)$.*

Above $K_G^{\mathbb{C}}$ stands for the *complex* Grothendieck constant; see Notes and Remarks for a precise definition and for estimates. Both propositions can be understood as statements about comparing different norms on tensor products of operator spaces. (The identification (11.24) gives one hint why this may be the case.)

The existence of large nonsignaling-to-classical violations is much easier to establish. We have

PROPOSITION 11.24. *In the class of boxes with binary outputs (i.e., $k=l=2$), the maximal nonsignaling-to-classical violation satisfies*

$$\max_V \omega_{\mathrm{NS}}(V)/\omega_{\mathrm{L}}(V) = \Omega\big(\min(\sqrt{m},\sqrt{n})\big).$$

Moreover, the same bound holds for violations involving correlation matrices.

PROOF. Combine Propositions 11.15 and 11.17. One way to take care of fine points is to use Urysohn's inequality to deduce that $w(\mathsf{NSC}_{m,n}) = \Omega(\sqrt{mn})$ and then compare it to the upper bound $w(\mathsf{LC}_{m,n}) = O(\sqrt{m}+\sqrt{n})$ from (11.9). This leads to a nonsignaling-to-classical violation (of correct order) of some Bell correlation inequality and shows the second (and hence the first) statement. □

We conclude the section by introducing another concept which quantifies nonlocality and which is, in a sense, a generalization of the geometric distance between sets (of boxes). Given $P \in \mathsf{NSB}$ we define the *local fraction* (or *classical fraction*) of P as

$$p_{\mathrm{L}} = p_{\mathrm{L}}(P) := \max\{t \in [0,1] \ : \ P \in t\mathsf{LB} + (1-t)\mathsf{NSB}\}. \tag{11.34}$$

The quantity $p_{\mathrm{NL}} := 1 - p_{\mathrm{L}}$ is the *nonlocal fraction.* Similar parameters can be defined for other pairs in place of LB, NSB. For example, replacing in (11.34) LB by DB, the set of deterministic boxes (defined by (11.18)) leads to the notion of *fraction of determinism*.

Clearly $P \in \mathsf{LB}$ iff $p_{\mathrm{L}} = 1$. Therefore, by the Hahn–Banach separation theorem, whenever $p_{\mathrm{NL}} > 0$, then there exists a Bell functional V such that $V(P) > \omega_{\mathrm{L}}(V)$ (i.e., P violates some Bell inequality). However, the size of the violation cannot be immediately ascertained. What can be quantified, though, is the relationship between different types of violations. We have

PROPOSITION 11.25. *Let $P \in \mathsf{NSB}$ and let V be a Bell functional. Then*

$$\frac{V(P)}{\omega_{\mathrm{L}}(V)} - 1 \leqslant p_{\mathrm{NL}}\left(\frac{\omega_{\mathrm{NS}}(V)}{\omega_{\mathrm{L}}(V)} - 1\right). \tag{11.35}$$

PROOF. By definition, there is a local box P' and a nonsignaling box P'' such that $P = p_{\mathrm{L}}P' + p_{\mathrm{NL}}P''$ and consequently

$$V(P) = p_{\mathrm{L}}V(P') + p_{\mathrm{NL}}V(P'') \leqslant p_{\mathrm{L}}\omega_{\mathrm{L}}(V) + p_{\mathrm{NL}}\omega_{\mathrm{NS}}(V),$$

which is equivalent to the asserted inequality (11.35). □

The meaningful case in (11.35) is when $0 < p_{\mathrm{NL}} < 1$. We can then conclude that while P violates some Bell inequalities, the violation is always noticeably smaller than the nonsignaling violation $\omega_{\mathrm{NS}}(V)/\omega_{\mathrm{L}}(V)$, uniformly over all V for which that ratio is strictly greater than 1.

An interesting and somewhat surprising setting when one has a nontrivial lower bound on the local fraction is for (bipartite) quantum boxes with $\min\{m, n\} = 2$. We have then

THEOREM 11.26 (Not proved here). *Consider a two-player game setup with $n = 2$ (i.e., two input settings at Bob's site) and arbitrary (but fixed) m, k, l. Then*

$$\inf\{p_{\mathrm{L}}(P) \ : \ P \in \mathsf{QB}_{k,l|m,2}\} \geqslant c, \tag{11.36}$$

where $c > 0$ is a constant that depends only on k and l (but not on m nor on the dimensions of the underlying Hilbert spaces). The same is true about the fraction of determinism.

Theorem 11.26, in combination with Exercise 11.33, provides an alternative argument that the PR-box cannot be realized as a quantum box. The same reasoning works for any bipartite setup with $\min\{m, n\} = 2$ and any box which yields the optimal nonsignaling value for any Bell functional V such that $\omega_{\mathrm{NS}}(V) > \omega_{\mathrm{L}}(V)$ (so, while more involved and less sharp, the present argument is very general).

The assertion of Theorem 11.26 does not hold when both players have three or more settings. This is because in that case there exist the so-called pseudotelepathy quantum games, i.e., the games that can be won with probability 1 using quantum strategies, while no foolproof classical strategy is possible. Consequently, if P is the corresponding quantum box and V is the probability of winning, then $V(P) = 1 = \omega_{\mathrm{NS}}(V)$, while $\omega_{\mathrm{L}}(V) < 1$, and so it follows from (11.35) that $p_{\mathrm{L}} = 1 - p_{\mathrm{NL}} = 0$. An outline of one such game, the Mermin–Peres magic square game, is given in Exercise 11.35.

EXERCISE 11.30 (Linear vs. affine Bell inequalities). Show that definitions (11.30) and (11.31) yield the same value if we allow V to vary over all affine functionals and not just over linear functionals.

EXERCISE 11.31 (Violations, symmetrizations, and the geometric distance). Verify that $\omega_{\mathrm{L}}(V) = \|V\|_{K^\circ}$, where $K = \mathsf{LB}_{\oslash}$ is the "cylindrical" symmetrization of K, and similarly for QB and NSB. Deduce that the maximal quantum violation equals $d_g(\mathsf{LB}_{\oslash}, \mathsf{QB}_{\oslash})$. (See (4.6) and (4.1) for definitions.)

EXERCISE 11.32 (Violations and widths). (i) Let $\delta = \max_u \frac{w(\mathsf{QB},u)-w(\mathsf{QB},-u)}{w(\mathsf{LB},u)-w(\mathsf{LB},-u)}$ be the maximal ratio of widths of QB and LB (see Section 4.3.3). Show that the maximal quantum violation is contained between δ and $2\delta - 1$. (ii) State and prove an analogous statement for NSB.
Note: It follows that the ratio of widths is an alternative measure of violation equivalent (up to a factor of 2) to the one based on values. Observe that, by Exercise 11.31, we would have equality—and not just equivalence—if we used $\mathsf{LB}_{\oslash}$, $\mathsf{QB}_{\oslash}$ in place of LB, QB in the definition of δ.

EXERCISE 11.33 (Nonsignaling value of the CHSH game). Show that the nonsignaling value of the CHSH game is 2 and deduce that the maximal nonsignaling violation for $m = n = k = l = 2$ is 2.

EXERCISE 11.34 (Quantum box for the CHSH game). Give an *explicit* example of an ensemble $\big(\rho, (E_i^\xi), (F_j^\eta)\big)$ which induces—via (11.21)—a quantum box giving the optimal violation of the CHSH game.

TABLE 11.1. The magic square game.

$\sigma_x \otimes \mathrm{I}$	$\mathrm{I} \otimes \sigma_x$	$\sigma_x \otimes \sigma_x$
$-\sigma_x \otimes \sigma_z$	$-\sigma_z \otimes \sigma_x$	$\sigma_y \otimes \sigma_y$
$\mathrm{I} \otimes \sigma_z$	$\sigma_z \otimes \mathrm{I}$	$\sigma_z \otimes \sigma_z$

EXERCISE 11.35 (The magic square game).
(i) Verify that the self-adjoint operators on $\mathbb{C}^2 \otimes \mathbb{C}^2$ given in Table 11.1 have the following properties
(a) the operators in each row commute and the same is true for each column
(b) the composition of the entries in each row is I, while the composition of the entries in each column is $-$I.
(ii) Show that there is no 3×3 table consisting of numbers such that the product of the entries in each row is 1, while the product of the entries in each column is -1.
(iii) The Mermin–Peres magic square game is played as follows. The number of input settings is $m = n = 3$ and the outputs are strings of ± 1 of length 3. An additional restriction is that the product of elements of Alice's string must be 1, while the product of elements of Bob's string must be -1 (so, in effect, $k = l = 4$). If the input settings communicated to Alice and Bob were (i, j), Alice and Bob win if their output strings placed respectively in ith row and jth column coincide on the common ij-th entry, and lose otherwise. Show that
(a) there is no deterministic (and hence classical) winning strategy,
(b) the following is a winning quantum strategy. Alice and Bob share a 4-qubit quantum state $\varphi^+ \otimes \varphi^+$, where $\varphi^+ = \frac{1}{\sqrt{2}}(|00\rangle + |11\rangle)$ is a Bell state with the first qubit of each copy of φ^+ going to Alice and the second to Bob. Given input i, Alice measures her part of the state in a basis in which the (commuting) operators from the ith row are simultaneously diagonal, and answers the corresponding triple of eigenvalues. Given input j, Bob does the same thing using the jth column.

Notes and Remarks

Section 11.1. The argument that the proper mathematical home of Bell inequalities belongs to the operator space theory was most explicitly put forward in [**JPPG**$^+$**10**].

For a proof of Theorem 11.4, we refer the reader to [**Por81**] (Theorem 13.68) or [**Kir76**]. There is a huge gap between that Theorem and Lemma 11.1, which both yield subspaces of dimension $\Theta(\log n)$ and the optimal ratio $\|\cdot\|_{\mathrm{op}}/\|\cdot\|_{\mathrm{HS}} \equiv 1/\sqrt{n}$, and the subspaces given by Dvoretzky's theorem, which feature $\|\cdot\|_{\mathrm{op}}/\|\cdot\|_{\mathrm{HS}} \approx 2/\sqrt{n}$ and are of dimension $\Theta(n)$ (see Theorem 7.37; for sharpness, see Exercise 7.25). Accordingly, we suggest the following problem.

PROBLEM 11.27. *Given $\lambda \geqslant 1$, denote by $d(n, \lambda)$ the maximal dimension of a subspace $E \subset \mathsf{M}_n(\mathbb{R})$ such that, for any $M \in E$,*

$$\frac{1}{\sqrt{n}}\|M\|_{\mathrm{HS}} \leqslant \|M\|_\infty \leqslant \frac{\lambda}{\sqrt{n}}\|M\|_{\mathrm{HS}}.$$

It follows from Lemma 11.1 that, for $\lambda > 1$, $d(n, \lambda) = \Omega(\log n)$. Is this sharp for $\lambda \in (1, 2]$? (For $\lambda > 2$, we have $d(n, \lambda) = \Theta_\lambda(n)$.)

Note that while Lemma 11.1 addresses only the case when n is a power of 2, one can readily deduce that (for $\lambda > 1$ and for arbitrary n) $d(n,\lambda) \geqslant 2\log_2 n - C(\lambda)$. (Consider $E = F \otimes \mathrm{I}_m$ where $F \subset \mathsf{M}_{2^k}$ is the subspace from Lemma 11.1, for appropriate k, m with $2^k m \leqslant n < 2^k(m+1)$.) Note, however, that $d(n,1) = 1$ if (and only if) n is odd, see Exercise 11.1.

Section 11.2. Proposition 11.7 is probably folklore, and Proposition 11.8 is due to Tsirelson [**Cir80, Tsi85, Tsi93**]. Theorem 11.12 is known as Grothendieck's inequality [**Gro53a**] and its reformulation via correlation matrices is also due to Tsirelson. The paper [**Gro53a**] went largely unnoticed for 15 years until it was "brought to the mathematical mainstream" by Lindenstrauss and Pełczyński in [**LP68**]. In particular, the elementary formulation (11.7) comes from [**LP68**]. For a beautiful recent survey about Grothendieck's inequality, including historical background and far-reaching generalizations, see [**Pis12a**].

Concerning values of constants $K_G^{(m,n)}$ for specific m, n, we have $K_G^{(2,n)} = \sqrt{2}$ for all n and $K_G^{(3,n)} = \sqrt{2}$ for all n (the latter is stated without proof in [**FR94**] and attributed ultimately to Kemperman's interpretation of results of Garg [**Gar83**]; see also [**AII06**] and [**BM08**], on which Exercise 11.15 is based). The approach that was used to calculate $K_G^{(3)}$ in Exercise 11.15 can be in principle replicated for larger dimensions, but the computational complexity of the problem increases very fast. It was implemented in [**Li**] to show rigorously that $K_G^{(4)} = \sqrt{2}$ (there are two new Bell correlation inequalities that appear in the 4×4 context, but neither of them leads to a violation that is $\sqrt{2}$ or larger). See [**DD16**], [**Kin17**] for an analysis leading to $K_G^{(4,7)} = K_G^{(5,5)} = \sqrt{2}$. Other values of $K_G^{(m,n)}$ seem to be unknown; it seems that $(m,n) = (11,14)$ are the smallest known values of the parameters for which $K_G^{(m,n)} > \sqrt{2}$, see [**Ver08**]. Various aspects of this circle of ideas, including in particular the significance of the constant $\sqrt{2}$, are discussed in [**For10**] and [**FR94**]. The CHSH inequality was introduced in [**CHSH69**].

One may also define $K_G^{[n]}$ as the best constant such that (11.7) holds for every matrix (a_{ij}) of arbitrary size and every vector $x_i, y_j \in \mathbb{R}^n$. An easy observation is that $K_G^{(n)} \leqslant K_G^{[n]}$. While $K_G^{[2]} = \sqrt{2}$ [**Kri79**], the value of $K_G^{[3]}$ seems unknown; see [**BNV16, HQV⁺17**] for recent lower and upper bounds.

The Grothendieck constant introduced in the text is the real Grothendieck constant. It has a complex counterpart defined as the smallest constant $K_G^{\mathbb{C}}$ such that, for any complex matrix (m_{ij}) of arbitrary size $m \times n$ and any unit vectors x_i, y_j in a complex Hilbert space, we have

$$\left|\sum_{i,j} m_{ij}\langle x_i, y_j\rangle\right| \leqslant K_G^{\mathbb{C}} \max_{\xi\in\mathbb{T}^m,\eta\in\mathbb{T}^n} \left|\sum_{i,j} m_{ij}\xi_i\eta_j\right|, \tag{11.37}$$

where $\mathbb{T}$ denotes the set of complex numbers of unit modulus. The best estimates are $1.338... < K_G^{\mathbb{C}} < 1.405...$, which in particular imply $K_G^{\mathbb{C}} < K_G$ (see [**Pis12a**] for more information and references). Somewhat surprisingly, for 2×2 matrices the complex Grothendieck inequality holds with constant 1, see Exercise 11.17 (based on [**BM08**]). For larger dimensions the optimal values of the constants do not seem to be known.

The argument from Exercise 11.8 is from [**WW01a**]. The description of the extremal Bell correlation inequalities (extreme points of LC°, or equivalently faces of LC) has attracted a lot of attention, see the website [**@4**].

Section 11.3. For more information on quantum boxes and Bell inequalities we refer the readers to the surveys [**PV16**] and [**BCP**$^+$**14**]. Older valuable references include [**Pit89**] and [**WW01b**].

Some authors reserve the term "value of the game" to payoff functions that are nonnegative. (Of course, any finite payoff function can be made nonnegative via an offset, but that makes a difference when we calculate the ratios of values for different strategies, as we do.) A 2-output game for which the payoff function is of the form (11.17) for *some* (m_{ij}) (or, perhaps, slightly more generally, $m_{ij}\xi\eta + n_{ij}$, which allows in particular, talking about $\frac{1}{2}(\xi\eta + 1)$, the probability of winning the game) is called an XOR game. This is because when we think of the outputs as Boolean data $a, b \in \{0, 1\}$, the value of the game depends only on the "exclusive or" (XOR) value $a \oplus b$. XOR games can also be defined for more than two players; their study is essentially equivalent to that of correlation matrices. It should be noted that while for local correlation matrices and boxes the link to the projective tensor product works perfectly (as in Proposition 11.7 and (11.24)), the correspondence to operator space tensor products in the quantum setting is slightly less satisfactory once we leave the setting of XOR games. This is pointed out, e.g., in section IV.B of [**PV16**]): while we still can, with some work, come up with two-sided estimates, constants larger than 1 do appear. It would be very useful to come up with a natural construction (such as the use of cylindrical symmetrizations in Exercise 11.31) which allows us to bypass this complication.

It is known [**AIIS04**] that determining whether a box is local is NP-complete, even for the class of boxes with 2 outputs, and similarly for correlation matrices. This is established via a connection to the concept of the *cut polytope* associated to a graph $G = (V, E)$, which is a polytope in $\mathbb{R}^E$ defined as

$$\operatorname{conv}\{(\delta_S(e))_{e \in E} \ : \ S \subset V\},$$

where $\delta_S(e) = 1$ if the edge e has one endpoint in S and one endpoint in $V \backslash S$, and 0 otherwise. It can be checked that $\mathsf{LC}_{m,n}$ is affinely equivalent to the cut polytope of the complete bipartite graph $K_{m,n}$ (cf. the comments on contextuality at the end of these notes) and that $\mathsf{LB}_{2,2|m,n}$ is affinely equivalent to the cut polytope of the complete tripartite graph $K_{m,n,1}$. For more information on cut polytopes we refer the reader to [**DL97**].

It is unknown whether the set QB of quantum boxes is closed. A closely related question is known as Tsirelson's problem and has to do with how quantum physics models locality: we may define a set QB' as the set of boxes $p(\xi, \eta|i, j)$ of the form

$$p(\xi, \eta|i, j) = \langle \psi | \bar{E}_i^\xi \bar{F}_j^\eta | \psi \rangle,$$

where ψ is a unit vector in a Hilbert space $\mathcal{H}$, and, for every i and j, $(\bar{E}_i^\xi)_\xi$ and $(\bar{F}_j^\eta)_\eta$ are POVMs on $\mathcal{H}$ which satisfy the commutation condition $\bar{E}_i^\xi \bar{F}_j^\eta = \bar{F}_j^\eta \bar{E}_i^\xi$ for any i, j, ξ, η. (It is crucial here to allow $\mathcal{H}$ to be infinite-dimensional.) To check that $\mathsf{QB} \subset \mathsf{QB}'$, simply take $\mathcal{H} = \mathcal{H}_A \otimes \mathcal{H}_B$, $\bar{E}_i^\xi = E_i^\xi \otimes \mathrm{I}$ and $\bar{F}_j^\eta = \mathrm{I} \otimes F_j^\eta$. A natural question is whether QB or its closure $\overline{\mathsf{QB}}$ are equal to QB' (the set QB' can be checked to be closed, see Proposition 3.4 in [**Fri12**]). It was proved in a series of papers [**JNP**$^+$**11**, **Fri12**, **Oza13**] that the equality $\overline{\mathsf{QB}} = \mathsf{QB}'$ is equivalent to Connes' embedding problem on von Neumann algebras. On the other hand, Slofstra proved [**Slo16**] using techniques from group theory that $\mathsf{QB} \subsetneq \mathsf{QB}'$.

The I_{3322} inequalities appeared in [**Fro81**] and the terminology was introduced in [**CG04**]. PR-boxes are usually credited to [**PR94**], were they were studied in some detail, but they make an appearance in [**KT85, Tsi85**].

A concept in the spirit of local/nonlocal fractions was introduced in [**BKP06**]. Fraction of determinism appears in [**JHH$^+$15**], which also contains Theorem 11.26 and its proof.

Peres conjecture was stated in a somewhat vague form in [**Per99**]. A more rigorous mathematical formulation and interesting positive partial results can be found in a series of papers [**WW00, WW01a, WW01b**]. The example of a quantum box mentioned in Remark 11.21, based on a PPT state on $\mathbb{C}^3 \otimes \mathbb{C}^3$ and disproving the Peres conjecture for bipartite systems was given in [**VB14**]. An earlier example in the multipartite setting was given in [**Dür01**]. See the discussion in [**VB14**] and in section III.A of [**BCP$^+$14**] for more on the relationship between nonlocality and entanglement, and for many more references.

The fact that the multipartite analogue of Theorem 11.12 does not hold has been known for some time. In the present context, unboundedness of $K_G^{(n_1,n_2,n_3)}$ as n_1, n_2, n_3 tend to infinity was shown in [**PGWP$^+$08**]. Quantitative estimates were obtained later in [**Pis12b**]. Proposition 11.22, with a slightly worse power of logarithm, appeared in [**BV13**]; the version stated here is from [**PV16**]. Proposition 11.23 is from [**PY15**]. See also [**JP11**]; more references can be found in [**PV16**].

The form of the Mermin–Peres magic square game given in Exercise 11.35 follows largely [**Ara04**]. Another (more explicit but less transparent) exposition can be found in [**BBT05**]. Other demonstrations of pseudotelepathy are based on versions of the Kochen–Specker theorem [**KS67**] which involves the concept of *contextuality*. Contextuality, or rather *noncontextuality*, is a generalization of locality. For example, a two party scenario allows to perform measurements indexed by pairs $\{(i,j)\}$, where i and j identify respectively local POVMs of Alice and Bob; this can be represented by a complete bipartite graph $K_{m,n}$. By contrast, the more general scenario permits a general hypergraph: the observables are still represented by vertices, with the hyperedges corresponding to their subsets that can be performed (simultaneously or sequentially) without mutually affecting the outcomes of other observables in the subset. See [**BBT05**] for more details and examples and [**CSW14**] for sophisticated links to graph theory.

CHAPTER 12

POVMs and the distillability problem

This last chapter consists of two parts which are linked by the central role played by the concept of POVMs, but are otherwise largely independent. The first part deals with the norms that are associated with POVMs and which are intimately related to zonoids. This connection allows us to derive a sparsification result for POVMs. The second part also uses the language of POVMs, but is focused on the distillability problem, a major unsolved problem in quantum information theory.

12.1. POVMs and zonoids

12.1.1. Quantum state discrimination. What happens when a quantum system in a state ρ is measured with a POVM? We only focus on the case of a discrete POVM, $\mathrm{M} = (M_i)_{1 \leqslant i \leqslant N}$ (continuous POVMs could then be treated by approximation).

We know from Born's rule (3.13) that the outcome i is obtained with probability $\mathrm{Tr}(\rho M_i)$. This simple formula can be used to quantify the efficiency of a POVM to perform the task of state discrimination. State discrimination can be described as follows: a quantum system is prepared in an unknown state which is either ρ or σ (both hypotheses being *a priori* equally likely), and we have to guess the unknown state.

After measuring it with the POVM $\mathrm{M} = (M_i)_{1 \leqslant i \leqslant N}$, the outcome i occurs with probability $p_i = \mathrm{Tr}(\rho M_i)$ if the unknown state is ρ and with probability $q_i = \mathrm{Tr}(\sigma M_i)$ if the unknown state is σ. Consequently, the optimal strategy is as follows: when outcome i is observed, guess ρ if $p_i > q_i$ and guess σ if $p_i < q_i$ (and use any rule if $p_i = q_i$). The probability of failure is then

$$\mathbf{P}(\text{failure}) = \frac{1}{2}\sum_{i=1}^{N} \min(p_i, q_i) = \frac{1}{2} - \frac{1}{4}\sum_{i=1}^{N} |p_i - q_i| .$$

It is convenient to introduce the *distinguishability (semi-)norm* $\|\cdot\|_{\mathrm{M}}$ defined for $\Delta \in B^{\mathrm{sa}}(\mathcal{H})$ by

$$\|\Delta\|_{\mathrm{M}} = \sum_{i=1}^{N} |\mathrm{Tr}(\Delta M_i)| . \tag{12.1}$$

Note that $\|\cdot\|_{\mathrm{M}}$ is a norm if and only if $\mathrm{span}\{M_i \ : \ 1 \leqslant i \leqslant N\} = B^{\mathrm{sa}}(\mathcal{H})$, which requires in particular $N \geqslant (\dim \mathcal{H})^2$. Since $\mathbf{P}(\text{failure}) = \frac{1}{2} - \frac{1}{4}\|\rho - \sigma\|_{\mathrm{M}}$, this norm can be used to quantify the performance of POVMs for state discrimination.

EXERCISE 12.1 (The Helstrom bound). Show that, for any POVM M, we have $\|\cdot\|_{\mathrm{M}} \leqslant \|\cdot\|_1$. Conversely, show that for any pair of states $\rho, \sigma \in \mathrm{D}(\mathcal{H})$ there is a POVM M such that $\|\rho - \sigma\|_{\mathrm{M}} = \|\rho - \sigma\|_1$. This gives operational meaning to the trace norm distance between quantum states; the optimal inequality $\mathbf{P}(\text{failure}) \geqslant \frac{1}{2} - \frac{1}{4}\|\rho - \sigma\|_1$ is known as the *Helstrom bound* for quantum hypothesis testing.

12.1.2. Zonotope associated to a POVM. Given a POVM M, we denote by $B_{\mathrm{M}} = \{\|\cdot\|_{\mathrm{M}} \leqslant 1\}$ the unit ball for the distinguishability norm, and $K_{\mathrm{M}} = (B_{\mathrm{M}})^\circ$ its polar, i.e.,

$$K_{\mathrm{M}} = \{A \in B^{\mathrm{sa}}(\mathcal{H}) \ : \ \mathrm{Tr}(AB) \leqslant 1 \text{ whenever } \|B\|_{\mathrm{M}} \leqslant 1\}.$$

The set K_{M} is a compact convex set. Moreover K_{M} has nonempty interior if and only if $\|\cdot\|_{\mathrm{M}}$ is a norm. It follows from the inequality $\|\cdot\|_{\mathrm{M}} \leqslant \|\cdot\|_1$ that K_{M} is always included in the unit ball for the operator norm.

The following proposition characterizes the convex sets that can be obtained by means of this construction.

PROPOSITION 12.1. *Let $K \subset B^{\mathrm{sa}}(\mathcal{H})$ be a symmetric closed convex set. Then the following are equivalent.*
(i) K is a zonotope such that $K \subset \{\|\cdot\|_\infty \leqslant 1\}$ and $\pm \mathrm{I} \in K$.
(ii) There exists a POVM M on $\mathcal{H}$ such that $K = K_{\mathrm{M}}$.

Zonotopes were defined in Section 4.1.3 and briefly discussed in Section 7.2.6.4; the insight implicit in the above Proposition permits us to relate the ideas and the techniques outlined in those sections to the task of state discrimination.

PROOF OF PROPOSITION 12.1. For a POVM $\mathrm{M} = (M_i)_{1 \leqslant i \leqslant N}$, we claim that

$$K_{\mathrm{M}} = [-M_1, M_1] + \cdots + [-M_N, M_N]. \tag{12.2}$$

Indeed, denoting by L the right-hand side of (12.2), we have for every $A \in B^{\mathrm{sa}}(\mathcal{H})$

$$\|A\|_{L^\circ} = \sup\{\mathrm{Tr}(AB) \ : \ B \in L\} = \sum_{i=1}^{N} |\mathrm{Tr}(AM_i)| = \|A\|_{K_{\mathrm{M}}^\circ},$$

so that $L = K_{\mathrm{M}}$. Conversely, suppose that K is a zonotope as in (i). By definition, there are operators $(M_i)_{1 \leqslant i \leqslant N}$ such that

$$K = [-M_1, M_1] + \cdots + [-M_N, M_N].$$

The hypotheses imply that I is an extreme point of K. Any extreme point of K has the form $\pm M_1 \pm \cdots \pm M_N$, and therefore by changing M_i into $-M_i$ if necessary, we may assume that

$$\mathrm{I} = M_1 + \cdots + M_N.$$

For every $1 \leqslant i \leqslant N$, we have $\mathrm{I} - M_i \in K$ and thus $\|\mathrm{I} - M_i\|_\infty \leqslant 1$. Therefore M_i is positive, and $\mathrm{M} = (M_i)_{1 \leqslant i \leqslant N}$ is a POVM such that $K_{\mathrm{M}} = K$. □

12.1.3. Sparsification of POVMs. We are going to show that POVMs can be sparsified, i.e., approximated by POVMs with few outcomes. The terminology "approximation" refers here to the associated distinguishability norms: a POVM M is considered to be ε-close to a POVM M′ when their distinguishability norms satisfy inequalities of the form

$$(1 - \varepsilon)\|\cdot\|_{\mathrm{M}} \leqslant \|\cdot\|_{\mathrm{M}'} \leqslant (1 + \varepsilon)\|\cdot\|_{\mathrm{M}}. \tag{12.3}$$

As an immediate consequence of Theorem 7.48 about approximation of zonotopes by zonoids, we obtain a result about sparsification of POVMs: given any POVM M, we can produce a POVM M′ with relatively few outcomes which performs the task of state discrimination almost as well as M.

THEOREM 12.2. *There is a constant C such that the following holds: for every POVM $\mathrm{M} = (M_i)_{1 \leqslant i \leqslant N}$ on $\mathbb{C}^n$ and every $\varepsilon \in (0,1)$, there exists another POVM $\mathrm{M}' = (M'_j)_{1 \leqslant j \leqslant N'}$ with $N' \leqslant Cn^2 \log n/\varepsilon^2$ outcomes such that*

$$(1-\varepsilon)\|\cdot\|_{\mathrm{M}} \leqslant \|\cdot\|_{\mathrm{M}'}. \tag{12.4}$$

PROOF. Consider the convex set $K_{\mathrm{M}} \subset \mathsf{M}_n^{\mathrm{sa}}$, which is a zonoid by Proposition 12.1. By Theorem 7.48, there is a zonotope

$$Z = [-A_1, A_1] + \cdots + [-A_{N'}, A_{N'}]$$

with A_i being positive operators, $N' \leqslant Cn^2 \log n/\varepsilon^2$, such that $(1-\varepsilon)K_{\mathrm{M}} \subset Z \subset K_{\mathrm{M}}$. (The positivity of A_i follows from the last sentence in Theorem 7.48.) Define $A_0 = \mathrm{I} - (A_1 + \cdots + A_{N'})$. Note that A_0 is positive since $Z \subset K_{\mathrm{M}} \subset S_\infty^{n,\mathrm{sa}}$ (the unit ball for the operator norm). It follows that $\mathrm{M}' := (A_0, A_1, \ldots, A_{N'})$ is a POVM such that $K_{\mathrm{M}'} \supset Z \supset (1-\varepsilon)K_{\mathrm{M}}$, and therefore $\|\cdot\|_{\mathrm{M}'} \geqslant (1-\varepsilon)\|\cdot\|_{\mathrm{M}}$ as claimed. □

REMARK 12.3. The one-sided inequality (12.4) in Theorem 12.2 is the meaningful half of (12.3) since we want the sparsified POVM to be not weaker than the initial one. However, it is natural to wonder whether one can insist on a two-sided inequality as in (12.3). This seems to require an extra insight.

12.2. The distillability problem

In this section we discuss the distillability problem, one of the most important open problems connected to entanglement.

Consider a bipartite Hilbert space $\mathcal{H} = \mathcal{H}_A \otimes \mathcal{H}_B$ shared between two parties customarily called Alice and Bob. For any integer $n \geqslant 1$, the Hilbert space $\mathcal{H}^{\otimes n}$ is also considered as a bipartite Hilbert space by identifying it with $\mathcal{H}_A^{\otimes n} \otimes \mathcal{H}_B^{\otimes n}$. Whenever we mention separability, partial transpose, LOCC,... for states or channels on $\mathcal{H}^{\otimes n}$, it is always understood as relative to the $A:B$ bipartition.

12.2.1. State manipulation via LOCC channels.

Given bipartite states $\rho \in \mathrm{D}(\mathcal{H}_A \otimes \mathcal{H}_B)$ and $\sigma \in \mathrm{D}(\mathcal{H}'_A \otimes \mathcal{H}'_B)$, we write $\rho \rightsquigarrow \sigma$ if, for any $\varepsilon > 0$, there is an integer n and an LOCC quantum channel $\Phi : B((\mathcal{H}_A \otimes \mathcal{H}_B)^{\otimes n}) \to B(\mathcal{H}'_A \otimes \mathcal{H}'_B)$ such that

$$\left\|\Phi(\rho^{\otimes n}) - \sigma\right\|_1 \leqslant \varepsilon.$$

In words, this property is referred to as "σ can be *distilled* from (multiple copies of) ρ".

We are going to discuss this notion without giving a precise definition of LOCC quantum channels. We only need to know that the class of LOCC channels is stable under composition (which implies, together with the result from Exercise 2.31, that the relation $\rightsquigarrow$ is transitive), that (see Section 2.3.4.8)

$$\mathrm{conv}\{\text{product channels}\} \subset \{\text{LOCC channels}\} \subset \{\text{separable channels}\},$$

and that the *local filtering* operation is LOCC: given a state ρ on $\mathcal{H}_A \otimes \mathcal{H}_B$, POVMs $(P_i)_{i \in I}$ on $\mathcal{H}_A$ and $(Q_j)_{j \in J}$ on $\mathcal{H}_B$, and $S \subset I \times J$, then (provided $\operatorname{Tr} M > 0$) $\rho \rightsquigarrow \frac{M}{\operatorname{Tr} M}$, where

$$M = \sum_{i,j \in S} (P_i \otimes Q_j)\rho(P_i \otimes Q_j).$$

The idea behind the last scheme is informally as follows: given n copies of the state ρ, Alice and Bob can successively measure copies of ρ locally using the

POVMs (P_i) and (Q_j) until they obtain outcomes i and j such that $(i,j) \in S$, the post-measurement state being then $\frac{M}{\operatorname{Tr} M}$. (The protocol fails if none of the n copies gives an outcome in S, but the probability of failure tends to zero as n tends to infinity.) This is where classical communication ("CC" of LOCC) comes in: Alice and Bob need a mechanism for certifying that $i, j \in S$ and this generally can not be accomplished by "local" means unless S itself has a product structure.

The above hierarchy of channels parallels somewhat the hierarchy of boxes (see Section 11.3.2). For example, conv{product channels} can be thought of as "local operations with shared randomness."

EXERCISE 12.2 (Distillation preserves separability and PPT). If $\rho \rightsquigarrow \sigma$, show that σ is separable (resp., PPT) whenever ρ is separable (resp., PPT).

12.2.2. Distillable states. Recall the standard notation: the canonical basis of $\mathbb{C}^2$ is $(|0\rangle, |1\rangle)$ and we often drop the tensor product signs (for example, $|00\rangle$ should be understood as $|0\rangle \otimes |0\rangle$). Next, it is convenient to work with the family of Bell vectors $\{\varphi^+, \varphi^-, \psi^+, \psi^-\}$, which is the orthonormal basis of $\mathbb{C}^2 \otimes \mathbb{C}^2$ consisting of maximally entangled vectors

$$\varphi^{\pm} = \frac{1}{\sqrt{2}}(|00\rangle \pm |11\rangle) \quad \text{and} \quad \psi^{\pm} = \frac{1}{\sqrt{2}}(|01\rangle \pm |10\rangle).$$

The corresponding states are called the Bell states. A bipartite state $\rho \in \mathrm{D}(\mathcal{H})$ is said to be *distillable* if $\rho \rightsquigarrow |\psi^+\rangle\langle\psi^+|$. The motivation for this concept is that many quantum information protocols (e.g., quantum teleportation) use Bell states as a resource. Distillable states are exactly those which are useful for these protocols.

Note that the choice of the Bell vector ψ^+ in this definition is arbitrary: if x, y are any two maximally entangled vectors on $\mathbb{C}^d \otimes \mathbb{C}^d$, then there exist $U, V \in \mathsf{U}(d)$ such that $y = (U \otimes V)x$. Since the channel $\rho \mapsto (U \otimes V)\rho(U \otimes V)^\dagger$ is LOCC (as a product channel), we have $|x\rangle\langle x| \rightsquigarrow |y\rangle\langle y|$. We repeatedly use this fact and refer to it as "conjugating with local unitaries".

It is easy to check that PPT states are not distillable (see Exercise 12.2). The distillability problem asks whether the converse holds.

PROBLEM 12.4 (Distillability problem). *Is every non-PPT state distillable?*

The answer to Problem 12.4 is commonly believed to be negative.

12.2.3. The case of two qubits.

PROPOSITION 12.5. *Every entangled state on $\mathbb{C}^2 \otimes \mathbb{C}^2$ is distillable.*

Since in the $\mathbb{C}^2 \otimes \mathbb{C}^2$ setting "entangled" and "non-PPT" are equivalent by Theorem 2.15, Proposition 12.5 is indeed an instance of Problem 12.4. In the argument it will be convenient to use states that are diagonal in the basis of Bell vectors. For $a, b, c, d \geqslant 0$ such that $a + b + c + d = 1$, let us denote

$$\rho_{a,b,c,d} = a|\varphi^+\rangle\langle\varphi^+| + b|\varphi^-\rangle\langle\varphi^-| + c|\psi^+\rangle\langle\psi^+| + d|\psi^-\rangle\langle\psi^-|.$$

The heart of the protocol lies in the following two lemmas, whose proofs we postpone. To each state $\rho \in \mathrm{D}(\mathbb{C}^2 \otimes \mathbb{C}^2)$, we associate the quantity $s(\rho) = \max\{\langle\chi|\rho|\chi\rangle\}$, where the maximum is taken over all maximally entangled vectors $\chi \in \mathbb{C}^2 \otimes \mathbb{C}^2$. Given that $\langle\chi|\rho|\chi\rangle$ is the square of the fidelity between ρ and $|\chi\rangle\langle\chi|$ (cf. Exercise B.3), the functional $s(\cdot)$ measures proximity to the set of maximally entangled states. In particular, ρ is distillable if and only if there exists a sequence (σ_n) in $\mathrm{D}(\mathbb{C}^2 \otimes \mathbb{C}^2)$ such that $s(\sigma_n) \to 1$ and that, for every n, $\rho \rightsquigarrow \sigma_n$.

LEMMA 12.6. *We have* $\rho \rightsquigarrow \rho_{s(\rho),\frac{1-s(\rho)}{3},\frac{1-s(\rho)}{3},\frac{1-s(\rho)}{3}}$.

LEMMA 12.7. *Given* $a, b, c, d \geqslant 0$ *with* $a+b+c+d = 1$, *denote* $\alpha = (a^2+b^2)/N$, $\beta = 2ab/N$, $\gamma = (c^2+d^2)/N$ *and* $\delta = 2cd/N$, *where* $N = (a+b)^2 + (c+d)^2$. *Then*

$$\rho_{a,b,c,d} \rightsquigarrow \rho_{\alpha,\beta,\gamma,\delta}.$$

PROOF OF PROPOSITION 12.5. Let $\rho \in \mathrm{D}(\mathbb{C}^2 \otimes \mathbb{C}^2)$ be an entangled state. By Theorem 2.15, this means that ρ is not PPT. Consequently, there exists a unit vector $x \in \mathbb{C}^2 \otimes \mathbb{C}^2$ such that $\langle x|\rho^\Gamma|x\rangle < 0$. Conjugating with local unitaries, we may assume that the Schmidt decomposition of x is $\alpha|00\rangle + \beta|11\rangle$. Consider the operator $W = \alpha|0\rangle\langle 0| + \beta|1\rangle\langle 1|$, then $x = \sqrt{2}\,(\mathrm{I}\otimes W)|\varphi^+\rangle$. By local filtering,

$$\rho \rightsquigarrow \sigma := \frac{(\mathrm{I}\otimes W)\rho(\mathrm{I}\otimes W)}{\operatorname{Tr}(\mathrm{I}\otimes W)\rho(\mathrm{I}\otimes W)} \tag{12.5}$$

(note that $0 \leqslant W \leqslant \mathrm{I}$, so that W can be one of the operators in a POVM) and one checks that $\langle\varphi^+|\sigma^\Gamma|\varphi^+\rangle < 0$. Using the formula $\operatorname{Tr}(A^\Gamma B) = \operatorname{Tr}(AB^\Gamma)$, we obtain

$$0 > \operatorname{Tr}\left(\sigma(|\varphi^+\rangle\langle\varphi^+|)^\Gamma\right) = \operatorname{Tr}\left(\sigma\left(\frac{\mathrm{I}}{2} - |\psi^-\rangle\langle\psi^-|\right)\right) = \frac{1}{2} - \langle\psi^-|\sigma|\psi^-\rangle$$

and therefore $s(\sigma) > 1/2$.

The problem is thus reduced to showing that any state σ with $s(\sigma) > 1/2$ is distillable. By applying successively Lemmas 12.6 and 12.7, we obtain that $\sigma \rightsquigarrow \sigma'$ for some state σ' such that $s(\sigma') \geqslant \phi(s(\sigma))$, where ϕ is the function

$$\phi(t) = \frac{t^2 + \frac{1}{9}(1-t)^2}{\frac{1}{9}(1+2t)^2 + \frac{1}{9}(2-2t)^2} = \frac{1 - 2t + 10t^2}{5 - 4t + 8t^2}.$$

Since $\phi(t) > t$ for $t \in (1/2, 1)$, we have $\lim_{n\to\infty} \phi^n(s(\sigma)) = 1$. In other words, iterating the above procedure shows that $\sigma \rightsquigarrow \sigma''$, where σ'' is a state such that $s(\sigma'')$ is as close to 1 as we wish. It follows that σ is distillable. □

PROOF OF LEMMA 12.6. By conjugating with local unitaries, we may assume that $s(\rho) = \langle\psi^-|\rho|\psi^-\rangle$. The twirling channel $\Upsilon : B(\mathbb{C}^2 \otimes \mathbb{C}^2) \to B(\mathbb{C}^2 \otimes \mathbb{C}^2)$ is defined as $\Upsilon(\rho) = \mathbf{E}(U \otimes U)\rho(U \otimes U)^\dagger$ where $U \in \mathsf{U}(2)$ is Haar-distributed. This is an LOCC channel (it belongs to the convex hull of the set of product channels) and moreover (see Exercise 2.16),

$$\Upsilon(\rho) = s(\rho)|\psi^-\rangle\langle\psi^-| + \frac{1-s(\rho)}{3}\left(|\varphi^+\rangle\langle\varphi^+| + |\varphi^-\rangle\langle\varphi^-| + |\psi^+\rangle\langle\psi^+|\right).$$

The result follows since ψ^- can be transformed into φ^+ by local unitaries. □

PROOF OF LEMMA 12.7. We write ρ for $\rho_{a,b,c,d}$. It will be convenient to consider ρ as a state on $\mathcal{H}_A \otimes \mathcal{H}_B$ and $\rho \otimes \rho$ as a state on $\mathcal{H}_A \otimes \mathcal{H}_B \otimes \mathcal{H}'_A \otimes \mathcal{H}'_B$ (all the spaces $\mathcal{H}_A, \mathcal{H}_B, \mathcal{H}'_A, \mathcal{H}'_B$ being equal to $\mathbb{C}^2$). When an operator X on $\mathbb{C}^2 \otimes \mathbb{C}^2$ is thought of as acting on $\mathcal{H}_A \otimes \mathcal{H}'_A$ (resp., $\mathcal{H}_B \otimes \mathcal{H}'_B$), we denote it by X_A (resp., by X_B). The same convention will be used for superoperators Ψ whose domain is $B(\mathbb{C}^2 \otimes \mathbb{C}^2)$.

Denote by $P = |00\rangle\langle 00| + |11\rangle\langle 11|$ and $Q = \mathrm{I} - P = |01\rangle\langle 01| + |10\rangle\langle 10|$ the complementary rank 2 projectors acting on the space $\mathbb{C}^2 \otimes \mathbb{C}^2$. Next, consider $\Pi = P_A \otimes P_B + Q_A \otimes Q_B$ as an operator acting on $\mathcal{H}_A \otimes \mathcal{H}_B \otimes \mathcal{H}'_A \otimes \mathcal{H}'_B$. A simple computation shows that Π is the orthogonal projection onto the subspace generated by the 8 vectors

$$\begin{array}{cccc} \varphi^+ \otimes \varphi^+, & \varphi^+ \otimes \varphi^-, & \varphi^- \otimes \varphi^+, & \varphi^- \otimes \varphi^-, \\ \psi^+ \otimes \psi^+, & \psi^+ \otimes \psi^-, & \psi^- \otimes \psi^+, & \psi^- \otimes \psi^-. \end{array}$$

Consider also the quantum channel $\Psi : B(\mathbb{C}^2 \otimes \mathbb{C}^2) \to B(\mathbb{C}^2)$ given by $\Psi(\rho) = \operatorname{Tr}_2 U\rho U^\dagger$, where Tr_2 denote the partial trace over the second factor, and U is the "CNOT" unitary transformation on $\mathbb{C}^2 \otimes \mathbb{C}^2$ defined by

$$U(|00\rangle) = |00\rangle, \quad U(|01\rangle) = |01\rangle, \quad U(|10\rangle) = |11\rangle, \quad U(|11\rangle) = |10\rangle.$$

A direct calculation shows that, for $\varepsilon, \eta = \pm$ and with the usual rules for sign multiplication,

$$(\Psi_A \otimes \Psi_B)(|\varphi^\varepsilon \otimes \varphi^\eta\rangle\langle\varphi^\varepsilon \otimes \varphi^\eta|) = |\varphi^{\varepsilon\eta}\rangle\langle\varphi^{\varepsilon\eta}|),$$

$$(\Psi_A \otimes \Psi_B)(|\psi^\varepsilon \otimes \psi^\eta\rangle\langle\psi^\varepsilon \otimes \psi^\eta|) = |\psi^{\varepsilon\eta}\rangle\langle\psi^{\varepsilon\eta}|).$$

(We emphasize that, in the above formulas, not all occurrences of the symbol $\otimes$ refer to the same bipartitions; for example in $\varphi^\varepsilon \otimes \varphi^\eta$ we have $\varphi^\varepsilon \in \mathcal{H}_A \otimes \mathcal{H}_B$ and $\varphi^\eta \in \mathcal{H}'_A \otimes \mathcal{H}'_B$.) It follows (using first local filtering, then the LOCC channel $\Psi_A \otimes \Psi_B$ and a tedious but straightforward computation) that

$$\rho \rightsquigarrow \frac{\Pi(\rho \otimes \rho)\Pi}{\operatorname{Tr} \Pi(\rho \otimes \rho)\Pi} \rightsquigarrow \rho_{\alpha,\beta,\gamma,\delta},$$

as asserted. □

12.2.4. Some reformulations of distillability. We start with a criterion for distillability.

LEMMA 12.8. *A state $\rho \in \mathrm{D}(\mathcal{H}_A \otimes \mathcal{H}_B)$ is distillable if and only if there exists an integer n and operators $A : \mathbb{C}^2 \to \mathcal{H}_A^{\otimes n}$, $B : \mathbb{C}^2 \to \mathcal{H}_B^{\otimes n}$ such that the operator $(A \otimes B)^\dagger \rho^{\otimes n} (A \otimes B)$ is non-PPT.*

PROOF. Assume that there exist n, A and B with the above properties. Then, by local filtering, we have $\rho \rightsquigarrow \sigma$, where

$$\sigma = \frac{(A \otimes B)^\dagger \rho^{\otimes n} (A \otimes B)}{\operatorname{Tr}((A \otimes B)^\dagger \rho^{\otimes n} (A \otimes B))}$$

is a non-PPT state on $\mathbb{C}^2 \otimes \mathbb{C}^2$. By Proposition 12.5, σ (and hence also ρ) is distillable.

Conversely, if ρ is distillable, there exists, for some n, an LOCC channel $\Phi : B((\mathcal{H}_A)^{\otimes n} \otimes (\mathcal{H}_B)^{\otimes n}) \to B(\mathbb{C}^2 \otimes \mathbb{C}^2)$ such that $\Phi(\rho^{\otimes n})$ is non-PPT. Since Φ is separable, it has the form

$$\Phi(X) = \sum_i (A_i \otimes B_i)^\dagger X (A_i \otimes B_i)$$

and therefore at least some couple (A_i, B_i) satisfies the desired conclusion. □

There is also a connection between distillability and 2-positivity. Fix an orthonormal basis (e_i) of $\mathcal{H}_A$ and denote $\chi = \sum e_i \otimes e_i \in \mathcal{H}_A \otimes \mathcal{H}_A$. We recall (see Section 2.3.2) that the Choi matrix associated to a completely positive map $\Phi \in \boldsymbol{CP}(\mathcal{H}_B, \mathcal{H}_A)$ is defined as $C(\Phi) = (\Phi \otimes \mathrm{Id}_{B(\mathcal{H}_A)})(|\chi\rangle\langle\chi|)$.

PROPOSITION 12.9. *Given a state* $\rho \in \mathrm{D}(\mathcal{H}_A \otimes \mathcal{H}_B)$, *let* $\Phi \in \boldsymbol{CP}(\mathcal{H}_B, \mathcal{H}_A)$ *be such that* $\rho = C(\Phi)$. *Denote by* $T \in \boldsymbol{P}(\mathcal{H}_A)$ *the transposition map. Then the following are equivalent:*

(1) ρ *is distillable,*
(2) *there exists an integer* n *such that the map* $(T\Phi)^{\otimes n}$ *is not* 2*-positive.*

PROOF. We apply the result of Exercise 2.48 (for $k = 2$) to the superoperator $(T\Phi)^{\otimes n}$. We note that $C(T\Phi) = (T\Phi \otimes \mathrm{Id})(|\chi\rangle\langle\chi|) = (T \otimes \mathrm{Id})(\rho) = \rho^\Gamma$.

It follows that $(T\Phi)^{\otimes n}$ is 2-positive iff the operator $(A \otimes B)^\dagger(\rho^\Gamma)^{\otimes n}(A \otimes B)$ is positive for any $A : \mathbb{C}^2 \to \mathcal{H}_A^{\otimes n}$ and $B : \mathbb{C}^2 \to \mathcal{H}_B^{\otimes n}$. This condition is also equivalent to the operator $(\bar{A} \otimes B)^\dagger \rho^{\otimes n}(\bar{A} \otimes B)$ being PPT, and the result is now immediate from Lemma 12.8. □

Problem 12.4 reduces therefore to the following.

PROBLEM 12.10. *Let* Φ *be a completely positive map such that* $(T\Phi)^{\otimes n}$ *is 2-positive for every* n *(where* T *denotes the transposition). Is* $T\Phi$ *necessarily completely positive?*

A remarkable result is the fact that in order to solve Problem 12.4 it is enough to search among Werner states.

PROPOSITION 12.11. *Fix* $d \geqslant 3$. *The following are equivalent:*
(i) Every non-PPT state on $\mathbb{C}^d \otimes \mathbb{C}^d$ *is distillable.*
(ii) Every entangled Werner state on $\mathbb{C}^d \otimes \mathbb{C}^d$ *is distillable.*

PROOF. Since PPT Werner states are separable (see Proposition 2.16), (i) implies (ii). Conversely, let $\rho \in \mathrm{D}(\mathbb{C}^d \otimes \mathbb{C}^d)$ be a non-PPT state. In other words, there is a unit vector $x \in \mathbb{C}^d \otimes \mathbb{C}^d$ such that $\langle x|\rho^\Gamma|x\rangle < 0$. By applying the same argument as in the proof of Proposition 12.5, we deduce that there is a state $\sigma \in \mathrm{D}(\mathbb{C}^d \otimes \mathbb{C}^d)$ such that $\rho \rightsquigarrow \sigma$ and $\langle\psi|\sigma^\Gamma|\psi\rangle < 0$, where $\psi = \frac{1}{\sqrt{d}}\sum_{i=1}^d e_i \otimes e_i$ is a maximally entangled vector. Equivalently (cf. Exercise 2.20), $\mathrm{Tr}(F\sigma) < 0$, where F is the flip operator on $\mathbb{C}^d \otimes \mathbb{C}^d$. Consider now $\Upsilon : B(\mathbb{C}^d \otimes \mathbb{C}^d) \to B(\mathbb{C}^d \otimes \mathbb{C}^d)$, the twirling quantum channel, defined by $\Upsilon(\rho) = \mathbf{E}(U \otimes U)\rho(U^\dagger \otimes U^\dagger)$ where U is Haar-distributed on the unitary group. This channel is an LOCC channel and maps any state σ to a Werner state $w = \Upsilon(\sigma)$ satisfying $\mathrm{Tr}(F\sigma) = \mathrm{Tr}(Fw)$ (see Exercise 2.16). It follows (see Proposition 2.16) that $\rho \rightsquigarrow w$ for some entangled Werner state w, so (ii) implies (i). □

A consequence of Lemma 12.8 and Proposition 12.11 is that Problem 12.4 can be reduced to the following question, where ψ denotes a maximally entangled vector on $\mathbb{C}^k \otimes \mathbb{C}^k$: *for every* $k \geqslant 3$ *and* $\varepsilon > 0$, *do there exist an integer* n *and vectors* $a, b, c, d \in (\mathbb{C}^k)^{\otimes n}$ *such that*

$$\Big\langle a \otimes b + c \otimes d \Big| \big(\mathrm{I} - (1+\varepsilon)|\psi\rangle\langle\psi|\big)^{\otimes n} \Big| a \otimes b + c \otimes d \Big\rangle < 0\,? \tag{12.6}$$

Notes and Remarks

Section 12.1. The distinguishability norms associated to POVMs were introduced in [**MWW09**]. The observation behind Exercise 12.1 is due to Helstrom [**Hel69**] and Holevo [**Hol73**]. The connection between POVMs and zonoids (Proposition 12.1) was noticed in [**AL15b**], where Theorem 12.2 was proved (and where improvements for specific examples of POVMs are also discussed). Volume and mean width estimates for norms associated to a family of POVMs on a bipartite state can also be found in [**AL15a**].

Section 12.2. For a precise description of the class of LOCC transformations we refer to [**HHHH09**, Section XI] and [**Wat**, Chapter 6] (see also [**CLM**$^+$**14**]).

A basic reference on the distillability problem is [**HH01**] (see also the survey [**Cla06**], and the website [**@5**]). The relationships between distillability, the PPT property, and teleportation have also been studied in [**HHH98, LP99, HHH99**].

The protocol described in Lemmas 12.6 and 12.7 appeared in [**BBP**$^+$**96**] (see also [**BDSW96**]). Proposition 12.5 appears in [**HHH97**] and Proposition 12.11 is from [**HH99**].

Proposition 12.9, and the equivalence between Problem 12.4 and Problem 12.10, are from [**DSS**$^+$**00**]. For numerical attempts to solve Problem 12.4 in its formulation (12.6), see [**DSS**$^+$**00, DCLB00**].

There is a quantitative version of the distillability problem, which asks for the asymptotic rate of Bell states production via LOCC channels from many copies of a given state; the supremum of achievable rates is called the distillable entanglement. Entanglement that is not distillable is often referred to as *bound entanglement*, and the states that exhibit it are called *bound entangled*.

If one uses operations preserving PPT instead of LOCC, then every non-PPT state can be "distilled" [**EVWW01**]. Note that some care is needed when analyzing this issue because the class in question is not closed under tensoring.

APPENDIX A

Gaussian measures and Gaussian variables

This appendix serves as a brief general reference for Gaussian random variables, both scalar and vector-valued. It addresses terminology, basic properties, and various elementary but useful identities and inequalities. More specialized properties are included elsewhere in this book, most notably in Chapter 6.

A.1. Gaussian random variables

The standard Gaussian distribution $N(0,1)$ is the probability measure on $\mathbb{R}$ (denoted by γ_1) with density $\frac{1}{\sqrt{2\pi}}\exp(-x^2/2)\,\mathrm{d}x$. The standard complex Gaussian distribution $N_{\mathbb{C}}(0,1)$ is the probability measure on $\mathbb{C}$ with density $\frac{1}{\pi}\exp(-|z|^2)\,\mathrm{d}z$. (Occasionally we will write $N_{\mathbb{R}}(0,1)$ for $N(0,1)$ to emphasize the distinction.) The word "standard" refers, in particular, to the unit variance normalization: if Z has distribution either $N(0,1)$ or $N_{\mathbb{C}}(0,1)$, then $\mathbf{E}\,|Z|^2 = 1$. We note also that if Z_1, Z_2 are independent random variables with distribution $N(0,1)$, then $\frac{1}{\sqrt{2}}(Z_1 + iZ_2)$ has distribution $N_{\mathbb{C}}(0,1)$.

If X has $N(0,1)$ distribution (resp., $N_{\mathbb{C}}(0,1)$ distribution) and $\sigma \geqslant 0$, the distribution of σX is denoted by $N(0,\sigma^2)$ (resp., by $N_{\mathbb{C}}(0,\sigma^2)$).

The moments of the Gaussian standard distributions can be computed explicitly: if Z has $N(0,1)$ distribution, then, for any $p \geqslant 0$,

$$\text{(A.1)} \qquad \mathbf{E}\,|Z|^p = \frac{2^{p/2}}{\sqrt{\pi}}\Gamma\left(\frac{p+1}{2}\right) \overset{p\to\infty}{\sim} \sqrt{2}\left(\frac{p}{e}\right)^{p/2}.$$

Similarly, if Z has $N_{\mathbb{C}}(0,1)$ distribution, then for any $p \geqslant 0$,

$$\text{(A.2)} \qquad \mathbf{E}\,|Z|^p = \Gamma\left(\frac{p}{2}+1\right)$$

(indeed, $|Z|^2$ follows an exponential distribution with parameter 1).

We also need some fine estimates on the cumulative distribution function of a standard Gaussian variable, denoted by

$$\text{(A.3)} \qquad \Phi(x) := \gamma_1\big((-\infty, x]\big).$$

For large x, we have $1 - \Phi(x) \sim (2\pi)^{-1/2}x^{-1}\exp(-x^2/2)$. This is refined by the Komatu inequalities which assert that for every $x \geqslant 0$,

$$\text{(A.4)} \qquad \frac{2}{x+\sqrt{x^2+4}} \leqslant e^{x^2/2}\int_x^\infty e^{-t^2/2}\,\mathrm{d}t \leqslant \frac{2}{x+\sqrt{x^2+2}}.$$

A further refinement is provided by the inequalities (where $x \geqslant 0$)

$$\text{(A.5)} \qquad \frac{\pi}{(\pi-1)x+\sqrt{x^2+2\pi}} \leqslant e^{x^2/2}\int_x^\infty e^{-t^2/2}\,\mathrm{d}t \leqslant \frac{4}{3x+\sqrt{x^2+8}}.$$

EXERCISE A.1 (A simple bound for the normal tail). Show the inequality (6.6): if Z is a standard normal variable (i.e., distributed according to the $N(0,1)$ law), then $\mathbf{P}(Z \geqslant t) = \frac{1}{2}\mathbf{P}(|Z| \geqslant t) \leqslant \frac{1}{2}e^{-t^2/2}$ for $t \geqslant 0$. This bound motivates the definition of *subgaussian processes*, see (6.19) and subsequent comments.

EXERCISE A.2 (Komatu inequalities). Prove the Komatu inequalities (A.4) by arguing as follows:
(i) If $f_-(x)$, $f(x)$ and $f_+(x)$ denote respectively the left, middle and right member of the inequality to be proved, show that for $x \geqslant 0$ we have $f_-' \geqslant xf_- - 1$, $f' = xf - 1$ and $f_+' \leqslant xf_+ - 1$.
(ii) Show (A.4). The same argument proves the upper bound in (A.5).

A.2. Gaussian vectors

A family of real-valued centered random variables (X_i) is *jointly Gaussian* if any linear combination of the variables has distribution $N(0,\sigma^2)$ for some σ. A jointly Gaussian family is also called a Gaussian process (see Section 6.1). A crucial property of jointly Gaussian families, or Gaussian processes, is that the joint distribution of (X_i) is uniquely determined by the covariance matrix $(a_{ij}) = (\mathbf{E}\, X_i X_j)$.

When V is a real (resp., complex) finite-dimensional space equipped with a Euclidean (resp., Hilbertian) norm, we call the *standard Gaussian vector* in V a V-valued random variable such that, in any orthonormal basis, the coordinates of V are independent standard real (resp., complex) random variables. More concretely, the distribution of a standard Gaussian vector in $\mathbb{R}^n$ (denoted by γ_n) has density

$$\frac{1}{(2\pi)^{n/2}} \exp(-|x|^2/2)\, \mathrm{d}x,$$

whereas the distribution of a standard Gaussian vector in $\mathbb{C}^n$ (denoted by $\gamma_n^{\mathbb{C}}$) has density

$$\frac{1}{\pi^n} \exp(-|x|^2)\, \mathrm{d}x. \tag{A.6}$$

In all these cases the respective distribution will be referred to as the *standard Gaussian measure* on the corresponding space V. Note that if $\mathbb{C}^n$ is identified with $\mathbb{R}^{2n}$, the distributions $\gamma_n^{\mathbb{C}}$ and γ_{2n} do not coincide: they differ by a scaling factor of $\sqrt{2}$.

While we are mostly interested in *standard* Gaussian vectors and measures, the joint distribution of any jointly Gaussian sequence $X_1, X_2, \ldots, X_n$ is referred to as a Gaussian measure on $\mathbb{R}^n$. Sequences or measures that are not centered are also considered. However, this does not add a lot to generality: any such measure is a pushforward of the standard Gaussian measure via a linear (or affine, as appropriate) map.

Let G be a standard Gaussian vector in $\mathbb{R}^n$. Rotational invariance of γ_n implies that the random variable $\frac{G}{|G|}$ is uniformly distributed on sphere S^{n-1}; moreover $|G|$ and $\frac{G}{|G|}$ are independent. This can be used to relate Gaussian averages and spherical averages. For any function $f : \mathbb{R}^n \to \mathbb{R}_+$ satisfying $f(tx) = tf(x)$ whenever $x \in \mathbb{R}^n$ and $t \geqslant 0$, we have

$$\int_{\mathbb{R}^n} f\, \mathrm{d}\gamma_n = \mathbf{E}\, f(G) = \kappa_n \int_{S^{n-1}} f\, \mathrm{d}\sigma, \tag{A.7}$$

where σ is the uniform measure on S^{n-1} and κ_n is the constant

$$\kappa_n := \mathbf{E}\,|G| = \frac{\sqrt{2}\Gamma((n+1)/2)}{\Gamma(n/2)}. \tag{A.8}$$

In particular, (A.7) can be applied when f is the gauge associated with a convex body. (See also Exercise A.6 .)

The constant κ_n appears in probability and statistics as the mean of $\chi(n)$, the chi distribution with n degrees of freedom. (See Exercise 5.34 for bounds for the median, which is necessarily smaller than κ_n by Proposition 5.34.) The first values are $\kappa_1 = \sqrt{2/\pi}$, $\kappa_2 = \sqrt{\pi/2}$, $\kappa_3 = 2\sqrt{2/\pi}$. Note also the formula $\kappa_n\kappa_{n+1} = n$. For large n, we have $\kappa_n \sim \sqrt{n}$. More precise estimates are gathered in the following proposition.

PROPOSITION A.1 (See Exercises A.4 and A.5; (iv) and (v) are not proved here). *Let κ_n be the constant defined in* (A.8). *Then*
(i) $\sqrt{n-1} \leqslant \kappa_n \leqslant \sqrt{n}$,
(ii) the sequence $\kappa_n/\sqrt{n}$ is increasing,
(iii) $\sqrt{n-\frac{1}{2}} \leqslant \kappa_n \leqslant \sqrt{n - \frac{n}{2n+1}}$,
(iv) the sequence $\sqrt{n} - \kappa_n$ is non-increasing,
(v) as n tends to infinity, we have $\kappa_n = \sqrt{n}(1 - 1/4n + 1/32n^2 + O(1/n^3))$.

The complex analogue of (A.7) is as follows: if $f : \mathbb{C}^n \to \mathbb{R}_+$ satisfies $f(tx) = tf(x)$ whenever $x \in \mathbb{C}^n$ and $t \geqslant 0$, we have

$$\int_{\mathbb{C}^n} f \,\mathrm{d}\gamma_n^{\mathbb{C}} = \kappa_n^{\mathbb{C}} \int_{S_{\mathbb{C}^n}} f \,\mathrm{d}\sigma \tag{A.9}$$

with $\kappa_n^{\mathbb{C}} = \kappa_{2n}/\sqrt{2}$.

EXERCISE A.3. Let $n \geqslant 2$. Prove the following result sometimes known as the Herschel–Maxwell theorem: *up to scaling, γ_n is the only rotationally-invariant probability measure on $\mathbb{R}^n$ which is also a product measure.*

EXERCISE A.4. Using the fact that the function $\log\Gamma$ is convex, show parts (i) and (ii) of Proposition A.1.

EXERCISE A.5. Prove part (iii) of Proposition A.1 by showing that the corresponding ratios are monotone along even and odd subsequences.

EXERCISE A.6. State and prove a variant of (A.7) for α-homogeneous functions, i.e., verifying $f(tx) = t^\alpha f(x)$ for $x \in \mathbb{R}^n$ and $t > 0$.

Notes and Remarks

The Komatu inequalities (A.4) appeared in [**Kom55**]. The upper bound in (A.5) was proved in [**Sam53, SW99**] and the lower bound in [**RW00**]. A survey paper on related inequalities is [**Due10**]. Part (iii) from Proposition A.1 is from [**Chu62**]. Part (iv) follows for example from the refined inequality $\kappa_n \geqslant \sqrt{n - \frac{1}{2} + \frac{1}{8(n+1)}}$ from [**Boy67**]. Another derivation appears as Lemma C.4 in [**FR13**].

For many characterizations of the Gaussian measure in the spirit of Exercise A.3, see [**Bry95**].

APPENDIX B

Classical groups and manifolds

This appendix contains an overview of the classical groups and manifolds that appear in this book, and of the natural structures, such as metrics and measures, which they carry. Most of the facts included here have been known for one hundred years or more, but the precise statements are often difficult to find in the literature, mostly because presentations of these topics usually focus on more general and more abstract settings. Again, more specialized features of these objects are studied elsewhere in this book, primarily in Chapter 5.

B.1. The unit sphere S^{n-1} or $S_{\mathbb{C}^d}$

We denote by $S^{n-1} = \{x \in \mathbb{R}^n \ : \ |x| = 1\}$ the unit sphere in $\mathbb{R}^n$. There are two natural distances on the sphere: the (intrinsic) geodesic distance ("as the crow flies") denoted by g and the extrinsic distance ("as the mole burrows"), i.e., the restriction to S^{n-1} of the Euclidean distance $|\cdot|$ on $\mathbb{R}^n$. Since they are related by the formula $|x-y| = 2\sin(g(x,y)/2)$, statements about $|\cdot|$ have immediate translations involving g and vice versa. Note also that, for any $x, y \in S^{n-1}$, we have

$$\frac{2}{\pi} g(x,y) \leqslant |x-y| \leqslant g(x,y). \tag{B.1}$$

We denote by σ the uniform measure on S^{n-1}, normalized so that $\sigma(S^{n-1}) = 1$. We note for the record that the *non-normalized* $(n-1)$-dimensional "surface area" of S^{n-1} equals

$$\mathrm{vol}_{n-1}\left(S^{n-1}\right) = \frac{2\pi^{n/2}}{\Gamma\left(\frac{n}{2}\right)}. \tag{B.2}$$

However, σ can be also induced by the Lebesgue measure vol_n on $\mathbb{R}^n$ as follows: for any Borel set $A \subset S^{n-1}$,

$$\sigma(A) = \frac{\mathrm{vol}_n \{tx \ : \ t \in [0,1], x \in A\}}{\mathrm{vol}_n B_2^n}.$$

We note for the record the formula for the volume of the unit ball B_2^n,

$$\mathrm{vol}\left(B_2^n\right) = \frac{\pi^{n/2}}{\Gamma\left(\frac{n}{2}+1\right)}. \tag{B.3}$$

If G is a standard Gaussian vector on $\mathbb{R}^n$, then $G/|G|$ is distributed according to σ. This is an efficient procedure to simulate the uniform measure on the sphere.

We denote by $S_{\mathbb{C}^d}$ the unit sphere in $\mathbb{C}^d$. Since $\mathbb{C}^d$ identifies with $\mathbb{R}^{2d}$ as a *real* vector space, and $S_{\mathbb{C}^d}$ with S^{2d-1} as a metric measure space, the preceding discussion is also valid for $S_{\mathbb{C}^d}$. Note also the formula, for $x, y \in S_{\mathbb{C}^d}$,

$$g(x,y) = \arccos \mathrm{Re}\,\langle x, y\rangle. \tag{B.4}$$

EXERCISE B.1. Show that $\mathrm{vol}(B_2^n)^{1/n} \sim \sqrt{2\pi e/n}$ as n tends to infinity, and that $\mathrm{vol}(B_2^n)^{1/n} \leqslant \sqrt{2\pi e/n}$ for every $n \geqslant 1$.

B.2. The projective space

We denote by $\mathsf{P}(\mathbb{C}^d)$ the complex *projective space* on $\mathbb{C}^d$ (more commonly denoted by $\mathbb{CP}^{d-1}$), i.e., the quotient of $S_{\mathbb{C}^d}$ under the identification of unit vectors φ, ψ which differ only by their phase; in other words, if $\varphi = e^{i\theta}\psi$ for some $\theta \in \mathbb{R}$. When $\psi \in S_{\mathbb{C}^d}$, we will occasionally denote by $[\psi]$ its class in $\mathsf{P}(\mathbb{C}^d)$. We equip $\mathsf{P}(\mathbb{C}^d)$ with the following metric (called *Fubini–Study metric*, or *Bures metric*),

$$d([\psi], [\chi]) = \arccos |\langle \psi, \chi \rangle|. \tag{B.5}$$

The quantity $|\langle \psi, \chi \rangle|$ is called the *overlap* of the vectors ψ and χ or, more properly, of $[\psi]$ and $[\chi]$.

We also introduce the *Segré variety* on the bipartite Hilbert space $\mathbb{C}^{d_1} \otimes \mathbb{C}^{d_2}$, defined as

$$\mathrm{Seg} = \{\varphi \otimes \psi \ : \ \varphi \in S_{\mathbb{C}^{d_1}}, \psi \in S_{\mathbb{C}^{d_2}}\}. \tag{B.6}$$

As defined in (B.6), Seg is a subset of the unit sphere $S_{\mathbb{C}^{d_1} \otimes \mathbb{C}^{d_2}}$ with real dimension $2(d_1 + d_2) - 3$. Alternatively, one could define the Segré variety as a subset of the projective space $\mathsf{P}(\mathbb{C}^{d_1} \otimes \mathbb{C}^{d_2})$. In that case it has complex dimension $d_1 + d_2 - 2$.

The real projective space $\mathsf{P}(\mathbb{R}^m)$ is defined and endowed with metric *mutatis mutandis* starting from the sphere S^{m-1}. However, the real setting generally appears in quantum theory only as a toy model. Note that the more standard (and more general) definition of the projective space $\mathsf{P}(V)$ associated to a vector space V over an arbitrary field $\mathbb{K}$ is by identification of vectors $u, v \in V \backslash \{0\}$ such that $u = kv$ for some $k \in \mathbb{K}\backslash\{0\}$. However, the equivalent approach starting from the sphere S^{m-1} or $S_{\mathbb{C}^d}$ better fits the standard setup of quantum theory.

EXERCISE B.2. Check that the Fubini–Study metric is obtained as the quotient metric from the geodesic metric on the unit sphere.

EXERCISE B.3 (Bures vs. Fubini–Study, fidelity vs. overlap). The Bures metric is usually defined for not-necessarily-pure states $\sigma, \tau \in \mathrm{D}$ by

$$d(\sigma, \tau) = \arccos F(\sigma, \tau), \tag{B.7}$$

where $F(\sigma, \tau) = \operatorname{Tr} \sqrt{\sqrt{\sigma}\,\tau\sqrt{\sigma}}$ is the *fidelity* between σ and τ. (Note that some texts define fidelity as the square of this quantity.) (i) Verify that if $\tau = |\chi\rangle\langle\chi|$, then $F(\sigma, \tau) = \sqrt{\langle\chi|\sigma|\chi\rangle}$. (ii) Deduce that if $\sigma = |\psi\rangle\langle\psi|$, $\tau = |\chi\rangle\langle\chi|$, then (B.5) and (B.7) yield the same value (in other words, the Fubini–Study metric is the restriction of the Bures metric to pure states and similarly for the fidelity vs. the overlap). (iii) Verify that $d(\sigma, \tau)$ is indeed a metric.

B.3. The orthogonal and unitary groups $\mathsf{O}(n)$, $\mathsf{U}(n)$

We denote by $\mathsf{O}(n) = \{O \in \mathsf{M}_n(\mathbb{R}) \ : \ OO^\dagger = \mathrm{I}\}$ the orthogonal group and by $\mathsf{U}(n) = \{U \in \mathsf{M}_n(\mathbb{C}) \ : \ UU^\dagger = \mathrm{I}\}$ the unitary group. Their dimensions, as real Riemannian manifolds, are $\dim \mathsf{O}(n) = n(n-1)/2$ and $\dim \mathsf{U}(n) = n^2$. We also recall the standard notation $\mathsf{SO}(n) = \{O \in \mathsf{O}(n) \ : \ \det(O) = 1\}$, $\mathsf{SU}(n) = \{U \in \mathsf{U}(n) \ : \ \det(U) = 1\}$, and $\mathsf{PSU}(n)$ for the quotient of $\mathsf{U}(n)$ under the relation $U \sim V \iff U = \lambda V$ for some $\lambda \in \mathbb{C}$. Note that $\mathsf{O}(n)$ is a disjoint union of two copies of $\mathsf{SO}(n)$, so all statements about $\mathsf{SO}(n)$ transfer *mutatis mutandis* to $\mathsf{O}(n)$. We also point out the classical isomorphism $\mathsf{PSU}(2) \leftrightarrow \mathsf{SO}(3)$ (see Exercise B.4).

It what follows G will stand for either $\mathsf{SO}(n)$, $\mathsf{SU}(n)$ or $\mathsf{U}(n)$. There are many metric structures one may consider on G. Each norm $\|\cdot\|$ on M_n induces two distances on G: the extrinsic distance (simply $\|U-V\|$, for $U, V \in \mathsf{G}$) and the geodesic distance (the length of a shortest path in G joining U to V, where length is measured with respect to $\|\cdot\|$). For $p \in [1, \infty]$, we will denote by g_p the geodesic distance induced by the Schatten p-norm.

Among these choices we single out the standard Riemannian metric g_2, which can be expressed for $U, V \in \mathsf{G}$ as

$$g_2(U,V) = \left(\sum_{i=1}^n \theta_i^2\right)^{1/2}, \tag{B.8}$$

where $e^{i\theta_1}, \ldots, e^{i\theta_d}$ are the eigenvalues of $U^{-1}V$, and $\theta_j \in [-\pi, \pi]$. (See Exercise B.5.)

PROPOSITION B.1 (Not proved here). *Let $1 \leqslant p \leqslant \infty$. Let $U, V \in \mathsf{G}$, and $A \in \mathsf{M}_n^{\mathrm{sa}}$ with $\|A\|_\infty \leqslant \pi$ such that $\exp(iA) = U^{-1}V$. Then the map $t \mapsto U\exp(itA)$, defined for $t \in [0,1]$, is a geodesic joining U to V for the distance g_p. If $\|U-V\|_\infty < 2$ and $1 < p < \infty$, this is the unique path of minimal length.*

The above result is very well-known for $p = 2$, but it is also valid, with the stated caveats, for other values of p. As a consequence of Proposition B.1, extrinsic and geodesic distances are easy to calculate and they are comparable, see Exercise B.5. Note that if $\|U-V\|_\infty = 2$, then A (which necessarily verifies $\|A\|_\infty = \pi$) is no longer uniquely determined and neither are the geodesics. See also Exercise B.6.

We point out that while Proposition B.1 appears to be stated in the complex setting (i.e., $\mathsf{G} = \mathsf{SU}(n)$ or $\mathsf{G} = \mathsf{U}(n)$), it makes sense just as well when $\mathsf{G} = \mathsf{SO}(n)$: the matrix $B = iA$ is then *real* skew-symmetric (see Exercise B.7). Moreover, it follows then that $\mathsf{SO}(n)$ is a geodesically convex submanifold of $\mathsf{U}(n)$, i.e., that the shortest curve in $\mathsf{U}(n)$ connecting any two points in $\mathsf{SO}(n)$ is entirely contained in $\mathsf{SO}(n)$ (or at least that there exists such a curve, if the shortest curve is not unique).

As compact groups, $\mathsf{O}(n)$ and $\mathsf{U}(n)$ carry a *Haar measure*: the unique probability measure which is invariant under right and/or left multiplication. The Haar measure can also be generated in a more concrete fashion. For example, start from a vector x_1 uniformly distributed on S^{n-1} (resp., $S_{\mathbf{C}^n}$), and construct inductively a random orthonormal basis $(x_1, \ldots, x_n)$ by choosing x_k uniformly on the unit sphere in the subspace $\{x_1, \ldots, x_{k-1}\}^\perp$. Then the random matrix with columns $(x_1, \ldots, x_n)$ is Haar-distributed on $\mathsf{O}(n)$ (resp., $\mathsf{U}(n)$). A slightly different scheme is outlined in Exercise B.14.

EXERCISE B.4 ($\mathsf{PSU}(2)$ and $\mathsf{SO}(3)$ are isomorphic). The group $\mathsf{U}(2)$ acts on the (real, 3-dimensional) hyperplane of trace zero matrices by the formula $X \mapsto UXU^\dagger$. This action preserves the Hilbert–Schmidt inner product. Check that this action induces an isomorphism between $\mathsf{PSU}(2)$ and $\mathsf{SO}(3)$.

EXERCISE B.5 (Equivalence of metrics on G). Let $1 \leqslant p \leqslant \infty$, G be either $\mathsf{SO}(n)$, $\mathsf{SU}(n)$ or $\mathsf{U}(n)$, and $U, V \in \mathsf{G}$.
(i) Denote by $e^{i\theta_1}, \ldots, e^{i\theta_n}$ the (complex) eigenvalues of $U^{-1}V$ with $|\theta_j| \leqslant \pi$. Show that

$$g_p(U,V) = \|(\theta_1, \ldots, \theta_n)\|_p,$$

where the norm on the right-hand side is the p-norm on $\mathbb{R}^n$.
(ii) Check that the geodesic and extrinsic metrics satisfy the inequalities

$$\frac{2}{\pi} g_p(U,V) \leqslant \|U-V\|_p \leqslant g_p(U,V)$$

for any $U, V \in \mathsf{G}$.

EXERCISE B.6 (All the geodesics in G). Show that it follows formally from Proposition B.1 that *all* paths of the form $t \mapsto We^{itA}$, $A \in \mathsf{M}_n^{\mathrm{sa}}$, $W \in \mathsf{G}$, and $t \in \mathbb{R}$ are geodesics in the sense that all their "sufficiently short" arcs are the shortest curves connecting their endpoints (unique if $1 < p < \infty$). Moreover, for $1 < p < \infty$, all such shortest curves are unique and hence all geodesics are of that form.

EXERCISE B.7 (Geodesical convexity of $\mathsf{SO}(n)$). Show that for any $U \in \mathsf{SO}(n)$ there is a real skew-symmetric matrix B with $\|B\|_\infty \leqslant \pi$ such that $U = e^B$ and $g_p(\mathrm{I}, U) = \|B\|_p$. Conclude that $\mathsf{SO}(n)$ is a geodesically convex submanifold of $\mathsf{U}(n)$ with respect to any metric g_p.

EXERCISE B.8 (Bi-Lipschitz estimates for the exponential map).
(i) Show that $\|\exp(iB) - \exp(iA)\|_{\mathrm{op}} \leqslant \|B - A\|_{\mathrm{op}}$ for every $A, B \in \mathsf{M}_n^{\mathrm{sa}}$.
(ii) Consider, for $\theta \in (0, \pi)$,

$$L(\theta) = \inf\left\{\frac{\|\exp(iB)-\exp(iA)\|_{\mathrm{op}}}{\|B-A\|_{\mathrm{op}}} \;:\; A, B \in \mathsf{M}_n^{\mathrm{sa}}, \|A\|_{\mathrm{op}} \leqslant \theta, \|B\|_{\mathrm{op}} \leqslant \theta, A \neq B\right\}.$$

Show that for $\theta \in (0, 2\pi/3)$ we have $L(\theta) \geqslant L(\theta/2)(1 - |1 - e^{i\theta/2}|)$. Conclude that (for example) $L(\pi/4) \geqslant 0.4$.

PROBLEM B.2. *What is the precise value of* $L(\theta)$ *in Exercise* B.8*? We did not find an answer in the literature (but we did not look very hard). Check that an easy upper bound is* $\sin(\theta)/\theta$.

B.4. The Grassmann manifolds $\mathsf{Gr}(k, \mathbb{R}^n)$, $\mathsf{Gr}(k, \mathbb{C}^n)$

Let V be a finite-dimensional real or complex vector space. For $0 < k < \dim V$, we denote by $\mathsf{Gr}(k, V)$ the family of all k-dimensional subspaces of V. The set $\mathsf{Gr}(k, V)$ is called the *Grassmann manifold* or the *Grassmannian.* Since its properties effectively depend only on the dimension of V, in what follows we consider only the concrete situations $\mathsf{Gr}(k, \mathbb{R}^n)$ and $\mathsf{Gr}(k, \mathbb{C}^n)$. (See, however, Exercise B.15.) Further, since the map $E \leftrightarrow E^\perp$ is a bijection between $\mathsf{Gr}(k, \mathbb{R}^n)$ and $\mathsf{Gr}(n-k, \mathbb{R}^n)$ preserving all the structures we will be interested in (and similarly for $\mathbb{C}^n$), the reader may always concentrate on the cases when $k \leqslant n/2$, which we will often tacitly assume.

Before discussing metrics on the Grassmann manifold we introduce the concept of principal angles. Given $E, F \in \mathsf{Gr}(k, \mathbb{R}^n)$ or $\mathsf{Gr}(k, \mathbb{C}^n)$, consider the singular value decomposition of the operator $P_E P_F$ (recall that P_E denotes the orthonormal projection onto E), which we will write in the form given by (2.10),

$$P_E P_F = \sum_{i=1}^{k} s_i |x_i\rangle\langle y_i| \tag{B.9}$$

with $s_i \in [0,1]$, $x_1, \ldots, x_k \in E$, and $y_1, \ldots, y_k \in F$ (the latter inclusions are automatic for x_i, y_i corresponding to coefficients $s_i \neq 0$ and can be arranged otherwise). The *principal angles* between E and F are the numbers $\theta_1, \ldots, \theta_k \in [0, \pi/2]$ defined by $\cos\theta_i = s_i$. The unit vectors $x_1,\dots,x_k$ and $y_1,\dots,y_k$ are called *principal vectors*. It is easily checked that we have $\langle x_i, x_j\rangle = \langle x_i, y_j\rangle = \langle y_i, y_j\rangle = 0$ for $i \neq j$ and that $s_i = \langle x_i, y_i\rangle$; the equality means that θ_i is actually the angle between x_i and y_i and, at the same time, the angle between $\mathbb{R}x_i$ and $\mathbb{R}y_i$ (or the Fubini–Study distance between $[x_i]$ and $[y_i]$—given by (B.5)—in the complex setting).

The principal angles quantify how close two subspaces are to each other. As we shall see, a natural Riemannian metric on Grassmann manifolds is as follows: if $E, F \in \mathsf{Gr}(k,\mathbb{R}^n)$ or $\mathsf{Gr}(k,\mathbb{C}^n)$, then

$$d(E,F) = \sqrt{2}\left(\sum_{i=1}^{k}\theta_i^2\right)^{1/2}, \tag{B.10}$$

where $\theta_1,\dots,\theta_k$ are the principal angles between E and F. The reader may wonder why we included the factor $\sqrt{2}$, which may appear redundant, both geometrically and esthetically. Indeed, as noted above, the natural metric on the projective space (Fubini–Study in the complex case), which corresponds to the case $k = 1$ of the Grassmannian, *does not* have that factor. However, as we shall see, there are sound functorial reasons for using the normalization (B.10): it shows up in two canonical constructions of the Grassmann manifold.

Another very natural way to define the distance in terms of principal angles is

$$d_\infty(E,F) = \max_{1\leqslant i\leqslant k}\theta_i. \tag{B.11}$$

However, the metric d_∞ is not Riemannian; an important (and obvious) inequality relating the two metrics is

$$d_\infty(E,F) \leqslant 2^{-1/2} d(E,F). \tag{B.12}$$

Fix $0 < k < n$ and consider the canonical action of $\mathsf{O}(n)$ on $\mathsf{Gr}(k,\mathbb{R}^n)$. Let $\mathbb{R}^k \subset \mathbb{R}^n$ be the canonical inclusion, so that $\mathbb{R}^k \in \mathsf{Gr}(k,\mathbb{R}^n)$. We now note that the stabilizer subgroup of $\mathsf{O}(n)$ that fixes $\mathbb{R}^k$ consists of block-diagonal matrices of the form

$$\begin{bmatrix} O_1 & 0 \\ 0 & O_2 \end{bmatrix},$$

where $O_1 \in \mathsf{O}(k)$ and $O_2 \in \mathsf{O}(n-k)$, and so it can be naturally identified with $\mathsf{O}(k) \times \mathsf{O}(n-k)$. Since the action of $\mathsf{O}(n)$ on $\mathsf{Gr}(k,\mathbb{R}^n)$ is transitive, it follows that $\mathsf{Gr}(k,\mathbb{R}^n)$ is a homogeneous space for $\mathsf{O}(n)$ and can be identified with the quotient space $\mathsf{O}(n)/(\mathsf{O}(k)\times\mathsf{O}(n-k))$. It follows in particular that the dimension of $\mathsf{Gr}(k,\mathbb{R}^n)$ equals $\dim\mathsf{O}(n) - \big(\dim\mathsf{O}(k) + \dim\mathsf{O}(n-k)\big) = k(n-k)$.

For a more concrete description of this correspondence, consider the map $\mathsf{O}(n) \to \mathsf{Gr}(k,\mathbb{R}^n)$ that associates to an orthogonal matrix O the subspace spanned by its first k columns, i.e., $O\mathbb{R}^k$. The preimage of $E \in \mathsf{Gr}(k,\mathbb{R}^n)$ under this map, i.e., the set $\{O \in \mathsf{O}(n) \quad O(\mathbb{R}^k) = E\}$ is a (left) coset of $\mathsf{O}(k)\times\mathsf{O}(n-k)$. Similarly, $\mathsf{Gr}(k,\mathbb{C}^n)$ identifies with the quotient space $\mathsf{U}(n)/(\mathsf{U}(k)\times\mathsf{U}(n-k))$. Note that $\mathsf{Gr}(k,\mathbb{C}^n)$ is a complex manifold (of complex dimension $k(n-k)$), although $\mathsf{U}(n)$ is not. As pointed out earlier, $\mathsf{Gr}(1,V)$ identifies with the projective space $\mathsf{P}(V)$, except that the metric (B.10) differs from the Fubini–Study metric (B.5) by a factor of $\sqrt{2}$ when $V = \mathbb{C}^n$. We explain the reasons for this factor further below, particularly in the paragraph containing (B.14). (The same formulas and the same caveats apply to the case $V = \mathbb{R}^n$.) On the other hand, the metric d_∞ defined by (B.11) coincides, for $k = 1$, with the Fubini–Study distance.

Whether we use the high-tech or simple-minded point of view, there is a canonical procedure that allows us to transfer metric structure(s) from $\mathsf{O}(n)$ to $\mathsf{Gr}(k,\mathbb{R}^n)$ (and from $\mathsf{U}(n)$ to $\mathsf{Gr}(k,\mathbb{C}^n)$). We will exemplify that procedure in the case of the (extrinsic) Schatten p-norm for $1 \leqslant p \leqslant \infty$. We set, for $E, F \in \mathsf{Gr}(k,\mathbb{R}^n)$,

$$\begin{aligned} \tilde{h}_p(E,F) &:= \min\{\|U-V\|_p \;:\; U,V \in \mathsf{O}(n),\, U\mathbb{R}^k = E,\, V\mathbb{R}^k = F\} \\ &= \min\{\|W - \mathrm{I}\|_p \;:\; W \in \mathsf{O}(n),\, WE = F\} \end{aligned} \tag{B.13}$$

and similarly for $E, F \in \mathsf{Gr}(k,\mathbb{C}^n)$. The definition ":=" works *mutatis mutandis* for any quotient map on (or, equivalently, for any family of "cosets" in) a metric space, but the second equality requires that space to be a group with invariant metric (see also Exercise B.9).

The same scheme can be applied to the geodesic metric g_p on $\mathsf{O}(n)$ or $\mathsf{U}(n)$. In particular, if $p = 2$, we obtain the standard Riemannian structure on $\mathsf{Gr}(k,\mathbb{R}^n)$ or $\mathsf{Gr}(k,\mathbb{C}^n)$ and the resulting metric is (B.10), while $p = \infty$ yields the metric d_∞ from (B.11) (see Exercise B.12). Moreover, it doesn't matter whether we first define the geodesic metric and then pass to a quotient, or whether we reverse the order of these operations (see Exercise B.10).

It is instructive to specify the calculations implicit in the above paragraph to the simplest nontrivial setting, that of the real projective space $\mathsf{P}(\mathbb{R}^2)$, or $\mathsf{Gr}(1,\mathbb{R}^2)$. If the angle between two lines $E, F \subset \mathbb{R}^2$ (i.e., their Fubini–Study distance (B.5)) is $\theta \in (0, \pi/2]$, then the eigenvalues of the rotation W mapping E to F are $e^{i\theta}$ and $e^{-i\theta}$ and so $W = e^{iA}$, where $A \in \mathsf{M}_2^{\mathrm{sa}}$ has eigenvalues θ and $-\theta$ (cf. the calculation in the hint to Exercise B.7). It follows that the intrinsic Riemannian distance induced by g_2 and the quotient map $\mathsf{O}(2) \to \mathsf{Gr}(1,\mathbb{R}^2)$ verifies

$$d(E,F) = g_2(W,\mathrm{I}) = \|A\|_2 = (\theta^2 + \theta^2)^{1/2} = \sqrt{2}\,\theta, \tag{B.14}$$

which explains the factor $\sqrt{2}$ appearing in (B.10). Observe that the second equality in (B.14) is a straightforward application of Proposition B.1; the first one requires noting that if $R \in \mathsf{O}(2)$ is the reflection swapping E and F, then (again by Proposition B.1) $g_2(R,\mathrm{I}) = \pi > \sqrt{2}\,\pi/2 \geqslant g_2(W,\mathrm{I})$.

And here is a slightly different approach to endowing a Grassmann manifold with a metric. The map $E \mapsto P_E$ allows one to consider (for example) $\mathsf{Gr}(k,\mathbb{R}^n)$ as a submanifold of $\mathsf{M}_n^{\mathrm{sa}}$, so any norm of M_n also induces two metrics (extrinsic vs. geodesic) on $\mathsf{Gr}(k,\mathbb{R}^n)$. As it turns out, the geodesic metric obtained from the Hilbert–Schmidt norm is again (B.10). For an analysis of this situation via principal angles, see Exercise B.13.

Finally, let us note that since $\mathsf{SO}(n)$ acts transitively on $\mathsf{Gr}(k,\mathbb{R}^n)$, the Grassmann manifold can be likewise represented as a quotient of $\mathsf{SO}(n)$, and similarly for $\mathsf{Gr}(k,\mathbb{C}^n)$ and $\mathsf{SU}(n)$, a point of view that can be occasionally useful (cf. the proof of Theorem 7.15). This circle of ideas is explored in Exercises B.16 and B.17.

Each Grassmann manifold carries a natural probability measure which can be constructed in two different but equivalent ways:
- as the normalized Riemannian volume induced by the metric (B.10)
- as the pushforward of the Haar measure on $\mathsf{O}(n)$ via the quotient map $\mathsf{O}(n) \to \mathsf{O}(n)/(\mathsf{O}(k) \times \mathsf{O}(n-k))$.

The latter construction can be described more tangibly as follows: fix $E \in \mathsf{Gr}(k,\mathbb{R}^n)$ and consider a Haar-distributed $O \in \mathsf{O}(n)$; then $O(E)$ is a random element in $\mathsf{Gr}(k,\mathbb{R}^n)$ whose distribution does not depend on the choice of E. Either way, we will call the resulting measure the *standard Haar measure* on $\mathsf{Gr}(k,\mathbb{R}^n)$. The same construction (using $\mathsf{U}(n)$ instead of $\mathsf{O}(n)$) defines similarly the standard Haar measure on $\mathsf{Gr}(k,\mathbb{C}^n)$. For an even more concrete realization of the standard Haar measure, see Exercise B.14.

Since $\mathsf{O}(n)$ consists of morphisms of the corresponding space that preserve the inner product, the Haar measure on $\mathsf{Gr}(k,\mathbb{R}^n)$ may be seen as depending on the choice of a Euclidean (i.e., inner product) structure on $\mathbb{R}^n$. Using another Euclidean structure on $\mathbb{R}^n$ leads to a different measure on $\mathsf{Gr}(k,\mathbb{R}^n)$, as illustrates Exercise B.15. The same caveat applies to the complex case.

To complete the discussion of Grassmann manifolds, we will mention briefly their "cousins," Stiefel manifolds. For $1 \leqslant k \leqslant n$, denote by

$$\mathsf{St}(k,\mathbb{R}^n) := \{(x_1,\dots,x_k) \in \mathbb{R}^n \ : \ \langle x_i, x_j\rangle = \delta_{ij}\}$$

the set of k-tuples of orthonormal vectors in $\mathbb{R}^n$. We have the canonical equivalences $\mathsf{St}(k,\mathbb{R}^n) \leftrightarrow \mathsf{O}(n)/\mathsf{O}(n-k)$ and $\mathsf{Gr}(k,\mathbb{R}^n) \leftrightarrow \mathsf{St}(k,\mathbb{R}^n)/\mathsf{O}(k)$. The complex version is defined similarly; as for Grassmann manifolds, Stiefel manifolds naturally inherit metrics and a Haar measure from the orthogonal group.

For simplicity, Exercises B.9–B.15 *are stated in the real case, but the statements are also valid in the complex case.*

EXERCISE B.9 (Induced metrics on spaces of cosets). Fix $0 < k < n$, $1 \leqslant p \leqslant \infty$, and denote $\mathsf{H} = \mathsf{O}(k)\times\mathsf{O}(n-k) \subset \mathsf{O}(n)$. Let $U_0\mathsf{H}$ and $V_0\mathsf{H}$ be two left cosets of H and let $U_1 \in U_0\mathsf{H}$. Show that $\min\{\|U_1 - V\|_p \ : \ V \in V_0\mathsf{H}\} = \min\{\|U_0 - V\|_p \ : \ V \in V_0\mathsf{H}\}$ and that a similar equality holds for the corresponding geodesic distance g_p.

EXERCISE B.10 (Geodesics in $\mathsf{Gr}(k,\mathbb{R}^n)$ and in $\mathsf{O}(n)$). Fix $0 < k < n$ and $1 < p < \infty$. Denote by $\tilde{g}_p$ the metric on $\mathsf{Gr}(k,\mathbb{R}^n)$ obtained as the quotient metric from the geodesic metric g_p on $\mathsf{O}(n)$. Show that any geodesic in $\big(\mathsf{Gr}(k,\mathbb{R}^n),\tilde{g}_p\big)$ can be lifted to a geodesic in $\big(\mathsf{O}(n),g_p\big)$, which is of the form given by Exercise B.6 and on which the quotient map acts as an isometry. If $p = 1$ or ∞, any two points in $\mathsf{Gr}(k,\mathbb{R}^n)$ can be connected by a geodesic with this property.

EXERCISE B.11 ($\mathsf{Gr}(k,\mathbb{R}^n)$ vs. $\mathsf{Gr}(n-k,\mathbb{R}^n)$). Let $E, F \in \mathsf{Gr}(k,\mathbb{R}^n)$. Show that the nonzero principal angles between E and F coincide with the nonzero principal angles between $E^\perp$ and $F^\perp$.

EXERCISE B.12 (Equivalence of metrics on $\mathsf{Gr}(k,\mathbb{R}^n)$). Let $\tilde{h}_p$ the metric on $\mathsf{Gr}(k,\mathbb{R}^n)$ given by (B.13) and let $\tilde{g}_p$ be the geodesic metric defined in Exercise B.10. Show that for $E, F \in \mathsf{Gr}(k,\mathbb{R}^n)$

$$\tilde{h}_p(E,F) = 2^{1/p}\|(2\sin\theta_1/2,\dots,2\sin\theta_k/2)\|_p, \quad \tilde{g}_p(E,F) = 2^{1/p}\|(\theta_1,\dots,\theta_k)\|_p,$$

where $\|\cdot\|_p$ is the ℓ_p-norm on $\mathbb{R}^k$ and $\theta_1,\dots,\theta_k$ are the principal angles between E and F. Conclude that $\frac{2\sqrt{2}}{\pi}\tilde{g}_p \leqslant \tilde{h}_p \leqslant \tilde{g}_p$ and that $\tilde{g}_\infty$ coincides with the metric d_∞ from (B.11).

EXERCISE B.13 (Equivalence of metrics on $\mathsf{Gr}(k,\mathbb{R}^n)$, take #2). Show that the metric on $\mathsf{Gr}(k,\mathbb{R}^n)$ induced from the Schatten p-norm on M_n via the embedding $E \mapsto P_E$ is equivalent to the metrics $\tilde{g}_p$ and $\tilde{h}_p$ from Exercise B.12. Show that the geodesic metric induced by it coincides with $\tilde{g}_p$.

EXERCISE B.14 (Simulating the Haar measure on $\mathsf{Gr}(k,\mathbb{R}^n)$). For $1 \leqslant k \leqslant n$, let $(x_i)_{1\leqslant i\leqslant k}$ be independent standard Gaussian vectors in $\mathbb{R}^n$. Show that the subspace span$\{x_i : 1 \leqslant i \leqslant k\}$ is almost surely k-dimensional and distributed with respect to the standard Haar measure on $\mathsf{Gr}(k, \mathbb{R}^n)$. Prove also that the same holds when (x_i) are uniformly distributed on the unit sphere.

EXERCISE B.15 (About the choice of the Euclidean structure). Given a k-dimensional subspace $E \subset \mathbb{R}^n$, show that there is a sequence of Euclidean structures on $\mathbb{R}^n$ such that the corresponding Haar measures on $\mathsf{Gr}(k, \mathbb{R}^n)$ converge towards the Dirac mass at E.

EXERCISE B.16. Does $\mathsf{Gr}(k, \mathbb{R}^n)$ identify with $\mathsf{SO}(n)/(\mathsf{SO}(k) \times \mathsf{SO}(n-k))$? Does $\mathsf{Gr}(k, \mathbb{C}^n)$ identify with $\mathsf{SU}(n)/(\mathsf{SU}(k) \times \mathsf{SU}(n-k))$?

EXERCISE B.17 (Another representation of $\mathsf{Gr}(k, \mathbb{R}^n)$ as a coset space). Since the stabilizer of $\mathbb{R}^k$ under the canonical action of $\mathsf{SO}(n)$ on $\mathsf{Gr}(k, \mathbb{R}^n)$ is $\mathsf{H} = \mathsf{SO}(n) \cap \big(\mathsf{O}(k) \times \mathsf{O}(n-k)\big)$ (and since the action is transitive), $\mathsf{Gr}(k, \mathbb{R}^n)$ can be likewise identified with $\mathsf{SO}(n)/\mathsf{H}$. Are the metrics induced this way by the Schatten p-norms the same as $\tilde{g}_p$'s and $\tilde{h}_p$'s? What about the analogous question for $\mathsf{Gr}(k, \mathbb{C}^n)$? Note that there are no subtleties as far as the induced probability measure is concerned: all reasonable constructions lead to the same object by uniqueness of the Haar measure.

B.5. The Lorentz group $\mathsf{O}(1, n-1)$

Just as the orthogonal group $\mathsf{O}(n)$ preserves the Euclidean norm on $\mathbb{R}^n$, the *Lorentz group* $\mathsf{O}(1, n-1)$ consists of linear transformations preserving the quadratic form $q(x) = x_0^2 - \sum_{k=1}^{n-1} x_k^2$, where $x = (x_0, x_1, \ldots, x_{n-1}) \in \mathbb{R}^n$. Let J be the diagonal matrix with diagonal entries $(1, -1, \ldots, -1)$, i.e., the matrix inducing q in the sense that $q(x) = \langle x|J|x\rangle$ for $x \in \mathbb{R}^n$. Then

$$M \in \mathsf{O}(1, n-1) \iff M^T J M = J. \tag{B.15}$$

This immediately shows that $M \in \mathsf{O}(1, n-1)$ verifies $\det M = \pm 1$ and motivates the definition of of the *proper Lorentz group*

$$\mathsf{SO}(1, n-1) := \{M \in \mathsf{O}(1, n-1) : \det M = 1\}. \tag{B.16}$$

Let $\mathcal{L}_n = \{x \in \mathbb{R}^n : x_0 \geqslant 0 \text{ and } q(x) \geqslant 0\}$ be the Lorentz cone. If $M \in \mathsf{O}(1, n-1)$, then clearly $M\big(\mathcal{L}_n \cup (-\mathcal{L}_n)\big) = \mathcal{L}_n \cup (-\mathcal{L}_n)$ and so there are two possibilities: either $M\big(\mathcal{L}_n\big) = \mathcal{L}_n$ or $M\big(\mathcal{L}_n\big) = -\mathcal{L}_n$. Again, this motivates the definition of the *orthochronous* subgroup of the Lorentz group (the transformations that preserve the direction of time, identified with the coordinate x_0)

$$\mathsf{O}^+(1, n-1) := \{M \in \mathsf{O}(1, n-1) : M\big(\mathcal{L}_n\big) = \mathcal{L}_n\} \tag{B.17}$$

and

$$\mathsf{SO}^+(1, n-1) := \mathsf{SO}(1, n-1) \cap \mathsf{O}^+(1, n-1), \tag{B.18}$$

the *restricted Lorentz group*. Actually, we will see later (Exercise C.2) that the condition $M\big(\mathcal{L}_n\big) = \mathcal{L}_n$ (i.e., M being a linear automorphism of $\mathcal{L}_n$) implies that M is a positive multiple of an element of $\mathsf{O}^+(1, n-1)$ and so an alternative definition of $\mathsf{O}^+(1, n-1)$ is $\{M \in \mathsf{SL}(n, \mathbb{R}) : M\big(\mathcal{L}_n\big) = \mathcal{L}_n\}$. More generally, the structure of the cone of linear maps M verifying $M\big(\mathcal{L}_n\big) \subset \mathcal{L}_n$ is studied in Appendix C.

The instance that is of most immediate physical significance is $n = 4$, with $\mathbb{R}^4$ being identified with the Minkowski spacetime of the theory of special relativity. Another special feature of the case $n = 4$ is that the Lorentz cone $\mathcal{L}_4$ is isomorphic to the positive semi-definite cone $\mathcal{PSD}(\mathbb{C}^2)$ (see Section 1.2.1) and so its group of automorphisms can be identified with the group of automorphisms of the latter cone described in Proposition 2.29. In particular, the fact that a linear map $\Phi : \mathsf{M}_d^{\mathrm{sa}} \to \mathsf{M}_d^{\mathrm{sa}}$ satisfying $\Phi(\mathcal{PSD}) = \mathcal{PSD}$ is either completely positive or co-completely positive corresponds—for $d = 2$—to the dichotomy $\mathsf{O}^+(1,3)$ vs. $\mathsf{O}(1,3)\backslash\mathsf{O}^+(1,3)$. When restricted to $\mathsf{SO}^+(1,3)$, that identification induces an isomorphism of that group with $\mathsf{PSL}(2,\mathbb{C})$, or the so-called *spinor map*, see Exercise B.19.

Exercise B.18 (Examples of automorphisms of the Lorentz cone).
(a) Show that $\mathsf{SO}^+(1,1) = \left\{ \begin{bmatrix} \cosh\theta & \sinh\theta \\ \sinh\theta & \cosh\theta \end{bmatrix} : \theta \in \mathbb{R} \right\}$.
(b) Deduce that if $c > 0$, then $\mathsf{SO}^+(1,1)$ acts transitively on the (branch of the) hyperbola $\{(x_0, x_1) \ : \ x_0 > 0,\ x_0^2 - x_1^2 = c\}$.

Exercise B.19 (A spinor map). Let $\Psi(\mathbf{x}) = \mathbf{x}\cdot\sigma = X$ be the map from (2.4)–(2.5) implementing the isomorphism between the cones $\mathcal{L}_4$ and $\mathcal{PSD}(\mathbb{C}^2)$.
(a) Show that if $V \in \mathsf{SL}(2,\mathbb{C})$, then $\Psi_V(\mathbf{x}) = \Psi^{-1}\big(V\Psi(\mathbf{x})V^\dagger\big)$ is an automorphism of $\mathcal{L}_4$ which belongs to $\mathsf{SO}^+(1,3)$, and that every element of $\mathsf{SO}^+(1,3)$ can be represented that way.
(b) Show that the map $\mathsf{SL}(2,\mathbb{C}) \ni V \mapsto \Psi_V \in \mathsf{SO}^+(1,3)$ is a group homomorphism whose kernel is $\{\mathrm{Id}, -\mathrm{Id}\}$ and deduce that it induces a group isomorphism between $\mathsf{PSL}(2,\mathbb{C}) = \mathsf{SL}(2,\mathbb{C})/\{\mathrm{Id}, -\mathrm{Id}\}$ and $\mathsf{SO}^+(1,3)$ (an example of the so-called "spinor map").

Notes and Remarks

Proposition B.1 appears in [**Sza98**]. For $p \in (1,\infty)$, $p \neq 2$, the assertion—but not the argument—is exactly the same as in the classical Riemannian case ($p = 2$). If $p \in (1,2)$, the Riemannian proof can be tweaked as it fits in the framework of Finsler geometry [**CS05**]. For $p \in (2,\infty)$, the metric structure induced by the p-Schatten norm does not satisfy the usual hypotheses of Finsler geometry and so a more specialized argument is needed.

Proposition B.1 can be extended to other bi-unitarily invariant norms (i.e., norms defined via singular values, see Exercise 1.47).

For more information and alternative definitions for principal angles, see the book [**GVL13**].

APPENDIX C

Extreme maps between Lorentz cones and the S-lemma

The focus of this appendix is the Lorentz cone

$$\mathcal{L}_n = \Big\{(x_0, x_1, \ldots, x_{n-1}) : x_0 \geqslant 0,\ \sum_{k=1}^{n-1} x_k^2 \leqslant x_0^2\Big\} \subset \mathbb{R}^n \tag{C.1}$$

and particularly the cone $\boldsymbol{P}(\mathcal{L}_n) := \{\Phi : \Phi(\mathcal{L}_n) \subset \mathcal{L}_n\}$ of linear maps that preserve it. We have the following.

Proposition C.1. *Let $\Phi : \mathbb{R}^n \to \mathbb{R}^n$ be a linear map which generates an extreme ray of $\boldsymbol{P}(\mathcal{L}_n)$. Then either Φ is an automorphism of $\mathcal{L}_n$ or Φ is of rank one, in which case $\Phi = |u\rangle\langle v|$ for some $u, v \in \partial\mathcal{L}_n\backslash\{0\}$. If $n > 2$, the converse implication also holds.*

In view of the isomorphism between the cones $\mathcal{PSD}(\mathbb{C}^2)$ and $\mathcal{L}_4$ (see (2.4)), Proposition 2.38—characterizing extreme rays of the cone of positivity-preserving linear maps on $\mathsf{M}_2^{\mathrm{sa}}$—is really a special case of Proposition C.1. Note that every element of $\partial\mathcal{PSD}(\mathbb{C}^2)\backslash\{0\}$ is of the form $|\varphi\rangle\langle\varphi|$, $\varphi \in \mathbb{C}^2\backslash\{0\}$, so $|\varphi\rangle\langle\varphi|$ and $|\psi\rangle\langle\psi|$ of Proposition 2.38 play the same role as u, v in Proposition C.1. However, the true reason why they appear in the statements is that they generate extreme rays respectively in $\mathcal{PSD}(\mathbb{C}^2)$ and $\mathcal{L}_n$ (cf. Corollary 1.10). The following simple observation completely characterizes extreme rays generated by rank one maps in a very general setting (we only need the "only if" part, which is easy).

Lemma C.2 (See Exercise C.1). *Let $\mathcal{C} \subset \mathbb{R}^n$ be a nondegenerate cone and let $\boldsymbol{P}(\mathcal{C})$ be the cone of linear maps preserving $\mathcal{C}$. A rank one map $\Phi : \mathbb{R}^n \to \mathbb{R}^n$ generates an extreme ray of $\boldsymbol{P}(\mathcal{C})$ iff it is of the form $\Phi = |u\rangle\langle v|$, with u and v generating extreme rays of respectively $\mathcal{C}$ and $\mathcal{C}^*$.*

While not as simple as the case of rank one maps, the structure of the set of automorphisms of $\mathcal{L}_n$ is very well understood: they are of the form $t\Phi$, where $t > 0$ and $\Phi \in \mathsf{O}^+(1, n-1)$ (see Appendix B.5), the *orthochronous subgroup* of the *Lorentz group* $\mathsf{O}(1, n-1)$ of transformations preserving the quadratic form $q(x) = x_0^2 - \sum_{k=1}^{n-1} x_k^2$. This follows easily from the so-called S-lemma, a well-known fact from control theory and quadratic/semi-definite programming. (This and similar issues are explored in Exercises C.2–C.3.) The same lemma underlies the proof of Proposition C.1. We first state the simplest version of the Lemma.

Lemma C.3 (S-lemma). *Let M, N be $n \times n$ symmetric real matrices. The following two statements are equivalent:*
(i) $\{x \in \mathbb{R}^n : \langle x|M|x\rangle \geqslant 0\} \cup \{x \in \mathbb{R}^n : \langle x|N|x\rangle \geqslant 0\} = \mathbb{R}^n$,
(ii) there exists $t \in [0, 1]$ such that the matrix $(1-t)M + tN$ is positive semi-definite.

We apply the S-lemma in the following form, which is an easy consequence of Lemma C.3 applied with $M = F$ and $N = -G$.

LEMMA C.4 (S-lemma reformulated). *Let F, G be $n \times n$ symmetric real matrices. Assume that there is an $\bar{x} \in \mathbb{R}^n$ such that $\langle \bar{x}|G|\bar{x}\rangle > 0$. Then the following two statements about such F, G are equivalent:*
(i) if $x \in \mathbb{R}^n$ verifies $\langle x|G|x\rangle \geqslant 0$, then $\langle x|F|x\rangle \geqslant 0$;
(ii) there exists $\mu \geqslant 0$ such that $F - \mu G$ is positive semi-definite.

We postpone the proof of the Lemma until the end of this appendix and show how it implies the Proposition.

PROOF OF PROPOSITION C.1. In view of Lemma C.2, we may assume that $\operatorname{rank} \Phi \geqslant 2$. Let J be the diagonal matrix with diagonal entries $(1, -1, \ldots, -1)$, i.e., the matrix inducing q in the sense that $q(x) = \langle x|J|x\rangle$ for $x \in \mathbb{R}^n$. The map Φ preserving $\mathcal{L}_n$ (and hence $-\mathcal{L}_n$) means that the hypothesis (i) of Lemma C.4 is satisfied with $G = J$ and $F = \Phi^* J \Phi$. Since clearly $-J$ is not positive definite, it follows that there is $\mu \geqslant 0$ and a positive semi-definite operator Q such that

$$\Phi^* J \Phi = \mu J + Q. \tag{C.2}$$

We now notice that since $\operatorname{rank} \Phi \geqslant 2$, there is $y = \Phi x \neq 0$ such that $y_0 = 0$. In particular, $\langle x|\Phi^* J \Phi|x\rangle = \langle y|J|y\rangle < 0$. Given that $\langle x|Q|x\rangle \geqslant 0$, it follows that μ cannot be 0. Next, if $Q = 0$, (C.2) means precisely that $\mu^{1/2}\Phi \in \mathsf{O}(1, n-1)$ and so Φ is an automorphism of $\mathcal{L}_n$.

To complete the argument, we will show that if $Q \neq 0$, then there is a rank one operator Δ such that $\Phi \pm \Delta \in \boldsymbol{P}(\mathcal{L}_n)$. Since Φ and Δ have different ranks, they are not proportional. Hence $\Phi + \Delta$ and $\Phi - \Delta$ do not belong to the ray generated by Φ, which implies that the ray is not extreme.

Let $|v\rangle\langle v|$, $v \neq 0$, be one of the terms appearing in the spectral decomposition of Q; then $Q = Q' + |v\rangle\langle v|$, where Q' is positive semi-definite. Next, let $u \in \mathbb{R}^n \backslash \{0\}$ be such that $\Phi^* J u = \delta v$, where δ is either 1 or 0. Such u exists: if Φ^* is invertible, then $u = J(\Phi^*)^{-1} v$ satisfies $\Phi^* J u = v$, while in the opposite case the nullspace of $\Phi^* J$ is nontrivial. We will show that, for some $\varepsilon > 0$,

$$\Phi + s|u\rangle\langle v| \in \boldsymbol{P}(\mathcal{L}_n) \text{ if } |s| \leqslant \varepsilon, \tag{C.3}$$

thus supplying the needed $\Delta = \varepsilon|u\rangle\langle v|$. We have, by (C.2) and by the choice of u,

$$\begin{aligned} (\Phi + s|u\rangle\langle v|)^* J (\Phi + s|u\rangle\langle v|) &= \mu J + Q + 2s\delta|v\rangle\langle v| + s^2|v\rangle\langle u|J|u\rangle\langle v| \\ &= \mu J + Q' + (1 + 2s\delta + s^2\langle u|J|u\rangle)|v\rangle\langle v|. \end{aligned} \tag{C.4}$$

Since clearly $1 + 2s\delta + s^2\langle u|J|u\rangle \geqslant 0$ if $|s|$ is sufficiently small, it follows that, for such s, $(\Phi + s|u\rangle\langle v|)^* J (\Phi + s|u\rangle\langle v|) - \mu J$ is positive semi-definite. Thus we can deduce from the easy part of Lemma C.4 that $\Phi + s|u\rangle\langle v| \in \boldsymbol{P}(\mathcal{L}_n)$, as needed. (To be precise, we need to exclude the possibility that $\Phi + s|u\rangle\langle v| \in -\boldsymbol{P}(\mathcal{L}_n)$, but this is simple.)

For the converse implication, Lemma C.2 takes care of rank one maps, so we just need to show that every automorphism Φ of $\mathcal{L}_n$ generates an extreme ray of $\boldsymbol{P}(\mathcal{L}_n)$ if $n > 2$. To that end, notice that the map $\Psi \mapsto \Phi \circ \Psi$ is a linear automorphism of the cone $\boldsymbol{P}(\mathcal{L}_n)$ sending Id to Φ. Since linear maps preserve faces and their character, the ray $\mathbb{R}_+ \Phi$ is extreme iff $\mathbb{R}_+ \operatorname{Id}$ is extreme. This means that either *all* automorphisms of $\mathcal{L}_n$ generate extreme rays of $\boldsymbol{P}(\mathcal{L}_n)$, or *none* of them do, and we just have to exclude the latter possibility.

Indeed, suppose that all extreme rays of $\boldsymbol{P}(\mathcal{L}_n)$ are generated by rank one maps. It then follows in particular (see Section 1.2.2) that $\mathrm{Id} = \sum_{i=1}^N |u_i\rangle\langle v_i|$ for some $u_i, v_i \in \partial\mathcal{L}_n$. Since $u, v \in \mathcal{L}_n$ implies that $\mathrm{Tr}\,(J|u\rangle\langle v|) = \langle v|J|u\rangle \geqslant 0$, we obtain

$$-1 \geqslant 2 - n = \mathrm{Tr}\, J = \mathrm{Tr}\Big(J \sum_{i=1}^N |u_i\rangle\langle v_i|\Big) = \sum_{i=1}^N \mathrm{Tr}\,(J|u_i\rangle\langle v_i|) \geqslant 0,$$

which yields a desired contradiction. (See Exercise C.4 for the discussion of the case $n = 2$.) □

PROOF OF LEMMA C.3. To show that $(i) \Rightarrow (ii)$, we argue by contradiction. Denote $M_t = (1-t)M + tN$ and assume that, for every $t \in [0,1]$, the smallest eigenvalue λ_t of M_t is strictly negative. For $t \in [0,1]$, let

$$\Lambda_t := \{x \in S^{n-1} : M_t x = \lambda_t x\}.$$

Note that $t \mapsto \lambda_t$ is continuous and $t \mapsto \Lambda_t$ is upper semicontinuous, i.e., $t_n \to t$, $x_n \in \Lambda_{t_n}$ and $x_n \to x$ imply $x \in \Lambda_t$, and of course all $\Lambda_t \neq \varnothing$.

Consider the sets $A = \{x \in \mathbb{R}^n : \langle x|M|x\rangle \geqslant 0\}$ and $B = \{x \in \mathbb{R}^n : \langle x|N|x\rangle \geqslant 0\}$. We have $A \cup B = \mathbb{R}^n$ by hypothesis. Since $M_0 = M$, it follows that $\Lambda_0 \cap A = \varnothing$ and so $\Lambda_0 \subset B$. Similarly, $\Lambda_1 \subset A$. Set

$$\tau = \sup\{t \in [0,1] : \Lambda_t \cap B \neq \varnothing\}.$$

We now note that $\Lambda_\tau \cap B \neq \varnothing$; this is immediate if $\tau = 0$ and follows from upper semicontinuity of $t \mapsto \Lambda_t$ if $\tau > 0$. For essentially the same reasons, $\Lambda_\tau \cap A \neq \varnothing$.

We now claim that $\Lambda_\tau \cap A \cap B \neq \varnothing$. This is clear if the eigenvalue λ_τ is simple (note that all three sets, Λ_τ, A and B, are symmetric by definition). On the other hand, if the multiplicity of λ_t equals $k > 1$, then Λ_τ is a $(k-1)$-dimensional sphere and hence is connected. Consequently, the closed nonempty sets $\Lambda_\tau \cap A$ and $\Lambda_\tau \cap B$, the union of which is Λ_τ, must have a nonempty intersection.

To conclude the argument, choose $x \in \Lambda_\tau \cap A \cap B \neq \varnothing$. Then, since $x \in \Lambda_\tau$,

$$\langle x|\Lambda_\tau|x\rangle = \lambda_t < 0.$$

On the other hand, since $x \in A \cap B$,

$$\langle x|\Lambda_\tau|x\rangle = (1-\tau)\langle x|M|x\rangle + \tau\langle x|N|x\rangle \geqslant 0,$$

a contradiction. □

EXERCISE C.1. Prove Lemma C.2.

EXERCISE C.2. Use the S-lemma to show that every linear automorphism of $\mathcal{L}_n$ is of the form $t\Phi$, where $t > 0$ and $\Phi \in \mathsf{O}^+(1, n-1)$. In other words, there exists $t > 0$ such that $\langle x|\Phi^* J\Phi|x\rangle = t^2\langle x|J|x\rangle$ for all $x \in \mathbb{R}^n$.

EXERCISE C.3. Show that maps of the form of $t\Phi$, where $t > 0$ and $\Phi \in \mathsf{SO}^+(1, n-1)$, act transitively on the interior of $\mathcal{L}_n$.

EXERCISE C.4. Show that the all extreme rays of the cone $\boldsymbol{P}(\mathcal{L}_2)$ consist of maps of rank one.

Notes and Remarks

The fact that statements similar to Proposition C.1 imply Størmer's theorem was apparently folklore for some time; it appears explicitly in [**MO15**]. Proposition C.1 was proved in [**LS75**] and then rediscovered (apparently independently) in [**Hil05**], where its relevance to the entanglement theory was also noted. The subsequent paper [**Hil07b**] by the same author contains a stronger result, a complete classification of elements of $\boldsymbol{P}(\mathcal{L}_n)$. Our proof of Proposition C.1 follows roughly that of [**Hil05**], but is substantially simpler. In turn, the argument from [**Hil07b**] was similar to, but simpler than [**LS75**]; all proofs seem to use either a variant of the S-lemma (Lemma C.4) or closely related facts. The papers [**Hil05, Hil07b**] actually characterize (for any $m, n \geqslant 2$) extreme rays of maps that satisfy $\Phi(\mathcal{L}_m) \subset \mathcal{L}_n$, but this slightly more general fact is easy to derive from Proposition C.1 combined with (for example) Exercise C.3.

APPENDIX D

Polarity and the Santaló point via duality of cones

The goal of this appendix is to explore the dependence of polarity on translation, which is otherwise not very transparent, by exploiting the duality of cones. We believe that this approach deserves to be better known. Besides recovering the characterization of the Santaló point of a convex body, we are able to easily explain other somewhat mysterious facts such as, for example, the polar of an ellipsoid with respect to any interior point being also an ellipsoid.

We start with a reformulation of Lemma 1.6 from Section 1.2.1 in a manner not appealing to the concept of scalar product. Let V be a real vector space and V^* its dual. To make the analogy with Lemma 1.6 more apparent, we will write $\langle x^*, x\rangle$ for the evaluation $x^*(x)$ whenever $x \in V$ and $x^* \in V^*$. If $\mathcal{C} \subset V$ is a closed convex cone, the dual cone $\mathcal{C}^* \subset V^*$ is now defined by (cf. (1.18))

$$\mathcal{C}^* := \{x \in V^* \ : \ \forall y \in \mathcal{C} \ \langle x, y\rangle \geqslant 0\}. \tag{D.1}$$

We then have

LEMMA D.1. *Let $e \in \mathcal{C}$ and $e^* \in \mathcal{C}^*$ be such that $\langle e^*, e\rangle = 1$. Let $H_e := \{y \in V^* \ : \ \langle y, e\rangle = 1\}$, $H_{e^*} := \{x \in V \ : \ \langle e^*, x\rangle = 1\}$, and let $\mathcal{C}^{\mathrm{b}} = \mathcal{C} \cap H_{e^*}$ and $(\mathcal{C}^*)^{\mathrm{b}} = \mathcal{C}^* \cap H_e$ be the corresponding bases of $\mathcal{C}$ and $\mathcal{C}^*$. Then*

$$(\mathcal{C}^*)^{\mathrm{b}} = \{y \in H_e \ : \ \forall x \in \mathcal{C}^{\mathrm{b}} \ \ \langle -(y - e^*), x - e\rangle \leqslant 1\}. \tag{D.2}$$

In other words, if we think of H_{e^*} as a vector space with the origin at e, of H_e as a vector space with the origin at e^* and as a dual of H_{e^*}, and of $\mathcal{C}^{\mathrm{b}}$ and $(\mathcal{C}^*)^{\mathrm{b}}$ as their respective subsets, then $(\mathcal{C}^*)^{\mathrm{b}} = -(\mathcal{C}^{\mathrm{b}})^\circ$.

The proof of Lemma D.1 fully parallels that of Lemma 1.6 and so we relegate it to Exercise D.1.

The formula in (D.2) suggest a definition of polarity in the *affine* context that is a tad different than the one usually used. Namely, if K and L are (say, closed and convex) subsets of two affine spaces whose underlying vector spaces are dual to each other, and if $a \in K$ and $b \in L$, then L is a polar of K *with respect to the pair* (a, b) if $L - b = (K - a)^\circ$ (in the sense indicated in Section 1.2.1).

This definition, and Lemma D.1, allow for a nice way to visualize polars of translates of a convex body. Indeed, let K be an n-dimensional closed convex set. We can represent K as a subset of $H = \{(x_0, x_1, \ldots, x_n) : x_0 = 1\} \subset \mathbb{R}^{n+1}$ and consider the cone $\mathcal{C} \subset \mathbb{R}^{n+1}$ generated by it (i.e., $\mathcal{C} = \overline{\mathbb{R}_+ K}$, with the closure not needed if K is compact). Then K is the base of $\mathcal{C}$ with respect to $e_0 = (1, 0, \ldots, 0)$ and automatically $e_0 \in \mathcal{C}^*$. Now, if $a \in K$, then Equation (D.2) shows that $(K - a)^\circ$ can be identified (up to a reflection with respect to e_0) with the base of $\mathcal{C}^*$ corresponding to a, that is, with the section $\mathcal{C}^* \cap \{y \in \mathbb{R}^{n+1} : \langle y, a\rangle = 1\}$ of the cone $\mathcal{C}^*$. This point of view is pictured in Figure D.1, where $\mathcal{C}$ and $\mathcal{C}^*$ are

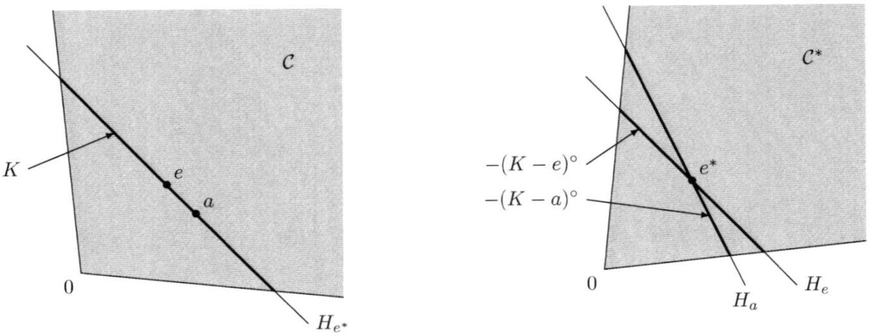

FIGURE D.1. If K is a base of $\mathcal{C}$, then the polars of K with respect to different points (defined in the way implicit in Lemma D.1) correspond to different sections of the cone $\mathcal{C}^*$. It is possible to superimpose the two pictures and even to assume that $e = e^*$, but that obscures the dependence of $(K-a)^\circ$ on a. The minus sign in front of $(K-\cdot)^\circ$ indicates a reflection inside $H_.$ with respect to e^*.

separately represented in two copies of $\mathbb{R}^{n+1}$ with e and e^* being two copies of e_0. (Note that while necessarily $e_0 \in \mathcal{C}^*$, it is *a priori* possible that $e_0 \notin \mathcal{C}$.)

Such approach has a number of nice immediate consequences, for example the fact that the polar of a not-necessarily-centered ellipsoid is an ellipsoid as long as 0 is an interior point (see Exercise D.3). Note, however, that we cannot directly compare (say, the volumes of) $(K-a)^\circ$ for different values of a since they do not live in the same hyperplane of $\mathbb{R}^{n+1}$. However, a simple trick permits such comparisons (cf. the comments following Theorem 4.17 in Section 4.3.4).

PROPOSITION D.2. *Let $K \subset \mathbb{R}^n$ be a convex body. Then there exists a unique interior point $s \in K$ such that $(K-s)^\circ$ has centroid at 0. Moreover, if $a \neq s$, then the volume of $(K-a)^\circ$ is strictly larger than the volume of $(K-s)^\circ$.*

The point s appearing in the statement of Proposition D.2 is called the *Santaló point* of K.

PROOF. We start with the construction outlined in the paragraph preceding the statement of the Proposition. Note that since K is a convex body (hence n-dimensional and compact), the cones $\mathcal{C}$ and $\mathcal{C}^*$ are both nondegenerate and e_0 is an interior point of $\mathcal{C}^*$ (see Lemma 1.7 and Exercise 1.32). We now consider the following auxiliary optimization problem: among the *solid* cones of the form

$$T_a = \{x \in \mathcal{C}^* : \langle x, a\rangle \leqslant 1\},$$

where a varies over the interior of K, find one for which $\mathrm{vol}_{n+1}(T_a)$ is the smallest. Note that the restrictions on a ensure that each T_a is indeed a (bounded) solid cone with the base $\{x \in \mathcal{C}^* : \langle x, a\rangle = 1\} =: B_a$ (this happens whenever a belongs to the interior of $\mathcal{C}$) and that e_0 belongs to B_a (this happens whenever $a \in \mathcal{C} \cap H$). The sets T_a and B_a are pictured in the first drawing in Figure D.2.

It is easy to see that $\inf_a \mathrm{vol}_{n+1}(T_a) > 0$, and that both the diameter and the volume of T_a tend to $+\infty$ as $a \to \partial K$. Since $\mathrm{vol}_{n+1}(T_a)$ is a continuous function of a, this implies that the infimum is attained. On the other hand, if $a \mapsto \mathrm{vol}_{n+1}(T_a)$ has a local extremum at s, then an elementary variational argument shows that

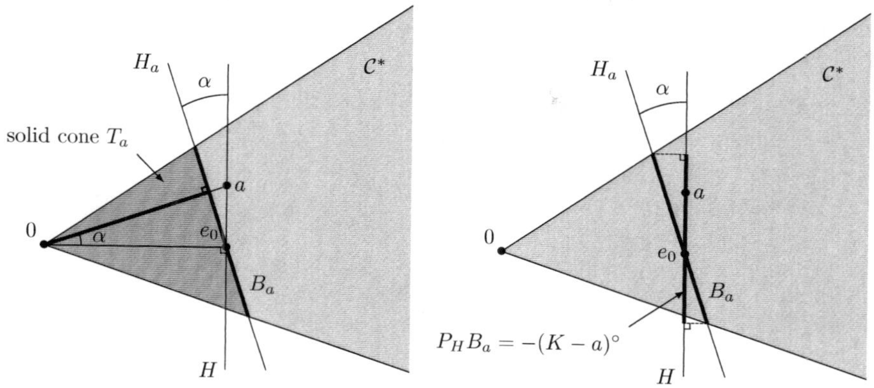

FIGURE D.2. The first drawing illustrates the calculation (D.3) of the volume of the solid cone T_a. The second drawing illustrates the calculation (D.4) of the volume of $(K-a)^\circ$, the polar of $K-a$ constructed inside H. The minus sign in front of $(K-a)^\circ$ indicates a reflection inside H with respect to e_0.

e_0 is the centroid of B_s (see Exercise D.4), which—according to Lemma D.1—is affinely equivalent to $(K-s)^\circ$, the polar of $K-s$ *inside* H. More precisely, one sees directly from (D.2) that, for every a, $(K-a)^\circ$ is (up to a reflection with respect to e_0) the orthogonal projection of B_a onto H, as pictured in the second drawing in Figure D.2.

Now comes a simple but crucial observation (illustrated in the two drawings of Figure D.2). On the one hand,

$$\mathrm{vol}_{n+1}(T_a) = \frac{1}{n+1}\,\mathrm{vol}_n(B_a) \times \frac{1}{|a|} \tag{D.3}$$

because $\frac{1}{|a|}$ equals the cosine of the angle between a and e_0 (denoted by α), and hence is the same as the height of the cone T_a. On the other hand, since $(K-a)^\circ$, the polar of $K-a$ constructed *inside* H is a reflection of $P_H(B_a)$, and since the angle α between B_a and H is the same as between a and e_0, it follows that

$$\mathrm{vol}_n((K-a)^\circ) = \mathrm{vol}_n(B_a) \times \frac{1}{|a|}. \tag{D.4}$$

This shows that $\mathrm{vol}_{n+1}(T_a)$ and $\mathrm{vol}_n((K-a)^\circ)$ differ only by a factor independent of a, and so they achieve their minima simultaneously. This concludes the argument, except for the uniqueness part (which is easy, see Exercise D.2). □

EXERCISE D.1. Prove Lemma D.1.

EXERCISE D.2. Let $K \subset \mathbb{R}^n$ be a convex body with 0 in the interior and such that K° has centroid at the origin. Then, for any point $a \neq 0$ in the interior of K, the centroid of $(K-a)^\circ$ is not 0.

EXERCISE D.3. This exercise supplies "soft" proofs of the facts derived previously via tedious calculations in Exercise 1.26. Let $\mathscr{E} \subset \mathbb{R}^n$ be an ellipsoid.
(i) Show that if $\mathscr{E}$ contains 0 in its interior, then $\mathscr{E}^\circ$ is also an ellipsoid.
(ii) Show that if $0 \in \partial\mathscr{E}$, then $\mathscr{E}^\circ$ is an elliptic paraboloid.
(iii) Show that, among translates of $\mathscr{E}$, the volume of the polar is minimal iff the

translate is 0-symmetric. Give a proof that does not use the uniqueness part of Proposition D.2.

EXERCISE D.4. Show that if (in the notation from the proof of Proposition D.2) the function $a \mapsto \mathrm{vol}_{n+1}(T_a)$ has a local extremum at $b \in K$, then e_0 is the centroid of B_b.

APPENDIX E

Hints to exercises

Chapter 0

Exercise 0.2. We may write $|x_1 \otimes x_2 + y_1 \otimes y_2\rangle\langle x_1 \otimes x_2 + y_1 \otimes y_2|$ as

$$|x_1\rangle\langle x_1| \otimes |x_2\rangle\langle x_2| + |y_1\rangle\langle y_1| \otimes |y_2\rangle\langle y_2| + \frac{1}{4}\sum_{k=0}^{3}(-1)^k |x_1 + \imath^k y_1\rangle\langle x_1 + \imath^k y_1| \otimes |x_2 + \imath^k y_2\rangle\langle x_2 + \imath^k y_2|.$$

Chapter 1

Exercise 1.1. If (x_i) are affinely dependent, then $\sum \mu_i x_i = 0$ for some (not identically zero) real numbers (μ_i) adding to zero. Then $x = \sum(\lambda_i + \varepsilon\mu_i)x_i$ and, for a well-chosen ε, this is a strictly shorter convex decomposition.

Exercise 1.2. By Carathéodory's Theorem 1.2, $\operatorname{conv} A$ is a continuous image of $\Delta_n \times A^{n+1}$.

Exercise 1.3. By the Hahn–Banach theorem, any boundary point of a convex body K admits a supporting hyperplane, whose intersection with K is an exposed face.

Exercise 1.4. If $y \in L\setminus\{x\}$, then x is an interior point of some segment $[y, z]$ with $z \in L$.

Exercise 1.5. (a) and (b) follow fairly directly from the definitions. (c) Consider a supporting hyperplane to K at a point in the relative interior of the face and apply Proposition 1.4 to the functional defining that hyperplane. (d) For the first assertion, take K to be the Minkowski sum of a disk and a segment (see Figure E.1), or of a disk and a square. For the second assertion, appeal to part (c). (e) Let L be the Minkowski sum of an n-dimensional cube and B_2^n. Consider a hyperplane supporting K which is parallel to one of the facets of the cube, and let F be the corresponding exposed facet. Show that F is a translate of that facet (hence an $(n-1)$-dimensional cube) and consider any k-dimensional face of F. (f) For sufficiency, use part (a) and part (b). For necessity, use part (c) to argue by induction with respect to the dimension. (g) Assume K is full-dimensional. If a supporting hyperplane does not isolate a point, then the boundary of K contains a segment.

Exercise 1.6. Proceed by induction with respect to $\dim K$. The base cases (dimension 0 or 1) are simple. For the inductive step, let $x \in K$ and assume first that x belongs to the relative interior of K. Next, note that every convex body

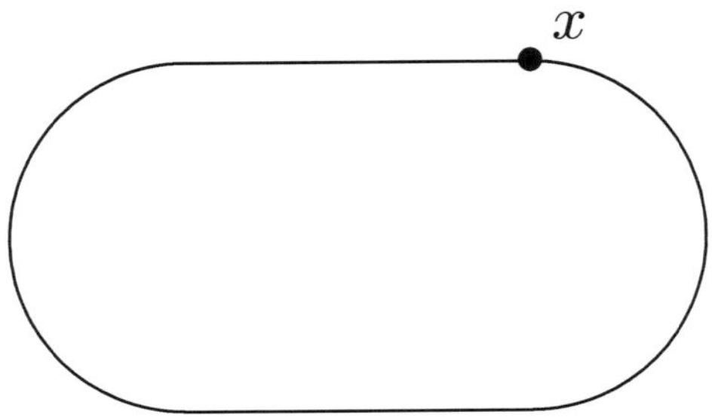

FIGURE E.1. An example of an extreme point which is not exposed.

admits at least one extreme point (for example, the smallest element with respect to the lexicographic order) and let y be one such point. There is a (unique) point $z \in \partial K$ (the relative boundary) such that x belongs to the segment $[z, y]$. Let H be a supporting hyperplane for K which contains z. We may apply the induction hypothesis to $K \cap H$ and produce a decomposition of z as a convex combination of extreme points of $K \cap H$ (hence of K, by Exercise 1.5(b)). Finally, if $x \in \partial K$, we may perform the dimension reduction immediately.

Exercise 1.7. For necessity, appeal to the spectral theorem. For sufficiency, use the following fact: *If ρ_1, ρ_2 are positive operators and $\rho = \rho_1 + \rho_2$, then the range of ρ contains the ranges of ρ_1 and ρ_2.* Alternatively, note that either all rank one projections $|\psi\rangle\langle\psi|$ are extreme or none of them are, and appeal to the Krein–Milman Theorem 1.3. (See also Section 2.1.3.)

Exercise 1.8. If $K = \operatorname{conv}\{x_1, \ldots, x_N\}$ and F is a face of K, then $F = \operatorname{conv}\{x_i : x_i \in F\}$.

Exercise 1.9. Prove that if F is an exposed face of a polytope P, and G is an exposed face of F, then G is an exposed face of P. Then use Exercise 1.5(f).

Exercise 1.10. The extreme points of B_1^n are the n vectors from the canonical basis in $\mathbb{R}^n$ and their opposites. The extreme points of B_∞^n are elements of $\{-1, 1\}^n$. For $1 < p < +\infty$, every boundary point is extreme (to show this, use the fact that the function $x \mapsto |x|^p$ is strictly convex; another "high level" argument is given in Exercise 1.11(iv)).

Exercise 1.11. (i) We may assume $0 < b \leqslant a$. Check that

$$\frac{\mathrm{d}}{\mathrm{d}t}\left(\alpha(t)a^p + \alpha(1/t)b^p\right) = (p-1)(a^p - b^p/t^p)((1+t)^{p-2} - |1-t|^{p-2})$$

so that the maximum is achieved for $t = b/a \leqslant 1$. (ii) Use (i) and the inequality $\sum_{i=1}^n \sup_{t>0}\{\cdots\} \geqslant \sup_{t>0} \sum_{i=1}^n \{\cdots\}$. For $p \geqslant 2$, the proof goes along the same lines except that the supremum in the variational formula is replaced by an infimum. (iii) To deduce (1.6) from (1.5), use the following inequalities

$$\text{(E.1)} \qquad \left(1 + (p-1)t^2\right)^{p/2} \leqslant 1 + \frac{p(p-1)}{2}t^2 \leqslant \frac{(1+t)^p + (1-t)^p}{2},$$

valid for $t \in [-1, 1]$ and applied with $t = \|y\|_p / \|x\|_p$ (we may assume that $\|y\|_p \leqslant \|x\|_p$). The second inequality in (E.1) can be proved by a Taylor expansion of the right-hand side. (iv) For $p \geqslant 2$, use (ii). For $p \leqslant 2$, use (iii) applied to the pair $(x + y, x - y)$.

Exercise 1.12. We may assume that K contains the origin in its interior. One possibility is to define $\Theta(x) = \Theta_K(x)$ as the (unique) element of minimal Euclidean norm in the set (denoted F) of points where $\langle \cdot, x \rangle$ is maximal on K. To see that this choice is Borel, define a sequence (K_m) of convex bodies approximating K from the inside by the relation $\|\cdot\|_{K_m} = \|\cdot\|_K + \frac{1}{m}|\cdot|$. One checks that (i) for each $x \in \mathbb{R}^n \backslash \{0\}$, the linear form $\langle \cdot, x \rangle$ achieves its maximum on K_m at a unique point, denoted $\phi_m(x)$; (ii) for each m, the map ϕ_m is continuous; and (iii) the sequence (ϕ_m) converges pointwise to Θ_K. To see the last point, write $\phi_m(x) = (1+|x_m|/m)^{-1}x_m$ for some $x_m \in \partial K$. If $y \in F$, then (by definition of ϕ_m) we have

$$(1+|y|/m)^{-1}\langle y, x\rangle \leqslant (1+|x_m|/m)^{-1}\langle x_m, x\rangle \leqslant (1+|x_m|/m)^{-1}\langle y, x\rangle,$$

which implies that $|\phi_m(x)| < |x_m| \leqslant |y|$. Deduce that $(\phi_m(x))$ must converge to the point of minimal Euclidean norm in F.

Exercise 1.13. If $h, h' : \mathbb{R}^n \to \mathbb{R}_+$ are positively homogeneous, then $\{h \leqslant 1\} = \{h' \leqslant 1\}$ implies $h = h'$. What may fail here is that $\sup_{x \in A}\langle x, \cdot \rangle$ may be negative.

Exercise 1.14. Use the fact that $K \subset RB_2^n \iff R^{-1}B_2^n \subset K^\circ$.

Exercise 1.15. $(K^\circ)^\circ$ is a closed convex set containing both K and 0, so one inclusion is clear. For the other inclusion, argue by contradiction using the Hahn–Banach separation theorem.

Exercise 1.17. If K does not contain 0 in the interior, then $\|\cdot\|_K$ takes the value $+\infty$ which forbids the application of Hahn–Banach theorem. For an illustration of the importance of the assumptions consider $K = L^\circ$, where $L = \{(x, y) \in \mathbb{R}^2 : (2-y)(2-x) \geqslant 1,\ x < 2\}$.

Exercise 1.18. (1.14) is simple and (1.15) can be deduced from it using the bipolar theorem. The example $K = -L = \{0, 0\} \cup (-\infty, 0) \times [-1, 1]$ shows that closedness is needed. The example $K = \{0, 2\}$, $L = [-1, 1]$ shows that convexity is needed. The example $K = [1, 2]$, $L = [3, 4]$ shows that containing the origin is needed. Finally, taking $K^\circ = (-\infty, 0) \times \{0\}$ and $L^\circ = \{(x, y) : x - 1 > 0,\ (x-1)(y-1) \geqslant 1\}$ shows that taking the closure is needed (it is clearly not needed if K° and L° are both compact).

Exercise 1.19. Consider $K = \operatorname{conv}\{V\} \subset \mathbb{R}^n$ containing 0 in the interior with V finite. For any extreme point $x \in K^\circ$ there is a subset $U \subset V$ such that $\operatorname{span} U = \mathbb{R}^n$ and x is the (unique) vector satisfying $\langle x, u \rangle = 1$ for every $u \in U$. It follows that K° has only finitely many extreme points.

Exercise 1.20. The hypotheses ensure that every supporting hyperplane to K is of the form $H_y = \{x : \langle y, x \rangle = 1\}$ for some $y \in \partial K^\circ$, so $\nu_K(F) \neq \varnothing$. To establish that $\nu_K(F)$ is an exposed face, show that if x_0 belongs to the relative interior of F and $H = \{y : \langle y, x_0 \rangle = 1\}$, then $\nu_K(F) = H \cap K^\circ$ (use the fact that for any $x \in F$ there exist $t \in (0, 1)$ and $x' \in F$ such that $x_0 = (1-t)x' + tx$). To establish injectivity, show that if F_1, F_2 are exposed faces of K with $F_2 \not\subset F_1$ and $F_1 = H_{y_0} \cap K$, then $y_0 \in \nu_K(F_1) \backslash \nu_K(F_2)$. With regards to the last property, $F \subset \nu_{K^\circ}(\nu_K(F))$ is easy; if we had a strict inclusion, injectivity and order reversing would imply the strict inclusion $\nu_K(\nu_{K^\circ}(\nu_K(F))) \subsetneq \nu_K(F)$, which is a contradiction since we just noted that the reverse inclusion always holds.

Exercise 1.21. Show that the interior of K is disjoint with F_y (always) and that, under our hypotheses, $F_y \neq \varnothing$. Deduce that $H_y = \{x : \langle y, x \rangle = 1\}$ is a supporting hyperplane and F_y an exposed face. For the second statement, if F is a maximal

exposed face of K, show that F coincides with F_y, where y is an extreme point of $\nu_K(F)$ (appeal to the Krein–Milman theorem and use maximality).
The same argument works in the general case with the caveat that, for some y, the set F_y (as defined by (1.17)) may be empty.

Exercise 1.23. The polars of the examples from Exercise 1.5(d) will work.

Exercise 1.24. Consider $K = \big(B_1^2 \cap \{x \leqslant 0\}\big) \cup \big(B_2^2 \cap \{x \geqslant 0\}\big)$, where x is the first coordinate in $\mathbb{R}^2$.

Exercise 1.25. If $a_1 \leqslant \cdots \leqslant a_{2n-1}$ denote the principal semi-axes of $\mathscr{E}$, produce an n-dimensional section which is a Euclidean ball of radius a_n by pairing each small semi-axis (a_k for $k < n$) with a large semi-axis (a_k for $k > n$).

Exercise 1.28. If $e \in \mathcal{C}^*$ is such that the functional $\langle e, \cdot \rangle$ doesn't vanish identically on $\mathcal{C}$, then it doesn't vanish identically on the relative interior of $\mathcal{C}$ and so, by Proposition 1.4 (applied with $K = \mathbb{R}_+$ and $F = \{0\}$), $\langle e, \cdot \rangle$ is strictly positive on the relative interior of $\mathcal{C}$. Show that this implies that the relative interior of $\mathcal{C}$ is contained in $\mathbb{R}_+\mathcal{C}^{\mathrm{b}}$ and deduce the assertion. For an example where closure is needed, take $\mathcal{C} = \mathbb{R}_+^2$ and $e = (1, 0)$.

Exercise 1.29. By the bipolar theorem, $\mathcal{C}^* = \mathcal{C}^\perp \iff \mathcal{C} = (\mathcal{C}^\perp)^* = \operatorname{span}(\mathcal{C})$, so whenever $\mathcal{C}$ is not a linear subspace, any vector $e \in \mathcal{C}^* \backslash \mathcal{C}$ induces a base.

Exercise 1.30. Try $\mathcal{C}_1 = \{(x, y, z) \in \mathbb{R}^3 \ : \ x \geqslant 0, y \geqslant 0, z \geqslant 0, xy \geqslant z^2\}$ and $\mathcal{C}_2 = \mathbb{R}^- \times \{0\} \times \{0\}$.

Exercise 1.31. We may assume after rotation that $y = te_0$ with $t > 1$. Note that $\mathcal{C}_{\sqrt{2}e_0}$ is the Lorentz cone $\mathcal{L}_n$ and is therefore self-dual. For the general case, define a linear map T_λ by $T_\lambda y = \lambda y$ and $T_\lambda x = x$ for $x \perp y$. For $\lambda = \sqrt{t^2 - 1}$, we have $\mathcal{C}_{te_0} = T_\lambda \mathcal{L}_n$ and therefore $\mathcal{C}_{te_0}^* = (T_\lambda^{-1})^T \mathcal{L}_n = T_{1/\lambda}\mathcal{L}_n = \mathcal{C}_{\sqrt{1+1/\lambda^2}e_0} = \mathcal{C}_{ue_0}$ for $u = t/\sqrt{t^2 - 1}$.

Exercise 1.32. Prove for example that (a) $\Rightarrow$ (c) $\Rightarrow$ (e) $\Rightarrow$ (f) $\Rightarrow$ (b) $\Rightarrow$ (g) $\Rightarrow$ (d) $\Rightarrow$ (a). The first implication is straightforward, the next two are Corollary 1.8. Other implications are simple. If $\mathcal{C}$ has a compact base, Lemma 1.6 implies that $\mathcal{C}^*$ has a $(n-1)$-dimensional base, so $\dim \mathcal{C}^* = n$. If $\mathcal{C}$ contains a line L, then $\mathcal{C}^* \subset L^\perp$ and $\operatorname{span} \mathcal{C}^* \subset L^\perp$.

Exercise 1.33. Let V be a maximal vector subspace contained in $\mathcal{C}$, then use Exercise 1.32.

Exercise 1.34. If $x \in \mathcal{C}$, then the map $\phi(z) = \langle z, x \rangle$ verifies $\phi(\mathcal{C}^*) \subset \mathbb{R}_+$ and $\{0\}$ is a face of $\mathbb{R}_+$.

Exercise 1.35. Show that if F' is a face of $\mathcal{C}$ then $\mathbb{R}_+ F' = F'$. Deduce that $F' \mapsto F' \cap \mathcal{C}^b$ is the inverse to the correspondence defined in the Proposition.

Exercise 1.36. Let $y \in C_1 \cap C_2$ define the common isolating hyperplane. The proof of the implication (a) $\Rightarrow$ (d) from Exercise 1.32 shows then that the corresponding bases of $\mathcal{C}_1^*$ and $\mathcal{C}_2^*$ are compact and hence so is their convex hull, which generates $\mathcal{C}_1^* + \mathcal{C}_2^*$ (cf. (1.23)).

Exercise 1.37. In (ii)–(iv) this is easy; in (v) consider t tending to $+\infty$ and to $-\infty$.

Exercise 1.38. The "if" direction is easy. For "only if", use induction on n. Let $x, y \in \mathbb{R}^n$ such that $x \prec_w y$. Assume for notational simplicity that $x = x^\downarrow$, $y = y^\downarrow$. Let $\delta = \min\{y_1 + \cdots + y_k - (x_1 + \cdots + x_k) \ : \ 1 \leqslant k \leqslant n\}$ ($\geqslant 0$) with the minimum achieved for $k = k_0$. Show that $(x_1 + \delta, x_2, \ldots, x_{k_0}) \prec (y_1, \ldots, y_{k_0})$ and (if $k_0 < n$) apply the induction hypothesis to the vectors $(x_{k_0+1}, \ldots, x_n)$ and $(y_{k_0+1}, \ldots, y_n)$.

Exercise 1.39. The statement and the proof are the same (simply replace everywhere "unitary" by "orthogonal").

Exercises 1.40 and 1.41. Apply Proposition 1.15.

Exercise 1.42. We may check strict concavity on lines; for A positive definite and $B \neq 0$ self-adjoint, we have $\log\det(A+tB) = \log\det(A) + \sum_i \log(1+t\lambda_i)$ where (λ_i) are the eigenvalues of $A^{-1/2}BA^{1/2}$, which is strictly concave wherever it is defined. Alternatively, use Klein's lemma and analyze the proof for equality conditions.

Exercise 1.43. This follows from the fact that, for any $X \in \mathsf{M}_n$, $\operatorname{diag} X = \operatorname{Ave} D_v X D_v$, where v varies over $\{-1,1\}^n$ endowed with normalized counting measure and D_v denotes the diagonal matrix made from the coordinates of a vector v.

Exercise 1.44. Extreme points of $S_1^{m,n}$ are of the form $|x\rangle\langle y|$, where x and y are unit vectors. Similarly, extreme points of $S_1^{m,\mathrm{sa}}$ are of the form $|x\rangle\langle x|$. Extreme points of $S_\infty^{m,n}$ are (if, say, $m \geqslant n$) the isometric embeddings of $\mathbb{R}^n$ into $\mathbb{R}^m$, in particular, for $m = n$ orthogonal matrices (resp., $\mathbb{C}^n$ into $\mathbb{C}^m$ unitary matrices). Extreme points of $S_\infty^{m,\mathrm{sa}}$ are reflections and have $m+1$ connected components (eigenvalues are ± 1), each of which can be identified with the Grassmann manifold $\mathsf{Gr}(k, \mathbb{R}^m)$ for the appropriate $k \in \{0, 1, \ldots, m\}$.

Exercise 1.45. If $X \in K = S_\infty^n$ (real or complex case), let $X = U\Sigma V^\dagger$ be the singular value decomposition with $U, V \in \mathsf{O}(n)$ (or $\mathsf{U}(n)$ in the complex case) and Σ a diagonal matrix with diagonal entries belonging to $[0,1]$. Consider the diagonal of Σ as an element of B_∞^n and apply Exercise 1.10 and Carathéodory's theorem in $\mathbb{R}^n$. Other instances of K are handled in similar way.
Applying Carathéodory's theorem directly leads to a convex combination of $m+1$ extreme points, where $m = \Theta(n^2)$ is the (real) dimension of the corresponding space of matrices.

Exercise 1.46. The set $S_1^{2,\mathrm{sa}}$ is a cylinder (whose base is the real version of the Bloch ball) and the set $S_\infty^{2,\mathrm{sa}}$ is a double-cone over a disk (see Figure E.2).

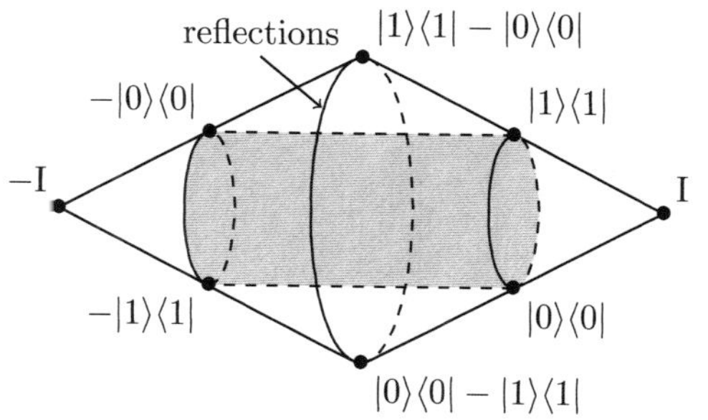

FIGURE E.2. Schatten unit balls in 2×2 real self-adjoint matrices.

Exercise 1.47. The delicate point is the triangle inequality. For $M, N \in \mathsf{M}_{m,n}$, consider $\tilde{M}, \tilde{N} \in \mathsf{M}_{m+n}^{\mathrm{sa}}$ as in Lemma 1.13. By mimicking the proof of Proposition 1.15, we obtain $\operatorname{spec}(\tilde{M} + \tilde{N}) \prec \operatorname{spec}(\tilde{M}) + \operatorname{spec}(\tilde{N})$, and therefore $s(M+N) \prec_w$

$s(M)+s(N)$. Using the result from Exercise 1.38, this implies that $\|s(M+N)\| \leqslant \|s(M)+s(N)\| \leqslant \|s(M)\|+\|s(N)\|$. For the second statement (and, say, $m=n$), consider the restriction of the norm to diagonal matrices with real entries.

Exercise 1.48. Mimic the explanation of the equality in (1.34), or use the bipolar theorem.

Exercise 1.50. Use Exercise 1.43. For the second statement, analyze the proofs for equality conditions.

Exercise 1.51. Since $S_p(\sigma) = H_p\big(\operatorname{spec}(\sigma)\big)$, it is enough to settle the commutative case. Calculate the derivative $\frac{\mathrm{d}}{\mathrm{d}p} H_p(\mathbf{q})$ and show that it equals $-\sum_i r_i \log(r_i/q_i)$ for some classical state $\mathbf{r}=(r_i)$ (depending on p). The quantity $\sum_i r_i \log(r_i/q_i)$ is called the Kullback–Leibler divergence (or relative entropy) between $\mathbf{r}$ and $\mathbf{q}$ and is always nonnegative by concavity of the logarithm.

Chapter 2

Exercise 2.1. Boundary states are states having 0 in their spectrum.

Exercise 2.2. Prove the statement by induction on d. Use the intermediate value theorem to show that the operator $\rho - \frac{1}{d}|\psi\rangle\langle\psi|$ is on the boundary of the $\mathcal{PSD}$ cone for some unit vector ψ.

Exercise 2.3. Use (2.6).

Exercise 2.4. (i) Each σ_a is a self-adjoint isometry, so its eigenvalues are ± 1. The assertion also follows formally from Exercise 2.3. (ii) It is enough to verify directly just one of the rules; the remaining ones follow then via simple algebra by repeatedly using (i).

Exercise 2.5. Hyperplanes in H_1 are described by the equation $\operatorname{Tr}(A\,\cdot\,) = t$ for some $A \in \mathsf{M}_d^{\mathrm{sa}}$ which is not a multiple of the identity (and which can be assumed to be of trace 0) and some $t \in \mathbb{R}$. For such $A \in \mathsf{M}_d^{\mathrm{sa}}$, we first note that

$$\max_{\rho \in \mathrm{D}(\mathbb{C}^d)} \operatorname{Tr}(A\rho) = \lambda_1(A),$$

so the value $t = \lambda_1$ corresponds to supporting hyperplanes (here $\lambda_1(A)$ denotes the largest eigenvalue of A). Let E be the eigenspace of A corresponding to the eigenvalue $\lambda_1(A)$. Given $\rho \in \mathrm{D}(\mathbb{C}^d)$, the condition $\operatorname{Tr}(A\rho) = \lambda_1(A)$ is equivalent to ρ having its range in E, and the result follows.

Exercise 2.6. This is an immediate consequence of the fact that $\mathrm{D}(\mathbb{C}^2)$ is linearly isometric to the unit ball of $\mathbb{R}^3$ (see Section 2.1.2).

Exercise 2.7. For $U \in \mathsf{U}(d)$, denote by Φ_U the map $\rho \mapsto U\rho U^\dagger$ and by Ψ_U the map $\rho \mapsto U\rho^T U^\dagger$. Check that the relations $\Phi_U \circ \Phi_V = \Phi_{UV}$, $\Phi_U \circ \Psi_V = \Psi_{UV}$, $\Psi_U \circ \Phi_V = \Psi_{U\overline{V}}$ and $\Psi_U \circ \Psi_V = \Phi_{U\overline{V}}$ hold for any $U, V \in \mathsf{U}(d)$.

Exercise 2.8. The statement is that isometries of the real projective space $\mathsf{P}(\mathbb{R}^n)$ are of the form $[\psi] \mapsto [O\psi]$ for some $O \in \mathsf{O}(n)$. This can be proved by induction on n since the set of points at largest distance from $[\psi]$ identifies with $\mathsf{P}(\psi^\perp)$.

Exercise 2.9. The hypothesis implies that the matrix of ρ in any orthonormal basis has real entries. Since this property remains true when one multiplies each basis element by a complex number with modulus 1, it follows that the matrix of ρ in any orthonormal basis is diagonal, and therefore $\rho = \rho_*$.

Exercise 2.10. For (i), work in the affine hyperplane of trace one self-adjoint operators, whose real dimension is $d^4 - 1$. For (ii), let $\mathrm{Seg} \subset S_{\mathbb{C}^d \otimes \mathbb{C}^d}$ be the set of product unit vectors (see (B.6)) and consider the map $\Psi : \Delta_{k-1} \times \mathrm{Seg}^k \to \mathrm{Sep}$ defined as $\Psi(\lambda, \psi_1, \dots, \psi_k) = \sum \lambda_i |\psi_i\rangle\langle\psi_i|$. Then prove that a necessary condition for the surjectivity of Ψ is that $\dim(\Delta_{k-1}) + k \dim(\mathrm{Seg}) \geqslant \dim(\mathrm{Sep})$ for an appropriate notion of dimension. One possible notion is the covering dimension of a compact metric space (X, d) defined as $\dim(X) = \liminf_{\varepsilon \to 0} \log N(X, \varepsilon)/\log(1/\varepsilon)$ where $N(X, \varepsilon)$ is the covering number defined in Section 5.1. We have $\dim(\Delta_{k-1}) = k$, $\dim \mathrm{Seg} = 4d - 3$ and $\dim \mathrm{Sep} = d^4 - 1$.

Exercise 2.11. Consider first the case $d_1 = d_2 = 2$ and let $E = \mathrm{span}(|00\rangle, |11\rangle) \subset \mathbb{C}^2 \otimes \mathbb{C}^2$. Since the only product vectors contained in E are $|00\rangle$ and $|11\rangle$, it follows that the intersection of $\mathrm{Sep}(\mathbb{C}^2 \otimes \mathbb{C}^2)$ with the hyperplane $\{A : \mathrm{Tr}(AE) = 1\}$ is the set of states of the form $\lambda|00\rangle\langle 00| + (1-\lambda)|11\rangle\langle 11|$ for $\lambda \in [0,1]$. This set is a 1-dimensional face. Deduce the case of arbitrary d_1, d_2.

Exercise 2.12. Expand all the objects with respect to the canonical bases, i.e., $|i\rangle, |i\rangle \otimes |j\rangle, |i\rangle\langle j|$ etc., as appropriate.

Exercise 2.13. Verify that $\mathrm{dist}(\psi, \mathrm{Seg}) = 2 - 2\lambda_1(\psi)$ and note that $\lambda_1(\psi)$ is minimal when ψ is a maximally entangled vector.

Exercise 2.14. The statement about the antisymmetric space follows from the relation $\mathrm{Asym}_d = \{\psi - F(\psi) : \psi \in \mathbb{C}^d \otimes \mathbb{C}^d\}$. For the symmetric space, what is clear is that $\mathrm{Sym}_d = \{\psi + F(\psi) : \psi \in \mathbb{C}^d \otimes \mathbb{C}^d\} = \mathrm{span}\{x \otimes y + y \otimes x\}$; then use the polarization formula $x \otimes y + y \otimes x = \frac{1}{2}(x+y)^{\otimes 2} - \frac{1}{2}(x-y)^{\otimes 2}$.

Exercise 2.15. (i) Write $P_E \otimes P_E$ as

$$\frac{1}{4}\left[(P_E + P_{E^\perp})^{\otimes 2} + (P_E + iP_{E^\perp})^{\otimes 2} + (P_E - P_{E^\perp})^{\otimes 2} + (P_E - iP_{E^\perp})^{\otimes 2}\right].$$

(ii) By Exercise 2.14, there are unit vectors $x, y \in \mathbb{C}^d$ such that $\langle \varphi, x \otimes x\rangle \neq 0$ and $\langle \psi, y \otimes y\rangle \neq 0$. Let $W \in \mathsf{U}(d)$ be such that $x = Wy$. By (i), $|y \otimes y\rangle\langle y \otimes y| \in \mathscr{A}$, and $V = (W \otimes W)(|y \otimes y\rangle\langle y \otimes y|)$ satisfies the desired conclusion. (iii) There are vectors $\chi = x \otimes y - y \otimes x$ and $\chi' = x' \otimes y' - y' \otimes x'$ (with $x, y, x', y' \in \mathbb{C}^d$) such that $\langle \psi, \chi\rangle \neq 0$ and $\langle \chi', \varphi\rangle \neq 0$. Denote $E = \mathrm{span}\{x, y\}$ and $E' = \mathrm{span}\{x', y'\}$ and show that necessarily $\dim E = \dim E' = 2$. Let $W \in \mathsf{U}(d)$ be such that $E' = WE$ and use $V = (W \otimes W)(P_E \otimes P_E)$. As before, $V \in \mathscr{A}$ by (i). To verify that $\langle \varphi|V|\psi\rangle \neq 0$ use the fact that $(P_E \otimes P_E)\varphi, \chi$ are all collinear (since $\dim \mathrm{Asym}_2 = 1$) and nonzero, and similarly for $V\varphi, (W \otimes W)\chi, \chi'$. (iv) First, by Exercise 2.14, both Sym_d and Asym_d are invariant under the $U \otimes U$ action of $\mathsf{U}(d)$ and hence $\mathscr{A}$-invariant. To show that they are $\mathscr{A}$-irreducible (and hence "$U \otimes U$-irreducible"), prove and use the following:

A semigroup $\mathscr{A} \subset B(\mathcal{H})$ acts irreducibly on $\mathcal{H}$ if and only if for any $\varphi, \psi \in \mathcal{H}\backslash\{0\}$ there exists $V \in \mathscr{A}$ such that $\langle \varphi|V|\psi\rangle \neq 0$.

Exercise 2.16. (i) Apply Proposition 2.9 to eigenspaces of ρ. (ii) Use (i) and the fact that VU is Haar-distributed for any fixed $V \in \mathsf{U}(d)$. (iii) Apply (ii) to $\rho = |x \otimes x\rangle\langle x \otimes x|$, where x is a fixed unit vector in $\mathbb{C}^d$.

Exercise 2.17. Convexity is easy. If $\rho = \sum \lambda_i \sigma_i \otimes \tau_i$ is separable (with $\lambda_i > 0$, $\sigma_i \in \mathrm{D}(\mathcal{H}_1)$ and $\tau_i \in \mathrm{D}(\mathcal{H}_2)$), then $\sum \lambda_i \sigma_i \otimes \tau_i^{\otimes l}$ is an l-extension of ρ. If ρ_k is a k-extension of ρ and $l < k$, taking partial trace over $k - l$ copies of $\mathcal{H}_2$ gives an l-extension.

Exercise 2.18. (i) Write $\rho = \sum \lambda_i |\chi_i\rangle\langle\chi_i|$ for $\lambda_i > 0$ and unit vectors $\chi_i \in \mathcal{H}_1 \otimes \mathcal{H}_2$. Necessarily $\mathrm{Tr}_{\mathcal{H}_2} |\chi_i\rangle\langle\chi_i| = |\psi\rangle\langle\psi|$ for all i and, by considering the Schmidt decomposition of χ_i, one sees that $\chi_i = \psi \otimes \varphi_i$ for some $\varphi_i \in \mathcal{H}_2$, hence the result. (ii) Let $\rho \in \mathrm{D}(\mathcal{H}_1 \otimes \mathcal{H}_2 \otimes \mathcal{H}_2)$ be a 2-extension of $|\psi\rangle\langle\psi|$. By (i), ρ has the form $|\psi\rangle\langle\psi| \otimes \sigma$ for some $\sigma \in \mathrm{D}(\mathcal{H}_2)$. Taking partial trace over the first copy of $\mathcal{H}_2$ shows that $|\psi\rangle\langle\psi|$ is a product state.

Exercise 2.19. If $\psi = \sum_{i=1}^d \lambda_i e_i \otimes f_i$ is the Schmidt decomposition, show that

$$|\psi\rangle\langle\psi|^\Gamma = \sum_{i,j=1}^{d} \lambda_i \lambda_j |e_j \otimes f_i\rangle\langle e_i \otimes f_j|$$

and that its spectrum is $\{\lambda_i^2 \ : \ 1 \leqslant i \leqslant d\} \cup \{\pm\lambda_i\lambda_j \ : \ 1 \leqslant i < j \leqslant d\}$.

Exercise 2.21. What are the operators ρ on $\mathcal{H}_1 \otimes \mathcal{H}_2$, for which we can be sure that $\rho^\Gamma = (V \otimes \mathrm{I})^\dagger \rho (V \otimes \mathrm{I})$? (Note that V depends on X.)

Exercise 2.22. Note that $\Gamma^2 = \mathrm{Id}$. Take $E = \{A \in B^{\mathrm{sa}}(\mathcal{H}_1 \otimes \mathcal{H}_2) \ : \ A^\Gamma = A\}$.

Exercise 2.23. $\rho_\beta^\Gamma = \frac{\beta}{d} F + (1-\beta)\frac{\mathrm{I}}{d^2}$, and therefore ρ_β is PPT if and only if $\beta \leqslant \frac{1}{d+1}$. It follows that ρ_β is entangled for $\beta > \frac{1}{d+1}$. Next, verify that $\rho_\beta = w_\lambda^\Gamma$, where w_λ is the Werner state (2.21) with $\lambda = (\beta(d^2-1)+d+1)/2d$. For $-\frac{1}{d^2-1} \leqslant \beta \leqslant \frac{1}{d+1}$, we have $\frac{1}{2} \leqslant \lambda \leqslant 1$, so w_λ is separable by Proposition 2.16. Since the partial transpose of a separable state is a separable state, the result follows.

Exercise 2.24. (i) For $\psi \in S_{\mathbb{C}^{d_1}}$ and $\varphi \in S_{\mathbb{C}^{d_2}}$, we have $|\psi \otimes \varphi\rangle\langle\psi \otimes \varphi|^R = |\psi \otimes \overline{\psi}\rangle\langle\overline{\varphi} \otimes \varphi|$; in particular $\||\psi \otimes \varphi\rangle\langle\psi \otimes \varphi|^R\|_1 = 1$. Using the triangle inequality for $\|\cdot\|_1$, it follows that $\|\rho^R\|_1 \leqslant 1$ for any separable state ρ. (ii) Let $\rho = |\chi\rangle\langle\chi|$ for $\chi \in S_{\mathbb{C}^{d_1} \otimes \mathbb{C}^{d_2}}$. Consider a Schmidt decomposition $\chi = \sum \lambda_i \psi_i \otimes \varphi_i$. We have

$$\rho^R = \sum_{i,j} \lambda_i \lambda_j |\psi_i \otimes \overline{\psi_j}\rangle\langle\varphi_i \otimes \overline{\varphi_j}|.$$

Since the families $(\psi_i \otimes \overline{\psi_j})_{i,j}$ and $(\varphi_i \otimes \overline{\varphi_j})_{i,j}$ consist of orthonormal vectors, it follows that $\|\rho^R\|_1 = \sum_{i,j} \lambda_i \lambda_j = (\sum \lambda_i)^2$, so $\|\rho^R\| > 1$ unless ρ is separable.

Exercise 2.25. Use the self-dual property (see Section 1.2.1 for more on this) of the positive semi-definite cone of operators: $A \geqslant 0$ if and only if $\mathrm{Tr}(AB) \geqslant 0$ for all $B \geqslant 0$.

Exercise 2.26. $\Phi \otimes \Psi = (\Phi \otimes \mathrm{Id}) \circ (\mathrm{Id} \otimes \Psi)$.

Exercise 2.27. Write the Choi matrix of Φ as the difference of two positive operators.

Exercise 2.28. When $\dim \mathcal{H}_1 \leqslant \dim \mathcal{H}_2$, this follows from the proof of Theorem 2.21. Otherwise, consider Φ^* to switch the roles of $\mathcal{H}_1$ and $\mathcal{H}_2$, and use Exercise 2.25 and the fact that Φ is completely positive if and only if Φ^* is completely positive.

Exercise 2.29. To show that the map Φ is not $(k+1)$-positive, consider the input operator $|\psi\rangle\langle\psi|$ for $\psi = \sum_{i=1}^{k+1} |i\rangle \otimes |i\rangle \in \mathbb{C}^n \otimes \mathbb{C}^{k+1}$. To establish k-positivity of Φ, write any $\psi \in \mathbb{C}^n \otimes \mathbb{C}^k$ as $\sum_{i=1}^k \chi_i \otimes \varphi_i$ with (φ_i) an orthonormal basis in $\mathbb{C}^k$, and argue that

$$(\Phi \otimes \mathrm{Id}_{\mathsf{M}_k})(|\psi\rangle\langle\psi|) \geqslant \sum_{i<j} |\chi_i \otimes \varphi_i - \chi_j \otimes \varphi_j\rangle\langle\chi_i \otimes \varphi_i - \chi_j \otimes \varphi_j| \geqslant 0.$$

Exercise 2.30. The unit ball in $\left(\mathsf{M}_d^{\mathrm{sa}}, \|\cdot\|_\infty\right)$ is an "order interval" $\{\tau : -\mathrm{I} \leqslant \tau \leqslant \mathrm{I}\}$ (where $\sigma \leqslant \tau$ means that $\tau - \sigma$ is positive semi-definite) and positive maps are exactly those that preserve this order.

Exercise 2.31. Use the preceding exercise and duality. Alternatively, use the fact that any $\tau \in \mathsf{M}_m^{\mathrm{sa}}$ can be written as $\tau_1 - \tau_2$ with τ_1, τ_2 positive and $\|\tau\|_1 = \|\tau_1\|_1 + \|\tau_2\|_1$.

Exercise 2.32. (i) The "only if" part follows from the preceding exercise (note that $\Phi \otimes \mathrm{Id}$ is trace-preserving if Φ is). In the opposite direction, if σ is positive and $\Phi(\sigma)$ is not, then $\|\sigma\|_1 = \operatorname{Tr}\sigma = \operatorname{Tr}\Phi(\sigma) < \|\Phi(\sigma)\|_1$. This takes care of $k = 1$, and the general case follows formally. (ii) The norm equals 2; note that the norm is necessarily attained on a pure state and use Exercise 2.19. Essentially the same argument gives k for the norm of $\Phi \otimes \mathrm{Id}$ on $\left(B^{\mathrm{sa}}(\mathbb{C}^m \otimes \mathbb{C}^k), \|\cdot\|_1\right)$. (iii) Use part (ii) and duality.

Exercise 2.33. The case $\operatorname{rank}\rho = n$ follows from Proposition 1.4 (the trace-preserving hypothesis is not needed). In the general case, argue by contradiction: let $E = \operatorname{range}(\sigma)$, $E' = \operatorname{range}(\Phi(\sigma))$, and assume that $r := \dim E > r' := \dim E'$. Next, use Propositions 2.1 and 1.4 to infer that $\Phi(\mathrm{D}(E)) \subset \mathrm{D}(E')$ and note that $r = \operatorname{Tr} P_E = \operatorname{Tr}\Phi(P_E) \leqslant r'\|\Phi(P_E)\|_\infty$, hence $\|\Phi(P_E)\|_\infty \geqslant \frac{r}{r'} > 1 = \|P_E\|_\infty$. Conclude by appealing to Exercise 2.30.

Exercise 2.34. (i) A channel is an affine map from the Bloch ball to itself; such a map is necessarily a contraction and preserves the center if and only if the channel is unital. We are allowed to compose the channel with maps $X \mapsto UXU^\dagger$ for $U \in \mathsf{U}(2)$, which correspond to rotations of the Bloch ball. This yields the desired form with $|a|, |b|, |c|$ being the singular values of the contraction. We may have a, b, c negative since we are only allowed proper rotations (from $\mathsf{SO}(3)$). (ii) follows from Theorem 2.21 after we compute explicitly the Choi matrix. For (iii), note that the inequalities for (a, b, c) obtained in part (ii) describe a tetrahedron whose vertices are $(1, 1, 1)$ (corresponding to the identity channel) and permutations of $(1, -1, -1)$ (corresponding to conjugations with Pauli matrices).

Exercise 2.35. Apply Carathéodory's theorem in the space of unital and trace-preserving superoperators.

Exercise 2.37. Check that $R(X) = \mathbf{E}\,UXU^\dagger$ with U Haar-distributed (see also Exercise 8.6), and that $D(X) = \mathbf{E}\,VXV^\dagger$ where V is a uniformly distributed among diagonal matrices with ± 1 on the diagonal.

Exercise 2.38. The condition is that $\operatorname{Tr} M_i = 1$, which implies in particular $N = \dim \mathcal{H}$. One checks then directly that $\Phi^*(\rho) = \sum M_i \langle i|\rho|i\rangle$.

Exercise 2.39. The implication (i) $\Rightarrow$ (ii) is immediate from (2.32). Assuming (ii), write $C(\Phi) = \sum |x_i \otimes y_i\rangle\langle x_i \otimes y_i|$ for $x_i \in \mathcal{H}^{out}, y_i \in \mathcal{H}^{in}$. Repeating the proof of Theorem 2.21 with this decomposition instead of (2.34) gives (iii). Finally, assuming (iii), there are positive semi-definite operators $A_i \in B(\mathcal{H}^{in})$ and $B_i \in B(\mathcal{H}^{out})$ such that $\Phi(Y) = \sum_i \operatorname{Tr}(B_i Y) A_i$ for any $Y \in B^{\mathrm{sa}}(\mathcal{H}^{in})$. Consequently, for any d and any positive operator $X \in \mathcal{B}(\mathcal{H}^{in} \otimes \mathbb{C}^d)$,

$$(\Phi \otimes \mathrm{Id}_{\mathsf{M}_d})(X) = \sum_i A_i \otimes \operatorname{Tr}_{\mathcal{H}^{in}}\left[\left(B_i^{1/2} \otimes \mathrm{I}\right) X \left(B_i^{1/2} \otimes \mathrm{I}\right)\right]$$

belongs to the separable cone, hence Φ is entanglement-breaking.

Exercise 2.40. If Φ is entanglement-breaking, write $\Phi \otimes \Psi = (\mathrm{Id} \otimes \Psi) \circ (\Phi \otimes \mathrm{Id})$ and use the fact that the product superoperator $\mathrm{Id} \otimes \Psi$ maps the separable cone to the separable cone.

Exercise 2.41. By (2.32), $C(\Phi)^\Gamma$ is positive whenever Φ is PPT-inducing. Conversely, assume that $C(\Phi)^\Gamma$ is positive. It is enough to show that $(\Phi \otimes \mathrm{Id}_{\mathsf{M}_d})(\rho)$ has a positive partial transpose for every pure state $\rho = |\psi\rangle\langle\psi|$, with $\psi \in \mathcal{H}^{in} \otimes \mathbb{C}^d$. Denoting $\chi = \sum e_i \otimes e_i \in \mathcal{H}^{in} \otimes \mathcal{H}^{in}$, we may write $\psi = (\mathrm{I} \otimes B)\chi$ for some $B \in B(\mathcal{H}^{in}, \mathbb{C}^d)$. It follows that

$$(\Phi \otimes \mathrm{Id})(\rho) = (\Phi \otimes \mathrm{Id})\left[(\mathrm{I} \otimes B)|\chi\rangle\langle\chi|(\mathrm{I} \otimes B^\dagger)\right] = (\mathrm{I} \otimes B)C(\Phi)(\mathrm{I} \otimes B^\dagger)$$

has a positive partial transpose.

Exercise 2.42. Prove that $A, B \geqslant 0$ implies $A \odot B \geqslant 0$ ($A \odot B$ is a submatrix of $A \otimes B$). Use also the fact that $\Theta_A \otimes \mathrm{Id}_{\mathsf{M}_k} = \Theta_{A \otimes J}$, where J is the matrix with all entries equal to 1.

Exercise 2.43. Observe that if $a \in \mathbb{C}^n$ and $D = D_a$ (i.e., the diagonal matrix with $D_{ii} = a_i$), then $DXD^\dagger = \Theta_{|a\rangle\langle a|}(X)$ for any $X \in \mathsf{M}_d$.

Exercise 2.44. The map Φ is completely positive, and trace-preserving because $\sum A_i^\dagger A_i = \mathrm{I}_{\mathbb{C}^2 \otimes \mathbb{C}^2}$. It is also obvious from the definition that Φ is a separable channel. Assume now that Φ can be written as a convex combination of product channels of the form $\sum \lambda_j \Psi_j \otimes \Xi_j$ with $\lambda_j > 0$ and $\sum \lambda_j = 1$. The pure product states $|0\rangle\langle 0| \otimes |0\rangle\langle 0|$ and $|1\rangle\langle 1| \otimes |1\rangle\langle 1|$ are mapped to themselves under Φ. It follows that for every j, $\Psi_j(|0\rangle\langle 0|) = \Xi_j(|0\rangle\langle 0|) = |0\rangle\langle 0|$ and $\Psi_j(|1\rangle\langle 1|) = \Xi_j(|1\rangle\langle 1|) = |1\rangle\langle 1|$. This leads to a contradiction since $\Phi(|0\rangle\langle 0| \otimes |1\rangle\langle 1|) = |0\rangle\langle 0| \otimes |0\rangle\langle 0|$.

Exercise 2.45. If $\{A_i^{(1)}\}$ are Kraus operators for Φ_1 and $\{A_j^{(2)}\}$ are Kraus operators for Φ_2, the family $\{A_i^{(1)} \otimes \mathrm{I}\} \cup \{\mathrm{I} \otimes A_j^{(2)}\}$ are Kraus operators for $\Phi_1 \oplus \Phi_2$.

Exercise 2.46. Prove that Φ is co-completely positive iff $T \circ \Phi$ is completely positive iff $\Phi \circ T$ is completely positive, where T denotes the transposition superoperator.

Exercise 2.47. Prove that, for $\Phi \in B(\mathsf{M}_m^{\mathrm{sa}}, \mathsf{M}_n^{\mathrm{sa}})$ and $\Psi \in B(\mathsf{M}_n^{\mathrm{sa}}, \mathsf{M}_m^{\mathrm{sa}})$, we have

$$\mathbf{Tr}(\Psi \circ \Phi) = \mathrm{Tr}(C(\Phi)FC(\Psi)F^*)$$

where $F : \mathbb{C}^m \otimes \mathbb{C}^n \to \mathbb{C}^m \otimes \mathbb{C}^n$ is the flip, and use self-duality of $\mathcal{PSD}$.

Exercise 2.48. A superoperator $\Phi : \mathsf{M}_m^{\mathrm{sa}} \to \mathsf{M}_n^{\mathrm{sa}}$ is k-positive if $\Phi \otimes \mathrm{Id}_{\mathsf{M}_k}$ is positive, i.e., if $\langle y|(\Phi \otimes \mathrm{Id})(|x\rangle\langle x|)|y\rangle \geqslant 0$ for any $x \in \mathbb{C}^m \otimes \mathbb{C}^k$ and $y \in \mathbb{C}^n \otimes \mathbb{C}^k$. Writing $x = \sum x_i \otimes |i\rangle$ and $y = \sum y_i \otimes |i\rangle$ for $x_i \in \mathbb{C}^m$ and $y_i \in \mathbb{C}^n$, this condition becomes

$$\sum_{i,j=1}^{k} \langle y_i|\Phi(|x_i\rangle\langle x_j|)|y_j\rangle \geqslant 0. \tag{E.2}$$

If we expand x_i as $x_i = \sum_l \langle e_l, x_i \rangle e_l$, where (e_l) is the basis in $\mathbb{C}^m$ used in the definition of the Choi matrix, (E.2) becomes

$$\sum_{i,j=1}^{k} \langle y_i \otimes \overline{x_i}|C(\Phi)|y_j \otimes \overline{x_j}\rangle \geqslant 0,$$

which is equivalent to $\langle \psi|C(\Phi)|\psi\rangle \geqslant 0$ for any $\psi \in \mathbb{C}^n \otimes \mathbb{C}^m$ with Schmidt rank at most k. This shows that (1) is equivalent to (2). The equivalence between (2) and (3) follows from the fact that a vector $x \in \mathbb{C}^m \otimes \mathbb{C}^n$ has Schmidt rank at most k iff it can be written as $x = (A \otimes B)y$, where $y \in \mathbb{C}^k \otimes \mathbb{C}^k$, $A \in \mathsf{M}_{k,m}$ and $B \in \mathsf{M}_{k,n}$.

Exercise 2.49. The "only if" part follows from Proposition 2.29. For the "if" part, argue first that if Φ is rank-preserving, then it maps the interior of $\mathcal{PSD}$ into itself. Next, if there was a positive definite operator τ that was not in the image of the interior of $\mathcal{PSD}$ under Φ, then some point of the segment connecting τ and $\Phi(\mathrm{I})$ would be of the form $\Phi(\sigma)$ for some $\sigma \in \partial\mathcal{PSD}$, in particular $\operatorname{rank}\sigma < n = \operatorname{rank}\Phi(\sigma)$. Infer that Φ is a bijection of the interior of $\mathcal{PSD}$ onto itself and conclude that it is an automorphism of $\mathcal{PSD}$.

Exercise 2.50. Start by showing that if Φ belongs to the interior of $\boldsymbol{P}$, then $\Phi(\mathrm{D}) \cap \partial\mathcal{PSD} = \varnothing$. Next, consider λ_n (the smallest eigenvalue) as a function on $\Phi(\mathrm{D})$.

Exercise 2.51. The condition is equivalent to $\langle\Phi(\rho),\sigma\rangle_{\mathrm{HS}} \geqslant \delta \operatorname{Tr}\rho \operatorname{Tr}\sigma$ for all $\rho,\sigma \in \mathcal{PSD}$.

Exercise 2.52. (a) In the language of the second proof of Theorem 2.36, if $\|R\|_\infty = 1$, then $R(S^2) \cap S^2$ consists of at least 2 points. On the other hand, there are nontrivial ellipsoids contained in B_2^3 that intersect S^2 only at one point. For a concrete example, consider $\rho \mapsto \frac{1}{2}(\rho + |0\rangle\langle 0|)$ for a state $\rho \in \mathrm{D}(\mathbb{C}^2)$ (b) Any unitary channel (even the identity channel!) will do.

Exercise 2.53. Same example as in Exercise 2.52 (a).

Exercise 2.54. If $\rho \in \mathrm{D}(\mathbb{C}^m \otimes \mathbb{C}^n)$ is an entangled state, let Φ be given by Theorem 2.34. Note that, for $\varepsilon > 0$ small enough, the map $\Phi' : X \mapsto \Phi(X) + \varepsilon(\operatorname{Tr} X)\mathrm{I}$ also satisfies the conclusions of Theorem 2.34. Finally, consider $\Psi : X \mapsto \Phi'(\mathrm{I})^{-1/2}\Phi'(X)\Phi'(\mathrm{I})^{-1/2}$.

Exercise 2.55. (i) Let $A = \Phi^*(\mathrm{I})$; then $\operatorname{Tr}\Phi(\rho) = \operatorname{Tr}(A\rho)$ for all $\rho \in \mathsf{M}_m^{\mathrm{sa}}$. (ii) We may assume that $A = \Phi^*(\mathrm{I})$ is positive definite and satisfies $\Phi^*(\mathrm{I}) \leqslant \mathrm{I}$. (iii) Set $B = (\mathrm{I} - A)^{1/2}$, define $\tilde{\Phi} : \mathsf{M}_m^{\mathrm{sa}} \to \mathsf{M}_{m+n}^{\mathrm{sa}}$ by $\tilde{\Phi}(\rho) = B\rho B \oplus \Phi(\rho)$, and verify that $\tilde{\Phi}$ is trace-preserving. (iv) Verify that $\tilde{\Phi}(\rho) \geqslant 0$ if and only if $\Phi(\rho) \geqslant 0$, and that the same is true for any extensions $\Phi \otimes \mathrm{Id}_{\mathcal{K}}$ and $\tilde{\Phi} \otimes \mathrm{Id}_{\mathcal{K}}$. (v) Deduce from (iv) that $\tilde{\Phi}$ preserves positivity and that it detects entanglement of a state ρ (in the sense of Theorem 2.34) iff Φ does.

Exercise 2.56. Let $\sigma \in \mathcal{BP}\backslash\mathcal{PSD}$, and $\tau \in \mathcal{BP}$ such that $\mathrm{E}(\sigma) \subset \mathrm{E}(\tau)$. We may apply Lemma C.4 with $F = -\tau$ and $G = -\sigma$ and conclude that $\mu\sigma - \tau \in \mathcal{PSD}$ for some $\mu \geqslant 0$ (in fact, $\mu > 0$). Since we may write $\mu\sigma = (\mu\sigma - \tau) + \tau$, the assumption that σ lies on an extreme ray forces τ to be proportional to σ.

Chapter 4

Exercise 4.1. Write $B = K \cap H = L \cap H$ and take a unit vector $u \perp H$. After applying to K and L linear transformations fixing H, we may assume that $u \in K$ and that $\max\{\langle u, x\rangle \ : \ x \in K\} = 1$, and similarly for L. Under these hypotheses, we have

$$\operatorname{conv}\{B \ \pm u\} \subset K \subset 2B \times [-u, u] \subset 3\operatorname{conv}\{B, \pm u\}$$

(same for L) which gives the result with $C = 9$. The result appears in [**Las08**] and we are not aware of a known better value for C.

Exercise 4.2. The inclusion $K \subset -n\Delta$ is shown by a variational argument. To prove the other inclusion we show that every point $a \notin (n+1)\Delta$ would form, together with one of the faces of Δ, a simplex of volume larger than that of Δ.

Exercise 4.3. Let (K_k) a sequence of convex bodies in $\mathbb{R}^n$. By Exercise 4.2, we may assume (applying invertible affine transformations if necessary) that $\Delta_n \subset K_k \subset (n+1)\Delta_n$. Then apply Ascoli's theorem to extract from $(\|\cdot\|_{K_k})$ a subsequence converging uniformly on S^{n-1}.

Exercise 4.4. We have $\frac{1}{2}(K-K) \subset K_{\cup} \subset K-K$ and $K-K$ does not depend on the choice of the origin.

Exercise 4.5. We know from (1.14) that $(K_{\oslash})^\circ = K^\circ \cap (-K^\circ)$. Then check that $x \in K^\circ \iff x - \frac{e}{|e|^2} \in -\mathcal{C}^*$.

Exercise 4.6. $\mu_K = \sum_{i=1}^p |u_i|\delta_{u_i/|u_i|}$ (replace δ_x by $\frac{1}{2}(\delta_x + \delta_{-x})$ to obtain an even measure).

Exercise 4.7. If P is a polygon whose edges are the segments $\pm S_1, \dots, \pm S_k$, then P is a translate of $S_1 + \cdots + S_k$. (This can also be checked by induction.) The result for zonoids follows by approximation.

Exercise 4.8. Prove that every face of a zonotope is a zonotope.

Exercise 4.9. No. For every partition of S^{n-1} as $A_1 \cup A_2$, the convex bodies K_1, K_2 defined for $x \in \mathbb{R}^n$ by

$$\|x\|_{K_i} = \int_{A_i} |\langle x, \theta\rangle| \, d\sigma(\theta)$$

are such that $K_1 + K_2$ is a multiple of a Euclidean ball.

Exercise 4.10. For the first statement, use Exercise 1.2. For the second statement, try $K = [0,1] \subset \mathbb{R}$ and $K' = \{(x,y) \in \mathbb{R}^2 \ : \ x \geqslant 0, \ y \geqslant 0, \ xy \geqslant 1\}$ (cf. the hint to Exercise 1.30).

Exercise 4.11. Straightforward from the definitions.

Exercise 4.12. Start by noticing that if $\{x_i\} \subset K$ is a basis of V and $\{x'_j\}$ is a basis of V', then $\{\pm x_i \mathbin{\widehat{\otimes}} x'_j\} \subset K \mathbin{\widehat{\otimes}} V'$.

Exercise 4.13. Let $d = \dim(V_1 \mathbin{\widehat{\otimes}} V_2)$. If $0 \in V_1$ and $0 \in V_2$, then $V_1 \mathbin{\widehat{\otimes}} V_2 = V_1 \otimes V_2$ and $d = \dim V_1 \dim V_2$. If, say, $0 \in V_1$ and $0 \notin V_2$, then $d = \dim V_1(\dim V_2 + 1)$. If $0 \notin V_1$ and $0 \notin V_2$, then $d = (\dim V_1 + 1)(\dim V_2 + 1) - 1$. The first is easy; the second follows, e.g., from Exercise 4.12. For the third, consider first the case when $V_j = e_{n_j} + \mathbb{R}^{n_j - 1}$ and then appeal to Exercise 4.11 (or, alternatively, use the approach from the paragraph following (4.13)).

Exercise 4.14. The part that is not obvious is that $\operatorname{conv}\{x \otimes x' \ : \ x \in \mathcal{C}, x' \in \mathcal{C}'\}$ is closed. Consider first the case when $\mathcal{C}, \mathcal{C}'$ are pointed and hence admit (by Exercise 1.32) compact bases, which allows appealing to Exercise 4.10. Next, use Exercise 1.33 and Exercise 4.12.

Exercise 4.15. Use Exercise 4.10 and then (to show full-dimensionality) the polarization formula

$$\frac{1}{2}\big((x+y) \otimes (x'+y') + (x-y) \otimes (x'-y')\big) = x \otimes x' + y \otimes y'.$$

To show full-dimensionality in the affine setting use the same ideas as in Exercises 4.12 and 4.13 to establish that the relative interior of $K_1 \mathbin{\widehat{\otimes}} K_2$ is nonempty.

Exercise 4.16. (i) The unit ball in $\ell_1^k(X)$, where X is the normed space whose unit ball is K. ($\ell_1^k(X)$ is the space X^k equipped with the norm $\|(x_1, \dots, x_k)\| = \|x_1\|_K + \cdots + \|x_k\|_K$.) (ii) Use the formula $\operatorname{vol}(L) = \frac{1}{n!}\int_{\mathbb{R}^n} \exp(-\|x\|_L)\, dx$, valid for any symmetric convex body $L \subset \mathbb{R}^n$.

Exercise 4.17. It is clear that any extreme point must be of the claimed form. Conversely, given extreme points $x \in K$, $x' \in K'$, let ϕ and ϕ' be supporting functionals, i.e., $\phi \leqslant 1$ on K with $\phi(x) = 1$ (and similarly for ϕ'). Given a decomposition $x \otimes x' = \sum \lambda_i x_i \otimes x'_i$, show that we may assume that $\phi(x_i) = \phi'(x'_i) = 1$. Now, if $x_i \neq x$ for some i, consider a linear functional ψ such that $\psi(x) > \sup\{\psi(x_i) : x_i \neq x\}$, and obtain a contradiction by computing $(\psi \otimes \phi')(x \otimes x')$.

Exercise 4.18. Straightforward from the definitions.

Exercise 4.19. Calculate $\mathrm{Tr}(T\,\mathrm{I}_n)$ by using (4.16).

Exercise 4.20. If (x_i, c_i) is a resolution of identity, then for any $x \in \mathbb{R}^n$ we have $\sum c_i \langle x, x_i \rangle^2 = |x|^2$ and $\sum c_i = n$, thus $\max_i |\langle x, x_i \rangle| \geqslant |x|/\sqrt{n}$.

Exercise 4.21. If (x_i, c_i) is an unbiased resolution of identity, then for any $x \in \mathbb{R}^n$ we have $\sum c_i \langle x, x_i \rangle^2 = |x|^2$, $\sum c_i \langle x, x_i \rangle = 0$, $\langle x, x_i \rangle \geqslant -|x|$, and $\sum c_i = n$. All this together implies $\max_i \langle x, x_i \rangle \geqslant |x|/n$.

Exercise 4.22. Use Carathéodory's theorem.

Exercise 4.23. We have $\mathrm{John}(B_\infty^n) = B_2^n$ and $\mathrm{Löw}(B_\infty^n) = \sqrt{n} B_2^n$. If $\mathscr{E} \subset B_\infty^n \subset \alpha \mathscr{E}$ for some ellipsoid $\mathscr{E}$, the extremal volume property implies $\alpha \geqslant \sqrt{n}$.

Exercise 4.24. $\mathsf{G}'_{\mathrm{unc}}$ consists of all diagonal matrices. If Δ is a diagonal matrix and P a permutation matrix, what are ΔP and $P \Delta$?

Exercise 4.25. (i) B_p^n is permutationally symmetric. (ii) Isometries of $S_p^{m,n}$ include maps $X \mapsto UXV$ for U, V orthogonal/unitary matrices; it follows that S_p^n has enough symmetries. (iii) Any isometry of $S_p^{n,\mathrm{sa}}$ preserves $\mathbb{R}\,\mathrm{I}$ (indeed, $\pm n^{-1/p}\,\mathrm{I}$ can be characterized as isolated points in the set of elements of the largest (for $p > 2$) or smallest (for $p < 2$) Hilbert–Schmidt norm in $\partial S_p^{n,\mathrm{sa}}$) so $S_p^{n,\mathrm{sa}}$ does not have enough symmetries. (iv) Isometries include $X \mapsto \pm UXU^\dagger$ for U orthogonal/unitary; there are enough symmetries. (v) Isometries of the regular simplex are obtained from permutations of its vertices; it has enough symmetries. (vi) See Theorem 2.3; $\mathrm{D}(\mathbb{C}^d)$ has enough symmetries.

Exercise 4.26. (i) Choosing for K a regular p-gon fails since the isometry group is a dihedral group. However, it is possible to slightly modify K to obtain the required isometry group, for example

$$K = \mathrm{conv}(\{R^k(1,0) \ : \ 0 \leqslant k \leqslant p-1\} \cup \{R^k(1,\varepsilon) \ : \ 0 \leqslant k \leqslant p-1\})$$

for $\varepsilon > 0$ small enough. (ii) Consider $L = K \mathbin{\widehat{\otimes}} B_1^n$ where K is the convex body from (i). We claim that the isometries of L have the form $\sum_{i=1}^n U_i \otimes |e_{\sigma(i)}\rangle\langle e_i|$ for some $\sigma \in \mathfrak{S}_n$ and $U_1, \dots, U_n \in \mathrm{Iso}(K)$, (e_i) being the canonical basis of $\mathbb{R}^n$. Indeed an isometry of L induces an isometry on the set $M = \{R^k(1,\varepsilon) \otimes e_i \ : \ 0 \leqslant k \leqslant p-1, 1 \leqslant i \leqslant n\}$ (the set of points in K farthest from the origin) and one checks that it must be of the announced form. It follows that L does not have enough symmetries (since $U \otimes \mathrm{I}$ commutes with $\mathrm{Iso}(L)$ for any $U \in \mathsf{SO}(2)$). On the other hand, there is no invariant subspace (if $x = \sum x_i \otimes e_i \in \mathbb{R}^2 \otimes \mathbb{R}^n$, one checks that for any j the vector $x_j \otimes e_j$ belongs to the $\mathrm{span}\{Vx \ : \ V \in \mathrm{Iso}(L)\}$, and therefore the orbit of x spans $\mathbb{R}^2 \otimes \mathbb{R}^n$ whenever $x \neq 0$).

Exercise 4.27. Isometries of $K \mathbin{\widehat{\otimes}} L$ include the maps $A \otimes B$ for $A \in \mathrm{Iso}(K)$ and $B \in \mathrm{Iso}(L)$. We claim that a linear map $S \in B(\mathbb{R}^m \otimes \mathbb{R}^n)$ which commutes with all such maps is a multiple of identity; this follows from the fact that, for every $y, y' \in \mathbb{R}^n$, the map $S_{y,y'}$ defined by the relation $\langle S_{y,y'}(x), x' \rangle = \langle S(x \otimes y), x' \otimes y' \rangle$

(for $x, x' \in \mathbb{R}^m$) commutes with $\mathrm{Iso}(K)$, and similarly with the role of both factors exchanged.

Exercise 4.28. We have $\mathrm{John}(\sqrt{n}B_1^n) = B_2^n$. The John ellipsoid of $\sqrt{n}B_1^n \hat{\otimes} \sqrt{n}B_1^n$ (which identifies with $nB_1^{n^2}$) is $\mathscr{E} = B_2^n \otimes_2 B_2^n$. The John ellipsoid of $B_2^n \hat{\otimes} B_2^n$ (which identifies with $S_1^{n,n}$) is $\frac{1}{\sqrt{n}}\mathscr{E}$.

Exercise 4.30. By "globally equivalent" we mean that the validity of *all* instances of (4.22) implies the validity of all instances of(4.21), and *vice versa.* To derive (4.21) from (4.22), appeal to the arithmetic mean/geometric mean inequality. To recover (4.22), apply (4.21) to $K/\mathrm{vol}(K)^{1/n}$ and $L/\mathrm{vol}(L)^{1/n}$ with $t = \mathrm{vol}(K)^{1/n}/(\mathrm{vol}(K)^{1/n} + \mathrm{vol}(L)^{1/n})$.

Exercise 4.31. Given convex bodies $K_1, K_2 \in \mathbb{R}^n$, consider $K = \mathrm{conv}(K_1 \times \{0\}, K_2 \times \{1\}) \subset \mathbb{R}^{n+1}$. Then $K \cap (\mathbb{R}^n \times \{\lambda\})$ corresponds to $\lambda K_2 + (1-\lambda)K_1$.

Exercise 4.32. Use the formula $\det(A \otimes B) = \det(A)^n \det(B)^m$ for $A \in \mathsf{M}_m$ and $B \in \mathsf{M}_n$.

Exercise 4.34. If $f = \exp(\varphi)$ is the density of μ, take $(1+\varphi/s)_+^s$ as the density of μ_s.

Exercise 4.35. To show that 1. implies 2., define

$$K = \bigcup_{x \in \mathrm{supp}\,\mu} \{x\} \times f(x)^{1/s} L,$$

where L is any convex body of volume 1 in $\mathbb{R}^s$. Conversely, apply the Brunn–Minkowski inequality in $\mathbb{R}^s$ to deduce that the function $x \mapsto \mathrm{vol}_s((\{x\} \times \mathbb{R}^s) \cap K)^{1/s}$ is concave.

Exercise 4.36. Is μ is log-concave, take (μ_s) as in Lemma 4.12 and show (4.28) for μ_s instead of μ by using Lemma 4.13 and (4.21) applied in $\mathbb{R}^{n+s}$. Conversely, apply (4.28) with K and L being balls of radius tending to 0 to prove that μ is log-concave. Note that the density f satisfies $f(x) = \lim_{\varepsilon \to 0} \mu(B(x,\varepsilon))/\mathrm{vol}(B(x,\varepsilon))$ for almost all $x \in \mathbb{R}^n$ (see Chapter 3, Theorem 1.4 in [**SS05**]).

Exercise 4.37. If the origin is not an interior point of K, then $w(K^\circ) = \infty$. Otherwise, for every $u \in S^{n-1}$ we have $1 \leqslant (w(K,u)w(K^\circ,u))^{1/2}$. Integrate over u and use the Cauchy–Schwarz inequality.

Exercise 4.38. Inradii and outradii are easy to compute. To compute $\mathrm{vol}(B_1^n)$, note that it is the union of 2^n essentially disjoint simplices, each with volume $1/n!$.

Exercise 4.39. Show and use $B_\infty^n \subset n^{1/p} B_p^n \subset nB_1^n$ and $\big(\mathrm{vol}(nB_1^n)/\mathrm{vol}(B_\infty^n)\big)^{1/n} = n/(n!)^{1/n} \sim e$.

Exercise 4.40. Integrate in Cartesian and polar coordinates, and appeal to (4.26).

Exercise 4.41. Observe that if $x \neq y$, then $B(x,r) \cap B(y,r) \subset B(\frac{x+y}{2}, r')$ for some $r' < r$.

Exercise 4.42. Consider rectangles of height 1 and width ε with $\varepsilon \to 0$.

Exercise 4.43. K contains a segment I with length $\ell = \mathrm{diam}\, K$, so $\kappa_n w(K) = w_G(K) \geqslant w_G(I) = \frac{1}{2}\kappa_1 \ell$. It is an interesting question whether we always have $w(K) \geqslant \frac{\kappa_1}{\kappa_n} \mathrm{outrad}(K)$. In other words, we are asking whether among all sets of given outradius R the segment of length $2R$ has the minimal mean width, which doesn't readily follow from the known results on sets for which—under certain constraints—the mean width is extremal (see, e.g., [**Bal91, Sch99, Bar98**]). The above question is equivalent to the following inequality (see Appendix A.2 for definitions): *If* $X_1, X_2, \ldots, X_N$ *are jointly Gaussian* $N(0,1)$*-distributed random variables such that, for some positive scalars* $t_1, t_2, \ldots, t_N$ *we have* $\sum_k t_k X_k = 0$*, then* $\mathbf{E} \max_k X_k \geqslant \mathbf{E}\,|X_1|$.

Exercise 4.44. Show the inequality for symmetric polygons by induction on the number of edges, then use symmetrization $K \mapsto K - K$ and approximation.

Exercise 4.45. If G is a standard Gaussian vector in $\mathbb{R}^n$, then $\mathbf{E}\, w(K, P_E G) \leqslant \mathbf{E}\, w(K, G)$ by Jensen's inequality.

Exercise 4.46. We can assume A linear. Use the classical fact ([**Hal82**], Problem 177) that any $A \in \mathsf{M}_n$ with $\|A\|_\infty \leqslant 1$ can be written as the top left block of an orthonormal matrix $O \in \mathsf{M}_{2n}$ to reduce to the case where A is an orthogonal projection, which is covered by Exercise 4.45.
The same assertion holds in fact for *any* contraction (i.e., not necessarily affine), see Proposition 6.6 and the comments following it. However, this change of generality makes the result much more subtle.

Exercise 4.47. By translation invariance, we can assume that $0 \in K \cap L$, so that the functionals $w(K, \cdot)$ and $w(L, \cdot)$ are nonnegative. Then $w(K \cup L, \cdot) = \max(w(K, \cdot), w(L, \cdot)) \leqslant w(K, \cdot) + w(L, \cdot)$.

Exercise 4.48. By modifying the proof of Proposition 4.16 show that

$$\left(\int_{S^{n-1}} \|\theta\|_K^p \,\mathrm{d}\sigma(\theta) \right)^{1/p} \mathrm{vrad}(K) \geqslant 1 \quad \text{for any } p > 0,$$

then let $p \to 0$. The inequality and the argument appear in Appendix A of [**Sza05**], but were likely known earlier.

Exercise 4.49. (i) When the measure μ is purely atomic with N atoms, the result can be proved by induction on N, the case $N = 2$ being exactly the Brunn–Minkowski inequality (4.22). The continuous case can then be derived by approximation. Minkowski integrals of convex bodies are defined via their support functions, so that inequality (4.37) makes sense whenever the map $t \mapsto w(K_t, \theta)$ is measurable for any $\theta \in \mathbb{R}^n$. (ii) In that case, $\mathrm{vol}(K_t) = \mathrm{vol}(K)$ for any $t \in \mathsf{O}(n)$. By invariance of the Haar measure (see Appendix B.3), the convex body $L := \int_{\mathsf{O}(n)} t(K)\,\mathrm{d}\mu(t)$ is necessarily a Euclidean ball centered at the origin. By computing the width of L in a fixed direction, we obtain that L is a Euclidean ball of radius $w(K)$, showing the result.

Exercise 4.50. We have $\mathrm{vrad}(K) \leqslant \mathrm{vrad}(K^\circ)^{-1} \leqslant w(K)$.

Exercise 4.51. In the symmetric case, combine the results from Proposition 4.15, Proposition 4.16 and Theorem 4.17. In the general case, sufficient conditions are that (i) 0 is the center of the largest Euclidean ball contained in K and (ii) 0 is the centroid of K (for Santaló's inequality to hold). These conditions are both satisfied whenever 0 is the unique fixed point under $\mathrm{Iso}(K)$.

Exercise 4.53. By Fubini theorem, we have

$$\mathrm{vol}(K) \leqslant \mathrm{vol}_F(P_F K) \max_{x \in F} \mathrm{vol}_E(K \cap (E + x))$$

and the convexity and symmetry of K imply that the maximum is achieved for $x = 0$.

Exercise 4.54. Apply Lemma 4.20 to the convex body $K \times K \subset \mathbb{R}^{2n}$ and to the pair of orthogonal subspaces $E = \{(x, x) \,:\, x \in \mathbb{R}^n\}$ and $F = \{(x, -x) \,:\, x \in \mathbb{R}^n\}$.

Exercise 4.55. The lower inequality follows from (4.21). For the upper inequality, assume $h = 1$ and apply Lemma 4.20 to the convex body

$$L = \{(\lambda x, (1-\lambda)y, \lambda) \; : \; x \in K, \; y \in K, \; \lambda \in [0,1]\} \subset \mathbb{R}^{2n+1}$$

and to the pair of subspaces $E = \{(x, x, 1/2) \; : \; x \in \mathbb{R}^n\}$, $F = \{(x, -x, t) \; : \; x \in \mathbb{R}^n, \; t \in \mathbb{R}\}$. We have $\mathrm{vol}(L) = \frac{n!^2}{(2n+1)!}\mathrm{vol}(K)^2$, $\mathrm{vol}(K \cap E) = 2^{-n/2}\mathrm{vol}(K)$ and $\mathrm{vol}(P_F K) = 2^{-n/2}\mathrm{vol}(K_{\diamond})$.

Exercise 4.56. Apply Lemma 4.20 to the convex body $L = (K \times \{1\})_{\diamond} \subset \mathbb{R}^{n+1}$ with E being the line generated by $(0, \ldots, 0, 1)$.

Exercise 4.57. Use Theorem 4.21.

Exercise 4.58. Consider $f, g, h : \mathbb{R}^n \to [0, \infty]$ verifying (4.46). For $s \in \mathbb{R}$, define $f_s : \mathbb{R}^{n-1} \to [0, \infty]$ by $f_s(z) = f(s, z)$ and similarly for g_s, h_s. If $t = \lambda u + (1-\lambda)v$, check that h_t, f_u, and g_v verify the $(n-1)$-dimensional instance of (4.46). Deduce that $\tilde{f}(s) = \int_{\mathbb{R}^{n-1}} f_s(z)\, dz$ and similarly defined $\tilde{g}, \tilde{h}$ verify the 1-dimensional instance of (4.46), and conclude by appealing to that instance.

Exercise 4.59. (i) Let $\alpha = f(0)$. Write $\int x^2 f(x)\, dx = 2\int_0^\infty 2t\mathbf{P}(Y \geqslant t)\, dt$, where Y is a random variable with density f. Show that the log-concavity hypothesis implies that $\mathbf{P}(X \geqslant t) \leqslant \mathbf{P}(Y \geqslant t) \leqslant \mathbf{P}(Z \geqslant t)$, where X is uniformly distributed on $[-1/2\alpha, 1/2\alpha]$ and Z has a symmetric exponential distribution with density $\alpha \exp(-2\alpha|t|)\, dt$. (ii) Reduce to $\lambda = 1$ by considering $L = \lambda^{-1}K$. Assume that $H = u^{\perp}$ for a unit vector u. For $\lambda = 1$, the function

$$f : t \mapsto (\mathrm{vol}_n K)^{-1} \mathrm{vol}_{n-1}\Big(K \cap \{\langle \cdot, u\rangle = t\}\Big)$$

satisfies the hypotheses from (i); log-concavity is given by Lemma 4.13 and Exercise 4.34.

Exercise 4.60. Use the inclusions $\frac{1}{\sqrt{2}}K \subset B_2^k \subset K$ for $K = B_2^k \times B_2^{n-k}$ and $L \subset B_2^n \subset \sqrt{2}L$ for $L = K^\circ = \mathrm{conv}\{B_2^k \times \{0\}, \{0\} \times B_2^{n-k}\}$, which correspond to equality cases/extreme cases of Lemmas 4.19 and 4.20.

Chapter 5

Exercise 5.1. To obtain some of the strict inequalities, consider $K = \{0, 1\} \subset [0, 1]$.

Exercise 5.2. This is an immediate consequence of Cauchy's integral formula for surface area (see [**Sch14**], Chapter 5.3). Alternatively, an elementary argument can be given as follows: consider the map $\phi : \mathbb{R}^n \to K$ which maps x to the closest point to x in K. It is easy to check that (i) ϕ is a contraction, (ii) ϕ maps ∂L onto ∂K. It follows that ϕ decreases the surface area.

Exercise 5.3. Let $K = \mathrm{conv}(C(x,t))$. We have $\mathrm{outrad}\, K = \sin t$. Let L be the n-dimensional half-ball with center x and radius $\sin t$, such that $K \subset L$ (see Figure 5.2). Comparing the areas of K and L using Exercise 5.2 gives the result. To prove the second part, check the inequality $\cos u \leqslant e^{-u^2/2}$ for $|u| < \pi/2$ (take logarithm of both sides and then differentiate).

Exercise 5.4. Use Exercise 5.2 with $L = B_2^n$ and $K = B_2^n \backslash \mathrm{conv}(C(x,t))$. This gives $\mathrm{area}(S^{n-1})V(t) \geqslant \sin(t)^{n-1}\mathrm{vol}_{n-1}(B_2^{n-1})$, which is equivalent to the lower bound in (5.4). To get the upper bound, compare the solid cap with the circumscribed solid cone whose base is the same as that of the cap. For the strengthened lower bound, consider an inscribed cone. See Figure E.3.

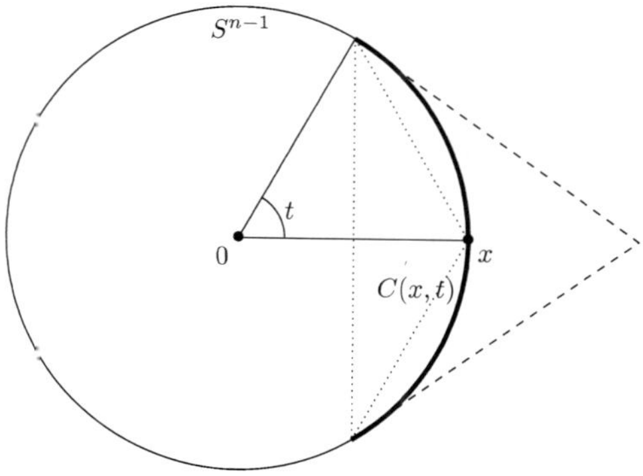

FIGURE E.3. Upper bound (dashed) and lower bounds (dotted) on the volume of a spherical cap.

Exercise 5.5. The problem is equivalent to showing that $r \mapsto r\frac{V'(r)}{V(r)}$ is nonincreasing. After some elementary manipulations, the inequality to verify becomes

$$\frac{1}{r}\int_0^r \left(\frac{\sin(ut)}{\sin(ur)}\right)^{n-2} dt \leqslant \frac{1}{r}\int_0^r \left(\frac{\sin t}{\sin r}\right)^{n-2} dt$$

for $r \in (0, \pi)$ and $u \in (0, 1)$. It can then be checked that the inequality $\frac{\sin(ut)}{\sin(ur)} < \frac{\sin t}{\sin r}$ holds pointwise if $0 < t < r < \pi$. The argument actually shows strict concavity in the "nontrivial" range (i.e., when $e^t \leqslant \pi$) for $n > 2$.
For the second part, note that Proposition (5.2) implies the inequality $V(\lambda r)/V(r) \geqslant V(\lambda s)/V(s)$ for $\lambda > 1$ and $r \leqslant s$, and we recover (5.6) when r tends to 0.

Exercise 5.6. If $n = 2$ and ε is slightly smaller than 1, we need at least 4 arcs to cover S^1.

Exercise 5.7. Argue by contradiction using the Hahn–Banach separation theorem; this is mostly planar geometry.

Exercise 5.8. Use Proposition 5.4 with $\theta + \eta = \varepsilon$ and $\eta = \varepsilon/n$, and (5.6). This choice gives $C = e$, but (as in the original Rogers's argument) optimizing over η leads to $C = 1$, at the expense of additional lower order terms.

Exercise 5.9. We know from Exercise 5.7 that $N(S^{n-1}, g, \varepsilon) \geqslant n + 1$ for any $\varepsilon < \pi/2$.

Exercise 5.10. Write $\mathcal{N} = \{[\psi] : \psi \in \mathcal{N}_1\}$ for some set $\mathcal{N}_1 \subset S_{\mathbb{C}^d}$. Take now $\mathcal{T}$ to be an ε-net in the unit circle $\mathbb{T} := \{\zeta \in \mathbb{C} : |\zeta| = 1\}$ and check that the set $\mathcal{N}_2 = \{\zeta\psi : \zeta \in \mathcal{T}, \psi \in \mathcal{N}_1\}$ is a 2ε-net in $(S_{\mathbb{C}^d}, g)$ (in fact even a $\sqrt{2}\varepsilon$-net). Note that $\operatorname{card}\mathcal{N}_2 = \operatorname{card}\mathcal{T} \operatorname{card}\mathcal{N}$. Since we can ensure that $\operatorname{card}\mathcal{T} = \lceil \pi/\varepsilon \rceil$, the result follows from the bound (5.7). For the upper bound, argue similarly that $P(S_{\mathbb{C}^d}, \varepsilon) \geqslant P(\mathbf{P}(\mathbb{C}^d), \varepsilon) \times P(\mathbb{T}, \varepsilon)$, and then appeal to (5.1).

Exercise 5.11. Rough two-sided estimates of the form $(C\varepsilon)^{2d-2}$ follow from Exercise 5.10 and (5.2), but the precise value requires a careful integration. First show that the question is a special case (with $n = 2d$) of the problem of calculating the spherical volume of the ε-neighborhood (in the geodesic distance) of S^1 considered as a subset of S^{n-1}. Next, observe that the (non-normalized) volume of that neighborhood equals $\mathrm{vol}_1(S^1)\,\mathrm{vol}_{n-3}(S^{n-3})\int_0^\varepsilon \cos t \sin^{n-3} t\,dt$. Conclude by evaluating the integral and using repeatedly the formula (B.2).

Exercise 5.12. (i) Expand $\|x_1 + \cdots + x_N\|^2$. (ii) Prove by induction on n that at most $2n$ nonzero vectors in $\mathbb{R}^n$ can have pairwise nonpositive inner products.

Exercise 5.13. (i) Consider the Gram matrix $G = (\langle x_i, x_j\rangle)_{1\leqslant i,j\leqslant N}$ and write $N = \|G\|_1 \leqslant \sqrt{n}\|G\|_2 \leqslant \sqrt{n}(N + N^2t^2)^{1/2}$. (ii) Observe that the vectors $(x_i^{\otimes k})$ span a space (the symmetric subspace of $(\mathbb{R}^n)^{\otimes k}$) of dimension at most $\binom{n+k-1}{k-1} \leqslant e^k(1+n/k)^k$. Then choose $k = \log(n)/2\log(1/t)$ and apply (i) to the vectors $(x_i^{\otimes k})$. See Theorem 9.3 in [**Alo03**]. (iii) Consider a maximal set of points verifying the condition from the exercise with $t = 1/r$.

Exercise 5.14. This is even simpler than the case of the sphere. Let $(x_i)_{1\leqslant i\leqslant N}$ be chosen uniformly and independently on $\{-1,1\}^n$ and let $A = \bigcup_{i=1}^N B(x_i,\varepsilon)$. Then, by (5.13), $\mathbf{E}\,\mathrm{card}\,A^c \leqslant 2^n(1-V(\varepsilon)/2^n)^N \leqslant \exp(n\log 2 - N2^{n(H(\varepsilon)-1)}/(n+1))$. This is less than 1 (and therefore the event $\{A = \{-1,1\}^n\}$ has positive probability) provided $N > n(n+1)\log(2)\cdot 2^{n(1-H(\varepsilon))}$. The matching lower bound on covering numbers is (5.2).

Exercise 5.15. We have $V_q(t) = \sum_{k=0}^{tn}\binom{n}{k}(q-1)^k \leqslant \sum_{k=0}^{n}\binom{n}{k}(q-1)^k\alpha^{k-tn} = q^{nH_q(t)}$ for $\alpha = t/((1-t)(q-1)) \leqslant 1$. For the lower bound, just keep the last term and write $q^{-nH_q(t)}V_q(t) \geqslant q^{-nH_q}\binom{n}{tn}(q-1)^{tn} = \binom{n}{tn}t^{tn}(1-t)^{(1-t)n} \geqslant \frac{1}{n+1}$. The last inequality follows from the fact that $\binom{n}{k}t^k(1-t)^{n-k}$ is maximal for $k = tn$.

Exercise 5.16. Consider $L = B_1^3$ and let K be a facet of L, then $N'(K,L) = 1 < N(K,L)$. To obtain an example with K symmetric, let K be a rhombus made of two opposite faces of L. Then $N'(K,L) = 2$ but two central sections of L (which are hexagons) cannot cover K. If we insist on having K with nonempty interior, that example can be modified by slightly enlarging K and L.

Exercise 5.17. If $x \in K$, then $K \cap (x + \varepsilon L) \subset x + (\varepsilon L \cap (K - K))$.

Exercise 5.18. If $K_\cap = K \cap (-K)$, then $N(K,K,\varepsilon) \leqslant N(K,K_\cap,\varepsilon) \leqslant (2+4/\varepsilon)^n$. The argument from Lemma 5.9 applies then *mutatis mutandis*.

Exercise 5.19. One may be tempted to say that the statement follows from Exercise 5.18 by duality, but this fails since K has centroid at the origin iff K° has Santaló point at the origin. However, a variant of the preceding hint gives a correct argument: by Lemmas 5.8 and Proposition 4.18, we have $N(\partial K, -K, \varepsilon) \leqslant N(\partial K, K_\cap, \varepsilon) \leqslant (2+4/\varepsilon)^n =: N$. Let now $x_1,\ldots,x_N$ in ∂K such that the sets $x_i - \varepsilon K$ cover ∂K. For each i, let f_i be a linear form such that $f_i(x_i) = 1$ and $f_i \leqslant 1$ on K, and let that $Q = \bigcap_{1\leqslant i\leqslant N}\{f_i \leqslant 1\}$. Show that $y \in \partial K$ satisfies $f_i(y) \geqslant 1-\varepsilon$ for some i, and conclude that $(1-\varepsilon)Q \subset K$.

Exercise 5.20. Use relations from Section 1.1.4 to show that $(K_\cap)^\circ \subset K^\circ - K^\circ$ and, subsequently, Theorem 4.21 to deduce that $\mathrm{vrad}(K_\cap) < 4\,\mathrm{vrad}(K^\circ)$. Next, use the hypothesis to conclude that $\mathrm{vrad}(K_\cap) > \frac{c}{4\kappa}\mathrm{vrad}(K)$, where c is the constant from Theorem 4.17. Then argue as in Exercises 5.18 and 5.19.

Exercise 5.21. If T is a linear map such that $\mathscr{F} = T(B_2^n)$, then $B_2^n = T(\mathscr{F}^\circ)$.

Exercise 5.22. (ii) Let $L = [0, \varepsilon/\sqrt{n}]^n$. The cubes $\{x + L : x \in \mathcal{N}\}$ have disjoint interiors and lie inside $B_2^n + L$, so $\operatorname{card} \mathcal{N} \leqslant \frac{\operatorname{vol}(B_2^n + L)}{\operatorname{vol} L} = (\varepsilon/\sqrt{n})^{-n} \operatorname{vol}(B_2^n + L)$. Then use Urysohn's inequality to upper-bound the volume radius of $B_2^n + L$.

Exercise 5.23. By the results of Exercise B.8, if $M = \mathsf{SO}(n)$ (resp., $M = \mathsf{U}(n)$, $M = \mathsf{SU}(n)$), then $N(\frac{\pi}{4}K, \|\cdot\|_\infty, 2.5\varepsilon) \leqslant N(M, \|\cdot\|_\infty, \varepsilon) \leqslant N(\pi K, \|\cdot\|_\infty, \varepsilon)$, where K denotes the operator norm unit ball in the space of real skew-symmetric (resp., complex skew-Hermitian, complex skew-Hermitian with zero trace) matrices. The result follows then from Lemma 5.8.

Exercise 5.24. Let $\theta_1, \dots, \theta_k \in [0, \pi/2]$ be the principal angles between $E, F \in \mathsf{Gr}(k, \mathbb{R}^n)$, as defined in (B.9). By Exercise B.12, the distance between E and F equals $\|(\pm 2 \sin \theta_i/2)\|_p \leqslant \sqrt{2}(2k)^{1/p}$.

Exercise 5.25. Use (5.2). Modulo the values of the constants C, c, the estimates are reciprocals of the bounds from (5.20).

Exercise 5.26. First observe that the extremal cases are when K is a cap and $L = K_\varepsilon^c$. Then prove, as a consequence of Proposition 5.2, that the function $t \mapsto V(t)V(\pi - t - \varepsilon)$ increases on the interval $[0, \frac{\pi}{2} - \frac{\varepsilon}{2}]$ and decreases on the interval $[\frac{\pi}{2} - \frac{\varepsilon}{2}, \pi - \varepsilon]$.

Exercise 5.27. Consider $f(x) = \operatorname{dist}(x, A)$.

Exercise 5.28. We may arrange by translation that K and L are inside RB_2^n, so that the functions $w(K, \cdot)$ and $w(L, \cdot)$ are R-Lipschitz on S^{n-1}. Then use the union bound and Lévy's lemma.

Exercise 5.29. Realize the normalized uniform measure on $\sqrt{N}S^{N-1}$ as the distribution of $\alpha_N G_N$, where G_N is a standard Gaussian vector in $\mathbb{R}^N$ and $\alpha_N = \sqrt{N}/|G_N|$, and use the law of large numbers to conclude that α_N tends almost surely to 1.

Exercise 5.30. By approximation, it is enough to show that $\gamma_n(A) > \gamma_1((-\infty, a])$ implies $\gamma_n(A_\varepsilon) > \gamma_1((-\infty, a + \varepsilon])$. Consider the orthogonal projection (restricted to the sphere) $\pi_{N,n} : \sqrt{N}S^{N-1} \to \mathbb{R}^n$. For N large enough, we know from Theorem 5.22 that the set $T := \pi_{N,n}^{-1}(A)$ has larger measure than the cap $C := \pi_{N,1}^{-1}((-\infty, a])$. It follows that $\sigma(T_\varepsilon) \geqslant \sigma(C_\varepsilon)$. Finally observe that $\pi_{N,n}^{-1}(A_\varepsilon) \supset T_\varepsilon$ while $C_\varepsilon = \pi_{N,1}^{-1}((-\infty, a + \varepsilon_N])$ for some ε_N tending to ε as N tends to infinity; this follows from the (geodesic) radius of C being $\sqrt{N} \arccos\left(-\frac{a}{\sqrt{N}}\right)$ and a similar formula for the radius of C_ε.

Exercise 5.31. Show that $\log \Phi$ is concave by computing its second derivative and appealing to (A.4). Alternatively, use Proposition 5.2 and Poincaré's lemma, and the fact that the function $u \mapsto \log(\arccos u)$ is concave near 0.

Exercise 5.32. The nontrivial part is to show that, for fixed n and $\delta > 0$, and for r large enough, we have $\Phi^{-1}\big(\gamma_n(rB_2^n)\big) > (1 - \delta)r$ or, equivalently, $\gamma_n\big((rB_2^n)^c\big) < \gamma_1\big([(1 - \delta)r, \infty)\big)$. Now choose a finite set $T \subset S^{n-1}$ such that $(rB_2^n)^c \subset \{x \in \mathbb{R}^n : \max_{u \in T}\langle x, u\rangle > (1 - \delta/2)r\}$ and use the fact that $\gamma_1\big([\theta r, \infty)\big) \gg \gamma_1\big([r, \infty)\big)$ as $r \to +\infty$ (with $\theta \in (0, 1)$ fixed). The last fact follows, e.g., from (A.4).

Exercise 5.33. To recover Ehrhard's inequality for convex bodies $A, B \subset \mathbb{R}^n$, consider the convex body $K = \operatorname{conv}\{(\{1\} \times A) \cup (\{0\} \times B)\} \subset \mathbb{R} \times \mathbb{R}^n$.

Exercise 5.34. We have $h(t) = C\exp\left(-(\frac{n}{2}-\frac{1}{3})(\log(t^3)-t^3)\right)$ for some constant C (depending on n). The same argument shows that the median of the gamma distribution with parameter p is greater than $p-\frac{1}{3}$.

Exercise 5.35. Use the exponential Markov inequality and optimize over $s \geqslant 0$.

Exercise 5.36. We may assume $b-a=1$. Write X as the convex combination $X=(b-X)a+(X-a)b$, then use Jensen's inequality and the convexity of the exponential function to reduce the problem to the inequality $b\exp(sa)-a\exp(sb) \leqslant \exp(s^2/8)$. The latter follows since $g'' \leqslant \frac{1}{4}$, where $g(s)=\log\left(b\exp(sa)-a\exp(sb)\right)$.

Exercise 5.37. Use the exponential Markov inequality for $\pm X$ with $s=\frac{\varepsilon}{2(1\pm\varepsilon)}$, and the bound $1+\varepsilon \leqslant \exp(\varepsilon-(\varepsilon^2-\varepsilon^3)/2)$. Checking the last inequality is a tedious but elementary computation.

Exercise 5.38. Argue as in the proof of Lemma 6.16 and Exercise 6.17, then set $Y_0=\lambda^{1/2}(Y-a)$.

Exercise 5.39. Easy.

Exercise 5.40. Reduce the problem to $\lambda=1$, then choose $\delta=\sqrt{\log(\kappa A)}$.

Exercise 5.41. Assume $\lambda=1$. For the median, check that if $0 \leqslant t \leqslant \sqrt{\log(2A)}$, then $2A^2\exp(-t^2/2) \geqslant \frac{1}{2}$. For the 3rd quartile, check that $3\sqrt{2}A^2\exp(-t^2/2) \geqslant \frac{3}{4}$ for $0 \leqslant t \leqslant \sqrt{\log(4A)}$, but that similar inequality holds with $3\sqrt{2}A^2$ replaced by $4A^2$ only if $A \geqslant 3^{2/3}/4$. For other quantiles, recalculate the bound on $|M-a|$ and then show analogous inequalities. The only verification that is not straightforward is establishing the bound $4A^2\exp(-\lambda t^2/2)$ when M is the mean under the original hypothesis $A \geqslant \frac{1}{2}$ (i.e., without assuming that $A \geqslant e^{-1/3}$); it can be accomplished by identifying a family of extremal c.d.f. of Y that are of the form

$$F(t)=\begin{cases}0 & \text{if} \quad t<0,\\ 1-p & \text{if} \quad 0 \leqslant t \leqslant t_0,\\ 1-Ae^{-t^2} & \text{if} \quad t \geqslant t_0,\end{cases}$$

where $t_0=\sqrt{\log(A/p)}$ is the solution to $p=Ae^{-t^2}$, and then using the calculation from the proof of Lemma 6.16 together with some numerics.

Exercise 5.42. The bound is $A(\kappa A)^{\alpha/(1-\alpha)}\exp(-\alpha\lambda t^2)$.

Exercise 5.43. To show (5.38), note that if $A=\{f \leqslant M\}$ and $B=\{f>M+\varepsilon\}$, then $\operatorname{dist}(A,B) \geqslant \varepsilon/L$. For the first assertion and $M=M_{1/4}$, the 1st quartile, consider $A=\{f \leqslant M_{1/4}\}$ and $B=\{f \geqslant M_f\}$, and similarly for the other quantiles.

Exercise 5.44. If we try to maximize $\mathbf{E}f$ among 1-Lipschitz functions with median 0, we may assume $f \geqslant 0$ (replace f by its positive part f^+). Writing $\mathbf{E}f=\int_0^1 \sigma(\{f \geqslant t\})\,dt$, we know from the solution of the isoperimetric problem that the extremal case is when $f_0(x)=\operatorname{dist}(x,A)$ for some half-sphere A (the distance being the geodesic distance). Consequently, $\mathbf{E}f_0=\int_0^{\pi/2} V(t)\,dt \leqslant \sqrt{\pi/8n}$ from (5.5).

Exercise 5.45. It follows from the solution of the isoperimetric problem (and from the formula $\mathbf{E}f^2=\int_0^\infty 2t\sigma(|f| \geqslant t)\,dt$) that among 1-Lipschitz functions with median 0, $\mathbf{E}f^2$ is maximal for $f_0(x)=\arcsin\langle x,u\rangle$ for some $u \in S^{n-1}$. For any Lipschitz function f, we have therefore $\mathbf{Var}\,f \leqslant \mathbf{E}(f-M_f)^2 \leqslant \mathbf{Var}\,f_0$. We compute

$$\mathbf{E}f_0^2=\int_0^{\pi/2} 2t\sigma(|f_0|>t)\,dt=4\int_0^{\pi/2} tV(\pi/2-t)\,dt \leqslant \frac{2}{n},$$

where we used (5.5). An example with variance $1/n$ is the function $x \mapsto \langle x, u \rangle$. *Note*: The estimate $2/n$ is not quite sharp; it follows from the Poincaré inequality (5.54) that the variance is at most $\frac{1}{n-1}$, since $n-1$ is the first nontrivial eigenvalue of $-\Delta$ (the Laplacian) on S^{n-1}. Numerical evidence suggests that the optimal bound is of the form $\frac{1}{B}$, where $n-1+\frac{1}{3n} < B < n-1+\frac{1}{3(n-1)}$.

Exercise 5.46. Let $m = \mathbf{E}f$. It follows from Exercise 5.45 that $\mathbf{Var}\, f \leqslant 2/n$. Consequently, $m \leqslant q \leqslant \sqrt{m^2+2/n}$ and $q - m \leqslant \sqrt{2/n}$. In what follows, use the values from Table 5.2. The first inequality is then immediate since $m \leqslant q$. The second one is trivial if $t \leqslant \sqrt{2/n}$ or if $t > q$. If $t \in (\sqrt{2/n}, q]$ (which implies $t > q - m$), then

$$\mathbf{P}(f \leqslant q-t) = \mathbf{P}(f \leqslant m-(t-(q-m))) \leqslant e^{-n(t-(q-m))^2/2} \leqslant e \cdot e^{-nt^2/2}.$$

To derive the last inequality, use $t \leqslant q \leqslant \sqrt{m^2+2/n}$. A very ambitious reader may try to come up with a better estimate based on the sharper bound $\mathbf{Var}\, f \leqslant \frac{1}{n-1}$ (see the hint for Exercise 5.45).

Exercise 5.48. Apply the hypothesis to $f(x) = \min\{\operatorname{dist}(x, A), \varepsilon\}$.

Exercise 5.49. Consider $A = X \backslash B_\varepsilon$.

Exercise 5.50. The concavity of g is a consequence of Ehrhard's inequality (Theorem 5.23). Since $g(M_f) = 0$, we conclude that the inequality $g(t) \leqslant \alpha(t - M_f)$ holds for some $\alpha > 0$ and every real t. This is equivalent to the statement $\gamma_n(\{f \leqslant t\}) \leqslant \mathbf{P}(Z \leqslant t)$ where Z is an $N(M_f, \alpha^{-2})$ random variable. The conclusion follows since stochastic domination allows comparison of the expectations.

Exercise 5.51. The distribution of $(X_k)_{1 \leqslant k \leqslant N}$ is the image of γ_N under an affine map.

Exercise 5.52. Consider the function $\tilde{f}$ defined on S^{n-1} by $\tilde{f}(x) = \inf\{f(y) + Lg(x, y) \,:\, y \in \Omega\}$. Show that $\tilde{f}$ is L-Lipschitz, coincides with f on Ω and that M_f is a central value for $\tilde{f}$. Then apply Corollary 5.32.

Exercise 5.53. Use the fact that for $B \subset Y$ and $\varepsilon > 0$, $\phi^{-1}(B_\varepsilon) \supset \phi^{-1}(B)_\varepsilon$. The statement about the median is an immediate consequence of (5.39). For the statement about the expectation, restrict the supremum on the right-hand side to functions of the form $g \circ \phi$. If ϕ is L-Lipschitz, one needs to replace A_ε by $A_{\varepsilon/L}$ in (5.39), $\mu(f - \mathbf{E}f > t)$ by $\mu(f - \mathbf{E}f > t/L)$ in (5.40), and similarly for the median.

Exercise 5.54. If $n = 1$, the function $x \mapsto \Phi(x)$ (the c.d.f. of the $N(0,1)$ distribution, see (A.3)) pushes forward γ_1 to the Lebesgue measure on $[0,1]$ and is Lipschitz with constant $(2\pi)^{-1/2}$, which allows us to transfer the results on Gaussian concentration to $[0,1]$. For general $n \in \mathbb{N}$, consider the surjection $\phi : \mathbb{R}^n \to [0,1]^n$ given by $\phi((x_j)) = (\Phi(x_j))$.

Exercise 5.55. (i) $x \mapsto \sup\{t : \nu(F(x, .) \geqslant t) \geqslant 1/2\}$ is 1-Lipschitz, and similarly for the other term. (ii) If $f(x, y) > M_\phi + t$, then either $\phi(x) > M_\phi + t/2$ or $f(x, y) > \phi(x) + t/2$. (iii) $(\frac{1}{2})^2 = \frac{1}{4}$. (iv) Argue as in (ii).

Exercise 5.56. (ii) For a 1-Lipschitz function $f : X_1 \times X_2 \to \mathbb{R}$, $x_1 \in X_1$, and $x_2 \in X_2$, introduce the functions $f_{x_2}(x_1) = f(x_1, x_2)$ and $g(x_2) = \int f_{x_2}\, d\mu_1$, and show that they are 1-Lipschitz.

Exercise 5.58. Let $\mathcal{B}$ be an orthonormal basis of $\mathsf{M}_{k,n-k}$ considered as a real space. We claim that, for any $X \in \mathsf{M}_{k,n-k}$,

$$\frac{1}{2}\sum_{Y\in\mathcal{B}} \|XY^\dagger - YX^\dagger\|_{\mathrm{HS}}^2 + \|X^\dagger Y - Y^\dagger X\|_{\mathrm{HS}}^2 = \alpha\|X\|_{\mathrm{HS}}^2, \tag{E.3}$$

where $\alpha = n-2$ in the real case and $\alpha = 2n$ in the complex case. The sum in (E.3) does not depend on the choice of the orthonormal basis since it is the trace of a quadratic form. Write the singular value decomposition of X as $X = \sum s_j |e_j\rangle\langle f_j|$ where $(e_1,\dots,e_k)$ and $(f_1,\dots,f_{n-k})$ are orthonormal bases. For $1 \leqslant a \leqslant k$ and $1 \leqslant b \leqslant n-k$, set $E_{ab} := |e_a\rangle\langle f_b|$ and $F_{ab} := i|e_a\rangle\langle f_b|$ and consider the orthonormal basis of $\mathsf{M}_{k,n-k}$ formed by (E_{ab}) (in the real case) or $(E_{ab}) \cup (F_{ab})$ (in the complex case). We compute

$$\frac{1}{2}\|XE_{ab}^\dagger - E_{ab}X^\dagger\|_{\mathrm{HS}}^2 + \frac{1}{2}\|X^\dagger E_{ab} - E_{ab}^\dagger X\|_{\mathrm{HS}}^2 = \begin{cases} s_a^2 + s_b^2 & \text{if } a \neq b,\\ 0 & \text{if } a = b,\end{cases}$$

$$\frac{1}{2}\|XF_{ab}^\dagger - F_{ab}X^\dagger\|_{\mathrm{HS}}^2 + \frac{1}{2}\|X^\dagger F_{ab} - F_{ab}^\dagger X\|_{\mathrm{HS}}^2 = \begin{cases} s_a^2 + s_b^2 & \text{if } a \neq b,\\ 4s_a^2 & \text{if } a = b,\end{cases}$$

and (E.3) follows by summing over a, b. In the above formulas it is tacitly assumed that $s_j = 0$ for $j > \min\{k, n-k\}$.

Exercise 5.59. For $\mathsf{U}(n)$, note that $\mathfrak{u}_n$ (the space of skew-Hermitian matrices) contains a central element $u_1 := i\,\mathrm{I}/\sqrt{n}$, and so it follows from (5.44) and (5.45) that $\mathrm{Ric}_{\mathrm{I}}(u_1) = 0$. In the case of $\mathsf{SO}(n)$, consider the orthonormal basis of $\mathfrak{so}_n$ of matrices of the form $S_{ij} = \frac{1}{\sqrt{2}}(|i\rangle\langle j| - |j\rangle\langle i|)$ and reduce to the case $X = S_{12}$. The argument for $\mathsf{SU}(n)$ is similar; note that $u_1 = i\,\mathrm{I}/\sqrt{n} \notin \mathfrak{su}_n$. For details of the computations for both $\mathsf{SO}(n)$ and $\mathsf{SU}(n)$, see Proposition E.15 in [**AGZ10**].

Exercise 5.60. Test the log-Sobolev inequality on the function $x \mapsto \exp(\lambda x)$ for some $\lambda \neq 0$. Alternatively, consider the function $F : x \mapsto x$ in (5.49) and let $t \to \infty$.

Exercise 5.61. (i) There is a contraction $\phi : S^1 \to [0,\pi]$ which pushes forward σ to the normalized Lebesgue measure. (ii) Consider the Fourier series of the function f from (5.54). (iii) Consider $f(x) = \cos(\pi x)$.

Exercise 5.62. Use Jensen's inequality in the form $|P_t f|^p \leqslant P_t(|f|^p)$, and the relation $\int P_t g \,\mathrm{d}\gamma_n = \int g \,\mathrm{d}\gamma_n$ (justify!) applied for $g = |f|^p$. Note that the argument is much easier when $p = 2$, the contractivity following right away from (5.57).

Exercise 5.63. Use the fact that $\mathbf{E}\exp(\lambda Z) = \exp(\lambda^2/2)$ when Z has an $N(0,1)$ distribution. The result is $P_t f_\lambda = \exp(\lambda^2(1-e^{-2t})/2) f_{\lambda e^{-t}}$. Since $\|f_\lambda\|_{L_p(\gamma_n)} = e^{p\lambda^2/2}$, the statement about sharpness follows by taking $\lambda \to \infty$.

Exercise 5.64. Write $A_{s/n} = \{y \in \{-1,1\}^n : (1+\varepsilon)y_1 + y_2 + \cdots + y_n \leqslant m + s + \varepsilon\}$ for small ε, use Hoeffding's inequality (5.43), and let $\varepsilon \to 0$.

Exercise 5.65. If $\varepsilon < 1/n$, then $A_\varepsilon = A$, and so in that case we may have $\mu(A_\varepsilon) = \frac{1}{2}$. Positive results follow from (5.59) and from Exercise 5.64.

Exercise 5.66. Try $N = 9$, and consider a Hamming ball of radius 1 plus any 4 of the 6 elements of its boundary.

Exercise 5.67. The second assertion of Theorem 5.54 can be restated as follows: *If $K, L \subset \mathbb{R}^n$ satisfy $\mathrm{dist}(K,L) \geqslant t$ and one of K, L is convex, then $\mu(K)\mu(L) \leqslant e^{-t^2/2}$.*

Exercise 5.68. Consider the supremum of all 1-Lipschitz affine functions that are smaller than f on K.

Exercise 5.69. We have for $y \in \{-1,1\}^n$

$$f(y) = \frac{1}{\sqrt{2}} \inf\{|y - z| \ : \ z \in \{-1,1\}^n, \ \sum z_i \leqslant 0\}$$

(this formula is valid for n even and has to be slightly modified for n odd), so f is $\frac{1}{\sqrt{2}}$-Lipschitz. The bound on the probability follows from the central limit theorem.

Exercise 5.70. Write $\mathbf{E}\,|f(G)|^p = \int_0^\infty pt^{p-1}\mathbf{P}(|f(G)| > t)\,dt$ and use the Gaussian isoperimetric inequality in the form given in Theorem 5.24.

Exercise 5.71. First note that clearly $\|X\|_{L_2}^2 = \sum_i |a_i|^2$. Next, for $p \geqslant 2$, use Proposition 5.58 and the fact that $\|\varepsilon_i\|_{\psi_2} = 1$ (this gives $B_p = O(\sqrt{p})$, which is the correct order of magnitude). The case $p = 1$ (and hence $p \in (1,2)$) follows then from the inequality $\mathbf{E}\,|X| \geqslant (\mathbf{E}\,X^2)^{3/2}/(\mathbf{E}\,X^4)^{1/2}$. An alternative approach is to appeal to Theorem 5.56 to upper-bound higher moments of X (or to Theorem 5.54 and to the fact that, for any nonnegative variable W, we always have $\mathbf{E}\,W \geqslant \frac{1}{2}M_W$).

Exercise 5.72. By change of variables, reduce the problem to comparing the moments of a norm (or a seminorm) $\|\cdot\|$ on $\mathbb{R}^n$ calculated with respect to the normalized counting measure μ on $\{-1,1\}^n$. Next, follow the last strategy from the hint to Exercise 5.71 combined with Theorem 5.54. The only difference is that while previously we got "for free" the fact that the Lipschitz constant of the linear function $(t_i) \mapsto \sum_i a_i t_i$ was exactly the same as $\|\sum_i a_i\varepsilon_i\|_{L_2}$, this is no longer automatically true for the function for $(t_i) \mapsto \|(t_i)\|$. However, the Lipschitz constant and the median of $\|\cdot\|$ can still be related: if $K = \{(t_i) \ : \ \|(t_i)\| \leqslant M_{\|\cdot\|}\}$, then the Euclidean inradius of K cannot be too small. This follows from the scalar case: if the Euclidean inradius of K *was* small, then K would be contained in a narrow band $\{t \ : \ |\langle t, a\rangle| \leqslant 1\}$ and, consequently, the median of function $(t_i) \mapsto \left|\sum_i a_i t_i\right|$ would be at most 1, much smaller than its L_2-norm (equal to $|a|$), contradicting the argument from the hint to Exercise 5.71.

Exercise 5.73. (i) We have $\mathbf{E}\,X^n = 0$ if n is odd; compare both Taylor series using the inequality $k^k k!/(2k)! \leqslant (e/4)^k$. (ii) Use Jensen's inequality.

Exercise 5.74. Use the bound on the Laplace transform obtained in Exercise 5.73(ii) to upper-bound the moments.

Exercise 5.75. The equality $\|Z\|_{\psi_2} = \sqrt{2/\pi}$ is equivalent to $\|Z\|_p \leqslant \sqrt{p}\,\|Z\|_1$ for $p \geqslant 1$. Unless p is small, this follows from Stirling's formula (on which the asymptotic formula (5.33) is based). For small p one can verify the inequality numerically. The inequality $\|\cdot\|_{\psi_1} \leqslant \|\cdot\|_{\psi_2}$ follows similarly from $\|T\|_p \geqslant \sqrt{p}$ for $p \geqslant 2$. (This is a very simple minded approach, we will be grateful to a reader who supplies a nice rigorous argument.)

Exercise 5.76. (iii) Choose λ to be the minimum of $1/2\|a\|_\infty$ and $t/4\sum a_i^2$.

Chapter 6

Exercise 6.1. Let $T = [0,1] = \Omega$ and let $f : [0,1] \mapsto \mathbb{R}_+$ be an arbitrary function. Define $X_t(t) = f(t)$ and $X_t(\omega) = 0$ if $\omega \neq t$.

Exercise 6.2. Define $t_N > 0$ by the formula $\exp(t_N^2/2) = N/\log^{3/2} N$ and check using (A.4) that $\mathbf{P}(M \leqslant t_N) = O(1/N^c)$ for some constant $c > 0$, where $M =$

$\max\{X_k : 1 \leqslant k \leqslant N\}$. Conclude that $\mathbf{E}\, M \geqslant \sqrt{2\log N} - O\left(\frac{\log\log N}{\sqrt{\log N}}\right)$ (handle $\mathbf{E}\, M^+$ and $\mathbf{E}\, M^-$ separately). See [**DLS14**] for more precise bounds.

Exercise 6.3. The suggested inequality follows from the formula

$$\mathbf{E}Y = \int_0^\infty \mathbf{P}(Y > t)\, \mathrm{d}t,$$

valid for any nonnegative random variable Y.

Exercise 6.4. By Carathéodory's theorem (Theorem 1.2), K equals the union of a family of simplices, each of which has vertices of the form $x_{k_1}, \ldots, x_{k_n}, x_{k_{n+1}}$. Next, upper-bound the number of such simplices (simple combinatorics) and the volume of each simplex (Hadamard's inequality). *Note*: If $N \gg n$, however, this argument does not yield the logarithmic dependence on N/n from Remark 6.4.

Exercise 6.5. Proceed as in Proposition 6.3 using Lemma 6.2 instead of Lemma 6.1. In the nontrivial range $2\log(2N) \leqslant n$, use the inequality $\kappa_n > \sqrt{n-1}$.

Exercise 6.6. We compute the Gaussian mean width $w_G(\cdot) = \kappa_n w(\cdot)$. By linearity of expectation, we have $w_G(B_\infty^n) = n\,\mathbf{E}\,|Z| = n\sqrt{2/\pi}$, where Z is an $N(0,1)$ random variable. It follows from Lemmas 6.1 and 6.2 that $(1-o(1))\sqrt{2\log n} \leqslant w_G(\Delta_{n-1}) \leqslant w_G(B_1^n) \leqslant \sqrt{2\log n}$.

Exercise 6.7. With the assumption that $\mathbf{E}\, X_k^2 = \mathbf{E}\, Y_k^2$ this is immediate by integration (take $\lambda_k = t$ for all k). Without this assumption, let Z be an $N(0,1)$ random variable independent of $(X_k), (Y_k)$. For $0 < t < 1$ and R large enough, define new processes $(\bar{X}_k)$ and $(\bar{Y}_k)$ by $\bar{X}_k = tX_k + \alpha_k Z$ and $\bar{Y}_k = Y_k + \beta_k Z$, where the positive numbers α_k, β_k are adjusted so that $\mathbf{E}\, \bar{X}_k^2 = \mathbf{E}\, \bar{Y}_k^2 = R^2$. Check that, for R large enough, the second part of Slepian's lemma can be applied to $(\bar{X}_k)$ and $(\bar{Y}_k)$. Check also that $\mathbf{E} \sup \bar{X}_k = R + t\,\mathbf{E} \sup X_k + O(1/R)$ and similarly for $(\bar{Y}_k)$, so that letting $R \to \infty$ and then $t \to 1$ yields (6.7).

Exercise 6.8. Any centered Gaussian measure is the pushforward of the standard Gaussian measure by a linear transformation.

Exercise 6.9. Without loss of generality, $L = \{(x_1, \ldots, x_n) \in \mathbb{R}^n : |x_1| \leqslant t\}$ for some $t > 0$. Define $f(s) = \gamma_{n-1}\left(\{y \in \mathbb{R}^{n-1} : (s,y) \in K\}\right)$, an even function of s. By (4.28), the function $\log f$ is concave, and therefore decreasing on $[0, +\infty)$. It follows (differentiate) that

$$\int_0^t f \,\mathrm{d}\gamma_1 \geqslant 2\gamma_1([0,t]) \int_0^\infty f \,\mathrm{d}\gamma_1,$$

which is equivalent to the statement $\gamma_n(K \cap L) \geqslant \gamma_n(L)\gamma_n(K)$.

We now prove Proposition 6.9. Without loss of generality we may assume that $X_k = \langle G, x_k\rangle$, where G is a standard Gaussian vector in $\mathbb{R}^d$ (for some $d \leqslant N$), and $x_1, \ldots, x_N \in \mathbb{R}^d$. We apply (6.13) to $L = \{x \in \mathbb{R}^d : |\langle x, x_1\rangle| \leqslant t_1\}$ and $K = \{x \in \mathbb{R}^d : |\langle x, x_k\rangle| \leqslant t_k \text{ for } 2 \leqslant k \leqslant N\}$ in order to obtain

$$\mathbf{P}\left(|X_k| \leqslant t_k \text{ for } 1 \leqslant k \leqslant N\right) \geqslant \mathbf{P}\left(|X_1| \leqslant t_1)\mathbf{P}(|X_k| \leqslant t_k \text{ for } 2 \leqslant k \leqslant N\right).$$

The result follows by induction on N.

Exercise 6.10. Slepian's lemma (6.7) supplies candidates for the extremal configuration. The worst case is when $\mathbf{X}$ contains only two elements.

Exercise 6.11. We have $w_G(K) = w_G(K_1) + \cdots + w_G(K_n) = \Theta(n)$ regardless of the choice of the sequence (d_j). Now choose $d_j = 2^j$. Given $1 \leqslant j \leqslant k \leqslant n$, let

$\mathcal{N}_{j,k}$ be a minimal $2^{j-\frac{3k}{2}}$-net in K_j and $\mathcal{M}_k = \mathcal{N}_{k,1} \times \cdots \times \mathcal{N}_{k,k} \times \{0\} \times \cdots \times \{0\}$. Check that $\mathcal{M}_k$ is an $O(2^{-k/2})$-net of K and that $\log \operatorname{card} \mathcal{M}_k = O(2^k)$.

Exercise 6.12. The only case that is not straightforward is applying the volumetric bound (5.8) when $\varepsilon \gg 1$: the upper bound requires subtle estimates on $\operatorname{vol}(B_1^n + \delta B_2^n)$, where $\delta = \frac{\varepsilon}{2\sqrt{n}}$. However, it is not hard to see that, in any case, the upper bound is not tight: replace B_1^n by the smaller set $B_2^n/\sqrt{n}$ and deduce that the ratio of volumes is greater than $e^{n/\varepsilon}$. The approach via Sudakov's inequality is *much* simpler and yields, for that range of ε, optimal or nearly optimal results.

Exercise 6.13. If $\operatorname{inrad}(K) \leqslant r$, then K is contained in a symmetric band of width $2r$. For the second part, we use Markov's inequality to write $\gamma_n(\|\cdot\|_K^\circ > \varepsilon) \leqslant w(K)/\varepsilon \leqslant .317$, which implies $\gamma_n(\varepsilon K^\circ) \geqslant .683$ and therefore $N(B_2^n, \varepsilon K^\circ) = 1$.

Exercise 6.14. Apply Proposition 5.34 to the convex function $\|\cdot\|_{L^\circ}$.

Exercise 6.15. Since we are covering a Euclidean ball with translates of another body, it is the dual Sudakov inequality (Proposition 6.11) that is relevant. Use the value of the appropriate mean width from Table 4.1.

Exercise 6.16. The optimal θ equals $1 + \sqrt{2}$.

Exercise 6.17. The worst case is when the random variables Y_i are non-negative and disjointly supported.

Exercise 6.18. Use the union bound to estimate $\mathbf{P}(\max_i Y_i > t)$ and argue as in the proof of Lemma 6.16.

Exercise 6.19. This is again similar to the proof of Lemma 6.16. First use (6.6) and the union bound to estimate $\mathbf{P}(\max_k X_k > t)$; this leads (as $n \to \infty$) to an expression involving Riemann zeta function $\zeta(s)$. Then just use the fact that if $s \geqslant 2$, then $\zeta(s) \leqslant \zeta(2) = \frac{\pi^2}{6}$. The best bound for $\mathbf{E} \max_k X_k$ that can be obtained by this line of argument is about 1.724. On the other hand, it is not hard to see that $\mathbf{E} \max_k X_k > \sqrt{2}$ for n large enough. The true value of $\mathbf{E} \sup_k X_k$ seems to be between 1.45 and 1.5. To get a lower bound on the Dudley integral, note that for $k \leqslant n$ the elements $X_1, \dots, X_k$ are ε-separated with $\varepsilon = \sqrt{2/(1 + \log k)}$.

Exercise 6.20. Use Lemma 6.1 and (6.17).

Exercise 6.21. For $k \leqslant l \leqslant n$, we compute $\|X_k - X_l\|_2 = \sqrt{2 - 2\sqrt{k/l}}$. Since the family $(X_{2^j})_{j \leqslant \log n}$ is $\sqrt{2 - \sqrt{2}}$-separated, the lower bound follows from Sudakov's inequality. For the upper bound, use Dudley's inequality and the fact that, for $\alpha > 1$, the family $(X_{\lfloor \alpha^j \rfloor}) \cup (X_{\lceil \alpha^j \rceil})$ gives a $\sqrt{2 - 2/\sqrt{\alpha}}$-net with at most $2 \log n / \log \alpha$ elements.

Exercise 6.22. Define a sequence (a_k) by $a_k = \inf\{\eta > 0 \ : \ N(T, \eta) \leqslant 2^{2^k}\}$ for $k \geqslant 1$, a_0 being the radius of T. The right-hand side of (6.27) is exactly $\sum_{k=0}^{\infty} 2^{k/2} a_k$. To compare with the left-hand side, use the bound $2^{2^k} \leqslant N(T, \eta) \leqslant 2^{2^{k+1}}$ for $\eta \in [a_{k+1}, a_k]$, $k \geqslant 1$.

Exercise 6.23. Consider the sets $T_k = \{X_1, \dots, X_{2^{2^k}-1}, X_n\}$.

Exercise 6.25. If $\|X - Y\|_{L_\infty} \leqslant \varepsilon$, then $f(Y) \geqslant g(X)$.

Exercise 6.26. The direction that is not entirely straightforward is showing that $d_\infty(X, Y)$ does not exceed the infimum in (6.32). If $\tilde{X} = F_X^{-1} : (0, 1) \to \mathbb{R}$ (the inverse function) exists, then, when considered as a random variable with respect to the Lebesgue measure, its law is the same as that of X. With care, such $\tilde{X}$

can be defined also if F_X is not strictly increasing and/or discontinuous. Given X, Y, what is $\|\tilde{X} - \tilde{Y}\|_\infty$? This argument shows also that the infimum in (6.30) is attained.

Exercise 6.27. Case 1° (the bounded case). If $\|Y_n\|_\infty \leqslant M$ for some finite M and all n, then also $\|Z\|_\infty \leqslant M$. Now approximate f on $[-M, M]$ by a Lipschitz function and apply (6.31). Case 2° (the general case). Let $\varepsilon > 0$ and choose M so that $\mathbf{P}(|Z| > M) < \varepsilon$. Then, for all sufficiently large n, $\mathbf{P}(|Y_n| > M + 1) < \varepsilon$. Apply Case 1° to Y_n's and Z truncated at the level $M + 1$, and then let $\varepsilon \to 0$. (The last step uses the hypothesis that f is bounded.)

Exercise 6.28. See the hint to Exercise 6.27; note that under the present hypotheses Case 1° always holds. For an example, consider Z with distribution $N(0, 1)$, $Y_n = Z + \frac{1}{n}$, and $f(x) = \frac{e^{x^2/2}}{1+x^2}$.

Exercise 6.29. The measures $(\frac{1}{2} + \frac{1}{n})\delta_0 + (\frac{1}{2} - \frac{1}{n})\delta_1$ converge weakly but do not converge in ∞-Wasserstein distance, as n tends to infinity.

Exercise 6.30. The function $A \mapsto \lambda_k(A)$ is 1-Lipschitz with respect to the operator norm. It is remarkable that a similar inequality (with an additional multiplicative constant $C < 3$ on the right-hand side) holds for normal matrices [**BDM83, BDK89**].

Exercise 6.31. For (2), use the fact that the image of a standard Gaussian vector under the orthogonal projection onto a subspace is the standard Gaussian vector in that subspace.

Exercise 6.32. Show that if a random matrix $X \in \mathsf{M}_n^{\mathrm{sa}}$ is unitarily invariant, then $U\,\mathrm{Diag}(X)U^\dagger$ (where U is a Haar-distributed random unitary matrix independent of X) has the same distribution as X.

Exercise 6.33. If $\mathcal{N}$ is an ε-net in $(S_{\mathbb{C}^n}, |\cdot|)$, show (argue as in the proof of Lemma 5.9) that for any $A \in \mathsf{M}_n$,

$$\|A\|_\infty = \sup_{x,y \in S_{\mathbb{C}^n}} \big|\langle x|A|y\rangle\big| \leqslant \frac{1}{1-2\varepsilon} \sup_{x,y \in \mathcal{N}} \big|\langle x|A|y\rangle\big|.$$

Then use Proposition 6.3.

Exercise 6.34. Use Exercise 6.30 with A being a GUE(n) matrix and with $B = A - \frac{\operatorname{Tr} A}{n}\mathrm{I}$.

Exercise 6.35. Show that the function $(z - x_+)(z - x_-)$ admits an analytic square root $g_\lambda : \mathbb{C}\backslash[x_-, x_+] \to \mathbb{C}$ such that $g_\lambda(x) > 0$ for $x \in (x_+, \infty)$, $g_\lambda(x) < 0$ for $x \in (-\infty, x_-)$, and $\lim_{y \to 0^\pm} g_\lambda(x + iy) = \pm i\sqrt{(x - x_-)(x_+ - x)}$ for $x \in [x_-, x_+]$. It follows that if $M := \int_{x_-}^{x_+} f_\lambda$, then $\int_\gamma \frac{g_\lambda(z)}{z}\,\mathrm{d}z = 2iM$ for any closed path γ which circles $[x_-, x_+]$ once in the clockwise direction, but does not wind around 0. To evaluate the path integral over γ we choose $R > x_+$ and set $\Gamma(t) = Re^{it}, 0 \leqslant t \leqslant 2\pi$, and note that

(i) $\int_\gamma \frac{g_\lambda(z)}{z}\,\mathrm{d}z + \int_\Gamma \frac{g_\lambda(z)}{z}\,\mathrm{d}z = 2\pi i g_\lambda(0)$ by the Cauchy integral formula, or by the residue theorem,

(ii) $\int_\Gamma \frac{g_\lambda(z)}{z}\,\mathrm{d}z$ can be related to the constant term of the Laurent expansion of g_λ, which in turn can be found by subtracting the dominant (as $z \to \infty$) term z and considering the limit of $g_\lambda(x) - x$ as $x \to +\infty$.

An alternative argument is to perform successive substitutions $x = y + (1+\lambda)$, $y = 2\sqrt{\lambda}u$, $u = \cos t$, and to recognize the resulting expression as an integral involving the Poisson kernel $P_r(t)$ for $r = \sqrt{\lambda}$.

Either approach allows to find also the expected value and the variance of $\mathrm{MP}(\lambda)$, the calculation being in both cases simpler than the one sketched above.

Exercise 6.36. The equality is easily verified. To extend Theorem 6.28 to $\lambda < 1$, let W_1 and W_2 be as in the paragraph following (6.39), and note that for $s \geqslant n$, $\mu_{\mathrm{sp}}(W_2) = (1 - n/s)\delta_0 + n/s\, \mu_{\mathrm{sp}}(W_1)$.

Exercise 6.37. Couple W and X, defined as (6.39) and (6.40), by realizing ψ_i as $G_i/|G_i|$. Using Exercise 6.30, it follows that $d_\infty(\mu_{\mathrm{sp}}(n^{-1}W), \mu_{\mathrm{sp}}(X)) \leqslant \sup\{|1 - n^{-1}|G_i|^2| \ : \ 1 \leqslant i \leqslant s\}$. This tends to zero in probability, by Corollary 5.27.

Exercise 6.38. By Theorem 6.28, the eigenvalue distribution of $BB^\dagger$ approaches a MP(1) distribution, therefore the singular value distribution of B approaches the law of $\sqrt{X^2}$, where $X \sim \mu_{\mathrm{SC}}$.

Exercise 6.39. For (a), use Lemma 6.20. The argument from (b) does not justify changing the order of the limits. A separate question (to which the authors do not know the answer) is whether we do actually have *uniform* convergence of $W_{n,s}/n$ to $X_{s/n}$ in ∞-Wasserstein distance as $n, s \to \infty$.

Exercise 6.40. If $W = BB^\dagger$, then $2\operatorname{Tr} W = \sum_{ij} 2|\operatorname{Re} B_{ij}|^2 + 2|\operatorname{Im} B_{ij}|^2$ is the sum of $2ns$ squared independent $N(0,1)$ variables.

Exercise 6.41. Let $\psi \in S_{\mathbf{C}^n}$ be uniformly distributed, $A = |\psi\rangle\langle\psi| - \mathrm{I}/n$, and let B be a $\mathrm{GUE}_0(n)$ random matrix. By symmetry, the covariances of A and B (considered as $\mathsf{M}_n^{\mathrm{sa}}$-valued random vectors) are proportional, i.e., there exists $\beta > 0$ such that $\mathbf{E}\operatorname{Tr}(AM)^2 = \beta^2\, \mathbf{E}\operatorname{Tr}(BM)^2$ for every $M \in \mathsf{M}_n^{\mathrm{sa}}$. We compute that $\beta = 1/\sqrt{n(n-1)}$, and the result follows from the multivariate central limit theorem.

Exercise 6.42. Use Proposition 6.34 and Proposition A.1(ii).

Exercise 6.43. (i) This is more transparent if we think of $S_{\mathbf{C}^n \otimes \mathbf{C}^s}$ as the Hilbert–Schmidt sphere $S_{\mathrm{HS}} \subset \mathsf{M}_{n,s}$, and identify ρ and $AA^\dagger$, with A uniformly distributed on S_{HS}. The function becomes $f(A) = |A^\dagger y|$ and is 1-Lipschitz with respect to $\|\cdot\|_\infty$, hence with respect to $\|\cdot\|_{\mathrm{HS}}$. To apply Exercise 5.46 we identify S_{HS} with S^{2ns-1}. (ii) Given $x \in S_{\mathbf{C}^n}$, let $y \in \mathcal{N}$ with $|x - y| \leqslant \delta$ and write

$$|\langle x|\Delta|x\rangle| \leqslant |\langle y|\Delta|y\rangle| + |\langle x - y|\Delta|y\rangle| + |\langle x|\Delta|x - y\rangle| \leqslant |\langle y|\Delta|y\rangle| + 2\delta\|\Delta\|_\infty,$$

then take the supremum over x. (iii) Choose for example $\delta = 1/4$. Using Lemma 5.3, the union bound and (ii), we have

$$\mathbf{P}(\|\Delta\| \geqslant 48/\sqrt{ns}) \leqslant 8^{2n}\mathbf{P}(|f^2 - 1/n| \geqslant 24/\sqrt{ns}) \leqslant 8^{2n}\mathbf{P}(|f - 1/\sqrt{n}| \geqslant 4/\sqrt{s}).$$

(The last inequality is valid whenever $s \geqslant n$.) By (i), it follows that

$$\mathbf{P}(\|\Delta\| \geqslant 48/\sqrt{ns}) \leqslant 64^n(1+e)\exp(-16n),$$

which tends to 0 as n tends to infinity.

Exercise 6.44. Combine the results from Exercise 2.19 and Exercise 6.38.

Exercise 6.45. (i) Expand $\mathbf{E}\operatorname{Tr} GG^\dagger GG^\dagger = \sum_{i,k=1}^n \sum_{j,l=1}^s \mathbf{E}[G_{ij}\overline{G_{kj}}G_{kl}\overline{G_{il}}]$ and notice that, by independence, $\mathbf{E}[G_{ij}\overline{G_{kj}}G_{kl}\overline{G_{il}}] = \mathbf{1}_{\{i=k\}} + \mathbf{1}_{\{j=l\}}$ (using the value $\mathbf{E}|Z|^4 = 2$ for $N \sim N_{\mathbf{C}}(0,1)$). The second computation is similar. (ii) Write $GG^\dagger$ as the product of independent random variables $\frac{GG^\dagger}{\operatorname{Tr} GG^\dagger} \times \operatorname{Tr} GG^\dagger$ and use (i).

Exercise 6.46. Notice that, for fixed $t \in \mathbb{R}^n$, $\mathbf{E}\max_{u\in L}\langle Bt, u\rangle = |t| w_G(L)$, and similarly with K and L switched.

Exercise 6.47. The inequality from (i) can be rewritten as

$$\left(|x||y| - |x'||y'|\right)^2 + 2\left(|x||x'| - \langle x, x'\rangle\right)\left(|y||y'| - \langle y, y'\rangle\right) \geqslant 0.$$

Part (ii) is proved similarly. For (iii), this fails already in dimension 1: if $x = x' = 1$ and $y = y' = e^{i\varepsilon}$, then as ε tends to zero, $|x \otimes x' - y \otimes y'| \sim 2\varepsilon$ while $|(x, x') - (y, y')| \sim \sqrt{2}\varepsilon$.

Exercise 6.48. If A is a GOE(n) matrix and G is a standard Gaussian vector in $\mathbb{R}^n$, consider the processes $X_t = \langle t, At\rangle = \operatorname{Tr}\left(A|t\rangle\langle t|\right)$ and $Y_t = \langle G, t\rangle$, both indexed by $t \in S^{n-1}$. Check that, for $s, t \in S^{n-1}$,

$$\|X_s - X_t\|_{L_2}^2 = 2\||s\rangle\langle s| - |t\rangle\langle t|\|_2^2 \leqslant 4|s-t|^2 = 4\|Y_s - Y_t\|_{L_2}^2$$

and conclude from Slepian's lemma that $\mathbf{E}\,\lambda_1(A) \leqslant 2\kappa_n$. The reason for the factor 2 in the first equality is that A is a standard Gaussian vector in the space $\mathsf{M}_n^{\mathrm{sa}}$ times $\sqrt{2}$. The argument for the inequality is a special case of that from Exercise 6.47, but using the bra-ket notation makes it easier to rewrite it when A is a GUE(n) matrix, in which case we get $\mathbf{E}\,\lambda_1(A) \leqslant \sqrt{2}\kappa_{2n}$

Exercise 6.49. Let G_1, G_2, G_3 be standard Gaussian vectors in $\mathbb{R}^m$, $\mathbb{R}^n$, $\mathbb{R}^m \otimes \mathbb{R}^n$ respectively. Compare the processes $X_{(t,u)} = \langle G_3, t \otimes u\rangle$ and $Y_{(t,u)} = r_L\langle G_1, t\rangle + r_K\langle G_2, u\rangle$ (indexed by $(t, u) \in K \times L$) via Slepian's lemma. To deduce the inequality for the usual mean width, use Proposition A.1(ii).

Exercise 6.50. Here is an outline of the complex case, the real case being similar. Proceed inductively as follows. Choose a (random) unitary matrix $V_0 \in \mathsf{U}(s)$ with the property that the matrix BV_0 has a zero entry at position $(1, j)$ for $j > 1$, while the $(1, 1)$-entry α is positive (note that α follows a $\chi(s)$ distribution). Then choose a (random) unitary matrix $U_0 \in \mathsf{U}(n)$ with the properties that $U|1\rangle = |1\rangle$ and that the matrix U_0BV_0 has a zero entry at position $(i, 1)$ for $i > 1$, while the $(2, 1)$-entry β is positive (note that β follows a $\chi(n-1)$ distribution). Repeat the procedure with the $(n-1) \times (s-1)$ bottom right block of U_0BV_0, which has independent $N_{\mathbb{C}}(0, 1)$ entries and is independent of α, β.

Once the Lemma is proved, the second part of the exercise follows formally from the facts that (a) B has the same distribution as WBX where $W \in \mathsf{U}(n)$ and $X \in \mathsf{U}(s)$ are Haar-distributed and independent of B and (b) if U is a random or deterministic unitary matrix and W is Haar-distributed and independent of U, then UW is Haar-distributed and independent of U.

Exercise 6.51. (i) Write $R = UAV$ as in Lemma 6.39, use Jensen's inequality to obtain $\mathbf{E}\,\|A\| \geqslant \|\mathbf{E}\,R\| = \|M\|$. (ii) Write $\|M\| \geqslant |Mx|/|x|$ where x is the vector $(1, \ldots, 1, 0, \ldots, 0)$ with k occurrences of "1", and use the lower bounds $\kappa_{s+1-i} \geqslant \sqrt{s-k}$ and $\kappa_{n+1-i} \geqslant \sqrt{n-k}$ for $2 \leqslant i \leqslant k$. The same argument applies to the complex case (with $\kappa_j^{\mathbb{C}}$ in place of κ_j).

Exercise 6.52. It is enough to show that the relation $\langle\Omega|P(a_i + a_i^\dagger)|\Omega\rangle = \int P\,\mathrm{d}\mu_{\mathrm{SC}}$ holds for every polynomial P. We reduce to the case $P(X) = X^n$ and check by expansion that $\langle\Omega|(a_i + a_i^\dagger)^n|\Omega\rangle$ is the number of Dyck paths (see Section 2.1.1 in [**AGZ10**]) of length $2n$, which is the nth Catalan number, and also equals $\int x^n\,\mathrm{d}\mu_{\mathrm{SC}}(x)$, see (6.34).

Chapter 7

Exercise 7.1. First, even if K is not symmetric, $\ell_K(-T) = \ell_K(T)$ due to symmetry of G. The only part that is not straightforward is (ii). By homogeneity we may assume $\|T\|_{\mathrm{op}} = 1$, and using (i) we may also assume that T is an extreme point of S_∞^n (the operator norm unit ball). This means that $T \in \mathsf{O}(n)$ (see Exercise 1.44), and then it follows from the rotational invariance of the Gaussian measure that $\ell_K(ST) = \ell_K(S)$. Note also that the second inequality in (v) is (1.13).

Exercise 7.2. No. Choosing T being a rank 1 operator, and S a rotation, one would get from Proposition 7.1(v) that all 1-dimensional projections of K° have the same length. This is true only if K is a Euclidean ball.

Exercise 7.3. (i) Note that $\|\!|\cdot|\!\|_{B_2^n}$ is the Euclidean norm associated to the inner product (7.2), and so $\mathbf{K}(B_2^n)$ is the norm of $\tilde{R}_1$ as an element of $B(\mathscr{H}_{k,n})$, which equals 1 since it is an orthonormal projection. (ii) First prove that $\mathbf{K}(K) = \mathbf{K}(TK)$ for any $T \in \mathsf{GL}(n,\mathbb{R})$; this follows from the formulas $\|\!|\Theta|\!\|_{TK} = \|\!|\Theta \circ T^{-1}|\!\|_K$ and $\tilde{R}_1(\Theta \circ T^{-1}) = \tilde{R}_1(\Theta) \circ T^{-1}$ for $\Theta \in \mathscr{H}_{k,n}$. Then show $\mathbf{K}(K) \leqslant d_g(K, B_2^n)$ using (i). (iii) Use Exercise 4.20.

Exercise 7.4. Use (7.4) and the fact that $\tilde{R}_1$ is self-adjoint in $\mathscr{H}_{k,n}$.

Exercise 7.5. Let $f : \mathbb{R}^k \to \mathbb{R}$ be the indicator function of $\mathbb{R}_+^k$, and z be the vector $(1,\dots,1)/\sqrt{k}$. We compute, for $x = (x_1,\dots,x_k) \in \mathbb{R}^k$,

$$R_1 f(x) = [\mathbf{E}\, f(G)\langle G, z\rangle]\,\langle x, z\rangle = \frac{1}{\sqrt{2\pi}} 2^{-k}(x_1 + \cdots + x_k).$$

It follows that

$$\tilde{R}_1(\Theta)(x_1,\dots,x_k) = \frac{1}{\sqrt{2\pi}}\, 2^{-k} \sum_{\varepsilon \in \{-1,1\}^k} \langle x, \varepsilon\rangle e_\varepsilon.$$

Since $\mathbf{E}\,|\langle G, \varepsilon\rangle| = \sqrt{k}\sqrt{2/\pi}$ for any $\varepsilon \in \{-1,1\}^k$, we obtain $\|\!|\tilde{R}_1(\Theta)|\!\|_{B_1^N} = \frac{1}{\pi}\sqrt{k}$, while $\|\!|\Theta|\!\|_{B_1^N} = 1$. For the last equality, appeal to Exercise 7.4.

Exercise 7.6. The version on S is: if $f : S \to \mathbb{C}$ is a holomorphic function on S such that $|f| \leqslant \lambda$ on S and $|f| \leqslant 1$ on $\mathbb{R}$, then $|f'(0)| \leqslant e \log \lambda$. Reduce to the case $f(0) = 0$ by considering $z \mapsto (f(z) - f(-z))/2$. Use the Hadamard three-lines lemma to conclude that $|f(z)| \leqslant \lambda^{|\operatorname{Im} z|}$. Write $f(z) = zg(z)$ and use the maximum principle (with $0 < t < 1$) to show that $|g(z)| \leqslant \lambda^t/t$ for $|\operatorname{Im} z| \leqslant t$. The optimal choice $t = 1/\log\lambda$ gives $|f'(0)| \leqslant e\log\lambda$.

Exercise 7.7. (i) We have $T_\alpha = \{f \leqslant M_f\}_\alpha \cap \{f \geqslant M_f\}_\alpha$; use Corollary 5.14. (ii) For $x \in S^{n-1}$, use $w((B_\beta)^c, x) \leqslant \cos\beta$ when $x \in B$ and $w((B_\beta)^c, x) \leqslant 1$ otherwise. (iii) Check numerically that $\varepsilon - \alpha \geqslant (1 - \sqrt{\log(2)/6}) \geqslant 0.66\varepsilon$ and $\frac{1+\cos 0.66\varepsilon}{2} \leqslant \sqrt{1 - \varepsilon^2/6}$ for $\varepsilon \in (0,1)$. Apply (ii) with $B = T_\alpha$ and $\beta = \varepsilon - \alpha$ to get

$$w_G(A) = \kappa_n w(A) \leqslant \sqrt{n}\frac{1+\cos\beta}{2} \leqslant \sqrt{n - n\varepsilon^2/6} \leqslant \sqrt{n - (k+1)} \leqslant \kappa_{n-k}.$$

Exercise 7.8. Let E be a random $c\varepsilon^2 n$-dimensional subspace. Since g is 1-Lipschitz and circled with mean μ_f, we can choose $c > 0$ such that $\operatorname{osc}(g, S_E, \mu_f) \leqslant \varepsilon/3$ with high probability, by Theorem 7.15. Moreover, we can write

$$h(x) \leqslant \frac{2\pi}{n} + \max\{|f(e^{i2k\pi/n}x) - f(x)| \;:\; 1 \leqslant k \leqslant n\}.$$

Using the union bound and Lévy's lemma shows that $M_h = O(\sqrt{\log n}/\sqrt{n})$, where M_h is a median of h. We can choose C such that $M_h \leqslant \varepsilon/3$. Another application of Theorem 7.15 (the function h is 2-Lipschitz and circled) gives that $\operatorname{osc}(h, S_E, M_h) \leqslant \varepsilon/3$ with high probability, for some choice of c. We conclude by using the triangle inequality in the form $\operatorname{osc}(f, S_E, \mu_f) \leqslant \operatorname{osc}(g, S_E, \mu_f) + M_h + \operatorname{osc}(h, S_E, M_h) \leqslant \varepsilon$.

Exercise 7.9. We have $\operatorname{inrad}(K)\operatorname{inrad}(K^\circ) \geqslant 1/A$ and $w(K)w(K^\circ) \geqslant 1$ (see Exercise 4.37). The second statement follows from Exercise 4.20.

Exercise 7.10. Without loss of generality we may assume that $\operatorname{inrad}(K \cap E) \geqslant 1$ and $\operatorname{outrad}(K \cap E) < A$. For $x \in E$ and $y \in E^\perp$, define $T_\lambda(x + y) = x + \lambda y$. As λ tends to $+\infty$, the inradius of $T_\lambda K$ tends to 1 and $w_G((T_\lambda K)^\circ)$ tends to $w_G((K \cap E)^\circ \cap E) > A^{-1}\kappa_k$. Therefore, for λ large enough, one has $k_*(T_\lambda K) \geqslant (\kappa_k/\kappa_n)^2 n/A^2 \geqslant (k-1)A^{-2}$. Compare also with Exercise B.15.

Exercise 7.11. (i) Let A be the maximum of $||\cdot||$ on S^{n-1}. Prove that $A \leqslant 1+\beta+\delta A$, yielding the upper bound in (7.14); the lower bound follows. (ii) Adjust the values of δ, α, β such that (7.14) implies $1 - \varepsilon \leqslant \|x\| \leqslant 1 + \varepsilon$ for any $x \in S^{n-1}$; then use Lemma 5.3 and the union bound.

Exercise 7.12. (i) Let $\mathbb{R}^n = \bigoplus E_i$ be a decomposition of $\mathbb{R}^n$ as the direct sum of $N = \lceil n/k \rceil$ subspaces, with $\dim E_i \leqslant k$, and $O \in \mathsf{O}(n)$ Haar-distributed. Using the union bound, show that the decomposition $\mathbb{R}^n = \bigoplus O(E_i)$ has the desired property with positive probability. (ii) If x_i is the projection of x onto the i-th subspace in a decomposition from (i), write $||x|| \leqslant \sum ||x_i|| \leqslant 2M \sum |x_i| \leqslant 2M\sqrt{N}|x|$. (iii) Use (ii) and the fact that $||x|| = b|x|$ for some $x \neq 0$.

Exercise 7.13. Let $K_r \subset \mathbb{R}^2$ be a disk of radius 1 centered at $(r, 0)$. Then $\lim_{r\to 1} \dim_V(K_r) = \infty$, or otherwise one would find a polytope P with $K_1 \subset P \subset 4K_1$, which is not possible.

Exercise 7.14. The n^2-dimensional convex body $B_1^n \times \cdots \times B_1^n$ has $(2n)^n$ vertices and $n2^n$ facets.

Exercise 7.15. Mimic the proof of Theorem 7.29, replacing the use of Lemma 7.28 by the inequalities $\dim_F(K, A)a(K)^2 \geqslant (n-1)/2A^2$ and $\dim_V(K, B)a(K)^2 \geqslant (n-1)/2B^2$.

Exercise 7.16. If the codimension of E is k, then E nontrivially intersects $\mathbb{R}^{k+1}$ (seen as a subspace of $\mathbb{R}^n$), on which $\|\cdot\|_1 \leqslant \sqrt{k+1}|\cdot|$.

Exercise 7.17. For $p < \infty$, mimic the proof of Theorem 7.31. For $p = \infty$, use Lemma 6.16.

Exercise 7.18. (i) is equivalent to the existence of a linear map $A : \mathbb{R}^k \to \mathbb{R}^n$ such that $(1+\varepsilon)^{-1}|x| \leqslant \|A(x)\|_\infty \leqslant |x|$ for any $x \in S^{k-1}$. The map A has the form $x \mapsto (\langle x, x_1\rangle, \ldots, \langle x, x_n\rangle)$ for $x_1, \ldots, x_n \in \mathbb{R}^k$. We have $|x_i| \leqslant 1$ and may assume $|x_i| = 1$ by replacing x_i with $x_i/|x_i|$. On the other hand, since $K \subset L$ is equivalent to the inequality $w(K, \cdot) \leqslant w(L, \cdot)$ between widths, the inclusion (7.22) means precisely that $(1+\varepsilon)^{-1}|x| \leqslant \|A(x)\|_\infty$ for $x \in \mathbb{R}^k$, hence the equivalence.

Exercise 7.19. (i) Denote $S = (T^{-1})^*$. We have $w_G((TB_p^n)^\circ) = \mathbf{E}\,\|T^{-1}G\|_p$, where G is a standard Gaussian vector in $\mathbb{R}^n$. The ith component of the random vector $T^{-1}G$ has variance $\sigma_i^2 = |Se_i|^2$ and therefore $\mathbf{E}\,\|T^{-1}G\|_p \leqslant \left(\mathbf{E}\,\|T^{-1}G\|_p^p\right)^{1/p} = m_p(\sum \sigma_i^p)^{1/p} \leqslant n^{1/p} m_p \max \sigma_i$, where m_p denotes the L_p-norm of an $N(0,1)$ Gaussian variable. On the other hand,

$$\operatorname{inrad}(T(B_p^n)) \leqslant \operatorname{inrad}(T(B_\infty^n)) = \operatorname{outrad}(SB_1^n)^{-1} = (\max \sigma_i)^{-1}.$$

It follows (cf. (A.1)) that $k_*(TB_p^n) \leqslant Cpn^{2/p}$.

Exercise 7.20. (i) Since $\|\cdot\| \geqslant (1+2\varepsilon)|\cdot|$, we have $(1+2\varepsilon) \leqslant A(1+\varepsilon)$, so $A \geqslant 1+\varepsilon/2$. Similarly, $\|\cdot\| \leqslant 2|\cdot|$ implies $A \leqslant 2$ and therefore $A \geqslant 1+\varepsilon A/4$. To get (ii), subtract $|x|$ from (7.24).

Exercise 7.21. We have $\operatorname{inrad}(K) = 1/\sqrt{m}$ and $\mathbf{E}\,\|G\|_K = m\kappa_n$ if G is a standard Gaussian vector in $\mathbb{R}^{mn}$.

Exercise 7.22. By Theorem 7.31, there is a subspace $E \subset \mathbb{R}^N$ of dimension $n = cN\varepsilon^2$ such that $P := E \cap B_1^N$ is $(1+\varepsilon)$-Euclidean. The polytope P has at most 2^N facets (since taking sections of polytopes cannot increase the number of facets). The polytope P also has at most 3^N vertices, since every vertex of P is the intersection of E with some face of B_1^N, and B_1^N has 3^N faces.

Exercise 7.23. (i) Let $B : \Omega \to \mathsf{M}_{n,s}$ be a standard Gaussian vector and let $W = W_{n,s} := BB^\dagger$ be the corresponding Wishart matrix. Consider first $p \in [1,\infty)$. As in the proof of Theorem 7.37, the problem is reduced to showing that $\mathbf{E}\,\|B\|_p = \mathbf{E}\left(\operatorname{Tr} W^{p/2}\right)^{1/p} \sim \alpha_p\, n^{1/p+1/2}$ or, equivalently, that

$$\mathbf{E}\left(n^{-1}\operatorname{Tr}(n^{-1}W)^{p/2}\right)^{1/p} = \mathbf{E}\left(\int |x|^{p/2}\,\mathrm{d}\mu_{\mathrm{sp}}(n^{-1}W)\right)^{1/p} \sim \alpha_p.$$

(Above and in what follows all expected values $\mathbf{E}$ are calculated on the probability space Ω, and all integrals are over $\mathbb{R}$, often with respect to empirical spectral measures depending on $\omega \in \Omega$.) Recalling that $\alpha_p = \left(\int |x|^{p/2}\,\mathrm{d}\mu_{\mathrm{MP}(\lambda)}\right)^{1/p}$, we see that we need to exploit the convergence $\mu_{\mathrm{sp}}(n^{-1}W) \to \mu_{\mathrm{MP}(\lambda)}$ explained in Section 6.2.3.2. However, there are a few technical points that need to be resolved. First, it is not enough to work with the weak convergence of measures since (by definition) $\nu_n \to \nu$ weakly iff $\int f\mathrm{d}\nu_n \to \int f\mathrm{d}\nu$ for every bounded continuous function, and $f(x) = |x|^{p/2}$ is not bounded. To address this problem, appeal to ∞-Wasserstein convergence and argue as in Exercise 6.28 (i.e., using Theorem 6.28 and Lemma 6.20) to conclude that $n^{-1}\operatorname{Tr}(n^{-1}W)^{p/2} \to \int |x|^{p/2}\,\mathrm{d}\mu_{\mathrm{MP}(\lambda)} = \alpha_p^p$ in probability, and similarly after raising all quantities to the power $1/p$.

Next, as every student of real analysis knows, the convergence $X_n \to Y$ in probability does not generally imply convergence in mean $\mathbf{E}\,X_n \to \mathbf{E}\,Y$: one only knows from Fatou's lemma that $\liminf_n \mathbf{E}\,X_n \geqslant \mathbf{E}\,Y$. However, we do have convergence in mean under some tightness assumptions, for example when the second moments $\mathbf{E}\,X_n^2$ are uniformly bounded. (Prove this if it sounds unfamiliar.) In our setting, we have

$$X_n = \left(n^{-1}\operatorname{Tr}(n^{-1}W)^{p/2}\right)^{1/p} \leqslant \|n^{-1}W\|_\infty^{1/2} = \|n^{-1/2}B\|_\infty.$$

To conclude, verify that Proposition 6.33 (or Corollary 6.38 in the real case) implies $\mathbf{E}\,\|n^{-1/2}B\|_\infty^2 \lesssim \lambda$. This is a simple instance of upper-bounding L_p-norms in presence of ψ_2 estimates explained in Section 5.2.6; actually it easily follows that $\mathbf{E}\,\|n^{-1/2}B\|_\infty^2 \sim (1+\sqrt{\lambda})^2$.

The case $p = \infty$ is easier since the quantities in question are more tangible; it follows from Proposition 6.31 (or Corollary 6.38) and Exercise 6.51. Note that the lower bound also follows formally from the case $p < \infty$ by using the facts that $\|\cdot\|_\infty \geqslant n^{-1/p}\|\cdot\|_p$ and $\lim_{p\to\infty}\alpha_p = 1+\sqrt{\lambda}$, while the upper bound is implicit in the last calculation above.

(ii) Argue in a similar way by using the analogous results from Section 6.2.2 concerning GUE/GOE matrices.

Exercise 7.24. The bounds on the mean width appear in (7.25). The bounds on the volume radius follow from the inequalities $\mathrm{vrad}(S_p^{m,n}) \leqslant w(S_p^{m,n})$ (Urysohn's inequality) $\mathrm{vrad}(S_p^{m,n})\,\mathrm{vrad}(S_q^{m,n}) \geqslant c$ (the inverse Santaló inequality). The constants C, c are independent of $p \in [1,2]$ (in addition to being dimension independent).

Exercise 7.25. (i) Let $\mathcal{M}$ and $\mathcal{N}$ be $\varepsilon/4$-nets in $(S^{m-1}, |\cdot|)$ and $(S^{n-1}, |\cdot|)$ respectively, and consider $P = \mathrm{conv}\{|x\rangle\langle y| \ : \ x \in \mathcal{M}, y \in \mathcal{N}\}$ (cf. the proof of Lemma 9.2). Use Lemma 5.3 to upper-bound the size of the nets. (ii) If $d_{BM}(E \cap S_\infty^{m,n}, B_2^k) \leqslant 2$, (i) implies that $d_{BM}(E \cap P^\circ, B_2^k) \leqslant 4$. Since $E \cap P^\circ$ is a 4-Euclidean polytope with C_0^{m+n} faces, Remark 7.34 implies that $k = O(n)$, as needed. (iii) If $d_{BM}(E \cap S_p^{m,n}, B_2^k) \leqslant 2$, then by (1.31) $d_{BM}(E \cap S_\infty^{m,n}, B_2^k) \leqslant 2m^{1/p}$. By Remark 7.22, $k_*(E \cap S_\infty^{m,n}) \geqslant km^{-2/p}/4$. This implies (Theorem 7.19) that $S_\infty^{m,n}$ has a 2-Euclidean section of dimension $ckm^{-2/p}$, hence we conclude from (ii) that $k \leqslant Cnm^{2/p}$.

Exercise 7.26. Identifying $\mathbb{C}^n$ with $\mathbb{R}^{2n}$, the ellipsoid John(K) is circled (as a consequence of its uniqueness, it inherits all the symmetries from K), and therefore we may assume that K is in John position. It suffices to check that Lemma 7.41 transfers *mutatis mutandis* to the complex case.

Exercise 7.27. (i) By the result from Exercise 7.9, we have either $k_*(K) \geqslant \sqrt{n}$ or $k_*(K^\circ) \geqslant \sqrt{n}$. Assuming the latter without loss of generality, it follows from Corollary 7.24 that there exists a subspace F of dimension $c\sqrt{n}$ such that $P_F K$ is 2-Euclidean. Conclude by applying Corollary 7.40 to $K \cap F$. (ii) Yes, since we can choose a position for which the Haar measure on $\mathsf{Gr}(k, \mathbb{R}^n)$ concentrates near E, see Exercise B.15.

Exercise 7.28. (a) Without loss of generality, one may assume that John$(K) = B_2^n$. Set $A := \mathrm{vrad}(K) = (\mathrm{vol}(K)/\mathrm{vol}(B_2^n))^{1/n}$. From Lemma 5.8, we obtain that $N(K, B_2^n) \leqslant \mathrm{vol}(K + B_2^n)/\mathrm{vol}(B_2^n) \leqslant \mathrm{vol}(2K)/\mathrm{vol}(B_2^n) \leqslant (2A)^n$. It follows that $K \cap E$ is covered by $(2A)^n$ translates of $B_2^n \cap E$, hence $\mathrm{vol}(K \cap E) \leqslant (2A)^n \mathrm{vol}(B_2 \cap E)$, which is the claimed estimate. (b) Consider $K = B_\infty^n \times B_2^N$ and check that $\mathrm{vrad}(K)$ is bounded by an absolute constant whenever $N \geqslant Cn \log n$, whereas $\mathrm{vrad}(B_\infty^n) = \Theta(\sqrt{n})$.

Exercise 7.29. The arguments in parts (i) and (ii) of the Exercise are identical, the key observation being that the intersection of two (or three) events with large probability also has large probability. For the first statement, use the fact that if E is Haar-distributed on $\mathsf{Gr}(k, \mathbb{R}^{2k})$, so is $E^\perp$. For the second statement, fix an orthogonal decomposition $\mathbb{R}^{3k} = F_1 \oplus F_2 \oplus F_3$ and consider $E_i = O(F_i)$ for $O \in \mathsf{O}(3k)$ Haar-distributed.

Exercise 7.30. Follows from Theorem 7.44 by duality.

Exercise 7.31. Both sets equal $E \cap (K + G)$.

Exercise 7.32. Use the example from Exercise 7.14, and Proposition 5.6.

Exercise 7.33. (i) It follows from Lemma 4.20 that $\mathrm{vol}(K) \geqslant 2^{-n} \mathrm{vol}(K \cap (E_1 \oplus E_2))\,\mathrm{vol}(K_3)$ and that $\mathrm{vol}(K \cap (E_1 \oplus E_2)) \geqslant 2^{-n} \mathrm{vol}(K_1)\,\mathrm{vol}(K_2)$. To obtain inequalities for K°, proceed similarly using (1.12) and (1.13). (ii) Use (4.55). (iii) Follows easily from part (ii).

Exercise 7.34. Apply Corollary 7.24 with $K = B_\infty^n$. Using the fact that $k_*(B_1^n) = \Omega(n)$, deduce that there exists a subspace $E \subset \mathbb{R}^n$ of dimension $\Omega(n\varepsilon^2)$ such that $P_E B_\infty^n$ is $(1+\varepsilon)$-Euclidean. Then note that $P_E B_\infty^n$ can be written as the Minkowski sum of n segments. Observe that an isomorphic version of the statement follows from Exercise 7.30.

Chapter 8

Exercise 8.1. Show that $\mathbf{P}(\mathrm{Seg} \cap U(\mathrm{Seg}) \neq \varnothing) = 0$ by arguing as in the proof of Theorem 8.1.

Exercise 8.2. This follows by restricting the minimum to product states, since $S_p(\rho \otimes \sigma) = S_p(\rho) + S_p(\sigma)$.

Exercise 8.3. (i) Use concavity of S_p. (ii) Define $\widetilde{\Phi}$ and $\widetilde{\Psi}$ as $\widetilde{\Phi}(\rho) = \Phi(\rho) \otimes \tau$ and $\widetilde{\Psi}(\rho) = \sigma \otimes \Psi(\rho)$, where σ (resp., τ) is a state minimizing the output entropy of Φ (resp., Ψ). Let $\Xi = \widetilde{\Phi} \oplus \widetilde{\Psi}$; then $S_p^{\min}(\Xi^{\otimes 2}) = S_p^{\min}(\widetilde{\Phi} \otimes \widetilde{\Psi}) < S_p^{\min}(\widetilde{\Phi}) + S_p^{\min}(\widetilde{\Psi}) = 2S_p^{\min}(\Xi)$.

Exercise 8.4. The right-hand side of (8.11) is achieved on extreme points, i.e., $\pm|\psi\rangle\langle\psi|$ for $\psi \in S_{\mathbb{C}^2}$. An immediate computation shows that $\|\Phi(\pm|\psi\rangle\langle\psi|)\|_p = 2^{1/p}/2$. On the other hand, if $\psi \perp \varphi$, then $\|\Phi(|\psi\rangle\langle\varphi|)\|_p = \||\psi\rangle\langle\varphi|\|_p = 1 > 2^{1/p}/2$.

Exercise 8.5. (i) Observe that nonzero eigenvalues of anti-symmetric matrices come in pairs. (ii) Argue as in the proof of Proposition 8.6, using (i) instead of Lemma 8.7.

Exercise 8.6. (i) The Choi matrix of R is $C(R) = \frac{1}{d}\mathrm{I}_{\mathbb{C}^d \otimes \mathbb{C}^d}$. (ii) Use direct computation, or argue that $\mathsf{G} = \{B^j A^k \ : \ 1 \leqslant j \leqslant d,\ 1 \leqslant k \leqslant d\}$ is a group (with the counting measure as Haar measure) which generates M_d as a vector space and therefore the argument used in the proof of Proposition 2.18 yields $\frac{1}{d^2}\sum_{G \in \mathsf{G}} G X G^\dagger = \operatorname{Tr} X \frac{\mathrm{I}}{d}$.

Exercise 8.7. The idea is to follow carefully the proof of Lemma 8.12 to come up with an exact calculation instead of an estimate.

Recall that the argument shows that L_k, the Lipschitz constant of the function $\psi \mapsto E(\psi)$, is *the same* as that of the function f from (8.17) and, in particular, independent of d (as long as $d \geqslant k$, which we assume). Next, compute w, the tangent (to S^{k-1}) component of the gradient of f; the supremum of the Euclidean norm of w will be equal to L_k. By direct calculation, show that $|w|^2 = 4F$ with

$$F = \sum_j p_j \big(\log(1/p_j)\big)^2 - H^2, \tag{E.4}$$

where $p_j = x_j^2$ (in the notation of (8.17)) and $H := \sum_j p_j \log(1/p_j)$. To find the maximum of F over the set $\{(p_j) \ : \ p_j \geqslant 0,\ \sum_j p_j = 1\}$, use Lagrange multipliers and deduce that the extremal sequences (p_j) take only two values, namely such that $\log(1/p_j) = (1 + H) \pm \alpha$, for some $\alpha > 0$. By analyzing the constraints, show that the maximum of the objective function F equals to $\alpha^2 - 1$, and that it is achieved when the smaller value of p_j is repeated $k-1$ times, in which case $\alpha = \alpha_k$ is the positive root of the equation

$$e^{2\alpha}(\alpha - 1)/(\alpha + 1) = k - 1, \tag{E.5}$$

which implies that $\alpha_k \sim \frac{1}{2}\log k$ (as $k \to \infty$). Since the argument shows that $L_k = 2\sqrt{\alpha_k^2 - 1}$, deduce the conclusion.

For any given value of k, L_k can be found numerically by solving equation (E.5); numerical evidence suggests that $\log k \leqslant L_k \leqslant \log(2k)$ for all $k \geqslant 2$.

Exercise 8.8. Fix an orthonormal basis $(\varphi_j)_{1\leqslant j\leqslant s}$ of F and define $y \in \mathbb{R}^n$ by $y_i = \big(\sum_{j=1}^s |\langle \varphi_j, i\rangle|^2\big)^{1/2}$. Then check that

$$\sup_{x\in F\,:\,|x|=1} \|x\|_\infty = \|y\|_\infty \geqslant \frac{1}{\sqrt{n}}|y| = \sqrt{\frac{s}{n}}.$$

Exercise 8.9. (i) Use Exercise 8.8 to show that $\mathcal{W}$ contains a unit vector ψ with largest Schmidt coefficient greater than $\sqrt{\alpha}$. This uses the identification of bipartite states with matrices (see Section 2.2.2) and the fact that the operator norm of a matrix is at least as large as the absolute value of the largest matrix element. (The latter seems very rough, but works; it is conceivable that refining the argument at this point could lead to closing the gap between the lower and the upper bound on the dimension of "very entangled subspaces", at least for some ranges of the parameters.) Then appeal to concavity of entropy to show that under this constraint the von Neumann entropy is maximized when the spectrum of $\rho = \operatorname{Tr}_{\mathbb{C}^m} |\psi\rangle\langle\psi|$ is $(\alpha, (1-\alpha)/(k-1), \ldots, (1-\alpha)/(k-1))$. (ii) This is a tedious but straightforward consequence of part (i); use the fact that if $y = \phi(x) = \Theta\big(x(1+\log x)\big)$ for $x \geqslant 1$, then $\phi^{-1}(y) = \Theta\big(y/(1+\log y)\big)$.

Exercise 8.10. Let $\theta : (\mathbb{R}^d, |\cdot|) \to (E, \|\cdot\|_{\mathrm{HS}})$ be an isometry. For $i, j \in \{1, \ldots, N\}$, consider the linear form $\phi_{i,j} : (\mathbb{R}^d)^k \to \mathbb{R}$ defined by

$$\phi_{i,j}(x_1, \ldots, x_k) = \langle i|\theta(x_1)\ldots\theta(x_k)|j\rangle.$$

Show that $\|\phi_{i,j}(x_1, \ldots, x_k)\| \leqslant N^{-k/2}|x_1|\cdots|x_k|$ and that $\sum_{i,j=1}^N |||\phi_{i,j}|||^2 = \frac{d^k}{N^{k-1}}$, so that $|||\phi_{i,j}||| \geqslant d^{k/2}/N^{(k+1)/2}$ for some indices (i,j). Then use Proposition 8.25(iii).

Chapter 9

Exercise 9.1. Use Proposition 6.3.

Exercise 9.2. Consider $A = |0\rangle\langle 0| - |1\rangle\langle 1|$ and $\mathcal{N} = \{\psi \in S_{\mathbb{C}^2} \ : \ \|\psi\|_\infty \leqslant \cos\alpha\}$. Then $\mathcal{N}$ is an α-net in $(S_{\mathbb{C}^d}, g)$ and $|\langle\psi|A|\psi\rangle| \leqslant \cos(2\alpha)$ for any $\psi \in \mathcal{N}$.

Exercise 9.3. For the first part, mimic the argument used in the proof of Proposition 8.28. The second part is straightforward, see (6.15).

Exercise 9.4. This is a reformulation of the statement from Lemma 4.3.

Exercise 9.5. The statement about the mean width is proved similarly as in the qubit case. For the lower bound, one may notice that since $\mathrm{Sep}((\mathbb{C}^2)^{\otimes k}))$ is a section of $\mathrm{Sep}((\mathbb{C}^d)^{\otimes k})$, its Gaussian mean width is smaller than that of the latter set. For the volume, to be able to generalize the argument from the proof of Theorem 9.11 one needs to find $\text{Löw}(\mathrm{D}(\mathbb{C}^d)_{\otimes})$. To that end, use Proposition 4.8 to show that it has the form $\mathscr{E}_{a,b} = (aP + bQ)B_{\mathrm{HS}}$, where P is the projection onto the hyperplane of trace zero matrices and $Q = \mathrm{I} - P$. Check that $\mathrm{D}(\mathbb{C}^d)_{\otimes} \subset \mathscr{E}_{a,b} \iff a^{-2}(1-1/d) + b^{-2}/d \leqslant 1$. Minimizing $\operatorname{vol}\mathscr{E}_{a,b} = a^{d^2-1}b\operatorname{vol}(B_{\mathrm{HS}})$ under this constraint gives $a = \sqrt{d/(d+1)}$ and $b = \sqrt{d}$.

Exercise 9.6. For the bound on $w(\mathrm{PPT}^\circ)$, argue as in the proof of Theorem 9.13, but in the last step use Exercise 5.28. In the displayed formula, the first inequality is Urysohn's inequality. For the second one, use the bound on $w(\mathrm{PPT}^{\circ})$ and appeal to the dual Urysohn inequality (Proposition 4.16).

Exercise 9.7. This follows from the fact that the measure μ_{d^2,d^2} on $\mathrm{D}(\mathbb{C}^d \otimes \mathbb{C}^d)$ is proportional to the Lebesgue measure (see the discussion following (6.47)), and from Proposition 6.34.

Exercise 9.8. (i) The operator whose norm has to be estimated is $A = \sum_i B_i$ with $B_i = \sum_j A_{ij} \otimes |i\rangle\langle j|$. Since $B_{i_1}^\dagger B_{i_2} = 0$ for $i_1 \neq i_2$, it follows that

$$\|A\|^2 = \|A^\dagger A\| = \|\sum_i B_i^\dagger B_i\| \leqslant \sum_i \|B_i^\dagger B_i\|.$$

Next, using $B_i B_i^\dagger = \left(\sum_j A_{ij} A_{ij}^\dagger\right) \otimes |i\rangle\langle i|$, conclude that $\|B_i^\dagger B_i\| = \|B_i B_i^\dagger\| = \|\sum_j A_{ij} A_{ij}^\dagger\| \leqslant \sum_j \|A_{ij} A_{ij}^\dagger\| = \sum_j \|A_{ij}\|^2$, as needed. (ii) It is enough to prove (see the comment following Theorem 9.15) that every matrix $A \in B^{\mathrm{sa}}(\mathbb{C}^{d_1} \otimes \mathbb{C}^{d_2})$ with $\|A\|_{\mathrm{HS}} \leqslant 1$ satisfies $\mathrm{I} + A \in \mathcal{SEP}$. By Theorem 2.34 and Remark 2.35, it suffices to show that for every unital positive map $\Phi : \mathsf{M}_{d_1} \to \mathsf{M}_{d_2}$, we have $\mathrm{I} + (\Phi \otimes \mathrm{Id})(A) \geqslant 0$. Writing A as $\sum A_{ij} \otimes |i\rangle\langle j|$, we have

$$\|(\Phi \otimes \mathrm{Id})(A)\|_\infty^2 = \|\sum \Phi(A_{i,j}) \otimes |i\rangle\langle j|\|_\infty^2 \leqslant \sum \|\Phi(A_{i,j})\|_\infty^2 \leqslant \sum \|A_{i,j}\|_\infty^2 \leqslant 1,$$

from which the result follows. In the chain of inequalities we used successively (i), Exercise 2.30, and $\|\cdot\|_{\mathrm{op}} \leqslant \|\cdot\|_{\mathrm{HS}}$.

Exercise 9.9. Note that $P(\mathrm{D}(\mathbb{C}^2))$ is the shifted Bloch ball (a 3-dimensional real Euclidean ball with radius $1/\sqrt{2}$). For the last inequality, argue as in the proof of Proposition 8.28.

Exercise 9.10. Use Stirling's formula.

Exercise 9.11. (i) The fact that the projection contains the section is a general obvious fact. For $\rho = |1\rangle\langle 1| \otimes |1\rangle\langle 1|$, we have $P_H \rho = \frac{1}{mn} \mathrm{I}_{\mathbb{C}^m \otimes \mathbb{C}^n} + |1\rangle\langle 1| \otimes (|1\rangle\langle 1| - \frac{1}{n} \mathrm{I}_{\mathbb{C}^n})$, which is not positive. (ii) Use (1.13). (iii) We have $P_F \rho = \frac{I}{m} \otimes \mathrm{Tr}_{\mathbb{C}^m} \rho$ for every state ρ.

Exercise 9.12. Use the fact that D has enough symmetries. The argument suggested in the hint to Exercise 9.14 also works.

Exercise 9.13. We have $\mathrm{vrad}(P) \geqslant \frac{1}{4} \mathrm{vrad}(\mathrm{D}(\mathbb{C}^d)) \geqslant c/\sqrt{d}$ (see Table 9.1). If P has N vertices, Proposition 6.3 implies that $\mathrm{vrad}(P) = O(\sqrt{\log N}/d)$, and the result follows. We point out that the result can also be proved by arguing as in the proof of Proposition 9.31.

Exercise 9.14. Argue that the smallest $\lambda > 0$ such that $(-1) \bullet \mathrm{Sep} \subset \lambda \bullet \mathrm{Sep}$ equals $d^2 - 1$ by considering a pure product state.

Exercise 9.15. Denote by $E \subsetneq \mathbb{C}^d$ the range of $\Phi(\mathrm{I})$ (which is a positive operator) and consider $\tilde{\Phi} : X \mapsto \Phi(X) + P_{E^\perp} X P_{E^\perp}$. The map $\tilde{\Phi}$ is clearly positivity-preserving and has the property that, for any state ρ on $\mathbb{C}^d \otimes \mathbb{C}^d$, we have

$$(\Phi \otimes \mathrm{Id})(\rho) \in \mathcal{PSD} \iff (\tilde{\Phi} \otimes \mathrm{Id})(\rho) \in \mathcal{PSD}.$$

(The key point in inferring the latter is that positivity of Φ implies then that, for any $X \in \mathsf{M}_d$, the range of $\Phi(X)$ is contained in E.) Finally, define $\Psi : X \mapsto \tilde{\Phi}(\mathrm{I})^{-1/2} \tilde{\Phi}(X) \tilde{\Phi}(\mathrm{I})^{-1/2}$.

Exercise 9.16. (i) is an immediate consequence of Exercise 7.15 applied with $B = 4$. For (ii), proceed exactly as in the proof of Theorem 9.34.

Chapter 10

Exercise 10.1. Check that $\frac{1}{\|x\|_\infty}\sum_{i=1}^k x_i^\downarrow \leqslant \min(k, n-k) \leqslant \frac{2n}{\|y\|_1}\sum_{i=1}^k y_i^\downarrow$.

Exercise 10.2. First remark that the unitary invariance of A implies that A and $VAV^\dagger$ have the same distribution when V is a random unitary matrix independent of A (V is not assumed to be Haar-distributed). Now, let W be a (random) unitary matrix W such that $\operatorname{Diag}(\operatorname{spec} A) = WAW^\dagger$ (W is not unique but can be chosen as a measurable function of A). If U is a Haar-distributed random matrix independent of A, then UW is independent of A (this follows from the translation-invariance of the Haar measure). Finally, we may apply the initial remark with $V = UW$.

Exercise 10.3. Argue as in the proof of Proposition 10.4 using the event $E = \{\|B\|_1 \geqslant c_1 n\}$, and Lemma 10.2.

Exercise 10.4. Consider the random vector Z defined by $\mathbf{P}(Z = e_i) = \mathbf{P}(Z = -e_i) = \frac{1}{2n}$, where (e_i) is the canonical basis of $\mathbb{R}^n$. We show separately that (i) $\mathbf{E}\,\|X\|_K \leqslant C_1\,\mathbf{E}\,\|Z\|_K$ and (ii) $\mathbf{E}\,\|Z\|_K \leqslant C_2\,\mathbf{E}\,\|Y\|_K$. Writing X as a positive combination of $\pm e_i$ gives (i) with $C_1 = 2n\,\mathbf{E}\,\|X\|_1$. For (ii), denote $\mathcal{A} = \{\mathbf{E}[Y\mathbf{1}_A] : A \text{ measurable}\}$. Note that for any $x \in \operatorname{conv}\mathcal{A}$, we have $\|x\|_K \leqslant \mathbf{E}\,\|Y\|_K$. The convex hull of $\mathcal{A}$ has nonempty interior (otherwise Y would be almost surely contained in some hyperplane) and therefore contains the $2n$ vectors $\pm\varepsilon e_i$ ($1 \leqslant i \leqslant n$) for some $\varepsilon > 0$. It follows that $\varepsilon\,\mathbf{E}\,\|Z\|_K \leqslant \mathbf{E}\,\|Y\|_K$.

Exercise 10.5. This is obvious since $\mu_{d^2,s}$ has a density with respect to the Lebesgue measure.

Exercise 10.6. Consider the set $L_s = \{(\psi_1, \dots, \psi_s) \in (\mathbb{C}^d \otimes \mathbb{C}^d)^s : \sum_{i=1}^s |\psi_i\rangle\langle\psi_i| \in \mathcal{SEP}\}$. Since $\mathcal{SEP}$ is convex, we have the following fact: if $(\psi_1, \dots, \psi_{s-1}, \varphi) \in L_s$ and $(\psi_1, \dots, \psi_{s-1}, \chi) \in L_s$, then $(\psi_1, \dots, \psi_{s-1}, \frac{1}{\sqrt{2}}\varphi, \frac{1}{\sqrt{2}}\chi) \in L_{s+1}$. It follows that whenever $(\psi_1, \dots, \psi_s)$ is a Lebesgue point for L_s, then $(\psi_1, \dots, \psi_{s-1}, \frac{1}{\sqrt{2}}\psi_s, \frac{1}{\sqrt{2}}\psi_s)$ is a Lebesgue point for L_{s+1}. (A point x is a Lebesgue point for $A \in \mathbb{R}^n$ if the ratio $\operatorname{vol}(A \cap B(x, \varepsilon))/\operatorname{vol} B(x, \varepsilon)$ goes to 1 as ε goes to 0.) The result follows from the fact that almost every point of L_s is a Lebesgue point (see Chapter 3, Corollary 1.5 in **[SS05]**).

Exercise 10.7. Realize ρ as $\operatorname{Tr}_{\mathbb{C}^s} |\psi\rangle\langle\psi|$ for ψ uniformly distributed on $S_{\mathbb{C}^d \otimes \mathbb{C}^d \otimes \mathbb{C}^s}$ (i) For $d_1 \leqslant d_2$, identify $\mathbb{C}^{d_1}$ as a subspace of $\mathbb{C}^{d_2}$ and let $P : \mathbb{C}^{d_2} \to \mathbb{C}^{d_1}$ be the orthogonal projection. Show that the map

$$\rho \mapsto \frac{(P \otimes P)\rho(P \otimes P)}{\operatorname{Tr}(P \otimes P)\rho(P \otimes P)}$$

pushes forward $\mu_{d_2^2,s}$ onto $\mu_{d_1^2,s}$, and preserves separability. (ii) Identify $\mathbb{C}^{2d}$ with $\mathbb{C}^2 \otimes \mathbb{C}^d$. Let $\Phi : B(\mathbb{C}^{2d} \otimes \mathbb{C}^{2d}) \mapsto B(\mathbb{C}^d \otimes \mathbb{C}^d)$ be the partial trace over $\mathbb{C}^d \otimes \mathbb{C}^d$. Then Φ pushes forward $\mu_{4d^2,s}$ onto $\mu_{d^2,4s}$, and preserves separability.

Exercise 10.8. We use the same notation as in Exercise 10.7. Theorem 10.12 applied for $\varepsilon = 1/2$ gives, with $\delta := 2\exp(-\alpha^N)$ for some $\alpha > 1$:

- if k, N are such that $2^{N-2k} \leqslant \frac{1}{2}s_0(2^k)$, then $\pi_{2^k,2^{N-2k}} \leqslant \delta$,
- if k, N are such that $2^{N-2k} \geqslant \frac{3}{2}s_0(2^k)$, then $\pi_{2^k,2^{N-2k}} \geqslant 1-\delta$.

Set $p_k := \pi_{2^k,2^{N-2k}}$. Exercise 10.7(ii) implies that (p_k) is non-increasing. Define k_N as the smallest k such that $p_k < 1-\delta$. It is clear from the estimates (10.10) that $k_N \sim N/5$. Our definition of k_N implies that $2^{N-2k_N} < \frac{3}{2}s_0(2^{k_N})$, and therefore

$2^{N-2k_N-2} \leqslant \frac{1}{2}s_0(2^{k_N})$, so that $\pi_{2^{k_N},2^{N-2k_N-2}} \leqslant \delta$. By Exercise 10.7(i), this implies that $p_{k_N+1} \leqslant \delta$ and the Corollary follows.

Exercise 10.9. (i) We have $\operatorname{Tr}(\rho^2) = \frac{1}{d^2} + \operatorname{Tr}(W\rho)$. The value of $\mathbf{E}\operatorname{Tr}(\rho^2)$ was computed in Exercise 6.45. To obtain concentration, use the fact that $\operatorname{Tr}\rho^2$ is related to the Schatten 4-norm of M when $\rho = MM^\dagger$, with M distributed on the Hilbert–Schmidt sphere in $\mathsf{M}_{d^2,s}$. (ii) Let Π be the orthogonal projection onto the subspace $\mathbb{C}x \otimes \mathbb{C}^s$. The function $|\Pi\psi| = \sqrt{\langle x|\rho|x\rangle}$ is 1-Lipschitz as a function of ψ and satisfies $\mathbf{E}\,|\Pi\psi|^2 = 1/d^2$; use Exercise 5.46. (iii) Use Lemma 9.4 and the union bound.

Exercise 10.10. It follows from (the proof of) Carathéodory's theorem (see Exercise 1.1) that the infimum in (10.15) can be restricted to convex combinations of length at most d^4. Then use a compactness argument.

Exercise 10.11. The inradius of PPT is the same as that of Sep (see Table 9.1), so the argument that led to (10.14) carries over to the present setting. For the bound in (i), the relevant range of s is $\Theta(d^2)$.

Chapter 11

Exercise 11.1. If $n \geqslant 3$ is odd, argue as in the comment following Lemma 11.1. If $n = 2k$, identify $\mathbb{R}^n$ with $\mathbb{R}^2 \otimes \mathbb{R}^k$ and consider $E = F \otimes \mathrm{I}_k$, where $F \subset \mathsf{M}_2(\mathbb{R})$ is the subspace spanned by the two real Pauli matrices.

Exercise 11.2. This can be seen directly from the definition. Alternatively, we may use the description from Proposition 11.8. Let $\lambda \in (0,1)$ and $(a_{ij}), (a'_{ij}) \in \mathsf{QC}_{m,n}$. We have $a_{ij} = \langle x_i, y_j\rangle$ and $a'_{ij} = \langle x'_i, y'_j\rangle$. Defining $\tilde{x}_i = \sqrt{\lambda}\,x_i \oplus \sqrt{1-\lambda}\,x'_i$ and $\tilde{y}_j = \sqrt{\lambda}\,y_j \oplus \sqrt{1-\lambda}\,y'_j$ leads to $\lambda a_{ij} + (1-\lambda)a'_{ij} = \langle \tilde{x}_i, \tilde{y}_j\rangle$. We then argue as in the end of the proof of Proposition 11.8 to ensure that vectors live in $\mathbb{R}^{\min(m,n)}$.

Exercise 11.3. For vectors x_i, y_j of norm at most 1, the unit vectors $x'_i = x_i + \sqrt{1-|x_i|^2}\,u$ and $y'_j = y_j + \sqrt{1-|y_j|^2}\,v$ satisfy $\langle x_i, y_j\rangle = \langle x'_i, y'_j\rangle$ provided u, v are unit vectors in $\{y_j \,:\, 1 \leqslant j \leqslant n\}^\perp \cap \{x_i \,:\, 1 \leqslant i \leqslant m\}^\perp$ such that $u \perp v$.

Exercise 11.4. When considered as elements of $\mathbb{R}^4$, the 8 distinct matrices $A^{\xi,\eta} = (\xi_i\eta_j)_{i,j=1}^2$ are either opposite or orthogonal. A less explicit argument goes as follows: use Proposition 11.7, the fact that B_∞^2 is congruent to $\sqrt{2}\,B_1^2$, and that $B_1^m \mathbin{\widehat{\otimes}} B_1^n$ identifies with B_1^{mn} (cf. Exercise 11.8).

Exercise 11.5. Given $\xi \in \{-1,1\}^m$ and $\eta \in \{-1,1\}^n$, let $I = \{i \,:\, \xi_i = 1\}$ and $J = \{j \,:\, \eta_j = 1\}$ and split the overall sum $\sum b_{ij}\xi_i\eta_j$ into 4 sums according to whether $i \in I$ or not, $j \in J$ or not; then use the triangle inequality.

Exercise 11.6. For the first statement, note that $\{-1,1\}^k$ is exactly the set of extreme points of $B_\infty^k = (B_1^k)^\circ$. The second statement is even more straightforward from Proposition 11.8: $\{(x_i)_{i=1}^k \,:\, x_i \in \mathcal{H}, |x_i| \leqslant 1\}$ is exactly the unit ball of $\ell_\infty^k(\mathcal{H}) = \big(\ell_1^k(\mathcal{H})\big)^*$.

Exercise 11.7. Choose $\sigma = \tau = \sigma_z$, $\tilde{X}_i = X_i \otimes |0\rangle\langle 0|$, and $\tilde{Y}_j = Y_j \otimes |0\rangle\langle 0|$.

Exercise 11.8. First observe that Proposition 11.7 generalizes to the present context, with the same proof: $\mathsf{LC}_{2,\dots,2}$ identifies with $(B_\infty^2)^{\otimes k}$. Next use the facts that B_∞^2 is congruent to $\sqrt{2}B_1^2$, and that $(B_1^2)^{\widehat{\otimes} k}$ identifies with $B_1^{2^k}$. It follows that $\mathsf{LC}_{2,\dots,2}$ is congruent to $2^{k/2}B_1^{2^k}$, a polytope with 2^{k+1} vertices and 2^{2^k} facets.

Exercise 11.9. The answers are most conveniently deduced from the characterizations given by Propositions 11.7 and 11.8. The outradius is in both cases easily seen to be $\sqrt{mn}$. It is a little more delicate to establish that the inradii are 1. For the lower bound on the inradius of $\mathsf{LC}_{m,n} = B_\infty^m \hat{\otimes} B_\infty^n$, note that it is in Löwner position by Lemma 4.9 and then appeal to Exercise 4.20. For the remaining conclusions, use $\mathsf{LC}_{m,n} \subset \mathsf{QC}_{m,n} \subset B_\infty^{mn}$.

Exercise 11.10. Since $\mathsf{LC}_{m,n} = B_\infty^m \hat{\otimes} B_\infty^n$, this follows from Exercise 4.27 and the fact that a cube has enough symmetries (Exercise 4.25). More concretely, symmetries of LC are generated by permutations and sign flips of rows and columns. Since these operations are also symmetries for QC, it follows that QC has likewise enough symmetries.

Exercise 11.11. Taking into account Remark 11.9, it is enough to check that for every self-adjoint operator X_1, X_2, Y_1, Y_2 with $X_1^2 = X_2^2 = \mathrm{I}$ and $Y_1^2 = Y_2^2 = \mathrm{I}$, we have $\operatorname{Tr} \rho B \leqslant 2\sqrt{2}$, where $B = X_1 \otimes Y_1 + X_1 \otimes Y_2 + X_2 \otimes Y_1 - X_2 \otimes Y_2$. To that end, show that $B^2 = 4\,\mathrm{I} - (X_1X_2 - X_2X_1) \otimes (Y_1Y_2 - Y_2Y_1)$ and conclude that $\|B^2\|_{\mathrm{op}} \leqslant 8$. For an example giving violation $\sqrt{2}$, appeal to Proposition 11.8 and consider the case where $x_1, y_1, x_2, y_2 \in \mathbb{R}^2$ are unit vectors separated by successive $45°$ angles.

Here is an alternative argument which allows us to arrive at an example without guessing. First, observe that

$$\sup\{\varphi_{CHSH}(A) \;:\; A \in \mathsf{QC}_{2,2}\} = \frac{1}{2} \sup\{|y_1 + y_2| + |y_1 - y_2| \;:\; y_j \in \mathcal{H},\ |y_j| \leqslant 1\}$$

(cf. Exercise 11.6). Next, note that for such y_1, y_2,

$$|y_1 + y_2| + |y_1 - y_2| \leqslant \sqrt{2}\big(|y_1 + y_2|^2 + |y_1 - y_2|^2\big)^{1/2} = 2\big(|y_1|^2 + |y_2|^2\big)^{1/2} \leqslant 2\sqrt{2}$$

and verify when equalities occur.

Exercise 11.12. By Exercise 11.4 and its hint, every normal to a facet is proportional to the sum of four vertices of that facet, which in turn are of the form $A^{\xi,\eta}$. All such sums can then be listed and classified: there are 8 that exhibit the CHSH pattern and another 8 with only one non-zero entry. Alternatively, one may notice that every such sum is a matrix of Hilbert-Schmidt norm 4, whose entries are even integers that sum up to ± 4. Finally, the functionals corresponding to matrices with only one non-zero entry cannot distinguish between classical and quantum correlations.

Exercise 11.13. If $m > n$, the set $\mathsf{LC}_{n,n}$ can be seen in a canonical way as a section of $\mathsf{LC}_{m,n}$, which in turn is a section of $\mathsf{LC}_{m,m}$, and similarly for $\mathsf{QC}_{n,n}$, $\mathsf{QC}_{m,n}$ and $\mathsf{QC}_{m,m}$. The fact that $\mathbf{K}_G^{(2)} \geqslant \sqrt{2}$ follows from Exercise 11.11, and the opposite inequality by combining Exercises 11.11 and 11.12.

Exercise 11.14. We have $\mathsf{LC}_{2,n} = B_\infty^2 \hat{\otimes} B_\infty^n$. Since B_∞^2 is congruent to $\sqrt{2}B_1^2$, it follows that $\frac{1}{\sqrt{2}}\mathsf{LC}_{2,n}$ is congruent to $B_1^2 \hat{\otimes} B_\infty^n$, which identifies with $B_\infty^n \oplus_1 B_\infty^n :=$ $\operatorname{conv}(\{(x,0) \;:\; x \in B_\infty^n\} \cup \{(0,x) \;:\; x \in B_\infty^n\})$. The facets of $B_\infty^n \oplus_1 B_\infty^n$ are of the form $\operatorname{conv}(F \times \{0\}, \{0\} \times G)$, where F, G are facets of B_∞^n. (This can be easily seen by identifying $(B_\infty^n \oplus_1 B_\infty^n)^\circ$ with $B_1^n \times B_1^n$.) It follows that $\mathsf{LC}_{2,n}$ has $(2n)^2$ facets: $4n$ facets express the fact that each entry of a correlation matrix belongs to $[-1,1]$, and $8\binom{n}{2} = 4n^2 - 4n$ are equivalent to the CHSH inequality.

Exercise 11.15. Fix $1 \leqslant i, j \leqslant 3$ and denote $E \subset \mathbb{R}^3 \times \mathbb{R}^3$ the subspace of matrices for which the ith row and the jth column are zero. It is clear from the definition that $P_E \mathsf{LC}_{3,3} = \mathsf{LC}_{2,2}$, where E is identified with $\mathbb{R}^2 \times \mathbb{R}^2$. It follows that whenever $\{\phi(\cdot) \leqslant 1\}$ is a facet-defining inequality for $\mathsf{LC}_{2,2}$, then $\{\phi(P_E(\cdot)) \leqslant 1\}$ is a facet-defining inequality for $\mathsf{LC}_{3,3}$. A careful counting (cf. Exercise 11.12) shows that this construction produces 18 facets of the kind $\pm a_{ij} \leqslant 1$ and $9 \times 8 = 72$ facets defined by inequalities equivalent to CHSH up to symmetries. The information that $\mathsf{LC}_{3,3}$ has 90 facets implies that $\mathsf{LC}_{3,3}$ is the intersection of the half-spaces associated to these $18 + 72 = 90$ facets. Since $P_E \mathsf{QC}_{3,3} = \mathsf{QC}_{2,2} \subset \sqrt{2}\mathsf{LC}_{2,2}$, it follows that $\mathsf{QC}_{3,3} \subset \sqrt{2}\mathsf{LC}_{3,3}$.

Exercise 11.16. If $M = (m_{ij})$, then

$$\|M : \ell_\infty^2(\mathbb{C}) \to \ell_1^2(\mathbb{C})\| = \max\Big\{ \sum_{i=1}^{m} \Big| \sum_{j=1}^{n} m_{ij} z_j \Big| \ : \ z_j \in \mathbb{C},\ |z_j| \leqslant 1,\ j = 1, \ldots, m \Big\}.$$

Since, as a real normed space, $(\mathbb{C}, |\cdot|)$ coincides with $(\mathbb{R}^2, |\cdot|)$, it remains to appeal to Exercise 11.6. (Note that we are concerned here with the case $m = n = 2$, but a similar argument works if $\min\{m, n\} = 2$.)

Exercise 11.17. Let $a, b, c, d \in \mathbb{C}$ and let $\phi : \mathbb{C} \to \mathbb{R}_+$ be defined by $\phi(z) = |az + b| + |cz + d|$. Then ϕ is convex and, in particular, its maximal value over the (closed) unit disk is attained on its boundary $\mathbb{T}$. Next, note that for $\eta_1, \eta_2 \in \mathbb{T}$ we have

$$|a\eta_1 + b\eta_2| + |c\eta_1 + d\eta_2| = \phi(\eta_1 \overline{\eta_2})$$

and, similarly, for $y_1, y_2 \in \mathbb{C}^2$ with $|y_1| = |y_2| = 1$,

$$|ay_1 + by_2| + |cy_1 + dy_2| = \phi(\langle y_1 | y_2 \rangle).$$

By the first observation, the maxima of these two expressions (over $\eta_1, \eta_2 \in \mathbb{T}$ and over unit vectors $y_1, y_2 \in \mathbb{C}^2$ respectively) coincide and it remains to notice that these maxima represent the expressions on the two sides of the inequality (11.37) for $[m_{ij}] = \begin{bmatrix} a & b \\ c & d \end{bmatrix}$.

Exercise 11.18. The polytope $\mathsf{LC}_{n,n}$ is a symmetric polytope with 2^{2n-1} vertices and dimension n^2 (see Proposition 11.7), so the result follows. For the "moreover" part, combine Exercise 7.15 and, if needed, Theorem 11.12. Note that, from general principles (see Exercise 4.20), $a(\mathsf{LC}_{n,n}) \leqslant n$ and $a(\mathsf{QC}_{n,n}) \leqslant n$ (in fact we have equality by Exercise 11.9).

Exercise 11.19. Via Santaló inequality and its reverse, Proposition 11.15 implies that $\operatorname{vrad}(\mathsf{LC}_{n,n}^\circ) = \Theta(1/\sqrt{n})$. Since $\operatorname{outrad}(\mathsf{LC}_{n,n}^\circ) = \operatorname{inrad}(\mathsf{LC}_{n,n}) = 1$ (see Exercise 11.9), Proposition 6.3 implies that $\mathsf{LC}_{n,n}^\circ$ has $\exp(\Omega(n))$ vertices, or equivalently that $\mathsf{LC}_{n,n}$ has $\exp(\Omega(n))$ facets.

Exercise 11.20. (a) The value of the game is $\sum_{i,j} \pi(i,j) m_{ij}\, \xi_i \eta_j$, where $(\pi(i,j))$ is the distribution on the set of inputs. If $\pi(i_0, j_0) < \frac{1}{4}$, choose ξ, η so that $(\xi_i \eta_j)$ agrees with (m_{ij}) except for the (i_0, j_0)th entry. (b) First, replacing (ξ, η) by $(-\xi, -\eta)$ does not change the outcome, so for each such pair of strategies only the sum of their probabilities matters. Next, there are four pairs of that kind that saturate (11.4) and (11.12), with each pair leading to a mismatch in exactly one of the four entries of the 2×2 matrices $(\xi_i \eta_j)$ and (m_{ij}). If one of these four pairs entered into the random strategy with a weight strictly larger than $\frac{1}{4}$, the referee could use as

the setting (i,j) the index of the corresponding mismatched entry.
The combination of (a) and (b) describes the von Neumann–Nash-type equilibrium for the CHSH game.

Exercise 11.21. Alice and Bob have a quantum strategy which gives the value of at least $\frac{\sqrt{2}}{2}$ independently of the distribution $(\pi(i,j))$ on the set of inputs; moreover, if that distribution is not uniform, they have a quantum strategy yielding a value strictly larger than $\frac{\sqrt{2}}{2}$. For the universal strategy, use the same x_i, y_j as those implicit in the hint to Exercise 11.11; it follows from the argument there that, when expressed in terms of x_i, y_j, such strategy is unique up to isometries of the Hilbert space in question. If $(\pi(i,j))$ is not uniform, then either $|\pi(1,1)y_1 + \pi(1,2)y_2| + |\pi(2,1)y_1 - \pi(2,2)y_2|$ or $|\pi(1,1)x_1 + \pi(2,1)x_2| + |\pi(1,2)x_1 - \pi(2,2)x_2|$ is strictly larger than $2\sqrt{2}$.

Exercise 11.22. Extreme points of the set $\mathrm{K}_{k,m}$ defined in (11.23) are deterministic distributions that are of the form $p(\xi|i) = \delta_{\xi,f(i)}$ for some function f. It follows from the Krein–Milman theorem that any conditional probability distribution is a convex combination of deterministic distributions. Since $\mathsf{LB} = \mathrm{K}_{k,m} \widehat{\otimes} \mathrm{K}_{l,n}$, the result follows.

Exercise 11.23. Consider $\lambda \in (0,1)$ and two boxes $P, \bar{P} \in \mathsf{QB}$. Represent $P = \{p(\xi,\eta|i,j)\}$ as $\{\operatorname{Tr} \rho(E_i^\xi \otimes F_j^\eta)\}$ and $\bar{P} = \{\bar{p}(\xi,\eta|i,j)\}$ as $\{\operatorname{Tr} \bar{\rho}(\bar{E}_i^\xi \otimes \bar{F}_j^\eta)\}$, where the operators $E_i^\xi, F_j^\eta, \bar{E}_i^\xi, \bar{F}_j^\eta$ act respectively on the Hilbert spaces $\mathcal{H}_A, \mathcal{H}_B, \bar{\mathcal{H}}_A, \bar{\mathcal{H}}_B$. Verify that

$$\lambda p(\xi,\eta|i,j) + (1-\lambda)\bar{p}(\xi,\eta|i,j) = \operatorname{Tr}\Big(\sigma\Big((E_i^\xi \oplus \bar{E}_i^\xi) \otimes (F_j^\eta \oplus \bar{F}_j^\eta)\Big)\Big),$$

where $\sigma = \lambda\rho \oplus (1-\lambda)\bar{\rho}$ is a state acting on the diagonal subspace $\mathcal{H}_A \otimes \mathcal{H}_B \oplus \bar{\mathcal{H}}_A \otimes \bar{\mathcal{H}}_B \subset (\mathcal{H}_A \oplus \bar{\mathcal{H}}_A) \otimes (\mathcal{H}_B \oplus \bar{\mathcal{H}}_B)$.

Exercise 11.24. Replace ρ by its appropriate purification (see Section 3.4), i.e., represent $\rho \in \mathcal{H}_A \otimes \mathcal{H}_B$ as $\rho = \operatorname{Tr}_{\mathcal{H}_C} |\psi\rangle\langle\psi|$ for some $\psi \in \mathcal{H}_A \otimes \mathcal{H}_B \otimes \mathcal{H}_C$. Then write $\operatorname{Tr} \rho(E_i^\xi \otimes F_j^\eta) = \langle\psi|E_i^\xi \otimes \overline{F}_j^\eta|\psi\rangle$, where $\overline{F}_j^\eta = F_j^\eta \otimes \mathrm{I}_{\mathcal{H}_C}$.

Exercise 11.25. (i) By Exercise 11.23, it is enough to show that $\mathsf{RB} \subset \mathsf{QB}$, which is easy. Note that a product box $P \in \mathsf{RB}$ can be represented in a trivial way: take $\mathcal{H}_A = \mathcal{H}_B = \mathbb{C}$, $\rho = \mathrm{I}_{\mathbb{C}\otimes\mathbb{C}}$ and $E_i^\xi = p(\xi|i)\,\mathrm{I}_{\mathbb{C}}$, $F_j^\eta = p(\eta|j)\,\mathrm{I}_{\mathbb{C}}$. (ii) Consider a local box of the form (11.20). By Carathéodory's theorem, we may assume that the index set Λ is finite. To obtain a representation as a quantum box with a separable state, consider $\mathcal{H}_A = \mathcal{H}_B = \mathbb{C}^\Lambda$ and let $(|\lambda\rangle)_{\lambda\in\Lambda}$ be the canonical basis in $\mathbb{C}^\Lambda$. Define $\rho = \sum_\lambda \mu(\lambda)|\lambda\rangle\langle\lambda| \otimes |\lambda\rangle\langle\lambda|$, $E_i^\xi = \sum_\lambda p(\xi|i,\lambda)|\lambda\rangle\langle\lambda|$ and $F_j^\eta = \sum_\lambda p(\eta|j,\lambda)|\lambda\rangle\langle\lambda|$. One checks then that the representation (11.21) holds. Note that this construction is essentially the argument used in Exercise 11.23 to prove convexity of QB, specified to the present (simpler) setting.

Exercise 11.26. Since LB is convex, it suffices to prove the result when ρ is a product state, in which case it is almost immediate.

Exercise 11.27. Use (11.24) in combination with Exercises 4.13 and 4.15. Note that the affine space $V_{k,m}$ generated by $\mathrm{K}_{k,m}$ does not contain 0 and similarly for $\mathrm{K}_{l,n}$.

Exercise 11.28. If $p_A(\cdot|i) \in \mathrm{K}_{k,m}$ and $p_B(\cdot|j) \in \mathrm{K}_{l,n}$, the dimension of the set of boxes $P = \{p(\xi,\eta|i,j)\}$ verifying (11.25) for inputs i,j and for that particular choice of p_A, p_B is $(k-1)(l-1)$. Consequently, $\dim \mathsf{NSB} \leqslant mn(k-1)(l-1) + \dim \mathrm{K}_{k,m} +$

$\dim \mathrm{K}_{l,n}$, which coincides with the value of $\dim \mathsf{LB}$ calculated in Exercise 11.27. Since $\mathsf{LB} \subset \mathsf{QB} \subset \mathsf{NSB}$, all dimensions must be the same. They are all convex sets with nonempty interior in the affine space $V_{k,m} \hat{\otimes} V_{l,n}$ analyzed in Exercise 11.27.

Exercise 11.29. Let $P = \{\operatorname{Tr} \rho(E_i^\xi \otimes F_j^\eta)\} \in \mathsf{QB}$ and $\bar{P} = \{\operatorname{Tr} \rho_*(E_i^\xi \otimes F_j^\eta)\}$, where ρ_* is the maximally mixed state. Since ρ_* is an interior point of Sep, it follows from Exercise 11.26 that the intersection of the segment $[\bar{P}, P]$ with LB is a segment of nonzero length, in particular P belongs to the affine subspace generated by LB. Since $P \in \mathsf{QB}$ was arbitrary, we conclude that QB is contained in that subspace and, in particular, $\dim \mathsf{QB} \leqslant \dim \mathsf{LB}$. (The converse inequality is trivial.)

Exercise 11.30. If $H \subset \mathbb{R}^N$ is an affine subspace not containing 0 and if V is an affine functional on $\mathbb{R}^N$, then there exists $v \in \mathbb{R}^N$ such that $\langle v, x\rangle = V(x)$ for $x \in H$.

Exercise 11.31. The first part is straightforward from the definitions. For the second part, note that we cannot have $\mathsf{LB}_{\wp} \subset b\mathsf{QB}_{\wp}$ if $|b| < 1$, and then appeal to the first part.

Exercise 11.32. (i) By Exercise 11.30, we can use affine functionals to exhibit violations. Given such functional V, the largest violation among functionals of the form $V_s = s + V$ (where $s \in \mathbb{R}$) occurs when $V_s(\mathsf{LB})$ is an interval of the form $[-a, a]$. Hence if V yields the maximal quantum violation, then

$$[-a, a] = V(\mathsf{LB}) \subset V(\mathsf{QB}) \subset [-a\omega_{\mathrm{Q}}(V), a\omega_{\mathrm{Q}}(V)]$$

and the last two intervals share (at least) one of the endpoints. In particular, the ratio of the lengths of the intervals $V(\mathsf{QB})$ and $V(\mathsf{LB})$ is between $(1+\omega_{\mathrm{Q}}(V))/2$ and $\omega_{\mathrm{Q}}(V)$. (ii) Replace everywhere QB by NSB.

Exercise 11.33. First, the PR-box yields value 4 (in the normalization given by (11.29)). In the opposite direction, use the fact that, for each i, j, $p(\xi, \eta|i, j)$ is a joint density to deduce that $\left|\sum_{\xi,\eta} p(\xi, \eta|i, j)\right| \leqslant 1$. The second statement follows then from Exercise 11.15 and the proof of Proposition 11.19.

Exercise 11.34. Reverse engineer the proof of Proposition 11.8 starting from the configuration $x_1, y_1, x_2, y_2 \in \mathbb{R}^2$ from the hint to Exercise 11.11. This leads (for example) to ρ being the maximally entangled state on $\mathbb{C}^2 \otimes \mathbb{C}^2$, the isometries $X_1 = \sigma_x$, $X_2 = \sigma_z$ (the Pauli matrices), $Y_1 = 2^{-1/2}(\sigma_x + \sigma_z)$, $Y_2 = 2^{-1/2}(\sigma_x - \sigma_z)$ and, finally, to the POVMs consisting of spectral projections of X_i's and Y_j's (as in the formulas following (11.13)). The last step is somewhat tedious, but instructive.

Exercise 11.35. (i) The composition rules for Pauli matrices are in Exercise 2.4. (ii) Multiply all the numbers in the matrix. (iii)(a) Use part (ii); it follows that the probability of winning under any classical strategy is at most 8/9. (b) First, the product of the elements of Alice's output string must be an eigenvalue of the composition of the corresponding operators, and similarly for Bob, and therefore by (i)(b) their answers are valid. Next, we can compute (as in Section 3.8) the joint probability distribution of outcomes when Alice and Bob measure a single shared φ^+ in the eigenbasis of a Pauli matrix: for σ_x and σ_z both outcomes are always equal, and for σ_y both outcomes are always different. It follows that for each of the entries in Table 11.1, the outcomes of Alice's and Bob's measurements on $\phi_+ \otimes \phi_+$ always coincide.

Chapter 12

Exercise 12.1. For the first part, use the triangle inequality. For the second part, consider the POVM $(P, \mathrm{I}-P)$ where P is the projection onto the range of $(\rho-\sigma)^+$.

Exercise 12.2. Show that separability and PPT properties are preserved under the action of a separable channel.

Appendix A

Exercise A.1. Use simple calculus (differentiation) for small t and (for example) the upper Komatu inequality (A.4) for large t.

Exercise A.2. (i) Is elementary calculus. (ii) Let $\delta(x)$ be either $f_+(x) - f(x)$ or $f(x) - f_-(x)$. We have $\delta'(x) \leqslant x\delta(x)$. Since $\delta(0) \geqslant 0$, and δ vanishes at $+\infty$, the result follows (otherwise consider a local minimum of δ).

Exercise A.3. Let μ be such a probability measure. The Fourier transform $\widehat{\mu} : u \mapsto \int_{\mathbb{R}^n} \exp(i\langle x, u\rangle)\, \mathrm{d}\mu(x)$ satisfies $\widehat{\mu}(u+v) = \widehat{\mu}(u)\widehat{\mu}(v)$ for $u \perp v$. Moreover $\widehat{\mu}$ is radial. If $f(t)$ denotes the value of $\widehat{\mu}$ on the sphere of radius t^2, we have $f(t+u) = f(t)f(u)$. Since f is continuous and real-valued (μ is even by assumption), this implies $f(t) = \exp(-t\sigma^2/2)$ for some $\sigma \geqslant 0$, and therefore μ is a Gaussian measure.

Exercise A.4. We have $\kappa_n = \mathbf{E}\,|G| \leqslant (\mathbf{E}\,|G|^2)^{1/2} = \sqrt{n}$. For the lower bound, use the functional equation $\Gamma(t+1) = t\Gamma(t)$. Note also that $\kappa_{n+1}/\kappa_n = n/\kappa_n^2$.

Exercise A.5. If $\alpha_n = \sqrt{n - 1/2}$ and $\beta_n = \sqrt{n - \frac{n}{2n+1}}$, it is elementary to check that the sequences κ_{2n}/α_{2n}, $\kappa_{2n+1}/\alpha_{2n+1}$, β_{2n}/κ_{2n}, and $\beta_{2n+1}/\kappa_{2n+1}$ are non-increasing. The result follows since all these sequences converge to 1.

Exercise A.6. Express $\int_{\mathbb{R}^n} f\, \mathrm{d}\gamma_n$ in polar coordinates. The factor is $\kappa_n(\alpha) = \mathbf{E}\,|G|^\alpha = 2^{\alpha/2}\Gamma((n+\alpha)/2)/\Gamma(n/2)$ and, under some minimal regularity assumptions on f, the formula is valid $\alpha > -n$.

Appendix B

Exercise B.1. Use Stirling's formula and the bound $n! \geqslant (n/e)^n$.

Exercise B.2. This is immediate from (B.4).

Exercise B.3. (i)–(ii) Easy. (iii) Use the non-commutative Hölder inequality and the fact that $A^\dagger A$ and $AA^\dagger$ (and hence $\sqrt{A^\dagger A}$ and $\sqrt{AA^\dagger}$) have the same non-zero eigenvalues.

Exercise B.4. The argument is essentially included in the proof of Theorem 2.3.

Exercise B.5. (i) Follows from Proposition B.1 and from the formula $\int_0^1 \|\gamma'(t)\|\, \mathrm{d}t$ for the length of an absolutely continuous curve $\gamma : [0,1] \to \mathsf{G}$. (ii) The singular numbers of $U - V$ are the same as those of $\mathrm{I} - U^{-1}V$ and hence (in the notation of part (i)) equal $|1 - e^{i\theta_j}| = 2\sin(\theta_j/2)$.

Exercise B.6. By rescaling the parameter t we can achieve $\|A\|_\infty \leqslant \pi$. Next, if $s, t \in \mathbb{R}$ with $|s-t| < 1$, apply Proposition B.1 with $U = e^{isA}$ and $V = e^{itA}$. Finally, use the fact that the map $X \mapsto WX$ is an isometry with respect to any Schatten p-norm.

The last assertion follows from the uniqueness part of Proposition B.1.

Note that allowing right (or two-sided) cosets does not increase generality since $e^{itA}W = We^{itW^\dagger AW}$.

Exercise B.7. Use the formula $\exp\begin{pmatrix} 0 & -\theta \\ \theta & 0 \end{pmatrix} = \begin{pmatrix} \cos\theta & -\sin\theta \\ \sin\theta & \cos\theta \end{pmatrix}$ and the fact that $\mathbb{R}^n$ can be decomposed as an orthogonal direct sum of subspaces of dimension at most 2 that are invariant for U. For the last equality apply Exercise B.5(i).

Exercise B.8. (i) For an integer n, write $e^{iB} - e^{iA} = \sum_{k=0}^{n-1} e^{ikA/n}(e^{iB/n} - e^{iA/n})e^{i(n-1-k)B/n}$. It follows that $\|e^{iB} - e^{iA}\| \leqslant n\|e^{iB/n} - e^{iA/n}\|$. Conclude by using the bound $\|e^X - e^Y\| \leqslant \|X - Y\| \exp(\max(\|X\|, \|Y\|))$ which follows from the series expansion, and take $n \to \infty$. Alternatively, consider $\phi(t) = e^{i(1-t)B}e^{itA}$ and show that $\phi'(t) = ie^{i(1-t)B}(A - B)e^{itA}$. These arguments work for any unitarily invariant norm. (ii) The functional inequality for $L(\cdot)$ follows from

$$e^{iB} - e^{iA} = 2(e^{iB/2} - e^{iA/2}) + (e^{iB/2} - \mathrm{I})(e^{iB/2} - e^{iA/2}) + (e^{iB/2} - e^{iA/2})(e^{iA/2} - \mathrm{I}).$$

Iterating (and using the simple fact that $L(\theta)$ tends to 1 as θ goes to 0) gives that $L(\theta) \geqslant \prod_{k=1}^{\infty}(1 - |1 - e^{i\theta/2^k}|) = \prod_{k=2}^{\infty}(1 - 2\sin(\theta/2^k))$, which is easy to estimate numerically.

Exercise B.9. Let $V \in V_0\mathsf{H}$, then $V = V_0 h$ and $U_1 = U_0 h_1$ for some $h, h_1 \in \mathsf{H}$. Now note that $\|U_0 - V\|_p = \|U_0 - V_0 h\|_p = \|U_0 h_1 - V_0 h h_1\|_p = \|U_1 - V_0 h h_1\|_p$ and $V_0 h h_1 \in V_0\mathsf{H}$; taking the infimum over V shows one inequality and the other follows by symmetry. Similarly, if $[0,1] \ni t \mapsto U_0 e^{itA}$ is a geodesic connecting U_0 to $V \in V_0 H$, then $t \mapsto U_0 e^{itA} h_1 = U_1 e^{ith_1^\dagger A h_1}$ is a curve of the same length connecting U_1 to $V h_1 \in V_0\mathsf{H}$.

Exercise B.10. In the notation from Exercise B.9 and its hint, we may assume that $g_p(U_0, V_0)$ equals the distance between $U_0\mathsf{H}$ and $V_0\mathsf{H}$ (in the sense of g_p; note that the distance is attained, for example by compactness). Next, consider the geodesic connecting U_0 to V_0, whose length is equal to that distance, and deduce from Exercise B.9 that the quotient map $\mathsf{O}(n) \to \mathsf{O}(n)/\mathsf{H}$ is an isometry when restricted to that geodesic.

Exercise B.11. In the notation of (B.9) let $E_i = \operatorname{span}\{x_i, y_i\}$. The subspaces $E_1, \ldots, E_k$ are pairwise orthogonal and invariant under P_E and P_F; they are 2-dimensional for $s_i < 1$ (which is equivalent to $x_i \neq y_i$) and 1-dimensional otherwise. We now note that the eigenvalues of $|x_i\rangle\langle x_i| - |y_i\rangle\langle y_i|$ are $\pm\sin\theta_i$; since $P_E = \sum_i |x_i\rangle\langle x_i|$ and $P_F = \sum_i |y_i\rangle\langle y_i|$, the principal angles $\theta_i \neq 0$ can be retrieved from the eigenvalues of $P_E - P_F$. It remains to use the relation $P_E - P_F = P_{F^\perp} - P_{E^\perp}$.

Exercise B.12. Show first the inequalities "$\leqslant$". In the notation of (B.9) and of the hint to Exercise B.11, define $W_0 \in \mathsf{O}(n)$ to be a rotation on each 2-dimensional space E_j such that $W_0 x_j = y_j$ (i.e., a rotation by θ_j) and to be an identity on the orthogonal complement of the union of such E_j's. The nonzero singular values of $W_0 - \mathrm{I}$ are then $|e^{i\theta_j} - 1| = 2\sin\theta_j/2$, each repeated twice, which combined with (B.13) shows the needed upper bound on $\tilde{h}_p(E, F)$. For an upper bound on the geodesic distance $\tilde{g}_p(E, F)$ consider a family $W(t)$, $t \in [0,1]$, where $W(t)$ acts as a rotation by $t\theta_j$ on E_j and calculate the length of the path $t \mapsto W(t)$ with respect to the Schatten p-norm. (Alternatively, you may note that $W(t) = e^{itA}$ for the appropriate $A \in \mathsf{M}_n^{\mathrm{sa}}$ and refer to the calculation from Exercise B.5.)

For the opposite inequality, show the following claim: *If $W \in \mathsf{O}(n)$ verifies $WE = F$, then the singular values of $W - \mathrm{I}$ dominate those of $W_0 - \mathrm{I}$, in the sense that*

$s_j(W_0 - \mathrm{I}) \leqslant s_j(W - \mathrm{I})$ *for* $1 \leqslant j \leqslant n$. The lower bound on $\tilde{h}_p(E,F)$ follows then immediately from (B.13); to get the lower bound on $\tilde{g}_p(E,F)$, observe that, by Exercise B.10, the optimal geodesic is of the form $W(t) = U_0 e^{itA}$, $t \in [0,1]$, and that its length—which is $\|A\|_p$ by Exercise B.5(i)—depends in a straightforward way on the singular values of $W(1) - \mathrm{I}$.
To show the claim, fix $s \in (0,1)$ and let $\theta \in (0, \pi/2)$ be such that $s = \cos\theta$. Next, consider $F_s = \mathrm{span}\{y_j : s_j \leqslant s\}$. If $y \in F_s$ is a unit vector, show that $|y - x| \geqslant |e^{i\theta} - 1|$ for all $x \in S^{n-1} \cap E$ and deduce that whenever $WE = F$, then there are at least $\dim F_s$ singular values of $(V - \mathrm{I})_{|E}$ that are at least $|e^{i\theta} - 1|$. Since, by Exercise B.11, the same is true for $(V - \mathrm{I})_{|E^\perp}$, the claim follows.

Exercise B.13. It follows from the argument sketched in the hint to Exercise B.11 that the (nonzero) eigenvalues of $P_E - P_F$ are $\pm \sin\theta_i$ and so $\|P_E - P_F\|_p = 2^{1/p}\|(\sin\theta_1, \ldots, \sin\theta_k)\|_p$. Comparing with the formulas from Exercise B.12 gives $\frac{2}{\pi}\tilde{g}_p(E,F) \leqslant \|P_E - P_F\|_p \leqslant \tilde{g}_p(E,F)$ and $\frac{1}{\sqrt{2}}\tilde{h}_p(E,F) \leqslant \|P_E - P_F\|_p \leqslant \tilde{h}_p(E,F)$. Finally, since $\|P_E - P_F\|_p$ and $\tilde{g}_p$ differ only in terms of the 3rd order and higher, they both induce the same geodesic distance.

Exercise B.14. Let $(x_1, \ldots, x_n)$ be independent Gaussian vectors in $\mathbb{R}^n$. Since the set of singular matrices has Lebesgue measure 0 in M_n, these vectors are almost surely linearly independent. Moreover, the orthonormal matrix obtained by applying the Gram–Schmidt procedure to the matrix with columns $x_1, \ldots, x_n$ is Haar-distributed on $\mathsf{O}(n)$. It follows that the subspace $\mathrm{span}(x_1, \ldots, x_k)$ is Haar-distributed on $\mathsf{Gr}(k, \mathbb{R}^n)$.

Exercise B.15. Let $g_1, \ldots, g_k$ (resp., $h_1, \ldots, h_k$) be independent standard Gaussian vectors in E (resp., in $E^\perp$). For every $\varepsilon > 0$, the random subspace $\mathrm{span}\{g_i + \varepsilon h_i : 1 \leqslant i \leqslant k\}$ is distributed with respect to some "Haar measure" on $\mathsf{Gr}(k, \mathbb{R}^n)$ and converges to E almost surely as ε goes to 0.

Exercise B.16. The answer to both questions is "no". The reason is that $\mathsf{SO}(k) \times \mathsf{SO}(n-k)$ is a proper subgroup of the stabilizer of $\mathbb{R}^k$ under the canonical action of $\mathsf{SO}(n)$ on $\mathsf{Gr}(k, \mathbb{R}^n)$, and similarly in the complex case. In the complex case, even the dimensions do not add up, we have $\dim \mathsf{SU}(n) - \big(\dim \mathsf{SU}(k) + \dim \mathsf{SU}(n-k)\big) = 2k(n-k) + 1 > 2k(n-k) = \dim \mathsf{Gr}(k, \mathbb{C}^n)$ (note that these are *real* dimensions).

A more complete answer is that $\mathsf{SO}(n)/(\mathsf{SO}(k) \times \mathsf{SO}(n-k))$ identifies with the set of *oriented* k-dimensional subspaces of $\mathbb{R}^n$ and is, in a canonical way, a two-fold cover of $\mathsf{Gr}(k, \mathbb{R}^n)$. (A particular example of this phenomenon is $S^{n-1} = \mathsf{SO}(n)/\mathsf{SO}(n-1)$ being a two-fold cover of $\mathsf{Gr}(1, \mathbb{R}^n) = \mathsf{P}(\mathbb{R}^n)$.) Similarly, the set $\mathsf{SU}(n)/(\mathsf{SU}(k) \times \mathsf{SU}(n-k))$ identifies, in a way, with the set of "signed" k-dimensional subspaces of $\mathbb{C}^n$. See also Exercise B.17.

Exercise B.17. There are two fine points: first, the cosets of H are subsets of the cosets of $\mathsf{O}(k) \times \mathsf{O}(n-k)$ and so the distances (extrinsic or geodesic) between the former may be larger than between the latter. Next, geodesics connecting cosets (as in Exercise B.10) may *a priori* turn out to be longer if we insist that they are entirely contained in $\big(\mathsf{SO}(n), g_p\big)$ (as opposed to the larger space $\big(\mathsf{O}(n), g_p\big)$). Similar issues arise in the complex case.
To address these concerns, check that W and $W(t)$ suggested in the hint to Exercise B.12 are minimizers that work *simultaneously* for $\mathsf{O}(n)$ and for $\mathsf{SO}(n)$ (resp., for $\mathsf{U}(n)$ and for $\mathsf{SU}(n)$).

Exercise B.19. (a) (i) By Proposition 2.29, automorphisms of $\mathcal{L}_4$ are maps of the form $\mathbf{x} \mapsto \Psi^{-1}(V\Psi(\mathbf{x})V^\dagger)$ or $\mathbf{x} \mapsto \Psi^{-1}(V\Psi(\mathbf{x})^T V^\dagger)$. (ii) Use (2.6) to show that automorphisms from (i) preserve q and hence belong to $\mathsf{O}^+(1,3)$ iff $|\det V| = 1$. (iii) Check that if $V = \mathrm{I}$, then the two maps from (i) belong respectively to $\mathsf{SO}^+(1,3)$ and $\mathsf{O}^+(1,3)\backslash\mathsf{SO}^+(1,3)$, and appeal to connectedness of $\mathsf{SL}(2,\mathbb{C})$.
(b) $V \mapsto \Psi_V$ being a homomorphism is straightforward; for the part about the kernel note that if $\mathbf{x} = \Psi^{-1}(|\xi\rangle\langle\xi|)$ for some $\xi \in \mathbb{C}^2$, then $\Phi_V(\mathbf{x}) = \mathbf{x}$ can be rewritten as $V|\xi\rangle\langle\xi|V^\dagger = |\xi\rangle\langle\xi|$; this means that $\Psi_V = \mathrm{I}_{\mathbb{R}^4}$ implies that every $\xi \in \mathbb{C}^2\backslash\{0\}$ is an eigenvector of V, which is only possible if V is a multiple of I.

Appendix C

Exercise C.1. Start by showing that $\Phi = |u\rangle\langle v| \neq 0$ belongs to $\boldsymbol{P}(\mathcal{C})$ iff $u \in \mathcal{C}$ and $v \in \mathcal{C}^*$ (or $u \in -\mathcal{C}$, $v \in -\mathcal{C}^*$); necessity easily follows. For sufficiency, start by observing that if u and v generate extreme rays in the respective cones and if, for some Δ, $\Phi \pm \Delta \in \boldsymbol{P}(\mathcal{C})$, then $\Delta(\mathcal{C}) \subset \mathbb{R}_+ u$ and $\Delta^*(\mathcal{C}^*) \subset \mathbb{R}_+ v$. To show that (for example) $\Delta(x) \in \mathbb{R}_+ u$ for $x \in \mathcal{C}$, consider separately the cases $\Phi(x) = 0$ and $\Phi(x) \neq 0$.

Exercise C.2. If Ψ is the automorphism in question, then $\Psi^* J \Psi = \mu J + Q$ and similarly $(\Psi^{-1})^* J(\Psi^{-1}) = \nu J + Q'$ with $\mu, \nu \geqslant 0$ and Q, Q' positive semi-definite (justify). Next, show that this implies that $(1 - \mu\nu)J = \nu Q + \Psi^* Q' \Psi$, which is only possible if $\mu\nu = 1$ and $Q = Q' = 0$.

Exercise C.3. If $n = 4$, this follows from Corollary 2.30, modulo identifying completely positive automorphisms of $\mathcal{PSD}(\mathbb{C}^2)$ with $\mathsf{SO}^+(1,3)$ (see the hint to Exercise B.19(a)). Deduce the conclusion for $n > 4$: when looking for an automorphism Ψ such that $\Psi(u) = v$, consider any 4-dimensional subspace $E \subset \mathbb{R}^n$ containing e_0, u and v, and define Ψ separately on E and $E^\perp$.
A similar line of argument allows us to derive the full statement ($n \geqslant 2$) from Exercise B.18.

Exercise C.4. "Reverse engineer" the failure of the proof of the converse implication from Proposition C.1 when for $n = 2$. Alternatively, notice that $\mathcal{L}_2$ is isomorphic to the positive quadrant $\mathbb{R}^2_+$ and that the structure of the cone $\boldsymbol{P}(\mathbb{R}^n_+)$ is particularly simple.

Appendix D

Exercise D.1. One fine point is in verifying that the bases are nontrivial and that they generate the respective cones, but this is assured by the hypothesis $\langle e^*, e\rangle = 1$ (cf. Exercise 1.28).

Exercise D.2. Start by noticing that $\|x\|_{(K-a)^\circ} \leqslant \|x\|_{K^\circ}$ if $\langle x, a\rangle \geqslant 0$ while $\|x\|_{(K-a)^\circ} \geqslant \|x\|_{K^\circ}$ if $\langle x, a\rangle \leqslant 0$ (this may be more obvious if instead of the gauges we consider the support functions $w(\cdot,\cdot)$, see Section 4.3.3), and that for some x (e.g., $x = \pm a$) the inequalities are strict. Deduce that $K^\circ \cap H^+ \subsetneq (K-a)^\circ \cap H^+$, where $H^+ = \{x \in \mathbb{R}^n : \langle x, a\rangle \geqslant 0\}$, with the reverse inclusion for the other halfspace, and show that this implies $\int_{(K-a)^\circ} \langle x, b\rangle \, dx > 0$.

Exercise D.3. (i)&(ii) By linear invariance, we may assume that $\mathscr{E}$ is a translate of B_2^n. Identify it with the base of the Lorentz cone $\mathcal{L}_{n+1}$ and apply Lemma D.1. (iii) This is immediate if we use the full force of Proposition D.2. For a proof that does not use the uniqueness part, note that if K is centrally symmetric and has centroid at the origin, then it is 0-symmetric. Apply this observation to $K = (\mathscr{E} - a)^\circ$ and use the bipolar theorem.

Exercise D.4. Let u be such that the segment $[b - u, b + u]$ lies in the interior of K. We now consider $a = a(t) := b + tu$ for $t \in [-1, 1]$ and the corresponding solid cones T_a. The first-order variation as $t \to 0$ is, for some constant $C(b, u) > 0$,

$$\mathrm{vol}_{n+1}(T_a) = \mathrm{vol}_{n+1}(T_b) + C(b, u)\, t \int_{B_b} \langle u, x - e_0 \rangle \, \mathrm{d}x + o(|t|).$$

If b is a local extremum, it follows that $\int_{B_b} x \, \mathrm{d}x = e_0$.

APPENDIX F

Notation

We list below mathematical symbols that appear in the book, particularly those that are subfield-specific or not generally accepted throughout mathematics, or just potentially ambiguous. We grouped them by theme/subfield; since any such classification is necessarily imperfect, it may sometimes be necessary to check more than one category. Within each category, we tried—to the extent it was possible—to arrange the symbols in the alphabetic order. The numbers following each brief description refer to the pages on which the corresponding symbol is defined, or at least appears in a context.

General notation

$\langle x\|, \|x\rangle$	Dirac bra-ket notation, 4
$\langle x\|y\rangle$	scalar product, alternative notation to $\langle x, y\rangle$, 5
$\|x\rangle\langle y\|$	ket-bra, the rank one operator mapping z to $\langle y, z\rangle \cdot x$, 5
$\|\cdot\|$	Euclidean or Hilbertian norm, or modulus of a scalar, 3
$\|\alpha\|$	weight of a multi-index $\alpha \in \mathbb{N}^n$, 135
$\mathbf{1}_A$	indicator function of a set A, 101
$\lesssim, \gtrsim, \simeq$	Landau notation (alternative form), 3
$\operatorname{card} A$	cardinality of a set A, 111
$\log$	natural logarithm, 27
$O(\cdot), \Omega(\cdot), \Theta(\cdot)$	Landau notation, 3
$o(\cdot), \sim, \ll$	asymptotic notation, 3
$\mathfrak{S}_m$	group of permutations of $\{1, 2, \ldots, m\}$, 27
$\mathrm{vol}, \mathrm{vol}_n, \mathrm{vol}_E$	Lebesgue measure on $\mathbb{R}^n$, on the subspace E, 4

Convex geometry

$\\|\cdot\\|_K$	gauge of a convex body K, 11
$\\|\cdot\\|_p$	p-norm on $\mathbb{R}^n$, 12
K°	polar of a set $K \subset \mathbb{R}^n$, 15
$\mathcal{C}^*$	cone dual to a cone $\mathcal{C}$, 19
$\mathcal{C}^{\mathrm{b}}$	base of a cone $\mathcal{C}$, 19
$K_\cap$	intersection symmetrization of a convex body K, 80
$K_\cup$	union symmetrization of a convex body K, 80
$K_{\oslash}$	cylindrical symmetrization of a convex body K, 81
$\langle\cdot,\cdot\rangle_{\mathscr{E}}$	scalar product associated with an ellipsoid $\mathscr{E}$, 18
B_p^n	unit ball of ℓ_p^n, 12
B_X	unit ball of a normed space X, 11
$\operatorname{conv} A$	convex hull of a set A, 12
Δ_n	n-dimensional simplex, 12

$h_K(\cdot)$	support function of a convex body K, 94
$\operatorname{inrad}(K)$	inradius of a convex body K, 96
$\operatorname{Iso}(K)$	group of isometries preserving K, 89
$\operatorname{John}(K)$	John ellipsoid of a convex body K, 84
ℓ_p^n	space $\mathbb{R}^n$ equipped with the p-norm, 12
$\mathcal{L}_n$	Lorentz cone in $\mathbb{R}^n$, 18
$\text{Löw}(K)$	Löwner ellipsoid of a convex body K, 84
ν_K	map which implements duality of faces, 17
$\operatorname{outrad}(K)$	outradius of a convex body K, 96
$S^{n-1}, S_{\mathbb{C}^n}, S_{\mathcal{H}}$	unit sphere in $\mathbb{R}^n$, in $\mathbb{C}^n$, in Hilbert space $\mathcal{H}$, 4, 311
$\operatorname{vrad}(K)$	volume radius of a convex body K, 92
$w(K,\cdot)$	support function of a convex body K, 94
$w(K)$	mean width of a convex body K, 95
$w_G(K)$	Gaussian mean width of a convex body K, 95

Linear algebra

$x^{\downarrow}$	non-increasing rearrangement of a vector $x \in \mathbb{R}^n$, 22
$\prec$	majorization, 22
$\prec_w$	submajorization, 23
$A^{\dagger}$	adjoint of a matrix (or operator) A, 4
$\lvert A\rvert$	absolute value of an operator A (equals $(A^{\dagger}A)^{1/2}$), 23
$\lVert\cdot\rVert_p$	Schatten p-norm on matrices, 23
$\lVert\cdot\rVert_{\mathrm{HS}}$	Hilbert–Schmidt norm (equals $\lVert\cdot\rVert_2$), 23
$\lVert\cdot\rVert_{\mathrm{op}}$	operator norm (equals $\lVert\cdot\rVert_\infty$), 23
$[\psi]$	equivalence class of a unit vector ψ in the projective space, 312
$B(\mathcal{H})$	bounded linear operators on a Hilbert space $\mathcal{H}$, 4
$B^{\mathrm{sa}}(\mathcal{H})$	bounded linear self-adjoint operators on a Hilbert space $\mathcal{H}$, 4
$d(A)$	vector formed by diagonal entries of a matrix A, 24
$\operatorname{diag} A$	matrix obtained from A by setting non-diagonal entries to zero, 25
$\operatorname{Diag}(v), D_v$	for a vector $v = (v_i)$, the diagonal matrix whose ii-th entry is v_i, 265, 333
E_{ij}	operator $\lvert e_i\rangle\langle f_j\rvert$, where (e_i), (f_j) are specified bases, 47
$\mathsf{Gr}(k,\mathbb{R}^n), \mathsf{Gr}(k,\mathbb{C}^n)$	Grassmann manifolds, 314
$\overline{\mathcal{H}}$	conjugate of a Hilbert space $\mathcal{H}$, 4
H_1	often (but not always) the hyperplane of trace one matrices in $\mathsf{M}_d^{\mathrm{sa}}$, 31
I	identity matrix or identity operator, 8
Id	identity superoperator, 8
J	diagonal matrix with diagonal entries $(1,-1,\dots,-1)$, 318
$\lambda_j(A)$	eigenvalues of a matrix A, usually arranged in the nonincreasing order if A is Hermitian, 160
$\lambda_j(\psi)$	Schmidt coefficients of a vector $\psi \in \mathcal{H}_1 \otimes \mathcal{H}_2$, 36
$\mathsf{M}_{m,n}$	space of $m \times n$ (real or complex) matrices, 7
M_n	equals $\mathsf{M}_{n,n}$, 7
$\mathsf{M}_n^{\mathrm{sa}}$	space of self-adjoint matrices (subspace of M_n), 7

$\mathsf{M}_n^{\mathrm{sa},0}$	subspace of $\mathsf{M}_n^{\mathrm{sa}}$ consisting of trace zero matrices, 265
$\mathsf{O}(n)$	orthogonal group, 312
$\mathsf{O}(1, n-1)$	Lorentz group, 318
$\mathsf{O}^+(1, n-1)$	orthochronous subgroup of the Lorentz group, 318
$\boldsymbol{P}(\mathcal{C})$	cone of linear maps preserving the cone $\mathcal{C}$, or preserving the order induced by the cone $\mathcal{C}$, 321
P_E	orthogonal projection onto the subspace E, 5
$\mathcal{PSD}$	cone of positive-semidefinite matrices, 18
$\mathsf{PSU}(n)$	projective special unitary group, 312
$q(\cdot)$	quadratic form of the Minkowski spacetime, 32, 318
$\mathbb{R}^{n,0}$	the hyperplane of $\mathbb{R}^n$ consisting of vectors whose coordinates add up to 0, 263
$s_j(A)$	singular values (arranged in non-increasing order) of a matrix A, 24
$s(A)$	the vector $\big(s_j(A)\big)$ of singular values of a matrix A, 24
S_{HS}	unit sphere for the Hilbert–Schmidt norm $\|\cdot\|_{\mathrm{HS}}$, 225
$S_p^{m,n}$	unit ball for $\|\cdot\|_p$ in $\mathsf{M}_{m,n}$, 26
$S_p^{m,\mathrm{sa}}$	unit ball for $\|\cdot\|_p$ in $\mathsf{M}_n^{\mathrm{sa}}$, 26
SVD	singular value decomposition, 36
$\mathsf{SO}(n)$	special orthogonal group, 312
$\mathsf{SO}(1, n-1)$	proper Lorentz group, 318
$\mathsf{SO}^+(1, n-1)$	restricted Lorentz group, 318
$\mathrm{spec}(A)$	spectrum (arranged in non-increasing order) of a self-adjoint matrix A, 24
$\mathsf{SU}(n)$	special unitary group, 312
T	transposition with respect to a specified basis, 41
$\mathsf{U}(n)$	unitary group, 312

Probability

$\|\cdot\|_{\psi_1}$	subexponential norm, 139
$\|\cdot\|_{\psi_2}$	subgaussian norm, 139
$\boxplus$	free additive convolution, 177
d_∞	∞-Wasserstein distance, 161
$\mathbf{E}f$	expected value of the random variable f, also referred to as the mean, the expectation, or the first moment, 117
$\mathrm{Ent}_\mu(f)$	continuous entropy of f (with respect to μ), 132
$F_X(\cdot)$	cumulative distribution function of a random variable X, 161
$\Phi(\cdot)$	cumulative distribution function of an $N(0,1)$ variable, 307
G	a standard Gaussian vector, 308
$\gamma_n, \gamma_n^{\mathbb{C}}$	standard Gaussian measure on $\mathbb{R}^n$, $\mathbb{C}^n$, 308
$\mathrm{GUE}(n)$	Gaussian Unitary Ensemble, 162
$\mathrm{GUE}_0(n)$	Gaussian Unitary Ensemble conditioned to have trace 0, 163
GOE	Gaussian Orthogonal Ensemble, 163
$H(\mathbf{p})$	Shannon entropy of a probability mass function $\mathbf{p}$, 28
$\chi(n), \chi^2(n)$	chi, chi-squared distribution with n degrees of freedom, 175
i.i.d.	independent, identically distributed, 160

κ_n	expected Euclidean norm of a standard Gaussian vector in $\mathbb{R}^n$, 309
$\kappa_n^{\mathbb{C}}$	expected Euclidean norm of a standard Gaussian vector in $\mathbb{C}^n$, 309
M_f	median of the random variable f, 117
$\mu_{\mathrm{sp}}(A)$	empirical spectral distribution of a self-adjoint matrix A, 160
$\mu_{\mathrm{MP}(\lambda)}$	Marčenko–Pastur distribution with parameter λ, 167
μ_{SC}	semicircular distribution, 163
$\operatorname{osc}(f, A, \mu)$	oscillation of f around μ on the subset A, 186
(P_t)	Ornstein–Uhlenbeck semigroup, 135
$\mathrm{Wishart}(n, s)$	Wishart distribution with parameters n, s, 166

Geometry and asymptotic geometric analysis

$\otimes_2$	Euclidean/Hilbertian tensor product, 18
$\widehat{\otimes}$	projective tensor product, 82
$\check{\otimes}$	injective tensor product, 83
A_ε	ε-enlargement of a set A, 117
$\lvert\lvert\lvert \cdot \rvert\rvert\rvert_K$	norm on $\mathscr{H}_{k,n}$ associated with K, 183
$a(K)$	asphericity of a convex body K, 193
$c(X)$	minimum of Ricci curvatures of the manifold X, 130
$C(x, \theta)$	spherical cap of angular radius θ with center at x, 109
$d(X, Y)$	Banach–Mazur distance between normed spaces X and Y, 103
$d_{BM}(K, L)$	Banach–Mazur distance between convex bodies K and L, 79
$d_g(K, L)$	geometric distance between convex bodies K and L, 79
diam	diameter of a set in a metric space, 116
$\dim_F(K)$	facial dimension of a convex body K, 193
$\dim_V(K)$	verticial dimension of a convex body K, 193
g	geodesic distance on the sphere, 311
$\mathscr{H}_k$	the space $L_2(\mathbb{R}^k, \gamma_k)$, 182
$\mathscr{H}_{k,n}$	$\mathscr{H}_k \otimes \mathbb{R}^n$, or $\mathbb{R}^n$-valued functions on $(\mathbb{R}^k, \gamma_k)$, 182
K_G	(real) Grothendieck constant, 281
$K_G^{\mathbb{C}}$	complex Grothendieck constant, 295
$K_G^{(n)}, K_G^{(m,n)}, K_G^{[n]}$	other variants of Grothendieck constant, 281, 295
$k_*(K)$	Dvoretzky dimension of a convex body K, 190
$\mathbf{K}(K)$	K-convexity constant of a convex body K, 183
ℓ_K	ℓ-norm associated to a convex body K, 181
$N(K, \varepsilon), N(K, d, \varepsilon)$	covering number (metric space), 107
$N(K, L), N(K, L, \varepsilon)$	covering number (convex bodies), 114
$\mathrm{LS}(X, \mu)$	logarithmic Sobolev constant of the space X, 132
LSI	logarithmic Sobolev inequality, 132
$P(K, \varepsilon), P(K, d, \varepsilon)$	packing number, 107
$\mathsf{P}(V)$	projective space associated to a vector space V, 312
$\mathrm{P}(X, \mu)$	Poincaré constant of the space X, 134
$R, R_1, \tilde{R}_1$	Rademacher projection, 183, 185
Ric_p	Ricci curvature at point p, 130
sec	sectional curvature, 130
σ	normalized Lebesgue measure on a Euclidean sphere, 4, 311

$V(\theta)$	measure of the spherical cap of angular radius θ, 109
$\mathrm{vr}(K)$	volume ratio of a convex body K, 201

Quantum information theory

$\bullet$	homotheties with respect to the maximally mixed state, 236
ρ^Γ	partial transposition of an operator ρ, 42
$\\|\cdot\\|_\diamond$	diamond norm, 52
$\\|\cdot\\|_{1\to p}$	$1\to p$ norm of a quantum channel, 217
$\\|\cdot\\|_{\mathrm{M}}$	distinguishability norm associated with a POVM M, 299
$\rho \rightsquigarrow \sigma$	state σ can be obtain from copies of ρ by an LOCC protocol, 301
Asym_d	antisymmetric subspace of $\mathbb{C}^d\otimes\mathbb{C}^d$, 40
$\mathcal{BP}$	cone of block-positive operators, 56
$C(\Phi)$	Choi matrix of a superoperator Φ, 48, 48
co-$\mathcal{PSD}$	cone of co-positive semidefinite operators, 56
$\boldsymbol{CP}$	cone of completely positive superoperators, 49
$\mathrm{D}(\mathcal{H})$	set of states on a Hilbert space $\mathcal{H}$, 9, 31
$\boldsymbol{DEC}$	cone of decomposable superoperators, 57
$E(\psi)$	entropy of entanglement of a pure state ψ, 215
$E_p(\psi)$	p-entropy of entanglement of a pure state ψ, 215, 229
$E_F(\rho)$	entanglement of formation of a state ρ, 271
$\boldsymbol{EB}$	cone of entanglement-breaking superoperators, 57
k-Ext	set of k-extendible states, 41
F	flip operator, 39
$g(\psi)$	geometric measure of entanglement of a pure state ψ, 229
$g_{\min}(\mathcal{H})$	extremal geometric measure of entanglement for $\mathcal{H}$, 229
Γ	partial transposition, 42
LB	set of local boxes, 286
LC	set of local correlations, 277
NSB	set of non-signaling boxes, 286
NSC	set of non-signaling correlations, 288
$\boldsymbol{P}$	cone of positivity-preserving superoperators, 56
$p_{\mathrm{L}}, p_{\mathrm{NL}}$	local, non-local fractions, 292
$\varphi^+,\varphi^-,\psi^+,\psi^-$	Bell vectors, 70
Φ_V	completely positive map $X\mapsto VXV^\dagger$, 59
π_a	antisymmetric state, 40
π_s	symmetric state, 40
PPT	set of states with positive partial transpose, 43
$\mathcal{PPT}$	cone of PPT operators, 55
$\boldsymbol{PPT}$	cone of PPT-inducing superoperators, 57
QB	set of quantum boxes, 286
QC	set of quantum correlations, 277
$R(\rho)$	robustness of a state ρ, 247
ρ_*	maximally mixed state, 32
$s_0(d)$	threshold for separability, 269
$S(\rho)$	von Neumann entropy of a state ρ, 27
$S_p(\rho)$	p-Rényi entropy of a state ρ, 28
$S^{\min}(\Phi), S_p^{\min}(\Phi)$	minimum output (p-)entropy of a quantum channel Φ, 216

Seg	Segré variety, 312
$\mathrm{Sep}(\mathcal{H})$	set of separable states on a multipartite Hilbert space $\mathcal{H}$, 37
$\mathcal{SEP}$	cone of separable operators, 38
$\sigma_x, \sigma_y, \sigma_z$	Pauli matrices, 32
Sym_d	symmetric subspace of $\mathbb{C}^d \otimes \mathbb{C}^d$, 39
$\mathrm{Tr}_1 \rho$, $\mathrm{Tr}_{\mathcal{A}} \rho$	partial trace of ρ with respect to subsystem 1 or $\mathcal{A}$, 35
w_λ	Werner state with parameter λ, 40
$\omega_{\mathrm{L}}(V)$	local value of a Bell expression V, 289
$\omega_{\mathrm{NS}}(V)$	non-signaling value of a Bell expression V, 289
$\omega_{\mathrm{Q}}(V)$	quantum value of a Bell expression V, 289

Bibliography

[AAGM15] Shiri Artstein-Avidan, Apostolos Giannopoulos, and Vitali D. Milman, *Asymptotic geometric analysis. Part I*, Mathematical Surveys and Monographs, vol. 202, American Mathematical Society, Providence, RI, 2015. MR3331351 87, 105, 143, 146, 147, 186, 207, 208, 209

[AAKM04] S. Artstein-Avidan, B. Klartag, and V. Milman, *The Santaló point of a function, and a functional form of the Santaló inequality*, Mathematika **51** (2004), no. 1-2, 33–48 (2005), DOI 10.1112/S0025579300015497. MR2220210 ↑105

[AAM06] S. Artstein-Avidan and V. D. Milman, *Logarithmic reduction of the level of randomness in some probabilistic geometric constructions*, J. Funct. Anal. **235** (2006), no. 1, 297–329, DOI 10.1016/j.jfa.2005.11.003. MR2216448 ↑207

[AAS15] Shiri Artstein-Avidan and Boaz A. Slomka, *A note on Santaló inequality for the polarity transform and its reverse*, Proc. Amer. Math. Soc. **143** (2015), no. 4, 1693–1704, DOI 10.1090/S0002-9939-2014-12390-2. MR3314082 ↑105

[AdRBV98] Juan Arias-de-Reyna, Keith Ball, and Rafael Villa, *Concentration of the distance in finite-dimensional normed spaces*, Mathematika **45** (1998), no. 2, 245–252, DOI 10.1112/S0025579300014182. MR1695717 ↑144

[AGMJV16] David Alonso-Gutiérrez, Bernardo González Merino, C. Hugo Jiménez, and Rafael Villa, *Rogers-Shephard inequality for log-concave functions*, J. Funct. Anal. **271** (2016), no. 11, 3269–3299, DOI 10.1016/j.jfa.2016.09.005. MR3554706 ↑105

[AGZ10] Greg W. Anderson, Alice Guionnet, and Ofer Zeitouni, *An introduction to random matrices*, Cambridge Studies in Advanced Mathematics, vol. 118, Cambridge University Press, Cambridge, 2010. MR2760897 ↑179, 245, 350, 356

[AII06] David Avis, Hiroshi Imai and Tsuyoshi Ito, *On the relationship between convex bodies related to correlation experiments with dichomotic observables*, J. Phys. A: Math. Gen. **39** (2006), 11283. 295

[AIIS04] David Avis, Hiroshi Imai, Tsuyoshi Ito, and Yuuya Sasaki, *Deriving tight Bell inequalities for 2 parties with many 2-valued observables from facets of cut polytopes*, arXiv eprint quant-ph/0404014 (2004). 296

[AJR15] Srinivasan Arunachalam, Nathaniel Johnston, and Vincent Russo, *Is absolute separability determined by the partial transpose?*, Quantum Inf. Comput. **15** (2015), no. 7-8, 694–720. MR3362609 ↑64

[AL15a] Guillaume Aubrun and Cécilia Lancien, *Locally restricted measurements on a multipartite quantum system: data hiding is generic*, Quantum Inf. Comput. **15** (2015), no. 5-6, 513–540. MR3362370 ↑306

[AL15b] Guillaume Aubrun and Cécilia Lancien, *Zonoids and sparsification of quantum measurements*, Positivity **20** (2016), no. 1, 1–23, DOI 10.1007/s11117-015-0337-5. MR3462036 ↑305

[Alo03] Noga Alon, *Problems and results in extremal combinatorics I*, Discrete Math. **273** (2003), no. 1-3, 31–53, DOI 10.1016/S0012-365X(03)00227-9. EuroComb'01 (Barcelona). MR2025940 ↑346

[AMS04] S. Artstein, V. Milman, and S. J. Szarek, *Duality of metric entropy*, Ann. of Math. (2) **159** (2004), no. 3, 1313–1328, DOI 10.4007/annals.2004.159.1313. MR2113023 ↑143

[AMSTJ04] S. Artstein, V. Milman, S. Szarek, and N. Tomczak-Jaegermann, *On convexified packing and entropy duality*, Geom. Funct. Anal. **14** (2004), no. 5, 1134–1141, DOI 10.1007/s00039-004-0486-3. MR2105957 ↑143

[AN12] Guillaume Aubrun and Ion Nechita, *Realigning random states*, J. Math. Phys. **53** (2012), no. 10, 102210, 16, DOI 10.1063/1.4759115. MR3050579 ↑273

[Ara04] P. K. Aravind, *Quantum mysteries revisited again*, Amer. J. Phys. **72** (2004), no. 10, 1303–1307, DOI 10.1119/1.1773173. MR2086837 ↑297

[Arv09] William Arveson, *Maximal vectors in Hilbert space and quantum entanglement*, J. Funct. Anal. **256** (2009), no. 5, 1476–1510, DOI 10.1016/j.jfa.2008.08.004. MR2490227 ↑233

[AS06] Guillaume Aubrun and Stanisław J Szarek, *Tensor products of convex sets and the volume of separable states on n qudits*, Phys. Rev. A **73** (2006), no. 2, 022109. 104, 233, 260, 261

[AS10] Erik Alfsen and Fred Shultz, *Unique decompositions, faces, and automorphisms of separable states*, J. Math. Phys. **51** (2010), no. 5, 052201, 13. MR2666955 (2011h:81012) 64

[AS15] Guillaume Aubrun and Stanisław Szarek, *Two proofs of Størmer's theorem*, arXiv:1512.03293 (2015). 64, 65

[AS17] Guillaume Aubrun and Stanisław Szarek, *Dvoretzky's theorem and the complexity of entanglement detection*, Discrete Anal. (2017), Paper No. 1, 20. MR3631615 ↑208, 261

[ASW10] Guillaume Aubrun, Stanisław Szarek, and Elisabeth Werner, *Nonadditivity of Rényi entropy and Dvoretzky's theorem*, J. Math. Phys. **51** (2010), no. 2, 022102, 7, DOI 10.1063/1.3271044. MR2605015 ↑232

[ASW11] Guillaume Aubrun, Stanisław Szarek, and Elisabeth Werner, *Hastings's additivity counterexample via Dvoretzky's theorem*, Comm. Math. Phys. **305** (2011), no. 1, 85–97, DOI 10.1007/s00220-010-1172-y. MR2802300 ↑144, 208, 233

[ASY12] Guillaume Aubrun, Stanisław J. Szarek, and Deping Ye, *Phase transitions for random states and a semicircle law for the partial transpose*, Phys. Rev. A **85** (2012). 272, 273

[ASY14] Guillaume Aubrun, Stanisław J. Szarek, and Deping Ye, *Entanglement thresholds for random induced states*, Comm. Pure Appl. Math. **67** (2014), no. 1, 129–171, DOI 10.1002/cpa.21460. MR3139428 ↑64, 179, 260, 270, 272

[Aub05] Guillaume Aubrun, *A sharp small deviation inequality for the largest eigenvalue of a random matrix*, Séminaire de Probabilités XXXVIII, Lecture Notes in Math., vol. 1857, Springer, Berlin, 2005, pp. 320–337, DOI 10.1007/978-3-540-31449-3_22. MR2126983 ↑179

[Aub09] Guillaume Aubrun, *On almost randomizing channels with a short Kraus decomposition*, Comm. Math. Phys. **288** (2009), no. 3, 1103–1116, DOI 10.1007/s00220-008-0695-y. MR2504867 ↑233

[Aub12] Guillaume Aubrun, *Partial transposition of random states and non-centered semicircular distributions*, Random Matrices Theory Appl. **1** (2012), no. 2, DOI 10.1142/S2010326312500013. MR2934718 ↑180, 273

[Aud09] Koenraad M. R. Audenaert, *A note on the $p \to q$ norms of 2-positive maps*, Linear Algebra Appl. **430** (2009), no. 4, 1436–1440, DOI 10.1016/j.laa.2008.09.040. MR2489405 ↑232

[Azu67] Kazuoki Azuma, *Weighted sums of certain dependent random variables*, Tôhoku Math. J. (2) **19** (1967), 357–367, DOI 10.2748/tmj/1178243286. MR0221571 ↑144

[BAG97] G. Ben Arous and A. Guionnet, *Large deviations for Wigner's law and Voiculescu's non-commutative entropy*, Probab. Theory Related Fields **108** (1997), no. 4, 517–542, DOI 10.1007/s004400050119. MR1465640 ↑245

[Bak94] Dominique Bakry, *L'hypercontractivité et son utilisation en théorie des semigroupes* (French), Lectures on probability theory (Saint-Flour, 1992), Lecture Notes in Math., vol. 1581, Springer, Berlin, 1994, pp. 1–114, DOI 10.1007/BFb0073872. MR1307413 ↑145

[Bal86] K. M. Ball, *Isometric problems in ℓ_p and sections of convex sets*, Ph.D. thesis, University of Cambridge, 1986. 105

[Bal89] Keith Ball, *Volumes of sections of cubes and related problems*, Geometric aspects of functional analysis (1987–88), Lecture Notes in Math., vol. 1376, Springer, Berlin, 1989, pp. 251–260, DOI 10.1007/BFb0090058. MR1008726 ↑106, 209

[Bal91] Keith Ball, *Volume ratios and a reverse isoperimetric inequality*, J. London Math. Soc. (2) **44** (1991), no. 2, 351–359, DOI 10.1112/jlms/s2-44.2.351. MR1136445 ↑209, 342

[Bal92a] Keith Ball, *Ellipsoids of maximal volume in convex bodies*, Geom. Dedicata **41** (1992), no. 2, 241–250, DOI 10.1007/BF00182424. MR1153987 ↑104

[Bal92b] Keith Ball, *A lower bound for the optimal density of lattice packings*, Internat. Math. Res. Notices (1992), no. 10, 217–221. MR1191572 (93k:11061) 142

[Bal97] Keith Ball, *An elementary introduction to modern convex geometry*, Flavors of geometry, Math. Sci. Res. Inst. Publ., vol. 31, Cambridge Univ. Press, Cambridge, 1997, pp. 1–58, DOI 10.2977/prims/1195164788. MR1491097 ↑87, 104, 209

[Bar98] Franck Barthe, *An extremal property of the mean width of the simplex*, Math. Ann. **310** (1998), no. 4, 685–693, DOI 10.1007/s002080050166. MR1619740 ↑342

[Bar02] Alexander Barvinok, *A course in convexity*, Graduate Studies in Mathematics, vol. 54, American Mathematical Society, Providence, RI, 2002. MR1940576 ↑29

[Bar14] Alexander Barvinok, *Thrifty approximations of convex bodies by polytopes*, Int. Math. Res. Not. IMRN **16** (2014), 4341–4356. MR3250035 ↑143

[BBP+96] Charles H Bennett, Gilles Brassard, Sandu Popescu, Benjamin Schumacher, John A Smolin, and William K Wootters, *Purification of noisy entanglement and faithful teleportation via noisy channels*, Phys. Rev. Lett. **76** (1996), no. 5, 722. 306

[BBT05] Gilles Brassard, Anne Broadbent, and Alain Tapp, *Quantum pseudo-telepathy*, Found. Phys. **35** (2005), no. 11, 1877–1907, DOI 10.1007/s10701-005-7353-4. MR2192453 ↑297

[BC02] Károly Bezdek and Robert Connelly, *Pushing disks apart—the Kneser-Poulsen conjecture in the plane*, J. Reine Angew. Math. **553** (2002), 221–236, DOI 10.1515/crll.2002.101. MR1944813 ↑178

[BCL94] Keith Ball, Eric A. Carlen, and Elliott H. Lieb, *Sharp uniform convexity and smoothness inequalities for trace norms*, Invent. Math. **115** (1994), no. 3, 463–482, DOI 10.1007/BF01231769. MR1262940 ↑29

[BCN12] Serban Belinschi, Benoît Collins, and Ion Nechita, *Eigenvectors and eigenvalues in a random subspace of a tensor product*, Invent. Math. **190** (2012), no. 3, 647–697, DOI 10.1007/s00222-012-0386-3. MR2995183 ↑233

[BCN16] Serban T. Belinschi, Benoît Collins, and Ion Nechita, *Almost one bit violation for the additivity of the minimum output entropy*, Comm. Math. Phys. **341** (2016), no. 3, 885–909, DOI 10.1007/s00220-015-2561-z. MR3452274 ↑233

[BCP+14] Nicolas Brunner, Daniel Cavalcanti, Stefano Pironio, Valerio Scarani, and Stephanie Wehner, *Bell nonlocality*, Rev. Mod. Phys. **86** (2014), 419–478. 296, 297

[BDG+77] G. Bennett, L. E. Dor, V. Goodman, W. B. Johnson, and C. M. Newman, *On uncomplemented subspaces of L_p, $1 < p < 2$*, Israel J. Math. **26** (1977), no. 2, 178–187, DOI 10.1007/BF03007667. MR0435822 ↑208

[BDK89] Rajendra Bhatia, Chandler Davis, and Paul Koosis, *An extremal problem in Fourier analysis with applications to operator theory*, J. Funct. Anal. **82** (1989), no. 1, 138–150, DOI 10.1016/0022-1236(89)90095-5. MR976316 ↑354

[BDM83] Rajendra Bhatia, Chandler Davis, and Alan McIntosh, *Perturbation of spectral subspaces and solution of linear operator equations*, Linear Algebra Appl. **52/53** (1983), 45–67, DOI 10.1016/0024-3795(83)80007-X. MR709344 ↑354

[BDM+99] Charles H. Bennett, David P. DiVincenzo, Tal Mor, Peter W. Shor, John A. Smolin, and Barbara M. Terhal, *Unextendible product bases and bound entanglement*, Phys. Rev. Lett. **82** (1999), no. 26, 5385–5388, DOI 10.1103/PhysRevLett.82.5385. MR1789694 ↑63

[BDMS13] Afonso S. Bandeira, Edgar Dobriban, Dustin G. Mixon, and William F. Sawin, *Certifying the restricted isometry property is hard*, IEEE Trans. Inform. Theory **59** (2013), no. 6, 3448–3450, DOI 10.1109/TIT.2013.2248414. MR3061257 ↑210

[BDSW96] Charles H. Bennett, David P. DiVincenzo, John A. Smolin, and William K. Wootters, *Mixed-state entanglement and quantum error correction*, Phys. Rev. A (3) **54** (1996), no. 5, 3824–3851, DOI 10.1103/PhysRevA.54.3824. MR1418618 ↑306

[BÉ85] D. Bakry and Michel Émery, *Diffusions hypercontractives* (French), Séminaire de probabilités, XIX, 1983/84, Lecture Notes in Math., vol. 1123, Springer, Berlin, 1985, pp. 177–206, DOI 10.1007/BFb0075847. MR889476 ↑145

[Bec75] William Beckner, *Inequalities in Fourier analysis*, Ann. of Math. (2) **102** (1975), no. 1, 159–182, DOI 10.2307/1970980. MR0385456 ↑146

[Bel64] J. S. Bell, *On the Einstein Podolsky Rosen paradox*, Physics **1** (1964), 195–200. 276

[Ben84] Yoav Benyamini, *Two-point symmetrization, the isoperimetric inequality on the sphere and some applications*, Texas functional analysis seminar 1983–1984 (Austin, Tex.), Longhorn Notes, Univ. Texas Press, Austin, TX, 1984, pp. 53–76. MR832231 ↑144

[Bez08] K. Bezdek, *From the Kneser-Poulsen conjecture to ball-polyhedra*, European J. Combin. **29** (2008), no. 8, 1820–1830, DOI 10.1016/j.ejc.2008.01.011. MR2463159 ↑178

[BF88] I. Bárány and Z. Füredi, *Approximation of the sphere by polytopes having few vertices*, Proc. Amer. Math. Soc. **102** (1988), no. 3, 651–659, DOI 10.2307/2047241. MR928998 ↑178

[BGK$^+$01] Andreas Brieden, Peter Gritzmann, Ravindran Kannan, Victor Klee, László Lovász, and Miklós Simonovits, *Deterministic and randomized polynomial-time approximation of radii*, Mathematika **48** (2001), no. 1-2, 63–105 (2003), DOI 10.1112/S0025579300014364. MR1996363 ↑142

[BGL14] Dominique Bakry, Ivan Gentil, and Michel Ledoux, *Analysis and geometry of Markov diffusion operators*, Grundlehren der Mathematischen Wissenschaften [Fundamental Principles of Mathematical Sciences], vol. 348, Springer, Cham, 2014. MR3155209 136, 144, 145

[BGM71] Marcel Berger, Paul Gauduchon, and Edmond Mazet, *Le spectre d'une variété riemannienne* (French), Lecture Notes in Math., Vol. 194, Springer-Verlag, Berlin-New York, 1971. MR0282313 145

[BGVV14] Silouanos Brazitikos, Apostolos Giannopoulos, Petros Valettas, and Beatrice-Helen Vritsiou, *Geometry of isotropic convex bodies*, Mathematical Surveys and Monographs, vol. 196, American Mathematical Society, Providence, RI, 2014. MR3185453 ↑105

[BH10] Fernando G. S. L. Brandão and Michał Horodecki, *On Hastings' counterexamples to the minimum output entropy additivity conjecture*, Open Syst. Inf. Dyn. **17** (2010), no. 1, 31–52, DOI 10.1142/S1230161210000047. MR2654951 ↑233

[BH13] Fernando G. S. L. Brandão and Aram W. Harrow, *Product-state approximations to quantum ground states (extended abstract)*, STOC'13—Proceedings of the 2013 ACM Symposium on Theory of Computing, ACM, New York, 2013, pp. 871–880, DOI 10.1145/2488608.2488719. MR3210849 ↑29

[Bha97] Rajendra Bhatia, *Matrix analysis*, Graduate Texts in Mathematics, vol. 169, Springer-Verlag, New York, 1997. MR1477662 ↑29

[BHH$^+$14] Piotr Badziąg, Karol Horodecki, Michał Horodecki, Justin Jenkinson, and Stanisław J. Szarek, *Bound entangled states with extremal properties*, Phys. Rev. A **90** (2014), 012301. 261

[Bil99] Patrick Billingsley, *Convergence of probability measures*, 2nd ed., Wiley Series in Probability and Statistics: Probability and Statistics, John Wiley & Sons, Inc., New York, 1999. A Wiley-Interscience Publication. MR1700749 ↑179

[BKP06] Jonathan Barrett, Adrian Kent, and Stefano Pironio, *Maximally nonlocal and monogamous quantum correlations*, Phys. Rev. Lett. **97** (2006), 170409. 297

[BL75] H.J. Brascamp and E.H. Lieb, *Some inequalities for Gaussian measures and the long range order of one-dimensional plasma*, pp. 1–14, Clarendon Press, Oxford, 1975. 105

[BL76] Herm Jan Brascamp and Elliott H. Lieb, *On extensions of the Brunn-Minkowski and Prékopa-Leindler theorems, including inequalities for log concave functions, and with an application to the diffusion equation*, J. Funct. Anal. **22** (1976), no. 4, 366–389. MR0450480 105

[BL01] H. Barnum and N. Linden, *Monotones and invariants for multi-particle quantum states*, J. Phys. A **34** (2001), no. 35, 6787–6805, DOI 10.1088/0305-4470/34/35/305. MR1862793 ↑233

[BLM89] J. Bourgain, J. Lindenstrauss, and V. Milman, *Approximation of zonoids by zonotopes*, Acta Math. **162** (1989), no. 1-2, 73–141, DOI 10.1007/BF02392835. MR981200 ↑210

[BLM13] Stéphane Boucheron, Gábor Lugosi, and Pascal Massart, *Concentration inequalities: a nonasymptotic theory of independence*, Oxford University Press, Oxford, 2013. MR3185193 ↑118, 119, 143, 144, 146

[BLPS99] Wojciech Banaszczyk, Alexander E. Litvak, Alain Pajor, and Stanisław J. Szarek, *The flatness theorem for nonsymmetric convex bodies via the local theory of Banach spaces*, Math. Oper. Res. **24** (1999), no. 3, 728–750, DOI 10.1287/moor.24.3.728. MR1854250 ↑103, 207, 208

[BM87] J. Bourgain and V. D. Milman, *New volume ratio properties for convex symmetric bodies in* $\mathbf{R}^n$, Invent. Math. **88** (1987), no. 2, 319–340, DOI 10.1007/BF01388911. MR880954 ↑105, 209

[BM08] Bhaskar Bagchi and Gadadhar Misra, *On Grothendieck constants*, unpublished, 2008. 295

[BMW09] Michael J. Bremner, Caterina Mora, and Andreas Winter, *Are random pure states useful for quantum computation?*, Phys. Rev. Lett. **102** (2009), no. 19, 190502, 4, DOI 10.1103/PhysRevLett.102.190502. MR2507884 ↑234

[BN02] Alexander Barg and Dmitry Yu. Nogin, *Bounds on packings of spheres in the Grassmann manifold*, IEEE Trans. Inform. Theory **48** (2002), no. 9, 2450–2454, DOI 10.1109/TIT.2002.801469. MR1929456 ↑143

[BN05] Brian M. Kurkoski, Paul H. Siegel, and Jack K. Wolf, *Correction to: "Joint message-passing decoding of LDPC codes and partial-response channels" [IEEE Trans. Inform. Theory 48 (2002), no. 6, 1410–1422; MR1908983 (2003e:94051)]*, IEEE Trans. Inform. Theory **49** (2003), no. 8, 2076, DOI 10.1109/TIT.2003.814494. MR2004715 ↑143

[BN06a] A. M. Barg and D. Yu. Nogin, *A spectral approach to linear programming bounds for codes* (Russian, with Russian summary), Problemy Peredachi Informatsii **42** (2006), no. 2, 12–25, DOI 10.1134/S0032946006020025; English transl., Probl. Inf. Transm. **42** (2006), no. 2, 77–89. MR2232886 ↑142

[BN06b] Alexander Barg and Dmitry Nogin, *A bound on Grassmannian codes*, J. Combin. Theory Ser. A **113** (2006), no. 8, 1629–1635, DOI 10.1016/j.jcta.2006.03.025. MR2269543 ↑143

[BN13] Teodor Banica and Ion Nechita, *Asymptotic eigenvalue distributions of block-transposed Wishart matrices*, J. Theoret. Probab. **26** (2013), no. 3, 855–869, DOI 10.1007/s10959-012-0409-4. MR3090554 ↑180

[BNV16] S. Brierley, M. Navascués, and T. Vértesi, *Convex separation from convex optimization for large-scale problems*, arXiv:1609.05011 (2016). 295

[Bob97] S. G. Bobkov, *An isoperimetric inequality on the discrete cube, and an elementary proof of the isoperimetric inequality in Gauss space*, Ann. Probab. **25** (1997), no. 1, 206–214, DOI 10.1214/aop/1024404285. MR1428506 ↑145

[Bom90a] Jan Boman, *Smoothness of sums of convex sets with real analytic boundaries*, Math. Scand. **66** (1990), no. 2, 225–230, DOI 10.7146/math.scand.a-12306. MR1075139 ↑104

[Bom90b] Jan Boman, *The sum of two plane convex* C^∞ *sets is not always* C^5, Math. Scand. **66** (1990), no. 2, 216–224, DOI 10.7146/math.scand.a-12305. MR1075138 ↑104

[Bon70] Aline Bonami, *Étude des coefficients de Fourier des fonctions de* $L^p(G)$ (French, with English summary), Ann. Inst. Fourier (Grenoble) **20** (1970), no. fasc. 2, 335–402 (1971). MR0283496 ↑146

[Bor75a] C. Borell, *Convex set functions in d-space*, Period. Math. Hungar. **6** (1975), no. 2, 111–136, DOI 10.1007/BF02018814. MR0404559 ↑104

[Bor75b] Christer Borell, *The Brunn-Minkowski inequality in Gauss space*, Invent. Math. **30** (1975), no. 2, 207–216, DOI 10.1007/BF01425510. MR0399402 ↑144

[Bor03] Christer Borell, *The Ehrhard inequality* (English, with English and French summaries), C. R. Math. Acad. Sci. Paris **337** (2003), no. 10, 663–666, DOI 10.1016/j.crma.2003.09.031. MR2030108 ↑144

[Bör04] Károly Böröczky Jr., *Finite packing and covering*, Cambridge Tracts in Mathematics, vol. 154, Cambridge University Press, Cambridge, 2004. MR2078625 ↑142

[Bou84] J. Bourgain, *On martingales transforms in finite-dimensional lattices with an appendix on the K-convexity constant*, Math. Nachr. **119** (1984), 41–53, DOI 10.1002/mana.19841190104. MR774175 ↑207

[Boy67] A. V. Boyd, *Note on a paper by Uppuluri*, Pacific J. Math. **22** (1967), 9–10. MR0216037 ↑309

[BP01] Imre Bárány and Attila Pór, *On 0-1 polytopes with many facets*, Adv. Math. **161** (2001), no. 2, 209–228, DOI 10.1006/aima.2001.1991. MR1851645 ↑281

[Bry95] Włodzimierz Bryc, *The normal distribution*, Lecture Notes in Statistics, vol. 100, Springer-Verlag, New York, 1995. Characterizations with applications. MR1335228 ↑309

[BS] Andrew Blasius and Stanisław Szarek, *Sharp two-sided bounds for the medians of gamma and chi-squared distributions*, in preparation. 124

[BS88] J. Bourgain and S. J. Szarek, *The Banach-Mazur distance to the cube and the Dvoretzky-Rogers factorization*, Israel J. Math. **62** (1988), no. 2, 169–180, DOI 10.1007/BF02787120. MR947820 ↑208

[BS10] Salman Beigi and Peter W. Shor, *Approximating the set of separable states using the positive partial transpose test*, J. Math. Phys. **51** (2010), no. 4, 042202, 10, DOI 10.1063/1.3364793. MR2662470 ↑261

[BTN01a] Aharon Ben-Tal and Arkadi Nemirovski, *Lectures on modern convex optimization*, MPS/SIAM Series on Optimization, Society for Industrial and Applied Mathematics (SIAM), Philadelphia, PA; Mathematical Programming Society (MPS), Philadelphia, PA, 2001. Analysis, algorithms, and engineering applications. MR1857264 ↑29

[BTN01b] Aharon Ben-Tal and Arkadi Nemirovski, *On polyhedral approximations of the second-order cone*, Math. Oper. Res. **26** (2001), no. 2, 193–205, DOI 10.1287/moor.26.2.193.10561. MR1895823 ↑210

[BV04] Stephen Boyd and Lieven Vandenberghe, *Convex optimization*, Cambridge University Press, Cambridge, 2004. MR2061575 ↑29

[BV13] Jop Briët and Thomas Vidick, *Explicit lower and upper bounds on the entangled value of multiplayer XOR games*, Comm. Math. Phys. **321** (2013), no. 1, 181–207, DOI 10.1007/s00220-012-1642-5. MR3089669 ↑297

[BW03] Károly Böröczky Jr. and Gergely Wintsche, *Covering the sphere by equal spherical balls*, Discrete and computational geometry, Algorithms Combin., vol. 25, Springer, Berlin, 2003, pp. 235–251, DOI 10.1007/978-3-642-55566-4_10. MR2038476 ↑142

[BY88] Z. D. Bai and Y. Q. Yin, *Convergence to the semicircle law*, Ann. Probab. **16** (1988), no. 2, 863–875. MR929083 ↑180

[BZ88] Yu. D. Burago and V. A. Zalgaller, *Geometric inequalities*, Grundlehren der Mathematischen Wissenschaften [Fundamental Principles of Mathematical Sciences], vol. 285, Springer-Verlag, Berlin, 1988. Translated from the Russian by A. B. Sosinskiĭ; Springer Series in Soviet Mathematics. MR936419 ↑144

[BŻ06] Ingemar Bengtsson and Karol Życzkowski, *Geometry of quantum states*, Cambridge University Press, Cambridge, 2006. An introduction to quantum entanglement. MR2230995 ↑63

[CĐ13] Lin Chen and Dragomir Ž. Đoković, *Dimensions, lengths, and separability in finite-dimensional quantum systems*, J. Math. Phys. **54** (2013), no. 2, 022201, 13, DOI 10.1063/1.4790405. MR3076372 ↑63

[CDJ+08] Jianxin Chen, Runyao Duan, Zhengfeng Ji, Mingsheng Ying, and Jun Yu, *Existence of universal entangler*, J. Math. Phys. **49** (2008), no. 1, 012103, 7, DOI 10.1063/1.2829895. MR2385250 ↑232

[Ceĭ76] I. I. Ceĭtlin, *Extremal points of the unit ball of certain operator spaces* (Russian), Mat. Zametki **20** (1976), no. 4, 521–527. MR0440345 ↑104

[CFG+16] Umut Caglar, Matthieu Fradelizi, Olivier Guédon, Joseph Lehec, Carsten Schütt, and Elisabeth M. Werner, *Functional versions of L_p-affine surface area and entropy inequalities*, Int. Math. Res. Not. IMRN **4** (2016), 1223–1250, DOI 10.1093/imrn/rnv151. MR3493447 ↑105

[CFN15] Benoît Collins, Motohisa Fukuda, and Ion Nechita, *On the convergence of output sets of quantum channels*, J. Operator Theory **73** (2015), no. 2, 333–360, DOI 10.7900/jot.2013dec04.2008. MR3346125 ↑233

[CFR59] H. S. M. Coxeter, L. Few, and C. A. Rogers, *Covering space with equal spheres*, Mathematika **6** (1959), 147–157, DOI 10.1112/S0025579300002059. MR0124821 ↑112, 142

[CG04] Daniel Collins and Nicolas Gisin, *A relevant two qubit Bell inequality inequivalent to the CHSH inequality*, J. Phys. A **37** (2004), no. 5, 1775–1787, DOI 10.1088/0305-4470/37/5/021. MR2044191 ↑297

[CGLP12] Djalil Chafaï, Olivier Guédon, Guillaume Lecué, and Alain Pajor, *Interactions between compressed sensing random matrices and high dimensional geometry*, Panoramas et Synthèses [Panoramas and Syntheses], vol. 37, Société Mathématique de France, Paris, 2012. MR3113826 ↑146

[Cha67] G. D. Chakerian, *Inequalities for the difference body of a convex body*, Proc. Amer. Math. Soc. **18** (1967), 879–884, DOI 10.2307/2035131. MR0218972 ↑105

[Che78] S. Chevet, *Séries de variables aléatoires gaussiennes à valeurs dans $E\hat{\otimes}_\varepsilon F$. Application aux produits d'espaces de Wiener abstraits* (French), Séminaire sur la Géométrie des Espaces de Banach (1977–1978), École Polytech., Palaiseau, 1978, pp. Exp. No. 19, 15. MR520217 ↑180

[CHL+08] Toby Cubitt, Aram W. Harrow, Debbie Leung, Ashley Montanaro, and Andreas Winter, *Counterexamples to additivity of minimum output p-Rényi entropy for p close to 0*, Comm. Math. Phys. **284** (2008), no. 1, 281–290, DOI 10.1007/s00220-008-0625-z. MR2443306 ↑232

[CHLL97] Gérard Cohen, Iiro Honkala, Simon Litsyn, and Antoine Lobstein, *Covering codes*, North-Holland Mathematical Library, vol. 54, North-Holland Publishing Co., Amsterdam, 1997. MR1453577 ↑142

[Cho75a] Man Duen Choi, *Completely positive linear maps on complex matrices*, Linear Algebra Appl. **10** (1975), 285–290. MR0376726 ↑64

[Cho75b] Man Duen Choi, *Positive semidefinite biquadratic forms*, Linear Algebra Appl. **12** (1975), no. 2, 95–100. MR0379365 ↑63

[CHS96] John H. Conway, Ronald H. Hardin, and Neil J. A. Sloane, *Packing lines, planes, etc.: packings in Grassmannian spaces*, Experiment. Math. **5** (1996), no. 2, 139–159. MR1418961 ↑143

[CHSH69] John F. Clauser, Michael A. Horne, Abner Shimony, and Richard A. Holt, *Proposed experiment to test local hidden-variable theories*, Phys. Rev. Lett. **23** (1969), 880–884. 295

[Chu62] J. T. Chu, *Mathematical Notes: A Modified Wallis Product and Some Applications*, Amer. Math. Monthly **69** (1962), no. 5, 402–404, DOI 10.2307/2312135. MR1531681 ↑309

[Cir80] B. S. Cirel′son, *Quantum generalizations of Bell's inequality*, Lett. Math. Phys. **4** (1980), no. 2, 93–100, DOI 10.1007/BF00417500. MR577178 ↑295

[CL06] Eric Carlen and Elliott H. Lieb, *Some matrix rearrangement inequalities*, Ann. Mat. Pura Appl. (4) **185** (2006), no. suppl., S315–S324, DOI 10.1007/s10231-004-0147-z. MR2187765 ↑29

[Cla36] James A. Clarkson, *Uniformly convex spaces*, Trans. Amer. Math. Soc. **40** (1936), no. 3, 396–414, DOI 10.2307/1989630. MR1501880 ↑29

[Cla06] Lieven Clarisse, *The distillability problem revisited*, Quantum Inf. Comput. **6** (2006), no. 6, 539–560. MR2253931 ↑306

[CLM+14] Eric Chitambar, Debbie Leung, Laura Mančinska, Maris Ozols, and Andreas Winter, *Everything you always wanted to know about LOCC (but were afraid to ask)*, Comm. Math. Phys. **328** (2014), no. 1, 303–326, DOI 10.1007/s00220-014-1953-9. MR3196987 ↑306

[CM14] Benoît Collins and Camille Male, *The strong asymptotic freeness of Haar and deterministic matrices* (English, with English and French summaries), Ann. Sci. Éc. Norm. Supér. (4) **47** (2014), no. 1, 147–163, DOI 10.24033/asens.2211. MR3205602 ↑180

[CM15] Fabio Cavalletti and Andrea Mondino, *Sharp and rigid isoperimetric inequalities in metric-measure spaces with lower Ricci curvature bounds*, Invent. Math. **208** (2017), no. 3, 803–849, DOI 10.1007/s00222-016-0700-6. MR3648975 ↑144

[CN10] Benoît Collins and Ion Nechita, *Random quantum channels I: graphical calculus and the Bell state phenomenon*, Comm. Math. Phys. **297** (2010), no. 2, 345–370, DOI 10.1007/s00220-010-1012-0. MR2651902 ↑233

[CN11] Benoît Collins and Ion Nechita, *Random quantum channels II: entanglement of random subspaces, Rényi entropy estimates and additivity problems*, Adv. Math. **226** (2011), no. 2, 1181–1201, DOI 10.1016/j.aim.2010.08.002. MR2737781 ↑233

[CN16] Benoît Collins and Ion Nechita, *Random matrix techniques in quantum information theory*, J. Math. Phys. **57** (2016), no. 1, 015215, 34, DOI 10.1063/1.4936880. MR3432743 ↑180, 233

[CNY12] Benoit Collins, Ion Nechita, and Deping Ye, *The absolute positive partial transpose property for random induced states*, Random Matrices Theory Appl. **1** (2012), no. 3, 1250002, 22, DOI 10.1142/S2010326312500025. MR2967961 ↑273

[Col06] Andrea Colesanti, *Functional inequalities related to the Rogers-Shephard inequality*, Mathematika **53** (2006), no. 1, 81–101 (2007), DOI 10.1112/S0025579300000048. MR2304054 ↑105

[Col16] Benoît Collins, *Haagerup's inequality and additivity violation of the Minimum Output Entropy*, Houston J. Math., to appear. 233

[CP88] Bernd Carl and Alain Pajor, *Gel′fand numbers of operators with values in a Hilbert space*, Invent. Math. **94** (1988), no. 3, 479–504, DOI 10.1007/BF01394273. MR969241 ↑178

[CR86] Jeesen Chen and Herman Rubin, *Bounds for the difference between median and mean of gamma and Poisson distributions*, Statist. Probab. Lett. **4** (1986), no. 6, 281–283, DOI 10.1016/0167-7152(86)90044-1. MR858317 ↑124

[CS90] Bernd Carl and Irmtraud Stephani, *Entropy, compactness and the approximation of operators*, Cambridge Tracts in Mathematics, vol. 98, Cambridge University Press, Cambridge, 1990. MR1098497 ↑143

[CS99] J. H. Conway and N. J. A. Sloane, *Sphere packings, lattices and groups*, 3rd ed., Grundlehren der Mathematischen Wissenschaften [Fundamental Principles of Mathematical Sciences], vol. 290, Springer-Verlag, New York, 1999. With additional contributions by E. Bannai, R. E. Borcherds, J. Leech, S. P. Norton, A. M. Odlyzko, R. A. Parker, L. Queen and B. B. Venkov. MR1662447 ↑142

[CS05] Shiing-Shen Chern and Zhongmin Shen, *Riemann-Finsler geometry*, Nankai Tracts in Mathematics, vol. 6, World Scientific Publishing Co. Pte. Ltd., Hackensack, NJ, 2005. MR2169595 ↑319

[CSW14] Adán Cabello, Simone Severini, and Andreas Winter, *Graph-theoretic approach to quantum correlations*, Phys. Rev. Lett. **112** (2014), 040401. 297

[CW03] Kai Chen and Ling-An Wu, *A matrix realignment method for recognizing entanglement*, Quantum Inf. Comput. **3** (2003), no. 3, 193–202. MR1985541 ↑63

[Dav57] Chandler Davis, *All convex invariant functions of hermitian matrices*, Arch. Math. **8** (1957), 276–278, DOI 10.1007/BF01898787. MR0090572 ↑29

[DCLB00] W. Dür, J. I. Cirac, M. Lewenstein, and D. Bruß, *Distillability and partial transposition in bipartite systems*, Phys. Rev. A (3) **61** (2000), no. 6, 062313, 8, DOI 10.1103/PhysRevA.61.062313. MR1767474 ↑306

[DD16] M. Deza and M. Dutour Sikirić, *Enumeration of the facets of cut polytopes over some highly symmetric graphs*, Int. Trans. Oper. Res. **23** (2016), no. 5, 853-860; arXiv:1501.05407. 295

[Dem97] Amir Dembo, *Information inequalities and concentration of measure*, Ann. Probab. **25** (1997), no. 2, 927–939, DOI 10.1214/aop/1024404424. MR1434131 ↑146

[DF87] Persi Diaconis and David Freedman, *A dozen de Finetti-style results in search of a theory* (English, with French summary), Ann. Inst. H. Poincaré Probab. Statist. **23** (1987), no. 2, suppl., 397–423. MR898502 ↑144

[DF93] Andreas Defant and Klaus Floret, *Tensor norms and operator ideals*, North-Holland Mathematics Studies, vol. 176, North-Holland Publishing Co., Amsterdam, 1993. MR1209438 ↑104

[DL97] Michel Marie Deza and Monique Laurent, *Geometry of cuts and metrics*, Algorithms Combin., vol. 15, Springer-Verlag, Berlin, 1997. MR1460488 ↑296

[DLS14] Anirban DasGupta, S. N. Lahiri, and Jordan Stoyanov, *Sharp fixed n bounds and asymptotic expansions for the mean and the median of a Gaussian sample maximum, and applications to the Donoho-Jin model*, Stat. Methodol. **20** (2014), 40–62, DOI 10.1016/j.stamet.2014.01.002. MR3205720 ↑178, 352

[Dmi90] V. A. Dmitrovskiĭ, *On the integrability of the maximum and the local properties of Gaussian fields*, Probability theory and mathematical statistics, Vol. I (Vilnius, 1989), "Mokslas", Vilnius, 1990, pp. 271–284. MR1153819 ↑144

[DPS04] Andrew C. Doherty, Pablo A. Parrilo, and Federico M. Spedalieri, *Complete family of separability criteria*, Phys. Rev. A **69** (2004), 022308. 63

[DR47] H. Davenport and C. A. Rogers, *Hlawka's theorem in the geometry of numbers*, Duke Math. J. **14** (1947), 367–375. MR0021018 ↑142

[DR50] A. Dvoretzky and C. A. Rogers, *Absolute and unconditional convergence in normed linear spaces*, Proc. Nat. Acad. Sci. U. S. A. **36** (1950), 192–197. MR0033975 ↑208

[DS85] Stephen Dilworth and Stanisław Szarek, *The cotype constant and an almost Euclidean decomposition for finite-dimensional normed spaces*, Israel J. Math. **52** (1985), no. 1-2, 82–96, DOI 10.1007/BF02776082. MR815604 ↑209

[DS01] Kenneth R. Davidson and Stanisław J. Szarek, *Local operator theory, random matrices and Banach spaces*, Handbook of the geometry of Banach spaces, Vol. I, North-Holland, Amsterdam, 2001, pp. 317–366, DOI 10.1016/S1874-5849(01)80010-3. MR1863695 ↑144, 180

[DSS⁺00] David P. DiVincenzo, Peter W. Shor, John A. Smolin, Barbara M. Terhal, and Ashish V. Thapliyal, *Evidence for bound entangled states with negative partial transpose*, Phys. Rev. A (3) **61** (2000), no. 6, 062312, 13, DOI 10.1103/PhysRevA.61.062312. MR1767473 ↑306

[Dud67] R. M. Dudley, *The sizes of compact subsets of Hilbert space and continuity of Gaussian processes*, J. Funct. Anal. **1** (1967), 290–330. MR0220340 ↑179

[Due10] Lutz Duembgen, *Bounding standard Gaussian tail probabilities*, Tech. report, University of Bern, Institute of Mathematical Statistics and Actuarial Science, 2010. 309

[Dum07] Ilya Dumer, *Covering spheres with spheres*, Discrete Comput. Geom. **38** (2007), no. 4, 665–679, DOI 10.1007/s00454-007-9000-7. MR2365829 ↑110, 142

[Dür01] W. Dür, *Multipartite bound entangled states that violate Bell's inequality*, Phys. Rev. Lett. **87** (2001), no. 23, 230402, 4. MR1870610 ↑297

[Dvo61] Aryeh Dvoretzky, *Some results on convex bodies and Banach spaces*, Proc. Internat. Sympos. Linear Spaces (Jerusalem, 1960), Jerusalem Academic Press, Jerusalem; Pergamon, Oxford, 1961, pp. 123–160. MR0139079 ↑208

[EC04] Fida El Chami, *Spectra of the Laplace operator on Grassmann manifolds*, Int. J. Pure Appl. Math. **12** (2004), no. 4, 395–418. MR2058707 ↑145

[Ehr83] Antoine Ehrhard, *Symétrisation dans l'espace de Gauss* (French), Math. Scand. **53** (1983), no. 2, 281–301, DOI 10.7146/math.scand.a-12035. MR745081 ↑144

[EPR35] A. Einstein, B. Podolsky, and N. Rosen, *Can quantum-mechanical description of physical reality be considered complete?*, Phys. Rev. **47** (1935), 777–780. 276

[ES70] P. Erdős and A. H. Stone, *On the sum of two Borel sets*, Proc. Amer. Math. Soc. **25** (1970), 304–306, DOI 10.2307/2037209. MR0260958 ↑104

[EVWW01] Tilo Eggeling, Karl Gerd H. Vollbrecht, Reinhard F. Werner, and Michael M. Wolf, *Distillability via protocols respecting the positivity of partial transpose*, Phys. Rev. Lett. **87** (2001), 257902. 306

[Fer75] X. Fernique, *Regularité des trajectoires des fonctions aléatoires gaussiennes* (French), École d'Été de Probabilités de Saint-Flour, IV-1974, Springer, Berlin, 1975, pp. 1–96. Lecture Notes in Math., vol. 480. MR0413238 ↑178, 179

[Fer97] Xavier Fernique, *Fonctions aléatoires gaussiennes, vecteurs aléatoires gaussiens* (French), Université de Montréal, Centre de Recherches Mathématiques, Montreal, QC, 1997. MR1472975 ↑144, 179

[FF81] P. Frankl and Z. Füredi, *A short proof for a theorem of Harper about Hamming-spheres*, Discrete Math. **34** (1981), no. 3, 311–313, DOI 10.1016/0012-365X(81)90009-1. MR613409 ↑146

[FHS13] Omar Fawzi, Patrick Hayden, and Pranab Sen, *From low-distortion norm embeddings to explicit uncertainty relations and efficient information locking*, STOC'11—Proceedings of the 43rd ACM Symposium on Theory of Computing, ACM, New York, 2011, pp. 773–782, DOI 10.1145/1993636.1993738. MR2932028 ↑208

[Fig76] T. Figiel, *A short proof of Dvoretzky's theorem on almost spherical sections of convex bodies*, Compositio Math. **33** (1976), no. 3, 297–301. MR0487392 ↑208

[FK94] S. K. Foong and S. Kanno, *Proof of Page's conjecture on the average entropy of a subsystem*, Phys. Rev. Lett. **72** (1994), no. 8, 1148–1151, DOI 10.1103/PhysRevLett.72.1148. MR1259245 ↑233

[FK10] Motohisa Fukuda and Christopher King, *Entanglement of random subspaces via the Hastings bound*, J. Math. Phys. **51** (2010), no. 4, 042201, 19, DOI 10.1063/1.3309418. MR2662469 ↑233

[FKM10] Motohisa Fukuda, Christopher King, and David K. Moser, *Comments on Hastings' additivity counterexamples*, Comm. Math. Phys. **296** (2010), no. 1, 111–143, DOI 10.1007/s00220-010-0996-9. MR2606630 ↑233

[FLM77] T. Figiel, J. Lindenstrauss, and V. D. Milman, *The dimension of almost spherical sections of convex bodies*, Acta Math. **139** (1977), no. 1-2, 53–94, DOI 10.1007/BF02392234. MR0445274 ↑144, 208

[FLPS11] Shmuel Friedland, Chi-Kwong Li, Yiu-Tung Poon, and Nung-Sing Sze, *The automorphism group of separable states in quantum information theory*, J. Math. Phys. **52** (2011), no. 4, 042203, 8, DOI 10.1063/1.3578015. MR2964164 ↑64

[FN15] Motohisa Fukuda and Ion Nechita, *Additivity rates and PPT property for random quantum channels*, Ann. Math. Blaise Pascal **22** (2015), no. 1, 1–72. MR3361563 ↑180

[Fol99] Gerald B. Folland, *Real analysis*, 2nd ed., Pure and Applied Mathematics (New York), John Wiley & Sons, Inc., New York, 1999. Modern techniques and their applications; A Wiley-Interscience Publication. MR1681462 ↑15

[For10] Dominique Fortin, *Hadamard's matrices, Grothendieck's constant, and root two*, Optimization and optimal control, Springer Optim. Appl., vol. 39, Springer, New York, 2010, pp. 423–447, DOI 10.1007/978-0-387-89496-6_20. MR2732733 ↑295

[FR94] P. C. Fishburn and J. A. Reeds, *Bell inequalities, Grothendieck's constant, and root two*, SIAM J. Discrete Math. **7** (1994), no. 1, 48–56, DOI 10.1137/S0895480191219350. MR1259009 ↑295

[FR13] Simon Foucart and Holger Rauhut, *A mathematical introduction to compressive sensing*, Applied and Numerical Harmonic Analysis, Birkhäuser/Springer, New York, 2013. MR3100033 ↑208, 309

[Fra99] Matthieu Fradelizi, *Hyperplane sections of convex bodies in isotropic position*, Beiträge Algebra Geom. **40** (1999), no. 1, 163–183. MR1678528 ↑103, 105

[Fre14] Daniel J. Fresen, *Explicit Euclidean embeddings in permutation invariant normed spaces*, Adv. Math. **266** (2014), 1–16, DOI 10.1016/j.aim.2014.07.017. MR3262353 ↑210

[Fri12] Tobias Fritz, *Tsirelson's problem and Kirchberg's conjecture*, Rev. Math. Phys. **24** (2012), no. 5, 1250012, 67, DOI 10.1142/S0129055X12500122. MR2928100 ↑296

[Fro81] M. Froissart, *Constructive generalization of Bell's inequalities* (English, with Italian and Russian summaries), Nuovo Cimento B (11) **64** (1981), no. 2, 241–251, DOI 10.1007/BF02903286. MR637011 ↑297

[FŚ13] Motohisa Fukuda and Piotr Śniady, *Partial transpose of random quantum states: exact formulas and meanders*, J. Math. Phys. **54** (2013), no. 4, 042202, 23, DOI 10.1063/1.4799440. MR3088232 ↑180

[FT97] Gábor Fejes Tóth, *Packing and covering*, Handbook of discrete and computational geometry, CRC Press Ser. Discrete Math. Appl., CRC, Boca Raton, FL, 1997, pp. 19–41. MR1730158 ↑142

[FTJ79] T. Figiel and Nicole Tomczak-Jaegermann, *Projections onto Hilbertian subspaces of Banach spaces*, Israel J. Math. **33** (1979), no. 2, 155–171, DOI 10.1007/BF02760556. MR571251 ↑207

[Fuk14] Motohisa Fukuda, *Revisiting additivity violation of quantum channels*, Comm. Math. Phys. **332** (2014), no. 2, 713–728, DOI 10.1007/s00220-014-2101-2. MR3257660 ↑233

[FW07] Motohisa Fukuda and Michael M. Wolf, *Simplifying additivity problems using direct sum constructions*, J. Math. Phys. **48** (2007), no. 7, 072101, 7, DOI 10.1063/1.2746128. MR2337661 ↑232

[Gal95] Janos Galambos, *Advanced probability theory*, 2nd ed., Probability: Pure and Applied, vol. 10, Marcel Dekker, Inc., New York, 1995. MR1350792 ↑179

[Gar83] Anupam Garg, *Detector error and Einstein-Podolsky-Rosen correlations*, Phys. Rev. D (3) **28** (1983), no. 4, 785–790, DOI 10.1103/PhysRevD.28.785. MR712443 ↑295

[Gar02] R. J. Gardner, *The Brunn-Minkowski inequality*, Bull. Amer. Math. Soc. (N.S.) **39** (2002), no. 3, 355–405, DOI 10.1090/S0273-0979-02-00941-2. MR1898210 ↑104, 105

[GB02] Leonid Gurvits and Howard Barnum, *Largest separable balls around the maximally mixed bipartite quantum state*, Phys. Rev. A **66** (2002), no. 6, 062311. ↑260

[GB03] Leonid Gurvits and Howard Barnum, *Separable balls around the maximally mixed multipartite quantum states*, Phys. Rev. A **68** (2003), no. 4, 042312. ↑261

[GB05] Leonid Gurvits and Howard Barnum, *Better bound on the exponent of the radius of the multipartite separable ball*, Phys. Rev. A **72** (2005), no. 3, 032322. ↑261

[Gem80] Stuart Geman, *A limit theorem for the norm of random matrices*, Ann. Probab. **8** (1980), no. 2, 252–261. MR566592 ↑179

[GFE09] D. Gross, S. T. Flammia, and J. Eisert, *Most quantum states are too entangled to be useful as computational resources*, Phys. Rev. Lett. **102** (2009), no. 19, 190501, 4, DOI 10.1103/PhysRevLett.102.190501. MR2507883 ↑234

[GG71] D. J. H. Garling and Y. Gordon, *Relations between some constants associated with finite dimensional Banach spaces*, Israel J. Math. **9** (1971), 346–361, DOI 10.1007/BF02771685. MR0412775 ↑104

[GG84] A. Yu. Garnaev and E. D. Gluskin, *The widths of a Euclidean ball* (Russian), Dokl. Akad. Nauk SSSR **277** (1984), no. 5, 1048–1052. MR759962 ↑208

[GGHE08] O. Gittsovich, O. Gühne, P. Hyllus, and J. Eisert, *Unifying several separability conditions using the covariance matrix criterion*, Phys. Rev. A **78** (2008), 052319. ↑64

[GHP10] Andrzej Grudka, Michał Horodecki, and Łukasz Pankowski, *Constructive counterexamples to the additivity of the minimum output Rényi entropy of quantum channels for all $p > 2$*, J. Phys. A **43** (2010), no. 42, 425304, 7, DOI 10.1088/1751-8113/43/42/425304. MR2726721 ↑233

[Gia96] A. A. Giannopoulos, *A proportional Dvoretzky-Rogers factorization result*, Proc. Amer. Math. Soc. **124** (1996), no. 1, 233–241, DOI 10.1090/S0002-9939-96-03071-7. MR1301496 ↑209

[GLMP04] Y. Gordon, A. E. Litvak, M. Meyer, and A. Pajor, *John's decomposition in the general case and applications*, J. Differential Geom. **68** (2004), no. 1, 99–119. MR2152910 ↑103

[GLR10] Venkatesan Guruswami, James R. Lee, and Alexander Razborov, *Almost Euclidean subspaces of ℓ_1^N via expander codes*, Combinatorica **30** (2010), no. 1, 47–68, DOI 10.1007/s00493-010-2463-9. MR2663548 ↑207

[Glu81] E. D. Gluskin, *The diameter of the Minkowski compactum is roughly equal to n* (Russian), Funktsional. Anal. i Prilozhen. **15** (1981), no. 1, 72–73. MR609798 ↑103

[Glu88] E. D. Gluskin, *Extremal properties of orthogonal parallelepipeds and their applications to the geometry of Banach spaces* (Russian), Mat. Sb. (N.S.) **136(178)** (1988), no. 1, 85–96; English transl., Math. USSR-Sb. **64** (1989), no. 1, 85–96. MR945901 ↑178

[GLW08] Venkatesan Guruswami, James R. Lee, and Avi Wigderson, *Euclidean sections of ℓ_1^N with sublinear randomness and error-correction over the reals*, Approximation, randomization and combinatorial optimization, Lecture Notes in Comput. Sci., vol. 5171, Springer, Berlin, 2008, pp. 444–454, DOI 10.1007/978-3-540-85363-3_35. MR2538806 ↑207

[GM00] A. A. Giannopoulos and V. D. Milman, *Concentration property on probability spaces*, Adv. Math. **156** (2000), no. 1, 77–106, DOI 10.1006/aima.2000.1949. MR1800254 ↑144

[GMW14] Whan Ghang, Zane Martin, and Steven Waruhiu, *The sharp log-Sobolev inequality on a compact interval*, Involve **7** (2014), no. 2, 181–186, DOI 10.2140/involve.2014.7.181. MR3133718 ↑145

[Gor85] Yehoram Gordon, *Some inequalities for Gaussian processes and applications*, Israel J. Math. **50** (1985), no. 4, 265–289, DOI 10.1007/BF02759761. MR800188 ↑180

[Gor88] Y. Gordon, *On Milman's inequality and random subspaces which escape through a mesh in $\mathbf{R}^n$*, Geometric aspects of functional analysis (1986/87), Lecture Notes in Math., vol. 1317, Springer, Berlin, 1988, pp. 84–106, DOI 10.1007/BFb0081737. MR950977 ↑180, 208, 209

[Gra14] Loukas Grafakos, *Classical Fourier analysis*, 3rd ed., Graduate Texts in Mathematics, vol. 249, Springer, New York, 2014. MR3243734 ↑158

[Gro53a] A. Grothendieck, *Résumé de la théorie métrique des produits tensoriels topologiques* (French), Bol. Soc. Mat. São Paulo **8** (1953), 1–79. MR0094682 ↑295

[Gro53b] A. Grothendieck, *Sur certaines classes de suites dans les espaces de Banach et le théorème de Dvoretzky-Rogers* (French), Bol. Soc. Mat. São Paulo **8** (1953), 81–110 (1956). MR0094683 ↑208

[Gro75] Leonard Gross, *Logarithmic Sobolev inequalities*, Amer. J. Math. **97** (1975), no. 4, 1061–1083, DOI 10.2307/2373688. MR0420249 ↑145

[Gro80] Misha Gromov, *Paul Levy's isoperimetric inequality*, preprint IHES (1980). 144

[Gro87] M. Gromov, *Monotonicity of the volume of intersection of balls*, Geometrical aspects of functional analysis (1985/86), Lecture Notes in Math., vol. 1267, Springer, Berlin, 1987, pp. 1–4, DOI 10.1007/BFb0078131. MR907680 ↑178

[Grü03] Branko Grünbaum, *Convex polytopes*, 2nd ed., Graduate Texts in Mathematics, vol. 221, Springer-Verlag, New York, 2003. Prepared and with a preface by Volker Kaibel, Victor Klee and Günter M. Ziegler. MR1976856 ↑29

[Gru07] Peter M. Gruber, *Convex and discrete geometry*, Grundlehren der Mathematischen Wissenschaften [Fundamental Principles of Mathematical Sciences], vol. 336, Springer, Berlin, 2007. MR2335496 ↑142

[Gur03] Leonid Gurvits, *Classical deterministic complexity of Edmond's problem and quantum entanglement*, Proceedings of the Thirty-Fifth Annual ACM Symposium on Theory of Computing, ACM, New York, 2003, pp. 10–19, DOI 10.1145/780542.780545. MR2121068 ↑63, 64

[GVL13] Gene H. Golub and Charles F. Van Loan, *Matrix computations*, 4th ed., Johns Hopkins Studies in the Mathematical Sciences, Johns Hopkins University Press, Baltimore, MD, 2013. MR3024913 ↑319

[GW93] Paul Goodey and Wolfgang Weil, *Zonoids and generalisations*, Handbook of convex geometry, Vol. A, B, North-Holland, Amsterdam, 1993, pp. 1297–1326. MR1243010 ↑103

[GZ03] A. Guionnet and B. Zegarlinski, *Lectures on logarithmic Sobolev inequalities*, Séminaire de Probabilités, XXXVI, Lecture Notes in Math., vol. 1801, Springer, Berlin, 2003, pp. 1–134, DOI 10.1007/978-3-540-36107-7_1. MR1971582 ↑144

[Haa81] Uffe Haagerup, *The best constants in the Khintchine inequality*, Studia Math. **70** (1981), no. 3, 231–283 (1982). MR654838 ↑147

[Hal82] Paul Richard Halmos, *A Hilbert space problem book*, 2nd ed., Graduate Texts in Mathematics, vol. 19, Springer-Verlag, New York-Berlin, 1982. Encyclopedia of Mathematics and its Applications, 17. MR675952 ↑343

[Hal07] Majdi Ben Halima, *Branching rules for unitary groups and spectra of invariant differential operators on complex Grassmannians*, J. Algebra **318** (2007), no. 2, 520–552, DOI 10.1016/j.jalgebra.2007.08.010. MR2371957 ↑145

[Hal15] Brian Hall, *Lie groups, Lie algebras, and representations*, 2nd ed., Graduate Texts in Mathematics, vol. 222, Springer, Cham, 2015. An elementary introduction. MR3331229 ↑145

[Han56] Olof Hanner, *On the uniform convexity of L^p and l^p*, Ark. Mat. **3** (1956), 239–244, DOI 10.1007/BF02589410. MR0077087 ↑29

[Har66] L. H. Harper, *Optimal numberings and isoperimetric problems on graphs*, J. Combin. Theory **1** (1966), 385–393. MR0200192 ↑146

[Har13] Aram W Harrow, *The church of the symmetric subspace*, arXiv:1308.6595 (2013). 63

[Has09] Matthew B Hastings, *Superadditivity of communication capacity using entangled inputs*, Nature Physics **5** (2009), no. 4, 255–257. 144, 232, 233

[Hel69] Carl W. Helstrom, *Quantum detection and estimation theory*, J. Statist. Phys. **1** (1969), 231–252, DOI 10.1007/BF01007479. MR0250623 ↑305

[Hen80] Douglas Hensley, *Slicing convex bodies—bounds for slice area in terms of the body's covariance*, Proc. Amer. Math. Soc. **79** (1980), no. 4, 619–625, DOI 10.2307/2042510. MR572315 ↑105

[Hen12] Martin Henk, *Löwner-John ellipsoids*, Doc. Math. **Extra vol.: Optimization stories** (2012), 95–106. MR2991474 ↑104

[HH99] Michał Horodecki and Paweł Horodecki, *Reduction criterion of separability and limits for a class of distillation protocols*, Phys. Rev. A **59** (1999), 4206–4216. 306

[HH01] Paweł Horodecki and Ryszard Horodecki, *Distillation and bound entanglement*, Quantum Inf. Comput. **1** (2001), no. 1, 45–75. MR1910010 ↑306

[HHH96] Michał Horodecki, Paweł Horodecki, and Ryszard Horodecki, *Separability of mixed states: necessary and sufficient conditions*, Phys. Lett. A **223** (1996), no. 1-2, 1–8, DOI 10.1016/S0375-9601(96)00706-2. MR1421501 ↑63, 64

[HHH97] Michał Horodecki, Paweł Horodecki, and Ryszard Horodecki, *Inseparable two spin-$\frac{1}{2}$ density matrices can be distilled to a singlet form*, Phys. Rev. Lett. **78** (1997), 574–577. 306

[HHH98] Michał Horodecki, Paweł Horodecki, and Ryszard Horodecki, *Mixed-state entanglement and distillation: Is there a "bound" entanglement in nature?*, Phys. Rev. Lett. **80** (1998), no. 24, 5239–5242, DOI 10.1103/PhysRevLett.80.5239. MR1627438 ↑306

[HHH99] Michał Horodecki, Paweł Horodecki, and Ryszard Horodecki, *General teleportation channel, singlet fraction, and quasidistillation*, Phys. Rev. A (3) **60** (1999), no. 3, 1888–1898, DOI 10.1103/PhysRevA.60.1888. MR1721040 ↑306

[HHHH09] Ryszard Horodecki, Paweł Horodecki, Michał Horodecki, and Karol Horodecki, *Quantum entanglement*, Rev. Modern Phys. **81** (2009), no. 2, 865–942, DOI 10.1103/RevModPhys.81.865. MR2515619 ↑63, 74, 306

[Hil05] Roland Hildebrand, *Cones of ball-ball separable elements*, arXiv eprint quant-ph/0503194 (2005). 324

[Hil06] Roland Hildebrand, *Separable balls around the maximally mixed state for a 3-qubit system*, arXiv eprint quant-ph/0601201 (2006). 234, 261

[Hil07a] Roland Hildebrand, *Entangled states close to the maximally mixed state*, Phys. Rev. A (3) **75** (2007), no. 6, 062330, 10, DOI 10.1103/PhysRevA.75.062330. MR2328762 ↑234, 261

[Hil07b] Roland Hildebrand, *Positive maps of second-order cones*, Linear Multilinear Algebra **55** (2007), no. 6, 575–597, DOI 10.1080/03081080701251280. MR2360836 ↑324

[HK11] Kil-Chan Ha and Seung-Hyeok Kye, *Entanglement witnesses arising from exposed positive linear maps*, Open Syst. Inf. Dyn. **18** (2011), no. 4, 323–337, DOI 10.1142/S1230161211000224. MR2875744 ↑261

[HLSW04] Patrick Hayden, Debbie Leung, Peter W. Shor, and Andreas Winter, *Randomizing quantum states: constructions and applications*, Comm. Math. Phys. **250** (2004), no. 2, 371–391, DOI 10.1007/s00220-004-1087-6. MR2094521 ↑233

[HLW06] Patrick Hayden, Debbie W. Leung, and Andreas Winter, *Aspects of generic entanglement*, Comm. Math. Phys. **265** (2006), no. 1, 95–117, DOI 10.1007/s00220-006-1535-6. MR2217298 ↑233, 272, 273

[HNW15] Aram W. Harrow, Anand Natarajan, and Xiaodi Wu, *An improved semidefinite programming hierarchy for testing entanglement*, Comm. Math. Phys. **352** (2017), no. 3, 881–904, DOI 10.1007/s00220-017-2859-0. MR3631393 ↑29

[Hol73] A. S. Holevo, *Statistical decision theory for quantum systems*, J. Multivariate Anal. **3** (1973), 337–394, DOI 10.1016/0047-259X(73)90028-6. MR0337245 ↑305

[Hol06] Alexander S. Holevo, *The additivity problem in quantum information theory*, International Congress of Mathematicians. Vol. III, Eur. Math. Soc., Zürich, 2006, pp. 999–1018. MR2275716 ↑232

[Hol12] Alexander S. Holevo, *Quantum systems, channels, information*, De Gruyter Studies in Mathematical Physics, vol. 16, De Gruyter, Berlin, 2012. A mathematical introduction. MR2986302 ↑63

[Hor97] Paweł Horodecki, *Separability criterion and inseparable mixed states with positive partial transposition*, Phys. Lett. A **232** (1997), no. 5, 333–339, DOI 10.1016/S0375-9601(97)00416-7. MR1467418 ↑63

[HP98] Fumio Hiai and Dénes Petz, *Eigenvalue density of the Wishart matrix and large deviations*, Infin. Dimens. Anal. Quantum Probab. Relat. Top. **1** (1998), no. 4, 633–646, DOI 10.1142/S021902579800034X. MR1665279 ↑245

[HP00] Fumio Hiai and Dénes Petz, *The semicircle law, free random variables and entropy*, Mathematical Surveys and Monographs, vol. 77, American Mathematical Society, Providence, RI, 2000. MR1746976 ↑180

[HQV+17] F. Hirsch, M.T. Quintino, T. Vértesi, M. Navascués, and N. Brunner, *Better local hidden variable models for two-qubit Werner states and an upper bound on the Grothendieck constant $K_G(3)$*, Quantum **1** (2017), 3, . 295

[HS05] Daniel Hug and Rolf Schneider, *Large typical cells in Poisson-Delaunay mosaics*, Rev. Roumaine Math. Pures Appl. **50** (2005), no. 5-6, 657–670. MR2204143 ↑178

[HSR03] Michael Horodecki, Peter W. Shor, and Mary Beth Ruskai, *Entanglement breaking channels*, Rev. Math. Phys. **15** (2003), no. 6, 629–641, DOI 10.1142/S0129055X03001709. MR2001114 ↑64

[HT03] Uffe Haagerup and Steen Thorbjørnsen, *Random matrices with complex Gaussian entries*, Expo. Math. **21** (2003), no. 4, 293–337, DOI 10.1016/S0723-0869(03)80036-1. MR2022002 ↑170, 180

[HT05] Uffe Haagerup and Steen Thorbjørnsen, *A new application of random matrices: $\mathrm{Ext}(C^*_{\mathrm{red}}(F_2))$ is not a group*, Ann. of Math. (2) **162** (2005), no. 2, 711–775, DOI 10.4007/annals.2005.162.711. MR2183281 ↑180

[Hun72] Walter Hunziker, *A note on symmetry operations in quantum mechanics*. Helv. Phys. Acta **45** (1972), no. 2, 233-236. 63

[HW08] Patrick Hayden and Andreas Winter, *Counterexamples to the maximal p-norm multiplicity conjecture for all $p > 1$*, Comm. Math. Phys. **284** (2008), no. 1, 263–280, DOI 10.1007/s00220-008-0624-0. MR2443305 ↑232, 233

[HW17] Han Huang and Feng Wei, *Upper bound for the Dvoretzky dimension in Milman-Schechtman theorem*, Geometric aspects of functional analysis, Lecture Notes in Math., vol. 2169, Springer, Cham, 2017, 181-186. 208

[Ide13] Martin Idel, *On the stucture of positive maps*, Master's thesis, Technische Universität München, 2013. 64

[Ide16] Martin Idel and Michael M. Wolf, *Sinkhorn normal form for unitary matrices*, Linear Algebra Appl. **471** (2015), 76–84, DOI 10.1016/j.laa.2014.12.031. MR3314325 ↑64

[Ind00] Piotr Indyk, *Dimensionality reduction techniques for proximity problems*, Proceedings of the Eleventh Annual ACM-SIAM Symposium on Discrete Algorithms (San Francisco, CA, 2000), ACM, New York, 2000, pp. 371–378. MR1754879 ↑207

[Ind07] Piotr Indyk, *Uncertainty principles, extractors, and explicit embeddings of l_2 into l_1*, STOC'07—Proceedings of the 39th Annual ACM Symposium on Theory of Computing, ACM, New York, 2007, pp. 615–620, DOI 10.1145/1250790.1250881. MR2402488 ↑207

[IS10] Piotr Indyk and Stanisław Szarek, *Almost-Euclidean subspaces of ℓ_1^N via tensor products: a simple approach to randomness reduction*, Approximation, randomization, and combinatorial optimization, Lecture Notes in Comput. Sci., vol. 6302, Springer, Berlin, 2010, pp. 632–641, DOI 10.1007/978-3-642-15369-3_47. MR2755868 ↑207, 210

[Jam72] A. Jamiołkowski, *Linear transformations which preserve trace and positive semidefiniteness of operators*, Rep. Mathematical Phys. **3** (1972), no. 4, 275–278. MR0342537 ↑64

[Jan97] Svante Janson, *Gaussian Hilbert spaces*, Cambridge Tracts in Mathematics, vol. 129, Cambridge University Press, Cambridge, 1997. MR1474726 ↑146

[Jen13] Justin Jenkinson, *Convex geometric connections to information theory*, Ph.D. thesis, Case Western Reserve University, 2013, `http://rave.ohiolink.edu/etdc/view?acc_num=case1365179413`. 109, 260

[JHH+15] P. Joshi, K. Horodecki, M. Horodecki, P. Horodecki, R. Horodecki, Ben Li, S. J. Szarek, and T. Szarek, *Bound on Bell inequalities by fraction of determinism and reverse triangle inequality*, Phys. Rev. A (3) **92** (2015), no. 3, 032329, 11, DOI 10.1103/PhysRevA.92.032329. MR3432598 ↑297

[JHK+08] Eylee Jung, Mi-Ra Hwang, Hungsoo Kim, Min-Soo Kim, DaeKil Park, Jin-Woo Son, and Sayatnova Tamaryan, *Reduced state uniquely defines the groverian measure of the original pure state*, Phys. Rev. A **77** (2008), 062317. 233, 234

[JL84] William B. Johnson and Joram Lindenstrauss, *Extensions of Lipschitz mappings into a Hilbert space.*, Contemp. Math. **26** (1984), 189–206 (English). 210

[JLN14] Maria Anastasia Jivulescu, Nicolae Lupa, and Ion Nechita, *On the reduction criterion for random quantum states*, J. Math. Phys. **55** (2014), no. 11, 112203, 27, DOI 10.1063/1.4901548. MR3390496 ↑273

[JLN15] Maria Anastasia Jivulescu, Nicolae Lupa, and Ion Nechita, *Thresholds for reduction-related entanglement criteria in quantum information theory*, Quantum Inf. Comput. **15** (2015), no. 13-14, 1165–1184. MR3443352 ↑273

[JM78] Naresh C. Jain and Michael B. Marcus, *Continuity of sub-Gaussian processes*, Probability on Banach spaces, Adv. Probab. Related Topics, vol. 4, Dekker, New York, 1978, pp. 81–196. MR515431 ↑179

[JNP+11] M. Junge, M. Navascues, C. Palazuelos, D. Perez-Garcia, V. B. Scholz, and R. F. Werner, *Connes embedding problem and Tsirelson's problem*, J. Math. Phys. **52** (2011), no. 1, 012102, 12, DOI 10.1063/1.3514538. MR2790067 ↑296

[Joh48] Fritz John, *Extremum problems with inequalities as subsidiary conditions*, Studies and Essays Presented to R. Courant on his 60th Birthday, January 8, 1948, Interscience Publishers, Inc., New York, N. Y., 1948, pp. 187–204. MR0030135 ↑104

[JP11] M. Junge and C. Palazuelos, *Large violation of Bell inequalities with low entanglement*, Comm. Math. Phys. **306** (2011), no. 3, 695–746, DOI 10.1007/s00220-011-1296-8. MR2825506 ↑297

[JPPG+10] M. Junge, C. Palazuelos, D. Pérez-García, I. Villanueva, and M. M. Wolf, *Operator space theory: a natural framework for Bell inequalities*, Phys. Rev. Lett. **104** (2010), no. 17, 170405, 4, DOI 10.1103/PhysRevLett.104.170405. MR2653547 ↑294

[JS] Justin Jenkinson and Stanisław Szarek, *Optimal constants in concentration inequalities on the sphere*, in preparation. 109, 144

[JS91] William B. Johnson and Gideon Schechtman, *Remarks on Talagrand's deviation inequality for Rademacher functions*, Functional analysis (Austin, TX, 1987/1989), Lecture Notes in Math., vol. 1470, Springer, Berlin, 1991, pp. 72–77, DOI 10.1007/BFb0090214. MR1126739 ↑146

[Kad65] Richard V. Kadison, *Transformations of states in operator theory and dynamics*, Topology **3** (1965), no. suppl. 2, 177–198, DOI 10.1016/0040-9383(65)90075-3. MR0169073 ↑63

[Kah85] Jean-Pierre Kahane, *Some random series of functions*, 2nd ed., Cambridge Studies in Advanced Mathematics, vol. 5, Cambridge University Press, Cambridge, 1985. MR833073 ↑147

[Kah86] Jean-Pierre Kahane, *Une inégalité du type de Slepian et Gordon sur les processus gaussiens* (French, with English summary), Israel J. Math. **55** (1986), no. 1, 109–110, DOI 10.1007/BF02772698. MR858463 ↑178

[Kar11] Zohar S. Karnin, *Deterministic construction of a high dimensional ℓ_p section in ℓ_1^n for any $p < 2$*, STOC'11—Proceedings of the 43rd ACM Symposium on Theory of Computing, ACM, New York, 2011, pp. 645–654, DOI 10.1145/1993636.1993722. MR2932015 ↑210

[Kaš77] B. S. Kašin, *The widths of certain finite-dimensional sets and classes of smooth functions* (Russian), Izv. Akad. Nauk SSSR Ser. Mat. **41** (1977), no. 2, 334–351, 478. MR0481792 ↑208, 209

[Kat75] G. O. H. Katona, *The Hamming-sphere has minimum boundary*, Studia Sci. Math. Hungar. **10** (1975), no. 1-2, 131–140. MR0453266 ↑146

[KCKL00] B. Kraus, J. I. Cirac, S. Karnas, and M. Lewenstein, *Separability in $2 \times N$ composite quantum systems*, Phys. Rev. A (3) **61** (2000), no. 6, 062302, 10, DOI 10.1103/PhysRevA.61.062302. MR1767463 ↑65

[Kec95] Alexander S. Kechris, *Classical descriptive set theory*, Graduate Texts in Mathematics, vol. 156, Springer-Verlag, New York, 1995. MR1321597 ↑104

[Kha67] C. G. Khatri, *On certain inequalities for normal distributions and their applications to simultaneous confidence bounds*, Ann. Math. Statist. **38** (1967), 1853–1867, DOI 10.1214/aoms/1177698618. MR0220392 ↑178

[Kin17] Sebastian Kinnewig, *Bell Inequalities and Grothendieck's Constant*, Bachelor Thesis, Leibniz Universität Hannover, Germany, June 2017. 295

[Kir76] A. A. Kirillov, *Elements of the theory of representations*, Springer-Verlag, Berlin-New York, 1976. Translated from the Russian by Edwin Hewitt; Grundlehren der Mathematischen Wissenschaften, Band 220. MR0412321 ↑294

[Kis87] Christer O. Kiselman, *Smoothness of vector sums of plane convex sets* (English, with Esperanto summary), Math. Scand. **60** (1987), no. 2, 239–252, DOI 10.7146/math.scand.a-12183. MR914337 ↑104

[KL78] G. A. Kabatjanskiĭ and V. I. Levenšteĭn, *Bounds for packings on the sphere and in space* (Russian), Problemy Peredachi Informatsii **14** (1978), no. 1, 3–25. MR0514023 ↑142

[KL09] Robert L. Kosut and Daniel A. Lidar, *Quantum error correction via convex optimization*, Quantum Inf. Process. **8** (2009), no. 5, 443–459, DOI 10.1007/s11128-009-0120-2. MR2540487 ↑29

[Kla06] B. Klartag, *On convex perturbations with a bounded isotropic constant*, Geom. Funct. Anal. **16** (2006), no. 6, 1274–1290, DOI 10.1007/s00039-006-0588-1. MR2276540 ↑105

[Kle32] O. Klein, *Zur Berechnung von Potentialkurven für zweiatomige Moleküle mit Hilfe von Spektraltermen*, Zeitschrift für Physik **76** (1932), no. 3-4, 226–235 (German). 29

[KM05] B. Klartag and V. D. Milman, *Geometry of log-concave functions and measures*, Geom. Dedicata **112** (2005), 169–182, DOI 10.1007/s10711-004-2462-3. MR2163897 ↑105

[KMP98] H. König, M. Meyer, and A. Pajor, *The isotropy constants of the Schatten classes are bounded*, Math. Ann. **312** (1998), no. 4, 773–783, DOI 10.1007/s002080050245. MR1660231 ↑105

[Kol05] Alexander Koldobsky, *Fourier analysis in convex geometry*, Mathematical Surveys and Monographs, vol. 116, American Mathematical Society, Providence, RI, 2005. MR2132704 ↑106

[Kom55] Yûsaku Komatu, *Elementary inequalities for Mills' ratio*, Rep. Statist. Appl. Res. Un. Jap. Sci. Engrs. **4** (1955), 69–70. MR0079844 ↑309

[KP88] G. A. Kabatyanskiĭ and V. I. Panchenko, *Packings and coverings of the Hamming space by unit balls* (Russian), Dokl. Akad. Nauk SSSR **303** (1988), no. 3, 550–552; English transl., Soviet Math. Dokl. **38** (1989), no. 3, 564–566. MR980783 ↑142

[Kra71] K. Kraus, *General state changes in quantum theory*, Ann. Physics **64** (1971), 311–335, DOI 10.1016/0003-4916(71)90108-4. MR0292434 ↑64

[Kra83] Karl Kraus, *States, effects, and operations*, Lecture Notes in Physics, vol. 190, Springer-Verlag, Berlin, 1983. Fundamental notions of quantum theory; Lecture notes edited by A. Böhm, J. D. Dollard and W. H. Wootters. MR725167 ↑64

[Kri79] J.-L. Krivine, *Constantes de Grothendieck et fonctions de type positif sur les sphères* (French), Adv. in Math. **31** (1979), no. 1, 16–30, DOI 10.1016/0001-8708(79)90017-3. MR521464 ↑295

[KS67] Simon Kochen and E. P. Specker, *The problem of hidden variables in quantum mechanics*, J. Math. Mech. **17** (1967), 59–87. MR0219280 ↑297

[KS03] Boris S. Kashin and Stanisław J. Szarek, *The Knaster problem and the geometry of high-dimensional cubes* (English, with English and French summaries), C. R. Math. Acad. Sci. Paris **336** (2003), no. 11, 931–936, DOI 10.1016/S1631-073X(03)00226-7. MR1994597 ↑208

[KT85] L. A. Khalfin and B. S. Tsirelson, *Quantum and quasiclassical analogs of Bell inequalities*, Symposium on the foundations of modern physics (Joensuu, 1985), World Sci. Publishing, Singapore, 1985, pp. 441–460. MR843870 ↑297

[KTJ09] Hermann König and Nicole Tomczak-Jaegermann, *Projecting l_∞ onto classical spaces*, Constr. Approx. **29** (2009), no. 2, 277–292, DOI 10.1007/s00365-008-9025-z. MR2481592 ↑210

[Kup92] Greg Kuperberg, *A low-technology estimate in convex geometry*, Int. Math. Res. Not. IMRN **9** (1992), 181–183, DOI 10.1155/S1073792892000205. MR1185832 ↑105

[Kup08] Greg Kuperberg, *From the Mahler conjecture to Gauss linking integrals*, Geom. Funct. Anal. **18** (2008), no. 3, 870–892, DOI 10.1007/s00039-008-0669-4. MR2438998 ↑105

[KV07] B. Klartag and R. Vershynin, *Small ball probability and Dvoretzky's theorem*, Israel J. Math. **157** (2007), 193–207, DOI 10.1007/s11856-006-0007-1. MR2342445 ↑209

[KVSW09] Dmitry S. Kaliuzhnyi-Verbovetskyi, Ilya M. Spitkovsky, and Hugo J. Woerdeman, *Matrices with normal defect one*, Oper. Matrices **3** (2009), no. 3, 401–438, DOI 10.7153/oam-03-24. MR2571564 ↑65

[Kwa76] S. Kwapień, *A theorem on the Rademacher series with vector valued coefficients*, Probability in Banach spaces (Proc. First Internat. Conf., Oberwolfach, 1975), Springer, Berlin, 1976, pp. 157–158. Lecture Notes in Math., Vol. 526. MR0451333 ↑147

[Kwa94] Stanisław Kwapień, *A remark on the median and the expectation of convex functions of Gaussian vectors*, Probability in Banach spaces, 9 (Sandjberg, 1993), Progr. Probab., vol. 35, Birkhäuser Boston, Boston, MA, 1994, pp. 271–272. MR1308523 ↑144

[Lan16] Cécilia Lancien, *k-extendibility of high-dimensional bipartite quantum states*, Random Matrices Theory Appl. **5** (2016), no. 3, 1650011, 58, DOI 10.1142/S2010326316500118. MR3533708 ↑260, 273

[Las08] Marek Lassak, *Banach-Mazur distance of central sections of a centrally symmetric convex body*, Beiträge Algebra Geom. **49** (2008), no. 1, 243–246. MR2410579 ↑339

[Lat96] Rafał Latała, *A note on the Ehrhard inequality*, Studia Math. **118** (1996), no. 2, 169–174. MR1389763 ↑144

[Lat97] Rafał Latała, *Estimation of moments of sums of independent real random variables*, Ann. Probab. **25** (1997), no. 3, 1502–1513, DOI 10.1214/aop/1024404522. MR1457628 ↑146

[Lat02] R. Latała, *On some inequalities for Gaussian measures*, Proceedings of the International Congress of Mathematicians, Vol. II (Beijing, 2002), Higher Ed. Press, Beijing, 2002, pp. 813–822. MR1957087 ↑144

[Lat06] Rafał Latała, *Estimates of moments and tails of Gaussian chaoses*, Ann. Probab. **34** (2006), no. 6, 2315–2331, DOI 10.1214/009117906000000421. MR2294983 ↑146

[Lea91] Imre Leader, *Discrete isoperimetric inequalities*, Probabilistic combinatorics and its applications (San Francisco, CA, 1991), Proc. Sympos. Appl. Math., vol. 44, Amer. Math. Soc., Providence, RI, 1991, pp. 57–80, DOI 10.1090/psapm/044/1141923. MR1141923 ↑146

[Led96] Michel Ledoux, *Isoperimetry and Gaussian analysis*, Lectures on probability theory and statistics (Saint-Flour, 1994), Lecture Notes in Math., vol. 1648, Springer, Berlin, 1996, pp. 165–294, DOI 10.1007/BFb0095676. MR1600888 ↑144

[Led01] Michel Ledoux, *The concentration of measure phenomenon*, Mathematical Surveys and Monographs, vol. 89, American Mathematical Society, Providence, RI, 2001. MR1849347 ↑117, 119, 127, 143, 144, 145

[Led03] Michel Ledoux, *A remark on hypercontractivity and tail inequalities for the largest eigenvalues of random matrices*, Séminaire de Probabilités XXXVII, Lecture Notes in Math., vol. 1832, Springer, Berlin, 2003, pp. 360–369, DOI 10.1007/978-3-540-40004-2_14. MR2053053 ↑179

[Led97] Michel Ledoux, *On Talagrand's deviation inequalities for product measures*, ESAIM Probab. Statist. **1** (1995/97), 63–87, DOI 10.1051/ps:1997103. MR1399224 ↑146

[Lei72] L. Leindler, *On a certain converse of Hölder's inequality. II*, Acta Sci. Math. (Szeged) **33** (1972), no. 3-4, 217–223. MR2199372 ↑105

[Lév22] Paul Lévy, *Leçons d'analyse fonctionnelle*, Gauthier–Villars, Paris, 1922. 144

[Lév51] Paul Lévy, *Problèmes concrets d'analyse fonctionnelle. Avec un complément sur les fonctionnelles analytiques par F. Pellegrino* (French), Gauthier-Villars, Paris, 1951. 2d ed. MR0041346 ↑144

[Li] Ben Li, *in preparation*, Ph.D. thesis, Case Western Reserve University. 295

[LJL15] Gao Li, Marius Junge, and Nicholas LaRacuente, *Capacity Bounds via Operator Space Methods*, arXiv:1509.07294 (2015). 232

[LKCH00] M. Lewenstein, B. Kraus, J. I. Cirac, and P. Horodecki, *Optimization of entanglement witnesses*, Phys. Rev. A **62** (2000), 052310. 64

[LLR83] M. R. Leadbetter, Georg Lindgren, and Holger Rootzén, *Extremes and related properties of random sequences and processes*, Springer Series in Statistics, Springer-Verlag, New York-Berlin, 1983. MR691492 ↑178

[LM17] Rafał Latała and Dariusz Matlak, *Royen's proof of the Gaussian correlation inequality*, Geometric aspects of functional analysis, Israel Seminar (GAFA) 2014-2016, 265-275, lecture Notes in Math. 2169, Springer, 2017. 179

[LMO06] Jon Magne Leinaas, Jan Myrheim, and Eirik Ovrum, *Geometrical aspects of entanglement*, Phys. Rev. A (3) **74** (2006), no. 1, 012313, 13, DOI 10.1103/PhysRevA.74.012313. MR2255390 ↑65

[LN16] Kasper Green Larsen and Jelani Nelson, *The Johnson-Lindenstrauss lemma is optimal for linear dimensionality reduction*, Proceedings of the 43rd International Colloquium on Automata, Languages and Programming (ICALP 2016), 2016. 210

[LO94] Rafał Latała and Krzysztof Oleszkiewicz, *On the best constant in the Khinchin-Kahane inequality*, Studia Math. **109** (1994), no. 1, 101–104. MR1267715 ↑147

[LO99] Rafał Latała and Krzysztof Oleszkiewicz, *Gaussian measures of dilatations of convex symmetric sets*, Ann. Probab. **27** (1999), no. 4, 1922–1938, DOI 10.1214/aop/1022677554. MR1742894 ↑207

[LP68] J. Lindenstrauss and A. Pełczyński, *Absolutely summing operators in L_p-spaces and their applications*, Studia Math. **29** (1968), 275–326. MR0231188 ↑295

[LP99] N. Linden and S. Popescu, *Bound entanglement and teleportation*, Phys. Rev. A (3) **59** (1999), no. 1, 137–140, DOI 10.1103/PhysRevA.59.137. MR1670566 ↑306

[LQ04] Daniel Li and Hervé Queffélec, *Introduction à l'étude des espaces de Banach* (French), Cours Spécialisés [Specialized Courses], vol. 12, Société Mathématique de France, Paris, 2004. Analyse et probabilités. [Analysis and probability theory]. MR2124356 ↑207

[LR10] Michel Ledoux and Brian Rider, *Small deviations for beta ensembles*, Electron. J. Probab. **15** (2010), no. 41, 1319–1343, DOI 10.1214/EJP.v15-798. MR2678393 ↑179

[LS75] Raphael Loewy and Hans Schneider, *Positive operators on the n-dimensional ice cream cone*, J. Math. Anal. Appl. **49** (1975), 375–392, DOI 10.1016/0022-247X(75)90186-9. MR0407654 ↑324

[LS93] L. J. Landau and R. F. Streater, *On Birkhoff's theorem for doubly stochastic completely positive maps of matrix algebras*, Linear Algebra Appl. **193** (1993), 107–127, DOI 10.1016/0024-3795(93)90274-R. MR1240275 ↑64

[LS08] Shachar Lovett and Sasha Sodin, *Almost Euclidean sections of the N-dimensional cross-polytope using $O(N)$ random bits*, Commun. Contemp. Math. **10** (2008), no. 4, 477–489, DOI 10.1142/S0219199708002879. MR2444845 ↑207

[LS13] M. S. Leifer and Robert W. Spekkens, *Towards a formulation of quantum theory as a causally neutral theory of Bayesian inference*, Phys. Rev. A **88** (2013), 052130. 64

[LT91] Michel Ledoux and Michel Talagrand, *Probability in Banach spaces*, Ergebnisse der Mathematik und ihrer Grenzgebiete (3) [Results in Mathematics and Related Areas (3)], vol. 23, Springer-Verlag, Berlin, 1991. Isoperimetry and processes. MR1102015 ↑179

[LT92] Chi-Kwong Li and Nam-Kiu Tsing, *Linear preserver problems: a brief introduction and some special techniques*, Linear Algebra Appl. **162/164** (1992), 217–235, DOI 10.1016/0024-3795(92)90377-M. Directions in matrix theory (Auburn, AL, 1990). MR1148401 ↑64

[Mat02] Jiří Matoušek, *Lectures on discrete geometry*, Graduate Texts in Mathematics, vol. 212, Springer-Verlag, New York, 2002. MR1899299 ↑146

[Mau79] Bernard Maurey, *Construction de suites symétriques* (French, with English summary), C. R. Acad. Sci. Paris Sér. A-B **288** (1979), no. 14, A679–A681. MR533901 ↑147

[Mau91] B. Maurey, *Some deviation inequalities*, Geom. Funct. Anal. **1** (1991), no. 2, 188–197, DOI 10.1007/BF01896377. MR1097258 ↑146

[Mau03] Bernard Maurey, *Type, cotype and K-convexity*, Handbook of the geometry of Banach spaces, Vol. 2, North-Holland, Amsterdam, 2003, pp. 1299–1332, DOI 10.1016/S1874-5849(03)80037-2. MR1999197 ↑207, 209

[McC06] Robert J. McCann, *Stable rotating binary stars and fluid in a tube*, Houston J. Math. **32** (2006), no. 2, 603–631. MR2219334 ↑179

[McD89] Colin McDiarmid, *On the method of bounded differences*, Surveys in combinatorics, 1989 (Norwich, 1989), London Math. Soc. Lecture Note Ser., vol. 141, Cambridge Univ. Press, Cambridge, 1989, pp. 148–188. MR1036755 ↑144

[McD98] Colin McDiarmid, *Concentration*, Probabilistic methods for algorithmic discrete mathematics, Algorithms Combin., vol. 16, Springer, Berlin, 1998, pp. 195–248, DOI 10.1007/978-3-662-12788-9_6. MR1678578 ↑146

[Mec] Elizabeth Meckes, *The random matrix theory of the classical compact groups*, Cambridge University Press, in preparation. 179

[Mec03] Mark William Meckes, *Random phenomena in finite-dimensional normed spaces*, ProQuest LLC, Ann Arbor, MI, 2003. Thesis (Ph.D.)–Case Western Reserve University. MR2704690 ↑146

[Mec04] Mark W. Meckes, *Concentration of norms and eigenvalues of random matrices*, J. Funct. Anal. **211** (2004), no. 2, 508–524, DOI 10.1016/S0022-1236(03)00198-8. MR2057479 ↑146

[Mer07] N. David Mermin, *Quantum computer science*, Cambridge University Press, Cambridge, 2007. MR2341010 ↑75

[Mil71] V. D. Milman, *A new proof of A. Dvoretzky's theorem on cross-sections of convex bodies* (Russian), Funkcional. Anal. i Priložen. **5** (1971), no. 4, 28–37. MR0293374 ↑208

[Mil85] V. D. Milman, *Random subspaces of proportional dimension of finite-dimensional normed spaces: approach through the isoperimetric inequality*, Banach spaces (Columbia, Mo., 1984), Lecture Notes in Math., vol. 1166, Springer, Berlin, 1985, pp. 106–115, DOI 10.1007/BFb0074700. MR827766 ↑209

[Mil86] Vitali D. Milman, *Inégalité de Brunn-Minkowski inverse et applications à la théorie locale des espaces normés* (English, with French summary), C. R. Acad. Sci. Paris Sér. I Math. **302** (1986), no. 1, 25–28. MR827101 ↑143, 209

[Mil87] V. D. Milman, *Some remarks on Urysohn's inequality and volume ratio of cotype 2-spaces*, Geometrical aspects of functional analysis (1985/86), Lecture Notes in Math., vol. 1267, Springer, Berlin, 1987, pp. 75–81, DOI 10.1007/BFb0078137. MR907686 ↑209

[Mil88] V. D. Milman, *A few observations on the connections between local theory and some other fields*, Geometric aspects of functional analysis (1986/87), Lecture Notes in Math., vol. 1317, Springer, Berlin, 1988, pp. 283–289, DOI 10.1007/BFb0081748. MR950988 ↑208

[Mil15] Emanuel Milman, *Sharp isoperimetric inequalities and model spaces for the curvature-dimension-diameter condition*, J. Eur. Math. Soc. (JEMS) **17** (2015), no. 5, 1041–1078, DOI 10.4171/JEMS/526. MR3346688 ↑144

[Min11] Hermann Minkowski, *Gesammelte Abhandlungen von Hermann Minkowski. Unter Mitwirkung von Andreas Speiser und Hermann Weyl, herausgegeben von David Hilbert. Band I, II.*, Leipzig u. Berlin: B. G. Teubner. Erster Band. Mit einem Bildnis Hermann Minkowskis und 6 Figuren im Text. xxxvi, 371 S.; Zweiter Band. Mit einem Bildnis Hermann Minkowskis, 34 Figuren in Text und einer Doppeltafel. iv, 466 S. gr. 8° (1911)., 1911. 29

[MM13] Elizabeth S. Meckes and Mark W. Meckes, *Spectral measures of powers of random matrices*, Electron. Commun. Probab. **18** (2013), no. 78, 13, DOI 10.1214/ECP.v18-2551. MR3109633 ↑134

[MO15] Marek Miller and Robert Olkiewicz, *Topology of the cone of positive maps on qubit systems*, J. Phys. A **48** (2015), no. 25, 255203, 9, DOI 10.1088/1751-8113/48/25/255203. MR3355261 ↑65, 324

[MO16] Marek Miller and Robert Olkiewicz, *Extremal positive maps on $M_3(\mathbb{C})$ and idempotent matrices*, Open Syst. Inf. Dyn. **23** (2016), no. 1, 1650001, 13, DOI 10.1142/S1230161216500013. MR3478740 ↑65

[Mon12] Ashley Montanaro, *Some applications of hypercontractive inequalities in quantum information theory*, J. Math. Phys. **53** (2012), no. 12, 122206, 15, DOI 10.1063/1.4769269. MR3058179 ↑136, 146

[Mon13] Ashley Montanaro, *Weak multiplicativity for random quantum channels*, Comm. Math. Phys. **319** (2013), no. 2, 535–555, DOI 10.1007/s00220-013-1680-7. MR3037588 ↑233

[MP67] V. A. Marčenko and L. A. Pastur, *Distribution of eigenvalues in certain sets of random matrices* (Russian), Mat. Sb. (N.S.) **72 (114)** (1967), 507–536. MR0208649 ↑179

[MP86] Vitali D. Milman and Gilles Pisier, *Banach spaces with a weak cotype 2 property*, Israel J. Math. **54** (1986), no. 2, 139–158, DOI 10.1007/BF02764939. MR852475 ↑209

[MP00] V. D. Milman and A. Pajor, *Entropy and asymptotic geometry of non-symmetric convex bodies*, Adv. Math. **152** (2000), no. 2, 314–335, DOI 10.1006/aima.1999.1903. MR1764107 ↑105

[MS86] Vitali D. Milman and Gideon Schechtman, *Asymptotic theory of finite-dimensional normed spaces*, Lecture Notes in Math., vol. 1200, Springer-Verlag, Berlin, 1986. With an appendix by M. Gromov. MR856576 ↑144, 147, 207

[MS97] V. D. Milman and G. Schechtman, *Global versus local asymptotic theories of finite-dimensional normed spaces*, Duke Math. J. **90** (1997), no. 1, 73–93, DOI 10.1215/S0012-7094-97-09003-7. MR1478544 ↑208

[MS12] Mark W. Meckes and Stanisław J. Szarek, *Concentration for noncommutative polynomials in random matrices*, Proc. Amer. Math. Soc. **140** (2012), no. 5, 1803–1813, DOI 10.1090/S0002-9939-2011-11262-0. MR2869165 ↑180

[MTJ87] V. D. Milman and N. Tomczak-Jaegermann, *Sudakov type inequalities for convex bodies in* $\mathbf{R}^n$, Geometrical aspects of functional analysis (1985/86), Lecture Notes in Math., vol. 1267, Springer, Berlin, 1987, pp. 113–121, DOI 10.1007/BFb0078140. MR907689 ↑179

[MWW09] William Matthews, Stephanie Wehner, and Andreas Winter, *Distinguishability of quantum states under restricted families of measurements with an application to quantum data hiding*, Comm. Math. Phys. **291** (2009), no. 3, 813–843, DOI 10.1007/s00220-009-0890-5. MR2534793 ↑305

[Naz12] Fedor Nazarov, *The Hörmander proof of the Bourgain-Milman theorem*, Geometric aspects of functional analysis, Lecture Notes in Math., vol. 2050, Springer, Heidelberg, 2012, pp. 335–343, DOI 10.1007/978-3-642-29849-3_20. MR2985302 ↑105

[NC00] Michael A. Nielsen and Isaac L. Chuang, *Quantum computation and quantum information*, Cambridge University Press, Cambridge, 2000. MR1796805 ↑63, 232

[Nel73] Edward Nelson, *The free Markoff field*, J. Funct. Anal. **12** (1973), 211–227. MR0343816 ↑145

[Nem07] Arkadi Nemirovski, *Advances in convex optimization: conic programming*, International Congress of Mathematicians. Vol. I, Eur. Math. Soc., Zürich, 2007, pp. 413–444, DOI 10.4171/022-1/17. MR2334199 ↑29

[NS06] Alexandru Nica and Roland Speicher, *Lectures on the combinatorics of free probability*, London Mathematical Society Lecture Note Series, vol. 335, Cambridge University Press, Cambridge, 2006. MR2266879 ↑177, 180

[O'D14] Ryan O'Donnell, *Analysis of Boolean functions*, Cambridge University Press, New York, 2014. MR3443800 ↑136, 146

[Oza13] Narutaka Ozawa, *About the Connes embedding conjecture: algebraic approaches*, Jpn. J. Math. **8** (2013), no. 1, 147–183, DOI 10.1007/s11537-013-1280-5. MR3067294 ↑296

[Pag93] Don N. Page, *Average entropy of a subsystem*, Phys. Rev. Lett. **71** (1993), no. 9, 1291–1294, DOI 10.1103/PhysRevLett.71.1291. MR1232812 ↑233

[Paj99] Alain Pajor, *Metric entropy of the Grassmann manifold*, Convex geometric analysis (Berkeley, CA, 1996), Math. Sci. Res. Inst. Publ., vol. 34, Cambridge Univ. Press, Cambridge, 1999, pp. 181–188. MR1665590 ↑143

[Par04] K. R. Parthasarathy, *On the maximal dimension of a completely entangled subspace for finite level quantum systems*, Proc. Indian Acad. Sci. Math. Sci. **114** (2004), no. 4, 365–374, DOI 10.1007/BF02829441. MR2067699 ↑232

[Peł80] Aleksander Pełczyński, *Geometry of finite-dimensional Banach spaces and operator ideals*, Notes in Banach spaces, Univ. Texas Press, Austin, Tex., 1980, pp. 81–181. MR606222 ↑208

[Per96] Asher Peres, *Separability criterion for density matrices*, Phys. Rev. Lett. **77** (1996), no. 8, 1413–1415, DOI 10.1103/PhysRevLett.77.1413. MR1401726 ↑63

[Per99] Asher Peres, *All the Bell inequalities*, Found. Phys. **29** (1999), no. 4, 589–614, DOI 10.1023/A:1018816310000. Invited papers dedicated to Daniel Greenberger, Part II. MR1714162 ↑297

[Pet01] Dénes Petz, *Entropy, von Neumann and the von Neumann entropy*, John von Neumann and the foundations of quantum physics (Budapest, 1999), Vienna Circ. Inst. Yearb., vol. 8, Kluwer Acad. Publ., Dordrecht, 2001, pp. 83–96. MR2042743 ↑30

[Pet06] Peter Petersen, *Riemannian geometry*, 2nd ed., Graduate Texts in Mathematics, vol. 171, Springer, New York, 2006. MR2243772 ↑131, 132

[PGWP+08] D. Pérez-García, M. M. Wolf, C. Palazuelos, I. Villanueva, and M. Junge, *Unbounded violation of tripartite Bell inequalities*, Comm. Math. Phys. **279** (2008), no. 2, 455–486, DOI 10.1007/s00220-008-0418-4. MR2383595 ↑297

[Pic68] James Pickands III, *Moment convergence of sample extremes*, Ann. Math. Statist. **39** (1968), 881–889, DOI 10.1214/aoms/1177698320. MR0224231 ↑178

[Pis80] G. Pisier, *Un théorème sur les opérateurs linéaires entre espaces de Banach qui se factorisent par un espace de Hilbert* (French), Ann. Sci. École Norm. Sup. (4) **13** (1980), no. 1, 23–43. MR584081 ↑207

[Pis81] G. Pisier, *Remarques sur un résultat non publié de B. Maurey* (French), Seminar on Functional Analysis, 1980–1981, École Polytech., Palaiseau, 1981, pp. Exp. No. V, 13. MR659306 ↑207

[Pis86] Gilles Pisier, *Probabilistic methods in the geometry of Banach spaces*, Probability and analysis (Varenna, 1985), Lecture Notes in Math., vol. 1206, Springer, Berlin, 1986, pp. 167–241, DOI 10.1007/BFb0076302. MR864714 ↑144

[Pis89a] Gilles Pisier, *A new approach to several results of V. Milman*, J. Reine Angew. Math. **393** (1989), 115–131, DOI 10.1515/crll.1989.393.115. MR972362 ↑209

[Pis89b] Gilles Pisier, *The volume of convex bodies and Banach space geometry*, Cambridge Tracts in Mathematics, vol. 94, Cambridge University Press, Cambridge, 1989. MR1036275 ↑143, 207, 209

[Pis12a] Gilles Pisier, *Grothendieck's theorem, past and present*, Bull. Amer. Math. Soc. (N.S.) **49** (2012), no. 2, 237–323, DOI 10.1090/S0273-0979-2011-01348-9. MR2888168 ↑295

[Pis12b] Gilles Pisier, *Tripartite Bell inequality, random matrices and trilinear forms*, arXiv:1203.2509 (2012). 297

[Pit89] Itamar Pitowsky, *Quantum probability—quantum logic*, Lecture Notes in Physics, vol. 321, Springer-Verlag, Berlin, 1989. MR984603 ↑296

[Por81] Ian R. Porteous, *Topological geometry*, 2nd ed., Cambridge University Press, Cambridge-New York, 1981. MR606198 ↑294

[PR94] Sandu Popescu and Daniel Rohrlich, *Quantum nonlocality as an axiom*, Found. Phys. **24** (1994), no. 3, 379–385, DOI 10.1007/BF02058098. MR1265577 ↑297

[Pré71] András Prékopa, *Logarithmic concave measures with application to stochastic programming*, Acta Sci. Math. (Szeged) **32** (1971), 301–316. MR0315079 ↑105

[Pré73] András Prékopa, *On logarithmic concave measures and functions*, Acta Sci. Math. (Szeged) **34** (1973), 335–343. MR0404557 ↑105

[PT86] Alain Pajor and Nicole Tomczak-Jaegermann, *Subspaces of small codimension of finite-dimensional Banach spaces*, Proc. Amer. Math. Soc. **97** (1986), no. 4, 637–642, DOI 10.2307/2045920. MR845980 ↑209

[PTJ85] Alain Pajor and Nicole Tomczak-Jaegermann, *Remarques sur les nombres d'entropie d'un opérateur et de son transposé* (French, with English summary), C. R. Acad. Sci. Paris Sér. I Math. **301** (1985), no. 15, 743–746. MR817602 ↑179

[PTJ90] Alain Pajor and Nicole Tomczak-Jaegermann, *Gel′fand numbers and Euclidean sections of large dimensions*, Probability in Banach spaces 6 (Sandbjerg, 1986), Progr. Probab., vol. 20, Birkhäuser Boston, Boston, MA, 1990, pp. 252–264. MR1056714 208

[PV07] Martin B. Plenio and Shashank Virmani, *An introduction to entanglement measures*, Quantum Inf. Comput. **7** (2007), no. 1-2, 1–51. MR2302673 ↑232, 272

[PV16] Carlos Palazuelos and Thomas Vidick, *Survey on nonlocal games and operator space theory*, J. Math. Phys. **57** (2016), no. 1, 015220, 41, DOI 10.1063/1.4938052. MR3446943 ↑275, 281, 296, 297

[PY15] C. Palazuelos and Z. Yin, *Large bipartite Bell violations with dichotomic measurements*, Phys. Rev. A **92** (2015), 052313. 297

[Ran55] R. A. Rankin, *The closest packing of spherical caps in n dimensions*, Proc. Glasgow Math. Assoc. **2** (1955), 139–144. MR0074013 ↑142

[Rei08] Michael Reimpell, *Quantum information and convex optimization*, Ph.D. thesis, Technische Universität Braunschweig, 2008. 29

[Roc70] R. Tyrrell Rockafellar, *Convex analysis*, Princeton Mathematical Series, No. 28, Princeton University Press, Princeton, N.J., 1970. MR0274683 ↑14, 29

[Rog47] C. A. Rogers, *Existence theorems in the geometry of numbers*, Ann. of Math. (2) **48** (1947), 994–1002, DOI 10.2307/1969390. MR0022863 ↑142

[Rog57] C. A. Rogers, *A note on coverings*, Mathematika **4** (1957), 1–6, DOI 10.1112/S0025579300001030. MR0090824 ↑142

[Rog63] C. A. Rogers, *Covering a sphere with spheres*, Mathematika **10** (1963), 157–164, DOI 10.1112/S0025579300004083. MR0166687 ↑142

[Rog64] C. A. Rogers, *Packing and covering*, Cambridge Tracts in Mathematics and Mathematical Physics, No. 54, Cambridge University Press, New York, 1964. MR0172183 ↑142

[Rot86] O. S. Rothaus, *Hypercontractivity and the Bakry-Emery criterion for compact Lie groups*, J. Funct. Anal. **65** (1986), no. 3, 358–367, DOI 10.1016/0022-1236(86)90025-X. MR826433 ↑145

[Rot06] Ron Roth, *Introduction to coding theory*, Cambridge University Press, 2006. 142

[Roy14] Thomas Royen, *A simple proof of the Gaussian correlation conjecture extended to some multivariate gamma distributions*, Far East J. Theor. Stat. **48** (2014), no. 2, 139–145. MR3289621 ↑179

[RP11] Eleanor Rieffel and Wolfgang Polak, *Quantum computing*, Scientific and Engineering Computation, MIT Press, Cambridge, MA, 2011. A gentle introduction. MR2791092 ↑75

[RS58] C. A. Rogers and G. C. Shephard, *Convex bodies associated with a given convex body*, J. London Math. Soc. **33** (1958), 270–281, DOI 10.1112/jlms/s1-33.3.270. MR0101508 ↑105

[RSW02] Mary Beth Ruskai, Stanisław Szarek, and Elisabeth Werner, *An analysis of completely positive trace-preserving maps on $\mathcal{M}_2$*, Linear Algebra Appl. **347** (2002), 159–187, DOI 10.1016/S0024-3795(01)00547-X. MR1899888 ↑64

[Rud97] M. Rudelson, *Contact points of convex bodies*, Israel J. Math. **101** (1997), 93–124, DOI 10.1007/BF02760924. MR1484871 ↑209

[Rud00] M. Rudelson, *Distances between non-symmetric convex bodies and the MM^*-estimate*, Positivity **4** (2000), no. 2, 161–178, DOI 10.1023/A:1009842406728. MR1755679 ↑103, 208

[Rud05] Oliver Rudolph, *Further results on the cross norm criterion for separability*, Quantum Inf. Process. **4** (2005), no. 3, 219–239, DOI 10.1007/s11128-005-5664-1. MR2187343 ↑63

[RW00] Mary Beth Ruskai and Elisabeth Werner, *Study of a class of regularizations of $1/|X|$ using Gaussian integrals*, SIAM J. Math. Anal. **32** (2000), no. 2, 435–463, DOI 10.1137/S0036141099353758. MR1781466 ↑309

[RW09] Mary Beth Ruskai and Elisabeth M. Werner, *Bipartite states of low rank are almost surely entangled*, J. Phys. A **42** (2009), no. 9, 095303, 15, DOI 10.1088/1751-8113/42/9/095303. MR2525543 ↑272

[RZ14] Dmitry Ryabogin and Artem Zvavitch, *Analytic methods in convex geometry*, Analytical and probabilistic methods in the geometry of convex bodies, IMPAN Lect. Notes, vol. 2, Polish Acad. Sci. Inst. Math., Warsaw, 2014, pp. 87–183. MR3329057 ↑105

[Sam53] M. R. Sampford, *Some inequalities on Mill's ratio and related functions*, Ann. Math. Statistics **24** (1953), 130–132. MR0054890 ↑309

[SBŻ06] Stanisław J. Szarek, Ingemar Bengtsson, and Karol Życzkowski, *On the structure of the body of states with positive partial transpose*, J. Phys. A **39** (2006), no. 5, L119–L126, DOI 10.1088/0305-4470/39/5/L02. MR2200422 ↑272

[SC74] V. N. Sudakov and B. S. Cirel′son, *Extremal properties of half-spaces for spherically invariant measures* (Russian), Zap. Naučn. Sem. Leningrad. Otdel. Mat. Inst. Steklov. (LOMI) **41** (1974), 14–24, 165. MR0365680 ↑144

[SC94] L. Saloff-Coste, *Precise estimates on the rate at which certain diffusions tend to equilibrium*, Math. Z. **217** (1994), no. 4, 641–677, DOI 10.1007/BF02571965. MR1306030 ↑145

[Sch48] Erhard Schmidt, *Die Brunn-Minkowskische Ungleichung und ihr Spiegelbild sowie die isoperimetrische Eigenschaft der Kugel in der euklidischen und nichteuklidischen Geometrie. I* (German), Math. Nachr. **1** (1948), 81–157, DOI 10.1002/mana.19480010202. MR0028600 ↑144

[Sch50] Robert Schatten, *A Theory of Cross-Spaces*, Annals of Mathematics Studies, no. 26, Princeton University Press, Princeton, N. J., 1950. MR0036935 ↑29

[Sch65] Hans Schneider, *Positive operators and an inertia theorem*, Numer. Math. **7** (1965), 11–17, DOI 10.1007/BF01397969. MR0173678 ↑64

[Sch70] Robert Schatten, *Norm ideals of completely continuous operators*, Second printing. Ergebnisse der Mathematik und ihrer Grenzgebiete, Band 27, Springer-Verlag, Berlin-New York, 1970. MR0257800 ↑29

[Sch82] Gideon Schechtman, *Lévy type inequality for a class of finite metric spaces*, Martingale theory in harmonic analysis and Banach spaces (Cleveland, Ohio, 1981), Lecture Notes in Math., vol. 939, Springer, Berlin-New York, 1982, pp. 211–215. MR668548 ↑147

[Sch84] Carsten Schütt, *Entropy numbers of diagonal operators between symmetric Banach spaces*, J. Approx. Theory **40** (1984), no. 2, 121–128, DOI 10.1016/0021-9045(84)90021-2. MR732693 ↑156, 157

[Sch87] Gideon Schechtman, *More on embedding subspaces of L_p in l_r^n*, Compositio Math. **61** (1987), no. 2, 159–169. MR882972 ↑210

[Sch89] Gideon Schechtman, *A remark concerning the dependence on ϵ in Dvoretzky's theorem*, Geometric aspects of functional analysis (1987–88), Lecture Notes in Math., vol. 1376, Springer, Berlin, 1989, pp. 274–277, DOI 10.1007/BFb0090061. MR1008729 ↑208

[Sch99] Michael Schmuckenschläger, *An extremal property of the regular simplex*, Convex geometric analysis (Berkeley, CA, 1996), Math. Sci. Res. Inst. Publ., vol. 34, Cambridge Univ. Press, Cambridge, 1999, pp. 199–202. MR1665592 ↑342

[Sch03] Gideon Schechtman, *Concentration results and applications*, Handbook of the geometry of Banach spaces, Vol. 2, North-Holland, Amsterdam, 2003, pp. 1603–1634, DOI 10.1016/S1874-5849(03)80044-X. MR1999604 ↑143, 144

[Sch07] G. Schechtman, *The random version of Dvoretzky's theorem in ℓ_∞^n*, Geometric aspects of functional analysis, Lecture Notes in Math., vol. 1910, Springer, Berlin, 2007, pp. 265–270, DOI 10.1007/978-3-540-72053-9_15. MR2349612 ↑208

[Sch14] Rolf Schneider, *Convex bodies: the Brunn-Minkowski theory*, Second expanded edition, Encyclopedia of Mathematics and its Applications, vol. 151, Cambridge University Press, Cambridge, 2014. MR3155183 ↑103, 104, 344

[SCM16] Gniewomir Sarbicki, Dariusz Chruściński, and Marek Mozrzymas, *Generalising Wigner's theorem*, J. Phys. A **49** (2016), no. 30, 305302, 7, DOI 10.1088/1751-8113/49/30/305302. MR3519269 ↑63

[See66] R. T. Seeley, *Spherical harmonics*, Amer. Math. Monthly **73** (1966), no. 4, 115–121, DOI 10.2307/2313760. MR0201695 ↑145

[Sen96] Siddhartha Sen, *Average entropy of a quantum subsystem*, Phys. Rev. Lett. **77** (1996), no. 1, 1. 233

[Sha48] C. E. Shannon, *A mathematical theory of communication*, Bell System Tech. J. **27** (1948), 379–423, 623–656, DOI 10.1002/j.1538-7305.1948.tb01338.x. MR0026286 ↑30

[Sha08] R. Shankar, *Principles of quantum mechanics*, 2nd ed., Springer, New York, 2008. Corrected reprint of the second (1994) edition. MR2722363 ↑75

[Shi95] Abner Shimony, *Degree of entanglement*, Fundamental problems in quantum theory (Baltimore, MD, 1994), Ann. New York Acad. Sci., vol. 755, New York Acad. Sci., New York, 1995, pp. 675–679, DOI 10.1111/j.1749-6632.1995.tb39008.x. MR1478449 ↑233

[Sho04] Peter W. Shor, *Equivalence of additivity questions in quantum information theory*, Comm. Math. Phys. **246** (2004), no. 3, 453–472, DOI 10.1007/s00220-003-0981-7. MR2053939 ↑232

[Šid67] Zbyněk Šidák, *Rectangular confidence regions for the means of multivariate normal distributions*, J. Amer. Statist. Assoc. **62** (1967), 626–633. MR0216666 ↑178

[Šid68] Zbyněk Šidák, *On multivariate normal probabilities of rectangles: Their dependence on correlations*, Ann. Math. Statist. **39** (1968), 1425–1434, DOI 10.1214/aoms/1177698122. MR0230403 ↑178

[Sil85] Jack W. Silverstein, *The smallest eigenvalue of a large-dimensional Wishart matrix*, Ann. Probab. **13** (1985), no. 4, 1364–1368. MR806232 ↑179, 180

[Sim76] Barry Simon, *Quantum dynamics: from automorphism to Hamiltonian*, Studies in Mathematical Physics. Essays in Honor of Valentine Bargmann (1976), 327–349. 63

[Sin64] Richard Sinkhorn, *A relationship between arbitrary positive matrices and doubly stochastic matrices*, Ann. Math. Statist. **35** (1964), 876–879, DOI 10.1214/aoms/1177703591. MR0161868 ↑64

[Sko16] Łukasz Skowronek, *There is no direct generalization of positive partial transpose criterion to the three-by-three case*, J. Math. Phys. **57** (2016), no. 11, 112201, 19, DOI 10.1063/1.4966984. MR3572696 ↑261

[Sla12] Paul B. Slater, *A concise formula for generalized two-qubit Hilbert-Schmidt separability probabilities*, J. Phys. A **46** (2013), no. 44, 445302, 13, DOI 10.1088/1751-8113/46/44/445302. MR3120911 ↑260

[Sle62] David Slepian, *The one-sided barrier problem for Gaussian noise*, Bell System Tech. J. **41** (1962), 463–501, DOI 10.1002/j.1538-7305.1962.tb02419.x. MR0133183 ↑178

[Slo16] William Slofstra, *Tsirelson's problem and an embedding theorem for groups arising from non-local games*, arXiv:1606.03140 (2016). 297

[Som09] Hans-Jürgen Sommers, *Contribution to discussion: On conditions for block positivity,* Mini-Workshop: Geometry of Quantum Entanglement, December 6–12, 2009. Organized by Andreas Buchleitner, Stanisław Szarek, Elisabeth Werner and Karol Życzkowski, Oberwolfach Rep. **6** (2009), no. 4, 2993–3031, MR2724314 260

[Spi93] Jonathan E. Spingarn, *An inequality for sections and projections of a convex set*, Proc. Amer. Math. Soc. **118** (1993), no. 4, 1219–1224, DOI 10.2307/2160081. MR1184087 ↑105

[SR95] Jorge Sánchez-Ruiz, *Simple proof of Page's conjecture on the average entropy of a subsystem*, Physical Review E **52** (1995), no. 5, 5653. 233

[SS98] P. W. Shor and N. J. A. Sloane, *A family of optimal packings in Grassmannian manifolds*, J. Algebraic Combin. **7** (1998), no. 2, 157–163, DOI 10.1023/A:1008608404829. MR1609881 ↑143

[SS05] Elias M. Stein and Rami Shakarchi, *Real analysis*, Princeton Lectures in Analysis, vol. 3, Princeton University Press, Princeton, NJ, 2005. Measure theory, integration, and Hilbert spaces. MR2129625 ↑342, 364

[ST80] Stanisław Szarek and Nicole Tomczak-Jaegermann, *On nearly Euclidean decomposition for some classes of Banach spaces*, Compositio Math. **40** (1980), no. 3, 367–385. MR571056 ↑209

[Sti55] W. Forrest Stinespring, *Positive functions on C^*-algebras*, Proc. Amer. Math. Soc. **6** (1955), 211–216, DOI 10.2307/2032342. MR0069403 ↑64

[Stø63] Erling Størmer, *Positive linear maps of operator algebras*, Acta Math. **110** (1963), 233–278, DOI 10.1007/BF02391860. MR0156216 ↑63, 64

[Stø13] Erling Størmer, *Positive linear maps of operator algebras*, Springer Monographs in Mathematics, Springer, Heidelberg, 2013. MR3012443 ↑65

[Stø16] Erling Størmer, *Positive maps which map the set of rank k projections onto itself*, Positivity **21** (2017), no. 1, 509–511, DOI 10.1007/s11117-016-0432-2. MR3613010 ↑63

[Sud71] V. N. Sudakov, *Gaussian random processes, and measures of solid angles in Hilbert space* (Russian), Dokl. Akad. Nauk SSSR **197** (1971), 43–45. MR0288832 ↑179

[SV96] Stanisław J. Szarek and Dan Voiculescu, *Volumes of restricted Minkowski sums and the free analogue of the entropy power inequality*, Comm. Math. Phys. **178** (1996), no. 3, 563–570. MR1395205 ↑104

[SV00] S. J. Szarek and D. Voiculescu, *Shannon's entropy power inequality via restricted Minkowski sums*, Geometric aspects of functional analysis, Lecture Notes in Math., vol. 1745, Springer, Berlin, 2000, pp. 257–262, DOI 10.1007/BFb0107219. MR1796724 ↑104

[Sve81] George Svetlichny, *On the foundations of experimental statistical sciences*, Found. Phys. **11** (1981), no. 9-10, 741–782, DOI 10.1007/BF00726947. MR660364 ↑104

[SW] Stanisław Szarek and Paweł Wolff, *Radii of Euclidean sections of L_p-balls*, in preparation. 197

[SW83] Rolf Schneider and Wolfgang Weil, *Zonoids and related topics*, Convexity and its applications, Birkhäuser, Basel, 1983, pp. 296–317. MR731116 ↑103

[SW99] Stanisław J. Szarek and Elisabeth Werner, *A nonsymmetric correlation inequality for Gaussian measure*, J. Multivariate Anal. **68** (1999), no. 2, 193–211, DOI 10.1006/jmva.1998.1784. MR1677442 ↑309

[Swe] Michael Swearingin, *Ph.D. thesis, Case Western Reserve University, in preparation.* 110

[SWŻ08] Stanisław J. Szarek, Elisabeth Werner, and Karol Życzkowski, *Geometry of sets of quantum maps: a generic positive map acting on a high-dimensional system is not completely positive*, J. Math. Phys. **49** (2008), no. 3, 032113, 21, DOI 10.1063/1.2841325. MR2406781 ↑106, 260, 261

[SWŻ11] Stanisław J. Szarek, Elisabeth Werner, and Karol Życzkowski, *How often is a random quantum state k-entangled?*, J. Phys. A **44** (2011), no. 4, 045303, 15. MR2754722 (2012e:81028) 260, 261

[Sza] Stanisław Szarek, *Coarse approximation of convex bodies by polytopes and the complexity of banach–mazur compacta*, in preparation. 143

[Sza74] Andrzej Szankowski, *On Dvoretzky's theorem on almost spherical sections of convex bodies*, Israel J. Math. **17** (1974), 325–338, DOI 10.1007/BF02756881. MR0350388 ↑208

[Sza76] S. J. Szarek, *On the best constants in the Khinchin inequality*, Studia Math. **58** (1976), no. 2, 197–208. MR0430667 ↑147, 282

[Sza78] Stanisław Jerzy Szarek, *On Kashin's almost Euclidean orthogonal decomposition of l_n^1* (English, with Russian summary), Bull. Acad. Polon. Sci. Sér. Sci. Math. Astronom. Phys. **26** (1978), no. 8, 691–694. MR518996 ↑209

[Sza82] Stanisław J. Szarek, *Nets of Grassmann manifold and orthogonal group*, Proceedings of research workshop on Banach space theory (Iowa City, Iowa, 1981), Univ. Iowa, Iowa City, IA, 1982, pp. 169–185. MR724113 ↑143

[Sza83] Stanisław J. Szarek, *The finite-dimensional basis problem with an appendix on nets of Grassmann manifolds*, Acta Math. **151** (1983), no. 3-4, 153–179, DOI 10.1007/BF02393205. MR723008 ↑143

[Sza90] Stanisław J. Szarek, *Spaces with large distance to l_∞^n and random matrices*, Amer. J. Math. **112** (1990), no. 6, 899–942, DOI 10.2307/2374731. MR1081810 ↑103

[Sza98] Stanisław J. Szarek, *Metric entropy of homogeneous spaces*, Quantum probability (Gdańsk, 1997), Banach Center Publ., vol. 43, Polish Acad. Sci. Inst. Math., Warsaw, 1998, pp. 395–410. MR1649741 ↑143, 319

[Sza05] Stanisław J. Szarek, *Volume of separable states is super-doubly-exponentially small in the number of qubits*, Phys. Rev. A (3) **72** (2005), no. 3, 032304, 10, DOI 10.1103/PhysRevA.72.032304. MR2189159 ↑104, 165, 260, 343

[Sza10] Stanisław J. Szarek, *On norms of completely positive maps*, Topics in operator theory. Volume 1. Operators, matrices and analytic functions, Oper. Theory Adv. Appl., vol. 202, Birkhäuser Verlag, Basel, 2010, pp. 535–538, DOI 10.1007/978-3-0346-0158-0_31. MR2723298 ↑232

[Tak08] Leon A. Takhtajan, *Quantum mechanics for mathematicians*, Graduate Studies in Mathematics, vol. 95, American Mathematical Society, Providence, RI, 2008. MR2433906 ↑75

[Tal87] Michel Talagrand, *Regularity of Gaussian processes*, Acta Math. **159** (1987), no. 1-2, 99–149, DOI 10.1007/BF02392556. MR906527 ↑179

[Tal88] Michel Talagrand, *An isoperimetric theorem on the cube and the Kintchine-Kahane inequalities*, Proc. Amer. Math. Soc. **104** (1988), no. 3, 905–909, DOI 10.2307/2046814. MR964871 ↑146

[Tal90] Michel Talagrand, *Embedding subspaces of L_1 into l_1^N*, Proc. Amer. Math. Soc. **108** (1990), no. 2, 363–369, DOI 10.2307/2048283. MR994792 ↑210

[Tal95] Michel Talagrand, *Concentration of measure and isoperimetric inequalities in product spaces*, Publ. Math. Inst. Hautes Études Sci. **81** (1995), 73–205. MR1361756 ↑146

[Tal96a] Michel Talagrand, *New concentration inequalities in product spaces*, Invent. Math. **126** (1996), no. 3, 505–563, DOI 10.1007/s002220050108. MR1419006 ↑146

[Tal96b] Michel Talagrand, *A new look at independence*, Ann. Probab. **24** (1996), no. 1, 1–34, DOI 10.1214/aop/1042644705. MR1387624 ↑146

[Tal01] Michel Talagrand, *Majorizing measures without measures*, Ann. Probab. **29** (2001), no. 1, 411–417, DOI 10.1214/aop/1008956336. MR1825156 ↑179

[Tal05] Michel Talagrand, *The generic chaining: upper and lower bounds of stochastic processes*, Springer Monographs in Mathematics, Springer-Verlag, Berlin, 2005. MR2133757 ↑179

[Tal11] Michel Talagrand, *Mean field models for spin glasses. Volume I*, Ergebnisse der Mathematik und ihrer Grenzgebiete. 3. Folge. A Series of Modern Surveys in Mathematics [Results in Mathematics and Related Areas. 3rd Series. A Series of Modern Surveys in Mathematics], vol. 54, Springer-Verlag, Berlin, 2011. Basic examples. MR2731561 ↑178

[Tal14] Michel Talagrand, *Upper and lower bounds for stochastic processes*, Ergebnisse der Mathematik und ihrer Grenzgebiete. 3. Folge. A Series of Modern Surveys in Mathematics [Results in Mathematics and Related Areas. 3rd Series. A Series of Modern Surveys in Mathematics], vol. 60, Springer, Heidelberg, 2014, Modern methods and classical problems. MR3184689 179

[Tao12] Terence Tao, *Topics in random matrix theory*, Graduate Studies in Mathematics, vol. 132, American Mathematical Society, Providence, RI, 2012. MR2906465 ↑179

[TH00] Barbara M. Terhal and Paweł Horodecki, *Schmidt number for density matrices*, Phys. Rev. A (3) **61** (2000), no. 4, 040301, 4, DOI 10.1103/PhysRevA.61.040301. MR1775394 ↑63

[Tik14] Konstantin E. Tikhomirov, *The randomized Dvoretzky's theorem in* l_∞^n *and the* χ*-distribution*, Geometric aspects of functional analysis, Lecture Notes in Math., vol. 2116, Springer, Cham, 2014, pp. 455–463, DOI 10.1007/978-3-319-09477-9_31. MR3364705 ↑208

[TJ89] Nicole Tomczak-Jaegermann, *Banach-Mazur distances and finite-dimensional operator ideals*, Pitman Monographs and Surveys in Pure and Applied Mathematics, vol. 38, Longman Scientific & Technical, Harlow; copublished in the United States with John Wiley & Sons, Inc., New York, 1989. MR993774 ↑104, 207

[TK04] Gr. Tsagas and K. Kalogeridis, *The spectrum of the Laplace operator for the manifold* $\mathrm{SO}(2p+2q+1)/\mathrm{SO}(2p)\times\mathrm{SO}(2q+1)$, Conference "Applied Differential Geometry: General Relativity"—Workshop "Global Analysis, Differential Geometry, Lie Algebras", BSG Proc., vol. 10, Geom. Balkan Press, Bucharest, 2004, pp. 188–196. MR2125114 ↑145

[Tom85] Jun Tomiyama, *On the geometry of positive maps in matrix algebras. II*, Linear Algebra Appl. **69** (1985), 169–177, DOI 10.1016/0024-3795(85)90074-6. MR798371 ↑64

[Tro12] Joel A. Tropp, *User-friendly tail bounds for sums of random matrices*, Found. Comput. Math. **12** (2012), no. 4, 389–434, DOI 10.1007/s10208-011-9099-z. MR2946459 ↑146, 255

[Tsi85] B. S. Tsirelson, *Quantum analogues of Bell's inequalities. The case of two spatially divided domains,* Problems of the theory of probability distributions, IX (Russian), Zap. Nauchn. Sem. Leningrad. Otdel. Mat. Inst. Steklov. (LOMI) **142** (1985), 174–194, 200. MR788202 ↑295, 297

[Tsi93] B. S. Tsirelson, *Some results and problems on quantum Bell-type inequalities*, Hadronic J. Suppl. **8** (1993), no. 4, 329–345. MR1254597 ↑295

[Tsu81] Chiaki Tsukamoto, *Spectra of Laplace-Beltrami operators on* $\mathrm{SO}(n+2)/\mathrm{SO}(2)\times\mathrm{SO}(n)$ *and* $\mathrm{Sp}(n+1)/\mathrm{Sp}(1)\times\mathrm{Sp}(n)$, Osaka J. Math. **18** (1981), no. 2, 407–426. MR628842 ↑145

[TVZ82] M. A. Tsfasman, S. G. Vlăduţ, and Th. Zink, *Modular curves, Shimura curves, and Goppa codes, better than Varshamov-Gilbert bound*, Math. Nachr. **109** (1982), 21–28, DOI 10.1002/mana.19821090103. MR705893 ↑143

[TW94] Craig A. Tracy and Harold Widom, *Level-spacing distributions and the Airy kernel*, Comm. Math. Phys. **159** (1994), no. 1, 151–174. MR1257246 ↑179

[TW96] Craig A. Tracy and Harold Widom, *On orthogonal and symplectic matrix ensembles*, Comm. Math. Phys. **177** (1996), no. 3, 727–754. MR1385083 ↑179

[Vaa79] Jeffrey D. Vaaler, *A geometric inequality with applications to linear forms*, Pacific J. Math. **83** (1979), no. 2, 543–553. MR557952 ↑106

[VADM01] Frank Verstraete, Koenraad Audenaert, and Bart De Moor, *Maximally entangled mixed states of two qubits*, Phys. Rev. A **64** (2001), 012316. 64

[VB14] Tamás Vértesi and Nicolas Brunner, *Disproving the Peres conjecture by showing Bell nonlocality from bound entanglement*, Nat. Commun. **5** (2014), article. 297

[VDD01] Frank Verstraete, Jeroen Dehaene, and Bart DeMoor, *Local filtering operations on two qubits*, Phys. Rev. A **64** (2001), 010101. 65

[Vem04] Santosh S. Vempala, *The random projection method*, DIMACS Series in Discrete Mathematics and Theoretical Computer Science, vol. 65, American Mathematical Society, Providence, RI, 2004. MR2073630 ↑124

[Ver] Roman Vershynin, *High-Dimensional Probability. An Introduction with Applications in Data Science*, Cambridge University Press, in preparation. 143, 207

[Ver08] T. Vértesi, *More efficient Bell inequalities for Werner states*, Phys. Rev. A **78** (2008), 032112. 295

[Ver12] Roman Vershynin, *Introduction to the non-asymptotic analysis of random matrices*, Compressed sensing, Cambridge Univ. Press, Cambridge, 2012, pp. 210–268. MR2963170 ↑146

[Vil09] Cédric Villani, *Optimal transport, Old and New*, Grundlehren der Mathematischen Wissenschaften [Fundamental Principles of Mathematical Sciences], vol. 338, Springer-Verlag, Berlin, 2009. MR2459454 ↑179

[Voi85] Dan Voiculescu, *Symmetries of some reduced free product C^*-algebras*, Operator algebras and their connections with topology and ergodic theory (Buşteni, 1983), Lecture Notes in Math., vol. 1132, Springer, Berlin, 1985, pp. 556–588, DOI 10.1007/BFb0074909. MR799593 ↑180

[Voi90] Dan Voiculescu, *Circular and semicircular systems and free product factors*, Operator algebras, unitary representations, enveloping algebras, and invariant theory (Paris, 1989), Progr. Math., vol. 92, Birkhäuser Boston, Boston, MA, 1990, pp. 45–60. MR1103585 ↑180

[Voi91] Dan Voiculescu, *Limit laws for random matrices and free products*, Invent. Math. **104** (1991), no. 1, 201–220, DOI 10.1007/BF01245072. MR1094052 ↑145, 180

[von27] John von Neumann, *Thermodynamik quantenmechanischer Gesamtheiten.*, Nachr. Ges. Wiss. Göttingen, Math.-Phys. Kl. **1927** (1927), 276–291 (German). 30

[von32] Johann von Neumann, *Mathematische Grundlagen der Quantenmechanik* (German), Unveränderter Nachdruck der ersten Auflage von 1932. Die Grundlehren der mathematischen Wissenschaften, Band 38, Springer-Verlag, Berlin-New York, 1968. MR0223138 ↑30

[VT99] Guifré Vidal and Rolf Tarrach, *Robustness of entanglement*, Phys. Rev. A (3) **59** (1999), no. 1, 141–155, DOI 10.1103/PhysRevA.59.141. MR1670562 ↑260

[VW01] K. G. H. Vollbrecht and R. F. Werner, *Entanglement measures under symmetry*, Phys. Rev. A **64** (2001), 062307. 63

[Wal02] Nolan R. Wallach, *An unentangled Gleason's theorem*, Quantum computation and information (Washington, DC, 2000), Contemp. Math., vol. 305, American Mathematical Society, Providence, RI, 2002, pp. 291–298, DOI 10.1090/conm/305/05226. MR1947343 ↑232

[Wat] John Watrous, *Theory of quantum information*, book in preparation, see `https://cs.uwaterloo.ca/~watrous/TQI/`. 63, 64, 260, 306

[Wat05] John Watrous, *Notes on super-operator norms induced by Schatten norms*, Quantum Inf. Comput. **5** (2005), no. 1, 58–68. MR2123899 ↑218, 232

[Wer89] Reinhard F. Werner, *Quantum states with Einstein-Podolsky-Rosen correlations admitting a hidden-variable model*, Phys. Rev. A **40** (1989), 4277–4281. 63, 281

[WG03] Tzu-Chieh Wei and Paul M. Goldbart, *Geometric measure of entanglement and applications to bipartite and multipartite quantum states*, Phys. Rev. A **68** (2003), 042307. 233

[WH02] R. F. Werner and A. S. Holevo, *Counterexample to an additivity conjecture for output purity of quantum channels*, J. Math. Phys. **43** (2002), no. 9, 4353–4357, DOI 10.1063/1.1498491. MR1924444 ↑232

[Wig55] Eugene P. Wigner, *Characteristic vectors of bordered matrices with infinite dimensions*, Ann. of Math. (2) **62** (1955), 548–564, DOI 10.2307/1970079. MR0077805 ↑179

[Wig58] Eugene P. Wigner, *On the distribution of the roots of certain symmetric matrices*, Ann. of Math. (2) **67** (1958), 325–327, DOI 10.2307/1970008. MR0095527 ↑179

[Wig59] Eugene P. Wigner, *Group theory and its application to the quantum mechanics of atomic spectra*, Expanded and improved ed. Translated from the German by J. J. Griffin. Pure and Applied Physics. Vol. 5, Academic Press, New York-London, 1959. MR0106711 ↑63

[Wil17] Mark M. Wilde, *Quantum information theory*, second ed., Cambridge University Press, Cambridge, 2017. 30, 63, 232

[Win16] Andreas Winter, *Tight uniform continuity bounds for quantum entropies: conditional entropy, relative entropy distance and energy constraints*, Comm. Math. Phys. **347** (2016), no. 1, 291–313, DOI 10.1007/s00220-016-2609-8. MR3543185 ↑233

[Wor76] S. L. Woronowicz, *Positive maps of low dimensional matrix algebras*, Rep. Math. Phys. **10** (1976), no. 2, 165–183, DOI 10.1016/0034-4877(76)90038-0. MR573218 ↑63

[WS08] Jonathan Walgate and A. J. Scott, *Generic local distinguishability and completely entangled subspaces*, J. Phys. A **41** (2008), no. 37, 375305, 15, DOI 10.1088/1751-8113/41/37/375305. MR2430582 ↑232

[WW00] R. F. Werner and M. M. Wolf, *Bell's inequalities for states with positive partial transpose*, Phys. Rev. A (3) **61** (2000), no. 6, 062102, 4, DOI 10.1103/PhysRevA.61.062102. MR1767459 ↑297

[WW01a] R. F. Werner and M. M. Wolf, *All-multipartite Bell-correlation inequalities for two dichotomic observables per site*, Phys. Rev. A **64** (2001), 032112. 296, 297

[WW01b] Reinhard F. Werner and Michael M. Wolf, *Bell inequalities and entanglement*, Quantum Inf. Comput. **1** (2001), no. 3, 1–25. MR1907485 ↑296, 297

[You14] Pierre Youssef, *Restricted invertibility and the Banach-Mazur distance to the cube*, Mathematika **60** (2014), no. 1, 201–218, DOI 10.1112/S0025579313000144. MR3164527 ↑209

[ŻHSL98] Karol Życzkowski, Paweł Horodecki, Anna Sanpera, and Maciej Lewenstein, *Volume of the set of separable states*, Phys. Rev. A (3) **58** (1998), no. 2, 883–892, DOI 10.1103/PhysRevA.58.883. MR1638209 ↑260

[Zie00] Günter M. Ziegler, *Lectures on* 0/1*-polytopes*, Polytopes—combinatorics and computation (Oberwolfach, 1997), DMV Sem., vol. 29, Birkhäuser, Basel, 2000, pp. 1–41. MR1785291 ↑281

[ŻS01] Karol Życzkowski and Hans-Jürgen Sommers, *Induced measures in the space of mixed quantum states*, J. Phys. A **34** (2001), no. 35, 7111–7125, DOI 10.1088/0305-4470/34/35/335. MR1863143 ↑180

[ŻS03] Karol Życzkowski and Hans-Jürgen Sommers, *Hilbert-Schmidt volume of the set of mixed quantum states*, J. Phys. A **36** (2003), no. 39, 10115–10130, DOI 10.1088/0305-4470/36/39/310. MR2024516 ↑260

Websites

[@1] `http://www2.stetson.edu/~efriedma/packing.html` 108

[@2] `http://mathworld.wolfram.com/GumbelDistribution.html` 178

[@3] `http://www.encyclopediaofmath.org/index.php?title=Banach-Mazur_compactum&oldid=22053` (an article originated by A. A. Giannopoulos) 103, 209

[@4] `http://qig.itp.uni-hannover.de/qiproblems/1` 296

[@5] `http://qig.itp.uni-hannover.de/qiproblems/2` 306

Index

In addition to pointing to definitions of concepts that appear throughout this book, the index is designed to direct the reader to fundamental or major results about such concepts and to other facts, which have—in the authors' opinion—a reference value. This includes sharp versions of well-known inequalities, proofs of standard results that are new or not widely known, or tables listing values of various geometric parameters for classical objects. The index *is not* meant to be an exhaustive catalogue of all occurrences of a given notion or phrase in the book.

absolutely separable state, 38
additivity problem, 216, 228
adjoint
  map, 49
  of an operator, 4
almost randomizing channels, 220
anti-unitary, 34
antisymmetric
  subspace, 40
asphericity, 193
asymmetry of a convex set, 143
asymptotic freeness, 177

Banach–Mazur
  compactum, 79
  distance, 79, 103
Bell correlation inequality, 280
Bell inequality for boxes, 289
Bell polytope, 277
Bell states, 39, 70, 302
  nonseparability, 44
Bell vectors, 70, 302
Bell violations, 280, 290
  arbitrarily large, 291
Bernstein's inequalities, 141, 186
bipartite Hilbert space, 6
bipolar theorem, 15
bistochastic channel, 50
bistochastic matrix, 23
Blaschke–Santaló inequality, 98
blessing of dimensionality, 205
Bloch ball, 32
Bloch sphere, 32
block matrix, 8
block-positive matrix, 56
Bonami–Beckner inequality, 146
Boolean cube, 113
Borel selection theorem, 14
Born rule, 67
bound entanglement, 306
box, 285
bra-ket notation, 4
Brouwer's fixed-point theorem, 60
Brunn–Minkowski inequality, 92
  restricted, 104
  reverse, 104, 209
Bures metric, 312
Busemann–Petty problem, 106

canonical, 4
Carathéodory's theorem, 12
Catalan numbers, 163
central value, 124
chaining argument, 158
channel, 50
Chevet–Gordon inequalities, 173
chi distribution, 175
  mean, 309
  median, 124
Choi matrix, 48
Choi's isomorphism, 48
Choi's theorem, 49
CHSH game, 284
CHSH inequality, 280
circled
  body, 11
  function, 187
classical box, 286

classical correlation, 277
classical-quantum (c-q) channel, 53
Clifford algebras, 275
CNOT, 304
co-completely positive map, 57
co-positive semi-definite, 56
column vector, 5
completely depolarizing channel, 53
completely positive cone, 49, 57
  duality, 57
completely positive map, 49
  norm of, 217
completely randomizing channel, 53, 220
complexification, 6
computational basis, 5, 67
concentration of measure, 117
  on standard spaces, 118
  subgaussian, 117
cone, 18
  base duality, 20
  base of, 19
  dual, 19
  self-dual, 19
conjugate
  of a Hilbert space, 4
  of a matrix, 7
contact point, 87
contextuality, 297
contraction principle, 127, 134
convex body, 11
  C-Euclidean, 189
  polytopal approximation, 193, 194
convex hull, 12
convex roof, 271
Copenhagen interpretation, 67
correlation conjecture, 154
correlation polytope, 277
covering, 107
  density, 142
  number, 107, 114
creation operators, 177
curse of dimensionality, 205
cut polytope, 296

Davis convexity theorem, 25
decomposable map, 57
decomposable matrix, 56
density matrix, 9, 71
deterministic box, 286
deterministic strategy, 284, 286
diamond norm, 52
difference body, 81
direct sum of channels, 55
discrete cube, 113
distillability problem, 302
  and 2-positivity, 305
  and Werner states, 305
distillable state, 302
distinguishability, 299
ℓ_1-distortion, 197
doubly stochastic channel, 50
dropping the complex structure, 7
Dudley's inequality, 157
Dvoretzky dimension, 190
Dvoretzky's theorem, 200
Dvoretzky–Milman theorem
  and the escape phenomenon, 189
  for ℓ_p^n-spaces, 195
  for convex bodies, 190
  for Lipschitz functions, 187
  for projections, 192
  for Schatten spaces, 198
  isometric, 275
Dvoretzky–Rogers lemma, 200

Earth Mover's distance, 161
Ehrhard symmetrization, 124
Ehrhard's inequality, 122
ellipsoids, 18
  polars of, 18
  tensor product of, 18
empirical spectral distribution, 160
ε-enlargements, 117
enough symmetries, 89
entangled state, 37
k-entangled state, 41
entangled subspaces, 214
  extremely entangled, 225
  very entangled, 223
entanglement of formation, 271
entanglement witness, 60
entanglement-breaking channel, 53
entropy of entanglement, 215
  p-entropy, 215
escape phenomenon, 176
explicit constructions, 205
exponential Markov inequality, 124
exposed face, 13
exposed point, 13
exposed ray, 21
k-extendible state, 41
extension of a map, 49
extreme point, 13
extreme ray, 21
extrinsic distance, 311

face, 13
facet, 13
facial dimension, 193
fidelity, 312
Figiel–Lindenstrauss–Milman inequality, 194
Finsler geometry, 319
flip operator, 39
Fock space, 177
fraction
  classical, 292

local, 292
nonlocal, 292
of determinism, 292
free additive convolution, 177
free Poisson distribution, 167
free probability, 176
Frobenius norm, 7
Fubini–Study metric, 312
full cone, 21

gauge, 11
Gaussian distribution, 307
tail estimates, 307
Gaussian mean width, 95
Gaussian processes, 149, 308
and the mean width, 150
comparison inequalities, 153
stationary, 159
Gaussian Unitary Ensemble, 162
generic chaining, 159
geodesic, 314
geodesic distance, 311
geodesically convex, 313
geometric distance, 79
geometric measure of entanglement, 229
Ginibre formula, 163
GOE, 163
large deviations, 165
Gordon's lemma, 153
Grassmann manifold, 314
ε-nets, 116
Gromov's comparison theorem, 130
Grothendieck constant, 281
complex, 295
other variants, 281, 295
Grothendieck's inequality, 280
GUE, 162
convergence to semicircle law, 164
eigenvalue distribution, 163
large deviations, 164
norm, 165
small deviations, 165
GUE_0, 163
Gumbel distribution, 178
Gurvits–Barnum theorem, 246

Haar measure, 313
Hamming distance, 113
Hanner's inequalities, 14
Harper's isoperimetric inequality, 137
Hastings's counterexample, 228
Heisenberg–Weyl operators, 222
Helstrom bound, 300
Herbst's argument, 132
Hermitian matrix, 7
Herschel–Maxwell theorem, 309
hidden variable, 75, 286
Hilbert–Schmidt norm/inner product, 7
Hoeffding's
inequality, 129
lemma, 124
matrix inequality, 255
α-homogeneous functions, 309
homogeneous space, 315
Horodecki's entanglement witness theorem, 61
hypercontractivity, 135
hyperplane conjecture, 101

injective tensor product, 83
inradius, 96
intrinsic distance, 311
irreducible, 89
isoperimetric inequality
Gaussian, 122
in $\mathbb{R}^n$, 92
on the discrete cube, 137
on the sphere, 119
isotropic convex body, 101
isotropic states, 39

Jamiołkowski isomorphism, 47
John ellipsoid, 84
John position, 84
Johnson–Lindenstrauss lemma, 205
jointly Gaussian variables, 308

K-convexity constant, 183, 185, 207
bounds, 183, 184
duality, 186
for B_1^n, the cube, 186
Kadison's theorem, 34
Kashin decomposition of ℓ_1^n, 202
Khatri–Šidák lemma, 153
Klein's lemma, 27
Knaster problem, 208
Kneser–Poulsen conjecture, 178
Kochen–Specker theorem, 297
Komatu inequalities, 307
Kraus decomposition, 49
Kraus rank, 49
Krein–Milman theorem, 13

ℓ-norm, 181
ℓ-position, 181
Löwner position, 84
Lévy distance, 161
Laplace transform
bilateral, 132
method, 124
law of the iterated logarithm, 160
Lévy's lemma, 120
for central values, 125
for the mean, 121
local version, 127
linear programming bound, 112
L-Lipschitz function, 120
extension, 227

local, 74
  box, 286
  correlation, 277
  filtering, 301
  polytope, 277
  unitaries, 46
local strategy, 284
  with shared randomness, 286
LOCC channel, 54, 301
log-concave measure, 93
log-Sobolev
  constants, 132, 134
  inequality, 132
  tensorization property, 133
Lorentz cone, 18
  automorphisms, 323
Lorentz group, 318
  proper, 318
  restricted, 318
Low M^*-estimate, 202
Löwner ellipsoid, 84
ℓ_p-norm, 12
ℓ_p product metric, 128

M-ellipsoid, 143
M-position, 143
magic square game, 293
Mahler conjecture, 105
majorization, 22
majorizing measure, 179
Marčenko–Pastur distribution, 167
maximally entangled, 39, 229
maximally mixed state, 20, 32
mean width, 95
  and Gaussian processes, 150
  of a union of sets, 121
  of classical bodies, 96
  of QIT bodies, 235
median, 117
  of a $\chi^2(n)$ variable, 124
  of a convex function, 126
Mermin–Peres game, 293
metric entropy, 107
  of ℓ_p^n-balls, 156, 157
  of classical manifolds, 116
Milman–Pajor inequality, 98
minimum output entropy, 216
Minkowski compactum, 79
Minkowski functional, 11
Minkowski operations, 81
Minkowski–Hlawka theorem, 142
mixed state, 31, 69
mixed-unitary channel, 52
MM^*-estimate, 184, 207
multipartite Hilbert space, 6
multiplicativity problem, 217, 218

ε-nets, 107
  of classical manifolds, 116
  of product spaces, 114
  of the discrete cube, 113
  of the projective space, 112
  of the sphere, 110, 111
non-commutative Hölder inequality, 25
nondegenerate cone, 21
nonlocal
  boxes, 287
  correlations, 277
  fraction, 292
nonsignaling
  box, polytope, correlation, 287
  principle, 287
  violations, 290, 292

operational, 29
Ornstein–Uhlenbeck semigroup, 135
orthochronous subgroup, 318
orthogonal group, 312
  ε-nets, 116
  geodesics, 313
oscillation, 186
outer product, 5
outradius, 96
overlap, 37, 229, 312

packing, 107
  density, 142
  number, 107
  on the discrete cube, 113
  on the sphere, 112
partial trace, 35, 70
partial transposition, 41
Pauli matrices, 32
  composition rules for, 33
Peres conjecture, 281, 291, 297
Peres–Horodecki criterion, 43
permutationally symmetric
  basis, 90
  body, norm, space, 90
phase of a vector in $\mathbb{C}^d$, 312
Poincaré's
  constants, 134
  inequality, 134
  lemma, 122
pointed cone, 21
polar
  of a convex body, 15
  of a linear image, 15
  of a translate, 326
  of sections, projections, 16
  of unions, intersections, 16
polarity, 15
  in the complex setting, 27
polytope, 12
positive cone, 56
  duality, 57
positive map, 49
  Sinkhorn's normal form, 59

n-positive map, 49
positive orthant, 18
positive semi-definite cone, 18
  automorphisms, 58
  extreme rays, 22
positivity-preserving map, 49
POVM, 53, 74
  associated zonotope, 300
  sparsification, 300
PPT cone, 55
PPT criterion, 44
PPT state, 43
PPT-inducing map, 54
Prékopa–Leindler inequality, 101
precognition, 287
principal angles, 314
probabilistic method, 205
projective measurement, 73
projective space, 31, 68, 312
  nets, 112
  volume of balls, 112
projective tensor product, 82
projective unitary group, 312
proper face, 13
pseudotelepathy, 293, 297
pure state, 31
  separable, 37
purification, 71
purity, 173
pushforward, 127

q-c-q channel, 53
quantum box, 286
quantum channel, 50, 72
  as a subspace, 216
quantum correlation, 277
quantum game, 284
quantum map, 47
quantum marginal, 70, 287
quantum operation, 47
quantum strategy, 285, 286
quantum violations
  for boxes, 289, 291
  for correlations, 280
quantum-classical (q-c) channel, 53
quartercircular distribution, 169
qubit, 6
quotient metric, 316
quotient of a subspace theorem, 204

R-transform, 177
random covering, 111
random induced states, 170
  convergence, 171
  density, 172
  large deviations, 171
random strategy, 284, 286
random subspace, 186
randomness reduction, 206

realignment, 45
regular simplex, 12
Rényi entropies, 28
  monotonicity, 29
resolution of identity, 87
  unbiased, 87
Ricci curvature, 129
  bounds, 131
robustness, 247
  bipartite, 247
  multipartite, 249
Rogers–Shephard inequalities, 100
row vector, 5

S-lemma, 321
  and automorphisms of $\mathcal{L}_n$, 323
Santaló inequality, 98
  reverse, 98
Santaló point, 326
Schatten
  p-norms, 23
  spaces, 24
Schmidt coefficients, 36
  and Courant–Fischer formulas, 37
Schmidt decomposition, 36
Schmidt rank, 36
Schur channel, 54
sectional curvature, 130
Segré variety, 213, 312
self-adjoint
  matrix, 7
  operator, 4
self-adjointness preserving map, 48
semicircle law, 163
semicircular distribution, 163
separable cone, 38
  duality, 57
separable map, 54
separable state, 37
ε-separated set, 107
set of PPT states, 43
  volume and mean width, 235, 245
set of quantum states, 9, 31
  centroid, 35
  facial structure, 33
  polytopal approximation, 253
  symmetries, 34
  volume and mean width, 235, 236
set of separable states, 37
  MM^*-estimate for, 240
  centroid, 46
  dimension, 37
  extreme points, 37
  facial structure, 38
  inradius, 235
  polytopal approximation, 253
  symmetries, 46
  volume and mean width, 235, 244

Shannon entropy
  continuous, 132
  discrete, 28
simplex, 12
simplicial order, 137
singular value decomposition, 36
singular values, 24, 36
Slepian's lemma, 153
Slepian–Gordon lemma, 153
spherical cap, 109
  volume, 109
Spingarn inequality, 99
spinor map, 32, 319
standard Gaussian
  measure, 94, 308
  vector, 95, 308
star-shaped set, 114
state
  classical, 8
  quantum, 8
Steiner symmetrization, 93
Stiefel manifold, 317
Stinespring representation, 51, 72
Stinespring theorem, 51
strictly convex set, 14
Størmer's theorem, 44
  proofs, 62
subexponential variable (ψ_1), 139
subgaussian process, 157
subgaussian variable (ψ_2), 139
submajorized, 23
Sudakov minoration, 154
  dual, 155
super-positive map, 53
superoperator, 8, 47
support function, 94
supporting hyperplane, 13
symmetric basis, 90
symmetric convex body, 11
symmetric subspace, 39
symmetrizations, 80

Talagrand's convex concentration inequality, 138
tensor product
  Hilbertian, 6
  injective, 83
  projective, 82
threshold for entanglement, 269
threshold for PPT, 272
trace duality, 7
trace norm, 24
trace-preserving map, 50
Tracy–Widom
  distribution, 179
  effect, 165
transposition, 41
Tsirelson's bound, 280
twirling channel, 40, 305

unconditional
  basis, 90
  body, norm, space, 90
  direct sum, 146
uniform convexity, 14
  for Schatten p-norm, 29
unital map, 50
unitarily invariant
  function, 25
  norm, 27
  random matrix, 265
unitary channel, 52
unitary evolution, 71
unitary group, 312
  ε-nets, 116
  geodesics, 313
universal entanglers, 215
Urysohn's inequality, 95
  dual, 96
  reverse, 184

verticial dimension, 193
volume of polytopes, 152
volume radius, 92
  superadditivity, 93
volume ratio, 201
  and Kashin decomposition, 202
von Neumann entropy, 27
  Lipschitz constant, 222, 224

∞-Wasserstein distance, 161
wave function, 67
weak convergence
  of measures, 179, 359
  of random variables, 162
Werner states, 40
  separability, 45
Wigner's semicircle law, 164
Wigner's theorem, 34
Wishart matrix, 166
  convergence to MP(λ), 167
  convergence to semicircle law, 168
  large deviations, 170
  norm, 169, 174
  partial transposition, 168
Woronowicz theorem, 44

XOR game, 296

zonoids, 82
  approximation by zonotopes, 205, 210
zonotopes, 82
  and POVMs, 300

Published Titles in This Series

223 **Guillaume Aubrun and Stanisław J. Szarek,** Alice and Bob Meet Banach, 2017

219 **Richard Evan Schwartz,** The Projective Heat Map, 2017

218 **Tushar Das, David Simmons, and Mariusz Urbański,** Geometry and Dynamics in Gromov Hyperbolic Metric Spaces, 2017

217 **Benoit Fresse,** Homotopy of Operads and Grothendieck–Teichmüller Groups, 2017

216 **Frederick W. Gehring, Gaven J. Martin, and Bruce P. Palka,** An Introduction to the Theory of Higher-Dimensional Quasiconformal Mappings, 2017

215 **Robert Bieri and Ralph Strebel,** On Groups of PL-homeomorphisms of the Real Line, 2016

214 **Jared Speck,** Shock Formation in Small-Data Solutions to 3D Quasilinear Wave Equations, 2016

213 **Harold G. Diamond and Wen-Bin Zhang (Cheung Man Ping),** Beurling Generalized Numbers, 2016

212 **Pandelis Dodos and Vassilis Kanellopoulos,** Ramsey Theory for Product Spaces, 2016

211 **Charlotte Hardouin, Jacques Sauloy, and Michael F. Singer,** Galois Theories of Linear Difference Equations: An Introduction, 2016

210 **Jason P. Bell, Dragos Ghioca, and Thomas J. Tucker,** The Dynamical Mordell–Lang Conjecture, 2016

209 **Steve Y. Oudot,** Persistence Theory: From Quiver Representations to Data Analysis, 2015

208 **Peter S. Ozsváth, András I. Stipsicz, and Zoltán Szabó,** Grid Homology for Knots and Links, 2015

207 **Vladimir I. Bogachev, Nicolai V. Krylov, Michael Röckner, and Stanislav V. Shaposhnikov,** Fokker–Planck–Kolmogorov Equations, 2015

206 **Bennett Chow, Sun-Chin Chu, David Glickenstein, Christine Guenther, James Isenberg, Tom Ivey, Dan Knopf, Peng Lu, Feng Luo, and Lei Ni,** The Ricci Flow: Techniques and Applications: Part IV: Long-Time Solutions and Related Topics, 2015

205 **Pavel Etingof, Shlomo Gelaki, Dmitri Nikshych, and Victor Ostrik,** Tensor Categories, 2015

204 **Victor M. Buchstaber and Taras E. Panov,** Toric Topology, 2015

203 **Donald Yau and Mark W. Johnson,** A Foundation for PROPs, Algebras, and Modules, 2015

202 **Shiri Artstein-Avidan, Apostolos Giannopoulos, and Vitali D. Milman,** Asymptotic Geometric Analysis, Part I, 2015

201 **Christopher L. Douglas, John Francis, André G. Henriques, and Michael A. Hill, Editors,** Topological Modular Forms, 2014

200 **Nikolai Nadirashvili, Vladimir Tkachev, and Serge Vlăduţ,** Nonlinear Elliptic Equations and Nonassociative Algebras, 2014

199 **Dmitry S. Kaliuzhnyi-Verbovetskyi and Victor Vinnikov,** Foundations of Free Noncommutative Function Theory, 2014

198 **Jörg Jahnel,** Brauer Groups, Tamagawa Measures, and Rational Points on Algebraic Varieties, 2014

197 **Richard Evan Schwartz,** The Octagonal PETs, 2014

196 **Silouanos Brazitikos, Apostolos Giannopoulos, Petros Valettas, and Beatrice-Helen Vritsiou,** Geometry of Isotropic Convex Bodies, 2014

195 **Ching-Li Chai, Brian Conrad, and Frans Oort,** Complex Multiplication and Lifting Problems, 2014

194 **Samuel Herrmann, Peter Imkeller, Ilya Pavlyukevich, and Dierk Peithmann,** Stochastic Resonance, 2014